chapter 15   pages
173, 176, 177, 178, 187

LIQUID WASTE OF INDUSTRY

Theories, Practices, and Treatment

Nelson Leonard Nemerow received his B.S. in chemical engineering from Syracuse University and his M.S. and Ph.D. degrees from Rutgers University. He was a research associate at Rutgers, an assistant professor and then associate professor at North Carolina State College before his return to the Syracuse campus, where his current assignment is Professor of Civil Engineering. He is also chairman of a faculty committee on environmental engineering.

In addition to his teaching activities, Dr. Nemerow acts as an engineering consultant for industry, states, and municipalities. He is a Director of Princeton Aqua Science, a conceptual design and equipment manufacturing concern. Dr. Nemerow is a member of the American Academy of Sanitary Engineers, Sigma Xi, Theta Tau Society, Federated Water Pollution Control Association, New York and South Carolina State Societies of Professional Engineers, and the American Society of Civil Engineers.

LIQUID WASTE OF INDUSTRY

Theories, Practices, and Treatment

Nelson L. Nemerow

Syracuse University

ADDISON-WESLEY PUBLISHING COMPANY

Reading, Massachusetts

Menlo Park, California · London · Amsterdam · Don Mills, Ontario · Sydney

Cover photo by L. a.j. Forté

Copyright © 1971 by Addison-Wesley Publishing Company, Inc.
Philippines copyright 1971 by Addison-Wesley Publishing Company, Inc.

All rights reserved. No part of this publication may be reproduced, stored in a retrieval system, or transmitted, in any form or by any means, electronic, mechanical, photocopying, recording, or otherwise, without the prior written permission of the publisher. Printed in the United States of America. Published simultaneously in Canada. Library of Congress Catalog Card No. 72-106639.

ISBN 0-201-05264-4
DEFGHIJKLM-CO-79876

PREFACE

This book is intended to meet the needs of many people: the college professor who teaches environmental engineering, the consulting engineer who seeks a solution to his client's problem, the municipal engineer who must understand the waste problem well enough to explain it to city officials and point out remedies, the industrial plant engineer who wants to prevent his company from polluting the water that receives his plant's wastes, the Environmental Protection Agency technical personnel charged with administering the Water Quality Act of 1970, and engineers at state and regional levels who are faced with the immediate and readily visible problem of pollution abatement.

A book to serve such diverse interests, in order to be useful, must naturally attack the subject from several viewpoints. The experience gained in using *Theories and Practices of Industrial Waste Treatment* (Addison-Wesley, 1963) over a period of five years proved invaluable to me when it came to writing this present text, which is divided into the following four sections.

Section 1 contains basic facts which the industrial waste engineer needs to know: effect of wastes on the surrounding environment, ways to protect the stream from further pollution, how to calculate the final treatment required before disposal of wastes into a receiving stream, how to sample the stream to ascertain the waste treatment required or the efficiency of existing treatment, and finally how administrative decisions in pollution-abatement problems are influenced by the economics of waste treatment.

Section 2 delves into the theories of waste treatment, and talks about how wastes can be reduced by proper operation of manufacturing plants. Since no waste problem is exactly similar to any other waste problem, students of this subject must have a coherent picture of the entire field of industrial waste treatment in order to decide which treatment process best suits the needs of a particular project. Section 2, therefore, differs from the conventional text on waste-water treatment in that it discusses not only removal of suspended and colloidal solids, but also the subjects of neutralization, equalization and proportioning, and removal of inorganic dissolved salts. The theories are similar to those expressed by other authors, since theories have not changed very much. However, this section does present new ideas for the removal of dissolved organic solids.

Section 3 accentuates engineering practice, and presents concrete examples of problems and their solutions. Theories are highly idealistic and seldom work in practice exactly as set forth on paper; quite often, in this field, they never work at all. The reasons for this are numerous. Economics, public opinion, personality differences, local laws or customs, previous community experiences with certain industrial wastes, contradictory advice by consulting engineers, views of the local industrial development board, views of regulatory agencies—all these and many other factors help determine whether even well-conceived theories can be put into practice.

In each chapter in Section 3 I have reported on actual cases which I either know about or have executed personally. I have attempted to keep these case histories as contemporary as possible. The reader must realize, however, than it often takes as long as five years between the realization of a given waste problem and the initiation of work on it and the evaluation of the treatment plant used in solving it. Thus some cases described in the text originated several years ago, but I hope that they serve as typical examples and stand the test of time satisfactorily. The reader can follow step-by-step analyses and results, just as law students study legal cases. The basic question underlying this section is whether to treat industrial wastes separately or in conjunction with municipal sewage. In many instances theory calls for joint treatment but practice demands separate treatment. One must know the reasons for recommending overall treatment systems, as well as certain specific methods of waste treatment.

In these times of plant relocation due to business expansion and market changes, the separate chapter on site selection is especially useful. The more this country's land is developed with industries, cities, highways, parks, and reservoirs, the more important site selection becomes.

Section 4 gives separate treatises on all the major liquid industrial wastes—a subject which normally requires an entire book. I have classified all industries into five categories: apparel, food processing, materials, chemicals, and energy. I have found it desirable to divide the energy industry into two parts in order to specify the separate and different handling that must be accorded to radioactive wastes. I do not attempt to present a comprehensive study of each waste; that in itself would require a separate text for each. Rather I have given a condensed evaluation of the nature of each waste—its origin, characteristics, and more acceptable treatments. In addition, there is an extensive bibliography, which presents the most readily available reference material for each type of waste. The bibliographies are largely divided into two groups; those papers published prior to 1962 and those published between 1962 and 1968. This should be invaluable to the person who needs to do efficient and rapid research on recent and older studies of a particular waste.

Since no author could personally amass the extensive and varied data presented in this text, in writing this book I have borrowed heavily—and gratefully—from source material on the subject by other writers. To these authors—as well as to my teachers, Dr. William Rudolfs, Dr. Hovhannes Heukelekian, and Dr. Harold Orford—I am sincerely indebted. I am also indebted to the hundreds of original researchers and to the journals which published their works. For their permission to quote excerpts in the book, I express my appreciation to the following publications, as well as to numerous others: *Journal of the Water Pollution Control Federation*, Washington, D.C.; *Wastes Engineering*, New York City; *Industrial Water and Wastes*, Chicago, Illinois; *Water and Sewage Works*, Chicago, Illinois; and *Proceedings of the Purdue University Industrial Waste Conference*, Lafayette, Indiana.

I am grateful to my many friends and colleagues who, after using my book, *Theories and Practices of Industrial Waste Treatment*, offered their suggestions in a gracious manner. To my graduate students who were forced to study this book in great detail, let me say thank you for serving me and society so well. Without cooperative students on whom to try out one's ideas, it would be impossible to produce a meaningful textbook on this subject for graduate study.

Syracuse, New York   N.L.N.
June 1971

To the memory of my beloved
late father, Benjamin Nemerow

CONTENTS

PART 1   BASIC KNOWLEDGE AND PRACTICES

Chapter 1   Effect of Wastes on Streams and Waste-Water Treatment Plants
- 1.1   Effects on streams — 3
- 1.2   Effects on sewage plants — 7

Chapter 2   Stream Protection Measures
- 2.1   Standards of stream quality — 11
- 2.2   Stream quality control — 15

Chapter 3   Computation of Organic Waste Loads on Streams
- 3.1   Streeter-Phelps formulations — 26
- 3.2   Thomas method for determining pollution-load capacity of streams — 29
- 3.3   Churchill method of multiple linear correlation — 37

Chapter 4   Stream Sampling — 43

Chapter 5   Economics of Waste Treatment
- 5.1   Benefits of pollution abatement — 49
- 5.2   Measurement of benefits — 50
- 5.3   A proposed method for resource allocation — 51

PART 2   THEORIES

Chapter 6   Volume Reduction
- 6.1   Classification of wastes — 61
- 6.2   Conservation of waste water — 61
- 6.3   Changing production to decrease wastes — 62
- 6.4   Reusing both industrial and municipal effluents for raw water supplies — 62
- 6.5   Elimination of batch or slug discharges of process wastes — 65

Chapter 7   Strength Reduction
- 7.1   Process changes — 68
- 7.2   Equipment modifications — 69
- 7.3   Segregation of wastes — 69
- 7.4   Equalization of wastes — 70
- 7.5   By-product recovery — 70
- 7.6   Proportioning wastes — 72
- 7.7   Monitoring waste streams — 72

Chapter 8   Neutralization
- 8.1   Mixing wastes — 73
- 8.2   Limestone treatment for acid wastes — 74
- 8.3   Lime-slurry treatment for acid wastes — 74
- 8.4   Caustic-soda treatment for acid wastes — 74
- 8.5   Using waste boiler-flue gas — 76
- 8.6   Carbon-dioxide treatment for alkaline wastes — 76
- 8.7   Producing carbon dioxide in alkaline wastes — 76
- 8.8   Sulfuric-acid treatment for alkaline wastes — 77
- 8.9   Acid-waste utilization in industrial process — 77

Chapter 9   Equalization and Proportioning
- 9.1   Equalization — 79
- 9.2   Proportioning — 80

Chapter 10	Removal of Suspended Solids			13.12	Well injection	124
10.1	Sedimentation	83		13.13	Foam phase separation	128
10.2	Flotation	87		13.14	Brush aeration	129
10.3	Screening	91		13.15	Subsurface disposal	129
				13.16	The bio-disc system	130
Chapter 11	Removal of Colloidal Solids			Chapter 14	Treatment and Disposal of Sludge Solids	
11.1	Characteristics of colloids	97		14.1	Anaerobic and aerobic digestion	135
11.2	Chemical coagulation	98		14.2	Vacuum filtration	137
11.3	Coagulation by neutralization of the electrical charges	98		14.3	Elutriation	138
11.4	Removal of colloids by adsorption	101		14.4	Drying beds	138
				14.5	Sludge lagooning	141
Chapter 12	Removal of Inorganic Dissolved Solids			14.6	The wet combustion process	141
				14.7	Atomized suspension	143
12.1	Evaporation	103		14.8	Drying and incineration	143
12.2	Dialysis	104		14.9	Centrifuging	144
12.3	Ion exchange	105		14.10	Sludge barging	147
12.4	Algae	106		14.11	Sanitary landfill	147
12.5	Reverse osmosis	108		14.12	Sludge pumping	147
12.6	Miscellaneous methods	108		14.13	Miscellaneous methods	148
Chapter 13	Removal of Organic Dissolved Solids			PART 3	APPLICATIONS	
13.1	Lagooning	110		Chapter 15	Joint Treatment of Raw Industrial Wastes with Domestic Sewage	
13.2	Activated-sludge treatment	113				
13.3	Modified aeration	114		15.1	Industrial use of municipal sewage plants	154
13.4	Dispersed-growth aeration	114				
13.5	Contact stabilization	116		15.2	Municipal ordinances	156
13.6	High-rate aerobic treatment	117		15.3	Sewer-rental charges	157
13.7	Trickling filtration	119		15.4	Existing situation	161
13.8	Spray irrigation	122		15.5	Stream survey	162
13.9	Wet combustion	122		15.6	Composite waste sampling	169
13.10	Anaerobic digestion	122		15.7	Composite waste analyses	169
13.11	Mechanical aeration system	123		15.8	Laboratory pilot-plant studies	171

15.9	Literature survey	176		18.2	General stream and waste survey	207
15.10	Conclusions from study	176		18.3	Evaluation of survey results with regard to fish killings	208
15.11	Overall planning study conclusions	178		18.4	Preliminary conclusions and suggestions	212
15.12	Solids handling	183		18.5	Findings of the pickle-factory survey and detailed recommendations	213
15.13	Final design of the Gloversville-Johnstown joint sewage treatment plant	183		18.6	Effects of changes on factory and stream	214
15.14	Estimated costs and financing	185				
15.15	Application of the plan in practice	187				

Chapter 16 Joint Treatment of Partially Treated Industrial Wastes and Domestic Sewage

16.1	Ascertaining present plant capacity	189
16.2	Reducing the incoming load	189
16.3	Reevaluation of present plant and suggestions for additions	190

Chapter 17 Discharge of Completely Treated Wastes to Municipal Sewer Systems

17.1	The sampling program	194
17.2	Analyses of wastes	194
17.3	Plant-production study	195
17.4	Suggested in-plant changes to reduce waste	195
17.5	City waste-water treatment plant	199
17.6	Toxic limits for metals	200
17.7	Treatment of industrial wastes	201

Chapter 18 Discharge of Raw Wastes to Streams

18.1	The pickle-making process and its wastes	206

Chapter 19 Discharge of Partially Treated Industrial Waste Directly to Streams

19.1	Procedure	217
19.2	River studies	218
19.3	Pilot-plant results	221
19.4	Substitution of soluble sizing	224

Chapter 20 Discharge of Completely Treated Wastes to Streams

20.1	The problem	229
20.2	Stream studies	231
20.3	State decision	231
20.4	Poultry waste characteristics	232
20.5	The solution	234
20.6	Results	236

Chapter 21 Site Selection

21.1	Evaluation of cost-of-product basis	241
21.2	Tangible and intangible factors	241
21.3	The importance of long-term planning	243
21.4	Waste disposal as a critical factor	244

xii CONTENTS

	21.5	Water supply as a critical factor	245		23.8	Characteristics of brewery, distillery, and pharmaceutical wastes	330

21.5 Water supply as a critical factor 245
21.6 Site selection for atomic-energy plants 247

PART 4 MAJOR INDUSTRIAL WASTES

Chapter 22 The Apparel Industries
TEXTILE WASTES
- 22.1 Origin and characteristics of textile wastes 255
- 22.2 Treatment of textile wastes 258
- 22.3 Final waste treatment 265
  TANNERY WASTES
- 22.4 Origin and characteristics of tannery wastes 277
- 22.5 Treatment of tannery wastes 279
  LAUNDRY WASTES
- 22.6 Origin and characteristics of laundry wastes 291
- 22.7 Treatment of laundry wastes 292

Chapter 23 Food-Processing Industries
- 23.1 Introduction 297
  CANNERY WASTES
- 23.2 Origin of cannery wastes 299
- 23.3 Characteristics of cannery wastes 299
- 23.4 Treatment of cannery wastes 300
  DAIRY WASTES
- 23.5 Origin and characteristics of dairy wastes 316
- 23.6 Treatment of dairy wastes 317
  BREWERY, DISTILLERY, AND PHARMACEUTICAL WASTES
- 23.7 Origin of brewery, distillery, and pharmaceutical wastes 329
- 23.8 Characteristics of brewery, distillery, and pharmaceutical wastes 330
- 23.9 Treatment of brewery, distillery, and pharmaceutical wastes 331
  MEAT-PACKING, RENDERING, AND POULTRY-PLANT WASTES
- 23.10 Origin and characteristics of meat-packing wastes 340
- 23.11 Treatment of meat-packing wastes 342
- 23.12 Feedlot wastes 348
  BEET-SUGAR WASTES
- 23.13 Origin and characteristics of beet-sugar wastes 350
- 23.14 Treatment of beet-sugar wastes 351
  MISCELLANEOUS FOOD-PROCESSING WASTES
- 23.15 Coffee wastes 355
- 23.16 Rice wastes 358
- 23.17 Fish wastes 359
- 23.18 Pickle wastes 361
- 23.19 Soft-drink bottling wastes 361
- 23.20 Bakery wastes 363
- 23.21 Water-treatment-plant wastes 363

Chapter 24 The Materials Industries
WOOD FIBER INDUSTRIES
- 24.1 Pulp- and paper-mill wastes 365
- 24.2 Photographic wastes 397
  METAL INDUSTRIES
- 24.3 Steel-mill wastes 398
- 24.4 Other metal-plant wastes 412
- 24.5 Metal-plating wastes 415
- 24.6 Motor industry wastes 431

24.7	Iron-foundry wastes	434
	LIQUID MATERIALS INDUSTRIES	
24.8	Oil-field and refinery wastes	435
24.9	Fuel-oil wastes	453
24.10	Rubber wastes	453
24.11	Glass-industry wastes	459
24.12	Naval-stores wastes	459
	SPECIAL-MATERIALS INDUSTRIES	
24.13	Animal-glue manufacturing wastes	463
24.14	Wood-preservation wastes	463
24.15	Candle-manufacturing wastes	465
24.16	Plywood-plant glue wastes	466
Chapter 25	Chemical Industries	
25.1	Acid wastes	468
25.2	Cornstarch-industry wastes	473
25.3	Phosphate-industry wastes	476
25.4	Soap- and detergent-industry wastes	479
25.5	Explosives-industry wastes	484
25.6	Formaldehyde wastes	489
25.7	Pesticide wastes	491
25.8	Plastic and resin wastes	493
Chapter 26	Energy Industries	
26.1	Steam power plants	501
26.2	The coal industry	509
Chapter 27	Radioactive Wastes	
27.1	Origin of wastes	533
27.2	Power-plant wastes	533
27.3	Fuel-processing wastes	536
27.4	Treatment of radioactive wastes	540
27.5	Cost of radioactive-waste treatment	548
	Index	577

President Lyndon B. Johnson, when he signed into law the Water Quality Act of 1965, said:

> *No one has a right to use America's rivers and America's waterways that belong to all the people as a sewer. The banks of a river may belong to one man or one industry or one state, but the waters which flow between those banks should belong to all the people.*

In his January 1967 economic report to Congress, Johnson said:

> *A polluted environment erodes our health and well-being. It diminishes individual vitality; it is costly to industry and agriculture; it has debilitating effects on urban and regional development; it takes some of the joy out of life. The 89th Congress enacted important legislation to improve the quality of our environment. All 50 states have now signified their intention to establish water quality standards for their interstate and coastal waters. The Federal Government is assisting state and local governments through comprehensive water basin planning, and is providing financial help to states for the administration of water pollution control and to local areas for the construction of sewage treatment facilities. In addition, we are studying appropriate methods to encourage industry to control its discharge of pollutants.*

Again, in his 1968 Report to the Nation on Water Resources, Johnson said:

> *A nation that fails to plan intelligently for the development and protection of its precious waters will be condemned to wither because of its shortsightedness. The hard lessons of history are clear, written on the deserted sands and ruins of once-proud civilizations.*

Part 1 | BASIC KNOWLEDGE AND PRACTICES

CHAPTER 1

EFFECT OF WASTES ON STREAMS AND WASTE-WATER TREATMENT PLANTS

1.1 Effects on Streams

All industrial wastes affect, in some way, the normal life of a stream [11].* When the effect is sufficient to render the stream unacceptable for its "best usage," it is said to be polluted. Best usage means just what the words imply: use of water for drinking, bathing, fishing, and so forth. A more detailed description of these uses is given in Chapter 2.

Streams can assimilate a certain quantity of waste before reaching a polluted state. Generally speaking, the larger, swifter, and more remote streams that are not much used are able to tolerate a considerable amount of waste, but too much of any type of polluting material causes a nuisance. To call a stream polluted, therefore, generally means that the stream contains an excessive amount of a specific pollutant or pollutants. The following materials can cause pollution:

Inorganic salts	Heated water
Acids and/or alkalis	Color
Organic matter	Toxic chemicals
Suspended solids	Microorganisms
Floating solids and liquids	Radioactive materials
	Foam-producing matter

Inorganic salts, which are present in most industrial wastes as well as in nature itself, cause water to be "hard" and make a stream undesirable for industrial, municipal, and agricultural usage. We mention here just a few of the hundreds of difficulties arising from the use of hard water.

Salt-laden waters deposit scale on municipal water-distribution pipelines, increasing resistance to flow and lowering the overall capacity of the lines.

Hard waters interfere with dyeing in the textile industry, brewing in the beer industry, and quality of the product in the canning industry. Magnesium sulfate, a particularly bothersome constituent in hard waters, has a cathartic effect on people. The chloride ion increases the conductance of electrical insulating paper; iron causes spots and stains on white goods manufactured by textile mills and on high-grade papers produced by paper mills; and carbonates produce a hard scale on peas processed in canneries. Most types of hard water encrust boiler tubes, so that transfer of heat to the water from the fire chamber is impaired. This condition, called "boiler scale," results in lowered boiler efficiency and increased cost of operation.

Another disadvantage is that, under proper environmental conditions, inorganic salts, especially nitrogen and phosphorus, induce the growth of microscopic plant life (algae) in surface waters. Although algae are really a secondary form of pollution, they can be of extreme importance. Their advantage is that of adding dissolved oxygen to the stream; their disadvantage is the organic loading they contribute after dying. Too little attention is given by industrial waste engineers to these inorganic products of waste liquors. The role of phosphorus is diverse and complicated, but it is known that in the absence of phosphorus there is practically total elimination of algae life.

There is another facet of the problem worth noting: a total absence of salts is apt to result in corrosive and/or tasteless water, whereas a certain degree of hardness enhances the development of a protective film on surfaces and renders water more palatable. Producers of baked goods, for instance, feel that some concentration of calcium

---

*Numbers in brackets refer to the bibliographical references at the end of each chapter.

sulfate helps to achieve a golden brown crust on bread. It is therefore desirable that *some* inorganic salts be present in the water supply. The amount, rather than the presence, is the important factor.

A rather different form of pollution may exist along the coasts of southern California and Florida, as well as in parts of Texas and Arizona, where excessive withdrawal of ground water has allowed subterranean intrusion of salt water into previously fresh-water aquifers.

Acids and/or alkalis discharged by chemical and other industrial plants make a stream unsuitable not only for recreational uses such as swimming and boating, but also for propagation of fish and other aquatic life. High concentrations of sulfuric acid, sufficient to lower the pH to below 7.0 when free chlorine is not present, have been reported to cause eye irritation to swimmers, rapid corrosion of ships' hulls, and accelerated deterioration of fishermen's nets. The toxicity of sulfuric acid for aquatic life is a function of the resulting pH; i.e., a dose that would be lethal in soft water may be quite harmless in hard or highly buffered water. It is generally agreed that the pH of a stream must be not less than 4.5 and not more than 9.5 if fish are to survive. Yet stream pH values from as low as 2 to as high as 11 may occur near industrial sources of pollution.

Sodium hydroxide—to cite an example of an alkali—is highly soluble in water and affects the alkalinity and pH. It appears in wastes from many industries, including soap manufacturing, textile dyeing, rubber reclaiming and leather tanning. Streams containing as little as 25 parts of sodium hydroxide per million have been reported deadly to fish. Alkali in boiler-feed water can, by its caustic action, cause caustic embrittlement of pipes. Water-treatment plants are also adversely affected by these pollutants; for example, treatment plants using alum as a coagulant often find shock loads of acid or alkali interfering with floc formation.

Some miscellaneous processes affected by using waters of certain pH values are the rate of industrial fermentation, quality of dough in baking, flavor in soft drinks, yeast activity in brewing of beer, taste of canned fruits, especially tomatoes, cleaning of industrial metals, and gelatin and glue manufacture. A low pH may cause corrosion in air-conditioning equipment, and a pH greater than 9.5 enhances laundering.

Organic matter exhausts the oxygen resources of rivers and creates unpleasant tastes, odors, and general septic conditions. Fish and most aquatic life are stifled by lack of oxygen, and the oxygen level, combined with other stream conditions, determines the life or death of fish. It is generally conceded that the critical range for fish survival is 3 to 4 parts per million (ppm) of dissolved oxygen. We know that some species of fish may not survive in water containing 3 ppm of dissolved oxygen, while other species may not be affected even slightly by the same low oxygen level. For example, trout are sensitive fish, requiring oxygen concentrations of at least 5 ppm, whereas carp are scavenger fish, capable of surviving in waters containing as little as 1 ppm of oxygen. This oxygen shortage, caused by organic matter, is often considered to be the most objectionable single factor in a stream's pollution.

Certain organic chemicals, such as phenols, affect the taste of domestic water supplies. If rivers containing phenols permeate nearby wells, they cause objectionable medicinal tastes, and in addition there is the less-obvious organic matter, which may cause discomfort or diseases.

Suspended solids settle to the bottom or wash up on the banks and decompose, causing odors and depleting oxygen in the river water. Fish often die because of a sudden lowering of the oxygen content of a stream, and solids that settle to the bottom will cover their spawning grounds and inhibit propagation. Visible sludge creates unsightly conditions and destroys the use of a river for recreational purposes. These solids also increase the turbidity of the watercourse. Although each stream varies in the quantity of suspended solids it can safely carry away, most pollution-control authorities specify that suspended solids may be discharged into a stream only in amounts that will not impair the best usage of the stream.

Floating solids and liquids. These include oils, greases, and other materials which float on the surface; they not only make the river unsightly but also obstruct passage of light through the water, retarding the growth of vital plant food. Some specific objections to oil in streams are that it: (1) interferes with natural reaeration; (2) is toxic to certain species of fish and aquatic life; (3) creates a fire hazard when present on the water surface in sufficient amounts; (4) destroys vegetation along the shoreline, with consequent erosion; (5) renders boiler-feed and cooling water unusable; (6) causes trouble in conventional water-treatment processes by imparting tastes and odors to water and coating sand filters with a tenacious film; (7) creates an unsightly film on the surface of the water; and (8) lowers recreational, e.g. boating, potential.

Heated water. An increase in water temperature, brought about by discharging wastes such as condenser waters into streams, has various adverse effects. Stream waters which vary in temperature from one hour to the next are difficult to process effectively in municipal and industrial water-treatment plants, and heated stream waters are of decreased value for industrial cooling. Indeed, one industry may so increase the temperature of a stream that a neighboring industry downstream cannot use the water. Furthermore, warm water is lighter than cold, so that stratification develops, and this causes most fish life to retreat to stream bottoms. Since there may be less dissolved oxygen in warm water than in cold, aquatic life suffers, and less oxygen is available for natural biological degradation of any organic pollution discharged into these warm surface waters. Also, bacterial action increases in higher temperatures, resulting in accelerated depletion of the stream's oxygen resources.

Color, contributed by textile and paper mills, tanneries, slaughterhouses and other industries, is an indicator of pollution. Compounds present in waste waters absorb certain wavelengths of light and reflect the remainder, a fact generally conceded to account for color development of streams. Color interferes with the transmission of sunlight into the stream and therefore lessens photosynthetic action. It may also interfere with oxygen absorption from the atmosphere—although no positive proof of this exists.

Visible pollution often causes more trouble for industry than invisible pollution. Unseen pollution which does not create a nuisance will often be tolerated by state agencies, but the red and deep-brown colors of slaughterhouse wastes, the browns of paper-mill wastes, various intense colors of textile-mill wastes, and the yellows of plating-mill wastes will focus public indignation directly on those industries. It is only human to complain about visible pollution: property values decrease along a visibly polluted river, and fewer people will swim, boat, or fish in a stream highly colored by industrial wastes. Furthermore, municipal and industrial water plants have great difficulty, and scant success, in removing color from raw water.

Toxic chemicals. Both inorganic and organic chemicals, even in extremely low concentrations, may be poisonous to fresh-water fish and other, smaller, aquatic microorganisms. Many of these compounds are not removed by municipal treatment plants and have a cumulative effect on biological systems. Such insecticides as toxaphene, dieldrin, and dichlorobenzene have allegedly killed fish in farm ponds and streams. Insecticides used in cotton and tobacco dusting have their maximum effect following heavy rainfalls—i.e. they are more lethal in solution—but insecticides and rodenticides are hard to detect in a stream. However, newer techniques, e.g. electron-capture gas chromatography, can detect chlorinated hydrocarbon pesticides in concentrations of 0.001 micrograms per liter in one-liter samples of water.

New, highly complex, organic compounds produced by the chemical industry for textile and other companies have also proved extremely toxic to fish life. One example is acrylonitrile, a raw material used in the manufacture of certain new synthetic fibers.

Almost all salts, some even in low concentrations, are toxic to certain forms of aquatic life. Thus, chlorides are reportedly toxic to fresh-water fish in 400 ppm concentration, as are hexavalent chromium compounds in concentrations of 5 ppm. Copper concentrations as low as 0.1 to 0.5 ppm are toxic to bacteria and other microorganisms. Although oyster larvae, for setting, *require* a copper concentration of about 0.05 to 0.06 ppm, concentrations above 0.1 to 0.5 ppm are toxic to some species. All three salts are often found in watercourses.

Accidental or intermittent discharge of certain toxic materials may go unnoticed and yet may com-

Table 1.1 Limits set on contents of chemical elements or compounds in water supplies [1].

Characteristic	Natural mandatory limit, ppm	Recommended limit, ppm
Lead	0.1	
Fluoride	1.5	0.7–1.2
Arsenic	0.05	0.01
Selenium	0.05	
Chromium (hexavalent)	0.05	
Copper		1.0
Iron		0.3
Magnesium		125
Zinc		5
Chloride		250
Sulfate		250
Phenolic compounds, in terms of phenol		0.001
Total solids		
Desirable		500
Permitted		1000
Normal carbonate ($CaCO_3$)		120
Excess alkalinity over hardness ($CaCO_3$)		35
pH (25°C)		10.6
Alkyl benzene sulfonate		0.5
Carbon chloroform extract (CCE)		0.2
Cyanide (Cn)		0.01
Manganese (Mn)		0.05
Nitrate (NO_3)		45
Strontium 90		10 $\mu\mu$c/liter
Radium 226		3 $\mu\mu$c/liter
Gross β radiation concentration		1000 $\mu\mu$c/liter

pletely disrupt stream life. Building-floor and stormwater drains that lead directly to the stream may convey contamination because of an upset in an industrial process or ignorance of the consequences. For example, the flushing of a chemical delivery tank at the unloading dock may carry dissolved toxic material into the stream through a storm drain.

Complex inorganic phosphates, such as P_2O_5, at levels as low as 0.5 ppm, perceptibly interfere with normal coagulation and sedimentation processes in water-purification plants. Increased coagulant dosages and/or increased settling times are required [9] to solve the problem. Phenols in concentrations exceeding one part per billion have been found to be objectionable in a stream. Phenol reacts with chlorine and, even in extremely small quantities, gives the residual drinking water a noticeable medicinal taste. Table 1.1 lists water-quality limits recommended by the U.S. Public Health Service [1].

Microorganisms. A few industries, such as tanneries and slaughterhouses, sometimes discharge wastes containing bacteria. Vegetable and fruit canneries may also add bacterial contamination to streams. These bacteria are of two significant types: (a) Bacteria which assist in the degradation of the organic matter as the waste moves downstream. This process may aid in "seeding" a stream (deliberate inoculation with biological life for the purpose of degrading organic matter) and in accelerating the occurrence of oxygen sag in the water. (b) Bacteria which are pathogenic, not only to other bacteria, but also to humans. An example is the anthrax bacillus, originating in tanneries where hides from anthrax-infected animals have been processed.

Radioactive materials. The manufacture of fissionable materials, the increasing peacetime use of atomic energy, and the projected development of atomic-power facilities have introduced new complications in the field of sanitary engineering. The problem of disposing of radioactive wastes is unique [2], since the effects of radiation can be immediate or delayed, and radiation is an insidious contaminant with cumulative damaging effects on living cells. Certain highly active radioisotopes such as Sr^{90} and Cs^{137} continue to release energy over long periods of time (several generations of the human race). This radiation is not readily detectable by the methods usually employed to determine the presence of contaminants in the environment. Furthermore, the biological and hydrological characteristics of a stream may have a profound influence on the uptake of radioactivity.

At present, the maximum safe concentration of mixed fission products for lifetime consumption, according to the Atomic Energy Commission, is 1×10^{-7} microcuries per milliliter. Therefore, regulatory agencies, as well as the public, are concerned about preventing contamination of surface streams by radioactive wastes [13].

Foam-producing matter, such as is discharged by textile mills, pulp and paper mills, and chemical plants, gives an undesirable appearance to the receiving stream. It is an indicator of contamination and is often more objectionable in a stream than lack of oxygen. More court cases have been fought and won on evidence about the appearance of a stream than about the unseen contents of the water. (This in itself should serve as a warning to industries discharging foam-producing wastes.)

1.2 Effects on Sewage Plants

It is only natural for industry to presume that its wastes can best be disposed of in the domestic sewer system, and municipal officials often feel that it is their responsibility to accept any wastes flowing into their city's disposal system. However, city authorities should not accept any waste discharges into the domestic sewer system without first learning the facts about the characteristics of the wastes, the sewage system's ability to handle them, and the effects of the wastes upon *all* components of the city disposal system. Institution of a sewer ordinance, restricting the types or concentrations of waste admitted in the sewer leading to a treatment plant, is one means of protecting the system.

To remove pollution from industrial wastes, a sewage-treatment plant must have sufficient capacity and of the proper type. Theoretically, a sewage-treatment plant could be designed to handle any type of industrial waste, but present plants fall short of this ideal. Joint treatment of municipal and industrial waste waters which are amenable to treatment may offer greater *removal* efficiencies, but economics will usually be the deciding factor.

The pollutional characteristics of wastes having readily definable effects on sewers and treatment plants can be roughly classed as follows: (1) biochemical oxygen demand (BOD); (2) suspended solids; (3) floating and colored materials; (4) volume; and (5) other harmful constituents. Table 1.2 presents a comparison of domestic-sewage pollutional characteristics and those of some industrial wastes.

Table 1.2 General comparison of pollutional loads in industrial wastes versus domestic sewage [14].

Origin of waste	Population equivalent*	
	Biochemical oxygen demand	Suspended solids
Domestic sewage	1	1
Paper-mill waste	16–1330	6100
Tannery waste	24–48	40–80
Textile-mill waste	0.4–360	130–580
Cannery waste	8–800	3–440

*Persons per unit of daily production.

Biochemical oxygen demand (BOD) is usually exerted by dissolved and colloidal organic matter and imposes a load on the biological units of the treatment plant. Oxygen must be provided so that bacteria can grow and oxidize the organic matter. An added BOD load, caused by an increase in organic waste, requires more bacterial activity, more oxygen, and greater biological-unit capacity for its treatment. This calls for an increase in both capital outlay and daily operating expense.

However, not all dissolved or colloidal organic matter oxidizes at the same rate, with the same ease, or to the same degree. Sugars, for example, are more readily oxidized than starches, proteins, or fats. The rate of decomposition for industrial organic matter may therefore be faster or slower than that for sewage organic matter, and this difference must be considered in the design and operation of biological units. Before private industry embarks on a joint disposal venture with the city, the oxidizability of industrial wastes should be determined by the use of Warburg or other similar respirometer tests, which instantaneously measure the oxygen utilized and the carbon dioxide evolved by various solutions.

Figure 1.1 illustrates one possible effect of a given industrial waste on a sewage plant. In this instance the industrial waste, with its constant rate of degradation, tends to smooth out the rate of decomposition of the sewage so that the result shows less upsurge due to nitrogenation. Also, the rate of decomposition of the industrial waste tends to slow down the initial rapid rate of domestic sewage.

After much experimentation, Ettinger [3] believes that there is still some doubt whether activated-sludge biological units are able to handle slug waste discharges better than digesters, which have the advantages of inherent storage capacity and assumed complete mixing.

Suspended solids are found in considerable quantity in many industrial wastes, such as cannery and paper-mill effluents. They are screened and/or settled out of the sewage at the disposal plant. Solids removed by settling and separated from the flowing sewage are called *sludge*, which may then undergo an anaerobic decomposition known as digestion and be pumped to drying beds or vacuum filters for extraction of additional water. Certain settleable suspended solids from industrial wastes, e.g. fine grit and insoluble metal precipitates, may hinder sludge digestion.

Suspended solids in industrial waste may settle more rapidly or slowly than sewage suspended matter. If industrial solids settle faster than those of municipal sewage, sludge should be removed at shorter intervals to prevent excessive build-up. Quantities of stale sludge may be "scoured" (dislodged by physical means) off the bottom of the basin, with resultant increase of sludge in the effluent. A faster-

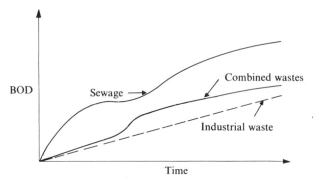

Fig. 1.1 Deoxygenation rates of sewage, a certain industrial waste, and a combination of the two.

settling industrial waste may accelerate the settling of sewage solids; a slower-settling one will require a longer detention period and larger basins and increase the likelihood of sludge decomposition, with accompanying nuisances, during slack sewage-flow periods. However, regardless of the settling rate, the quantity of sludge to be pumped to the sludge-disposal facilities at the treatment plant will be increased by the addition of such industrial waste. Since sludge digesters, drying beds, and filters are designed to handle a certain number of pounds of solids per unit of capacity, any increased demands on the system usually require larger sludge-handling devices and may ultimately necessitate an increase in the plant's capacity, with resulting higher capital and operating expenses.

The settling characteristics of industrial wastes, alone and combined with municipal waste, should be determined before any disposal agreement between industry and city. Sludge consistency, percentage of total suspended solids removed, and weight of suspended solids removed are the criteria for evaluating settling characteristics.

Floating materials and colored matter, such as oil, grease, and dyes from textile-finishing mills, are disagreeable and visible nuisances. Visible pollution retards the development of a community or area, since it discourages camping, boating, swimming, and fishing—recreations indispensable to the vitality of a physically and mentally healthy community—and industry is reluctant to locate on a stream which is visibly polluted. Lack of industry further depresses the growth of city, county, and state, for less tax money means less progress. It is therefore imperative that nuisances such as color and floating matter be removed by the sewage-treatment plant.

A modern treatment plant will remove normal grease loads in primary settling tanks, but abnormally high loads of predominantly emulsified greases from laundries, slaughterhouses, rendering plants, and so forth, passing through the primary units (screens, grit chambers, and settling basins) into the biological units, will clog flow-distributing devices and air nozzles. A lengthy shutdown of these units may result in stream pollution and sudden loss of fish life.

Color removal by the treatment units of sewage plants is a knotty problem, and too little effort has been made so far to find an effective solution. The author found that trickling-filter plants in North Carolina were removing between 34 and 44 per cent of the dye color in the influent [12]. An overloaded primary plant, on the other hand, *added* 12 per cent to the color as waste passed through the plant. A knowledge of the character and measurement of color is essential. Since most colored matter is in a dissolved state, it is not altered by conventional primary devices, although secondary treatment units, such as activated sludge and trickling filters, remove a certain percentage of some types of colored matter. Sewage-treatment plants are generally not designed to remove color, so any reduction in this constituent is a fortunate coincidence, but, because of the previously mentioned detriment to streams, municipal disposal plants should in the future give increased consideration to removal of color. If an industry defines the type and quantity of colored matter in its waste, engineers can then make some prediction concerning the effectiveness of color removal by the treatment designed for domestic sewage.

Volume. A sewage plant can handle any volume of flow if its units are sufficiently large. Unfortunately, most sewage plants are already in operation when a request comes to accept the flow of waste from some new industrial concern. The hydraulic capacity of all units must then be analyzed; sewer lines must be examined for carrying capacity, bar screens for horizontal flow velocity, settling basins for detention periods and surface and weir overflow rates, trickling filters for excessive hydraulic loadings, and so forth.

An industry with a relatively clean waste such as condenser water can usually discharge it, after cooling, directly into the receiving stream and thus avoid overloading the sewage-treatment plant. This expedient saves capital and operating expenses at the disposal plant. However, before seemingly clean waters are accepted for direct disposal, they must be carefully examined for dissolved solids. Even a small concentration of solids in a large volume of waste water will sometimes result in a significant total-solids load.

Other harmful constituents. Industrial wastes may contain harmful ingredients in addition to the polluting load. These wastes can cause malfunctioning of the sewer system and/or the disposal plant. Some nuisances and their accompanying effects are:

(a) Toxic metal ions (Cu^{++}, Cr^{+6}, Zn^{++}, CN^-), which interfere with biological oxidation by tying up enzymes required to oxidize organic matter.
(b) Feathers, which clog nozzles, overload digesters, and impede proper pump operation.
(c) Rags, which clog pumps and valves and interfere with proper operation of bar screens or comminutors.
(d) Acids and alkalis, which may corrode pipes, pumps, and treatment units, interfere with settling, upset the biological purification of sewage, release odors, and intensify color.
(e) Inflammables, which cause fires and may lead to explosions.
(f) Pieces of fat, which clog nozzles and pumps and overload digesters.
(g) Noxious gases, which present a direct danger to workers.
(h) Detergents, which cause foaming of aeration units.
(i) Phenols and other toxic organic material.

References

1. "Drinking water standards," U.S. Public Health Service, *Public Health Rept.* **61**, 371 (1946); (revised) U.S. Public Health Service Publication no. 956, Washington, D.C. (1962).

2. Dugan, P. R., R. M. Pfister, and M. L. Sprague, *Bibliography of Organic Pesticides—Publications Having Relevance to Public Health and Water Pollution Problems*, Prepared for the New York State Department of Health by Microbiology and Biochemical Center, Syracuse University Research Corporation, Syracuse, N.Y. (1963).
3. Ettinger, M. B., "Heavy metals in waste-recovery systems," Paper read at Interdepartmental Water Resources Seminar, March 1963, at Ohio State University, Columbus.
4. Gibbs, C. V., and R. H. Bòthel, "Potential of large metropolitan sewers for disposal of industrial wastes," *J. Water Pollution Control Federation* **37**, 1417 (1965).
5. Gorman, A. E., "Waste disposal as related to site selected," Preprint 3, American Institute of Chemical Engineers Meeting, December 12–16, 1955.
6. Huet, M., "Water quality criteria for fish life," Paper read at Third Seminar on Biological Problems in Water Pollution, 13–17 August, 1962, U.S. Public Health Service Publication no. 999-WP-25, Washington, D.C. (1965), p. 160.
7. Jones, E., *Fish and River Pollution*, Butterworths, London (1964).
8. *Modern pH and Chlorine Control*, 19th ed., W. A. Taylor & Co., Baltimore (1966), pp. 57–103.
9. Moss, H. V., "Continuing research related to detergents in water and sewage treatment," *Sewage Ind. Wastes* **29**, 1107 (1967).
10. National Technical Advisory Committee, *Interior Report to the Federal Water Pollution Control Administration on Water Quality Criteria, June 30, 1967*, U.S. Department of the Interior, Washington, D.C. (1967).
11. Nemerow, N. L., *Water Wastes of Industry*, Bulletin no. 5, Facts for Industry Series, Industrial Experiment Program, North Carolina State College, Raleigh (1956).
12. Nemerow, N. L., and T. A. Doby, "Color removal in waste water treatment plants," *Sewage Ind. Wastes* **30**, 1160 (1958).
13. Palange, R. C., G. G. Robeck, and C. Henderson, "Radioactivity as a factor in stream pollution," Preprint 190, American Institute of Chemical Engineers Meeting, December 12–16, 1955.
14. "Survey of the Ohio river," in *Industrial Wastes Guides*, Supplement D, U.S. Public Health Service, Washington, D.C. (1943).
15. Tarzwell, C. M., "Water quality criteria for aquatic life," in *Biological Problems in Water Pollution*, U.S. Department of Health, Education and Welfare, Cincinnati (1957), pp. 246–272.
16. Tarzwell, C. M., "Dissolved oxygen requirement for fishes," in *Oxygen Relationships in Streams*, Report W58-2, U.S. Public Health Service, Cincinnati (1958), pp. 15–24.
17. Wurtz, C. B., "Misunderstandings about heated discharges," *Ind. Water Eng.* **4**, 28 (1967).

CHAPTER 2

STREAM PROTECTION MEASURES

Streams serve people in many ways, and the carrying away of pollution certainly is one of the chief services performed. However, there are other more important uses of stream waters: drinking, bathing, fishing, irrigation, navigation, recreation, and power. A stream must therefore be protected, so that it can serve the best interests of the people using it.

2.1 Standards of Stream Quality

The methods of maintaining a stream in acceptable condition range from very flexible control, in which individuals make decisions about waste treatment, to rigid control by laws specifying stream or effluent standards. Because of the variations in procedure under the flexible control system, the method cannot be described in any detail. Normally, the state regulatory agency demands certain types and degrees of waste treatment based on the best interests of the persons living along the stream. An advantage of such a loose procedure is the freedom to make immediate decisions; a danger is the lack of representation of all the groups involved in the use of the stream. In addition, treatment may be different from one location to another, with resultant unfairness. However, nonuniformity can also be an advantage; for example, a plant located on a stream above the water-supply intake of a municipality would be expected to provide more complete treatment than a similar plant discharging waste into the same stream, but below the water-supply intake.

There are two schools of thought in the United States in regard to rigid protection: one group prefers "effluent standards" and the other "stream standards." The first system requires that, in all effluents from a certain type of industry, the waste discharged be kept below either a fixed percentage or a certain maximum concentration of polluting matter. A disadvantage to this approach is that there is normally no control over the total volume of polluting substance added to the stream each day. The large industry, although providing the same degree of waste treatment as the small one, may actually be responsible for a major portion of the pollution in the stream. It might be argued, however, that larger industries, by virtue of their value to the area, should be allocated a larger portion of the assimilating capacity of a stream.

The effluent-standard system is easier to control than the stream-standard system. No detailed stream analyses are needed to determine the exact amount of waste treatment required, and effluent standards can serve as a guide to a state in stream classification or during the organization of any pollution-abatement program. On the other hand, unless the effluent standards are upgraded, this system does not provide any effective protection for an overloaded stream. Standards for effluents are based more on economics and practicability of treatment than on absolute protection of the stream; the best usage of the stream is not the primary consideration. Rather the usage of the stream will depend on its condition after industrial-effluent standards have been satisfied. Upgrading and conservation of natural resources are somewhat neglected in favor of industrial economics.

The stream-standard system is based on establishing classifications or standards of quality for a stream and regulating any discharge into it to the extent necessary to maintain the established stream classification or quality. The primary motive of stream standards is to protect and preserve each stream for its best usage, on an equitable basis for both upstream and downstream users, although the upstream user often possesses a decided advantage because of location. A long and involved process of stream classification usually precedes any decision about waste treatment. Streams are classified in a manner set forth by state laws, sampled and analyzed for existing pollution, and surveyed for present and potential usages. The regulatory agency, after holding a public hearing to listen to comments from interested parties concerning the best usage, then decides on the highest usage of the stream, or section of stream. Naturally, the higher the classification, the cleaner the stream must be, with a resulting greater degree of waste treatment required. Formal notification is served on each polluter, giving a time limit for positive remedial action to maintain the stream in its classification. Implementation of the law becomes a matter of education, persuasion, public pressure, and sometimes action from the attorney general's office. It is left to each industry to decide the type and extent of treatment it must give to its waste to meet stream standards, although industries generally communicate directly with pollution-control authorities to determine what will be acceptable to them, since the pollution-control agency must review and approve the final construction plans for waste-treatment plants.

The main advantage of the stream-standard system is the prevention of excessive pollution, regardless of the type of industry or other factors such as the location of industries and municipalities. It also allows the public to establish goals for present and future water quality [10]. Loading is limited to what the stream can assimilate, and this may impose hardship on an industrial plant located at a critical spot along the stream. On the other hand, pollution abatement should be considered in decisions concerning the location of a plant, just as carefully as labor, transportation, market, and other conditions.

The difficulties in carrying to completion a classification system based on stream standards are:

1) Confusion when zones of different classifications are straddled by the waste: Does Waste 1 need to be treated to meet Class C or B stream standards?

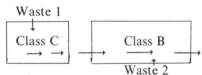

2) Controversy over the proportion of the stream to be reserved for future industrial, municipal, agricultural, and other uses.

3) Red tape and opposition from industry or the public to changes in the established classifications (upgrading or downgrading of a zone).

4) The need for a complex and thorough stream survey prior to classification. This can be costly, cumbersome, and result in delays.

On the brighter side is the security felt by industry, municipalities, and engineers, who know exactly the stream conditions for which they must design their plants. Industry will usually find a method of solving its problem *if* the extent of the situation and the precise degree of treatment required are known, although economics may still influence the action of the industry. Some states such as North Carolina have built in a degree of flexibility that would be unobtainable with effluent standards [5].

New York State has put a stream-standard method into practice [8] with the following classifications:

Fresh Water

Class AA—*Drinking water* after approved disinfection, with additional treatment, if necessary, to remove naturally present impurities.

Class A—*Drinking water* if subjected to approved treatment equal to a minimum of coagulation, sedimentation, filtration, and disinfection, plus additional treatment for natural impurities, if necessary.

Class B—*Bathing*, or any lesser use.

Class C—*Fishing*, or any lesser use.

Class D—*Agricultural, industrial cooling*, or *industrial process water,* or any lesser use.

Tidal Salt Waters

Class SA—*Shellfishing* for market purposes, and any other usages.

Class SB—*Bathing*, and any other usages except shellfishing for market purposes.

Class SC—*Fishing* and other usages, except bathing or shellfishing for market purposes.

Class SD—*Any usages* except fishing, bathing, and shellfishing.

Underground Waters

Class GA—*Drinking, culinary*, or *food processing*, and any other usages.

Class GB—*Industrial* or other water supply, and any other usages, except as used in GA.

2.1 STANDARDS OF STREAM QUALITY

Table 2.1 Water standards for various uses.

Water quality	Recreation and aesthetic	Public water supply		Fish and aquatic wildlife			Agriculture		
		Permissible	Desirable	Fresh water organisms	Wild life	Marine and estuarine organisms	Farm water supplies	Livestock	Irrigation
Color, units		75	<10	10% of light penetrating to bottom	10% of light penetrating 6 ft.				
Temperature, °F	<85°	<85°	<85°	83–96° for 6 hr.					55–85°
Fecal coliform, no/100 ml	2000–200	2000	20						4000
Alkalinity (CACO$_3$), ppm		30–500	30–500	>20	35–200	35–200			
Chloride, ppm		250	25						
Hexavalent chromium, ppm		0.05	Absent				0.05	0.05	5–20
Copper, ppm		1.0	Absent				1.0		0.2–5.0
Dissolved oxygen, ppm		>3.0	Near to saturation	>4.0	Bottom aerobic	>4.0			
Hardness (CACO$_3$), ppm		300–500	60–120						
Iron, ppm		0.3	Virtually absent				0.3		
Manganese, ppm		0.05	Absent				0.05		2.0–20
Nitrates, ppm		10.0(N) Ind. NO$_2$	Virtually absent				45.0		
pH	5.0–9.0	6.0–8.5		6–9	7.0–9.2	6.5–8.5	6.0–8.5		4.5–9.0
Sulfate, ppm		250	50						
Total dissolved solids, ppm		500	200				500–5000	10,000	0–5000
Carbon chloroform extractable, ppm		0.15	0.04				0.0001–0.02		
Pesticide, ppm		0.001–0.1	Absent	Varies with organism	Varies with organism	Varies with organism			
Phenol, ppm		0.001	Absent						
Gross beta radioactivity, μμc/liter		1000	100	1000	1000	1000	1000	1000	1000
Cyanide, ppm		0.20	Absent				0.20		
Turbidity, ppm			Virtually absent	10–50					

Table 2.2 Water quality standards for industrial use.

Industry and process	Color, ppm	Alkalinity, ppm (CaCO$_3$)	Chloride, ppm	Hardness, ppm (CaCO$_3$)	Iron, ppm	Manganese, ppm
Textiles						
Size	5			25	0.3	0.05
Scouring	5			25	0.1	0.01
Bleaching	5			25	0.1	0.01
Dyeing	5			25	0.1	0.01
Paper						
Mechanical	30		1000		0.3	0.1
Chemical						
Unbleached	30		200	100	1.0	0.5
Bleached	10		200	100	0.1	0.05
Chemicals						
Alkali and chlorine	10	80		140	0.1	0.1
Coal tar	5	50	30	180	0.1	0.1
Organic	5	125	25	170	0.1	0.1
Inorganic	5	70	30	250	0.1	0.1
Plastic and resin	2	1.0	0	0	0.005	0.005
Synthetic rubber	2	2	0	0	0.005	0.005
Pharmaceutical	2	2	0	0	0.005	0.005
Soaps and detergents	5	50	40	130	0.1	0.1
Paints	5	100	30	150	0.1	0.1
Gum and wood	20	200	500	900	0.3	0.2
Fertilizer	10	175	50	250	0.2	0.2
Explosives	8	100	30	150	0.1	0.1
Petroleum			300	350	1.0	
Iron and steel						
Hot-rolled						
Cold-rolled						
Miscellaneous						
Fruit and vegetable canning	5.0	250	250	250	0.2	0.2
Soft drinks	10	85			0.3	0.05
Leather tanning	5		250	150	50	
Cement		400	250		25	0.5

Water-quality surveillance programs making use of monitoring networks have been used to measure such contaminants and ingredients of the air and water as dissolved oxygen, chlorides, pH, conductivity, oxidation-reduction potential, temperature, and solar radiation. In an increasing number of cases (such as ORSANCO) these programs have resulted in an effective improvement in the receiving water quality.

Once a stream has been classified, the regulatory agency attempts to maintain that classification, or better. Under this system, the burden of proof rests on the engineer designing the waste-treatment works. He must show that the plant's waste will not alter the classification of the stream into which the plant discharges. The regulatory agency is responsible to the public for reporting on the existing quality of streams and pointing out sources of significant pollution.

Table 2.2 (*continued*)

NO$_3$, ppm	pH	SO$_4$, ppm	Total dissolved solids, ppm	Suspended solids, ppm	SiO$_2$, ppm	Ca, ppm	Mg, ppm	HCO$_3$, ppm
	6.5–10		100	5.0				
	3.0–10.5		100	5.0				
	2.0–10.5		100	5.0				
	3.5–10		100	5.0				
	6–10							
	6–10			10	50	20	12	
	6–10			10	50	20	12	
	6–8.5			10		40	8	100
	6.5–8.3	200	400	5		50	14	60
	6.5–8.7	75	250	5		50	12	128
	6.5–7.5	90	425	5		60	25	210
0	7.5–8.5	0	1.0	2.0	0.02	0	0	0.1
0	7.5–8.5	0	2.0	2.0	0.05	0	0	0.5
0	7.5–8.5	0	2.0	2.0	0.02	0	0	0.5
		150	300	10.0		30	12	60
	6.5	125	270	10		37	15	125
5	6.5–8.0	100	1000	30	50	100	50	250
5	6.5–8.0	150	300	10	25	40	20	210
2	6.8	150	200	5	20	20	10	120
	6.0–9.0		1000	10		75	30	
	5–9							
	5–9			10				
10	6.5–8.5	250	500	10	50	100		
	6.0–8.0	250				60		
	6.5–8.5	250	600	500	35			

Cleary warns that "to strive for stream cleanliness is one thing but to insist on striving to make all waters as pure as holy water introduces further delays with what is now the obvious problem—the gross and visible pollution" [2].

2.2 Stream Quality Control

The Water Quality Act of 1965 has amended the 1948 Federal Water Pollution Control Act, to provide for establishment of water-quality standards for interstate waters. Most states have decided to establish their own standards which the federal government is expected to accept in turn. The act specifically states that standards shall be such as to protect the public health or welfare and enhance the quality of water. In establishing these standards consideration will be given to the use and value of streams for public water supplies, propagation of fish and wildlife, recreational purposes, agricultural, industrial, and other legitimate uses. The act specifically states that "the discharge of matter into such interstate waters or portions thereof, which reduces the quality of such waters below the water quality standards established under this subsection (whether the matter causing or contributing to such reduction is discharged directly into such waters or reaches such waters after discharge into tributaries of such waters), is subject to abatement in accordance with provisions of..." the act. It also states that "Economic, health, esthetic, and conservation values which contribute to the social and economic welfare of an area must be taken into account in determining the most appropriate use or uses of a stream."

The federal government has established temporary quality guidelines for water used for (1) recreation, (2) public water supplies, (3) fish and wildlife, (4) agriculture, and (5) industry. A summary of the government's current limits is presented in Tables 2.1 and 2.2.

Illustrations of two current state quality standards are given by Grossman [4] for New York State in Table 2.3 and Rambow and Sylvester [10] for Washington State in Table 2.4. Current stream water quality standards are also presented for the New England Interstate Water Pollution Control Commission in Table 2.5 and for the Ohio River Sanitation Commission in Table 2.6. (These four tables appear at the end of the chapter.) The reader can study these governmental agency standards and notice the tendency for more stringent controls from both federal and state governments.

The state of Pennsylvania, practicing "effluent standards," has established (as an example) "raw waste" standards for pulp and paper mills (Table 2.7). These standards give the 5-day BOD and suspended solids per ton of product expected in wastes from well-run plants employing good housekeeping and recovery methods. The Sanitary Water Board has then applied the required effluent reduction for mills located on streams classified for primary or secondary treatment against the raw waste standard.

Four other procedures are becoming increasingly prevalent for improving stream water quality: (1) stream specialization, (2) stream aeration, (3) low-flow augmentation, and (4) pumped storage. These are described briefly below.

Stream specialization. This lets one stream become degraded so that others in the area are preserved in a relatively pristine state. The system is similar to that of stream classification except that it encourages the use of one stream as "an open sewer." Stream classification, on the other hand, is an attempt to label and use each stream for its best purpose and tends to upgrade all streams uniformly. Because social benefits are difficult to identify, all attempts to use streams in a discriminatory manner have been disappointing—most streams are found in a "great gray area."

Stream aeration. A novel approach to waste disposal for organic, decomposable-type wastes is to increase the amount of oxygen in the receiving streams by artificial means. Attempts are being made to use vertical pumps (similar to those used in deep ponds) to replenish the oxygen in the wastes. Some experiments have been carried out using pure oxygen rather than air, which appears to be rather costly at present. Although it apparently is less expensive to aerate the wastes than the streams which receive the wastes, a major exception occurs where a power dam is located across the stream. In this situation nearby industrial plants can, and have, used the draft tubes of large turbines to draw oxygen into the water flowing through them by gravity.

Low-flow stream augmentation. Storage of water is generally considered beneficial for handling large amounts of pollution. Dams and reservoirs built on the upland portion of the main stream or on one of the tributaries of the main stream can store water during high-flow periods to be released on a programmed schedule when the stream flow diminishes below critical values. However, this can be somewhat dangerous to the water quality, since low-oxygen water may be released during low stream flow, scouring of attached growth may occur during any sudden increase in volume, tributary flow may be temporarily retarded (leading to quality deterioration) during the flow increase in the main stream, and the temperature of the stream may rise if the released water is also used as cooling water in power plants.

Pumped storage is a relatively new and promising approach to drought control [13], which also has great potential for protecting streams from excessive

Table 2.3 New York State classes and standards for fresh surface waters.

Class and best use*	Water standards†				
	Minimum dissolved oxygen, ml/liter	Coliform bacteria median, no/100 ml	pH	Toxic wastes, deleterious substances, colored wastes, heated liquids, and taste- and odor-producing substances‡	Floating solids, settleable solids, oil, and sludge deposits
AA—Source of unfiltered public water supply and any other usage	5.0 (trout) 4.0 (nontrout)	Not to exceed 50	6.5–8.5	None in sufficient amounts or at such temperatures as to be injurious to fish life or make the waters unsafe or unsuitable	None attributable to sewage, industrial wastes or other wastes
A—Source of filtered public water supply and any other usage	5.0 (trout) 4.0 (nontrout)	Not to exceed 5000	6.5–8.5		
B—Bathing and any other usages except as a source of public water supply	5.0 (trout) 4.0 (nontrout)	Not to exceed 2400	6.5–8.5		None which are readily visible and attributable to sewage, industrial wastes or other wastes
C—Fishing and any other usages except public water supply and bathing	5.0 (trout) 4.0 (nontrout)	Not applicable	6.5–8.5	None in sufficient amounts or at such temperatures as to be injurious to fish life or impair the waters for any other best usage	
D—Natural drainage, agriculture, and industrial water supply	3.0	Not applicable	6.0–9.5	None in sufficient amounts or at such temperatures as to prevent fish survival or impair the waters for agricultural purposes or any other best usage	

*Class B and C waters and marine waters shall be substantially free of pollutants that: unduly affect the composition of bottom fauna; unduly affect the physical or chemical nature of the bottom; interfere with the propagation of fish. Class D and SD (marine) will be assigned only where a higher water use class cannot be attained after all appropriate waste-treatment methods are utilized. Any water falling below the standards of quality for a given class shall be considered unsatisfactory for the uses indicated for that class. Waters falling below the standards of quality for Class D, or SD (marine), shall be Class E, or SE (marine), respectively and considered to be in a nuisance condition.

†These Standards do not apply to conditions brought about by natural causes. Waste effluents discharging into public water supply and recreation waters must be effectively disinfected. All sewage-treatment plant effluents shall receive disinfection before discharge to a watercourse and/or coastal and marine waters. The degree of treatment and disinfection shall be as required by the state pollution control agency. The minimum average daily flow for seven consecutive days that can be expected to occur once in ten years shall be the minimum flow to which the standards apply.

‡Phenolic compounds cannot exceed 0.005 mg/liter; no odor-producing substances that cause the threshold-odor number to exceed 8 are permitted; radioactivity limits are to be approved by the appropriate state agency, with consideration of possible adverse effects in downstream waters from discharge of radioactive wastes, and limits in a particular watershed are to be resolved when necessary after consultation between states involved.

Table 2.4 Summary of surface water quality limit proposals for the state of Washington.* (After Rambow and Sylvester [10].)

Characteristic	Fresh water		Salt water	
	Goal	Standard	Goal	Standard
Alkalinity (phenolphthalein and total)[1]				
Ammonia nitrogen	0.3	0.5	0.0025	0.003
Arsenic	0.003	0.005	0.003	0.004
Bacteria[2]				
Barium	0.01	0.05	0.05	0.06
Bicarbonate[3]				
BOD	1.0	2.0	1.0	2.0
Boron	0.1	0.3	4.7	5.5
Bottom deposits from waste-water discharge	None	None	None	None
Cadmium	0.0005	0.001	0.00011	0.00013
Calcium[4]				
Carbonate[5]				
CCE (carbon chloroform extract)	0.00	0.10	0.05	0.10
Chloride	10	20	Natural	120% of natural
Chromium	Trace	0.01	0.00005	0.00006
COD[6]				
Coliforms (domestic sewage origin)	50/100 ml	240/100 ml	50/100 ml	240/100 ml
Color	5 units	5 units over natural	None	5 units
Conductivity	110% of natural	125% of natural	Natural	120% of natural
Copper	0.05	0.02 above background	0.05	0.06
Cyanide	0.005	0.01	None	0.01
Dissolved oxygen	95% saturation	85% saturation	95% saturation	85% saturation
Fecal streptococci[7]				
Floating solids	None	None	None	None
Fluoride	0.5	1.0	1.3	1.5
Hardness (as CaCO$_3$)	20 to 75	20 to 125	—	—
Hydroxide	None	None	None	None
Iron	0.0 above natural	0.1 above natural	0.01 above natural	0.2
Lead	Limit of detection	0.02	Limit of detection	0.004
Magnesium[8]				
Manganese	Trace	0.01	0.002	0.04
Nitrate	0.1 above natural	1.0 above natural	0.5	0.6
Nitrogen (total)	0.4 above natural	1.0 above natural	0.5	0.6
Threshold-odor number	1.0	3	1.0	3
Oil and tars	None	None	None	None
Pesticides[9]				
pH	7.0–8.0	6.5–8.5	7.5–8.4	7.5–8.4

Table 2.4 (*continued*)

Characteristic	Fresh water		Salt water	
	Goal	Standard	Goal	Standard
Phenol	Limit of detectability	0.0005	0.04	0.05
Phosphate (total)	0.03	0.15	0.3	0.4
Potassium	2.5	5.0	380	450
Radioactivity	None	USPHS DWS	None	USPHS DWS
Selenium	Limit of detectability	0.002	0.004	0.005
Silica[10]				
Silver	Limit of detectability	0.003	0.0003	0.0004
Sodium	10 over natural	35 over natural	10,500	12,500
Spent sulfite liquor[11]				
Sulfate	15	30	2700	3200
Surfactants	Trace (LAS)	0.10 (LAS)	Trace (LAS)	0.10 (LAS)
Temperature	Natural + 1°C	Natural + 2°C	Natural + 1°C	Natural + 2°C
Total dissolved solids[12]				
Toxicants, miscellaneous	None detectable	None detectable	None detectable	None detectable
Turbidity	5 units	Natural	3 units	5 units
Viruses[13]				
Zinc	Limit of detectability	Limit of detectability	0.01	0.012

*All values in mg/liter unless otherwise specified.
[1]No specific limits. A waste discharge is not to increase the natural total alkalinity by more than 10 per cent or to impart phenolphthalein (CO_3 and OH^-) alkalinity to a receiving water.
[2]No limit specified: refer to coliform organisms.
[3]No limit specified: relates to conductivity and pH.
[4]No limit specified: see hardness.
[5]Although carbonate itself in moderate concentrations is not particularly detrimental, it is associated with high pH values (greater than 8.3). Any carbonate discharge is not to be detectable below the point of discharge.
[6]Since the chemical oxygen demand is related to the BOD, DO, and CCE, it has no limit specified herein.
[7]Fecal bacteria are represented by the coliform group standard.
[8]Controlled by hardness content: no limit specified.
[9]Insufficient data.
[10]No standard proposed. Turbidity will include colloidal silica.
[11]Effect covered by other parameters.
[12]No standard proposed. Conductivity standards are related.
[13]None proposed at present.

Table 2.5 New England Interstate Water Pollution Control Commission: classification and standards of quality for interstate waters (as revised and adopted April 18, 1967).

	Standards of water quality		
Water-use class and description*	Dissolved oxygen	Sludge deposits, solid refuse, floating solids, oils, grease, and scum	Color and turbidity
A—Suitable for water supply and all other water uses; character uniformly excellent†	75% saturation, 16 hr/day; 5 mg/liter at any time	None allowable	None other than of natural origin
B—Suitable for bathing, other recreational purposes, agricultural uses; industrial processes and cooling; excellent fish and wildlife habitat; good aesthetic value; acceptable for public water supply with appropriate treatment	75% saturation, 16 hr/day; 5 mg/liter at any time	None allowable	None in such concentrations that would impair any usages specifically assigned to each class
C—Suitable for fish and wildlife habitat; recreational boating; industrial processes and cooling; under some conditions acceptable for public water supply with appropriate treatment; good aesthetic value	5 mg/liter, 16 hr/day; not less than 3 mg/liter at any time. (For cold-water fishery not less than 5 mg/liter at any time)	None§	
D—Suitable for navigation, power, certain industrial processes and cooling, and migration of fish; good aesthetic value	Minimum of 2 mg/liter at any time.	None§	

*Waters shall be free from chemical constituents in concentrations or combinations which would be harmful to human, animal, or aquatic life for the appropriate, most sensitive, and governing water-class use. In areas where fisheries are the governing considerations and approved limits have not been established, bioassays shall be performed as required by the appropriate agencies. For public drinking-water supplies the limits prescribed by the U.S. Public Health Service may be used where not superseded by more stringent signatory state requirements.
†Class A waters reserved for water supply may be subject to restricted use by state and local regulation.

Table 2.5 (*continued*)

	Standards of water quality			
Coliform bacteria, no./100 ml	Taste and odor	pH	Allowable temperature increase
Not to exceed a median of 100/100 ml nor more than 500 in more than 10% of samples collected	None other than of natural origin	As naturally occurs	None other than of natural origin
Not to exceed a median of 1000/ml nor more than 2400 in more than 20% of samples collected	None in such concentrations that would impair any usages specifically assigned to each class or cause taste and odor in edible fish	6.5–8.0	Only such increases that will not impair any usages specifically assigned to each class‡
		6.0–8.5	
None in such concentrations that would impair any usages specifically assigned to each class	None in such concentrations that would impair any usages specifically assigned to this class	6.0–9.0	None except where the increase will not exceed the recommended limits on the most sensitive water use and will in no case exceed 90° F

‡The temperature increase shall not raise the temperature of the receiving waters above 68°F for waters supporting cold-water fisheries and 83°F for waters supporting a warm-water fishery. In no case shall the temperature of the receiving water be raised more than 4°F.
§Sludge deposits, floating solids, oils, grease, and scum shall not be allowed except for such small amounts that may result from the discharge of appropriately treated sewage or industrial waste effluents.

Table 2.6 ORSANCO quality criteria.

ORSANCO resolution no. 16–66 (adopted May 12, 1966; amended September 8, 1966, and May 11, 1967)

Whereas: The assessment of scientific knowledge and judgments on water-quality criteria has been a continuing effort over the years by the Commission in consultation with its advisory committees; and

Whereas: The Commission now finds it appropriate to consolidate viewpoints and recommendations relating to such criteria;

Now, therefore, be it resolved: That the Ohio River Valley Water Sanitation Commission hereby adopts the following statement and specifications:

Criteria of quality are intended as guides for appraising the suitability of interstate surface waters in the Ohio Valley for various uses, and to aid decision-making in the establishment of waste-control measures for specific streams or portions thereof. Therefore, the criteria are not to be regarded as standards that are universally applicable to all streams. What is applicable to all streams at all places and at all times are certain minimum conditions, which will form part of every ORSANCO standard.

Standards for waters in the Ohio River Valley Water Sanitation District will be promulgated following investigation, due notice and hearing. Such standards will reflect an assessment of the public interest and equities in the use of the waters, as well as consideration of the practicability and physical and economic feasibility of their attainment.

The ORSANCO criteria embrace water-quality characteristics of fundamental significance, and which are routinely monitored and can be referenced to data that are generally available. The characteristics thus chosen may be regarded as primary indicators of water-quality, with the understanding that additional criteria may be added as circumstances dictate. Unless otherwise specified, the term average as used herein means an arithmetical average.

Minimum Conditions Applicable to All Waters at All Places and at All Times

1. Free from substances attributable to municipal, industrial or other discharges or agricultural practices that will settle to form putrescent or otherwise objectionable sludge deposits
2. Free from floating debris, oil, scum and other floating materials attributable to municipal, industrial or other discharges or agricultural practices in amounts sufficient to be unsightly or deleterious
3. Free from materials attributable to municipal, industrial or other discharges or agricultural practices producing color, odor or other conditions in such degree as to create a nuisance
4. Free from substances attributable to municipal, industrial or other discharges or agricultural practices in concentrations or combinations which are toxic or harmful to human, animal, plant or aquatic life

Stream-Quality Criteria

For public water supply and food-processing industry

The following criteria are for evaluation of stream quality at the point at which water is withdrawn for treatment and distribution as a potable supply:

1. *Bacteria:* Coliform group not to exceed 5000/100 ml as a monthly average value (either MPN or MF count); nor exceed this number in more than 20 per cent of the samples examined during any month; nor exceed 20,000 per 100 ml in more than five per cent of such samples
2. *Threshold-odor number:* Not to exceed 24 (at 60°C) as a daily average
3. *Dissolved solids:* Not to exceed 500 mg/l as a monthly average value, nor exceed 750 mg/l at any time. (For Ohio River water, values of specific conductance of 800 and 1200 micromhos/cm (at 25°C) may be considered equivalent to dissolved-solids concentrations of 500 and 750 mg/l)
4. *Radioactive substances:* Gross beta activity not to exceed 1000 picocuries per liter (pCi/l), nor shall activity from dissolved strontium 90 exceed 10 pCi/l, nor shall activity from dissolved alpha emitters exceed 3 pCi/l
5. *Chemical constituents:* Not to exceed the following specified concentrations at any time:

Table 2.6 (*continued*)

Constituent	Concentration (mg/l)
Arsenic	0.05
Barium	1.0
Cadmium	0.01
Chromium (hexavalent)	0.05
Cyanide	0.025
Fluoride	1.0
Lead	0.05
Selenium	0.01
Silver	0.05

For industrial water supply
The following criteria are applicable to stream water at the point at which the water is withdrawn for use (either with or without treatment) for industrial cooling and processing:
1. *Dissolved oxygen:* Not less than 2.0 mg/l as a daily-average value, nor less than 1.0 mg/l at any time
2. *pH:* Not less than 5.0 or greater than 9.0 at any time
3. *Temperature:* Not to exceed 95°F at any time
4. *Dissolved solids:* Not to exceed 750 mg/l as a monthly average value, nor exceed 1000 mg/l at any time. (For Ohio River water, values of specific conductance of 1200 and 1600 micromhos/cm (at 25°C) may be considered equivalent to dissolved-solids concentrations of 750 and 1000 mg/l)

For aquatic life
The following criteria are for evaluation of conditions for the maintenance of a well-balanced, warm-water fish population. They are applicable at any point in the stream except for areas immediately adjacent to outfalls. In such areas cognizance will be given to opportunities for the admixture of waste effluents with river water.
1. *Dissolved oxygen:* Not less than 5.0 mg/l during at least 16 hours of any 24-hour period, or less than 3.0 mg/l at any time
2. *pH:* No values below 5.0 or above 9.0; daily average (or median) values preferably between 6.5 and 8.5
3. *Temperature:* Not to exceed 93°F at any time during the months of May through November and not to exceed 73°F at any time during the months of December through April
4. *Toxic substances:* Not to exceed one-tenth of the 48-hour median tolerance limit, except that other limiting concentrations may be used in specific cases when justified on the basis of available evidence and approved by the appropriate regulatory agency

For recreation
The following criterion is for evaluation of conditions at any point in waters designated to be used for recreational purposes, including such water-contact activities as swimming and water skiing:
Bacteria: Coliform group not to exceed 1000/100 ml as a monthly average value (either MPN or MF count); nor to exceed this number in more than 20 per cent of the samples examined during any month or exceed 2400/100 ml (MPN or MF count) on any day

For agricultural use and stock watering
Criteria are the same as those shown for minimum conditions applicable to all waters at all places and at all times

Table 2.7 Pennsylvania raw-waste standards for pulp and paper mills.

Type of product or process	Population equivalent, per ton of product, based on 5-day BOD		Pounds of suspended solids per ton of product	
	3-day average*	8-hr average	3-day average	8-hr average
Group A				
Tissue paper	75	80	40	50
Glassine paper	25	30	15	20
Parchment paper	40	45	20	30
Miscellaneous papers	25	30	5	10
Flax papers—condenser	375	415	300	350
Group B (speciality group)				
Fiber paper	800	850	200	235
Asbestos paper	125	185	290	350
Felt paper	210	230	60	65
Insulating paper	2250	2500	325	350
Speciality papers	1000	1200	135	160
Group C (coarse paper)	90	120	35	50
Group D (integrated mills)				
Wood preparation	80	100	40	50
Pulp (sulfite)	3000	3500	35	40
Pulp (alkaline)	300	350	20	35
Pulp (groundwood)	115	130	80	85
Pulp (deinked unfilled stock)	500	650	375	500
Pulp (deinked filled stock)	400	500	600	800
Pulp (rag cooking)	1400	1550	475	500
Bleaching, (long-fiber stock, multi- or single-stage bleaching and short-fiber stock, single-stage bleaching)	60	70	3	6
Bleaching (short-fiber stock, multi-stage bleaching)	165	185	30	35
Paper-making	100	125	75	85

*All averages are for periods of consecutive operation.

contamination during critical low flows. The concept was originally developed as a means of storing electrical energy for subsequent reuse, thereby levelling demand. As Velz [13] points out, there are many river basins where conventional or even advanced treatment (tertiary) is not now, or will not be, adequate to meet desirable water-quality standards. Pollution control with the use of pumped storage for low-flow augmentation may provide an economical method of ensuring water quality, as well as having other benefits. Velz discusses four major advantages of pumped over conventional storage:

1) It minimizes the problem of locating reservoir sites on the main stream channel. Water may be pumped from a small on-channel pool to the pumped storage reservoir, located at a higher elevation and closer to the demand for augmentation. Simpler dams and elimination of costly flood spillways are obvious advantages.

2) The quality of stored water is improved, since it is pumped during maximum stream flow and stored away from the main stream.

3) There is a higher degree of flexibility and immediate response to deterioration of stream water.

4) Flow can be augmented incrementally along the course of a stream at or near points of greatest need, with high-quality water available at all times to augment the stream flow.

References

1. Bloodgood, D. E., "1953 Industrial wastes forum," *Sewage Ind. Wastes* **26**, 640 (1954).
2. Cleary, E. J., "Water pollution control—Gearing performance to promise," *Civil Eng.* **38**, 63 (1968).
3. Dappert, A. P., "Pollution control through the mechanism of classes and standards," *Sewage Ind. Wastes* **24**, 313 (1952).
4. Grossman, I., "Experiences with surface water quality standards," *J. Sanit. Eng. Div. Am. Soc. Civil Engrs.* **94**, (SAl) 13 (1968).
5. Hubbard, E. C., "Stream standards," *J. Water Pollution Control Federation* **37**, 308 (1965).
6. Ingols, R. S., "Surface water pollution and natural purification," *Municipal South*, p. 31 (January 1956).
7. Kittrell, F. W., "Effects of impoundments on dissolved oxygen resources," *Sewage Ind. Wastes* **31**, 1965 (1959).
8. New York State Water Pollution Control Board, "Classification and standards of water quality and purity," *J. Am. Water Works Assoc.* **42**, 1137 (1950).
9. "ORSANCO's success story," *Public Works*, **19**, 66 (1965).
10. Rambow, C. A., and R. O. Sylvester, "Methodology in establishing water quality standards," *J. Water Pollution Control Federation* **39**, 1155 (1967).
11. Sheets, J. L., *Evaluation of Pollution Abatement Benefits from Low-Flow Augmentation*, Department of Civil Engineering, University of Illinois, Urbana (July 1964) p. 91.
12. Streeter, H. W., "Standards of stream sanitation," *Sewage Ind. Wastes* **21**, 115 (1949).
13. Velz, C. J., J. D. Calvert, Jr., R. A. Deininger, W. L. Heilman, and J. Z. Reynolds, *J. Sanit. Eng. Div. Am. Soc. Civil Engrs.* SAl, 159 (1968).
14. Wendell, M., "Intergovernmental relations in water quality control," *J. Water Pollution Control Federation* **39**, 278 (1967).
15. Worley, J. L., F. J. Burgess, and W. W. Towne, "Identification of low flow augmentation requirements for water quality control by computer techniques," *J. Water Pollution Control Federation* **37**, 659, (1965).

CHAPTER 3

COMPUTATION OF ORGANIC WASTE LOADS ON STREAMS

Despite an industrial plant's efforts to reduce the volume and strength of its wastes to a minimum (to be described in Chapters 6 and 7), some waste will still remain to be disposed of. A detailed analysis is needed, to show the volume and character of the remaining waste and to determine how much treatment the waste requires, before an industry can satisfy either laws or public opinion.

The degree of treatment necessary depends primarily on the condition and best usage of the receiving stream. Overtreatment of wastes results in unnecessary, burdensome expense; undertreatment is only a waste of effort and money, since it does not abate the pollution problem. Thus one can readily comprehend the value of calculating as closely as possible the amount of pollution that can safely be discharged into a stream.

Although this text does not propose to cover the entire field of stream sanitation, the major methods of determining the required degree of waste treatment will be described here in some detail. The Streeter-Phelps formulation [6] has been used with various degrees of success over the past thirty years. Thomas simplified the Streeter-Phelps formulations in actual practice. Another method (Churchill) has come into common usage only within the last thirteen years. Other methods are available for determining stream reaeration and stream purification coefficients (such as Fair's factor), but these will not be discussed in this chapter because of space limitations.

3.1 Streeter-Phelps Formulations

We first present a list of the symbols used to denote the various parameters to be calculated:

K_1 = deoxygenation rate/day
Δ_t = time of travel (days)
L_A = ultimate upstream BOD (ppm or lb)
L_B = ultimate downstream BOD (ppm or lb)
K_2 = reaeration rate/day
$\overline{L}$ = average ultimate oxygen demand in stream section (ppm or lb)
$\overline{D}$ = average oxygen deficit in stream section (ppm or lb)
ΔD = change in oxygen deficit from upstream to downstream sampling points (ppm or lb)
t = time of stream flow from upstream to downstream points of sampling (days)
D_t = dissolved-oxygen deficit downstream (ppm or lb), at time t
D_A = dissolved-oxygen deficit upstream (ppm or lb).

The following formulas are used to compute the deoxygenation rate (K_1),* reaeration rate (K_2), and the dissolved-oxygen deficit (D_t) at a downstream location:

$$K_1 = \frac{1}{\Delta t} \log \frac{L_A}{L_B} = \text{deoxygenation rate}, \quad (1)$$

$$K_2 = K_1 \frac{\overline{L}}{\overline{D}} - \frac{\Delta D}{2.3 \Delta t \overline{D}} = \text{reaeration rate}, \quad (2)$$

$$D_t = \frac{K_1 L_A}{K_2 - K_1}[10^{-K_1 t} - 10^{-K_2 t}] + D_A \cdot 10^{-K_2 t}$$
$$= \text{dissolved oxygen deficit downstream}. \quad (3)$$

We now apply the Streeter-Phelps formulations to calculate the allowable organic loading for the stream situation sketched in Fig. 3.1. Using the stream data

---

*K_1 includes K_3 (deoxygenation due to bottom deposits), unless computed separately.

K_3 is the constant of proportionality which reflects the composition of the waste and the characteristics of the receiving water, as well as the quiescence of the stream at the point under consideration. It represents the amount of BOD which is removed by sedimentation. Therefore, in regions of considerable turbulence, K_3 is usually zero and under scouring conditions may even be negative. Since it is so difficult to compute, it is generally considered as an integral part of the deoxygenation rate, K_1.

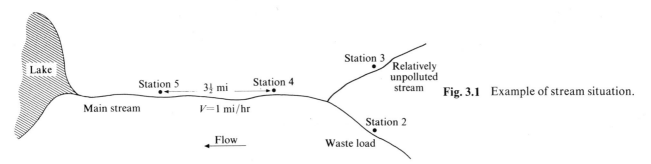

Fig. 3.1 Example of stream situation.

Table 3.1 Data for the stream illustrated in Fig. 3.1. Two samples, collected on different days, were taken at each station.

	Temperature, °C	Flow		5-day BOD			Dissolved oxygen			
		cfs*	mgd	ppm	lb/day		lb/day	Saturated, ppm	ppm	Deficit, ppm
Station 4	20	60.1	38.82	36	11,650		1325	9.2	4.1	5.1
	17	54	34.88	10.35	3,015		1512	9.8	5.2	4.6
					Total	14,665				
					Average	7,332				
Station 5	20	51.7	33.40	21.2	5,905		725	9.2	2.6	6.6
	16.5	66.6	43.02	5.83	2,095		1050	9.9	2.8	7.1
					Total	8,000				
					Average	4,000				

*Cubic feet per second.

given in Table 3.1, as representing the stream prior to any waste treatment, we obtain the following quantitative results:

$$K_1 = \frac{1}{3.5/24} \log \frac{7332 \times 1.46\dagger}{4000 \times 1.46\dagger},$$
$$= 1.8; \qquad (4)$$

$$\bar{L} = \frac{\text{BOD (Station 4)} + \text{BOD (Station 5)}}{2}$$
$$= \frac{7332 \times 1.46 + 4000 \times 1.46}{2}$$
$$= 8273 \text{ lb/day};$$

$$\bar{D} = \frac{\{6.6 \times 8.34 \times 33.4 + 5.1 \times 8.34 \times 38.8 + 7.1 \times 8.34 \times 43 + 4.6 \times 8.34 \times 34.9\}}{4}$$
$$= 1843 \text{ lb/day};$$

$$\Delta D = 1/2 \,(6.6 \times 8.34 \times 33.4 + 7.1 \times 8.34 \times 43)$$
$$\quad - 1/2 \,(5.1 \times 8.34 \times 38.8 + 4.6 \times 8.34 \times 34.9)$$
$$= 698 \text{ lb/day};$$

$$\Delta t = \frac{3.5}{24} = 0.146 \text{ day};$$

$$K_2 = 1.8 \cdot \frac{8273}{1843} - \frac{698}{2.3(0.146)(1843)}$$
$$= 8.08 - 1.12 = 6.96. \qquad (5)$$

†Multiplier used to convert 5-day 20°C BOD to ultimate first-stage BOD, assuming normal domestic sewage deoxygenation rates.

Using Fair's f formula, we have

$$f = \frac{K_2}{K_1} = \frac{6.96}{1.80} = 3.8.$$

According to Fair's classification, the result is characteristic of streams falling in his group D, i.e., "streams with normal velocity that can almost be considered as swift streams."

With these stream reaction rates (K_1 and K_2) and the initial condition at Station 4 (D_A and L_A), it is possible to plot the dissolved-oxygen deficit (sag curve) at any point in the stretch between Station 4 and Station 5, assuming that the reaction rates remain constant in the stretch. Figure 3.2 is a graphic plot of the actual sag curve.

If we assume that the waste load entering at Station 2 has undergone primary treatment resulting in a 30 per cent reduction of BOD and that the distance between Stations 2 and 4 does not change the effect of the pollution load or dissolved-oxygen deficit, we obtain the following BOD values:

90 (ppm BOD at Station 2)
 × 8.34 × 8.1 cfs
 × 0.65 (mgd/cfs) = 3951 BOD lb/day,

92 (ppm BOD at Station 2)
 × 8.34 × 8.22 cfs
 × 0.65 (mgd/cfs) = 4099 BOD lb/day,
   Total = 8050,
   Average = 4025 BOD lb/day,
   $\bar{L}$ = 4025 × 1.46
    = 5876 BOD lb/day.

Hence, at Station 4, we have
$$L_4 = 7332 \times 1.46 - 0.30\,(5876)$$
$$= 8942.3 \text{ lb/day}.$$

The dissolved-oxygen deficit at Station 4 is obtained by the following calculation:

$$D_4 = \frac{4.6 \times 8.34 \times 34.9 + 5.1 \times 8.34 \times 38.8}{2}$$

1494 lb/day,

and the dissolved-oxygen deficit downstream at Station 5 is obtained from Eq. (3):

$$D_5 = 1.8 \times \frac{8942}{6.96 - 1.80}[10^{-1.8(0.146)} - 10^{-6.96(0.146)}]$$
$$+ 1494 \times 10^{-6.96(0.146)}$$
$$= 3118\,[10^{-0.263} - 10^{-1.016}] + 1494 \times 10^{-1.016}$$
$$= 3118\,[0.5458 - 0.0964] + 1494\,(0.0964)$$
$$= 3118\,[0.4494] + 144$$
$$= 1400 + 144$$
$$= 1544 \text{ lb/day}.$$

At the lowest flow (33.4 mgd) and the highest temperature (20°C) observed at Station 5, we have 9.2 (ppm dissolved oxygen (DO) at saturation at Station 5) $-\dfrac{1544}{33.4 \times 8.34}$,

or

$$9.20 - 5.55 = 3.65 \text{ ppm DO};$$

and at 25°C we have

$$8.40 - 5.55 = 2.85 \text{ ppm DO}.$$

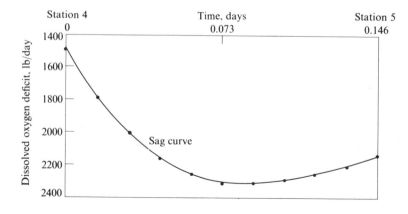

Fig. 3.2 Plot of sag curve.

Therefore, primary treatment would be insufficient to keep the stream at a minimum of 4 ppm DO during summer low flows and high temperatures.

The actual dissolved-oxygen deficit at the bottom of the sag at $t = 0.073$ days (see Fig. 3.2) can be computed as follows:

$$D_{0.075 \text{ days}} = 1.8 \times \frac{8942}{6.96-1.80}(10^{-1.86 \times 0.146}$$
$$-10^{-6.96 \times 0.146}) + 1494 \times 10^{-6.96 \times 0.146}$$

Cohen and O'Connell [2] have found the analog computer valuable in providing rapid information on the effect of variations in initial stream conditions. Because it is easy to modify the data fed in, many more conditions of potential interest can be simulated and thus investigated with little additional effort. It will be apparent to the reader that more rapid, automatic methods are essential whenever the Streeter-Phelps equations are used and especially when the engineer is attempting to ascertain the effects of industrial wastes under a variety of possible stream conditions.

3.2 Thomas Method for Determining Pollution-Load Capacity of Streams

Thomas [7] developed a useful simplification of the Streeter-Phelps equations for computing stream capacity. In his method the stream-reaction constants K_1 and K_2 are computed as in Section 3.1. However, he proposes using a nomograph to compute the dissolved-oxygen deficit at any time t downstream from a source of pollution load. Conversely, a pollution load producing a critical dissolved-oxygen deficit downstream can be calculated from the same nomograph. The nomograph which plots D/L_A versus $K_2 t$ for various ratios of K_2/K_1 is shown in Fig. 3.3. Thomas recognizes that Eq. (3) is unwieldy and in most practical applications can be solved only by tedious trial-and-error procedures. He believes that this disadvantage can be overcome by the use of his nomogram. Before the nomogram is used, K_1, K_2, D_A, and L_A must be computed. By means of a straight-edge, a straight line (isopleth) is drawn, connecting the value of D_A/L_A at the left with the point presenting the appropriate day multiplied by the reaeration constant ($K_2 t$) on the appropriate (K_2/K_1)-curve. The value of D/L_A is then read at the intersection with the isopleth. Finally, the value of the deficit at the end of the appropriate day is obtained by multiplying L_A and the intersection value.

The author has used the Thomas nomogram on many occasions and found it very convenient, accurate, and time-saving. The following problem illustrates the use of the method.

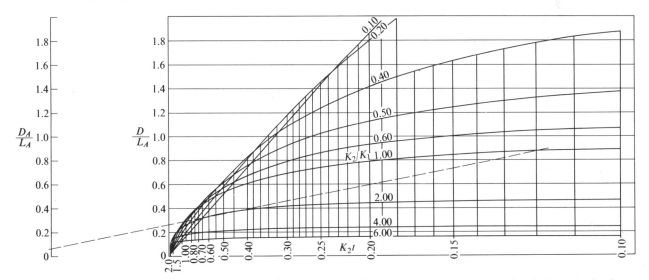

Fig. 3.3 Nomogram for the dissolved-oxygen sag. Oxygen deficits downstream from a point of pollution may be determined from initial BOD and DO and stream self-purification constants. (After Thomas [7].)

30 COMPUTATION OF ORGANIC WASTE LOADS ON STREAMS

Fig. 3.4 Section of Cape Fear River from ferry at Tarheel, N.C., to ferry near Carvers, N.C. (N.C. State Board of Health, Sanitary Eng. Div., 8-2-51.)

An industrial plant planning to discharge a 5-day 20°C BOD load of 5000 pounds was proposed at Station 1 in the Cape Fear River in North Carolina (Fig. 3.4). The North Carolina Stream Sanitation Committee was concerned about both the present condition of the stream stretch below Station 1 and the future condition, if and when the site was approved.

The river was sampled at six locations from Station 1 to Station 6, a distance of 38.1 miles. Ultimate BOD, deoxygenation, and reaeration rates were computed in each stretch of the river and used to obtain the oxygen-sag curve. A hypothetical load of 5000 pounds of 5-day BOD was imposed on the river at Station 1, and the Thomas method was used to draw the new oxygen-sag curve based on similar reaction rates.

The calculations are shown in Tables 3.2 through 3.9. The subscript B refers to laboratory bottle values and Str is the value at existing stream temperature. The sag curves (existing and projected), with present flow and 5-year minimum flow, are shown in Fig. 3.5.

Calculations of oxygen deficits in stream. We first present the calculations of the oxygen deficit existing in the stream at the time of the investigation. To obtain the K_2-value for the stream stretch between stations, we proceeded as follows, using the K_1-value (0.331) as the most representative, realistic, and appropriate value. (Negative K_1 values must be erroneous.)

Table 3.2 Cape Fear velocity measurements.

Floats: 4 miles/280 min = 0.86 miles/hr × 0.85* = 0.69 miles/hr

Station	Mile	Distance between stations, miles	Flow time hr	Flow time days	Total flow time, days	L at 20°C, ppm	K_B	DO, ppm	T, °C	L_R, ppm	$K_{B_{Str}}$	K_1	K_2
1	0	0	0	0	0	1.71	0.101	5.8	26.0	1.92	0.133	−0.95	−0.139
2	5.33	5.33	7.73	0.322	0.322	3.40	0.054	5.4	26.5	3.84	0.073	0.585	0.159
3	13.33	8.00	11.6	0.483	0.805	1.75	0.098	5.0	27.2	2.00	0.135	−2.21	−1.73
4	16.27	2.94	4.25	0.177	0.982	4.18	0.035	5.9	29	4.93	0.052	0.331	0.463
5	31.17	14.90	21.65	0.902	1.884	2.09	0.100	5.4	28.9	2.47	0.149	0.332	0.105
6	38.11	6.94	10.06	0.419	3.303	1.54	0.151	5.1	28.2	1.79	0.216		

*To correct for wind and other surface effects.

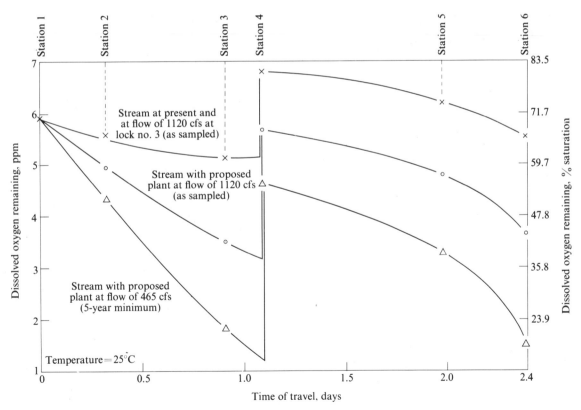

Fig. 3.5 Oxygen-sag curves of Cape Fear River below Tarheel, N.C. (N.C. State Board of Health, Sanitary Eng. Div., 8-2-51, N.L.N.)

Table 3.3 Calculations of deoxygenation rate (K_1) of stream.

$$K_1 = \frac{1}{t}\log\frac{L_A}{L_B}$$

From Station 1 to Station 2:
$$K_1 = \frac{1}{0.322}\log\frac{1.92}{3.84}$$
$$= \frac{1}{0.322}(0.282 - 0.584)$$
$$= \frac{1}{0.322}(-0.302)$$
$$= -0.95$$

From Station 2 to Station 3:
$$K_1 = \frac{1}{0.483}\log\frac{3.84}{2.00}$$
$$= \frac{0.282}{0.483} = 0.585$$

From Station 3 to Station 4:
$$K_1 = \frac{1}{0.177}\log\frac{2.00}{4.93}$$
$$= \frac{1}{0.177}(0.301 - 0.692)$$
$$= -\frac{0.391}{0.177}$$
$$= -2.21$$

From Station 4 to Station 5:
$$K_1 = \frac{1}{0.902}\log\frac{4.93}{2.47}$$
$$= \frac{0.299}{0.902} = 0.331$$

From Station 5 to Station 6:
$$K_1 = \frac{1}{0.419}\log\frac{2.47}{1.79}$$
$$= \frac{0.139}{0.419} = 0.332$$

Between Stations 1 and 2:

$$K_2 = \frac{\overline{L}}{\overline{D}}K_1 - \frac{\Delta D}{2.3\,\Delta t\,\overline{D}},$$

$$\overline{L} = \frac{11{,}600 + 25{,}800}{2} = 18{,}700,$$

$$\overline{D} = \frac{18{,}500 + 14{,}600}{2} = 16{,}550,$$

$$D = 18{,}550 - 14{,}600 = 3900,$$
$$t = 0.322;$$

hence

$$K_2 = 0.331\,\frac{18{,}700}{16{,}550} - \frac{3900}{2.3 \times 0.322 \times 16{,}550}$$
$$= 0.381 - \frac{3900}{12{,}275} = 3.81 - 0.318 = 0.063.$$

Again, using 0.331 as K_1 of the stream section, one obtains

K_2	K_1	K_2/K_1	D_A	L_A	D_A/L_A	$K_2 t$
0.067	0.331	0.202	2.42	1.92	1.26	0.067×0.322 = 0.022

From extension of the nomograph, we obtain

$$\frac{D}{L_A} = 1.46,$$

and therefore

$D = 1.46 \times 1.92 = 2.80$ (calculated deficit at Station 2)
(2.75 = observed deficit).

Between Stations 2 and 3:

K_2	K_1	K_2/K_1	D_A	L_A	D_A/L_A	$K_2 t$
0.159	0.331	0.48	2.75	3.84	0.716	0.159×0.483 = 0.077

From extension of the nomograph, we obtain

$$\frac{D}{L_A} = 0.9,$$

and therefore

$D = 0.9 \times 3.84 = 3.25$ (calculated deficit at Station 3)
(3.04 = observed deficit).

Between Stations 3 and 4:

$$K_2 = 0.331\,\frac{31{,}150}{22{,}550} - \frac{(-12{,}100)}{(2.3 \times 0.177 \times 22{,}550)}$$
$$= 0.457 + \frac{12{,}100}{9{,}180} = 0.457 + 1.320 = 1.777.$$

K_2	K_1	D_A	L_A	K_2/K_1	D_A/L_A	$K_2 t$
1.777	0.331	3.04	2.00	5.36	1.53	1.777 (0.177) = 0.314.

From nomograph, we obtain

$$\frac{D}{L_A} = 0.85,$$

and therefore

$D = 0.85 \times 2.00 = 1.70$ (calculated deficit at Station 4)
(1.87 = observed deficit).

Between Stations 4 and 5:

K_2	K_1	K_2/K_1	D_A	L_A	D_A/L_A	$K_2 t$
0.463	0.331	1.4	1.87	4.93	0.38	0.463 (0.902) = 0.418

From nomograph, we obtain

$$\frac{D}{L_A} = 0.45,$$

and therefore

$D = 0.45 (4.93) = 2.22$ (calculated deficit at Station 5)
(2.38 = observed deficit).

Between Stations 5 and 6:

K_2	K_1	K_2/K_1	D_A	L_A	D_A/L_A	$K_2 t$
0.105	0.322	0.316	2.38	2.47	0.96	0.105 (0.42) = 0.044

From extension of the nomograph, we obtain

$$\frac{D}{L_A} = 1.18,$$

and therefore

$D = 1.18 \times 2.47 = 2.91$ (calculated deficit at Station 6)
(2.79 observed deficit).

Imposition of 5000 pounds of 5-day BOD at Station 1. We next present calculations which establish what the new oxygen-sag curve would be after imposition of 5000 pounds of 5-day BOD on the river at Station 1. The Thomas nomograph was used, and it was assumed that reaction rates would be similar to those already observed in the river:

724 mgd × 5.8 + 1.5 × 0 = 725.5X   (initial river DO with industrial waste)

$$\frac{4200}{725.5} = X = 5.79$$

8.22 − 5.79 = 2.43 = oxygen deficit (D)
5000 lb/day = ppm × 8.34 × 1.5 mgd
ppm = 400 of 5-day BOD.

(a)   (b)

(724 mgd × 1.23 + 1.5 × 400) $\frac{1.92}{1.23}$ = 725.5X

(890 + 600) $\frac{1.92}{1.23}$ = 725X

ppm L $\frac{2325}{725.5}$ = X = 3.21

(a) = 5-day BOD of river at Station 1 at 20°C;
(b) = increase due to conversion to *L*-value and stream temperature.

Table 3.4 Calculations of stream oxygen deficits.

Station	Temperature, °C	DO, ppm Saturated	DO, ppm Observed	DO, ppm Deficit	Assumed flow, mgd	Downstream minus upstream oxygen deficit, lb/day
1	26	8.22	5.8	2.42	724	(807 × 8.34 × 2.75) − (724 × 8.34 × 2.42) 18,500 − 14,600 = +3,900
2	26.5	8.15	5.4	2.75	807	(1130 × 8.34 × 3.04) − (807 × 8.34 × 2.75) 28,600 − 18,500 = +10,100
3	27.2	8.04	5.0	3.04	1130	(1060 × 8.34 × 1.87) − (1130 × 8.34 × 3.04) 16,500 − 28,600 = −12,100
4	29	7.77	5.9	1.87	1060	(1060 × 8.34 × 2.38) − (1060 × 8.34 × 1.87) 21,000 − 16,500 = +4,500
5	28.9	7.78	5.4	2.38	1060	(1060 × 8.34 × 2.79) − (1060 × 8.34 × 2.38) 24,700 − 21,000 = +3,700
6	28.2	7.79	5.1	2.79	1060	

Table 3.5 Calculations of stream ultimate biochemical oxygen demand (BOD).

Station	$L_{20°C}$, ppm	L_R, ppm	$\overline{L}$, lb/day
1	1.71	1.92	$\dfrac{\{1.92 \times 724 \times 8.34 \\ + 3.84 \times 807 \times 8.34\}}{2}$
			$= \dfrac{11{,}600 + 25{,}800}{2}$
			$= 18{,}700$
2	3.40	3.84	$\dfrac{\{3.84 \times 807 \times 8.34 \\ + 2.00 \times 1130 \times 8.34\}}{2}$
			$= \dfrac{25{,}800 + 18{,}800}{2}$
			$= 21{,}800$
3	1.75	2.00	$\dfrac{\{2.00 \times 1130 \times 8.34 \\ + 4.93 \times 1060 \times 8.34\}}{2}$
			$= \dfrac{18{,}800 + 43{,}500}{2}$
			$= 31{,}150$
4	4.18	4.93	$\dfrac{\{4.93 \times 1060 \times 8.34 \\ + 2.47 \times 1060 \times 8.34\}}{2}$
			$= \dfrac{43{,}500 + 21{,}800}{2}$
			$= 32{,}650$
5	2.09	2.47	$\dfrac{\{2.47 \times 1060 \times 8.34 \\ + 1.79 \times 1060 \times 8.34\}}{2}$
			$= \dfrac{21{,}800 + 15{,}800}{2}$
			$= 18{,}800$
6	1.54	1.79	

Between Stations 1 and 2:

K_2	K_1	K_2/K_1	D_A	L_A	D_A/L_A	$K_2 t$
0.067	0.331	0.202	2.43	3.21	0.76	0.067×0.322 $= 0.022$

From nomograph, we obtain

$$\frac{D}{L_A} = 1.06,$$

and therefore

$$D = 1.06(3.21) = 3.40 \text{ (calculated deficit at Station 2)}.$$

Between Stations 2 and 3:

K_2	K_1	K_2/K_1	D_A
0.159	0.331	0.48	3.40

L_A	D_A/L_A	$K_2 t$
$\left(3.21 \times \dfrac{3.84}{1.92}\right)$ 6.42*	0.53	0.159×0.483 $= 0.077$

From nomograph, we obtain

$$\frac{D}{L_A} = 0.76,$$

and therefore

$$D = 0.76(6.42) = 4.88 \text{ (calculated deficit at Station 3)}.$$

Between Stations 3 and 4:

K_2	K_1	K_2/K_1	D_A
1.777	0.331	5.36	4.88

L_A	D_A/L_A	$K_2 t$
$\left(6.42 \times \dfrac{2.00}{3.84}\right)$ 3.34	1.46	1.777×0.177 $= 0.314$

From nomograph, we obtain

$$\frac{D}{L_A} = 0.81,$$

and therefore

$$D = 0.81(3.34) = 2.71 \text{ (calculated deficit at Station 4)}.$$

---

*$\dfrac{\text{New } L_{R^2}}{\text{New } L_{R^1}} = \dfrac{\text{Old } L_{R^2}}{\text{Old } L_{R^1}}$

and therefore

New L_{R^2} = New $L_{R^1} \times \dfrac{\text{Old } L_{R^2}}{\text{Old } L_{R^1}}$.

(R^1 and R^2 refer to station locations on the river, R.)

3.2 THOMAS METHOD FOR DETERMINING POLLUTION-LOAD CAPACITY OF STREAMS

Between Stations 4 and 5:

K_2	K_1	K_2/K_1	D_A
0.463	0.331	1.4	2.71

L_A
$\left(3.34 \times \dfrac{4.93}{2.00}\right)$
8.23

D_A/L_A
0.33

$K_2 t$
0.463×0.902
$= 0.418$

From nomograph, we obtain

$$\dfrac{D}{L_A} = 0.44,$$

and therefore

$D = 0.44(8.23) = 3.62$ (calculated deficit at Station 5).

Between Stations 5 and 6:

K_2	K_1	K_2/K_1	D_A
0.105	0.332	0.316	3.62

L_A
$\left(8.23 \times \dfrac{2.47}{4.93}\right)$
4.13

D_A/L_A
0.88

$K_2 t$
$0.105(0.42)$
$= 0.044$

From nomograph, we obtain

$$\dfrac{D}{L_A} = 1.16,$$

and therefore

$D = 1.16 \times 4.13 = 4.80$ (calculated deficit at Station 6).

Table 3.6 Calculation of stream average oxygen deficits.

Station	$\overline{D}$, lb/day
1	$\dfrac{18{,}500 + 14{,}600}{2} = \dfrac{33{,}100}{2} = 16{,}555$
2	$\dfrac{28{,}600 + 18{,}500}{2} = \dfrac{47{,}100}{2} = 23{,}550$
3	$\dfrac{16{,}500 + 28{,}600}{2} = \dfrac{45{,}100}{2} = 22{,}550$
4	$\dfrac{21{,}000 + 16{,}500}{2} = \dfrac{37{,}500}{2} = 18{,}750$
5	$\dfrac{24{,}700 + 21{,}000}{2} = \dfrac{45{,}700}{2} = 22{,}850$

Table 3.7 Calculations of stream reaeration rate (K_2).

Between Stations	
1 and 2	$K_2 = K_1 \cdot \dfrac{\overline{L}}{\overline{D}} - \dfrac{\Delta D}{2.3\,\Delta t\,\overline{D}}$ $= -0.95 \cdot \dfrac{18{,}700}{16{,}555} - \dfrac{(3900)}{2.3(0.322)16{,}555}$ $= -1.07 - \dfrac{3900}{12{,}265} = -1.07 - 0.318$ $= -1.390$
2 and 3	$K_2 = 0.585 \cdot \dfrac{21{,}800}{23{,}550} - \dfrac{10{,}100}{2.3 \times 0.483 \times 23{,}550}$ $= 0.541 - \dfrac{10{,}100}{26{,}200} = 0.541 - 0.382$ $= 0.159$
3 and 4	$K_2 = -2.21 \cdot \dfrac{31{,}150}{22{,}550} - \dfrac{-12{,}100}{2.3 \times 0.177 \times 22{,}550}$ $= -3.05 + \dfrac{12{,}100}{9160} = -3.05 + 1.32$ $= -1.73$
4 and 5	$K_2 = 0.331 \cdot \dfrac{32{,}650}{18{,}750} - \dfrac{4500}{2.3 \times 0.902 \times 18{,}750}$ $= +0.579 - \dfrac{4500}{38{,}850} = +0.579 - 0.116$ $= +0.463$
5 and 6	$K_2 = 0.332 \cdot \dfrac{18{,}800}{22{,}850} - \dfrac{3700}{2.3 \times 0.419 \times 22{,}850}$ $= +0.273 - 0.168 = 0.105$

Table 3.8 Summary of deficits and reaction rates used.

Between Stations	K_1	K_2	Downstream oxygen deficit	
			Calculated, ppm	Observed, ppm
1 and 2	0.331	0.067	2.80	2.75
2 and 3	0.331	0.159	3.25	3.04
3 and 4	0.331	1.777*	1.70	1.87
4 and 5	0.331	0.463	2.22	2.38
5 and 6	0.332	0.105	2.91	2.79

*City of Elizabethtown plus Lock No. 2 caused this high value.

Table 3.9 Oxygen left in water at Stations 1 to 6 at 25°C.

Station	DO saturated, ppm	Deficits (calculated), ppm			O_2 remaining, ppm			O_2 remaining, % saturation		
		No plant	With plant	Plant at low flow	No plant	With plant	Plant at low flow	No plant	With plant	Plant at low flow
1	8.38	2.42	2.43	2.45	5.96	5.95	5.93	71	71	70.5
2	8.38	2.80	3.40	4.07	5.58	4.98	4.31	66.7	59.5	51.4
3	8.38	3.25	4.88	6.54	5.13	3.50	1.84	61.3	41.8	22.0
4	8.38	1.70	2.71	3.71	6.68	5.67	4.67	79.8	67.7	55.8
5	8.38	2.22	3.62	5.16	6.16	4.76	3.22	73.6	57.0	38.4
6	8.38	2.91	4.80	6.97	5.47	3.58	1.41	65.3	42.8	16.8

Imposition of 5000 pounds 5-day BOD at 5-year minimum flow. We shall now examine the situation if the plant's waste load were imposed on the river at its 5-year minimum flow. Again we use the Thomas nomograph to calculate the oxygen deficit:

$$300 \times 5.8 + 1.5 + 0 = 301.5X$$

$$\frac{1740}{301.5} = X = 5.77 \text{ (initial DO at low flow)}$$

$8.22 - 5.77 = 2.45 =$ oxygen deficit (D) at low flow.

When $y = $ ppm L at low flow, then

$$(300 \times 1.23 + 1.5 \times 400)\frac{1.92}{1.23} = 301.5y$$

$$(369 + 600)\frac{1.92}{1.23} = 301.5y$$

$$\frac{1510}{301.5} = y = 5.02 \text{ ppm } L \text{ at low flow.}$$

Between Stations 1 and 2:

K_2	K_1	K_2/K_1	D_A	L_A	D_A/L_A	K_2t
0.067	0.331	0.202	2.45	5.02	0.488	0.067×0.322 = 0.022

From nomograph, we obtain

$$\frac{D}{L_A} = 0.81,$$

and therefore

$D = 0.81(5.02) = 4.07$ (calculated deficit at Station 2).

Between Stations 2 and 3:

K_2	K_1	K_2/K_1	D_A
1.59	0.331	0.48	4.07

L_A	D_A/L_A	K_2t
$\left(5.02 \times \dfrac{3.84}{1.92}\right)$ 10.05	0.405	0.159×0.483 = 0.077

From nomograph, we obtain

$$\frac{D}{L_A} = 0.65,$$

and therefore

$D = 0.65(10.05) = 6.54$ (calculated deficit at Station 3).

Between Stations 3 and 4:

K_2	K_1	K_2/K_1	D_A
1.777	0.331	5.36	6.54

L_A	D_A/L_A	K_2t
$\left(10.05 \times \dfrac{2.00}{3.84}\right)$ 5.23	1.25	1.777×0.177 = 0.314

From nomograph, we obtain

$$\frac{D}{L_A} = 0.71,$$

and therefore

$D = 0.71(5.23) = 3.71$ (calculated deficit at Station 4).

Between Stations 4 and 5:

K_2	K_1	K_2/K_1	D_A
0.463	0.331	1.4	3.71
L_A		D_A/L_A	$K_2 t$
$\left(5.23 \times \dfrac{4.93}{2.00}\right)$		0.288	0.463×0.902
12.9			$= 0.418$

From nomograph, we obtain

$$\frac{D}{L_A} = 0.4,$$

and therefore

$D = 0.4(12.9) = 5.16$ (calculated deficit at Station 5).

Between Stations 5 and 6:

K_2	K_1	K_2/K_1	D_A
0.105	0.332	0.316	5.16
L_A		D_A/L_A	$K_2 t$
$\left(12.9 \times \dfrac{2.47}{4.93}\right)$		0.8	0.105×0.42
6.45			$= 0.044$

From nomograph, we obtain

$$\frac{D}{L_A} = 1.08,$$

and therefore

$D = 1.08(6.45) = 6.97$ (calculated deficit at Station 6).

Drainage from Harrison Creek was evidently the cause of the existing oxygen sag in the river stretch between Station 1 and the dam at Station 4.

The critical oxygen concentration at 25°C in the river stretch sampled was 5.1 ppm just above the dam at U.S. Lock No. 2 (Station 4), as illustrated in Fig. 3.5. However, aeration of the water as it passed over this dam brought the oxygen level up to 6.8 ppm (the highest level obtained in the entire 38-mile reach). The pollution from Elizabethtown caused the oxygen to sag once again to 6.1 ppm at Station 5, and the relatively low reaeration rate in the stretch from Station 5 to Station 6 caused a continued sag to 5.75 ppm at the end of the reach at Station 6 (Table 3.9).

Rather drastic results occurred when the plant load was superimposed on the river at Station 1. The oxygen sagged from 5.9 ppm at Tarheel to about 3.2 ppm just above the dam at Station 4. Aeration at this point brought the level back to 5.7 ppm, from which point it sagged once again to 3.6 ppm in the stretch between Stations 4 and 6.

Thomas [7] also made available a formulation which allows the industrial-waste analyst to approximate L_A, the maximum BOD load that may be introduced into the stream without causing the oxygen concentration downstream to fall below a specified value:

$$\log L_A = \log D_c + \left[1 + \frac{K_1}{K_2 - K_1}\left(1 - \frac{D_A}{D_C}\right)^{0.418}\right]\log\frac{K_2}{K_1}.$$

By substituting in the above equation the values of the oxygen deficit to be maintained downstream, the stream-reaction constants (K_1 and K_2), and the initial oxygen deficit at the point of pollution, it is possible to calculate the maximum ultimate first-stage BOD which can be added to the stream.

When the plant waste load was imposed on the river at its 5-year minimum low flow, the oxygen sagged to about 1.4 ppm just above the dam. The greater the oxygen deficit, the greater the reaeration rate; thus, aeration at the dam brought the oxygen level to 4.7 ppm. From this point (Station 4), the oxygen level began to sag once again to a value of only 1.4 ppm at Station 6.

It was therefore recommended that the proposed plant be located just below U.S. Lock No. 2 (Station 4) where stream assets are at a maximum. With this location, some damage would be done to the river, but the oxygen level would not drop as low as if the plant were located at Station 1. Because the river already had an oxygen sag from Station 4 to Station 6, it was also recommended that a primary-treatment installation be a minimum condition of building the plant at this acceptable site. Furthermore, it was the author's opinion that this plant should not be allowed to locate at Station 1 unless it installed efficient secondary treatment.

3.3 Churchill Method of Multiple Linear Correlation

From 24 stream samples taken at appropriate points Churchill and Buckingham [1] found that a good correlation generally exists between BOD, DO, temperature, and stream flow. In other words, they found that the dissolved-oxygen sag in a stream depends upon only three variables: BOD, temperature, and flow. By means of the least-squares method [2], the line of regression can be computed, so as to predict the dissolved-oxygen sag for any desired BOD loading. This method eliminates the often questionable and always cumbersome procedure for determining times of flow between stations and the resulting stream-reaction rates (K_1, K_2, and K_3).

The author [4] found that the Churchill and Buckingham method provides a good correlation, if each stream sample is collected and observed under maximum or minimum conditions of one of the three stream variables. Only six samples were required in one study [4] to produce practical and dependable results. Additional samples may add some small degree of refinement to the results, but the refinement probably would not offset the effort of planning, collecting, and analyzing the samples and calculating the results.

The following data, which refer to the stream situation sketched in Fig. 3.6, illustrate the use of the Churchill method to obtain the line of best fit and the resulting oxygen sag. Data collected on four different days during various low- to medium-flow periods gave the sag values between Stations 3 and 5 shown in Table 3.10.

For our calculations, we use the following three normal equations, based on the principle of least squares [3], and the Doolittle method [5] of solving three simultaneous equations:

$$b_1 \Sigma X_1^2 + b_2 \Sigma X_1 X_2 + b_3 \Sigma X_1 X_3 = \Sigma X_1 Y, \quad (1)$$
$$b_1 \Sigma X_1 X_2 + b_2 \Sigma X_2^2 + b_3 \Sigma X_2 X_3 = \Sigma X_2 Y, \quad (2)$$
$$b_1 \Sigma X_1 X_3 + b_2 \Sigma X_2 X_3 + b_3 \Sigma X_3^2 = \Sigma X_3 Y. \quad (3)$$

The form of the equation for the dissolved oxygen drop is

$$Y = a + b_1 X_1 + b_2 X_2 + b_3 X_3,$$

where a, b_1, b_2, and b_3 are constants and

X_1 = BOD at sag (ppm)
X_2 = temperature at sag (°C)
X_3 = flow at sag (1000/cfs)
Y = DO drop (ppm).

Table 3.10 Multiple linear correlation of DO drop, BOD, temperature and stream discharge (four samples).

Date	Dissolved oxygen At Sta. 3, ppm	Dissolved oxygen At Sta. 5, ppm	Drop in O_2, ppm (Y)	BOD at sag, ppm (X_1)	Temp., °C (X_2)	Flow, 1000/cfs (X_3)	Y^2	YX_1	YX_2	YX_3	X_1^2	X_1X_2	X_1X_3	X_2^2	X_2X_3	X_3^2
6/18/58	10.0	7.9	2.1	8.4	11.5	7.14	4.41	17.64	24.15	15.00	70.50	96.60	59.98	132.25	82.11	50.98
7/1/58	8.2	7.2	1.0	2.6	20.5	11.71	1.00	2.60	20.50	11.71	6.76	53.30	30.45	420.25	240.05	137.12
7/22/58	7.2	2.6	4.6	21.2	20.0	19.35	21.15	97.52	92.00	89.0	449.44	424.00	410.22	400.00	387.00	374.42
8/4/58	8.3	2.8	5.5	5.83	16.5	15.02	30.25	32.06	90.75	82.61	33.99	96.19	87.57	272.25	247.83	225.60
Total			13.2	38.03	68.5	53.22	56.81	149.82	227.40	198.33	560.70	670.10	588.20	1224.75	956.95	788.12
			$\bar{Y}$	$\bar{X}_1$	$\bar{X}_2$	$\bar{X}_3$										
Means			3.30	9.51	17.12	13.31										
							$*n\bar{Y}^2$	$n\bar{Y}\bar{X}_1$	$n\bar{Y}\bar{X}_2$	$n\bar{Y}\bar{X}_3$	$n\bar{X}_1^2$	$n\bar{X}_1\bar{X}_2$	$n\bar{X}_1\bar{X}_3$	$n\bar{X}_2^2$	$n\bar{X}_2\bar{X}_3$	$n\bar{X}_3^2$
Corrected items							43.50	125.54	225.98	175.69	361.77	651.25	506.30	1172.31	911.47	708.62
Corrected sums							13.25	24.29	1.42	22.64	199.10	18.86	81.58	52.48	44.48	79.50

*n = number of terms.

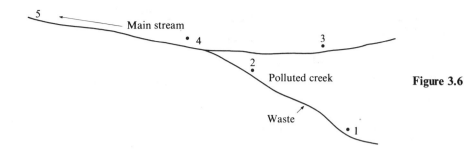

Figure 3.6

We now substitute the numerical data given in Table 3.10 and proceed step by step as follows. From Eq. (1) we obtain

$$198.93b_1 + 18.85b_2 + 81.90b_3 = 24.28. \quad (4)$$

Dividing Eq. (4) by 198.93 yields

$$b_1 + 0.0947b_2 + 0.4117b_3 = 0.1220. \quad (5)$$

Multiplying by 0.0934 gives

$$18.85b_1 + 1.785b_2 + 7.65b_3 = 2.27. \quad (6)$$

We next apply Eq. (2) and obtain

$$18.85b_1 + 52.44b_2 + 44.48b_3 = 1.42. \quad (7)$$

Subtracting Eq. (7) from (6) yields

$$-50.66b_2 - 36.72b_3 = 0.82. \quad (8)$$

Dividing Eq. (8) by -50.66, we obtain

$$b_2 + 0.727b_3 = -0.01760. \quad (9)$$

We now multiply Eq. (4) by -0.412 and have

$$-81.96b_1 - 7.76b_2 - 33.74b_3 = -10.00. \quad (10)$$

Next we multiply Eq. (8) by 0.725:

$$-36.72b_2 - 26.62b_3 = +0.6179. \quad (11)$$

Finally, we apply Eq. (3) and obtain

$$81.90b_1 + 44.48b_2 + 79.50b_3 = 22.64. \quad (12)$$

Adding Eqs. (10), (11), and (12), we have

$$19.14b_3 = 13.28, \quad (13)$$

$$b_3 = 0.6938. \quad (14)$$

From Eq. (9),

$$b_2 = -0.5198. \quad (15)$$

From Eq. (5),

$$b_1 = -0.1160. \quad (16)$$

As a check, we substitute all values in Eq. (12):

$$22.64 = 22.64,$$

$$a = \bar{Y} - (b_1\bar{X}_1 + b_2\bar{X}_2 + b_3\bar{X}_3),$$

$$a = 4.058.$$

The preceding computations yield the following equation for DO drop:

$$Y = a + b_1X_1 + b_2X_2 + b_3X_3,$$
$$Y = 4.058 - 0.1160X_1 - 0.5198X_2 + 0.6938X_3,$$

where Y = DO drop (ppm)
X_1 = BOD at sag (ppm)
X_2 = temperature at sag (°C)
X_3 = flow at sag (1000/cfs).

The results obtained from the dissolved-oxygen equation are summarized in Table 3.11 to show that this equation can predict with an acceptable degree of accuracy the sag occurring below a source of pollution. Another correlation procedure must be used to compute the allowable BOD loading at the source of

Table 3.11 Comparison of calculated oxygen drops with observed values of oxygen drop at the sag.

Date	Oxygen drop	
	Calculated from formula, ppm	Observed, ppm
6/18/58	2.07	2.10
7/1/58	1.22	1.00
7/22/58	4.66	4.60
8/4/58	5.23	5.50

pollution. The BOD equation can be derived from the same least-squares method by correlating the upstream BOD load with the temperature, discharge, and resulting BOD at the sag point in the stream. The necessary data for the development of the BOD equation are given in Table 3.12.

Using the three normal equations, (1), (2), and (3), we obtain from Eq. (1)

$$4.65b_1 - 15.06b_2 + 13.47b_3 = 4.99. \quad (17)$$

Dividing Eq. (17) by 4.65 yields

$$b_1 - 3.24b_2 + 2.90b_3 = 1.07. \quad (18)$$

We now use Eq. (2) and get

$$-15.06b_1 + 51.00b_2 - 39.23b_3 = -15.92. \quad (19)$$

If we multiply Eq. (17) by 3.24, we get

$$+15.06b_1 - 48.79b_2 + 43.64b_3 = 16.17. \quad (20)$$

Adding Eqs. (19) and (20) yields

$$2.21b_2 + 4.41b_3 = 0.25; \quad (21)$$

dividing Eq. (21) by 2.21 leads to

$$b_2 + 2.00b_3 = 0.113. \quad (22)$$

Multiplying Eq. (17) by -2.897, we have

$$-13.47b_1 + 43.63b_2 - 39.02b_3 = -14.46. \quad (23)$$

Multiplying (22) by -4.40 yields

$$-4.40b_2 - 8.80b_3 = -0.497. \quad (24)$$

Using Eq. (3), we obtain

$$13.47b_1 - 39.23b_2 + 44.87b_3 = 10.13. \quad (25)$$

We then add Eqs. (23), (24), and (25):

$$-2.95b_3 = -4.827,$$
$$b_3 = +1.64.$$

Substitution of this result in Eq. (22) yields

$$b_2 = -3.167,$$

and substitution in Eq. (25) gives

$$b_1 = -13.93.$$

We use Eq. (20) to check these results:

$$-209.79 = -209.92,$$
$$a = Y - b_1 X_1 - b_2 X_2 - b_3 X_3,$$
$$a = 114.96.$$

Table 3.12 Multiple linear correlation of BOD loads, temperature, and discharge from the source of pollution to the sag (Station 5)

Date	BOD at Sta. 5, ppm (Y)	Applied BOD load (1000 lb/day) at Sta. 2 + 3 (X_1)	At Station 5		Y^2	YX_1	YX_2	YX_3	X_1^2	$X_1 X_2$	$X_1 X_3$	X_2^2	$X_2 X_2$	X_3^2
			Temp., °C (X_2)	Discharge, 10 cfs (X_3)										
6/18/58	6.37	6.91	11.5	14.0	40.58	44.02	73.26	89.18	47.75	79.47	96.74	132.25	161.00	196.00
7/1/58	1.20	4.41	20.5	8.54	1.44	5.29	24.60	10.25	19.45	90.41	37.66	420.25	175.07	72.93
7/22/58	5.94	4.19	20.0	5.17	35.23	24.89	118.80	30.71	17.56	83.80	21.66	400.00	103.40	26.73
8/4/58	2.10	4.85	16.5	6.66	4.41	10.19	34.65	13.99	23.52	80.03	32.30	272.25	109.89	44.36
Sums	15.61	20.36	68.5	34.37	81.66	84.39	251.31	144.13	108.28	333.71	188.36	1224.75	549.36	340.02
	$\bar{Y}$	$\bar{X}_1$	$\bar{X}_2$	$\bar{X}_3$										
Means	3.90	5.09	17.13	8.59										
					$n\bar{Y}^2$	$n\bar{Y}\bar{X}_1$	$n\bar{Y}\bar{X}_2$	$n\bar{Y}\bar{X}_3$	$n\bar{X}_1^2$	$n\bar{X}_1\bar{X}_2$	$n\bar{X}_1\bar{X}_3$	$n\bar{X}_2^2$	$n\bar{X}_2\bar{X}_3$	$n\bar{X}_3^2$
Corrected items					60.79	79.40	267.23	134.0	103.63	348.77	174.89	1173.75	588.59	295.15
Corrected sums					20.87	4.99	−15.92	10.13	4.65	−15.06	13.47	51.00	−39.23	44.87

Therefore, the equation yielding the allowable BOD load becomes

$$Y = 114.96 - 13.93X_1 = 3.167X_2 + 1.64X_3,$$

where Y = BOD load at sag (Station 5) (1000 lb/day)
X_1 = combined BOD loads of upstream Stations 2 and 3 (1000 lb/day)
X_2 = temperature at sag (Station 5) (°C)
X_3 = flow at sag (Station 5) (10 cfs).

Applying the 30 per cent BOD reduction (primary treatment) to the pollution load added at Station 2 only, we have

0.70 × 3692* = 2584 lb/day at Station 2
  159 lb/day at Station 1 ⎱Natural
 1159 lb/day at Station 3 ⎰pollution
Total = 3902 lb/day.

If the above result (a somewhat reduced load) is inserted in the BOD equation at 20.5°C and 51.7 cfs at Station 5 (most critical temperature and flow), we should arrive at the BOD load remaining at the sag:

BOD = 114.96 − 13.93(3.902) − 3.167(20.5)
  + 1.64(5.17)
 = 4.17 (1000 lb/day)
 = 4170 lb/day, or 14.94 ppm at 51.7 cfs.

Using this BOD in the dissolved-oxygen-sag formula, we obtain the DO drop that would occur if the reduced effects from bottom deposits were ignored. (Less sewage pollution results in less bottom deposits, and this factor may change the deoxygenation rate of a flowing river.) We thus have

$$\text{DO drop} = 4.555 - 0.1371(14.94) - 0.5931(20.5)$$
$$+ 0.7639\left(\frac{1000}{51.7}\right)$$
$$= 5.095 \text{ ppm.}$$

At saturation (20°C),

DO = 9.2 ppm
Dissolved oxygen = 9.2 − 5.095 = 4.105 ppm.

Thus, Churchill's method (with only four samples at critical conditions) shows that primary treatment would be sufficient to sustain 4 ppm DO in the stream during critical conditions, providing the upstream station is fully saturated with dissolved oxygen.† The somewhat higher level of DO obtained by the Churchill method (3.65 according to the Streeter-Phelps formula) might be attributed to the omission of the slime effect when the BOD load is reduced. Slime growths tend to accelerate the removal of dissolved organic matter from the flowing water by increasing biological action. On the other hand, when sludge deposits exist or the slime growths become dense and voluminous, it is safe to conclude that the anaerobic decomposition products formed will require oxygen from the overlying water layers. The increased removal of organic matter in the slime counteracts to some degree the demand for oxygen from the water. The net result from slime growths and bottom deposits is accelerated local oxygen demand with subsequent (downstream) reduced oxygen demand.

*Average BOD added at Station 2.
†See result obtained by Streeter-Phelps formulations on p. 39.

References

1. Churchill, M. A., and R. A. Buckingham, "Statistical method for analysis of stream purification capacity," *Sewage Ind. Wastes* **28**, 517 (1956).
2. Cohen, J. B., and R. L. O'Connell, "The analog computer as an aid to stream self-purification computations," *J. Water Pollution Control Federation* **35**, 951 (1963).
3. Croxton, F. E., and D. J. Cowden, *Applied General Statistics*, Prentice-Hall, Englewood Cliffs, N. J. (1955), pp. 261–280.
3a. Fair, G. M., "The dissolved oxygen sag—An analysis," *Sewage Works J.* **11**, 451 (1939).
4. Simmons, J. D., N. L. Nemerow, and T. F. Armstrong, "Modified river sampling for computing dissolved oxygen sag," *Sewage Ind. Wastes* **29**, 936 (1957).
5. Steel, R. G. D., and J. H. Torrie, *Principles and Procedures of Statistics*, McGraw-Hill Book Co., New York (1960), p. 290.
6. Streeter, H. W., and E. B. Phelps, *A Study of the Pollution and Natural Purification of the Ohio River*, Bulletin no. 146, U.S. Public Health Service, Washington, D.C. (1925).
7. Thomas, H. A., "Pollution load capacity of streams," *Water Sewage Works*, **95**, 409 (1948).

Suggested Additional Reading

Camp, T. R., "Field estimates of oxygen balance parameters," *J. Sanit. Eng. Div. Am. Soc. Civil Engrs.* **92**, 115 (1966).

Fair, G. M., "The dissolved oxygen sag—An analysis," *Sewage Works J.* **11**, 445 (1939).

Kittrell, F. W., and O. W. Kochtitsky, "Shallow turbulent stream, self purification characteristics," *Sewage Works J.* **19**, 1032 (1947).

LeBosquet, M., Jr., and E. C. Tsivoglou, "Simplified dissolved oxygen computations," *Sewage Ind. Wastes* **22**, 1054 (1950).

Liebman, J. C., and Loucks, D. P., "A note on oxygen sag equations," *J. Water Pollution Control Federation* **38**, 1963 (1966).

Streeter, H. W., "A nomograph solution of the oxygen sag equation," *Sewage Ind. Wastes* **21**, 884 (1949).

CHAPTER 4

STREAM SAMPLING

Any decision on industrial waste treatment is only as dependable as the stream-sampling program on which it is based. Any time spent in planning a comprehensive stream-sampling program will be well rewarded when engineers design treatment plants which operate efficiently. Among the many factors to be considered when one designs such a sampling program [8] are:

Overall objectives of the program
Total number of samples
Points of collection
Method of collection
Data to be obtained
Frequency of sample collection
Time of year for sampling
Statistical handling of data
Care of samples prior to analysis

Overall objectives of the program. Programs may vary considerably in their objectives. In one instance the engineer may be concerned with the effect of an upstream industry on the water quality downstream; of special interest in this case might be the color of the receiving stream. In another instance he may be attempting to ascertain the dissolved-oxygen-sag characteristics of the stream during the summer season only. In still another case, he could be concerned that the stream characteristics comply with the classification standards established by the state pollution-control authorities. These are but a few examples. The importance of the other eight factors will depend to some degree on the overall objectives of the survey.

Total number of samples. The number of samples required depends on the objectives of the program and the amount of time and effort being devoted to the survey. The use of a few locations and enough samples to define the results in terms of statistical significance is usually much more reliable than using many stations with only a few samples from each. Also, samples are frequently taken over a long time interval, during which the condition of the watercourse is subject to variation. In many instances, an attempt to test all conditions by infrequent, random sampling produces no definite pattern and, in fact, may be misleading. It may be better to concentrate on well-defined, frequent, and intensive sampling.

A well-planned survey with a specific objective will require a minimum of samples. If, for example, someone wishes to determine the river characteristics—especially the dissolved-oxygen profile—during low-flow periods, two or three samples, collected at the proper time, will suffice. These samples must be collected, however, during extreme drought periods, if factors influencing the character of the stream are to be clearly established. The number of samples required often depends on the ability of the engineer to collect samples which include all significant factors. To illustrate, if a stream is known to contain phenols in addition to other organic matter but only in the fall season, the engineer must obviously collect one or more samples during the fall, despite the fact that the lowest flows and highest temperatures occur in July. Four to six river samples are the generally accepted minimum for reliable analysis and predictions. Factors such as sporadic influx of a toxic metal, flooding of the banks during certain seasons, or a peculiar flow pattern during droughts may warrant the collection of more than the minimum number of samples. It should be quite clear, however, that industrial processing is so varied, and usually unpredictable, that many samples of the receiving water under all conditions must be collected in order to evaluate truly the effect of a waste upon the stream.

Points of collection. Sampling points should be selected with great care and special consideration given to sources of pollution, dilution by branch streams, changes in surrounding topography, and slope of the river. Significant riverside features should also influence the choice of sampling points: a municipal water intake, a state park, an industrial area, a good fishing spot, a hotel, or a camping site would each have a definite bearing on the usage of the stream. Since the acceptable pollution limits for waters vary according to usage, samples should be collected and a record made of the condition of the stream just above and just below all such points of stream use or change.

As in the case history presented in Chapter 3, a minimum of four stream stations (sampling points) is recommended: (a) an upstream site, where the water is uncontaminated; (b) just below the source of pollution or dilution; (c) where the stream is in the worst condition due to a specific source of pollution (bottom of oxygen sag); (d) a point midway between bottom of oxygen sag and recovery of oxygen level. A rapid method of locating these points is to run a small boat up and down the stretch in question, using a dissolved-oxygen probe to locate sag and recovery points. This can be done from bridges but only with approximate location accuracy.

Whichever of the methods described in Chapter 3 is utilized to ascertain acceptable stream loadings, these four stations will be adequate for subsequent analysis. The necessity of additional sampling depends on various local conditions mentioned above or on unforeseen abnormalities in the characteristics of a stream, such as areas of immediate oxygen demand, sludge deposits, biological absorption, and algae growth and decay. Sampling stations should be located, as nearly as possible, at points of uniform cross-section, nonshifting bottom, minimum stream width (to facilitate sampling and increase accuracy of flow measurements), and average velocity. In addition, one should consider ease of approach to the station and ease of obtaining a representative sample. Bridges over the stream are of considerable assistance in collecting uniform samples and measuring cross-sections and stream velocity.

Sampling in tidal estuaries is a difficult and controversial problem, and engineers' answers to it have ranged from complete mathematical formulations to simple grab samples. The problem arises from the fact that pollution in tidal streams ebbs and floods with the tide, so that a portion often remains in the reach below the source of pollution for many days, rather than hours. One method of sampling which the author has found effective is to determine the tide cycle of the particular stream, then to collect samples on the high and low, as well as the mean, tide. This method provides the analyst with a consistent and overall picture of the tidal pollution situation.

Method of collection. Samples should be taken from a 0.6 depth in streams less than 2 feet deep (i.e., 1.2 feet in a 2-foot-deep stream). In deeper streams, it is necessary for the sampler to composit portions taken from depth levels of 0.2 and 0.8. When the stream flow remains quite steady, equal portions of each sample may be composited for analysis. The volume of samples depends on the number and type of analyses to be carried out on the individual and/or the composite samples. A standard-type dissolved-oxygen sampler is recommended for collecting most samples. Glass bottles with glass caps or polyethylene containers are most widely used. Any doubt about the cleanliness of the sample bottle can usually be dispelled by rinsing it first with some of the actual stream water, but special sterile bottles are required for bacteriological samples.

Data to be obtained. The scientific collection of data for stream analysis may be divided into three major categories: hydrologic factors, sources of pollution, and watercourse sampling. The type of data to be obtained depends on the objectives of the survey and the amount of time and money available for the investigation. For example, if the oxygen resources of a section of a stream are the main concern of the regulatory agency, dissolved oxygen, water temperature, and stream flow should be measured over as long and as critical a period as possible. If the survey is of a general nature, the stream analyst should undertake as many chemical, physical, and biological tests as possible, to assist him in later interpretation and evaluation of the data. The writer has found that many surveys containing only four measurements—rate of flow, temperature, BOD, and dissolved oxygen—can supply information sufficient for design of waste-treatment units. In addition, data on pH, color, and turbidity may indicate the general physical condition of the stream. Biological analyses are required when the stream water is used for drinking, bathing, or fishing. In this case, the coliform count is usually determined.

Frequency. Samples should be collected as frequently as is necessary to provide a representative total sample. The master sample should contain individual constituents of every variation expected. For example, if the pH is known to vary from 4 to 10, individual samples with pH values of 4, 5, 6, 7, 8, 9, and 10 should appear in the composite at least once during each sampling period. If the situation requires instantaneous analysis, more individual samples, with little or no compositing, are collected. The former method is practiced when one seeks to determine average existing stream conditions.

Time of year. The time of year is of utmost importance when there is a deadline for producing results. In stream studies dealing with industrial-waste treatment one is concerned primarily with critical conditions of pollution, which generally exist when the environment is at its warmest, the stream flow slowest, and the man-made pollution greatest. In most parts of the United States, under normal conditions, these critical situations occur in the summer months, so that the ideal time for stream survey is during the summer. But many studies must be undertaken in the spring and fall seasons, owing to stream conditions or the manpower situation. Indeed, because of the urgency of the problem or unusual conditions of runoff or pollution, investigations may be carried out in any season. However, the objective of every stream analyst and industrial waste engineer should be to collect his data during critical stream conditions of temperature, flow, and pollution load, when the probability of error is less.

It may be necessary to project stream analyses to conditions which might obtain during future critical periods. This is sometimes required even when a survey is made during the summer, because the intensity of the problem varies to some degree from year to year and even from one period of years to another.

Statistical handling of data. It is a well-known fact that data can be manipulated to emphasize that aspect of the survey which the analyst deems most important. This can be an ethical practice, and it does not preclude other conditions or phenomena which may exist. However, the engineer must have a working knowledge of statistics and mathematics in order to convey this information in the best form to the layman. For example, when studying the "most probable number" (MPN) of coliform bacteria present in a stream, the arithmetic mean would not clearly describe and emphasize this number, whereas the geometric mean, or mode, may well illustrate it accurately. In addition, considerable variation can, and often does, exist between the arithmetic mean and the mode in biological systems. The following example illustrates the difference between the arithmetic mean and mode for a series of coliform bacteria counts:

Sample number	Coliform count, MPN/100ml
1	0
2	0
3	0
4	9.0
5	3.6

$$\text{Arithmetic mean} = \frac{12.6}{5} = 2.52$$
$$\text{Mode} = 0$$

The rate of flow during critical conditions is of importance to the industrial waste engineer. If he incorporates in his figures the minimum flow ever recorded in a stream, he may end up with an unrealistic evaluation. On the other hand, the use of the mean summer or low-flow value can be dangerous, since lower flows than this occur quite often. Some state regulatory agencies use the minimum seven-day flow likely to recur once in ten years as the criterion for designing waste-treatment facilities.

Statistical handling of industrial-waste data is just as important as statistical handling of river data. The waste engineer should realize, for example, that figures on peak waste flows are significant under certain conditions, but the arithmetic mean BOD values are the figures required when one is designing facilities for treating these wastes. In computing treatment-plant efficiencies, the engineer can obtain a more complete picture of the plant's operation by using the standard deviation from the arithmetic mean than by using the mean value alone. Also, when comparing the efficiency of one treatment plant with another, he may find it desirable to use a coefficient of variation.

Care of samples prior to analysis. All samples should be analyzed as soon as possible after collection. The writer would recommend on-the-bank analysis whenever this is possible. With modern portable testing equipment available, there should be little reason (except convenience) for bringing samples back to the laboratory for every analysis. However, it is impractical to carry out detailed tests such as coliform counts and determinations of phenol concentration and suspended-solids quantity on the stream site. All samples subject to even the slightest chemical, physical, or biological change should be chilled immediately and kept at a temperature from 0°C to 10°C (4°C is optimum) until analyses are carried out. Dissolved-oxygen samples should be carried through the acidification stage on the stream site. Phenol samples should be preserved with copper sulfate. (For preserving samples for other analyses, see reference 7.) Plastic sampling bottles should be avoided when a reaction is possible between constituents of the waste, such as organic solvents, and plastic. Likewise, metal containers and caps should not be used to hold wastes on which metals are to be determined.

Porges [6] concerned himself with measuring the true concentration of dissolved oxygen in a stream water. He suggests: (1) whenever possible, dissolved-oxygen determinations should be made immediately after a sample is collected, with minimum exposure to light; (2) fixing a sample with acid-azide or processing it to the iodine stage and then leaving it exposed to sunlight may result in unreliable values; (3) for the particular waters sampled, untreated samples, samples processed to the iodine stage, and samples fixed by acid-azide, when stored not more than six hours on ice in the dark, all gave reasonably accurate dissolved-oxygen results; (4) icing of untreated samples under dark conditions merits consideration for stream surveys since the collection time is minimal and the samples are brought to the laboratory sooner, where analyses may be performed under optimum conditions; (5) samples should be exposed to a minimum of light during the time necessary for collection, transportation, and analysis. Porges' recommendations, though not always practicable, are significant, since the dissolved-oxygen measurement is one of the most vital in any stream survey.

Kittrell and West [5] suggest twelve "commandments" on stream survey procedures:

1. Develop a specific objective and define it clearly before undertaking a stream-pollution study

2. Review all available reports and records (essential to sound planning)

3. Do not assume that a previous occurrence will necessarily be duplicated but be alert to possible variations in the patterns of events; because of the complexities of a river system almost anything can happen, and frequently does

4. Make a thorough reconnaissance of the stream, sources of wastes, and water uses and prepare a detailed plan of operation before sending a crew into the field

5. Observe and take into account all the characteristics of the stream

6. Select stream sampling stations that will give accurate measurements of waste loads and of the orderly course of natural purification, not those that primarily save time or trouble in sampling

7. When waste discharges, stream flow, or other stream conditions vary diurnally, around-the-clock sampling is highly desirable, if not imperative; in any event, sampling times should be varied throughout the daylight hours as much as is feasible

8. Carefully evaluate wastes that may cause water quality degradation and make all analyses necessary to show damage, but avoid wasting energy and money on analyses that do not contribute to proof of pollution. One or two sets of preliminary samples will indicate whether certain analyses will be productive in marginal cases

9. Always obtain agreement with the laboratory chief on the number of samples, for the specified determinations, that can be processed daily by the personnel and facilities available

10. Maintain a continuing review of the analytical data produced by the laboratory to detect any need for revision of the original study plan

11. Always be alert for any significant clue to some important facet of the survey that could not be anticipated in the original planning of the study

12. Consider all possible interpretations of data and beware of pat answers that on first consideration appear to fit the findings

References

1. Baily, T. E., "Fluorescent-tracer studies of an estuary," *J. Water Pollution Control Federation* **38**, 1986 (1966).
2. Clark, R. N., "Discussion on sampling for effective evaluation of stream pollution," *Sewage Ind. Wastes* **22**, 683 (1950).
3. Gunnerson, C. G., "Hydrologic data collection in tidal estuaries," *Water Resources Res.* **2**, 491 (1967).
4. Haney, P. D., and J. Schmidt, *Representative Sampling and Analytical Methods in Stream Studies*, Technical Report W58-2, U.S. Public Health Service, Washington, D.C. (1958), pp. 133-142.
5. Kittrell, F. W., and A. W. West, "Stream survey procedures," *J. Water Pollution Control Federation* **39**, 627 (1967).
6. Porges, R., "Dissolved oxygen determination for field surveys," *J. Water Pollution Control Federation* **36**, 1247 (1964).
7. *Standard Methods for the Examination of Water, Sewage, and Wastes,* 10th ed., American Public Health Association, New York (1955).
8. Velz, C. J., "Sampling for effective evaluation of stream pollution," *Sewage Ind. Wastes* **22**, 666 (1950).
9. Weaver, L., "Stream surveillance programs," *J. Water Pollution Control Federation* **38**, 1334 (1966).

CHAPTER 5

ECONOMICS OF WASTE TREATMENT

At this point the author feels it appropriate to express his views on the choices open to an industry faced with the cost of pollution abatement. One often hears industry threaten to move to another part of the country or to close its doors completely, and, unfortunately, some of these threats have been carried out. The author believes rather that industry has a moral, legal, and economic responsibility to consider waste treatment as one of the variable costs of doing business, akin to labor, marketing, and raw materials. A company should seek to minimize its production costs by selecting a plant site where the total costs (labor, taxes, raw materials, water and other utilities, marketing, *and* waste treatment) are lowest, regardless of the existence or absence of a municipal plant, tax aid, or subsidy.

Industry should attempt to treat its waste at the lowest cost that will yield a satisfactory effluent for the particular receiving stream, which may necessitate considerable study, research, and pilot investigations. Planning ahead will provide time to make appropriate decisions. Conversely, lack of planning on minimizing waste-treatment costs may mean that a sudden demand for an immediate solution will cause industry to decide to cease production. We can expect industrial production costs to rise all over the U.S. as pollution-abatement costs are accepted as part of production costs. However, the advantages of particular locations and ingenuity in waste-treatment methods can have a considerable effect and industry is rapidly becoming aware of the importance of these factors.

Nevertheless, among the remarks most often made by industry when faced with abatement demands are: "We can't afford it," or "It'll put us out of business." The viewpoint of the federal government is typified by a statement by former Secretary of the Interior Stewart Udall, "It's not pollution abatement that we cannot afford—it's pollution." Thus, we have apparently reached an impasse in communication between these groups. The purpose of this chapter is to elucidate the actual benefits to industry of pollution abatement, to illustrate means of putting a dollar value on these benefits, and finally to relate the dollar value of benefits to the quality of stream water through the use of a new pollution index being developed.

Industrial waste treatment has been encouraged, promoted, and even dictated by state and federal governments on the premise that industrial wastes are similar to municipal wastes and constitute a public health menace. It has long been recognized, but never openly, that the public health hazard of industrial waste is unlike that of municipal wastes and is often nonexistent. However, there are benefits from waste treatment which exist but are difficult to describe in quantifiable terms. Since both the polluter and the public are aware that industrial waste treatment to prevent stream pollution is "a good thing," few have questioned the true reasons why their wastes should be treated.

Pollution of our watercourses has increased during the last ten years despite an increase in treatment plants. The U.S. Public Health Service reported that industrial waste-water effluents carry more than twice as much degradable organic matter into U.S. watercourses as the sewage effluents of all our municipalities combined. Population growth and increased industrial production have surpassed the elimination of wastes by proper treatment. We are faced with the dilemma of demanding expanded industrial waste treatment at more cost to industry without being able to present the benefits to be derived from such expenditure.

Most of the real benefits which result from industrial waste treatment are considered "irreducibles." This is primarily because no one has been willing or able to put a dollar value on them. If we can identify these benefits and indicate means to quantify them (whenever possible), both industry and the public will be in a position to examine the economics of a given waste-treatment situation. It then becomes important to devise a workable system for determining how pollution capacity resources in our streams should be equitably proportioned and distributed among the various competing consumers.

Streams are no longer just a means of waste conveyance to the oceans but valuable resources which can be used and reused for many purposes during their passage. Their pollution-carrying capacity must, however, be protected and preserved by all consumers and utilized in ways which are most beneficial to all society. Determination of the optimum beneficial uses of our rivers and streams and their true total costs must be a major concern of scientists and administrators.

Some of these pollution-carrying stream resources include tolerance levels for accepting quantities of:

Organic matter (decomposable)
Salt (chloride, sulfate, etc.)
Toxic materials
Color and turbidity
Suspended solids
Heat (temperature)
Algae nutrients
Radioactive matter
Grease and oils (floating matter)
Surface-active materials
Bacteria, protozoa, and viruses
Odors or odoriferous matter
Other contaminants such as pesticides and nondegradable organics (CCE)

J. G. Moore, Jr., the former Commissioner of the Federal Water Pollution Control Administration, reveals that the water pollution control policy is moving toward determining what degree of waste treatment is feasible and accepting that limit [5]. He gives the following principal reasons for this new policy:

Continued growth of population and industry, with the dual effect of generating both more wastes and greater demand on limited water resources

Growing awareness that the assimilative capacity of many lakes and streams would still be severely taxed by natural and other diffuse sources of pollution even if all point-sources of pollution were brought under complete control

The knowledge that feasible means are available, from both a technical and economic standpoint, to provide a high degree of treatment for industrial as well as municipal wastes

A growing conviction that effective water pollution control should be looked upon as one of the normal costs of running a government or a business

Increasing interest in the aesthetic value of clean water

Rising concern over the accelerated eutrophication* of lakes

Mounting public disgust over the wholesale killing of fish, birds, and other wildlife, and the growing belief that whatever effective water pollution control ultimately costs, it will be worth it

The increasing sophistication and effectiveness of conservation and other special interest groups in assessing both the strengths and weaknesses of specific water pollution control programs and policies.

The list could be continued. But these few points are enough to suggest that the current trend toward maximum feasible treatment of both municipal and industrial wastes is not likely to be reversed, now or in the future.

5.1 Benefits of Pollution Abatement

Most economists agree that there are three categories of benefit in any project: primary, secondary, and intangible. A notable exception is the system of categorizing benefits into "technological" and "pecuniary" types. Difficulties arise in this method when there is overlap from one type of benefit to another. The writer is inclined to prefer a system which separates benefits on the basis of the recipients and measurability.

Primary benefits may be defined as the accumulated worth of products and services originating directly from the project. Although many industrial firms claim that there are no primary benefits associated with pollution abatement, their proclamations may be intentionally misleading. What industries really wish to say is that the direct costs of waste treatment *to them* exceed any measurable good to them

---

*This term refers to an accelerated dying of bodies of water caused by the presence of excessive biological nutrient material.

which results from the treatment. If, on the other hand, more of these benefits were measurable, industry might be inclined to think and act differently.

Secondary benefits are often called "indirect benefits," since they tend to occur to those who do not use the output of the product and services directly. Many readily understood benefits of waste treatment fit into this category, such as a community's recreational use of clean water downstream after waste treatment by an industrial firm. The people of the community benefit *indirectly* from the industrial firm treating its waste. This situation can be referred to as a technical external economy for the community.

Intangible benefits are irreducible since no dollar value can be easily assigned to them, although it is readily apparent that they exist. For example, waste treatment might improve the morale of the community by virtue of its possession of a clean river: a sort of mental well-being exists among the inhabitants which, although real, defies quantification.

From a practical engineering standpoint the benefits of waste treatment are directly related to the value of the water and associated land downstream and should include (1) the lowered true cost of the water downstream, (2) the lessened damages for consumers utilizing contaminated water downstream, and (3) increased opportunities of associated land and water use downstream.

Commissioner Moore supports the contention that a minimum of secondary treatment for all domestic, commercial, and industrial wastes discharged to fresh water and for most wastes discharged to salt water should be required, regardless of the benefits:

1. In the vast majority of instances, secondary treatment is the minimum needed either to enhance or maintain the quality of the receiving waters.
2. Secondary treatment is economically and technologically feasible for municipal wastes; its equivalent is also feasible, through treatment or process changes or both, for industrial wastes. This fact has been abundantly demonstrated by both cities and industries.

What you have, in other words, is a virtually universal need and practicable means for meeting it. There is a lot of thrust in that combination.

Mr. Moore believes that using a true benefit-cost ratio varies from "the easy to the impossible." Where aesthetic values are concerned, he says:

> One of the major long-term benefits of water pollution prevention and control—to use the word "benefit" in its broad sense rather than in the technical sense of the word as used in the term benefit-cost analysis—is the aesthetic value of clean lakes and streams. It is not the function of benefit-cost analysis to set water quality goals or to provide economic justification for one level of water quality against another. The function of benefit-cost analysis or cost-effectiveness analysis is to determine the most practicable means of *achieving agreed-upon water quality goals*. Determining water quality *goals* is a matter of public policy. And in water quality management, as in a growing number of other areas, public policy is taking into account the indirect as well as the direct benefits of water pollution control.

5.2 Measurement of Benefits

Benefits even more than costs must indicate the recipient of the services. Whom are we benefiting: the local persons, the regional inhabitants, the entire country, or civilization as a whole? Our answer to this question will affect the decisions of administrations.

Any pollution abatement action undertaken on a river basin will have some reverberating effects on all the peoples of the earth. In most cases the effects will be greatly reduced as the distance increases from the point of abatement—like the ripples produced by a pebble thrown into a large, still lake. In some instances the ripples will be helpful or beneficial to some persons and harmful or costly to others. The total picture is complex indeed when we consider that each ripple represents many benefits and costs, some of which are intangible and most are of the secondary type. We must presume that the sum of benefits exceeds the total costs to all persons.

The question of whether the benefits exceed the costs of waste treatment on anything less than an entire civilization basis is one worth pondering. One would imagine that the smaller the area selected surrounding the abatement action the greater the excess of benefits over costs. This premise is based on the theory that people are basically selfish and will tend to do more good for themselves and immediate neighbors than harm; whereas they might do less good and more harm to persons situated far from the pollution abatement. Some would argue, on the other hand, that local people must share the major portion of the costs of pollution abatement (and thus deserve the major benefits thereof) while much of the benefit accrues to the larger segments of society. In other words, Industry A treats its wastes and incurs most of the total costs while Industry B 20 miles downstream receives most of the benefits from this treatment and pays little of the costs.

If all the true benefits (Table 5.1) could be quantified, the author believes that they would not only exceed the total costs, but also would fall largely upon persons living and working in the local river basin and using the facilities provided by the local governments. These premises cannot be verified, however, until one can truly measure all the benefits. It is suggested that engineers proceed on faith that this premise is valid until and unless disproven by factual evidence. In view of these statements the author has decided to limit the sphere of benefits of waste treatment to the area of local river basin influence.

5.3 A Proposed Method for Resource Allocation

Pollution cannot be continued without seriously affecting our country's progress as measured by the

Table 5.1 Specific benefits of industrial waste treatment.

Primary benefits

a) Savings in dollars to the industrial firm by reuse of treated effluent instead of fresh water
b) Savings in dollars resulting from compliance with regulatory agencies, i.e. avoidance of legal and expert fees and time of management involved in court cases
c) Savings in dollars from increased production efficiency, made possible by improved knowledge of the waste-producing processes and practices

Secondary benefits

a) Saving in dollars to downstream consumers from improved water quality and hence lowered operating and damage costs
b) Increase in employment, higher local payroll, and greater economic purchasing power of labor force used in construction and operation of waste-treatment facilities
c) Increased economic growth of the area due to the commitment of industry to waste treatment and potential for expansion at the existing plant
d) Increased economic growth of area with more clean water available for additional industrial operations, which in turn yield more employment and money for the area.
e) Increased value of adjacent properties as a result of a cleaner, more desirable, receiving stream
f) Increased population potential for the area since cleaner water will be available at a lower cost; the limiting factors of water cost and quantity have been pushed back further into the future

g) Increased recreational uses, such as fishing, boating, swimming, as a result of increased purity of water; recreational opportunities previously eliminated are available again

Intangible benefits

a) Good public relations and an improved industrial image after installation of pollution abatement devices
b) Improved mental health of citizens in the area confident of having adequate waste treatment and clean waters
c) Improved conservation practices, which will eventually yield payoffs in the form of more clean water for more people for more years
d) Renewal and preservation of scenic beauty and historical sites
e) Residential development potential for land areas nearby because of the presence of clean recreational waters
f) Elimination of relocation costs (of persons, groups, and establishments) because of impure waters
g) Removal of potential physical health hazards of using polluted water for recreation
h) Industrial capital investment assures permanence of the plant in the area thus lending confidence to other firms and citizens depending on the output produced by the industry
i) Technological progress, resulting from the conception, design, construction, and operation of industrial waste treatment facilities

gross national product and without placing rather severe mental and physical limitations on the existing population using stream resources. Two solutions of the pollution dilemma are open to us: the first involves regulation—policing and the courts—while the second allows the free market to solve the problem of supply and demand which is now becoming unbalanced.

Regulation with the aid of education and gentle persuasion has been practiced in the United States since the federal and state governments recognized the problem. Progress has been too slow; roadblocks have been placed in many critical areas; and politics have often replaced pollution abatement. Society seems reluctant to enforce the regulation of a resource which it does not really believe is exhaustible or limited. People are still slow to realize that pollution will finally overtake all our streams unless measures are taken to protect them now. Too many responsible citizens really believe that it will be time enough to correct the situation if and when our streams become unusable.

The economists advocate the use of the free market as a means of limiting the overdraft of water resources. In essence, they say, "Let the user pay the price of polluting a stream." If and when the price becomes too high he will be forced by economics to reduce his pollution and thus make more resources available for other consumers. There is little experience with this theory in the United States, although Kneese [4] and Fair [2] have reported and discussed the Ruhr River Valley Genossenschaften system in Germany.

However, there would be certain indisputable dangers involved in depending solely on the free market system in this country. The major one, ironically, is that aspect of the theory which makes it so desirable—making pollution-carrying resources available to those who can contribute most to the growth of the gross national product. In other words, those consumers most able to pay for the resources would receive preferential treatment. Herein lies the paradox: certain consumers unable to pay a market-clearing price would not be able to afford either to pollute or to treat wastes in competition with other consumers. Yet these very consumers might be important and useful for the good of the people. We are therefore faced with the rather common situation that what may be good for a firm, or an industry, or even the nation as measured by the gross national product may not be good for society as a whole. In fact, Galbraith [3] has recently written that our society should consider "quality of life as well as the Gross National Product" when making technical decisions. Since both regulation and marketing solutions have certain strong points in their favor, the obvious answer would be to devise a new system including these advantages and at the same time eliminating the major objections.

Establishment of the firm. In order to preserve our watercourses it is recommended that, where needed and desired, River Resource Allocation Boards be established. For example, a board might consist of a municipal official, an area farmer, a local conservationist, and a leading local manufacturer whose firm uses water. The board shall set out to determine the identity of all stream users and obtain from them (or determine for them) the separate costs of using the water and the measurable additive benefits of its use to the consumer and society.

The board—by a procedure described later—shall set the price for each unit of pollution-carrying resource used by the consumer. The product of the number of units used and the unit price will represent the actual cost to each consumer, to be paid as revenue to the board. Any revenue so collected will be used by the board in carrying out its stated objectives, which are more fully specified later. The board, in effect, shall act as a firm (and is hereafter referred to as the "firm") in its accounting and operational procedures. However, any profits which accrue will be "plowed back" into the firm to meet its stated objective of preserving the pollution-carrying resources.

Unit charges to consumers will be based on the financial condition of the firm. For example, when the stated objectives are being met readily and the firm is financially "flush," only a token unit charge will be necessary. On the other hand, when stream improvements are urgently required and/or the finances of the firm are low, relatively high unit resource charges will be called for. Submarginal consumers—such as recreational and farming users—may be partially subsidized by the firm if the latter has (a) the finances necessary and (b) the majority approval of the board.

The firm will also have the power to deny use of the pollution-carrying resource of the stream by (a) direct edict, (b) pricing the unit charges too high, or (c) denying subsidization to the user. This power symbolizes the will of the local firm and represents the best interests of the local society. For example, a slaughterhouse in the area may not be desired by the firm and it would be denied use of the stream resource by one of the three methods described above. In another case, the firm might recognize the benefits of a slaughterhouse to its community and even wish to subsidize part of its unit charge in order to enhance its operation.

In either case the will of the local people has been the deciding factor in the decision. This, the democratic way, is a most important aspect in the success of the proposed system. Nevertheless, the firm will not conflict with the objectives or the operations of any state, regional, or federal agency which exists to administer a law preserving minimum stream-quality criteria.

The firm would retain the services of legal, economic, and engineering experts to assist in its studies and decisions. This system is based on the theory that the local or regional river basin administration is the proper one to determine the highest level of water quality and the price of the resource. This is valid only where the regional consumers are willing to assume the major portion of the costs of using these resources. Likewise, state agencies should administer the minimum stream standards and approve abatement plans. However, the local firm can best determine the optimum use for its own resources that are available locally. Its case may have to be presented and defended to the state and federal governments in order to obtain extra monies for local use. The major argument for local control involves subsidies: only local representatives can carry out valid and equitable subsidies, since the firm members live and work in the area and represent the various interests of the region. Another strong point in favor of local control pertains to the direct effects of the wastes: wastes discharged into a stream affect primarily the people who reside, work, and travel in the vicinity of this stream.

Objectives of the firm.

1. To sell pollution-carrying resources in such a manner as to encourage consumers to utilize resources carefully.
2. To protect the low-level (minimum) resources in the stream by a system of charges that discourages the use of the last units of pollution-carrying resource.
3. To make certain that all potential consumers have an equal opportunity to obtain a portion of the resources at a reasonable price, by increasing the unit cost according to the number of units purchased or as fewer units remain.
4. To exert local influence over the use of the pollution-carrying resources in the form of prices and subsidies.
5. To minimize the cost of these resources by taking advantage of economies of scale to treat combined wastes where desirable or, by introducing other engineering systems, to make maximum quantities of resource available for sale.

Activities and procedures of the firm. The river resources allocation firm must conduct a market survey similar to that of any other industrial firm. The survey should reveal the following:

1. A listing of all of its potential customers.
2. The types and relative amounts of stream resource (product) desired by all consumers.
3. The amount of each type of stream resource which is available for sale by the firm. This will depend basically upon the decision of the firm to maintain a specific pollution index of the receiving stream.
4. An approximate value (reasonable market price) of each type of stream resource on a unit basis. This can be computed as a first approximation by ascertaining the total measurable benefits, ΣB,* of the use of each type of water resource in dollars and dividing this value by the number of units available for sale by the firm.

As an example, if the total measurable benefits of using the oxygen resource of a stream were $100,000

---

*ΣB = a summation of the dollar value of a unit of resource for the following:
  a) monies paid by consumers to purchase water service,
  b) damages done to this resource by existing pollution,
  c) losses incurred when opportunities of resource usage are foregone because of existing pollution.

For example, ΣBO_2 = revenue paid + damages + opportunities foregone; revenue paid = \$/gal/lb O_2/gal = \$/lb O_2.

per year and one million pounds of oxygen resources were available for sale each year, the unit charge for the first quantity of oxygen would be $0.10 per pound. Since most consumers cannot treat their wastes to reduce oxygen demand for this price, they would become interested purchasers.

A plan of increasing the price of the resource as it becomes used up seems ideal and in line with the economic principle of scarcity value. When the price of the resource reaches a level at which it is uneconomical for the consumer to buy any more, he will either curtail his production at that level or seek his own solution—such as waste treatment—at a cost somewhat lower than that charged by the firm. In either case, the firm's underlying objective of preserving a minimum stream quality will be achieved.

It is suggested that the firm hold an auction each year to assign as much of its pollution-carrying resource as possible. All potential consumers will be notified of the auction and told what resources will be sold and the lowest unit price anticipated for each resource. Let us assume that the meeting is attended by ten potential consumers interested in purchasing the firm's yearly one million pounds of oxygen resources, which were announced beforehand to be sold at the lowest price of $0.10 per pound. Since the unit price is so low, let us continue to assume that all ten consumers wish to purchase as many of these units as they can (up to their needs) at this price. The firm makes a decision that only the first 100,000 pounds may be bought at this price and that each consumer will be entitled to purchase an equal amount, i.e. 100,000/10 or 10,000 units of oxygen. Now, only 900,000 units are left for sale at a higher price of $150,00/900,000 or about $0.17 per pound. Not only did the pounds of product for sale decrease for the second allocation, but the total benefits to all users also increased, since damages would increase at the lower level of oxygen. Since both numerator and denominator reflect an increased unit price for the resource, the net effect is a much higher price.

Let us continue our assumptions, saying that one customer has sufficient units to satisfy his needs and that another cannot afford to purchase any more units at the next price. Each of the eight remaining consumers would be offered 100,000/8 or 12,500 units of oxygen at this price. Eventually, the last 100,000 units might cost $1,000,000/100,000 or $10 per pound. This last unit price would automatically be high enough to discourage any consumer from purchasing all of the remaining resources. It is suggested that these sales transactions be made final only after the firm has had a chance to review its sales and purchasers.

Difficulty arises in situations where two firms operate on the same basin, Firm A on the upper and Firm B on the lower section, and where Firm A sells all the stream resources available, leaving none for Firm B. One obvious remedy for this problem would be to merge the two firms into one whenever possible.

The firm will be faced with an important decision at the outset: what water-quality level should be maintained? In order to make this decision the firm will need to know the benefits of maintaining the stream at various water-quality levels. Total benefits (assumed to be equal to the expenditure for recreational use, the cost of damages for all uses and the opportunities foregone, in dollars), as mentioned earlier in the discussion, can be obtained by detailed surveys of the possible uses and the damages and wasted opportunities caused by the existing water quality. These total benefits can be obtained for various levels of water quality and expressed as pollution indices, determined from weighted averages of ratios of existing contaminant levels to allowable contaminant levels for each water use. At Syracuse University we are currently investigating this approach, which is designed to yield a plot of total benefits versus pollution index (Fig. 5.1).

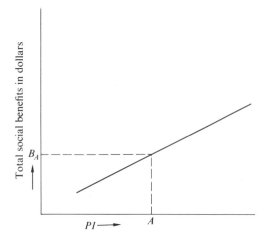

Fig. 5.1 Pollution benefits as related to water quality.

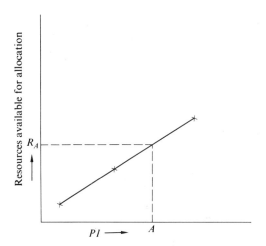

Fig. 5.2 Pollution benefits.

An increasing pollution index (PI) indicates increasing contamination. The firm may then decide to maintain the water quality at a pollution index of A, from which the total benefit (B) can be calculated as B_A. Resources (R) available for sale at a water quality comparable to PI_A can be measured at a first price of B_A/R_A obtained for marketing by the firm as shown in Fig. 5.2.

The higher the pollution index (relative contamination) the firm is willing to tolerate, the more stream resources, such as dissolved oxygen, it will have available for sale to polluters at a lower unit cost.

Summary of benefit analyses. Our society desperately needs dollar values for the true benefits of abating pollution of a given water resource to various levels of water quality. All benefits accruing to the river basin community must be evaluated. The total benefits must be related to water quality, which in turn determines the stream resources available for marketing among the many polluters of the stream. With this information a realistic price can be placed upon each unit of water resource. The firm or water resources board is provided with true benefit information for its natural resources maintained at various quality levels. This procedure enhances the job of state, regional, and federal governments and prevents pollution by giving polluters an opportunity to purchase excess natural resources or to treat wastes, whichever is more economical for them.

Initial estimates indicate that the cost of treating industrial wastes on a level comparable to secondary treatment of municipal wastes would be in the range of $2.6 to $4.6 billion during the 1969–73 federal fiscal period [7]. This includes $1.8 to $3.6 billion for new industrial treatment works and between $0.8 and $10.0 billion for replacing equipment. However, these estimates are based upon the minimal levels of control considered necessary to comply with water-quality standards. Should implementation of the standards call for higher levels of industrial waste reduction, these cost estimates could rise sharply. Meanwhile, industrial waste-abatement requirements could be met more efficiently through better in-plant controls and process changes and joint municipal and industrial treatment systems.

The estimated costs of operating and maintaining industrial waste-treatment facilities will range from $3.0 to $3.4 billion during 1969–73. As in the case of sewage-treatment works, these costs will continue to rise with increases in new treatment plants.

Manufacturing is the principal source of controllable water-borne wastes. In terms of the measurements of strength and volume usually quoted, wastes of manufacturing establishments are about *three times as great as those of the nation's sewered population* (Table 5.2). Moreover, the volume of industrial production, which gives rise to industrial wastes, is increasing by about 4.5 per cent a year or three times as fast as the population. Within industry as a whole, waste-load estimates, based on an estimate of the "average" quantity of pollutant per product unit, indicate that the chemical, paper, and kindred food industries generated about 90 per cent of the 5-day BOD (BOD_5) in untreated industrial waste-water.

Many industrial wastes differ markedly in chemical composition and toxicity from the wastes found in normal domestic sewage. Thus, the BOD_5 often is not an adequate indicator of the nature of industrial effluents. For example, industrial wastes frequently contain persistent organics which resist the secondary treatment procedures normally applied to domestic sewage and, in order to treat some industrial effluents, specific organic compounds must be stabilized or trace elements must be removed as part of the process. Obviously, these extra treatments increase the cost of joint municipal and industrial waste treatment.

An editorial in *Forbes Magazine* [1] reports the

Table 5.2 Comparison of annual totals of industrial and domestic wastes in the United States.

Source	Waste-water flow, billion gal	BOD, million lb	Settleable and suspended solids, million lb
All manufacturing	13,100	22,000	18,000
Sewered population*	5,300	7,300	8,800
	(120 gal)	(1/6 lb)	(1/5 lb)

*Annual totals calculated by multiplying individual daily figures (in parentheses) by 120 million persons × 365 days.

conclusions of a recent conference of 100 leaders of business, education, and government. Their main concern was why pollution abatement through industrial waste treatment remains hard to achieve. According to the editorial, seven difficulties stand in the way of any solution.

The problem of gathering adequate water-pollution data with regard to the movement and final disposition of wastes, the exact environmental effects, and the measurements of frequency and duration of specific pollutants

The difficulty of measuring costs against benefits

The difficulties in defining "purity," and in determining the degree of "acceptable" impurity that might be allowed

The difficulties in sorting out the relationships among separate chemical and organic pollutants, which may cancel one another out, add to each other's strength, or combine to produce even stronger new pollutants

The difficulties of informing the public of the costs... and of their need to cooperate

The difficulties in offering incentives for industry to take pollution abatement measures as long as states have different standards and enforcement measures

The difficulties of urging industries or governments to invest heavily in abatement measures before scientifically proven standards are determined.

It is readily apparent, from these difficulties and the previous discussion on resource allocation, that there should be a greater emphasis upon economics rather than on regulation alone to solve the problem of industrial pollution. One important aspect we might mention here is that of tax allowances.

New guidelines for tax depreciation of waste-treatment facilities, which provide broader classification of facilities and allow for more objective consideration by the tax examiner, were made available to industry in 1964; however, they still provide for rather long depreciation allowances or schedules. This makes waste treatment a somewhat costly investment, especially as no profit is realized from such expenditure. The size and nature of the industry will dictate in large measure not only the depreciation scheduling but also whether separate waste treatment can be economically feasible. Small companies may be induced for economic reasons to combine their wastes with municipal sewage (when the latter is adequately treated) for combined treatment, but if municipal treatment services are unavailable or incompatible with that of the industry, an insurmountable economic hardship to the industry can result. Therefore, certain tax improvements for waste treatment capital assets seem in order at both the federal and state levels. At present (1968) little positive action is being taken at the federal level to improve the tax situation, but some states are making an attempt. For example, New York State recently passed a bill which provides for (1) local tax forgiveness and (2) a 1-year depreciation write-off on state income taxes for the full value of the waste-treatment facility. Low-interest loans backed by the state might be an added inducement, especially to assist smaller industries in constructing pollution-abatement facilities.

References

1. "Cease pollution" (Editorial), *Forbes Magazine*, 1 June 1968.
2. Fair, G. M., "Pollution abatement in the Ruhr district," *J. Water Pollution Control Federation* **34**, 749 (1962).
3. Galbraith, J., *The New Industrial State*, Houghton Mifflin Co., New York (1967).
4. Kneese, A. V., *The Economics of Regional Water Quality Management*, The Johns Hopkins Press, Baltimore (1964).
5. Moore, J. G., "Water quality management in transition." *Civil Eng.* **38**, 30 (1968).
6. Nemerow, N. L., "Economics of industrial waste treatment," *Water Sewage Works J.* 238 (1968).
7. U.S. Department of the Interior, *The Cost of Clean Water*, Vol. 1, Summary Report, Jan. 10, 1968, Washington, D.C.

Part 2 | THEORIES

CHAPTER 6

VOLUME REDUCTION

In general, the first step in minimizing the effects of industrial wastes on receiving streams and treatment plants is to reduce the volume of such wastes. This may be accomplished by: (1) classification of wastes; (2) conservation of waste water; (3) changing production to decrease wastes; (4) reusing both industrial and municipal effluents as raw water supplies; or (5) elimination of batch or slug discharges of process wastes.

6.1 Classification of Wastes

If wastes are classified, so that manufacturing-process waters are separated from cooling waters, the volume of water requiring intensive treatment may be reduced considerably. Sometimes it is possible to classify and separate the process waters themselves, so that only the most polluted ones are treated and the relatively uncontaminated are discharged without treatment. The three main classes of waste are:

a) Wastes from manufacturing processes. These include waters used in forming paper on traveling wire machines, expended from plating solutions in metal fabrication, discharged from washing of milk cans in dairy plants, and so forth.

b) Waters used as cooling agents in industrial processes. The volume of these wastes varies from one industry to another, depending on the total Btu's to be removed from the process waters. One large refinery discharges a total of 150 million gallons per day (mgd), of which only 5 mgd is process waste; the remainder is only slightly contaminated cooling-water waste. Although cooling water can become contaminated by small leaks, corrosion products, or the effect of heat, these wastes contain little, if any, organic matter and are classed as nonpollutional from that standpoint.

c) Wastes from sanitary uses. These will normally range from 25 to 50 gallons per employee per day. The volume depends on many factors, including size of the plant, amount of waste-product materials washed from floors and the degree of cleanliness required of workers in the process operation.

Unfortunately, in most older plants, process, cooling, and sanitary waste waters are mixed in one pipeline; before 1930, industry paid little attention to segregating wastes to avoid stream pollution.

6.2 Conservation of Waste Water

Water conserved is waste saved. Conservation begins when an industry changes from an "open" to a "closed" system. For example, a paper mill which recycles white water (water passing through a wire screen upon which paper is formed) and thus reduces the volume of wash waters it uses is practicing water conservation. Concentrated recycled waste waters are often treated at the end of their period of usefulness, since usually it is impractical and uneconomical to treat the waste waters as they complete each cycle. The savings are twofold: both water costs and waste-treatment costs are lower. However, many changes to effect conservation are quite costly and their benefits must be balanced against the costs. If the net result is deemed economical, then new conservation practices can be installed with assurance.

A paperboard mill may discharge 10,000 gallons of waste water per ton of product, although there are many variations from one mill to the next. Paper mills may release as much as 100,000 gallons or as little as 1,000 gallons of waste water per ton of product. The latter figure is usually the result of a scarcity of water and/or an awareness of the stream-pollution problem and demonstrates what can be accomplished by effective waste elimination and conservation of water. One large textile mill reduced its water consumption by 50 per cent during a municipal water shortage, without any drop in production. The author observed that, despite the savings to the mill, water usage returned to its original level once the shortage was over. This incident further illustrates the "cheapness" of water in the public's mind.

Steel mills reuse cooling waters to quench ingots, and coal processors reuse water to remove dirt and other noncombustible materials from coal. Many industries have installed countercurrent washing to reduce water consumption. By the use of multiple vats, the plating industry utilizes make-up water, so that only the most exhausted waters are released as waste. Automation, in such forms as water-regulating devices, also aids in conservation of water. Introduction of conservation practices requires a complete engineering survey of existing water use and an inventory of all plant operations using water and producing wastes, so as to develop an accurate balance for peak and average operating conditions.

6.3 Changing Production to Decrease Wastes

This is an effective method of controlling the volume of wastes but is difficult to put into practice. It is hard to persuade production men to change their operations just to eliminate wastes. Normally, the operational phase of engineering is planned by the chemical, mechanical, or industrial engineer, whose primary objective is cost savings. The sanitary engineer, on the other hand, has the protection of public health and the conservation of a natural resource as his main considerations. Yet there is no reason why both objectives cannot be achieved.

Waste treatment at the source should be considered an integral part of production. If the chemical engineer argues that it would cost the company money to change its methods of manufacture in order to reduce pollution at the source, the sanitary engineer can do more than simply enter a plea for the improvement of mankind's environment. He can point out, for instance, that reduction in the amount of sodium sulfite used in dyeing, of sodium cyanide used in plating, and of other chemicals used directly in production has resulted in both lessening of wastes and saving of money. He can also mention the fact that balancing the quantities of acids and alkalis used in a plant often results in a neutral waste, with a saving of chemicals, money, and time spent in waste treatment. Rocheleau and Taylor [15] point out several measures that can be used to reduce wastes: improved process control, improved equipment design, use of different or better quality raw materials, good housekeeping, and preventative maintenance.

6.4 Reusing Both Industrial and Municipal Effluents for Raw Water Supplies

Practiced mainly in areas where water is scarce and/or expensive, this is proving a popular and economical method of conservation; of all the sources of water available to industry, sewage plant effluent is the most reliable at all seasons of the year and the only one that is actually increasing in quantity and improving in quality. Though there are many problems involved in the reuse of effluents for raw water supply, it must be remembered that *any* water supply poses problems to cities and industries. Since the problems of reusing sewage effluents are similar to those of reusing industrial effluents, they will be discussed here jointly.

Many industries and cities hesitate to reuse effluents for raw water supply. The reasons given [7] include lack of adequate information on the part of industrial managers, difficulty of negotiating contracts satisfactory to both municipalities and industrial users, certain technical problems such as hardness, color, and so forth, and an esthetic reluctance to accept effluents as a potential source of water for any purpose. Also, treatment plants are subject to shutdown and slug (sudden) discharges, both of which may make the supply undependable or of variable quality. In either case, industry may need an alternate source of water supply for these emergency situations. In addition, the "resistance to change in practice" factor cannot be overlooked as a major obstacle. However, as the cost of importing a raw water supply increases, it would seem logical to reuse waste-treatment plant effluents to increase the present water supply by replenishing the ground water. It cannot be denied that the ever-available treatment-plant effluent can produce a low-cost, steady water source through groundwater recharge. If any portion of a final industrial effluent can be reused, there will be less waste to treat and dispose of. Similarly, reuse of sewage effluent will reduce the quantity of pollution discharged by the municipality.

The greatest manufacturing use of water is for cooling purposes. Since the volume of this water requirement is usually great, industries located in areas where water is expensive should consider reuse of effluents. Even if the industry is fortunate enough to have a treated municipal water supply available, the cost will usually be excessive in comparison, which may have a generally beneficial effect. The reuse of municipal and industrial effluents saves water and brings revenue into the city. The design of waste-water treatment plants will be greatly influenced because the effluent must satisfy not only conventional stream requirements but those of industry as well.

Many cases are cited in the literature of industrial reuse of intermediate, untreated effluents, such as white waters from paper machines as spray and wash waters. The practice of reusing treated industrial effluents, however, is still in its infancy; there are more instances of industrial reuse of municipal effluents. For example, Wolman [17] has described the design and performance of a sewage-effluent treatment plant producing treated water at a rate of about 65 mgd for use in steel-mill processing operations. The plant employed a conventional coagulation treatment, using alum combined with chlorination; final water averaged 5 to 10 ppm turbidity, with little or no coliform bacterial contamination. The most serious problem encountered was the presence of a high concentration of chlorides. Operating costs, exclusive of interest and amortization but including pumping costs, were $1.75 per million gallons (though this figure does not include the cost of raw-sewage treatment). It is interesting to compare this with the usual municipal cost of collecting, treating, and distributing raw water of $50 to $250 per million gallons, excluding fixed charges. Even when one adds $15 to $50 per million gallons for treating the raw sewage, the reusable effluent is much more economical than water obtained by developing a separate source of raw water. Treatment-plant reuse facilities at the Sun Oil Toledo refinery have been evaluated [9] for use as make-up water in the cooling towers. The cost savings resulting from elimination of municipal fresh-water make-up were found to be $100,000 per year.

Keating and Calise [7] list five main differences between most sewage-plant effluents and typical surface- or well-water supplies: (1) higher color; (2) higher nitrogenous content; (3) higher BOD content; (4) higher total dissolved solids; and (5) the presence of phosphates, due to detergents. Industrial effluents may also possess these characteristic differences, as well as others such as higher temperature. Despite these contaminants, in many parts of the United States, the effluent from properly operated secondary sewage plants is actually superior to available surface- or well-water supplies.

The number and variety of return-flow and on-site reuse systems are increasing. Information from the latest national census shows that the overall reuse rate increased from 106 per cent (of water reused) to 136 per cent between 1954 and 1959 alone. Reuse in all industries other than steam-electric generation increased from 82 to 139 per cent during this same period. In 1959 the primary metal, chemical, paper, oil, and food industries were especially large reusers of water.

In 1957, El Paso Products Company founded a petrochemical complex near Odessa, Texas, designed to use sewage-plant effluent for cooling and boiler water. After pretreatment the only problem encountered was foaming (largely eliminated by the current changeover to "soft" detergents in domestic use). Reuse of sewage effluents often frees municipal or surface water for other valuable purposes. For example, reutilization of sewage for agricultural purposes in Israel will add a potential of 10 per cent to its total water supply. It was found that the soil structure is improved by the organics in sewage, but where industrial wastes, particularly heavy metals, are present further treatment beyond oxidation ponds is needed.

"Dry" cleaning of processing equipment instead of washing with water can greatly reduce the volume of waste water. However, this will still leave a solid waste for disposal rather than a liquid one. Hoak [6] presents a set of conservation techniques largely adapted from his experiences in steel mills:

1. Install meters in each department to make operators cost- and quantity-conscious
2. Regulate pressure to prevent needless waste
3. Use thermostatic controls to save water and increase efficiency
4. Install automatic valves to prevent loss through failure to close valves when water is no longer needed
5. Use spring-closing sanitary fixtures to prevent constant or intermittent flow of unused water

6. Descale heat exchangers to prevent loss of heat transfer and subsequent inefficient and excessive use of cooling water
7. Insulate pipes so that water is not left running to get it either cold or hot
8. Instigate leak surveys as a routine measure
9. Use centralized control to prevent wastages from improper connections
10. Recirculate cooling water, thereby saving up to 95 per cent of the water used in this process
11. Reuse, for example, blast-furnace cooling water for gas washing and clarified scale-pit water on blooming mills
12. Use high-pressure, low-volume rinse sprays for more efficiency and use a slight amount of detergent, wetting agent, or acid to improve the rinsing operation
13. Recondition waste water (often some slight in-plant treatment will provide water suitable for process use)

Eden and Truesdale [3] give typical analyses of effluents from three towns in the south of England (Table 6.1). Eden found that the total-solids content appears to increase by about 340 mg/liter between the water supply and the sewage effluent derived

Table 6.1 Typical analyses of sewage effluents after conventional primary and secondary treatment (after Eden and Truesdale [3]).

Constituent*	Source		
	Stevenage	Letchworth	Redbridge
Total solids	728	640	931
Suspended solids	15		51
Permanganate value	13	8.6	16
BOD	9	2	21
COD (chemical oxygen demand)	63	31	78
Organic carbon	20	13	
Surface-active matter			
Anionic (as Manoxol OT)	2.5	0.75	1.4
Nonionic (as Lissapol NX)			0.4
Ammonia (as N)	4.1	1.9	7.1
Nitrate (as N)	38	21	26
Nitrite (as N)	1.8	0.2	0.4
Chloride	69	69	98
Sulfate	85	61	212
Total phosphate (as P)	9.6	6.2	8.2
Total phenol			3.4
Sodium	144	124	
Potassium	26	21	
Total hardness	249	295	468
pH value	7.6	7.2	7.4
Turbidity (A.T.U.)†			66
Color (Hazen units)	50	43	36
Coliform bacteria (no./ml)	1300		3500

*Results are given in milligrams per liter, unless otherwise indicated.
†Absorptiometric turbidity units.

from it. The total-solids concentration is one of the chief limiting factors in reuse of any waste water; the number of times sewage can be reused for industrial water supply is controlled by the pickup of dissolved solids, which can be removed only by expensive treatment methods. Some discussion of the contaminants listed in Table 6.1 is relevant to potential reuse of sewage effluents for industrial water. Many industrial purposes would demand concentrations of suspended solids less than 2 mg/liter, but sewage effluents contain considerably more than this and even after tertiary treatment often contain at least 7 mg/liter. The organic constituents of sewage effluents are still largely unknown. Absorption has been suggested as a method for reducing most of the organic matter. At Lake Tahoe, for example, it has been possible to reduce the organic matter (as measured by COD) to less than 16 mg/liter by a combination of coagulation, filtration, and absorption. Detergents can also be removed in this manner to a theoretical minimum level of about 0.2 mg/liter. Additional removal of ammonia, nitrite, and nitrate is relatively expensive and difficult. Ammonia, which can be air-stripped at high pH values, is objectionable in concentrations of more than 0.1 mg/liter for drinking-water supplies which are to be chlorinated. Removal of phosphates is important whenever the water used by industry will be subjected to algae growth conditions. The Tahoe method will reduce the phosphate to less than 1.0 mg/liter; controlled activated-sludge and lime-precipitation methods are also effective. At high chlorine levels it has been found possible even to remove many viruses. Since sewage effluents contain many types of microorganisms, they should be sterilized even for industrial process use. In addition, color and hardness in sewage effluents may be harmful to certain industries.

6.5 Elimination of Batch or Slug Discharges of Process Wastes

In "wet" manufacturing of a product, one or more steps are sometimes repeated, which results in production of a significantly higher volume and strength of waste during that period. If this waste is discharged in a short period of time, it is usually referred to as a slug discharge. This type of waste, because of its concentrated contaminants and/or surge in volume, can be troublesome to both treatment plants and receiving streams. There are at least two methods of reducing the effects of these discharges: (1) the manufacturing firm alters its practice so as to increase the frequency and lessen the magnitude of batch discharges; (2) slug wastes are retained in holding basins from which they are allowed to flow continuously and uniformly over an extended (usually 24-hour) period.

References

1. Applebaum, S. B., "Industry does benefit from pollution control," *Water Wastes Eng.* **3**, 46 (1966).
2. Clarke, F. E., "Industrial re-use of water," *Ind. Eng. Chem.* **54**, 18 (1962).
3. Eden, G. E., and G. A. Truesdale, "Reclamation of water from sewage effluents," Paper read at Symposium on Conservation and Reclamation of Water, 28 November 1967, London, Reprint no. 519, Water Pollution Research Laboratory, London (1968).
4. "Flourishing on sewage-plant effluent," *Chem. Process.* **29**, 30 (1966).
5. Hershkovitz, S. Z., and F. Feinmesser, "Utilization of sewage for agricultural purposes," *Water Sewage Works* **113**, 181 (1967).
6. Hoak, R. D., "Water resources and the steel industry," *Iron Steel Engr.*, May 1964, p. 87.
7. Keating, R. J., and V. J. Calise, *The Treatment of Sewage Plant Effluent for Water Reuse in Process and Boiler Feed*, Technical Reprint T-129, Graver Water Conditioning Company, Union, N.J. (1954).
8. Marks, R. H., "Waste water treatment," *Power*, **111**, S32 (1967).
9. Mohler, E. F., Jr., H. F. Elkin, and L. R. Kumnick, "Experience with reuse and biooxidation of refinery wastewater in cooling tower systems," *J. Water Pollution Control Federation* **36**, 1380 (1964).
10. Morris, A. L., "Water renovation," *Ind. Water Eng.* **4**, 18 (1967).
11. National Association of Manufacturers and Chamber of Commerce of the United States, in cooperation with National Task Committee on Industrial Wastes, *Water in Industry*, Washington, D.C. (1967).
12. Rawn, A. M., and F. R. Bowerman, "Planned water reclamation," *J. Sewage Ind. Wastes* **29**, 1134 (1957).
13. Renn, C. E., "Serendipity at Hempstead—A study

in water management," *Ind. Water Eng.* **4**, 25 (1967).
14. Rice, J. K., "Water management to reduce wastes and recover water in plant effluents," *Chem. Eng.* **73**, 125 (1966).
15. Rocheleau, R. F., and E. F. Taylor, "An industry approach to pollution abatement," *J. Water Pollution Control Federation* **36**, 1185 (1964).
16. Unwin, H. D., "In plant wastewater management," *Ind. Water Eng.* **4**, 18 (1967).
17. Wolman, A., "Industrial water supply from processed sewage treatment plant effluent at Baltimore, Maryland," *Sewage Works J.* **20**, 15 (1948).

Suggested Additional Reading

Alexander, D. E., "Wastewater transformation at Amarillo. II. Industrial phase," *Sewage Ind. Wastes* **31**, 1107 (1959).

Berg, E. J., "Considerations in promoting the sale of sewage treatment plant effluent," *Sewage Ind. Wastes* **30**, 96 (1959).

Besselievre, E. B., "Industries recover valuable water and by-products from their wastes," *Wastes Eng.* **30**, 760, (1959).

Besselievre, E. B., "Industry must reuse effluents," *Wastes Eng.* **31**, 734 (1960).

Black, A. P., "Statement by Dr. A. P. Black," *J. Sanit. Eng. Div. Am. Soc. Civil Engrs.* **90** (**SA4**), 11 (1964).

Burrell, R., "Uses of effluent water in sewage treatment plants," *Sewage Works J.* **18**, 104 (1946).

California State Water Pollution Control Board, *Direct Utilization of Waste Waters*, Sacramento, Calif. (1955).

California State Water Pollution Control Board, "Industry utilizes sewage and waste effluent for processing operations," *Waste Eng.* **28**, 444 (1957).

Cecil, L. K., "Sewage treatment plant effluent for water reuse," *Water Sewage Works* **111**, 421 (1964).

Clarke, F. E., "Industrial re-use of water," *Ind. Eng. Chem.* **54**, 18 (1962).

Cohn, M. M., "A million tons of steel with sewage," *Wastes Eng.* **27**, 309 (1956).

Connell, C. H., "Utilization of waste waters," *Ind. Wastes* **2**, 148 (1957).

Connell, C. H., and E. J. Berg, "Industrial utilization of municipal wastewater," *Sewage Ind. Wastes* **31**, 212 (1959).

Connell, C. H., and M. C. Forbes, "Once used municipal water as industrial supply," *Water Sewage Works*, **111**, 397 (1964).

"Copper mining plant squeezes water dry," *Public Works*, **88**, 125 (1957).

Eliezer, R., R. Everett, and J. Weinstock, *Contaminant Removal from Sewage Plant Effluents by Foaming*, Advanced Waste Treatment Research Publication no. 5, U.S. Public Health Service, Cincinnati (December 1963).

Elkin, H. F., "Successful initial operation of water re-use at refinery," *Ind. Wastes* **1**, 75 (1955).

Gerster, J. A., *Cost of Purifying Municipal Waste Waters by Distillation*, Advanced Waste Treatment Research Publication no. 6, U.S. Public Health Service, Cincinnati (November 1963).

Geyer, J. C., "Reuse of sewage effluents for industrial water supply," in Proceedings of the Sixth Southern Municipal and Industrial Waste Conference, April 1957, at North Carolina State College, Raleigh.

Gloyna, E., J. Wolff, J. Geyer, and A. Wolman, "A report upon present and prospective means for improved re-use of water," Unpublished observations.

Hoak, R. D., "Industrial water conservation and re-use," *Tappi*, **44**, 40 (1961).

Hoak, R. D., "Water resources and the steel industry," *Iron Steel Eng.*, **41**, 1 (1964).

Hoppe, T. C., "Industry will reuse effluent in future waste economy drive," *Wastes Eng.* **31**, 596 (1960).

Hoot, R. A., "Plant effluent use at Fort Wayne," *Sewage Works J.* **20**, 908 (1948).

Howell, G. A., "Re-use of water in the steel industry," *Public Works*, **94**, 114 (1963).

Jenkins, S. H., "Re-use of water in industry. II. The composition of sewage and its potential use as a source of industrial water," *Water Sewage Works* **111**, 411 (1964).

Kabler, P. W., "Bacteria can be a nuisance," *Chem. Eng. Progr.* **59**, 23 (1963).

Keating, R. J., and V. J. Calise, "Treatment of sewage plant effluent for industrial re-use," *Sewage Ind. Wastes* **27**, 773 (1955).

Keefer, C. E., "Bethlehem makes steel with sewage," *Wastes Eng.* **27**, 310 (1956).

Middleton, F. M., "Advance treatment of waste waters for re-use," *Water Sewage Works* **111**, 401 (1964).

Morris, J. C., and W. J. Weber, *Preliminary Appraisal of Advanced Wastes Treatment Process*, Advanced Waste Treatment Research Publication no. 2, U.S. Department of Health, Education and Welfare, Washington, D.C. (1964).

Morris, J. C., and W. J. Weber, *Adsorption of Biochemically Resistant Materials from Solution*, Advanced Waste Treatment Research Publication no. 9, U.S. Department of Health, Education and Welfare, Washington, D.C. (1964).

Powell, S. T., "Some aspects of the requirements for the quality of water for industrial uses," *Sewage Works J.* **20**, 36 (1948).

Powell, S. T., "Adaptation of treated sewage for industrial use," *Ind. Eng. Chem.* **48**, 2168 (1956).

Randall, D. J., "Reclamation of process water," *Water Sewage Works* **111**, 414 (1964).

Scherer, C. H., "Sewage plant effluent is cheaper than city water," *Wastes Eng.* **30**, 124 (1959).

Scherer, C. H., "Wastewater transformation at Amarillo," *Sewage Ind. Wastes* **31**, 1103 (1959).

Sessler, R. E., "Waste water use in a soap and edible-oil plant," *Sewage Ind. Wastes* **27**, 1178 (1955).

Silman, H., "Re-use of water in industry. I. The re-use of water in the electroplating industry," *Chem. Ind.* **49**, 2046 (1962).

Stanbridge, H. H., "From pollution prevention to effluent re-use. Part I," *Water Sewage Works* **111**, 446 (1964).

Stanbridge, H. H., "From pollution prevention to effluent re-use. Part II," *Water Sewage Works* **111**, 494 (1964).

Stephan, D. G., "Water renovation, what it means to you," *Chem. Eng. Progr.* **59**, 19 (1963).

Stone, R., and J. C. Merrell, Jr., "Significance of minerals in waste-water," *Sewage Ind. Wastes* **30**, 928 (1958).

Tolman, S. L., "Reclaiming valuable water and bark," *Wastes Eng.* **30**, 21 (1959).

Veatch, N. T., "Industrial uses of reclaimed sewage effluents," *Sewage Works J.* **20**, 3, (1948).

Williamson, J. N., A. M. Heit, and C. Calmon, *Evaluation of Various Adsorbents and Coagulants for Waste Water Renovation*, Advanced Waste Treatment Research Program no. 12, U.S. Department of Health, Education and Welfare, Washington, D.C. (1964).

Wolman, A., "Industrial water supply from processed sewage treatment plant effluent at Baltimore, Maryland," *Sewage Works J.* **20**, 15 (1948).

CHAPTER 7

STRENGTH REDUCTION

Waste strength reduction is the second major objective for an industrial plant concerned with waste treatment. Any effort to find means of reducing the total pounds of polluting matter in industrial wastes will be well rewarded by the savings due to the reduced requirements for waste treatment. The strength of wastes may be reduced by: (1) process changes; (2) equipment modifications; (3) segregation of wastes; (4) equalization of wastes; (5) by-product recovery; (6) proportioning wastes; and (7) monitoring waste streams.

7.1 Process Changes

In reducing the strength of wastes through process changes, the sanitary engineer is concerned with wastes that are most troublesome from a pollutional standpoint. His problems and therefore his approach differ from those of the plant engineer or superintendent. Sometimes tremendous resistance by a plant superintendent must be overcome in order to effect a change in process. The superintendent possesses considerable security because he can do a familiar job well; why should he jeopardize his position merely to prevent stream pollution? The answer is obvious. Industry dies when its progress stops. No manufacturer can meet present-day market competition without continually, and critically, reviewing and analyzing his production techniques. In addition, pollution abatement can no longer be considered by industry as a "optional" act; on the contrary, it must be regarded as a vital step in preserving water resources for all users. Many industries have resolved waste problems through process changes. Two such examples of progressive management are the textile and metal-fabricating industries. On the other hand, the leather industry still generally uses lime and sulfides (major contaminants of tannery wastes), although it is known that amines and enzymes could be substituted. The lag between research and actual application is often extensive and is caused by many operational difficulties.

Textile-finishing mills were faced with the disposal of highly pollutional wastes from sizing, kiering, desizing, and dyeing processes. Starch had been traditionally used as a sizing agent before weaving and this starch, after hydrolysis and removal from the finished cloth, was the source of 30 to 50 per cent of the mill's total oxygen-demanding matter. The industry began to express an interest in cellulosic sizing agents, which would exhibit little or no BOD or toxic effect in streams. Several highly substituted cellulosic compounds, such as carboxymethyl cellulose, were developed and used in certain mills, with the result that the BOD contributed by desizing wastes was reduced almost in direct relation to the amount of cellulosic sizing compound used.

In the metal-plating industries [1], seven changes of process or materials have been suggested. Thus, to eliminate or reduce cyanide strengths: (1) change from copper–cyanide plating solutions to acid–copper solutions; (2) replace the $CuCN_2$ strike before the copper-plating bath with a nickel strike; (3) substitute a carbo-nitriding furnace, which uses a carburizing atmosphere and ammonia gas, for the usual molten cyanide bath. For other purposes: (4) use "shot blast" or other abrasive treatment on nonintricate parts instead of H_2SO_4, in pickling of steel; (5) substitute H_3PO_4 for H_2SO_4 in pickling; (6) use alkaline derusters instead of acid solutions to remove light rust which occurs during storage (the overall pH will be raised nearer to neutrality by this procedure, which will also alleviate corrosive effects on piping and sewer lines); (7) replace soluble oils, and other short-term rust-preventive oils applied to parts after cleaning, with "cold" cleaners. These cleaners can be used in both the wash solution and the rinse solution. They inhibit rust chemically rather than by a film of oil or grease. These process changes will become more understandable to the reader after the discussion of metal-plating wastes in Chapter 24.

A Pennsylvania coal-mining company modified its process to wash raw coal with acid mine waste rather than a public or private water supply. In this way, the mine drainage waste is neutralized while the coal is washed free from impurities. In one analysis, for example, the initial mine water had a pH of 3, an acidity of 4340 ppm as $CaCO_3$, and an iron content of 551 ppm; the waste water finally discharged from the process had a pH of 6.7 to 7.1 and an iron content of less than 1 ppm.

7.2 Equipment Modifications

Changes in equipment can effect a reduction in the strength of the waste, usually by reducing the amounts of contaminants entering the waste stream. Often quite slight changes can be made in present equipment to reduce waste. For instance, in pickle factories, screens placed over drain lines in cucumber tanks prevent the escape of seeds and pieces of cucumber which add to the strength and density of the waste. Similarly, traps on the discharge pipeline in poultry plants prevent emission of feathers and pieces of fat.

An outstanding example of waste strength reduction (with a more extensive modification of equipment) occurred in the dairy industry. Trebler [8] redesigned the large milk-cans used to collect farmers' milk. The new cans were constructed with smooth necks so that they could be drained faster and more completely. This prevented a large amount of milk waste from entering streams and sewage plants. Dairymen have also installed drip pans in assembly lines to collect milk which drains from the cans after they have been emptied into the sterilizers. The drip-pan contents are returned to the milk tanks daily.

In the chemical industry Hyde [4] described a chemical plant which effected a 23 per cent decrease in average BOD, through the installation of calandrias on open-bottom steam stills and by using refrigerated condensers ahead of vacuum jets, among other process modifications.

7.3 Segregation of Wastes

Segregation of wastes reduces the strength and/or the difficulty of treating the final waste from an industrial plant. It usually results in two wastes: one strong and small in volume and the other weaker, with almost the same volume as the original unsegregated waste. The small-volume strong waste can then be handled with methods specific to the problem it presents. In terms of volume reduction alone, segregation of cooling waters and storm waters from process waste will mean a saving in the size of the final treatment plant. Many dye wastes, for example, can be more economically and effectively treated in concentrated solutions. Although this type of segregation may increase the strength of the waste being treated, it will normally produce a final effluent containing less polluting matter.

Another type of segregation is the removal of one particular process waste from the other process wastes of an industrial plant, which renders the major part of the waste more amenable to treatment, as illustrated in the following examples.

A textile mill manufacturing finished cloth produced the wastes listed in Table 7.1. The combined waste was quite strong, difficult and expensive to treat, and very similar to laundry waste. However, when the liquid kiering waste was segregated from the other wastes, chemically neutralized, precipitated, and settled, the supernatant (that part which remained on the surface) could be treated chemically and biologically along with the other three wastes, because the strength of the resulting mixture was considerably less than that of the original combined waste. This type of segregation is also practiced in metal-finishing plants, which produce wastes containing both chromium and cyanide, as well as other metals. In almost all cases, it is necessary to segregate the cyanide-bearing wastes, make them alkaline, and oxidize them. The chromium wastes, on the other hand, have to be acidified and reduced. The two effluents can then be combined and precipitated in an alkaline solution to remove the metals. Without segregation, poisonous hydrogen cyanide gas would develop as a result of acidification. A recent patent [5] allows the separation of paint from waste water by precipitation with ferric chloride and/or ferric sulfate along with calcium hydroxide.

Segregation of certain wastes is of great advantage in all industries. It is dangerous, however, to arrive at a blanket conclusion that segregation of strong or dangerous wastes is always desirable. Just the reverse

Table 7.1 Wastes from a textile mill.

	Grey water*	White water*	Dye waste*	Kier waste*	Combined waste
pH	4.0	7.3	11.0	11.8	9.4
Total solids, ppm	2680	420	2880	18,880	1560
Suspended solids, ppm	224	67	148	218	156
Oxygen consumed, ppm	1560	31	556	4,900	460

* Defined in Chapter 22.

technique—complete equalization—may be necessary in certain circumstances.

7.4 Equalization of Wastes

Plants which have many products, from a diversity of processes, prefer to equalize their wastes. This requires holding wastes for a certain period of time, depending on the time taken for the repetitive processes in the plant. For example, if a manufactured item requires a series of operations that take eight hours, the plant needs an equalization basin designed to hold the wastes for that eight-hour period. The effluent from an equalization basin is much more consistent in its characteristics than is each separate influent to that same basin. Stabilization of pH and BOD and settling of solids and heavy metals are among the objectives of equalization. Stable effluents are treated more easily and efficiently than unstable ones by industrial and municipal treatment plants. Sometimes equalization may produce an effluent which warrants no further treatment. The graph in Fig. 7.1 illustrates one of the beneficial effects of equalization.

A large chemical corporation producing a predominantly acid waste has found it an advantage to equalize its wastes for a 24-hour period in an earthen holding basin. Following this equalization, a nearby plant, producing a highly alkaline waste, pumps *its* waste into the acid-waste effluent for neutralization. Considerably greater neutralizing power would be required for the acid waste were it not equalized, to iron out the peaks before neutralization.

A textile-finishing mill, which discharged its waste into a domestic secondary sewage-treatment plant, upset the efficiency of the plant. Although this waste represented only about 10 per cent of the total being

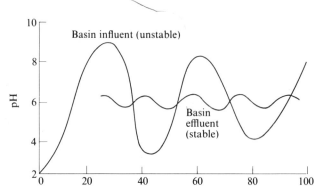

Figure 7.1

treated, it caused fluctuations, primarily in pH and BOD, which were responsible for the plant's difficulties. The solution was to build an equalization basin capable of detaining the wastes long enough to reduce the fluctuations in pH and BOD. In addition, the mill decided to deliver the equalized waste to the city treatment plant at three different rates of flow: the highest flow rate corresponded to the time when the greatest amount of sewage was reaching the plant, and vice versa. This gave a more constant dilution of the mill's waste with domestic sewage.

7.5 By-Product Recovery

This is the utopian aspect of industrial-waste treatment, the one phase of the entire problem which may lead to economic gain. Yet, many consultants deprecate this approach to the solution of waste problems. Their attitude is based mainly on statistics concerning the low percentage of successful by-products developed from waste salvage. However, any use of waste materials obviously eliminates at least some of the waste which eventually must be disposed of and the search for by-products should be encouraged, if only because it provides management with a clearer insight into processing and waste problems. All wastes contain by-products, the exhausted materials used in the process. Since some wastes are very difficult to treat at low cost, it is advisable for the industrial management concerned to consider the possibility of building a recovery plant which will produce a marketable by-product and at the same time solve a troublesome waste problem. There are many examples of positive results from adapting waste-treatment procedures to by-product recovery.

Metal-plating industries use ion exchangers to recover phosphoric acid, copper, nickel, and chromium from plating solutions. The de-ionized water, without any further treatment, is ideal for boiler-feed requirements. For final recovery of valuable chromium, copper, and nickel, companies use vacuum evaporation of the concentrated plating solutions. A nickel-wire plating plant, faced with a nickel shortage, made the plating waste alkaline with soda ash and precipitated nickel as the carbonate, then dried the sludge and treated it to recover the nickel. A silver-plating plant spends about $120,000 a year on waste treatment, of which $60,000 is returned as credit for silver recovered from the waste. The electrical industry recovers silver, gold, and (as by-products) water, valuable metals, and acids. Plants such as Scotscraft, Inc., report the recovery and reuse of by-product cyanide from plating wastes. A system of evaporation is used here to effect an overall plant saving in cost.

Specialty paper mills, with the aid of multiple-effect evaporators, recover caustic soda from cooking liquors. Chemical plants spray dilute waste acids into hot, lead-lined, brick-faced towers to concentrate the acids for reuse. Pharmaceutical houses recover the mold by drying the cake from vacuum filters or evaporating spent broth in multiple-effect evaporators. Distilleries screen the "slop" and thicken it for by-product use. Yeast factories evaporate a portion of their waste and sell the residue for cattle feed.

Even sewage plants have entered the by-product business. Methane gas from sewage digesters is commonly utilized for heat and power, and some cities make fertilizers and vitamin constituents from digested and dried sewage sludges. The sewage plant in Bradford, England, recovers grease by cracking with sulfuric acid and precipitating with alum and iron salts.

Classic examples of multiple usages of waste are the sulfite waste-liquor by-products from paper mills. They are used in fuel, road binder, cattle fodder, fertilizer, insulating compounds, as boiler-water additives and flotation agents, and in the production of alcohol and artificial vanillin. There are some 2000 U.S. patents for products made from waste sulfite liquor.

Packing houses and slaughterhouses recover waste blood, which is used as a binder in laminated wood products and in the manufacture of glue; they also sell waste greases to rendering plants.

The dairy industry treats skim milk with dilute acid to manufacture casein. Casein manufacturers in turn utilize their waste to precipitate albumin. The resulting albumin waste is used in the crystallization of milk sugar, and the residue from *this* process is utilized as poultry feed. Calcium and sodium lactate are also produced from skim milk, and dried and evaporated buttermilk is used for chicken feed. It is even rumored that chocolate ice cream originated as a by-product of the dairy industry.

Some companies, such as rendering plants, are in business primarily to develop by-products from other plants' waste products. Many rendering plants make feeds and fertilizers from chicken feet and feathers and recover grease, which is used to make soap.

Once a by-product is developed and put into production it is difficult to identify the new product with a waste-treatment process. For example, when sugar is extracted from sugar cane, a thick syrupy liquid known as blackstrap molasses is left. This molasses used to be so cheap that it was almost given away. Today it has many uses, one of the best-known being in the production of commercial alcohol. People have even found a use for the cane stalks: an insulating wallboard, called Celotex, is made from it.

These are only a few of the many ways in which industry can turn wastes into usable products. Although the problem of waste disposal usually persists, it is greatly lessened by the utilization of waste for by-products. In the final analysis both economic considerations and compliance with the requirements of pollution abatement play the major roles in any decisions involving by-product recovery.

7.6 Proportioning Wastes

By proportioning its discharge of concentrated wastes into the main sewer a plant can often reduce the strength of its total waste to the point where it will need a minimum of final treatment or will cause the least damage to the stream or treatment plant. It may prove less costly to proportion one small but concentrated waste into the main flow, according to the rate of the main flow, than to equalize the entire waste of the plant in order to reduce the strength.

7.7 Monitoring Waste Streams

Sophistication in plant control should include that of waste-water controls. Remote sensing devices that enable the operator to stop, reduce, or redirect the flow from any process when its concentration of contaminants exceeds certain limits are an excellent method of reducing waste strengths. In fact, accidental spills are often the sole cause of stream pollution or malfunctioning of treatment plants and these can be controlled, and often eliminated completely, if all significant sources of wastes are monitored.

References

1. Davis, L., "Industrial wastes control in the General Motors Corporation," *Sewage Ind. Wastes* **29**, 1024 (1957).
2. Dillon, K. E., "Waste disposal made profitable," *Chem. Eng.* **74**, 146 (1967).
3. "Factory recovers cyanides from plating wastes," *Water Works Wastes Eng.* **2**, 65 (1965).
4. Hyde, A. C., "Chemical plant waste treatment by ten methods," *J. Water Pollution Control Federation* **37**, 1486 (1965).
5. Koelsh-Folzer-Werke, A. G., "Separating paint from waste or circulating water containing paint," British Patent 1,016,673 (1966); *Chem. Abs.* **64**, 649423 (1966).
6. Rosengarten, G. M., "Union Carbide Corporation's water pollution control program," *Water Sewage Works* **114**, R181 (1967).
7. Sanders, M. E. "Implementation to meet the new water quality criteria," *Water Sewage Works* **114**, R-5 (1967).
8. Trebler, H. A., "Waste saving by improvements in milk plant equipment," in Proceedings of 1st Industrial Waste Conference, Purdue University, November 1944, pp. 6–21.

CHAPTER 8

NEUTRALIZATION

Excessively acid or alkaline wastes should not be discharged without treatment into a receiving stream. A stream even in the lowest classification—that is, one classified for waste disposal and/or navigation—is adversely affected by low or high pH values. This adverse condition is even more critical when *sudden* slugs of acids or alkalis are imposed upon the stream.

There are many acceptable methods for neutralizing overacidity or overalkalinity of waste waters, such as: (1) mixing wastes so that the net effect is a near-neutral pH; (2) passing acid wastes through beds of limestone; (3) mixing acid wastes with lime slurries or dolomitic lime slurries; (4) adding the proper proportions of concentrated solutions of caustic soda (NaOH) or soda ash (Na_2CO_3) to acid wastes; (5) blowing waste boiler-flue gas through alkaline wastes; (6) adding compressed CO_2 to alkaline wastes; (7) producing CO_2 in alkaline wastes; (8) adding sulfuric acid to alkaline wastes.

The material and method used should be selected on the basis of the overall cost, since material costs vary widely and equipment for utilizing various agents will differ with the method selected. The volume, kind, and quantity of acid or alkali to be neutralized are also factors in deciding which neutralizing agent to use.

In any lime neutralization treatment, the waste engineer should establish a minimum acceptable effluent pH and allow adequate reaction time for an acid effluent to reach this minimum pH. This will usually save considerable unnecessary expense [13]. In many cases, a mill can cut down on neutralization costs by providing sufficient detention time and sacrificing some efficiency in subsequent biological treatment (if used). During storage of alkaline wastes in contact with air, CO_2 will slowly dissolve in the waste and lower the pH. However, detention time alone, within feasible limits, will not effect as low a final pH as can be obtained by the use of neutralizing chemicals. Since biological treatment is more efficient at pH values nearer neutrality, prior neutralization by chemicals renders such treatment more effective.

8.1 Mixing Wastes

Mixing of wastes can be accomplished within a single plant operation or between neighboring industrial plants. Acid and alkaline wastes may be produced individually within one plant and proper mixing of these wastes at appropriate times can accomplish neutralization (Fig. 8.1), although this usually requires some storage of each waste to avoid slugs of either acid or alkali.

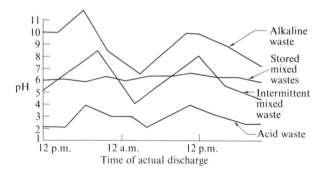

Figure 8.1

If one plant produces an alkaline waste which can be pumped conveniently to an area adjacent to a plant discharging an acid waste, an economical and feasible system of neutralization results for each plant. For example, a building-materials plant producing an alkaline (lime and magnesia) waste pumps the slurry, after some equalization, about one-half mile to mix with the effluent from a chemical plant producing an acid waste. The neutralized waste resulting from this combination is more readily treatable for final disposal and both plants thus solve problems in economics, politics, and engineering. In another instance, Hyde [8] reports the use of a 500,000-gallon reservoir ahead of an anaerobic digestion pond to mix various plant wastes prior to treatment. The resulting pH of the reservoir effluent ranges from 6.5 to 8.5.

8.2 Limestone Treatment for Acid Wastes

Passing acid wastes through beds of limestone was one of the original methods of neutralizing them [4, 15]. The wastes can be pumped up or down through the bed, depending on the head available and the cost involved, at a rate of about 1 gallon per minute (gpm) per square foot or less. Neutralization proceeds chemically according to the following typical reaction:

$$CaCO_3 + H_2SO_4 \rightarrow CaSO_4 + H_2CO_3.$$

The reaction will continue as long as excess limestone is available and in an active state. The first condition can be met simply by providing a sufficient quantity of limestone; the second condition is sometimes more difficult to maintain. A sulfuric acid solution must be diluted to an upper limit of about 5 per cent and applied at a rate less than 5 gpm/ft^2 to avoid fouling the bed. According to Jacobs [10], no attempt should be made to neutralize sulfuric acid above 0.3 per cent concentration or at a rate of feed less than 1 gpm/ft^2 because of the low solubility of calcium sulfate. Excessive acid will precipitate the calcium sulfate and cause subsequent coating and inactivation of the limestone.

Disposing of the used limestone beds can be a serious drawback to this method of neutralization, since the used limestone must be replaced by fresh at periodic intervals, the frequency of replacement depending on the quantity and quality of acid wastes being passed through a bed. When there are extremely high acid loads, foaming may occur, especially when organic matter is also present in the waste.

8.3 Lime-Slurry Treatment for Acid Wastes

Mixing acid wastes with lime slurries is an effective procedure for neutralization [5, 17–19]. The reaction is similar to that obtained with limestone beds. In this case, however, lime is used up continuously because it is converted to calcium sulfate and carried out in the waste. Though slow acting, lime possesses a high neutralizing power and its action can be hastened by heating or by oxygenating the mixture. It is relatively inexpensive, but in large quantities the cost can be an important item.

Hydrated lime is sometimes difficult to handle, since it has a tendency to arch, or bridge, over the outlet in storage bins and possesses poor flow properties, but it is particularly adaptable to neutralization problems involving small quantities of acid waste, as it can be stored in bags without the erection of special storage facilities.

In an actual case [2], neutralization of nitric and sulfuric acid wastes in concentrations up to about 1.5 per cent (in the case of sulfuric acid) was accomplished satisfactorily by using a burned dolomitic stone containing 47.5 per cent CaO, 34.3 per cent MgO, and 1.8 per cent $CaCO_3$. The concentration of acid was limited to the stated 1.5 per cent, at least in part, because of the absence of dilution water to vary the percentage. This stone provided the additional advantage of holding residual sulfation to a minimum, an impossibility with any of the high-calcium limes [9].

8.4 Caustic-Soda Treatment for Acid Wastes

Adding concentrated solutions of caustic soda or sodium carbonate to acid wastes in the proper proportions results in faster, but more costly, neutralization. Smaller volumes of the agent are required, since these neutralizers are more powerful than lime or limestone. Another advantage is that the reaction products are soluble and do not increase the hardness of receiving waters. Caustic soda is normally bled into the suction side of a pump discharging acid wastes. This method is suitable for small volumes, but for neutralizing large volumes of acid waste water, special proportioning equipment (see Chapter 9) should be provided, as well as a suitably sized storage tank for the caustic soda, with a multiple-speed pump for direct addition of the alkali to the flow of acid wastes.

We have now discussed four methods of neutralizing acid wastes. Before we move on to alkaline wastes, let us compare the basicity and costs of the acid-neutralizing methods and agents we have considered (Table 8.1).

Since the basicity factor, as shown in Table 8.1, is one of the vital factors in selecting a neutralizing agent, Hoak [7] provides not only a method for

Chemical	Cost, $/ton (approx.)	Basicity factor†	Cost, $/ton of basicity
NaOH (78% Na_2O)	106	0.687	154
Na_2CO_3 (58% Na_2O)	57	0.507	112
MgO	83	1.306	64
High-calcium hydrated lime	14	0.710	20
Dolomitic hydrated lime	14	0.912	15
High-calcium quicklime	11	0.941	12
Dolomitic quicklime	11	1.110	10
High-calcium limestone	4	0.489	8
Dolomitic limestone	4	0.564	7

Table 8.1 Cost comparison of various alkaline agents.* (After Hoak [6].)

*Based on 1954 cost quotations.
†A measure of the alkali available for neutralization (grams of equivalent CaO per gram).

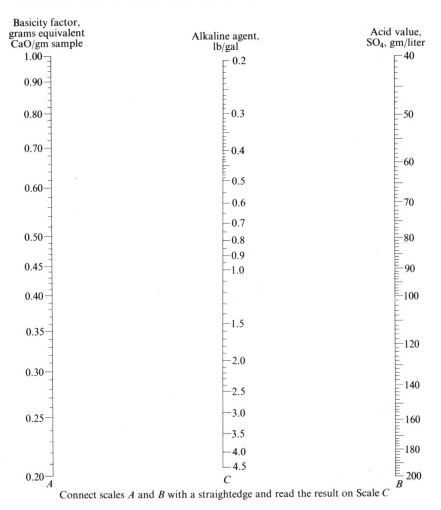

Fig. 8.2 Nomograph for treatment of acid wastes: a chart for determining the amount of alkaline agent needed. (After Hoak [7].)

Connect scales *A* and *B* with a straightedge and read the result on Scale *C*.

computing this factor but also a nomograph for calculating the pounds of neutralizing agent required per gallon of waste (Fig. 8.2). He determines the acid value by titrating a 5-ml sample of sulfuric-acid waste with an excess amount of $0.5\ N$ NaOH and back-titrating with $0.5\ N$ HCl to a phenolphthalein endpoint. The basicity factor of the lime (or neutralizing agent) is determined by titrating a 1-gm sample of alkaline agent with an excess of $0.5\ N$ HCl, boiling the sample for 15 minutes, and back-titrating with $0.5\ N$ NaOH to the phenolphthalein endpoint. The acid value (line B) and basicity factor (line A) are then connected in Hoak's nomograph to find the pounds of alkaline agent required per gallon of acid waste (line C).

When sodium hydroxide is used as a neutralizing agent for carbonic and sulfuric acid wastes, the following reactions take place:

$$Na_2CO_3 + \underbrace{CO_2 + H_2O}_{\text{carbonic acid waste}} \rightarrow 2NaHCO_3,$$

$$2NaOH + CO_2 \rightarrow Na_2CO_3 + H_2O;$$

$$NaOH + \underbrace{H_2SO_4}_{\text{sulfuric acid waste}} \rightarrow NaHSO_4 + HOH,$$

$$NaHSO_4 + NaOH \rightarrow Na_2SO_4 + HOH.$$

Both these neutralizations take place in two steps and the end-products depend on the final pH desired. For example, one treatment may require a final pH of only 6, and thus $NaHSO_4$ would make up the greater part of the products; another treatment may require a pH of 8, with most of the product being Na_2SO_4.

We shall now take up the subject of neutralization of alkaline wastes.

8.5 Using Waste Boiler-Flue Gas

Blowing waste boiler-flue gas through alkaline wastes is a relatively new and economical method for neutralizing them. Most of the experimental work has been carried out on textile wastes [1, 14, 18, 20, 21]. Well-burned stack gases contain approximately 14 per cent carbon dioxide. CO_2, dissolved in waste water, will form carbonic acid (a weak acid), which in turn reacts with caustic wastes to neutralize the excess alkalinity as follows:

$$\underset{\text{flue gas}}{CO_2} + \underset{\text{waste water}}{H_2O} \longrightarrow \underset{\text{carbonic acid}}{H_2CO_3},$$

$$\underset{\text{carbonic acid}}{H_2CO_3} + \underset{\substack{\text{caustic} \\ \text{soda in} \\ \text{waste water}}}{2NaOH} \longrightarrow \underset{\text{soda ash}}{Na_2CO_3} + 2H_2O,$$

$$\underset{\substack{\text{excess} \\ \text{carbonic} \\ \text{acid}}}{H_2CO_3} + \underset{\substack{\text{soda ash} \\ \text{in waste}}}{Na_2CO_3} \xrightarrow{H_2O} \underset{\substack{\text{sodium bi-} \\ \text{carbonate} \\ \text{in waste}}}{2NaHCO_3} + H_2O.$$

The equipment required usually consists of a blower placed in the stack, a gas pipeline to carry the gases to the waste-treatment site, a filter to remove sulfur and unburned carbon particles from gases, and a gas diffuser to disperse the stack gases in the waste water. Stack gases evolve hydrogen sulfide from waste waters which contain any appreciable quantity of sulfur, and this H_2S must be burned, absorbed, or vented positively to the upper atmosphere to prevent nuisance conditions.

8.6 Carbon-Dioxide Treatment for Alkaline Wastes

Bottled CO_2 is applied to waste waters in much the same way as compressed air is applied to activated-sludge basins. It neutralizes alkaline wastes on the same principle as boiler-feed gases (i.e., it forms a weak acid (carbonic acid) when dissolved in water) but with much less operating difficulty. The cost may be prohibitive, however, when the quantity of alkaline wastes is large. A textile mill [14] producing about 6 mgd of alkaline waste studied the practical aspects of this method and found that installation of the equipment necessary to provide bottled CO_2 would cost about $150,000 and the power and fuel to generate it about $275 per day: a considerable expense, even for so large a plant.

8.7 Producing Carbon Dioxide in Alkaline Wastes

Another way to produce carbon dioxide is to burn gas under water. This process is called submerged combustion and has been used in the disposal of nylon wastes [16] to neutralize the waste prior to biological treatment. In pilot-plant studies, the researchers [16] investigated submerged combustion on a continuous basis, using an evaporation vessel, a burner

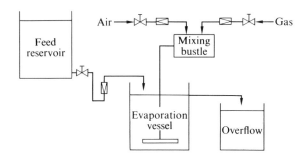

Fig. 8.3 Submerged combustion pilot unit as used by Remy and Lauria [16].

with flame jets submerged below the waste surface in the vessel, a bustle where air and natural gas were mixed to form a combustible mixture, and other equipment to measure air, gas, and waste flows and the weight of waste volatilized during each run. A schematic drawing of the submerged-combustion plant used is shown in Fig. 8.3. They concluded that submerged combustion, rather than aeration, should be used to treat part of the plant waste, for economic reasons. (The researchers in this case, however, were primarily concerned with stripping toxic materials from the waste, rather than with neutralizing it.) Krofchak [11] describes this method of neutralization and suggests its use for spent pickle liquors and spent electrolytes from nickel refining. CO_2 may also be produced by fermentation of an alkaline, organic waste; the resulting pH is thus lowered. Ebara-Infilco, Ltd., in 1965 patented such a process for fermenting alkaline beet-sugar wastes with yeast; the CO_2 produced can be used for neutralization and the excess yeast as forage.

8.8 Sulfuric-Acid Treatment for Alkaline Wastes

The addition of sulfuric acid to alkaline wastes is a fairly common, but rather expensive, means of neutralization. Sulfuric acid can cost as much as two or three cents per pound, although it may be as low as one cent per pound in large quantities. Storage and feeding equipment requirements are low as a result of its great acidity but it is difficult to handle because of its corrosiveness. The neutralization reaction which occurs when it is added to waste water is as follows:

$$2NaOH + H_2SO_4 \rightarrow Na_2SO_4 + 2H_2O.$$
waste — sulfuric — resulting
water — acid as — neutral
— neutralizer — salt

A titration curve of the alkaline waste neutralized with various amounts of H_2SO_4 is helpful to ascertain the quantities of acid required for neutralization to definite pH values and the relevant costs. Figure 8.4 represents the titration curve of an actual mixed alkaline waste in Niagara Falls, N.Y.

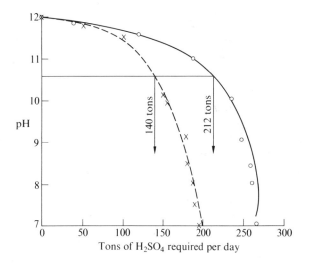

Fig. 8.4 Acid required to neutralize industrial wastes in sewer.

8.9 Acid-Waste Utilization in Industrial Processes

In some situations it may be possible to use acid wastes to effect a desired result in industrial processing—to wash, cool, or neutralize products. For example, Dillon [3] reports the use of acid mine drainage water for cleaning raw coal. Mine waste water occurs in large quantities in the coal industry. These waters are usually acid and contain sulfates of iron and aluminum; if they are used to wash raw coal, neutralization results, since coal contains calcium and magnesium carbonates. Dillon describes the treatment of 600 tons of raw coal per hour with an average of 225 gpm of mine waste water. The pH of the mine water is thereby raised from 3.0 to neutrality.

References

1. Beach, C. J., and M. G. Beach, "Treatment of alkaline dye wastes with flue gas," in Proceedings of 5th Southern Municipal and Industrial Waste Conference, April 1956, p. 162.
2. Dickerson, B. W., and R. M. Brooks, "Neutralization of acid wastes," *Ind. Eng. Chem.*, **42**, 599 (1950).
3. Dillon, K. E., "Waste disposal made profitable," *Chem. Eng.*, p. 146, 13 March 1967.
4. Gehm, H. W., "Neutralization with up-flow expanded limestone bed," *Sewage Works J.*, **16**, 104 (1944).
5. Hoak, R. D., "Neutralization studies on basicity of limestone and lime," *Sewage Works J.*, **16**, 855 (1944).
6. Hoak, R. D., "Acid iron wastes neutralization," *Sewage Ind. Wastes* **22**, 212 (1950).
7. Hoak, R. D., "A neutralization nomograph," *Ind. Wastes*, **3**, D-48 (1958).
8. Hyde, A. C., "Chemical plant waste treatment by ten methods," *J. Water Pollution Control Federation* **37**, 1486 (1965).
9. Jacobs, H. L., "Acid neutralization," *Chem. Eng. Progr.* **43**, 247 (1947).
10. Jacobs, H. L., "Neutralization of acid wastes," *Sewage Ind. Wastes* **23**, 900 (1951).
11. Krofchak, O., "Submerged combustion evaporation of acid wastes," *Ind. Water and Wastes*, **7**, 63 (1962).
12. Leidner, R. N., "Burns Harbor—Waste treatment planning for a new steel plant," *J. Water Pollution Control Federation* **38**, 1767 (1966).
13. Lewis, C. J., and L. J. Yost, "Lime in waste acid treatment," *Sewage Ind. Wastes* **22**, 893 (1950).
14. Nemerow, N. L. "Holding and aeration of cotton mill finishing wastes," in Proceedings of 5th Southern Municipal and Industrial Waste Conference, April 1956, p. 149.
15. Reidl, A. L., "Neutralization with up-flow limestone bed," *Sewage Works J.*, **19**, 1093 (1947).
16. Remy, E. D., and D. T. Lauria, "Disposal of nylon wastes," in Proceedings of 13th Industrial Waste Conference, May 1958, Purdue University Engineering Extension Series no. 96, p. 596.
17. Rudolfs, W., "Pretreatment of acid wastes," *Sewage Works J.*, **15**, 48 (1943).
18. Rudolfs, W., "Neutralization with lime," *Sewage Works J.*, **15**, 590 (1943).
19. Smith, F., "Neutralization of pickle liquor," *Sewage Works J.*, **15**, 157 (1943).
20. Steele, W. R., "Application of flue gas to the disposal of caustic textile wastes," in Proceedings of 3rd Southern Municipal and Industrial Waste Conference, March 1954, p. 190.
21. "Treatment of alkaline sulfur dye waste with flue gas," Research Report no. 8, *Proc. Am. Soc. Civil Engrs.* **82 (SA-5)**, 1078 (1956).

CHAPTER 9

EQUALIZATION AND PROPORTIONING

9.1 Equalization

Equalization is a method of retaining wastes in a basin so that the effluent discharged is fairly uniform in its sanitary characteristics (pH, color, turbidity, alkalinity, BOD, and so forth). A secondary but significant effect is that of lowering the concentration of effluent contaminants. This is accomplished not only by ironing out the slugs of high concentration of contaminants but also by physical, chemical, and biological reactions which may occur during retention in equalization basins. For example, the recent increases in industrial wastes reported by Fall [1] at Peoria have greatly varied the organic loading at the treatment plant. A retention pond serves to level out the effects of peak loadings on the plant while substantially lowering the BOD and suspended-solids load to the aeration unit. Air is sometimes injected into these basins to provide: (1) better mixing; (2) chemical oxidation of reduced compounds; (3) some degree of biological oxidation; and (4) agitation to prevent suspended solids from settling.

The size and shape of the basins vary with the quantity of waste and the pattern of its discharge from the factory. Most basins are rectangular or square, although Metzger [5] has recently found that triangular tanks produce satisfactory flow distribution. The capacity should be adequate to hold, and render homogeneous, all the wastes from the plant. Almost all industrial plants operate on a cycle basis; thus, if the cycle of operations is repeated every two hours, an equalization tank which can hold a two-hour flow will usually be sufficient. If the cycle is repeated only each 24 hours, the equalization basin must be big enough to hold a 24-hour flow of waste. Herion and Roughhead [3] report the use of 72-hour equalization for a pharmaceutical waste to ensure ample mixing. This period (three times the 24-hour cycle of operations) was selected as the proper detention time in order not to disrupt the biota of the activated-sludge

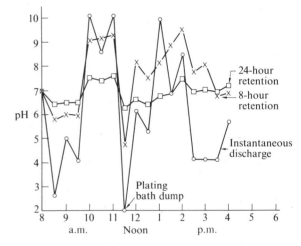

Fig. 9.1 Effect of equalization.

units. In a wool-finishing-mill waste containing dieldrin (a mothproofing insecticide) an equalization period of 44 days was necessary to yield a receiving stream concentration of less than 0.0005 mg/liter. Figure 9.1 compares the effects of 8-hour and 24-hour detention periods on the final pH of metal-plating wastes.

The mere holding of waste, however, is not sufficient to equalize it. Each unit volume of waste discharged must be adequately mixed with other unit volumes of waste discharged many hours previously. This mixing may be brought about in the following ways: (1) proper distribution and baffling; (2) mechanical agitation; (3) aeration; and (4) combinations of all three.

Proper distribution and baffling is the most economical, though usually the least efficient, method of mixing. Still, this method may suffice for many plants. Horizontal distribution of the waste is achieved by using either several inlet pipes, spaced at regular intervals across the width of the tank, or a perforated pipe

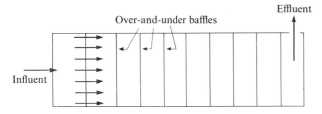

Fig. 9.2 Top view of an equalizing basin, with perforated inlet pipe and over-and-under baffles.

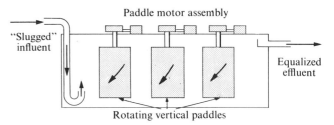

Fig. 9.3 Side view of an equalizing basin, with mechanical agitators instead of baffles.

across the entire width. Over-and-under baffles are advisable when the tank is wide, since they provide more efficient horizontal and vertical distribution (Fig. 9.2). Baffling is especially important when several different types of waste enter the basin at various locations across the width. The influent should be forced to the bottom of the basin so that the entrance velocity prevents suspended particles from sinking and remaining on the bottom.

Mechanical agitation eliminates most of the need for baffles and generally provides better mixing than baffles alone. One typical arrangement [6], shown in Fig. 9.3, utilizes three wooden gate-type agitators spaced equidistantly along the center line of the length of the tank. Agitators operated at a speed of 15 rpm by a 3-hp motor are usually adequate.

The design in Fig. 9.3 approximates the theoretically ideal tank, because of its relatively high efficiency at similar detention times, as a result of mechanical mixing, and also because it prepares varied chemical wastes for direct disposal or final treatment. If subsequent treatment is necessary, the process is made easier because the problem of wastes with rapidly changing characteristics varying from one extreme to the other is eliminated. Rudolfs and Millar [6] recommended this method of equalization when: (1) limited space is available; (2) removal of suspended solids is not desired; (3) there are rapid fluctuations in the characteristics of the wastes; and (4) facility of subsequent treatment is a goal.

This type of equipment is good not only for equalization but also for dilution, oxidation, reduction, or any other function in which one wants chemical compounds discharged at one time to react with compounds discharged before or after them, to produce a desired effect.

Aeration of equalizing basins is the most efficient way to mix wastes, but also the most expensive. To aerate an equalizing basin takes about half a cubic foot of air per gallon of waste. Aeration facilitates mixing and equalization of wastes, prevents or decreases accumulation of settled material in the tank, and provides preliminary chemical oxidation of reduced compounds, such as sulfur compounds. It is of special benefit in situations where wastes have varying character and quantity, excess of reduced compounds, and some settleable suspended solids.

9.2 Proportioning

Proportioning means the discharge of industrial wastes in proportion to the flow of municipal sewage in the sewers or to the stream flow in the receiving river. In most cases it is possible to combine equalization and proportioning in the same basin. The effluent from the equalization basin is metered into the sewer or stream according to a predetermined schedule. The objective of proportioning in sewers is to keep constant the percentage of industrial wastes to domestic-sewage flow entering the municipal sewage plant. This procedure has several purposes: (1) to protect municipal sewage treatment using chemicals from being impaired by a sudden overdose of chemicals contained in the industrial waste; (2) to protect biological-treatment devices from shock loads of industrial wastes, which may inactivate the bacteria; (3) to minimize fluctuations of sanitary standards in the treated effluent.

The rate of flow of industrial waste varies from instant to instant, as does the flow of domestic sewage, and both empty into the same sewage system. Therefore, the industrial waste must be equalized and retained, then proportioned to the sewer or stream according to the volume of domestic sewage or stream flow. To facilitate proportioning, an industry should construct a holding tank with a variable-speed pump to control the effluent discharge. Because the domestic-sewage treatment plant is usually located some distance from an industry, signalling the time and amount of flow is difficult and sometimes quite expensive. For this reason, many industries have separate pipelines through which they pump their wastes to the municipal treatment plant. The wastes are equalized separately at the site of the municipal plant and proportioned to the flow of incoming municipal waste water. Separate lines are not, of course, always possible or even necessary. One textile mill found that it could effectively proportion its waste to the variable domestic-sewage flow by adjusting the valve on the holding-tank effluent pump three times a day: 8 a.m., 12 noon, and 7 p.m.

There are two general methods of discharging industrial waste in proportion to the flow of domestic sewage at the municipal plant: manual control, related to a well-defined domestic-sewage flow pattern, and automatic control by electronics.

Manual control is lower in initial cost but less accurate. It involves determining the flow pattern of domestic sewage for each day in the week, over a period of months. Usually one does this by examining the flow records of the sewage plant or by studying the hourly water-consumption figures for the city. It is better to spend time on a careful investigation of the actual sewage flow than to make predictions based on miscellaneous, nonpertinent records. Actual investigative data should be used to support those records which are applicable to the case.

Automatic control of waste discharge according to sewage flow involves placing a metering device, which registers the amount of flow, at the most convenient main sewer connection. This device translates the rate of flow in the sewer to a recorder which is located near the industrial plant's holding tank. The pen on the recorder actuates either a mechanical (gear) or a pneumatic (air) control system for opening or closing the diaphragm of the pump. There are, of course, many variations of automatic flow-control systems. Although their initial cost is higher than that of manual control, they will usually return the investment many times by the savings in labor costs.

Some industrial and municipal sewage-plant superintendents think that the best time to release a high proportion of industrial waste to the sewer is at night, when the domestic-sewage flow is low. Whether night release is a good idea depends on the type of treatment used and the character of the industrial waste. If the treatment is primarily biological and the industrial wastes contain readily decomposable organic matter and no toxic elements, discharging the largest part of the industrial waste to the treatment plant at night is indeed advisable, since this ensures a relatively constant organic load delivered to the plant day and night.

One equipment-manufacturing company recommends a three-component system for automatic proportioning of wastes into sewers (Fig. 9.4). These three components are: (1) a kinematic manometer with integral pneumatic transmitter; (2) a remotely located indicator–program controller which receives air signals and has a precut time-pattern cam for continuously adjusting the set point of the pneumatic controller to give a waste-flow rate in accordance with

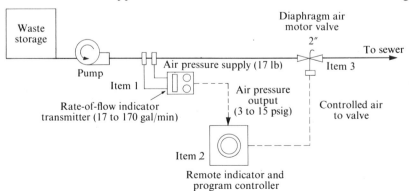

Fig. 9.4 Waste-metering system. (Courtesy Fischer and Porter Company.)

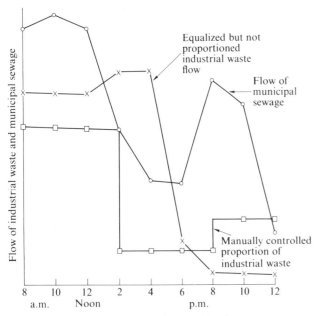

Fig. 9.5 Effect of proportioning.

the desired pattern; (3) a diaphragm-actuated, motor-controlled valve which is actuated by the air signal from the program controller. Practically speaking, the length of the pneumatic capillary tubing limits the physical separation between the sensing components, but this difficulty can be overcome with an electrical system.

The typical waste-flow proportioning system (Bubbler System) shown in Fig. 9.4, as supplied to me by Fischer and Porter Company, consists essentially of the three separate devices described above. Item 1, with a linear air-pressure output of 3 to 15 pounds per square inch, has a flow range of 17 to 170 gpm of an industrial waste (specific gravity assumed, 1.1). Item 2 is a remotely located indicator for receiving air signals from item 1, as explained in the text. Item 3 is an automatic valve capable of operating at a maximum pressure drop of 10 pounds at maximum flow rate. This valve is actuated by air signals from the program controller, item 2.

Another arrangement for proportioning industrial wastes, in a situation where pipelines are flowing only partly full or waste flows in open channels, is the use of a weir, flume, or Kennison nozzle in the main flow line to measure the flow. A float-operated transmitter (either electrical or pneumatic) is connected to this measuring device and the electrical or pneumatic signals are used to actuate a flow splitter in a proportioning weir tank (such as is provided by Proportioneers, Inc.).

The Belle, West Virginia, works of the Du Pont Nemours Company has been impounding its waste in two 2.5-million-gallon tanks and releasing it to the Kanawha River in proportion to the river flow for over 10 years [3a]. This has been necessary owing to the flashiness of the river flows. Figure 9.5 compares the effects on the flow at a municipal treatment plant of both equalization and proportioning.

References

1. Fall, E. B., "Retention pond improves activated sludge effluent quality," *J. Water Pollution Control Federation*, **37**, 1194 (1965).
2. Gibbs, C. V., and R. H. Bothel, "Potential of large metropolitan sewers for disposal of industrial wastes," *J. Water Pollution Control Federation* **37**, 1417 (1965).
3. Herion, R. W., and H. O. Roughhead, "Two treatment installations for pharmaceutical wastes," in Proceedings of 18th Industrial Waste Conference, 1964, Purdue University Engineering Extension Series, Bulletin no. 115, p. 218.
3a. Hyde, A. C., "Chemical plant waste treatment by ten methods," *J. Water Pollution Control Federation*, **37**, 1486 (1965).
4. *Manual for Sewage Plant Operators*, Texas Water and Sewage Works Association, Austin (1955), pp. 342–345.
5. Metzger, I., "Triangular tank for equalizing liquid wastes," *Water Sewage Works,* **114,** 9 (1967).
6. Rudolfs, W., and J. N. Millar, "A method for accelerated equalization of industrial wastes," *Sewage Works J.,* **18**, 686 (1946).
7. Wilroy, R. D., "Industrial wastes from scouring rug wool and the removal of dieldrin," in Proceedings of 18th Industrial Wastes Conference, 1964, Purdue University Engineering Extension Series, Bulletin no. 115, p. 413.

CHAPTER 10

REMOVAL OF SUSPENDED SOLIDS

10.1 Sedimentation

Although sedimentation is a method of treatment utilized in almost all domestic-sewage treatment plants, it should be considered for industrial-waste treatment only when the industrial waste is combined with domestic sewage or contains a high percentage of settleable suspended solids, such as are found in cannery, paper, sand-and-gravel, coal-washery, and certain other wastes. The efficiency of sedimentation tanks depends, in general, on the following factors:

Detention period
Waste-water characteristics
Tank depth
Floor surface area and overflow rate
Operation (cleanliness)
Temperature
Particle size
Inlet and outlet design
Velocity of particles
Density of particles
Container-wall effect
Number of basins (baffles)
Sludge removal
Pretreatment (grit removal)
Flow fluctuations
Wind velocity

Although settling tanks have been used for other purposes, such as grease flotation, equalization, and BOD reduction, they are primarily used for removing settleable suspended matter. Theoretically, a suspended particle in a waste-water solution will continue to settle at a fixed velocity relative to the solution, as long as the particle remains discrete; when it coalesces with other particles, its size, shape, and resulting density will change, as will its settling velocity. Coagulation, or self-flocculation, of particles causes an increase in velocity. In liquid wastes containing high percentages of suspended solids, greater reductions in the suspended solids will occur primarily because of increased flocculation. The fixed settling velocity will also be altered by changes in the temperature and density of the liquid solvent through which the particle is moving. Rising layers of warmer liquid can cause eddying and a disturbance in the settling of particles; an increased density in the lower layers of liquid can deter the particle from settling to the bottom. These factors can interfere with settling to such an extent that particles may be carried out of the tank with the effluent.

Depth of tank is also of great importance. The deeper the tank (all other factors being equal) the better the chance of preventing the deposited solids from being resuspended—e.g., by sudden scouring due to turbulence caused by unequal flow distribution or by exposure to wind or temperature effects—and thus being carried out with the effluent. This is especially important when sludge is stored in sedimentation basins for lengthy periods before pumping. If the solids are continuously removed from the bottom of settling tanks as soon as they land, shallower tanks can be built.

Surface area is another factor affecting tank efficiency, and engineers agree that floor area must be adequate to receive all the particles to be removed from the waste waters. However, many state health departments, when establishing acceptable dimensions for settling basins, do so on the basis of standard detention periods. In certain designs this method may not provide adequate floor area and complete settling is not achieved.

Figure 10.1 illustrates the effect of doubling the floor area and halving the depth of a settling basin, with volume and detention time remaining constant. Theoretically, the basin in Fig. 10.1 (b) will remove twice as many discrete particles as the basin in (a). Therefore, the engineer should strive to design settling basins which are as shallow as possible and contain ample floor area. However, tanks less than six feet deep have been found impractical from an operational standpoint, because they are subject to upsetting by scouring or velocity of currents. The floor area is increased most satisfactorily by extending the length of the basin.

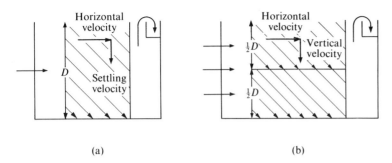

Fig. 10.1 Effect of doubling the floor area and halving the depth of a settling basin.

Since the percentage of particles reaching the bottom of the settling basin also depends on the rate of waste flow, an expression correlating horizontal flow with the floor or surface area has been devised. It is commonly referred to as the *overflow rate* and is expressed as gallons per square foot per day. Typical overflow rates vary from 200 to 800 gallons per square foot per day for primary sedimentation basins and from 1000 to 3000 gallons per square foot per day for final tanks, since particles in the final tanks usually settle more rapidly than those in primary basins. Exceptions include grit particles, which settle faster than the average particle in primary basins, and activated sludge floc, which tends to slow down the settling rate in secondary basins. Because of these discrepancies, both primary and secondary settling basins are often designed for the same overflow rates. Lower overflow rates for domestic-type wastes generally result in the removal of more suspended solids and BOD, as shown in Fig. 10.2 [10]. For further reading on theories of sedimentation the reader is referred to Eckenfelder [7] and O'Conner and Eckenfelder [13].

Unfortunately, actual settling velocities may vary from theoretical formulations. Turbulence and flocculation are the main causes of variation. Another factor is that velocities do not remain constant throughout a cross-sectional area of a tank. The settling velocity of discrete particles of diameter d in a quiescent viscous fluid is given by

$$V = \frac{4}{3} \cdot \frac{gd}{C_d} \left(\frac{\rho_s - 1}{\rho} \right)$$

where C_d is the drag coefficient between the fluid and the particle, g is the acceleration due to gravity, and ρ_s and ρ are the densities of the particle and the fluid. The drag coefficient does not remain constant but varies with the Reynolds number, R, which equals $\rho d V/\mu$. The correlation between C_d and R has been plotted in various textbooks, but a trial-and-error procedure is still required to obtain V.

Turbulence in sedimentation basins has both a positive and a negative effect on the settling velocity of a particle. It causes eddies, which carry some particles down and some up (as shown in Fig. 10.3), and thus it both helps flocculation and hinders sedimentation. Settling or rising velocities can be unequal, depending on the local circumstances causing the turbulence, such as increased horizontal velocity of water at the inlet.

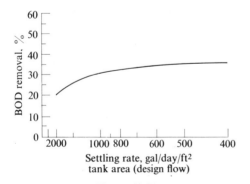

Figure 10.2

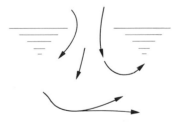

Fig. 10.3 Effect of turbulence on particle path.

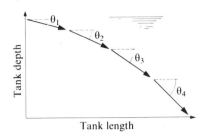

Fig. 10.4 Flocculation increases settling rate.

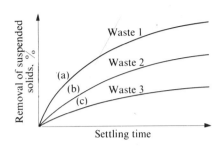

Fig. 10.5 (a) Fast and good settling characteristics typical of heavy suspended solids. (b) Medium and normal settling characteristics typical of homogeneous mixture of solids. (c) Slow and poor settling characteristics typical of highly colloidal and finely divided solids.

Other factors that induce eddying include wind, unequal distribution of flow, changes in temperature, and changes in density of the liquid at various depths. Eddying generally decreases the settling velocity and efficiency of operation, while flocculation generally increases the overall total of solids removed. The influence of flocculation is illustrated in Fig. 10.4, where θ is the angle of vertical settling of the average particle. Since shallow tanks appear to induce more flocculation, for this and other reasons they are preferred to deep tanks, provided scouring of settled particles is prevented. The average settling velocities of particles in industrial wastes vary appreciably (Fig. 10.5).

The percentage of suspended solids removed depends on the tank design, which in turn depends on the demands of the particular situation. In recent years, design engineers have been using either circular or square tanks instead of the conventional rectangular basins, for reasons of space and/or economics. Circular tanks require less form work, materials, and land space than rectangular basins for large flows and for any size of tank. However, they are less efficient, owing to (1) reduced length of effective settling zone and (2) short circuiting (waste water leaving the tank prior to theoretical detention time). The efficiency of circular tanks has been increased somewhat by the introduction of peripheral feed with center draw-off. This system eliminates the turbulence at the inlet.

The relative percentages of total transverse distance occupied by the inlet zones of circular and rectangular talks are shown in Fig. 10.6. Since the inlet zone of a circular tank occupies such a large portion of the horizontal particle path, special care must be used in designing inlet and outlet devices. The slightest disturbance in flow conditions will tend to disrupt the operation of a circular tank, but with long, narrow, rectangular tanks, the design of the inlet and outlet zones becomes less important.

Short circuiting means that effective sedimentation is not taking place in the entire volume of the settling tank; that is, a given entering volume of waste is hindered from spreading uniformly throughout the tank in a quiescent manner, so that it reaches the effluent weir before the theoretical detention time has been utilized. This is essentially true in all tanks, regardless of shape, but it seems to occur most readily in circular and square tanks, as illustrated in Fig. 10.7. To avoid short circuiting, some state regulatory agencies specify a minimum distance between the inlet and exit of the tank. It has also been demonstrated graphically by Camp [4] (see Fig. 10.8)

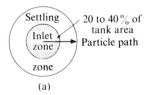

Fig. 10.6 Inlet zone of a circular tank (a) occupies 20 to 40 per cent of tank area. Inlet zone of a rectangular tank (b) occupies only 10 to 15 per cent of tank area.

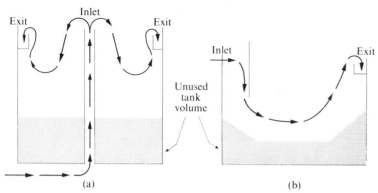

Fig. 10.7 (a) Circular tank. (b) Square tank.

that different shapes of sedimentation tanks cause different degrees of short circuiting. Villemonte et al. [17] recently showed that hydraulic efficiencies, predicted by basin dispersion curves, are related directly to the basin performance, measured by suspended-solids reduction.

In Fig. 10.8, the higher peaks occurring over shorter ranges of t/T indicate the absence of short circuiting. Curve A is a theoretical one for an ideal, instantaneous dispersion of a slug with entire tank contents. (Short circuiting approximates this.) Curve B is for a circular tank and indicates that some suspended contaminant reaches the outlet after about 15 per cent of the detention period, and the greatest concentration of matter reaches the outlet after about 50 per cent of the detention period. Curve C shows the situation in a wide rectangular tank, which approximates a square one. Curve D refers to a long, narrow, rectangular tank and indicates that no contaminant reaches the end of the tank until after 50 per cent of the detention period and most reaches the outlet after about 80 per cent of the detention period. Curve E is the dispersion curve of a round-the-end, long, baffled, rectangular chamber, with great length compared to width and depth. This type of tank gives a theoretical maximum contaminant content in the effluent after 100 per cent of the detention period, but little or none before this time. The student can readily appreciate, from a study of Fig. 10.8, the importance of proper design of sedimentation tanks. Preference should be given, wherever possible, to long, rectangular tanks with proper baffling.

A major objective of sedimentation is to produce sludge with the highest possible solids concentration. As the reader will see in Chapter 14, the volume and weight of sludge requiring final disposal is a major factor in waste treatment. A relatively new piece of equipment to achieve this objective is the Clarithickener, which combines sludge separation in circular settling tanks and thickening by means of slowly

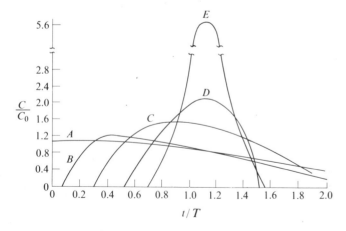

Fig. 10.8 Typical dispersion curves for various tanks (see text for explanation). The vertical axis shows the ratio of the actual concentration of contaminant (C) to the concentration of contaminant mixed with the entire tank volume (C_0); the horizontal axis shows the ratio of the actual time (t) a concentration takes to reach the end of the tank to (T), the total detention period (vol/rate). (After Camp [4].)

rotating picket-fence arms. Other methods of decreasing short circuiting include effective inlet and outlet design, properly located baffling, inboard weirs, and modification of existing sedimentation tanks to obtain better flow distribution.

Although the differences between domestic sewage and industrial wastes are often quite significant, some general statements made for domestic sewages hold true for all wastes. Normally, with detention periods of two hours, primary sedimentation basins remove 50 to 70 per cent of the suspended solids in the influent. Data collected from waste-water treatment-plant superintendents [3] are presented in Tables 10.1 and 10.2, to show design criteria and efficiencies of removal for rectangular and circular tanks.

10.2 Flotation

Flotation is the process of converting suspended substances and some colloidal, emulsified, and dissolved substances to floating matter [11]. The term "flotation" includes both violently agitated froth flotation, as used in the separation of ores in the mining industry, and quiescent flotation, which is now becoming popular as an efficient method for the removal of most suspensions from waste waters.

Small and difficult-to-settle particles in suspension can be flocculated and buoyed to the liquid surface by the lifting power of the many minute air bubbles which attach themselves to the suspended particles. Floated agglomerated sludges can be readily and continuously removed from the surface of the liquid by skimming. These skimmings are usually collected as a concentrated sludge and normally drain quite readily. A convenient practice is to detain the sludge float in a receiving tank for a few hours before draining the subnatant liquor from the bottom. The solids content of the float can be more than doubled by this concentration method; water is actually squeezed out of the float while the particles compact. Such a sludge float is usually quite stable and free from odors. Since the flotation process brings partially reduced chemical compounds into contact with oxygen in the form of tiny air bubbles, satisfaction of any immediate oxygen demand of the waste water is thereby aided.

Typical vacuum flotation units first aerate the waste with air diffusers or mechanical beaters. Aeration periods are brief, some as short as 30 seconds, and require only about 0.025 to 0.05 cubic feet of air per gallon of waste water. A brief de-aeration period is then provided at atmospheric pressure, to remove large bubbles. The waste, at this point nearly saturated with dissolved air, passes to an evacuation tank which is enclosed and maintained under a vacuum of about nine inches of mercury. This vacuum gives rise to bubbles, which cause flotation.

Pressure flotation differs from vacuum flotation in that air is injected into the waste under pressure, and bubbles of air are then formed when the waste is exposed to atmospheric pressure. Wastes are normally pressurized to about 30 to 40 pounds per square inch and retained at this pressure for approximately a minute. Some coagulant aids (alum and/or silica) and a small volume of air can be bled into the system at the suction end of the pump, where waste water enters the tank. Passage through the pump usually suffices to provide good mixing of the chemicals and air with the waste. When released to the atmosphere in the flotation tank, the tiny, rising bubbles trap suspended, colloidal, and (some) emulsified particles. The floated sludge is usually continuously skimmed and removed from the tank by sludge pumps.

Vrablik [19] makes a distinction between two methods of flotation: dissolved-air and dispersed-air. Dispersed-air flotation generates gas bubbles by the mechanical shear of propellers, diffusion of gas through porous media, or by homogenizing a gas and liquid stream. Dissolved-air flotation generates gas bubbles by precipitation from a solution supersaturated with the gas. These bubbles are much smaller than dispersed-air bubbles, generally not exceeding 80 microns* in diameter, while dispersed-air bubbles often reach 1000 microns in diameter.

To understand the theory of dissolved-air flotation, the student must investigate the gas, liquid, and solid phases as they are brought into intimate contact with each other. Henry's law indicates the relationship between the solubility of gas (in this case, dissolved air) and the total pressure:

$$C = kp,$$

where C is the concentration of gas in solution, k is Henry's law constant, and p is the absolute pressure above the solution at equilibrium.

---

*1 micron = 0.001 mm = 0.0000394 in.; 1 in. = 2.54 cm.

Table 10.1 Rectangular primary settling-tank data.*

Plant location	No. of tanks	Length, ft	Width, ft	Depth, ft	Length/width	Length/depth	Flow, mgd	Detention, hr	Overflow, gpd/ft²	Weir rate, gpd/ft	Raw suspended solids, mg/liter	Removal suspended solids, %	Raw BOD, mg/liter	Removal BOD, %
Hartford, Conn.	8	100	68	8.8	1.5	11.4	24.30	3.53	450	56,800	173	61	240	42
Detroit, Mich.	8	270	117	13	2.3	20.8	418.00	1.41	1650	408,000	184	44	153	39
Racine, Wis.	4	140	40	10.5	3.5	13.3	17.03	2.48	760	106,500	149	67	133	48
New York City, Bowery Bay	3	124	50	12	2.5	10.3	41.00	0.98	2210	284,000	152	39	169	22
New York City, Tallmans Island	3	124	50	11.6	2.5	10.7	31.00	1.25	1670	215,000	137	55	128	39
Fort Wayne, Ind.	3	100	33	13	3.3	7.7	18.70	1.25	1890	94,500	409	61	231	34
Rochester, N.Y.	2	37	12	8	3.1	4.6	0.81	1.56	914	41,200	233	21	260	21
Marshalltown, Iowa	3	80	16	8	5.0	10.0	1.22	1.51	950	13,550	436	58	414	42
Kenosha, Wis.	4	132	32	10.4	4.1	12.7	12.77	2.49	755	100,000	138	48	102	48
Jackson, Mich.	3	67.3	31	10	2.2	6.7	0.17	1.22	1470	118,000	193	16.1	134	22
Hammond, Ind.	6	120	16	13.25	7.5	9.0	20.70	1.32	1800	24,000	273	30	206	25
New York City, 26th Ward	4	162	67	12	2.4	13.5	41.00	2.16	930	35,500	139	31	127	28
New York City, Hunts Point	4	168	108.9	12	1.5	14.0	95.00	1.70	1300	97,000	140	48	113	30
Abington, Pa.	2	50	14	10	3.6	5.0	1.24	2.02	855	44,400	237	39	198	29
Portsmouth, Va.	4	100	15.25	10	6.5	10.0	7.36	1.49	1200	46,000	153	63	185	45
Canton, Ohio	3	124	32	10.6	3.9	11.7	17.00	1.33	1430	214,000	577	40	253	33
Niles, Mich.	6	75	14	9	5.4	8.3	2.30	1.86	362	27,200	250	69.2	106	57
Dallas, Tex.	2	180	50	12	3.6	15.0	19.40	2.00	1080	24,000	358	66	256	41
Richmond, Ind.	4	95	16	14.5	5.9	6.5	6.10	2.64	990	25,000	159	40	133	23
Lansing, Mich.	16	87.5	16	10	5.5	8.7	16.45	2.45	735	23,700	445	76	201	68
Winsted, Conn.	2	65	12	9	5.5	7.2	0.50	5.00	320	20,800	130	75	170	51
Waterbury, Conn.	3	212.5	33	10	6.4	21.2	13.94	2.71	660	14,500	144	54	166	33
Oklahoma City, Okla.	3	85	33	10	2.5	8.5	5.19	2.91	619	20,400	242	50	228	31
Tampa, Fla.	4	170	40	13	4.2	13.1	12.30	5.12	455	17,300	215	69	183	37
Roanoke, Va.	2	120	32	10.5	3.8	11.4	7.76	1.87	1010	120,000	230	67	190	51
Blackstone Valley, R.I.	2	230	68	10.8	3.4	21.1	12.21	4.97	390	62,000	212	62	333	12
East Hartford, Conn.	2	125	32	7.5	3.9	16.7	1.50	7.18	187	12,500	212	54	242	50
Milford, Conn.	2	55	16	9.75	3.5	5.1	9.70	4.40	400	21,800	150	79	130	72
Springfield, Mass.	4	115	50	14.5	2.3	7.9	17.5	3.36	761		160	49	145	26
Orrville, Ohio	2	43.8	16	10.4	2.7	4.2	0.73	3.65	515		342	64	415	18
New Haven, Conn.	3	145	31	11.5	4.7	12.6	14.7	1.90	1090		176	49		
Cleveland, Ohio (Easterly)	8	115	50	15	2.3	7.7	97.7	1.27	2120		240	37	149	35

*Data from plant superintendents. See reference 8, pp. 90–91.

Table 10.2 Circular primary tanks: long-term performance data.*

Location	Data period Years	No.	Average flow, mgd	No. of tanks	Diameter, ft	Sidewater depth, ft	Detention, hr	Overflow, gpd/ft²	Suspended solids Raw, mg/liter	Effluent, mg/liter	Removal, %	BOD Raw, mg/liter	Effluent, mg/liter	Removal, %	Sludge Solids, %	Volatile matter, %
Washington, D.C.	1944–45	2	136.3	12	106	14	1.88	1350	163	83	49	173	120	30.5	8.05	67.5
Winnipeg, Man.	1943–44	2	22.8	2	115	12	1.98	1100	348	159	55	310	231	25.5	9.0	70.5
Battle Creek, Mich.	1938–42	5	4.92	2	80	10	3.66	490	282	85	70	264	174	34.1	5.5	82.5
Buffalo, N.Y.	1939–41	3	135	4	160	15	1.6	1690	209	114	46	138	107	22.5	5.8	59
Albuquerque, N. Mex.	1939–46	7	5	1	80	12.2	2.21	995	254	91	61	282	150	44.5	3.9	81
Yakima, Wash.	1942	1	9.5	4	90	9	4.32	373	110	23	74	175	92	50	7.0	74.4
Appleton, Wis.	1938–45	7	4.8	2	70	10	2.90	623	276	63	77	284	141	50	5.6	58
Baltimore, Md.	1939–44	4	89.5	3	170	12	1.64	1360	214	83	61	281	204	27.5	3.9	82.7
Springfield, Ohio	1937–40	4	14.8	2	90	10	1.55	1160	166	63	62	90	43	52		
Mansfield, Ohio	1944–45	2	3	1	65	12	2.38	905	208	87	58	227	139	38.8	4.2	76
Cedar Rapids, Iowa	1936–44	9	4.21	1	70	11.5	1.95	1060	354	132	63	383	291	24	5.5	81.2
Austin, Tex.	1944–45	2	5.64	1	75	12	1.69	1275	263	95	64	285	152	46.3	4.0	83
Denver, Colo.	1939–43	5	46	4	140	9.7	2.34	750	187	44	77	212	108	49	5.4	76
Ypsilanti, Mich.	1943–45	3	1.66	2	40	9	2.5	660	226	87	62	141	95	33	8.2	71.4
Monroe, Mich.	1938–46	8	4.3	2	85	7.5	3.55	378	329	75	77	135	73	46	5.2	67.7

*Data from plant superintendents and/or annual reports. See reference 8, pp. 90–91.

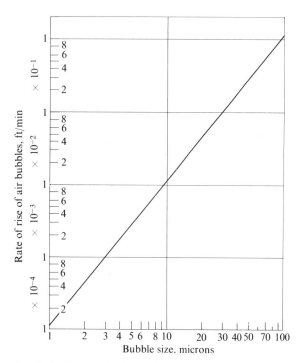

Fig. 10.9 Rate of rise of air bubbles in tap water (calculated by means of Stoke's law) as a function of bubble size. (After Vrablik [18].)

rise of air bubbles in the water. This can best be expressed by Stokes's law, which holds true for particles with a diameter of less than 130 microns:

$$V = kD,$$

where V is the rate of bubble rise (ft/min), k is Stokes's conversion factor (this includes all the factors which affect the rise or fall of bubbles, such as density or viscosity of the liquid, excluding the density of the bubble), and D is the diameter of the air bubble. The Stokes relationship is shown quantitatively in Fig. 10.9.

Typical results obtained from samples of several industrial wastes [11] treated by dissolved-air flotation show suspended solids and BOD reductions of 69 to 97.5 and 60 to 91.8 per cent, respectively (Table 10.3).

Since almost twice as much air can be dissolved in water, all other factors being equal, at 0°C than at 30°C, the temperature of waste water is a significant factor in the effectiveness of the flotation process. This relationship is shown in Fig. 10.10.

By attachment to, or inclusion in, a suspended-solids structure or liquid phase, the bulk density of the paired system may be less than the density of the parent system, causing the agglomeration to be floated to the top. The gas bubbles therefore render a buoyancy to the original suspended particle in accordance with Archimedes' principle: the resultant pressure of a fluid on an immersed body acts vertically upward through the center of gravity of the displaced fluid and is equal to the weight of the fluid displaced. The resultant upward force exerted by the fluid on the body is called *buoyancy* and this force is responsible for the floating of solids which were originally somewhat heavier than the surrounding fluid.

Since we are usually dealing with large volumes of water in waste treatment, detention time in flotation chambers becomes a critical factor. Detention time, in turn, is dependent primarily on the rate of

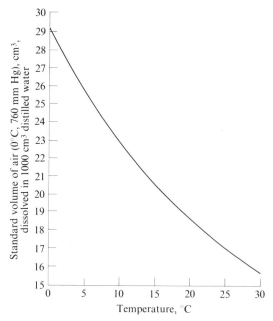

Fig. 10.10 Solubility of air in distilled water at various temperatures. (From *Handbook of Chemistry and Physics*, 36th edition, Chemical Rubber Publishing Company (1955), p. 1609.)

Table 10.3 Typical efficiencies of dissolved-gas flotation treatment of wastes [14].

Waste source	Suspended solids in influent, ppm	Reduction obtained, %	BOD in influent, ppm	Reduction obtained, %
Petroleum production	441	95.0		
Railroad maintenance	500	95.0		
Meat packing	1400	85.6	1225	67.3
Paper manufacturing	1180	97.5	210	62.6
Vegetable-oil processing	890	94.8	3048	91.6
Fruit-and-vegetable canning	1350	80.0	790	60.0
Soap manufacture	392	91.5	309	91.6
Cesspool pumpings	6448	96.2	3399	87.0
Primary sewage treatment	252	69.0	325	49.2
Glue manufacture	542	94.3	1822	91.8

Generally, air bubbles are negatively charged, the anions collecting mainly on the gas side of the interface, while the cations spread themselves out thinly on the water side of the interface. Since suspended particles or colloids may have a significant electrical charge, either attraction or repulsion will occur between these and the air bubbles.

Vrablik [18] made an extensive study of the three different processes by which flotation may be caused: (1) adhesion of a gas bubble to a suspended liquor or solid phase; (2) the trapping of gas bubbles in a floc structure as the gas bubble rises; (3) the absorption of a gas bubble in a floc structure as the floc structure is formed. These three phenomena are illustrated in Fig. 10.11.

An illustration of pressure flotation is shown in Fig. 10.12.

Finally, the engineer should be aware of both the advantages and disadvantages of flotation as a waste-treatment process [8]. The advantages are as follows:

1) Grease and light solids rising to the top and grit and heavy solids settling to the bottom are all removed in one unit

2) High overflow rates and short detention periods mean smaller tank sizes, resulting in decreased space requirements and possible savings in construction costs

3) Odor nuisances are minimized because of the short detention periods and, in pressure and aeration-type units, because of the presence of dissolved oxygen in the effluent

4) Thicker scum and sludge are obtained, in many cases, from a flotation unit than from gravity settling and skimming.

The disadvantages are as follows:

1) The additional equipment required results in higher operating costs

2) Flotation units generally do not give as effective treatment as gravity-settling units, although the efficiency varies with the waste

3) The pressure type has high power requirements, which increase operating cost

4) The vacuum type requires a relatively expensive airtight structure capable of withstanding a pressure of nine inches of mercury; any leakage to the atmosphere will adversely affect performance

5) More skilled maintenance is required for a flotation unit than for a gravity-settling unit.

Quigley and Hoffman [14] deserve credit for daring to refer to flocculation and dissolved-air flotation as "secondary treatment" when it follows sedimentation. They describe an effective dissolved-air flotation system for treating oil-refinery wastes. By recycling treated effluent and using lime as a coagulant they were able to obtain oil removals of 68 to 96 per cent.

10.3 Screening

Screening of industrial wastes is generally practiced on wastes containing larger suspended solids of variable sizes, e.g., from canneries, pulp and paper

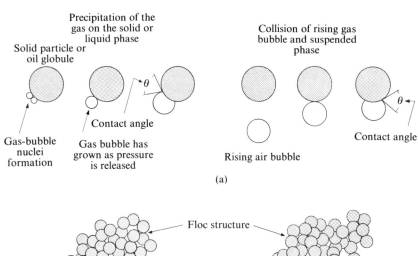

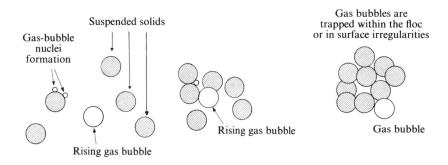

Fig. 10.11 Three methods of dissolved-air flotation. (a) Adhesion of a gas bubble to a suspended liquid or solid phase. (b) The trapping of gas bubbles in a floc structure as the gas bubbles rise. (c) The absorption and adsorption of gas bubbles in a floc structure as the floc structure is formed. (After Vrablik [18].)

mills, or poultry processing plants. It is an economical and effective means of rapid separation of these larger suspended solids from the remaining waste material. In many cases screening alone will reduce the suspended solids to a low enough concentration to be acceptable for discharge into a municipal sewer or a nearby stream. Often considerable BOD is also removed by the screening process, the percentage removed varying almost directly with the size of the screen and the amount of BOD associated with the

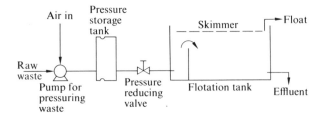

Fig. 10.12 Schematic drawing of pressure flotation system.

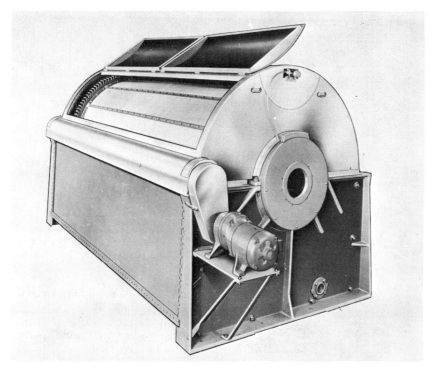

Fig. 10.13 North water filter. (Photograph courtesy Green Bay Foundry and Machine Works, Green Bay, Wisconsin.)

screenable solids. Screens are available in sizes ranging from coarse (10 or 20 mesh) to fine (120 to 320 mesh).

The North and Sweco screens are typical of two major types used in industry today. The former are generally rotary, self-cleaning, gravity-type units (Fig. 10.13). The latter are mostly circular, overhead-fed, vibratory units (Fig. 10.14).

The rotary, gravity-type, waste-disposal screens are manufactured in several sizes to handle almost any volume of waste liquid. In general, they vary from 3 to 5 feet in diameter and from 4 to 12 feet in length and weigh between one and five tons. These screens separate solid and liquid constituents from waste materials at a location where they gravitationally flow or can be pumped into the screened cylinder. The machine's large drum rotates at 4 rpm. The lift paddles within the drum pick up the solid material from the water and deposit it into a stationary, perforated hopper within the cylinder. The hopper holds a spiral screw conveyor that moves the solids to the rear end of the machine and out through the discharge spout. In the process, it compresses the wastes and squeezes out more liquid, which drains through the perforated hopper back into the cylinder. The water in the cylinder drains through the wire mesh and collects in a steel or wooden tank which is part of the machine. The fine wire mesh is at all times kept free from clogging by a continuous spray pipe with jet nozzles, located above the rotating cylinder. They have been used successfully in treating wastes from meat-packing, canning, grain-washing, tanning, malting, woolen, and sea-food plants.

The circular, vibratory screens have been quite effective in screening wastes from food-packing processes such as meat and poultry packing or fruit and vegetable canning. Vibration is designed usually to remove solids at the periphery of the screen, although Swallow [16] reports the use of a new center-discharge separator.

Microstraining, a particular screening device, was first introduced by Dr. P. L. Boucher in England in 1945 for water clarification and there are now about 70 water-treatment plants in the United States utilizing this process [2]. It involves the use of high-speed, continuously backwashed, rotating drum filters work-

Fig. 10.14 The 48-inch-diameter Sweco separator shown is screening lint from waste water at the Eastern Overall Company, in Baltimore, Maryland. The waste water is fed onto the 60-mesh market-grade screen at a rate of 300 gpm. The screened waste water is discharged to the sewer. (Photograph courtesy Sweco Inc., Los Angeles, California.)

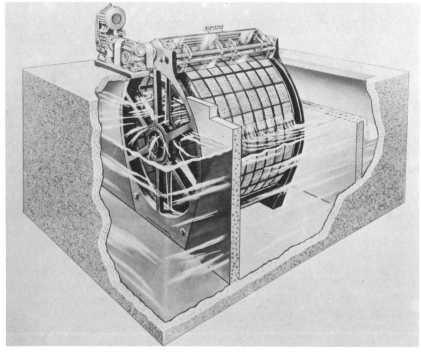

Fig. 10.15 Cutaway view of a $7\frac{1}{2}$-foot-diameter microstrainer. (Photograph courtesy Crane Company, King of Prussia, Pennsylvania.)

Table 10.4 Results obtained on humus tank effluent at Eastern Sewage Works, London, England, December 30, 1966–January 13, 1967.

Characteristic	Effluent from			
	Humus tank	Micro-strainer	Ozonizer	Sand filter
Suspended solids	51†	19	15	10
Total solids	931			928
BOD	21	13	11	9
COD	78	54	44	39
Permanganate value	16	10	6	5
Organic carbon		23	19	10
Surface-active matter				
Anionic (as Manoxol OT)	1.4	1.4	0.6	0.6
Nonionic (as Lissapol NX)	0.37		0.07	0.07
Ammonia (as N)	7.1	7.5	7.4	7.6
Nitrite (as N)	0.4	0.4	0.02	0.01
Oxidized nitrogen (as N)	26	26	26	27
Total phosphorus (as P)	8.2			7.4
Orthophosphate (as P)	6.6			7.0
Total hardness (as $CaCO_3$)				468
Chloride				98
Sulfate	212			213
Color (Hazen units)	36		4	7
Turbidity (ATU)‡	66		27	13
Total phenol	3.4			0.9
Temperature (°C)	8.1	8.0	7.9	7.7
Dissolved oxygen (% saturation)	52	52	99	94
Conductivity (μmho/cm^3)	1173	1175	1170	1150
Langlier index	−0.08			+0.12
pH	7.4	7.4	7.4	7.5
Pesticides (μg/l)				
α BHC		0.025	0.007	
γ BHC		0.035	0.030	
Aldrin		0.004	0.000	
Dieldrin		0.193	0.032	
pp DDT		0.031	0.030	

*After E.W.J. Diaper, *Water Wastes Eng.* **5**, 56 (1968).
†All results are given in milligrams per liter, unless otherwise specified.

ing in open gravity-flow conditions (see cutaway picture in Fig. 10.15). The principal filtering fabrics employed have apertures of 35 or 25 microns and are fitted on the drum periphery. Head loss is between 4 and 6 inches. Recent results in London, England, show that microstraining removes most of the suspended solids remaining after biological treatment (Table 10.4).

References

1. Bewtra, J. K., "Diagram for the settling of discrete particles in viscous fluids," *Water Sewage Works*, **114,** 60 (1967).
2. Boucher, P. L., "Micro-straining, microzon, and demicellization applied to public and industrial water supply," in Proceedings of Water Treatment Sympo-

sium, May 1965, Adelaide, S. Australia.
3. Bramer, H. C., and R. D. Hoak, "Measuring sedimentation-flocculation efficiencies," *Ind. Eng. Chem. Process Design Develop.* **5,** 316 (1966).
4. Camp, T. R., "Studies of sedimentation basin design," *Sewage Ind. Wastes* **25,** 1 (1953).
5. Clark, J. W., and W. Viessman, Jr., *Water Supply and Pollution Control*, International Textbook Company, Scranton, Pa. (1965), pp. 274–294.
6. Dobbins, W. E., "Advances in sewage treatment design," Paper read to the Sanitary Engineering Division of the A.S.C.E. (Metropolitan Section) Conference at Manhattan College, New York City (May 1961).
7. Eckenfelder, W. W., *Industrial Water Pollution Control*, McGraw-Hill Book Co., New York (1966).
8. Federation of Sewage and Industrial Wastes Association, *Sewage Treatment Design*, Manual of Practice no. 8 (American Society of Civil Engineers Manual of Engineering Practice no. 36) (1959), p. 78.
9. Fitch, B., "Current theory and thickener design," *Ind. Eng. Chem.* **10,** 18 (1966).
10. Great Lakes–Upper Mississippi River Board of State Sanitary Engineers, *Recommended Standards for Sewage Works*, Harrisburg, Pa., May 10, 1960.
11. Hess, R. W., et al., "1952 Industrial wastes forum," *Sewage Ind. Wastes* **25,** 709 (1953).
12. Katz, W. J., "Adsorption—Secret of success in separating solids by air flotation," *Ind. Wastes* **30,** 11 (1959).
13. Eckenfelder, W. W., *Industrial Water Pollution Control*, McGraw-Hill Book Co., New York (1966), Chap. 2, p. 28.
14. Quigley, R. E., and E. L. Hoffman, "Flotation of oily wastes," in Proceedings of 21st Industrial Wastes Conference, Purdue University, May 1966, p. 527.
15. Swallow, D. M., "Design and operation of the center-discharge separator," in Proceedings of the Seminar on Water Pollution Control, during 30th Exposition of Chemical Industries, New York, Nov. 30, 1965, p. 20.
16. Villemonte, J. R., "Hydraulic characteristics of circular sedimentation basins," in Proceedings of 17th Industrial Waste Conference, at Purdue University, 1962, p. 682.
17. Villemonte, J. R., et al., "Hydraulic and removal efficiencies in sedimentation basins," *J. Water Pollution Control Federation* **38,** 371 (1966).
18. Vrablik, E. R., "Fundamental principles of dissolved-air flotation of industrial wastes," in Proceedings of 14th Industrial Waste Conference, Purdue University Engineering Extension Series, Bulletin no. 104, May 1960, p. 743.

Suggested Additional Reading

Sedimentation
Camp, T. R., "Sedimentation and the design of settling tanks," *Trans. Am. Soc. Civil Engrs.*, **111,** 895 (1946).
Dobbins, W. E., "Effect of turbulence on sedimentation," *Trans. Am. Soc. Civil Engrs,* **109,** 629 (1944).
Federation of Sewage and Industrial Wastes Association, *Sewage Treatment Design*, Manual of Practice no. 8 (American Society of Civil Engineers Manual of Engineering Practice no. 36), (1959), pp. 90–91.
Hazen, A., "On sedimentation," *Trans. Am. Soc. Civil Engrs,* **53,** 45 (1904).
Rich, L. G., *Unit Operations in Sanitary Engineering*, John Wiley & Sons, New York (1961), Chapter 4, pp. 81–109.

Flotation
Beebe, A. H., "Soluble oil wastes treatment by pressure flotation," *Sewage Ind. Wastes* **25,** 1314 (1953).
D'Arcy, N. A., Jr., "Dissolved air flotation separates oil from waste water," *Oil Gas J.* **50,** 319 (1951).
Rich, L. G., *Unit Operations in Sanitary Engineering*, John Wiley & Sons, New York (1961), Chapter 5, pp. 110–35.

Screening and Microstraining
For full bibliography see: Boucher, P. L., *J. Inst. Public Health Engrs.* **60,** 294 (1961); and Reference 2 above.
Campbell, R. M., and M. B. Prescod, *J. Inst. Water Engrs* **19,** 101 (1965).
Boucher, P. L., "Micro-straining and ozonisation of water and waste water," in Proceedings of 22nd Industrial Waste Conference, Purdue University Engineering Extension Series, Bulletin no. 129, May 1967.
Diaper, E. W. J., "Micro-straining and ozonisation of water and waste water," *Water Wastes Eng.* **5,** 56 (1968).
Hazen, R. "Application of the microstrainer to water treatment in Great Britain," *J. Am. Water Works Assoc.* **45,** 723 (1953).

CHAPTER 11

REMOVAL OF COLLOIDAL SOLIDS

11.1 Characteristics of Colloids

A colloid may be defined as a particle held in suspension by its extremely small size (1 to 200 millimicrons), its state of hydration, and its surface electrical charge. There are two types of colloids: *lyophobic* and *lyophilic*. Because of the difference in their characteristics, they react differently to alterations in their environment. Table 11.1 will assist the student in understanding their properties. Colloids are often responsible for a relatively high percentage of the color, turbidity, and BOD of certain industrial wastes. Since it is important to remove colloids from waste waters before they can get into streams, one must understand their physical and chemical characteristics.

Colloids exhibit Brownian movement, a bombardment of the particles of the disperse phase by molecules of the dispersion medium. They are essentially nonsettleable because of their charge, small size, and low particle weight. They are dialyzable; that is, they can be separated from their crystalloid (low molecular weight) counterparts by straining through a semipermeable membrane. The colloids diffuse very slowly compared to soluble ions. Colloidal particles, in general, exhibit very low (if any) osmotic pressure because of their large size relative to the size of soluble ions. They also possess the characteristic of imbibition (the taking in of water by gels). In fact, it is by this very process that bacteria spores (often considered colloidal) take up water and germinate. Colloidal gels are very often used as ultrafilters, having pores sufficiently small to retain the dispersed phase of a colloidal system but large enough to allow the dispersion medium and its crystalloid solutes to pass through. For example, Perona *et al.* [10] in 1967 found that the formed membranes may be used to remove up to 90 per cent of the colored material and somewhat less of the COD and total dissolved solids of pulp-mill sulfite wastes. Colloidal systems show a wide range in viscosity or plasticity. Usually the lyophobic colloidal suspensions exhibit a viscosity only slightly higher than that of the pure dispersing medium (Fig. 11.1) and this concentration increases only very slightly when the concentration of the dispersed material is increased. On the other hand, lyophilic systems may reach very high values of viscosity. With these types of colloids, a parabolic, rather than a linear, relationship exists between viscosity and the concentration of dispersed phase, as shown in Fig. 11.1. Woodard and Etzel [21] have shown that, under certain conditions, one may change a lyophilic colloid in an industrial waste to a lyophobic one. In this case, lignin was altered by the addition of acetone and sodium hydroxide to render the colloid less stable and to enhance color removal.

Many colloidal systems, especially lyophilic (gel) systems, possess the property of elasticity ("springiness" or resistance). This property enables the gels to resist deformation and thereby recover their original shape and size once they have been deformed. If a concentrated beam of light is passed through a colloidal solution in which the dispersed phase has a different refractive index from that of the dispersion medium, its path is plainly visible as a milky turbidity when viewed perpendicularly. This is known as the Tyndall effect (see Table 11.1).

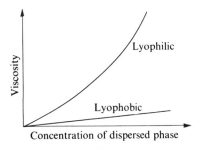

Fig. 11.1 Effect of colloidal type on viscosity.

Table 11.1 Types and characteristics of colloidal sols.

Characteristic	Lyophobic (hydrophobic)	Lyophylic (hydrophylic)
Physical state	Suspensoid	Emulsoid
Surface tension	The colloid is very similar to the medium	The colloid is of considerably less surface tension than the medium
Viscosity	The colloid suspension is very similar to the dispersing phase alone	Viscosity of colloid suspension alone is greatly increased
Tyndall effect	Very pronounced (ferric hydroxide is an exception)	Small or entirely absent
Ease of reconstitution	Not easily reconstituted after freezing or drying	Easily reconstituted
Reaction to electrolytes	Coagulated easily by electrolytes	Much less sensitive to the action of electrolytes, thus more is required for coagulation
Examples	Metal oxides, sulfides, silver halides, metals, silicon dioxide	Proteins, starches, gums, mucilages, and soaps

An important property of colloidal particles is the fact that they are generally electrically charged with respect to their surroundings. An electric current passing through a colloidal system causes the positive particles to migrate to the cathode and the negative ones to the anode.

11.2 Chemical Coagulation

The removal of oxygen-demanding and turbidity-producing colloidal solids from waste waters is often called intermediate treatment, since colloids are intermediate in size between suspended and dissolved solids. The most common and practical method of removing these solids is by chemical coagulation. This is a process of destabilizing colloids, aggregating them, and binding them together for ease of sedimentation. It involves the formation of chemical flocs that absorb, entrap, or otherwise bring together suspended matter, more particularly suspended matter that is so finely divided as to be colloidal.

The chemicals most commonly used are: alum, $Al_2(SO_4)_3 \cdot 18H_2O$; copperas, $FeSO_4 \cdot 7H_2O$; ferric sulfate, $Fe_2(SO_4)_3$; ferric chloride, $FeCl_3$; and chlorinated copperas, a mixture of ferric sulfate and chloride. Aluminum sulfate appears to be more effective in coagulating carbonaceous wastes, while iron sulfates are more effective when a considerable quantity of proteins is present in the waste. The use of organic polymers, which have the ability to act as either negatively or positively charged ions, has been increasing recently. Smaller dosages and the elimination of many storage problems are among the major advantages of these polymers. Dey [2] presents results obtained in various industries where water-soluble polymeric coagulation chemicals are used to achieve improved waste solids settling. Schaffer [15] found that these polymers were useful in maintaining higher solids concentrations in an anaerobic contact treatment process for meat-packing wastes.

The process of chemical coagulation involves complex equilibria among a number of variables including colloids of dispersed matter, water or another dispersing medium, and coagulating chemicals. Driving forces—such as the electrical phenomenon, surface effects, and viscous shear—cause the interaction of these three variables.

11.3 Coagulation by Neutralization of the Electrical Charges

This can be accomplished by:

1) Lowering the zeta potential of the colloids (Fig. 11.2). Zeta potential is the difference in

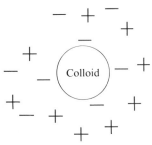

Fig. 11.2 Stable colloid.

electrical charge existing between the stable colloid and the dispersing medium.

2) Neutralizing the colloidal charge by flooding the medium with an excess of oppositely charged ions, usually hydrous oxide colloids formed by reaction of the coagulant with ions in the water. The coagulant colloids also become destabilized by the reaction with foreign, oppositely charged, colloids and produce hydrous oxide, which is a floc-forming material.

From the standpoint of electrical charges, there are two predominant types of colloid in waste waters:

1) Colloids naturally present, including several proteins, starches, hemicelluloses, polypeptides, and other substances, all possess negative charges (mostly lyophilic in nature).

2) Colloids artificially produced by coagulants, usually the hydroxides of iron and aluminum (mostly lyophobic in nature), are mainly positively charged ions.

In most scientific circles it is believed that the charge on colloidal particles is due mainly to the preferential adsorption of ions (H^+ or OH^-), from the dispersing medium. The charge may also be due, in part, to the direct ionization of a portion of its structural groups, such as NH_2^+ and COO^-.

Hydrous aluminum and iron oxides, as well as other metal sols, can acquire both positive and negative charges. Excess Fe^{+++} makes colloids positively charged. The following expression depicts a resultant positively charged colloid:

$$\begin{matrix} FeO \\ FeO \cdot x\ HOH \quad Fe^{+++} \leftarrow \\ O \end{matrix} \begin{cases} OH^- \\ OH^- \\ OH^- \end{cases}$$

Excess OH^- makes colloids negatively charged.

The following expression depicts a resultant negatively charged colloid:

$$\begin{matrix} FeO \\ FeO \cdot x\ H_2O \quad OH^- \leftarrow H^+ \\ O \end{matrix}$$

However, a colloid can acquire a charge by means other than adsorption. A protein dissolved in solution can be schematically illustrated as follows:

$$COO^- \text{—— [protein base molecule] ——} NH_2^+.$$

It may become necessary to add up all the positive NH_2^+ groups and the negative COO^- groups to ascertain the final ionic charge of the solution, because of the inherent charge brought about by direct ionization of the particle. The sol is thereby stabilized by inherent ionization of groups within the molecule itself.

Any alteration of the type and number of double-layer ions should reduce the zeta potential to such a point that the colloid will lose its stability. Stability is defined as the ability to resist precipitation and/or coagulation into a relatively large particle. A colloid is most stable when it possesses the greatest electrical charge and smallest size. The coagulating power of ions rises rapidly as the electrical charge increases, as is stated by the Schulze-Hardy rule. Table 11.2

Table 11.2 Valence and coagulant dosage.

Electrolyte	Anion or cation valence	Minimum concentration required, mmols/liter
	Anion	
KCl	1	103
KBr	1	138
KNO_3	1	131
K_2CrO_3	2	0.325
K_2SO_4	2	0.219
$K_3Fe(Cn)_6$	3	0.096
	Cation	
NaCl	1	51
KNO_3	1	50
K_2SO_4	1	63
$MgSO_4$	2	0.81
$ZnCl_2$	2	0.68
$BaCl_2$	2	0.69
$AlCl_3$	3	0.09

illustrates the minimum concentration of various chemical coagulants required for anions and cations to complete the reaction. Ratios of concentrations of electrolytes required for valences of 1, 2, or 3 are in the order of 729:11.4:1.

Electrolytes and colloids react readily to changes in the pH of the waste water. Most negatively charged particles, including the majority of contaminating colloids present in waste waters, coagulate at an optimum pH value of less than 7.0. Flocculent hydroxide colloids, on the other hand, are insoluble only at pH values above 7.0 and usually over 9.0. Lime is normally added to raise the pH, as well as to aid in precipitation of colloids.

Alum has a pH range of maximum insolubility between 5 and 7; the ferric ion coagulates only at pH values above 4; and the ferrous ion only above 9.5. Copperas ($FeSO_4 \cdot 7H_2O$) is a useful coagulant only in highly alkaline wastes. Lime, a coagulant in itself, is often added with iron salts to raise the pH to the isoelectric point of the coagulant. At this point, the colloid has its minimum electrical charge and is least stable. Since lime is quite insoluble at pH values of 9 and over, coagulation with lime and copperas together increases the pH range. Aeration of waste waters before addition of lime enhances coagulation by evolving lime (thus consuming carbon dioxide and supplying oxygen for converting iron to the oxide and hydroxide states).

Since the ferrous ion when oxidized to the ferric ion can also be used as a coagulant at low pH values, oxidation may be carried out by chlorination, as follows:

$$6Fe^{++}SO_4 \cdot 7H_2O + 3Cl_2 \rightleftarrows 6Fe^{+++} + 6SO_4^= + 6Cl^- + 42H_2O.$$
(Copperas)

Negative ions already present in waste waters extend the useful range of pH in the acid category and positive ions extend the useful pH range in the basic category. Thus, in soft waters, the negatively charged color colloids coagulate best in the acid pH range, and positively charged iron and aluminum ions are good precipitating chemicals in alkaline waters. Prechlorination of alum-treated wastes sometimes increases color removal. Finely divided clay, activated silica, bentonite, or other coagulant aids are often used for relatively clear waters. The addition of any of these produces an effect similar to that of seeding clouds with silver-iodide crystals: they provide nuclei

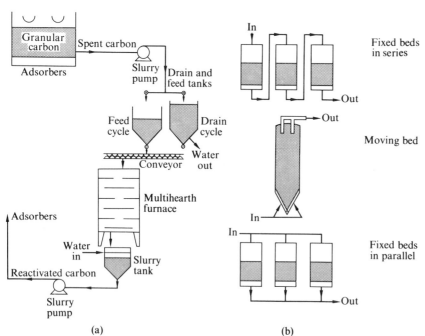

Fig. 11.3 (a) Granular carbon reactivation cycle. (b) Adsorber configuration for granular carbon waste treatment.

Table 11.3 Properties of coal-derived granular carbon for waste treatment.

Characteristic	Type SGL	Type CAL
Mesh size, U.S. Sieve Series	8 × 30	12 × 40
Effective size, mm	0.8–0.9	0.50–0.60
Uniformity coefficient	1.9 or less	1.7 or less
Mean particle diameter, mm	1.5	0.9
Real density, g/cm^3	2.1	2.1
Apparent density		
g/cm^3	0.48	0.44
lb/ft^3	30.0	27.5
Particle density wetted with water, g/cm^3	1.4–1.5	1.3–1.4
Total surface area (N$_2$ BET method), m^2/g	950–1050	1000–1100
Pore volume, cm^3/g	0.85	0.94

about which the precipitate can gather, agglomerate, and flocculate, with a resultant increase in density and settling rate.

Sometimes the presence of iron and manganese in waste waters will add to the effect of the cationic coagulants. An increase in the concentration of the coagulant shortens the time of the coagulation reaction considerably. Gentle agitation of the waste water also enhances coagulation, by increasing the number of collisions and thus bringing about more rapid floc formation.

11.4 Removal of Colloids by Adsorption

A large number of compounds which are not amenable to other types of treatment may be removed from wastes by adsorption. For example, pesticides, such as 2,4-D herbicides and carbamate insecticides may be removed by adsorption onto powdered activated carbon but not onto clay materials such as illite, kaolinite, and montmorillonite [16]. In addition, colloidal suspensions of DDT, chlorobenzene, and p-chlorobenzenesulfonic acid resulting in DDT production may be removed by using activated carbon [8]. Cooper and Hager [1] also suggest activated carbon for advanced waste treatment where reclamation is of paramount importance. They present three typical activated-carbon treatment systems and a granular-carbon reactivation system (Fig. 11.3). The granular carbon used in most reactivation systems in the world is made from bituminous coal. Cooper and Hager also present a summary of properties for two types of this coal (see Table 11.3) and claim this treatment is especially effective in removing biologically resistant (refractory) compounds.

References

1. Cooper, J. C., and D. G. Hager, "Water reclamation with activated carbon," *Chem. Eng. Progr.* **62**, 85 (1966).
2. Dey, R. F., "Use of organic polymers in treatment of industrial wastes," in Proceedings of 12th Ontario Industrial Waste Conference, June 1965, pp. 89–104.
3. Fair, G. M., and J. Geyer, *Elements of Water and Waste Water*, John Wiley & Sons, New York (1958), p. 616.
4. Hogg, R., T. W. Healy, and D. W. Fuerstenau, "Mutual coagulation of colloidal dispersions," *Trans. Faraday Soc.* **62**, 1638 (1966).
5. Johnson, R. L., F. J. Lowes, Jr., R. M. Smith, and T. J. Powers, *Evaluation of the Use of Activated Carbon and Chemical Regenerants in the Treatment of Waste Water*, Publication no. 999–13, U.S. Public Health Service, Washington, D.C. (1964).
6. Joyce, R. S., and V. A. Sukenik, *Feasibility of Granular Activated Carbon Adsorption for Wastewater Renovation*, Environmental Health Service Supply and Pollution Control Publication no. 999-WP-28, U.S. Public Health Service, Washington, D.C. (1965).
7. Kawamura, S., and T. Yoshitaro, "Applying colloid titration techniques to coagulant dosage control," *Water Sewage Works* **113**, 398 (1966).
8. Kul'skii, L. A., and A. G. Shabolina, "Adsorption of DDT from colloidal solutions on the iodine KAD activated carbon," *Chem. Abstr.* **67**, no. 14686y, 1967.
9. Middleton, A. E., "Activated silica solution applications," *Water Sewage Works* **100**, 251 (1963).
10. Perona, J. J., et al., "Hyperfiltration—Processing of pulp mill sulfite wastes with a membrane dynamically formed from feed constituents," *Environ. Sci. Technol.* **1**, 991 (1967).
11. Rudolfs, W., and J. L. Belmat, "A separation of sewage colloids with the aid of the electron microscope," *Sewage Ind. Wastes* **24**, 247 (1952).
12. Rudolfs, W., and H. W. Gehm, "Chemical coagulation of sewage," *Sewage Works J.* **8**, 195, 422, 537 and 547 (1936).
13. Rudolfs, W., and H. Gehm, "Colloids in sewage

treatment: 1. Occurrence and role. A critical review," *Sewage Works J.* **11**, 727 (1939).
14. Sawyer, C. N., and P. E. McCarty, *Chemistry for Sanitary Engineers*, McGraw-Hill Book Co., New York (1960).
15. Schaffer, R. B., "Polyelectrolytes in industrial waste treatment," *Water Sewage Works* **111**, 300R (1964).
16. Schwartz, H. G., Jr., "Adsorption of selected pesticides on activated carbon and mineral surfaces," *Environ. Sci. Technol.* **1**, 332 (1967).
17. Sennet, P., and J. P. Oliver, "Colloidal dispersions, electrokinetic effects and concept of zeta potential," *Ind. Eng. Chem.* **57**, 32 (1965).
18. Weber, W. J., "Adsorption," in Proceedings of Summer Institute for Water Pollution Control, Manhattan College, New York, 1967.
19. Weber, W. J., and J. C. Morris, *Adsorption of Biochemically Resistant Materials from Solution*, Publication no. 999-WP 33 W62-24, U.S. Public Health Service, Washington, D.C. (1966).
20. Williamson, J. N., A. M. Heit, and C. Calmon, *Evaluation of Various Adsorbents and Coagulants for Waste Water*, Publication no. 999-WP, U.S. Public Health Service, Washington, D.C. (1964).
21. Woodard, F., and J. Etzel, "Coacervation and chemical coagulation of lignin from pulpmill black liquor," *J. Water Pollution Control Federation* **37**, 990 (1965).

CHAPTER 12

REMOVAL OF INORGANIC DISSOLVED SOLIDS

The removal of dissolved minerals from waste waters has been given relatively little attention by waste-treatment engineers, because minerals have been considered less pollutional than other constituents, such as organic matter and suspended solids. However, as we learn more about the causes and effects of pollution, the importance of reducing the quantity of certain types of inorganic matter which sewage plants permit to enter streams is apparent. Chlorides, phosphates, nitrates, and certain metals are examples of the more common and significant inorganic dissolved solids. Among the methods employed mainly for removing inorganic matter from wastes are: (1) evaporation; (2) dialysis; (3) ion exchange; (4) algae; (5) reverse osmosis; and (6) miscellaneous methods. Other treatment methods which remove minerals incidentally but are aimed primarily at other contaminants are discussed in Chapters 10, 11, and 13. One should not overlook the minerals contributed by natural runoff from overland flow. The amount of dissolved solids which these natural flows contain often exceeds that contributed by waste waters from industry.

12.1 Evaporation

Evaporation is a process of bringing waste water to its boiling point and vaporizing pure water. The vapor is either used for power production, or condensed and used for heating, or simply wasted to the surrounding atmosphere. The mineral solids concentrate in the residue, which may be sufficiently concentrated for the solids either to be reusable in the production cycle or to be disposed of easily. This method of disposal is used for radioactive wastes, and paper mills have for a long time been evaporating their sulfate cooking liquors to a degree where they may be returned to the cookers for reuse.

Major factors in the selection of the evaporation method are: (1) *Economics:* does the value of the reusable residue outweigh the cost of fuel for evaporation? (2) *Initial dissolved solids:* are there enough solids in the waste, of a variable nature, to warrant evaporation? Generally, 10,000 ppm are required. (3) *Foreign matter:* is there foreign matter present which could cause scale formation or corrosion or interfere with heat transfer in evaporation? (4) *Pollution situation:* what effect will the minerals have on the receiving stream? For example, caustic soda kills fish, ammonium salts initiate troublesome algae growths and in some cases stimulate bacterial growth upon organic matter already present [1], salt interferes with water use by industries and municipalities, and so forth.

Today many evaporators are heated by steam condensing on metallic tubes, through which flows the waste to be concentrated or evaporated. The steam is at a low pressure, usually less than 50 pounds per square inch (psi) (absolute). Most evaporators operate with a slight vacuum on the vapor side, to lower the boiling point and to increase the rate of vapor removal from the evaporator. Vacuum systems are especially preferable to atmospheric evaporators when the decomposition of organic matter is involved. Care must be exercised, however, that the vacuum is not great enough to permit priming of the waste water into the vapor.

Evaporating a waste presents many problems, which include concentration changes during evaporation, foaming, temperature sensitivity, scale formation, and the materials used in evaporator construction. In industrial-waste concentration, scale formation usually presents the major obstacle. As crust is deposited on the heating surface, the overall heat-transfer coefficient decreases, causing the efficiency to drop until it is necessary to shut down and clean the tubes—a complicated process when the scale is hard and tenacious.

Chrome, nickel, and copper acid-type plating wastes may be reclaimed from the rinse tank by evaporation in glass-lined equipment, or other suitable evaporators, and the concentrated solution returned to the plating system [16]. Initial cost of equipment is high, so that the quantity and value of chemicals to be recovered, plus the estimated cost of operation of a treatment system if evaporative recovery were not practiced, are criteria one must use to justify purchasing such equipment.

Efficiency of evaporation is directly related to heat-transfer rate—expressed in British thermal units per hour (Btu/hr)—through the heating surface (tube wall). This rate is equal to the product of three factors: the overall heat-transfer coefficient, the heating surface area, and the overall change in temperature between the waste and the steam. It is expressed mathematically as

$$q = UA(t_s - t_w) = UA\,\Delta t,$$

where q is the rate of heat transfer (Btu/hr), U is the overall coefficient (Btu/ft²/hr/°F), A is the heating-surface area (ft²), t_s is the temperature of steam condensate (°F), t_w is the boiling temperature of waste (°F), and $\Delta t = t_s - t_w$ is the overall temperature change between steam and waste. Typical values of U for various types of evaporators are given in Table 12.1. These figures are estimated within broad ranges, by considering the viscosity of the waste, scale formation, and operating temperatures (greater temperature differentials yield higher coefficients). Tube wall thickness also influences U; the greater the thickness, the lower the value of U.

Table 12.1 Typical overall coefficients in evaporators [4].

Type of evaporator	Overall coefficient, Btu/ft²/hr/°F
Long-tube vertical	
Natural recirculation	200–600
Forced circulation	400–2000
Short-tube	
Horizontal tube	200–400
Calandria type	150–500
Coil	200–400
Agitated-film	
Newtonian liquid viscosity	
1 centipoise	400
100 centipoises	300
10,000 centipoises	120

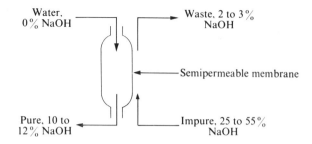

Fig. 12.1 Typical dialysis flow diagram.

12.2 Dialysis

Dialysis is the separation of solutes by means of their unequal diffusion through membranes [2, 6, 9, 12, 13, 27–30, 32, 34]. It is most useful in recovering pure solutions for reuse in manufacturing processes, for example, caustic soda in the textile industry [17]. Recovery involves separation of a crystalloid (NaOH) from a sol in which about 96 per cent of the impurities are in the form of hemicellulose and the rest include pectins, waxes, and dyes.

There are some eight to ten commercial dialyzers presently on the market. In our example, they all operate on the simple principle of passing a concentrated, impure caustic solution upward, countercurrent to a downstream water supply, from which it is separated by a semipermeable membrane (Fig. 12.1). The caustic soda permeates the membrane and goes into the water more rapidly than the other impurities contained in the waste. The concentration of caustic is always greater in the impure solution than in the water, and the water which flows through the membrane into the impure caustic solution tends to dilute it. The quantity of sodium hydroxide diffusing through the diaphragm depends upon the time, the area of the dialyzing surface, the mean concentration difference, and the temperature. These factors are expressed in the equation

$$Q = KAt(\Delta c),$$

where K is the overall diffusion coefficient, t is the time in minutes, A is area of dialyzing surface, and

$$\Delta c = \Delta c_{av} = \frac{(\Delta c_1 - \Delta c_2)}{2.3 \log_{10} \Delta c_1/\Delta c_2},$$

where Δc_1 and Δc_2 are the differences in concentration between the two solutions at the top and bottom of the diaphragm, respectively.

When one actually computes the weight of NaOH recovered, it becomes apparent that the quality and type of diaphragm are of paramount importance. This is evident from the following equation:

$$W = UA \, \Delta c_{\text{log mean}},$$

where W is the weight of material passing through the membrane in a unit of time (gm/min), U is the overall dialysis coefficient, and $\Delta c_{\text{log mean}}$ is the logarithmic mean concentration gradient across the membrane ($=\Delta c_{\text{av}}$). Also,

$$\frac{1}{U} = \frac{1}{U_1} + \frac{1}{U_2},$$

where U_1 is the combined film resistance (cm/min) and U_2 is the membrane resistance (cm/min). Each diaphragm shows a different membrane resistance (U_2). The restrictive characteristics of porous membranes are due to both a mechanical sieve action and a physicochemical interaction between solute, solvent, and membrane. Cellulose nitrate, parchment, and cellophane are the principal membranes in use today.

Smith and Eisemann [32] present an excellent evaluation of electrodialysis. Dialysis is an operation requiring very little operator attention and, although its main role is to conserve raw materials and to reduce plant waste, at the same time it aids in waste treatment. With the introduction of acid-resistant membranes, dialysis has been used successfully in the recovery of sulfuric acid in the copper, stainless-steel, and other industries. Some operations can recover as much as 70 to 75 per cent of the acid, but a recovery of as little as 20 per cent may justify the process. In dialysis the driving force of separation is natural diffusion because of concentration gradient; in electrodialysis this natural force is enhanced by the application of electrical energy. McRae [15] found that, for a secondary effluent containing 900 ppm of dissolved solids, electrodialysis could achieve 44 per cent reduction, with costs ranging from 10 to 15 cents per 1000 gallons. He finds this process useful for treating the wastes from dairies.

12.3 Ion Exchange

Ion exchange is basically a process of exchanging certain undesirable cations and anions of the waste water for sodium, hydrogen, or other ions in a resinous material. The resins, both natural and artificial, are commonly referred to as *zeolites*. The ion-exchange process was originally developed to reduce hardness in domestic water supplies, but has recently been used to treat industrial waste waters, such as metal-plating wastes. The softening reactions may be illustrated as follows [22]:

$$\left.\begin{matrix}\text{Ca}\\ \text{Mg}\end{matrix}\right\}\begin{matrix}(\text{HCO}_3)_2\\ \text{SO}_4\\ \text{Cl}_2\end{matrix} + \text{Na}_2\text{Z} \longrightarrow \left.\begin{matrix}\text{Ca}\\ \text{Mg}\end{matrix}\right\}\text{Z} + \begin{matrix}2\text{NaHCO}_3\\ \text{or}\\ \text{Na}_2\text{SO}_4\\ \text{or}\\ 2\text{NaCl}\end{matrix}$$

where Z is the symbol for the zeolite radical. When the ability of the zeolite bed to produce soft water is exhausted, the softener is temporarily cut out of service. It is then backwashed to cleanse and hydraulically regrade the bed, regenerated with a solution of common salt, which removes the calcium and magnesium in the form of their soluble chlorides and simultaneously restores the zeolite to its original condition, rinsed free of these and the excess salt, and finally returned to service. The reaction may be indicated as follows:

$$\left.\begin{matrix}\text{Ca}\\ \text{Mg}\end{matrix}\right\}\text{Z} + 2\,\text{NaCl} \to \left.\begin{matrix}\text{Ca}\\ \text{Mg}\end{matrix}\right\}\text{Cl}_2 + \text{Na}_2\text{Z}.$$

Ion exchange as a means of waste treatment is only a new application of a traditional method of water softening. If the proper approach is used, it offers great potential for material and water conservation. For instance, in the treatment of metal-plating wastes [16], rinse water is passed through beds of cationic and anionic resins selected for the particular application and the deionized water is then recycled through the rinse tank. This method may be applied on a continuous basis to the removal of contaminating metals [16] from chromic-acid solutions, permitting the return of pure chromic-acid solution to the process tank. In the case of nickel- and copper-plating solutions, both the contaminating metals and the metal to be plated are cationic, and therefore all will be extracted. Cation-exchange resins are suggested [21] for use in the steel industry to remove the iron from spent liquor and to recover sulfuric acid and iron oxide for further use. Unless the aim of the procedure is recovery of metals, ion exchange becomes simply a concentration method, and some treatment for the regenerated solution must be devised.

Walther [36] reports the use of a continuous ion-exchange unit, consisting of a stainless-steel loop divided into sections by butterfly valves, which successfully removed over 700 mg/liter of dissolved inorganic solids. The unit contains about 15 cubic feet of ion-exchange resin which moves around the loop in about three minutes. When the resin becomes saturated with hardness, it is removed from the loop and regenerated resin is exchanged. The spent resin is then regenerated and returned to the loop on the next cycle.

Organic matter and pH have a pronounced effect on the operation and efficiency of resin beds; the leaching of organic matter from certain resins may have a detrimental effect on the metals plated. Chemicals used for regenerating resin beds may also require special treatment before disposal.

General appraisal. Demineralization (ion exchange) is most useful when water of the highest quality is required, but it involves complex chemical reactions and therefore requires careful operation and supervision at all times. Furthermore, ion-exchange processes sometimes utilize chemicals which are hazardous to personnel and equipment. These are matters to think about before selecting an ion-exchanger system in preference to an evaporator; although evaporators, too, are uneconomical in certain instances, e.g. when the flow is light. Dialysis is normally economical and can compete in efficiency with both evaporation and ion exchange when the recovery of a pure compound is considered essential. The decision whether to use evaporation or demineralization can be intelligently made only after a thorough evaluation of the heat balance of the plant and expected operating conditions [26]. These factors, as well as operating costs, must be considered in relation to the capital investment needed for either system.

Table 12.2 Elemental composition of green algae (After Krauss [10].)

Element	Range of dry weight, %
Chlorella	
Carbon	51.4–72.6
Hydrogen	7.0–10.9
Oxygen	11.6–28.5
Scenedesmus	
Nitrogen	2.2–7.7
Phosphorus	1.1–2.0
Sulfur	0.28–0.39
Magnesium	0.36–0.80
Potassium	0.85–1.62
Calcium	0.005–0.08
Iron	0.04–0.55
Zinc	0.0006–0.005
Copper	0.001–0.004
Cobalt	0.000003–0.0003
Manganese	0.002–0.01

12.4 Algae

Algae require nine minor essential elements (Fe, Mn, Si, Zn, Cu, Co, Mo, B, and Va) and seven major essential elements (C, N, P, S, K, Mg, and Ca) for their optimum growth. The use of algae for removing minerals from waste waters is still in the experimental stages; most investigations have been carried out on sewage effluents. One such study carried out in the author's laboratory [7] involved a suburban housing-development treatment plant and utilized primary sedimentation, trickling filtration, and stabilization ponds. Although the sedimentation and filtration did not remove any phosphorus, the algae actively growing in the ponds caused a reduction of about 42 per cent of the phosphate content. Other mineral concentrations were not measured.

If this method is used to remove minerals such as phosphate over a period, algae must also be removed from the effluent before this is released into a stream used for water supplies and recreation. Golueke and Oswald [7] observed three steps in harvesting oxidation-pond algae: (1) collection and initial concentration, (2) dewatering or secondary concentration, and (3) final drying. They found chemical precipitation and centri-

Table 12.3 Occurrence of *Cyanophyceae* and *Chlorophyceae* in Massachusetts lakes and reservoirs [36].

Characteristic	Chemical analysis, ppm	Often above 1000/cm³		Below 100/cm³	
		Cyanophyceae	*Chlorophyceae*	*Cyanophyceae*	*Chlorophyceae*
Color	0–30	2	2	11	0
	30–60	2	2	3	1
	60–100	3	1	7	2
	>100	0	0	1	1
Chlorides (excess above normal)	0	2	1	3	1
	0.1–0.3	1	1	10	5
	0.4–2.5	1	0	9	6
	>2.5	3	3	0	0
Hardness	0–5	0	0	6	4
	5–10	2	1	10	5
	10–20	2	1	5	2
	>20	3	3	1	1
Albuminoid ammonia (dissolved)	0–0.10	0	0	4	3
	0.1–0.15	0	0	6	4
	0.15–0.20	2	2	7	3
	>0.20	5	3	5	2
Free ammonia	0–0.01	0	0	10	4
	0.01–0.03	0	0	8	5
	0.03–0.10	3	2	4	3
	>0.1	4	3	0	0
Nitrates	0–0.05	1	0	12	6
	0.05–0.10	3	2	10	6
	0.10–0.20	1	0	0	0
	>0.20	2	3	0	0

fugation to be most economical. The harvested algae can be sold as animal feed supplements. Oswald [25] describes *Chlorella* and *Scenedesmus* as the most active algae in stabilization ponds, because they are extremely hardy. Krauss [10] presents the elemental composition of these two algal types to validate their fixation of minerals (Table 12.2).

Table 12.2 shows the extent to which algae take up minerals from any solution in which they grow. In fact, the continued photosynthesis of algae depends directly on the ability of the culture medium (waste water) to supply these inorganic components over a long period, at a rate sufficient to support the growth potential of the algae. There is some evidence that the uptake (and the algal growth) depends on the availability, as well as the presence, of inorganic nutrients. Thus, insolubility and colloidal characteristics of the nutrients may hamper algal growth, but hardness in waste waters can contribute to it. A statistical study of Massachusetts lakes and reservoirs carried out in 1900 showed that the hard water supplies yielded more algae than the soft (Table 12.3). Bogan [3] capitalized on the ability of algae to utilize phosphorus in providing tertiary treatment of the sewage from Seattle, Washington, which utilized both algal activity and lime and which removed over 90 per cent of the phosphorus in the secondary sewage-plant effluent. Oxidation-pond usage has been increasing since the advent of lower-cost mechanical aeration.

12.5 Reverse Osmosis

Reverse osmosis is a method of applying sufficient pressure on a concentrated fluid to overcome osmotic pressure and thus force water through a semipermeable membrane. In this manner dissolved solids are separated from the water in which they were originally dissolved. New modifications of cellulose-acetate membranes have greatly increased the efficiency of this process. Although its use for industrial-waste treatment is at present conjectural, Okey and Stavenger [24] give total costs of $0.60 to $1.04 per 1000 gallons (for low flows of about 20,000 gallons per day of an unspecified industrial waste).

12.6 Miscellaneous Methods

Chemical precipitation or coagulation have been used to remove some inorganic matter from waste waters. For example, elevated pH values aid in the removal of heavy metals by precipitation of the hydroxide or carbonate, and, under some conditions, treatment of waste waters with calcium hydroxide is reasonably effective in the removal of nitrogen and phosphorus. Continued experimentation with various coagulants is warranted and is being carried out currently by the author.

Oxidation-reduction chemical reactions are used in certain cases to alter inorganic matter and thus enhance its removal. For example, chromate must be reduced, usually with ferrous sulfate or sulfur dioxide under acid conditions, to the trivalent form as a preliminary to precipitation with lime and subsequent removal as a chromic-hydroxide sludge (see reactions on page 419. Likewise, cyanides must be completely oxidized, usually with chloride under alkaline conditions, to split them up into harmless and volatile nitrogen gas and carbonate ions (see reaction on page 418).

References

1. Amberg, H. R., "The effect of nutrients upon the rate of stabilization of spent sulfite liquor in receiving waters," *Proc. Am. Soc. Civil Engrs.* **81**, 821 (1955).
2. Bassett, H. P., "Super filtration by dialysis," *Chem. Met. Eng.* **42**, 254 (1938).
3. Bogan, R. H., *Pilot Plant Evaluation of a Tertiary Stage Treatment Process for Removing Phosphorus from Sewage*, A report prepared for the city of Seattle (December 1959).
4. Brown, G. G., D. Katz, A. S. Foust, and R. Schneidewind, *Unit Operations*, John Wiley & Sons, New York (1950), p. 484.
5. Bryson, J. C., *Control of Algae through Phosphate Control*, Unpublished report, Syracuse University, Syracuse, N.Y., September 1961.
6. Eynon, D. J., "Operation of Cerini dialysers for recovery of caustic soda solutions containing hemicellulose," *J. Soc. Chem. Ind.* **52**, 173T (1933).
7. Golueke, C. G., and W. J. Oswald, "Harvesting and processing sewage grown planktonic algae," *J. Water Pollution Control Federation*, **37**, 471 (1965).
8. Keating, R. J., and R. Dvorin, "Dialysis for acid recovery," in Proceedings of Industrial Waste Conference, Purdue University, 1960, pp. 567–76.
9. Kirk, R. E., and D. F. Othmer, *Encyclopedia of Chemical Technology*, Interscience, New York (1950) p. 5.
10. Krauss, R. W., "Photosynthesis in the algae," *Ind. Eng. Chem.* **48**, 1449 (1956).
11. Kunin, R., and F. Y. McGarvey, "Status of ion exchange technology," *Ind. Eng. Chem.* **55**, 51 (1963).
12. Lee, J. A., "Caustic soda recovery in rayon industry," *Chem. Met. Eng.* **42**, 483 (1935).
13. Lovett, L. E., "Application of osmosis to recovery of caustic soda solutions containing hemicelluloses in rayon industry," *Trans. Electrochem. Soc.* **73**, 163 (1938).
14. McCabe, W. L., and J. C. Smith, *Unit Operations of Chemical Engineering*, McGraw-Hill Book Co., New York (1956) p. 530.
15. McRae, W. A., "Electrodialysis in wastewater reclamation," in Proceedings 2nd Water Quality Research Symposium, New York State Department of Health, April 14, 1965, pp. 97–119.
16. Merrill, G. R., A. R. Macommer, and H. R. Manusberger, *American Cotton Handbook*, 2nd ed., Textile Book Publishers, New York (1949).
17. Michalson, A. W., and C. W. Burhans, Jr., "Chemical waste disposal by ion exchange," *Ind. Water Wastes* **1**, 11 (1962).
18. Nemerow, N. L., and J. C. Bryson, "How efficient are oxidation ponds?" *Wastes Eng.* **34**, 133 (1963).
19. Nemerow, N. L., and W. R. Steele, "Dialysis of caustic

textile wastes," in Proceedings of 10th Industrial Waste Conference, Purdue University, May 1955, pp. 74–81.
20. *Textile Wastes—A Review*, New England Interstate Water Pollution Control Commission, December 1950.
21. "New process developed to recover acid and iron from spent pickle liquor," *Iron Steel Engr* **42**, 167 (1965).
22. Nordell, E., *Water Treatment*, Reinhold, New York (1951), p. 341.
23. Ohio River Valley Water Sanitation Commission, *Methods for Treating Metal Finishing Wastes*, January 1953, p. 58.
24. Okey, R. W., and P. L. Stavenger, "Membrane technology. A process report," *Ind. Water Eng.* **4**, 36 (1967).
25. Oswald, W. J., "Fundamental factors in oxidation pond design," in Conference on Biological Waste Treatment, at Manhattan College, New York, April 20–22, 1960, Paper no. 44.
26. Paulson, C. F., "Chromate recovery by ion-exchange," in Proceedings of 7th Industrial Waste Conference, Purdue University, 1952, p. 209.
27. Powell, S. T., *Water Conditioning for Industry*, McGraw-Hill Book Co., New York (1954), p. 214.
28. "Reverse osmosis. An old concept in new hardware," *Ind. Water Eng.* **4**, 20 (1967).
29. Roetman, E. T., "Viscose rayon manufacturing wastes and their treatment," *Water Works Sewerage* **91**, 295 (1944).
30. Roetman, E. T., "Stream pollution control at Front Royal, Virginia, rayon plant," *Southern Power Ind.* **62**, 86 (1944).
31. Rudolfs, W., "A survey of recent developments in the treatment of industrial wastes," *Sewage Works J.* **9**, 998 (1937).
32. Smith, J. D., and J. L. Eisemann, "Electrodialysis in waste water recycle," in Proceedings of 19th Industrial Waste Conference, Purdue University, 1964, pp. 738–760.
33. U.S. Department of Health, Education and Welfare, *Cost of Purifying Municipal Waste Waters by Distillation*, Publication no. AWTR-6, Washington, D.C. (1963).
34. U.S. Department of Health, Education and Welfare, *Advanced Waste Treatment Research*, AWTR-14 S Summary Report (PHS Publication NQ 999-WP-24), Washington, D.C. (1965).
35. Volbrath, H. B., "Applying dialysis to colloid-crystalloid separations," *Chem. Met. Eng.* **43**, 303 (1936).
36. Walther, A. T., "LaGrange tests ion exchange unit," *Water Sewage Works* **112**, 212 (1965).
37. Whipple, G. C., *Microscopy of Drinking Water*, John Wiley & Sons, New York (1948), pp. 214–215.

CHAPTER 13

REMOVAL OF ORGANIC DISSOLVED SOLIDS

The removal of dissolved organic matter from waste waters is one of the most important tasks of the waste engineer, and, unfortunately, also one of the most difficult. These solids are usually oxidized rapidly by microorganisms in the receiving stream, resulting in loss of dissolved oxygen and the accompanying ill effects of deoxygenated water. They are difficult to remove because of the extensive detention time required in biological processes and the elaborate and often expensive equipment required for other methods. In general, biological methods have proved most effective for this phase of waste treatment, since bacteria are adept at devouring organic matter in wastes, and the greater the bacterial efficiency the greater the reduction of dissolved organic matter. Microorganisms, however, are quite "temperamental" and sensitive to changes in environmental conditions, such as temperature, pH, oxygen tension (level of oxygen concentration), mixing, toxic elements or compounds, and character and quantity of food (organic matter) in the surrounding medium. It is the responsibility of the engineer to provide optimum environmental conditions for the proliferation of the particular biological species desired.

There are many varieties of biological treatment, each adapted to certain types of waste waters and local environmental conditions such as temperature and soil type. Some specific processes for treating organic matter are: (1) lagooning in oxidation ponds; (2) activated-sludge treatment; (3) modified aeration; (4) dispersed-growth aeration; (5) contact stabilization; (6) high-rate aerobic treatment (total oxidation); (7) trickling filtration; (8) spray irrigation; (9) wet combustion; (10) anaerobic digestion; (11) mechanical aeration system; (12) deep well injection; (13) foam phase separation; (14) brush aeration; (15) subsurface disposal; and (16) the Bio-Disc system.

13.1 Lagooning

Lagooning in oxidation ponds is a common means of both removing and oxidizing organic matter and waste waters as well. More research is needed on this method of treatment, which originally developed as an inexpensive procedure for ridding industry of its waste problem. An area adjacent to a plant was excavated, and waste waters either flowed or were pumped into the excavation at one end and out into a receiving stream at the other end. The depth of the lagoon depended on how much land was available, the storage period desired or required, and the condition of the receiving stream. Little attention was paid originally to the effect of depth on bacterial efficiency. In fact, reduction of dissolved organic matter was usually not anticipated or even desired, since it was presumed, and with good reason, that biological degradation of organic matter would lead to oxygen depletion and accompanying nuisances from odors. Thus, the lagoons served solely to settle sludge and equalize the flow. Now, modern techniques have led to new theories about the stabilization of organic matter in lagoons.

We now know that stabilization or oxidation of waste in ponds is the result of several natural self-purification phenomena. The first phase is sedimentation. Settleable solids are deposited in an area around the inlets to the ponds, the size of the area depending on the manner of feeding in the waste and location of the inlet. Some suspended and colloidal matter is precipitated by the action of soluble salts; decomposition of the resulting sediment by microorganisms changes the sludge into inert residues and soluble organic substances, which in turn are required by other microorganisms and algae for their metabolic processes.

Decomposition of organic material is the work of microorganisms, either aerobic (living in the presence of free oxygen) or anaerobic (living in absence of free oxygen). In a pond in which the pollution load is exceedingly high or which is deep enough to be void of oxygen near the bottom, both types of microorganism may be actively decomposing organic material at the same time. A third type of microorganism, the facultative anaerobic, is capable of growth under either aerobic or anaerobic conditions and aids in decomposing waste in the transition zone between aerobic and anaerobic conditions. It is desirable to maintain aerobic conditions, since aerobic microorganisms cause the most complete oxidation of organic matter. Anaerobic fermentation has proved effective for treating citrus, slaughterhouse, and certain paper-mill wastes, while aerobic bacteria have been most effective in oxidizing dairy, textile, and other highly-soluble organic wastes.

Table 13.1 gives a general scheme of the microbial degradation of the organic constituents in sewage. It also points out the difference between aerobic and anaerobic decomposition.

Algae are significant in stabilization ponds in that they complete nature's balanced plant-animal cycle. Whether seasonal or perennial, algae utilize CO_2, sulfates, nitrates, phosphates, water, and sunlight to synthesize their own organic cellular material and give off free oxygen as a waste product. This oxygen, dissolved in pond water, is available to bacteria and other microbes for their metabolic processes, which include respiration and degradation of organic material in the pond. Thus, we have a completed cycle in which: (a) microorganisms use oxygen dissolved in the water and (b) break down organic waste materials to produce (c) waste products such as CO_2, H_2O, nitrates, sulfates, and phosphates, which (d) algae use as raw materials in photosynthesis, thereby (e) replenishing the depleted oxygen supply and keeping conditions aerobic, so that the microorganisms can function at top efficiency (see Fig. 13.1). However, one drawback of algae should be mentioned, namely, when they die, they impose a secondary organic loading on the pond. Another disadvantage is

Table 13.1 Biological degradation of organic constituents in sewage.

Substance decomposing	Class of microbial enzymes	End-products	
		Anaerobic decomposition	Aerobic decomposition
Proteins	Proteinase*	Amino acids Ammonia Hydrogen sulfide Methane Carbon dioxide Hydrogen Alcohols Organic acids Phenols Indols	Ammonia, nitrites, nitrates Hydrogen sulfide, sulfuric acid Alcohols Organic acids Carbon dioxide Water
Carbohydrates	Carbohydrase*	Carbon dioxide Hydrogen Alcohols Fatty acids	Alcohols Fatty acids Carbon dioxide Water
Lipids (fats)	Lipase*	Fatty acids Carbon dioxide Hydrogen Alcohols	Fatty acids and glycerol Alcohols Carbon dioxide Water

*Class of enzymes only. Dozens of enzymes may be utilized in this degradation.

a seasonal one: algae are less effective in winter.

Ice and snow cover during winter months interferes with the stabilization process in the following manner:

1) It prevents sunlight from penetrating the pond, causing a reduction in the size and number of algae present. Algae are not necessarily killed by the absence of sunlight (those known as facultative chemo-organotrophs can carry on metabolic processes despite darkness), but they release little or no oxygen without sunlight.

2) It prevents mixing and reaeration by wind action.

3) It prevents reaeration by atmosphere–water dynamic equilibrium phenomena.

4) It usually results in anaerobic conditions if it continues over an extended period of time.

These factors tend to result in a lowered pond or lagoon efficiency during the winter.

Hermann and Gloyna [10] (disregarding the part played by minerals) describe the reaction in high-rate ponds in which sewage is oxidizing as:

$$C_{11}H_{29}O_7N + 14O_2 + H^+ \rightarrow 11CO_2 + 13H_2O + NH_4^+$$

The canning industry, one of the first to attempt lagooning, soon found it difficult to maintain aerobic conditions in basins; other industries experienced similar situations. As industries became aware that biological degradation occurs in lagoons, they made attempts to encourage and control the oxidation and began to refer to such lagoons as waste-oxidation basins.

Most modern oxidation basins have a maximum water depth of four feet and operate on a continuous-flow basis. Engineers try to maintain in the basin near-neutral pH, adequate oxygen concentration, and sufficient nutrient minerals for biological oxidation. Chemical neutralizers are used to alter pH values, oxygen concentrations are maintained by reducing detention times and using shallow basins, and mineral-salts nutrients may be added as needed, to accelerate biological activity. BOD removals range from as low as 10 per cent to as high as 60 to 90 per cent.

In an interesting full-scale study [26], the author treated an air-base oxidation pond, at 43° north latitude with ice cover during the winter, with an elevated loading of 130 pounds of BOD in the waste water per acre of pond area. The BOD reductions at these relatively high loadings ranged from 87.7 per cent in August to 53 per cent in January, with a yearly average of 69.3 per cent. In another pilot-plant study [25], the writer achieved BOD removals in excess of 80 per cent, using close-baffled four-foot-deep, or unbaffled eight-foot-deep, basins, during the critical summer period in central New York State, at elevated loadings of 312 to 467 pounds per acre per day. A photograph of the five parallel pilot-plant basins appears as Fig. 13.2.

Oswald [30] believes that in such heavily loaded ponds, particularly during periods when methane fermentation is either nonexistent or limited by temperature and when algal photosynthesis is not taking place in the surface layers, a buildup of organic acid occurs, with a subsequent lowering of the pH level and emission of hydrogen sulfide from the pond. The writer, however, did not experience these odors, even at the high loadings described above. Oswald offers the explanation that, if methane fermentation becomes established in the bottom deposits, high rates of BOD removal may be attained without

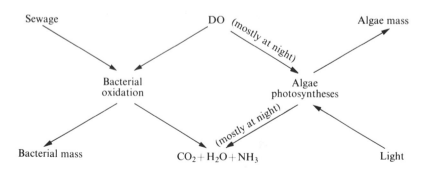

Fig. 13.1 The role of algae in stabilization ponds [37].

Fig. 13.2 Accelerated-oxidation pilot-plant basins [25].

appreciable odors. He also believes that ponds in which both photosynthetic oxygenation and methane fermentation occur (facultative ponds) must be restricted to about 50 pounds of BOD per acre per day, because conditions are at times unfavorable for either process. The author, at this point, does not necessarily agree with these findings. Furthermore, Oswald's later high-rate oxidation ponds for treating sewage in warmer climates have been loaded to over 600 pounds of BOD per acre per day or over, being aerated for an hour each midnight.

The reader is referred to the discussion in Section 12.4 of the necessity of preventing algae growth in bodies of water that are used for water supplies and recreational activities.

13.2 Activated-Sludge Treatment

The activated-sludge process has proved quite effective in the treatment of domestic sewage, as well as a few industrial wastes from large plants. In this process, biologically active growths are created, which are able to adsorb organic matter from the wastes and convert it by oxidation–enzyme systems to simple end-products like CO_2, H_2O, NO_3, and SO_4. Biological slimes develop naturally in aerated organic wastes which contain a considerable portion of matter in the colloidal and suspended state, but for the efficient removal of organic dissolved solids there must be high floc concentrations, to provide ample contact surface for accelerated biological activities. The flocs (zoogleal masses) are living masses of organisms, food, and slime material and are highly active centers of biological life—hence the term "activated sludge." They require food, oxygen, and living organisms in a delicately controlled environment.

Various degrees of efficiency are obtained by controlling the contact period and/or the concentration of active floc. The contact period can be regulated by careful design of the hydraulic systems of aeration basins, the average time of aeration being 6 hours for domestic sewage and 6 to 24 hours for various industrial wastes. The desired concentration of active floc is maintained by recirculating a specific volume of secondary settled sludge, normally about 20 per cent.

Higher sludge quantities lead to greater BOD removal and create a need for more air and food (organic matter) for proper balance. Also, "old," heavy sludge tends to become mineralized and devoid of oxygen, which results in a less-active floc. The reverse is true of a "young," light, sludge floc. The "age" of the growths, therefore, becomes an important consideration.

Busch [4] summarizes the situation by saying that for optimum activity the kinetics of activated sludge require: a young, flocculent sludge in the logarithmic stage of growth; maintenance of the logarithmic growth state by controlled sludge wastage; continuous loading of the organisms; and elimination of anaerobic conditions at any point in the oxidative treatment.

Hazeltine [9] has said of the present status of domestic-sewage activated-sludge treatment that BOD removals are usually above 90 per cent when the loadings are below 0.3 pound of BOD per pound of suspended solids in the waste under aeration. Efficiencies are difficult to predict when these loadings are increased to 0.5 pound per pound. Normally, the BOD loading is related to the aeration-tank capacity; about 30 to 35 pounds of BOD per 1000 cubic feet can be treated in plants with about 2000 ppm of suspended solids under aeration.

Sawyer [34] lists the limitations of the domestic-sewage activated-sludge process as follows: BOD loadings are limited to about 35 pounds per 1000 cubic feet of tank capacity, thus requiring relatively long detention time and resulting high capital investment; there is a high initial oxygen demand by the mixed liquors; there is a tendency to produce bulking sludge; the process cannot produce an intermediate quality of effluent; high sludge-recirculation ratios are required for high-BOD wastes; there are high solids loadings on final clarifiers; and large air requirements accompany the process.

The Kraus process [17] attempts to overcome some of the sludge-bulking problems of conventional activated-sludge plants by controlling the sludge volume index (a measure of the volume occupied by one gram of suspended solid). The process is similar to that of conventional activated-sludge treatment, employing separate reaeration for sludge, except that some digester sludge, digester supernatant, and activated sludge are aerated together for as much as 24 hours, in what he terms a nitrifying aeration tank.

BOD loadings as high as 170 pounds per 1000 cubic feet per day have recently been used, with removals near 90 per cent [18].

Von der Emde [7] notes that ciliated and flagellated protozoa, as well as bacteria, are normally prevalent in activated sludge. When the BOD loading is high or very low, flagellates replace the ciliates, regardless of the level of oxygen present. Where there are short aeration periods or when only traces of oxygen are maintained, bacteria only are observed in the sludge.

Many characteristics of industrial wastes—e.g. toxic metals, lack of nutrients required for biological oxidation, organic nondegradable matter, high temperature, and high or low pH values—give rise to problems requiring careful analysis. When the suitability of this process for a particular industrial waste is in question, laboratory and/or field pilot plant will yield the results necessary for decision making.

13.3 Modified Aeration

Modified, tapered, and step aeration are variations of the activated-sludge treatment. The objective is to supply the maximum of air to the sludge when it is in the optimum condition (sludge age) to oxidize adsorbed organic matter. The location of the aerator and the quantity of air supplied is varied, depending on sludge solids and organic matter to be oxidized. Lower volumes of air and shorter detention times are claimed for these processes, while the mechanisms and theories of operation are similar to those of activated sludge.

Step aeration attempts to eliminate the problems encountered with plain aeration by providing a two- to three-hour aeration only. Highly activated and concentrated sludge floc is returned to the aeration tank at the proper location (usually the inlet); this reduces bacterial lag, accelerates logarithmic bacterial growth, and provides abundant surfaces for adsorption of new cells. The chief advantage of this process is the flexibility it offers the operator. Figure 13.3 shows that one can obtain almost any desired ratio of primary effluent to sludge seed returned.

13.4 Dispersed-Growth Aeration

Dispersed-growth aeration is a process for oxidizing dissolved organic matter in the absence of flocculent

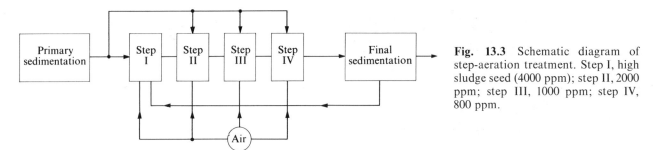

Fig. 13.3 Schematic diagram of step-aeration treatment. Step I, high sludge seed (4000 ppm); step II, 2000 ppm; step III, 1000 ppm; step IV, 800 ppm.

growths [11]. The bacteria (seed) for oxidizing are present in the supernatant liquor after wastes have been aerated and settled. A portion of this supernatant liquor is retained for seeding incoming wastes, while the settled sludge from the secondary settling tank is digested or treated by other sludge-treatment methods. This process has been successfully used to treat many types of dissolved organic wastes [13, 21–24]. Its advantage is that it eliminates certain problems associated with sludge seeding. With many industrial wastes, it is difficult to build up any significant sludge concentration; in such cases, dispersed-growth aeration (which is not dependent on sludge) finds ready acceptance. Dispersed-growth aeration does require more air to achieve the same BOD reduction as the activated-sludge process. However, when one considers that the initial BOD in dispersed-growth aeration is usually quite high, the amount of air required per pound of BOD removed is about the same as that used in the activated-sludge process, even though aeration periods to reach the same BOD reduction are normally quite lengthy (24 hours as compared to 6). Treatment by dispersed-growth aeration involves complete removal by oxidation, rather than by adsorption and partial oxidation, as in activated-sludge treatment.

Heukelekian [11] originally conceived this idea of seeding concentrated, soluble organic wastes with dispersed, instead of flocculent, growths, when he discovered that bacteria in culture mediums normally grow in the dispersed state or in small groups and that seeding is essential for high-rate biological activity. If a waste contains only soluble material, no flocculent growth should form. In his early work on penicillin and streptomycin wastes [12, 13], he made the following claims for the dispersed-growth aeration process:

1) It is better adapted than activated-sludge methods for the treatment of concentrated, soluble organic wastes because: (a) activated sludge has a tendency to bulk with concentrated organic wastes; (b) it is difficult to develop an activated sludge from a soluble waste.

2) Little sludge is formed with dispersed growths when soluble substrates are decomposed.

3) The percentage of BOD reduction decreases as the strengths of penicillin and streptomycin wastes are increased, but 80 per cent reduction may be expected with wastes up to 3000 ppm of BOD. Greater BOD reductions are possible when the BOD is less than 1000 ppm and the waste is aerated for 24 hours.

4) The effluent has a higher turbidity than the raw waste, and color is not removed.

5) The process may be used as a pretreatment unit for conventional biological-treatment processes.

6) The seed material can readily be developed and adapted from soil or sewage within a few days.

7) Optimum results are obtained with air rates of 2 to 3 cubic feet per gallon per hour; stronger wastes require higher air rates.

8) The initial pH of the waste does not seem a critical factor, since the pH increases during aeration. The BOD of raw waste with a pH of 6.4 is reduced as much as the BOD of the same waste adjusted to 7.2.

The author [28] also found this method of treatment suitable for rag and jute paper-mill wastes. In a basic study of the oxidation of glucose by dispersed-growth aeration [27], the authors found 5 million bacteria per milliliter when using nutrient broth as a medium. The two major types of bacteria found during the 24-hour aeration period were:

1) Dispersed, short, thick, round-ended rods;

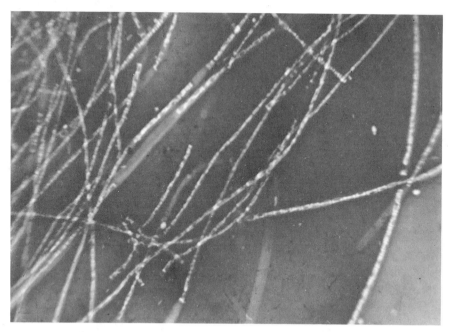

Fig. 13.4 Round-ended rods in a capsule of slime ($\times 620$).

approximate size, 2 to 2.5 microns $\times$ 1 micron. Some of these organisms appeared as fingerlike capsules. As the aeration period progressed, there was an apparent increase in the number of slime-enmeshed bacteria (Fig. 13.4).

2) Sphaerotilus-like organisms, often as unsheathed forms (Fig. 13.5); these were more abundant after 6 hours of aeration and reached an apparent maximum after 24 hours.

In studying the suitability of dispersed-growth aeration for industrial wastes containing both proteins and carbohydrates, Struzeski and Nemerow [38] found such wastes amenable to oxidation by this process. Biological oxidation was enhanced by an increase in temperature, as shown in Fig. 13.6, and initial pH values up to 9.5 did not hamper it. It was also found that, when soluble protein–carbohydrate wastes are to be treated by dispersed-growth aeration, units must be designed to allow ample detention time, since air rates above the critical level (1050 cubic feet of air per pound of BOD per day) do not increase the reduction of BOD.

13.5 Contact Stabilization

Biosorption is the commercial name of one equipment manufacturer's high-rate biological-oxidation process, used mainly for domestic sewage. It was originally developed at Austin, Texas, by Ullrich and Smith [40]. It is essentially a modification of the activated-sludge process and is similar in some respects to the step-aeration process, but generally requires less air and plant space than these other two methods. In the contact-stabilization process, raw waste is mixed by aeration with previously formed activated sludge from a stabilization–oxidation tank, or aerobic digester, for a short period of time (15 to 20 minutes). This activated-sludge—raw-waste mixture is then clarified by settling for about two hours, after which the settled sludge (consisting of activated-sludge floc with adsorbed impurities from the raw waste) goes through intense biological oxidation in the stabilization—oxidation basin for an aeration period of one to two hours. It then returns to the mixing tank and is again mixed with raw waste, so that it can absorb and adsorb added organic

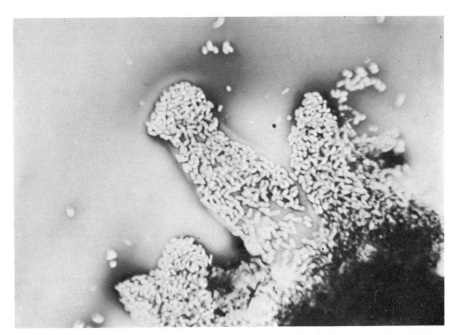

Fig. 13.5 *Sphaerotilus*-like organism, sheathed and unsheathed (× 620).

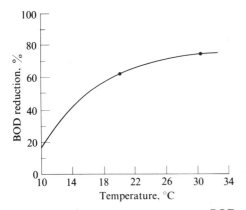

Fig. 13.6 Effect of temperature on average BOD reduction of a synthetic protein-glucose waste, using a dispersed-growth-aeration system, after 24 hours of aeration and no settling.

matter, and so on, in a continuous process. Excess or waste sludge can be taken from the system after either the clarifying or stabilizing steps, for anaerobic digestion or for dewatering on vacuum filters (see Fig. 13.7).

Ullrich and Smith claimed that this process requires less aeration-tank capacity than other processes, since the real aeration or reactivation takes place in the settled and concentrated sludge, not in the mixed liquor. Because the sewage and returned sludge is given only a brief mix, a small mixing compartment is needed. Pertinent pilot-plant and full-scale operating results are given in Table 13.2 [40, 41].

13.6 High-Rate Aerobic Treatment

High-rate aerobic treatment (total oxidation) has developed in the last ten years as a means of oxidizing organic wastes [39]. This process consists of comminution of the waste, long-period aeration (one to three days), final settling of the sludge, and return of the settled sludge to the aeration tank. There is no need for primary settling or sludge digestion, but the aeration system must be large, to provide the required aeration period. The total-oxidation process is particularly useful in small installations, since it does not require a great deal of supervision. Little difficulty occurs with bulking on the sludge, even though the

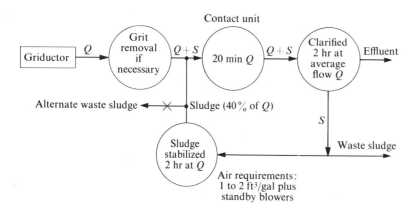

Fig. 13.7 Schematic arrangement of contact stabilization process.

settling period is relatively short at times. In fact, since the solids resulting from this process are mostly of low volatility and therefore high in ash, the settling rate is quite fast. Return of the sludge is continuous and very rapid in comparison with normal activated-sludge practice. By returning sludge at a high rate (100 to 300 per cent of flow), the system is kept completely aerobic at all times. The concentration of solids in the mixed liquor after a long period reaches a high level, and a portion of the sludge can then be wasted to reduce the concentration to 3000 to 5000 ppm. Lesperance [19] suggests that a waste sludge can be expected equal to 0.15 pound per pound of BOD removed. The small volume of wasted sludge is then stored and further concentrated until removed by tank car or other means to an area away from the plant.

The high-rate aerobic treatment, though it produces little waste sludge, has the disadvantages of requiring about three times as much air as conventional activated-sludge plants and of releasing some floc in the effluent. On the other hand, it needs very little operational maintenance and is well-suited for shock loadings from industrial operations.

Another version of this treatment is referred to as completely mixed systems [20]. It operates on the assumption that, if microorganisms are kept in a constant state of growth, they operate at maximum efficiency and are adapted to the particular character and concentration of the waste. This constant-growth state can be maintained only if: (1) the microorganisms and raw wastes are thoroughly and continuously mixed; (2) the organic concentration is held constant; and (3) the effluent is separated from the microorganisms at a constant rate that is equal to the waste-feed rate. Figure 13.8 depicts a typical complete-mixing activated-sludge system [20]. A loading of 60 pounds of oxidizable organics per 1000 cubic feet of aeration tank is possible with this type of treatment.

Table 13.2 Biosorption operating data. (After Ullrich and Smith [40, 41].)

Item	1951 (Pilot plant)		1955 (Full scale)	
	Mean	% removal	Mean	% removal
Influent BOD, ppm	264		307	
Effluent BOD, ppm	19.5	92.5	20	93.4
BOD loading, lb/1000 ft^3			144	
Detention time, hr	2.83		3	
Air required, ft^3/lb BOD removed			665	
Suspended-solids influent, ppm	226		226	
Suspended-solids effluent, ppm	13.9	93.8	18	92.1

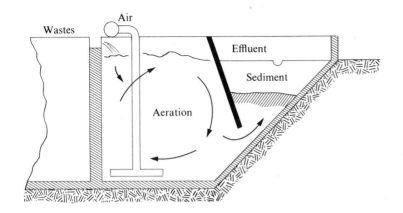

Fig. 13.8 Complete-mixing activated-sludge system. (After McKinney et al. [20].)

13.7 Trickling Filtration

Trickling filtration is a process by which biological units are coated with slime growths (zoogleal forms) from the bacteria in the wastes. These growths adsorb and oxidize dissolved and colloidal organic matter from the wastes applied to them. When the rate of application is excessive—10 to 30 million gallons per acre per day (mgad)—and continuous, the humus collected on the filter-bed surfaces is sloughed off continuously. Crushed stone, such as traprock, granite, and limestone, usually forms the surface material in the filter, although recently other materials, such as plastic rings, have proved very effective. Since smaller stones provide more surface per unit of volume, the contact material must be small, in order to support a large surface of active film, but not so small that its pores become filled by the growths or clogged by accumulated suspended matter or sloughed film. Crushed stone, $1\frac{1}{4}$ to 3 inches in diameter, is used, with the smallest stone at the top. The integral parts of a trickling-filter system are the distribution nozzles, contact surface, and underdrain units. The process may be summarized as follows:

1) An active surface film grows on the stone or contact surface.
2) Concentration of colloidal material and gelatinous matter occurs.
3) These adsorbed substances are attacked by bacteria and enzymes and reduced to simpler compounds, so that NH_3 is liberated and oxidized by chemical and bacterial means, giving a gradual reduction of NH_3 and an increase of NO_2 and NO_3 (Fig. 13.9).
4) A flocculent, humuslike residue or sludge, containing many protozoa and fungi, accumulates on the surface. When it gets too heavy it will slough off and resettle (a continuous process with biofilters). Part of the oxygen is supplied by spraying waste, blowing air into the filter, or allowing waste to drip into the filter. Another portion is supplied by convection due to the temperature difference between the incoming waste and the bed. The larger the surface, the greater the number of bacterial organisms that come into contact with the liquid to be purified; the greater the

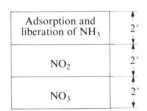

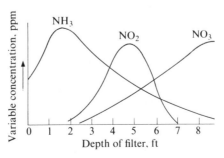

Fig. 13.9 Changes in nitrogen occurring in filter.

number of organisms, the higher the purification of the liquid. The smaller the pieces of rock in the surface media, the greater the purification; too-small particles, however, promote clogging. We can summarize by saying that trickling filters act as both strainers and oxidizers.

Zobell [44] was the first to point out the importance of providing solid contact surfaces to further the physiological activities of bacteria growing in dilute nutrient solutions, such as most industrial wastes are. The following phenomena, according to Zobell, cause this increase in biological oxidation:

1) Solid surfaces make possible the concentration of nutrients and enzymes by adsorption to the surface.

2) The interstices between bacterial cells and surfaces act as concentration points; they retard the diffusion of exoenzymes and metabolites away from the cell, thereby favoring both digestion and adsorption of foodstuffs.

3) The interstices between surfaces and cells produce optimum conditions for oxidation–reduction and other physicochemical reactions.

4) Surfaces function as attachment points for microorganisms which are obligatory periphytes.

A typical, standard-rate, stone-bed trickling filter provides about 100 square feet of surface material per square foot of ground on which the filter is constructed. Velz [42] proposed the performance equation for trickling filters as

$$\frac{L_D}{L} = 10^{-kD},$$

where L_D is the removable fraction of BOD remaining at depth D, L is total removal, k is the logarithmic extraction rate, and D is depth of the bed. The reader will note the similarity between this equation and the monomolecular rate of decomposition of organic matter in streams:

$$\frac{L_t}{L} = 10^{-kt}.$$

The student should realize that the contact time in a filter is relatively short, compared with an activated-sludge process. However, the organic matter (bacterial food) resides in the bed longer than computed from the detention time. Howland [14] has contributed to our knowledge of contact time in filters. Assuming that a sheet of water is flowing steadily down an inclined plane under laminar flow conditions, he expresses the contact time as

$$T = \left(\frac{3v}{gs}\right)^{1/3} \frac{l}{q^{2/3}},$$

where T is the time of flow down the inclined plane, l is the length of the plane, s is the sine of the angle which the plane makes with the horizontal, g is the acceleration due to gravity, v is the kinematic viscosity of water (μ/ρ), and q is the rate of flow per unit width of the plane.

Howland indicates that the amount of oxidizable organic matter removed in a filter depends directly on the length of time of the flow. He recommends a deep filter containing the smallest practical media, to achieve an optimum contact time and maximum efficiency. Although some researchers recommend shallower filters and larger stone to reduce both initial and operating costs, the tendency today appears to be to follow Howland's recommendations, because engineers want an increasing degree of removal of BOD. Still, because of clogging and head-loss difficulties, there is a limit to how deep a bed and how small a stone size one can utilize.

Ingram [15] suggests the following drawbacks of present-day trickling filters: they occupy too much space; they exhibit seasonal variation in efficiency; clogging and pooling present problems; there are limitations on hydraulic and organic loading; and there are limitations on the strength of sewage applied. He proposes a trickling-filtration process called controlled filtration, which utilizes deep filters (18 to 24 feet). He was able to achieve greater than 70 per cent BOD removal (the removal expected in high-rate filters loaded at a normal rate of 20 mgad and 1300 pounds of BOD per acre-foot per day) with a minimum hydraulic loading of twice—and an organic loading of $1\frac{1}{2}$ to $10\frac{1}{2}$ times—these normal standards. His experimental filter is shown in Fig. 13.10.

Behn [2] points out that deviations from the usual reaction rates sometimes occur because of temperature of the waste and degree of filter saturation. Rankin [33] has been concerned with recirculation of filter effluent and concludes, from a study of a

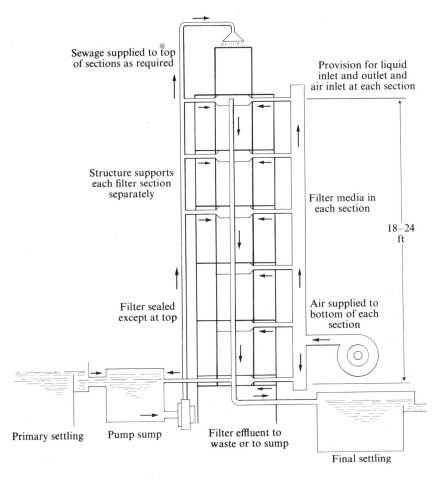

Fig. 13.10 Diagram of experimental controlled-filtration system. (After Ingram [15].)

number of treatment plants, that performance appears to depend primarily on the ratio of recirculation to raw waste-water flow, rather than on dosing rate, loading of the filter, or depth of filter (within the ranges studied). In smaller plants, with only one filter, single-stage filters appear to be most feasible, while for larger plants, where multiple filters are necessary or where stronger wastes are being treated, a two-stage series-parallel arrangement of filters yields a better effluent than single-stage filters with the same tank and filter capacity and the same volume of recirculated liquor. Diagrams of each of these systems are shown in Figs. 13.11 and 13.12.

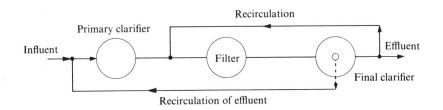

Fig. 13.11 Single-stage trickling filter (with recirculation).

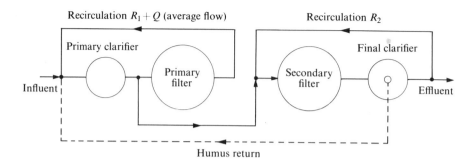

Fig. 13.12 Two-stage series-parallel biofiltration process.

13.8 Spray Irrigation

Spray irrigation is an adaptation of the familiar method of watering agricultural crops by portable sprinkling-irrigation systems; wastes are pumped through portable pipes to self-actuated sprinkler heads. Lightweight aluminum or galvanized piping, equipped with quick-assembly pipe joints, can be easily moved to areas to be irrigated and quickly assembled. Wastes are applied as a rain to the surface of the soil, with the objective of applying the maximum amount that can be absorbed without surface run-off or damage to the cover crops. A spray-irrigation system is composed of the following units: (1) the land on which to spray; (2) a vegetative cover crop to aid absorption and prevent erosion; (3) a mechanically operated screening unit; (4) a surge tank or pit; (5) auxiliary stationary screens; (6) a pump which develops the required sprinkler-nozzle pressure; (7) a main line; (8) lateral lines; and (9) self-actuated revolving sprinklers operating under 35 to 100 psi nozzle pressure.

With good cover crops (dense, low-growing grasses) and fairly level areas, waste to a depth of 3 to 4 inches can be applied at a rate of 0.4 to 0.6 inch per hour. The process is generally limited to spring, summer, and autumn. Anderson *et al.* [1], in a recent study with citrus wastes, found that aerobic conditions are maintained without odors to a depth of at least three feet.

13.9 Wet Combustion

Wet combustion is a process [43] of pumping organics-laden waste water and air into a reactor vessel at elevated pressure (1200 psi) (Fig. 13.13). The organic fractions undergo rapid oxidation, even though they are dissolved or suspended in the waste. This rapid oxidation gives off heat to the water by direct convection, and the water flashes into steam. Inorganic chemicals, which are present in many industrial wastes, can be recovered from the steam in a separate chamber. Heat from an external source is applied just to start the process; thereafter, it requires only 12 to 20 per cent of its own heat to maintain itself. The remaining 80 to 88 per cent can be utilized as process steam or to drive turbines for electrical or mechanical power. This process has a good potential where steam is essential and inexpensive enough to justify the cost of the equipment and where the inorganic chemicals in the waste are worth recovering and reusing. The wet combustion process can maintain itself only when the waste has a high percentage of organic material (usually about 5 per cent solids and 70 per cent organic).

13.10 Anaerobic Digestion

Anaerobic digestion is a process for oxidizing organic matter in closed vessels in the absence of air. The process has been highly successful in conditioning sewage sludge for final disposal. (Since digestion is primarily used for the treatment of sludge, rather than liquid wastes, the theory of its operation is described in more detail in Chapter 14.) It is also effective in reducing the BOD of soluble organic liquid wastes, such as yeast, cotton-kiering, slaughterhouse, dairy, and white-water (paper-mill) wastes. Generally, anaerobic processes are less effective than aerobic

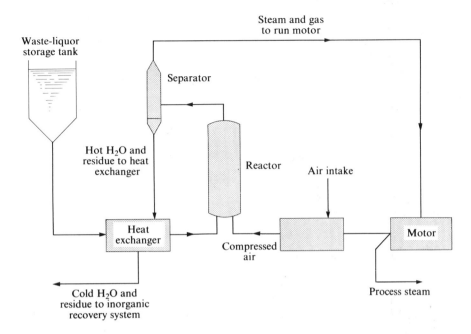

Fig. 13.13 Schematic arrangement of wet-combustion-process units.

processes, mainly because of the small amount of energy that results when anaerobic bacteria oxidize organic matter. Anaerobic processes are therefore slow and require low daily loadings and/or long detention periods. However, since little or no power need be added, operating costs are very low. Where liquid waste volumes are small and contain no toxic matter and there are high percentages of readily oxidized dissolved organic matter, this process has definite advantages over aerobic systems. The pH in digesters must be controlled to near the neutral point.

Buswell [5] proposed the following general equation for conversion of organic matter in industrial wastes to carbon dioxide and methane:

$$C_n H_a O_b + \left(n - \frac{a}{4} - \frac{b}{2}\right) H_2 O$$

$$= \left(\frac{n}{2} - \frac{a}{8} + \frac{b}{4}\right) CO_2 + \left(\frac{n}{2} + \frac{a}{8} - \frac{b}{4}\right) CH_4.$$

In the United States, anaerobic treatment plants have been built to treat yeast, butanol–acetone, brewery, chewing-gum, and meat-packing wastes. Pettet et al. [32], in a review of British practices, found that slaughterhouse waste appears to respond extremely well to anaerobic digestion, although up to 1959 there were no full-scale anaerobic-digestion plants in Great Britain. In the United States, BOD reductions of 60 to 92 per cent have been attained with all these wastes, at loadings of 0.003 to 0.191 pound of BOD per cubic foot of digester per day. Concentrations of organic matter ranged from 1565 to 17,000 ppm BOD.

13.11 Mechanical Aeration System

Cavitation is a typical process for mechanical aeration of wastes. The complete Cavitator assembly consists of a vertical-draft tube with openings for connection to the influent pipe and a rotor assembly of the multiblade type, supported by an adjustable ball thrust bearing mounted at the motor level. The rotor is mounted on a stainless-steel shaft and the entire unit, including the draft tube, is supported by a structural-steel bridge. A cross-section is shown in Fig. 13.14 [36]. As soon as the rotor exceeds a certain critical speed, air is drawn in from the atmosphere through the vertical hollow tube and dispersed

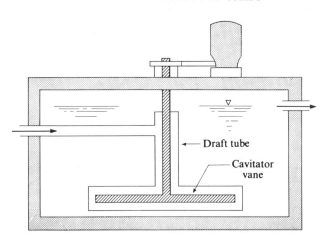

Fig. 13.14 Typical Cavitator system.

into the waste. The rotor creates a zone of cavitation in its turbulent trail and air moves in to fill the areas of rarefied underpressure. The amount of air which is being entrained depends on the size and shape of the rotor, the rpm, and the water depth. The manufacturers claim that their system utilizes at least 25 per cent of the available oxygen in the air, in contrast with conventional aeration equipment, which utilizes only 5 per cent. At least one waste-treatment plant (dealing with canning wastes and sewage) attained over 90 per cent BOD removal with an air supply of 110 cubic feet per pound of BOD per day. Operational costs [36] were $12.80 per day for an equivalent population of 12,000 persons. A recent modification of this system employs mechanical mixing by a rotor submerged (but near the surface of) the waste water. Power costs are thus reduced, with no apparent loss of aeration or mixing efficiency. This system promises to be the most economical one for secondary treatment of wastes with a highly dissolved organic content.

13.12 Well Injection

Disposal of wastes containing dissolved organic matter by injecting them into deep wells has been successful in areas of low or nonexistent stream flow, especially when wastes are malodorous or toxic and contain little or no suspended matter. Deep well injection has been used successfully to dispose of organic solutions from chemical, pharmaceutical, petrochemical, paper, and refinery wastes; in addition, many inorganic solutions may be disposed of in this manner. To be effective, the wastes must be placed in a geological formation which prevents the migration of the wastes to the surface or to ground-water supplies. The rock types most frequently used are the more porous ones such as limestones, sandstones, and dolomites, since the porosity may help develop a filter cake which plugs the well. Other factors, in addition to geology, to be considered are depth and diameter of well, injection pressures, and the volume and characteristics of the wastes. At the end of 1966, there were 78 industrial disposal wells in the United States, most of which are used for chemical and refinery wastes (86 per cent); most are less than 4000 feet deep (74 per cent), dispose of less than 400 gpm per well (87 per cent), and operate at less than 300 psi (57 per cent). Costs of injection disposal installations vary from $30,000 for a shallow (1800-foot) well not requiring pretreatment to over $1,400,000 for a very deep (12,000-foot) well with intricate pretreatment. Actual costs vary depending on depth, surface equipment, pretreatment, diameter of well, injection pressure, variability of composition of waste water, and availability of drilling equipment.

Donaldson [6] reports on a wide variety of industrial wastes being injected into formations ranging in age from Precambrian to modern-day. In the United States, up to 1964, more than 30 wells, ranging in depth from 300 to 12,000 feet, were being used for waste disposal into subsurface formations which include unconsolidated sand, sandstone, regular limestone, and fractured gneiss. Although subsurface injection offers an economical method of final disposal where receiving surface water is inadequate to carry the waste water away safely, circumstances can limit its effectiveness, e.g. the area lacks suitable underground formations for waste injection, the initial capital expense is excessive, or the pretreatment required may be too extensive and expensive.

The industrial waste engineer must work closely with a geologist familiar with the subsurface formations in the area in order to select the proper waste-disposal zone. A well is drilled and core samples are analyzed for specific characteristics such as permeability and reactivity with the waste. Tests are carried out to determine the injection pressure required at various waste-water flows. Certain procedures, such as fracturing and acidizing, may be used to improve the soil permeability and thus reduce the injection pressure required at various flows.

Schematic drawings of typical complete subsurface waste-disposal systems are shown in Figs. 13.15 and 13.16. Although cement tanks up to 50,000-gallon capacity are commonly employed within the basement of a factory, large, shallow, open ponds may be used where land is available and where some settling and oxidation is required as a pretreatment. The oil separator is required for petroleum refinery wastes, since oil tends to plug disposal formation and the oil can be recovered and reused. The usual separator consists of a tank with many internal baffles to cause the oil to separate and rise. If a clarifier is then used, heavier material such as dirt, resin flocs, and suspended grease can settle out. Mechanical equipment such as sludge rakes and surface skimmers can also be used

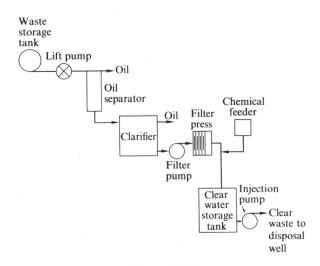

Figure 13.15

with this equipment. Since not all solids are completely removed by the treatment so far described, filters are then used to protect sand or sandstone formations from plugging. The screens are usually metal and coated with diatomaceous earth, but in some situations sand filters are preferred. If wastes contain slime that will form bacteria, algae, iron bacteria, sulfate-reducing bacteria, or fungi, a suitable bactericide (such as quaternary amines, formaldehyde, chlorinated hydrocarbons, chlorine, or copper sulfate) is used to control their detrimental effects. The clear-water storage tank is normally equipped with a float switch designed to operate the injection pump at certain liquid levels. The size and type of injection

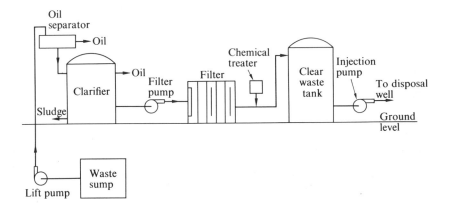

Fig. 13.16 Typical surface equipment for deep well waste injection. (From Bureau of Mines Information Circular no. 8212.)

Table 13.3 Summary of disposal systems.

Company	Type of waste	Injection rate, gpm	Injection pressure, psi	Subsurface depth of wells, feet
A	Brine; chlorinated hydrocarbons	200	500	12,045
B	Clear 4% solution Na$_2$SO$_4$	300	45	295
C	Masic waste, pH ±10	70	1000	6,160
D	Magnesium; calcium hydroxides	200	Vacuum	400
	Manufacturing waste, pH may change from 1 to 9 in 8 hr	400	Vacuum	4,150
E	Lachrymator waste from acrolein and glycerine units	700	150–170	1,960
F	Aqueous solution—phenols, mercaptans, and sulfides	215	30–90	1,795
G	Phenols; mercaptans; sulfides; brine	100	40–100	1,980
H	Phenols; chlorinated hydrocarbons	200	450	4,000
	Brine	200	150	4,000
	Phenols; mercaptans; sulfides	50	—*	4,000
I	Coke oven phenols; quench water	50	300	563
J	Organic wastes	60	500	1,472
K	Sulfuric acid waste	400	Vacuum	1,830
L	Detergents; solvents; salts	254	280	1,807
M	38% HCl solution	14		1,110
N	Stripping steam condensate; cooling tower blowdown	50	10–20	1,110
	Aqueous petroleum refinery effluent	400	50–70	1,110
O	Phenols; brine	75	400	7,650

*Information not available.

Table 13.3 (*continued*)

Formation age, type, and name	Total cost of system ($)	Date started	Problems	Solutions and remarks
Precambrian fractured gneiss (unnamed)	1,419,000	March 1962	Microorganisms in waste	
Sandstone	—*	June 1951	None	
Cambrian sandstone	250,000	Nov. 1960	Inadequate filtration	Larger filter planned
Permian salt bed (Hutchinson)	—*	—*	None	
Ordovician vugular limestone (Arbuckle)	500,000	Dec. 1957	Corrosion and water hammer	Heavier tubing planned
Pleistocene	135,000	1956	Sand incursion increased injection pressure	Back-washing every 4 months
Pleistocene	30,000	Sept. 1959	Sand incursion	Periodic back-washing
Pleistocene	—*	March 1960	Sand incursion	Periodic back-washing
Devonian vugular limestone (Dundee)	—*	1950	None	
Devonian vugular limestone (Dundee)	—*	1931	None	
Devonian vugular limestone (Dundee)	—*	1950	None	
Silurian sandstone (Sylvania)	25,000	Aug. 1956	High wellhead pressure	Acidizing and fracturing
Devonian sandy limestone (Dundee); Traverse, Dundee & Monroe	400,000	1954	None	
Permian sandstone	562,000	Jan. 1960	Microorganisms decreased injectivity	Formaldehyde
Ordovician vugular limestone (Arbuckle)	300,000	Feb. 1960	Mechanical failure of surface equipment	
Unconsolidated sand (Glorieta)	—*	April 1962	None	
Unconsolidated sand (Glorieta)	—*	1959	None	
Unconsolidated sand (Glorieta)	—*	1958	None	
Eocene sand and clay (Frio)	—*	1958	High injection pressure	Periodic acidizing

pump is controlled by wellhead pressure, waste-water flow, and waste-water characteristics such as pH and corrosiveness. The multiplex piston pump is most commonly used when wellhead pressures of greater than 150 psi are required, whereas single-stage centrifugal pumps are used at lower pressures.

To construct the well, first a 15-inch-diameter hole is drilled to 200 feet below the deepest fresh-water aquifer and a $10\frac{1}{2}$-inch (O.D.) casing is set and cemented to the surface. Next, a 9-inch-diameter hole is drilled to the bottom of the potential disposal formation, a 7-inch (O.D.) casing is set at the total depth of the hole, and cement is circulated in the annulus between the injection casing and the 9-inch hole to the surface (Fig. 13.17). This method has been proved to seal off water aquifers from the well and to protect other water resources. Table 13.3 summarizes Donaldson's findings for 20 separate installations, presenting much valuable information such as costs, associated problems, well depth, injection pressures, and formation type.

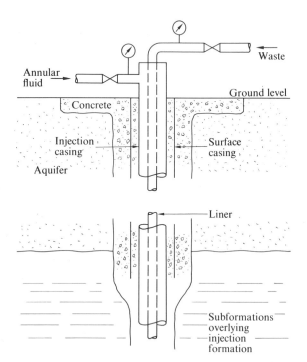

Fig. 13.17 Typical injection well. (From Bureau of Mines Information Circular no. 8212.)

For the disposal of acid wastes, there are five requirements for deep well disposal: (1) a satisfactory disposal horizon; (2) an horizon filled with salt water; (3) an horizon located at a sufficient depth; (4) a suitable cap rock; (5) a waste compatible with the natural water in the disposal horizon. The possible dangers of deep well disposal include: (1) contamination of potable water supplies either by lateral migration to existing unplugged dry holes or producing wells, or by vertical migration through the subsurface, or by vertical migration due to mechanical failure; and (2) possible movements along old fault planes. The representative cost for a 4000-foot well is $450,000, which includes the cost of well and equipment, above-ground pumping and equipment, and holding-tank and collection equipment.

The reader may use the following checklist in the design of deep-well disposal systems:

Factors to consider in subsurface disposal of industrial wastes by deep well injection
   A. State laws and legal aspects
      1) State recognition of this method
      2) Subsurface trespass
   B. Geology
      1) Employment of a geologist or well contractor
      2) Disposal formation
         a) Porosity
         b) Permeability
         c) Composition (sandstone or limestone)
   C. Waste characteristics
      1) Volume reduction
      2) Injection flow rate
      3) Injection pressure
      4) Corrosiveness
      5) Biological effects
   D. Surface equipment needs
   E. Wells
      1) Number
      2) Size
      3) Monitoring
   F. Economics

13.13 Foam Phase Separation

Figure 13.18 illustrates the equipment used for foam phase separation. A sparger producing small gas bubbles (usually air) causes these bubbles to rise through the liquid and adsorb surface-active solutes and suspended matter. When the bubbles reach the surface, a foam forms, which is forced out of the foamer, collapsed, and discharged as a concentrated waste.

If the following assumptions are made: (1) complete mixing in the foamer; (2) sufficient depth of liquid to reach maximum solute adsorption of the gas–liquid interface; (3) constant liquid density; (4) no bubble rupture in the foam phase; and (5) negligible volume of the liquid layer containing the surface excess of solute—one can arrive at a material-balance equation

$$C_F - C_B = 1000 \frac{G}{F} \Gamma_B S,$$

where C_F and C_B are feed and bottom product concentrations in mg/liter, G is the volumetric gas rate in liters/minute, Γ_B is solute surface excess corresponding to C_B in mg/cm², and S is the specific surface of bubbles in foam phase in cm²/cc.

At flows of air to liquid feed of $G/F \geq 3$, it is reported that the COD is reduced by 25 per cent and alkyl benzene sulfonate (ABS) concentrations are reduced by 50 to 75 per cent. At air rates of 1.5 liters/mg of ABS in secondary effluents, removals of 0.4 ppm ABS have been reported. The success of the process, in general, depends upon the foamability of the liquid waste, which is said to be of low order of magnitude. Bruner and Stephen [3] have calculated foam separation costs (not including the foamate disposal) as follows:

mgd	cents/1000 gal.
1	3.6
10	1.9
100	1.4

Schoen et al. [35] used this treatment successfully to separate radium from uranium-mill waste water. They found the pH of the waste water very important in selecting foaming agents. An increase in foaming agent will generally produce a similar increase in foam during treatment. Grieves and Crandall [8] also experimented with both iron and alum as coagulants, using bentonite as an aid, in foaming low-quality waters.

13.14 Brush Aeration

According to Pasveer [31] the brush aeration system

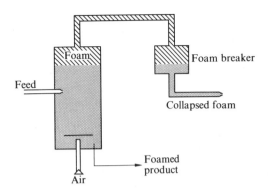

Fig. 13.18 A column foam fractionator.

was evolved between 1925 and 1930 by Dr. Kessener for use in the activated-sludge process. It is essentially an extended aeration process providing over 24 hours of aeration. Since the 1930s, it has found application, particularly in the Netherlands and in a few plants in Britain, and about a dozen plants were constructed in Canada and the United States prior to 1964 [29] (see Fig. 13.19). Most of these aeration systems are installed in "oxidation ditches." The design of the oxidation ditch combines an aeration tank and a holding tank in a single unit; the aeration rotor circulates the mixed waste through the whole ditch by means of the rotating cage, but aeration occurs only in the vicinity of the rotor. The rotor is fixed at both ends and set transversely across the aeration ditch and rotates in the direction of waste flow. Aeration is obtained by means of long, rectangular, angle irons welded to the rotating cage. Although the results have been obtained mostly with domestic sewage, it is apparently adaptable to any organic industrial waste.

13.15 Subsurface Disposal

Three other methods of disposing of dissolved organic wastes below the ground surface are injection, placement in underground cavities, and spreading. Since injection is discussed in some detail in Section 13.12 and placement in underground cavities is limited to either small volumes of wastes or particular situations of subsurface formation, they will only be mentioned here as possibilities. Koenig [16] reports, however, that, in 1956, 244 cavities were used for storage, mostly for hydrocarbons and mostly in salt mines.

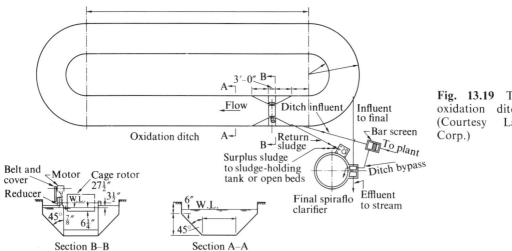

Fig. 13.19 Typical layout of an oxidation ditch treatment plant. (Courtesy Lakeside Engineering Corp.)

Spreading may be defined as the dispersal of liquid wastes on the ground in order to enhance their infiltration into it. Reclamation of waste waters by spreading on land, with subsequent withdrawal of ground water, has been extensively practiced, mostly for secondary sewage effluents. Infiltration rates govern the use of this method, while ultimate effects on underground water supplies govern its acceptability. Because of numerous physical limitations this method should be considered mainly for small volumes of concentrated, organic wastes in particularly suited soils. Koenig [16] gives comparative costs for this method of disposal and for injection, wet oxidation, and incineration (Fig. 13.20).

13.16 The Bio-Disc System

The Bio-Disc system was developed independently in West Germany (by Hartmann and Pöpel) and the United States (by Welch and Antonie).* It consists of a series of flat, parallel discs which are rotated while partially immersed in the waste being treated. Biological slime covers the surface of the discs and adsorbs and absorbs colloidal and dissolved organic matter present in the waste water. Excess slime generated by synthesis of the waste materials is sloughed off gradually into the mixed liquor and subsequently separated by settling (Fig. 13.21). The rotating discs carry a film of the waste water into the air where it absorbs the oxygen

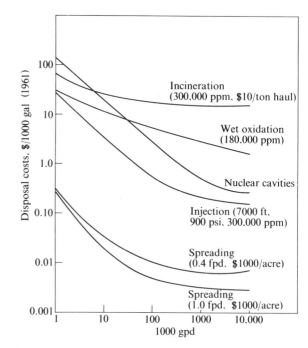

Fig. 13.20 A comparison of unit disposal costs (1961 figures). (After Koeing [16].)

---

*For references see the section on Bio-Disc in **Suggested Additional Reading**.

Fig. 13.21 RBC disc system. (Courtesy Allis Chalmers Company.)

necessary for aerobic biological activity of the slime. Disc rotation also provides contact between the slime and the waste water. Thus, the rotating discs provide: (1) mechanical support for a captive, microbial population; (2) a mechanism of aeration, the rate of which can be adjusted by changing the rotational speed; and (3) contact between the biological slime and the waste water, the intensity of which can be varied by changing the rotational speed.

Use of closely spaced parallel discs achieves a high concentration of active biological surface area. This high concentration of active organisms and the ability to achieve the required aeration rate by adjusting the rotational speed of the discs enables this process to give effective treatment to highly concentrated wastes. At a loading of 11 lb BOD/day/1000 ft² of surface area, 90 per cent BOD removal is obtained in 2000 ppm BOD dairy waste. Secondary treatment of domestic sewage is accomplished with a retention time of 1 hour or less, and 90 per cent BOD reduction is obtained at a loading of 5 lb BOD/day/1000 ft². Because a buoyant plastic material is used for the discs and negligible head loss is encountered through the RBC itself the power requirement for this process is very low. Its simplicity of construction and operation has demonstrated that minimal unskilled maintenance is all that is required for efficient operation.

The RBC process has gained wide acceptance in Europe. In the past ten years over 400 Bio-Disc plants have been constructed there ranging in size from 24,000 to 55,000 population equivalents for treatment of domestic and industrial wastes. This process is now being introduced commercially to the United States.

References

1. Anderson, D. R., W. D. Bishop, and H. L. Ludwig, "Percolation of citrus wastes through soil," in Proceedings of 21st Industrial Waste Conference, Purdue University Engineering Extension Series, Bulletin no. 121, 1966, p. 892.
2. Behn, V. C., "Trickling filter formulations," in Conference on Biological Treatment, at Manhattan College, New York, April 20–22, 1960, Paper no. 26.
3. Bruner, C. A., and B. G. Stephen, "Foam fractionation," *Ind. Eng. Chem.* **57**, 40 (1965).
4. Busch, A. W., and A. A. Kalinske, "The utilization of the kinetics of activated sludge in process and equipment of design," in J. McCabe and W. W. Eckenfelder (eds.), *Biological Treatment of Sewage and Industrial Wastes*, Reinhold, New York (1956), p. 277.
5. Buswell, A. M., and W. D. Hatfield, "Anaerobic fermentations," Bulletin no. 32, State of Illinois, Division of State Water Survey, Urbana, Ill. (1939).
6. Donaldson, E. C., *Subsurface Disposal of Industrial Wastes in the United States*, Information Circular 8212, Bureau of Mines, U.S. Department of the Interior, Washington, D.C. (1964).
7. Emde, W. von der, "Aspects of high rate activated sludge process," in Conference on Biological Waste

Treatment at Manhattan College, New York, April 20–22, 1960, Paper no. 35.
8. Grieves, R., and C. Crandall, "Water clarification by foam separation: Bentonite as a flotation aid," *Water Sewage Works* **113**, 432 (1966).
9. Haseltine, T. R., "A rational approach to the design of activated sludge plants," in J. McCabe and W. W. Eckenfelder (eds.), *Biological Treatment of Sewage and Industrial Wastes*, Reinhold, New York (1956), p. 257.
10. Hermann, E. R., and E. F. Gloyna, "Waste stabilization ponds," *Sewage Ind. Wastes* **30**, 511 and 646 (1958).
11. Heukelekian, H., "Aeration of soluble organic wastes with non-flocculent growths," *Ind. Eng. Chem.* **41**, 1412 (1949).
12. Heukelekian, H., "Treatment of streptomycin wastes," *Ind. Eng. Chem.* **41**, 1412 (1949).
13. Heukelekian, H., "Characteristics and treatment of penicillin wastes," *Ind. Eng. Chem.* **41**, 1535 (1949).
14. Howland, W. E., "Flow over porous media as in a trickling filter," in Proceedings of 12th Industrial Waste Conference, Purdue University, 1957, p. 435.
15. Ingram, W. T., "A new approach to trickling filter design," *Proc. Am. Soc. Civil Engrs*, **82**, Paper no. 999, (1956).
16. Koenig, L., *Ultimate Disposal of Advanced-Treatment Waste*, Environmental Health Series AWTR-8, U.S. Department of Health, Education and Welfare, Washington, D.C. (1964).
17. Kraus, L. S., "The use of digested sludge and digester overflow to control bulking of activated sludge," *Sewage Works J.* **17**, 1177 (1945).
18. Kraus, L. S., "Dual aeration as a rugged activated sludge process," *Sewage Ind. Wastes* **27**, 1347 (1955).
19. Lesperance, T. W., "Extended aeration and high rate treatment," *Water Works Wastes Eng.* **2**, 40 (1965).
20. McKinney, R. E., J. M. Symons, W. G. Shifrin, and M. Vezina, "Design and operation of a complete mixing activated sludge system," *Sewage Ind. Wastes* **30**, 287 (1958).
21. Nemerow, N. L., "Oxidation of enzyme desize and starch rinse textile wastes," *Sewage Ind. Wastes* **26**, 1231 (1954).
22. Nemerow, N. L., "Oxidation of cotton kier wastes," *Sewage Ind. Wastes* **25**, 1060 (1955).
23. Nemerow, N. L., "Holding and aeration of cotton mill finishing wastes," in Proceedings of 5th Southern Municipal and Industrial Waste Conference, Chapel Hill, N. C., April 1956, p. 149.
24. Nemerow, N. L., "Dispersed growth aeration of cotton finishing wastes: II. Effect of high pH and lowered air rate," *Am. Dyestuff Reptr.* **46**, 575 (1957).
25. Nemerow, N. L., "Accelerated waste oxidation pond studies," in Proceedings of Third Conference on Biological Waste Treatment, Manhattan College, New York, April 20–22, 1960.
26. Nemerow, N. L., and J. C. Bryson, *Hancock Air Force Base Waste Stabilization Research Report*, Syracuse University Reports to U.S. Air Force, March 1960.
27. Nemerow, N. L., and J. Ray, "Biochemical oxidation of glucose by dispersed growth aeration," Chapters 1–7 of *Biological Treatment of Sewage and Industrial Wastes*, Reinhold, New York (1956).
28. Nemerow, N. L., and W. Rudolfs, "Rag, rope, and jute wastes from specialty paper mills: V. Treatment by aeration," *Sewage Ind. Wastes* **24**, 1005 (1952).
29. Ontario Water Resources Commission, *Evaluation of the Oxidation Ditch as a Means of Wastewater Treatment in Ontario*, Research Publication No. 6, Ottawa (July 1964).
30. Oswald, W. J., "Fundamental factors in oxidation pond design," in Proceedings of Third Conference on Biological Waste Treatment, at Manhattan College, New York, April 20–22, 1960, Paper no. 44.
31. Pasveer, A., "New developments in the application of Kessener brushes in the activated-sludge treatment of trade-waste waters," in *Waste Treatment* (ed. P. Isaac), Pergamon Press, New York (1959), pp. 126–155.
32. Pettet, A. E. J., T. G. Tomlinson, and J. Hemens, "The treatment of strong organic wastes by anaerobic digestion," *J. Inst. Public Health Engrs* 170 (1959).
33. Rankin, R. S., "Performance of biofiltration plants by three methods," *Proc. Am. Soc. Civil Engrs* **79**, Separate No. 336 (1953).
34. Sawyer, C. N., "Activated sludge modifications," *J. Water Pollution Control Federation* **32**, 233 (1960).
35. Shoen, H. M., E. Rubin, and D. Ghosh, "Radium removal from uranium mill wastewater," *J. Water Pollution Control Federation* **34**, 1026 (1962).
36. Schulze, K. L., and H. S. Foth, "New low cost secondary treatment by new cavitation system," *Water Sewage Works* **102**, 74 (1955).
37. *Sewage Stabilization Ponds in the Dakotas*, Joint report by North and South Dakota Departments of

Health and the United States Department of Health, Education and Welfare, (1957).
38. Struzeski, E. J., and N. L. Nemerow, "Dispersed growth aeration of protein—glucose mixtures," in Proceedings of 12th Industrial Waste Conference, Purdue University, May 1957, p. 145.
39. Tapleshay, J. A., "Total oxidation treatment of organic wastes," *Sewage Ind. Wastes* **30**, 652 (1958).
40. Ullrich, A. H., and M. W. Smith, "The Biosorption process of sewage and waste treatment," *Sewage Ind. Wastes* **23**, 1248 (1951).
41. Ullrich, R. A., and M. W. Smith, "Operation experience with activated sludge—Biosorption at Austin, Texas," *Sewage Ind. Wastes* **29**, 400 (1957).
42. Velz, C. J., "A basic law for the performance of biological filters," *Sewage Works J.* **20**, 607 (1948).
43. "Wet combustion of wastes," *Power Eng.* **59**, 63 (1955).
44. Zobell, C. E., "The influence of solid surface upon the physiological activities of bacteria in sea water," *J. Bacteriol.* **33**, 86 (1937).

Suggested Additional Reading

Flower, W. A., "Spray irrigation—A positive approach to a perplexing problem," in Proceedings of 20th Industrial Waste Conference, Purdue University, 1965, p. 679.
Ling, J. T., "Pilot study of treating chemical wastes with an aerated lagoon," *J. Water Pollution Control Federation* **35**, 963 (1963).
Luley, H. G., "Spray irrigation of vegetable and fruit processing wastes," *J. Water Pollution Control Federation* **35**, 1252, (1963).
Oswald, W. J., and H. B. Gotaas, "Photosynthesis in sewage treatment," *Trans. Am. Soc. Civil Engrs.* **122**, 73 (1957).
Parker, C. D., "Food treatment waste treatment by lagoons and ditches at Shepparton, Victoria, Australia," in Proceedings of 21st Industrial Waste Conference, Purdue University, 1966, p. 284.

Deep Well Injection

Barraclough, J. T., "Waste injection into deep limestone in northwestern Florida," *Groundwater* **4**, 22 (1966).
Hundley, C. L., and J. T. Matulis, "Deep well disposal," in Proceedings of 17th Industrial Waste Conference, Purdue University, 1962, p. 175.
Koenig, L., "Advanced waste treatment," *Chem. Eng.* **70**, 210 (1963).

Powers, T. J., and G. W. Querio, "Check on deep-well disposal for specially troublesome waste," *Power* **105**, 94 (1961).
"Production waste goes underground at Holland-Suco. Mich.," *Water Sewage Works* **113**, 329 (1966).
Querio, C. W., and T. J. Powers, "Deep well disposal of industrial waste water," *J. Water Pollution Control Federation* **34**, 136 (1962).
Selm, R. P., "Deep well disposal of industrial wastes," in Proceedings of 14th Industrial Waste Conference, Purdue University, 1959.
Talbot, J. S., "Deep well method of industrial waste disposal," *Chem. Eng. Progr.* **60**, 1 (1964).
Warner, D. L., "Deep well waste injection—Reaction with aquifer water," *J. Sanit. Eng. Div. Am. Soc. Civil Engrs.* **92**, no. SA4, 95 (1966).
"Waste well goes down over two miles," *Eng. News-Record* **165**, 32 (1960).
Winar, R. M., "The disposal of wastewater underground," *Ind. Water Eng.* **4**, 21 (1967).

Foam Phase Separation

Advanced Waste Treatment Research, Publication no. AWTR-14, U.S. Public Health Service, Cincinnati, Ohio (1955).
Brown, D. J., "A photographic study of froth flotation," *Fuel Soc. J. Univ. Sheffield* **16**, 22 (1965).
Eldib, I. A., "Foam fractionation for removal of soluble organics from wastewater," *J. Water Pollution Control Federation* **33**, 914 (1961).
Gassett, R. B., O. J. Sproul, and P. F. Atkin, Jr., "Foam separation of ABS and other surfactants," *J. Water Pollution Control Federation* **37**, 460 (1965).
Grieves, R. B., C. J. Crondall, and R. K. Wood, *Air Water Pollution* **8**, 501 (1964).
Rubin, E., R. Everett, Jr., J. J. Weinstock, and H. M. Shoen, *Contaminant Removal from Sewage Effluents by Foaming*, Publication no. 999-WP-5, U.S. Public Health Service, Cincinnati, Ohio (1963).

Bio-Disc

Antonie, R. L., and F. M. Welch, "Preliminary results of a novel biological process for treating dairy wastes," in Proceedings of 24th Industrial Waste Conference, Purdue University, 1969.
Hartmann, H., "Investigation of the biological purification of sewage using the Bio-Disc filter," Stuttgarter Berichte zur Siedlungswasserwirtschaft no. 9, R. Oldenbourg, Munich (1960).

Pöpel, F., "Estimating construction and output of Bio-Disc filter plants," Stuttgarter Berichte zur Siedlungswasserwirtschaft no. 11, R. Oldenbourg, Munich (1964).

Welch, F. M., "Preliminary results of a new approach in the aerobic biological treatment of highly concentrated wastes," in Proceedings of 23rd Industrial Waste Conference, Purdue University, 1968.

CHAPTER 14

TREATMENT AND DISPOSAL OF SLUDGE SOLIDS

Of prime importance in the treatment of all liquid wastes is the removal of solids, both suspended and dissolved. Once these solids are removed from the liquids, however, their disposal becomes a major problem. Unfortunately, waste engineers spend more time and money removing the solids than finally treating and disposing of them, so that often a poor solids-disposal program will cause trouble in an otherwise properly designed and operated waste-treatment plant. When the solids-disposal system is poor, the solids tend to build up in the flow-through treatment units and overall removal efficiencies then begin to decrease. Therefore, proper sludge handling enhances the overall treatment of all wastes. The following list contains most of the methods commonly used to deal with sludge solids: (1) anaerobic and aerobic digestion; (2) vacuum filtration; (3) elutriation; (4) drying beds; (5) sludge lagooning; (6) wet combustion; (7) atomized suspension; (8) drying and incineration; (9) centrifuging; (10) sludge barging; (11) landfill; (12) sludge pumping; and (13) miscellaneous methods.

14.1 Anaerobic and Aerobic Digestion

Anaerobic digestion is a common method of readying sludge solids for final disposal. All solids settled out in primary, secondary, or other basins are pumped to an enclosed airtight digester, where they decompose in an anaerobic environment. The rate of their decomposition depends primarily on proper seeding, pH, character of the solids, temperature, and degree of mixing of raw solids with actively digesting seed material. Digestion serves the dual purpose of rendering the sludge solids readily drainable and converting a portion of the organic matter to gaseous end-products. It may reduce the volume of sludge by as much as 50 per cent. After digestion, the sludge is dried and/or burned, or used for fertilizer or landfill.

Two main groups of microorganisms, hydrolytic and methane, carry out digestion. Hydrolytic bacteria exist in great numbers in sewage and waste sludges and are capable of rapid rates of reproduction; they are saprophytic microorganisms that attack complex organic substances and convert them to simple organic compounds. Among these saprophytes are many acid-forming bacteria which produce fatty acids of low molecular weight, such as acetic and butyric, during degradation processes. In some cases, such acids are produced in quantities sufficient to lower the pH to a level where all biological activity is arrested.

Fortunately, methane bacteria, the other group of microorganisms, are capable of utilizing the acid and other end-products formed by the hydrolytic bacteria. Methane producers, however, are sensitive to pH changes and proliferate only within a narrow pH range of 6.5 to 8.0, with an optimum of 7.2 to 7.4; furthermore, they are few in number and reproduce slowly. Consequently, organic acids may form faster than they can be assimilated by the limited population of methane bacteria. As a result, the pH may be lowered and conditions made even more unfavorable for methane bacteria. When this happens, lime is usually added and the digestion process stopped until normal conditions return.

The proper environment for both types of bacteria requires a balance between population of organisms, food supply, temperature, pH, and food accessibility. The following factors are measures of the effectiveness of digestive action: gas production (both quantity and quality), solids balance (total, volatile, and fixed), BOD, acidity and pH, volatile acids, grease, sludge characteristics, and odor.

As mentioned before, fermentation (digestion) of organic matter proceeds in two stages: (1) hydrolytic action, converting organic matter to low-molecular-weight, organic acids and alcohols, and (2) evolution of carbon dioxide and the simultaneous reduction to methane (carbon dioxide is actually consumed). The following general equations represent the digestion of carbohydrates, fats, and proteins:

Carbohydrates:

$$(C_6H_{10}O_5)_x + x\,H_2O \rightarrow x(C_6H_{12}O_6),$$
$$C_6H_{12}O_6 \rightarrow 2C_2H_5OH + 2CO_2,$$
$$2CH_3CH_2OH \xrightarrow{+CO_2} 2CH_3COOH + CH_4,$$
$$CH_3COOH \rightarrow CH_4 + CO_2.$$

Fats:

$$\begin{array}{c}
H_2C-O-\overset{O}{\overset{\|}{C}}-R_1 \\
| \\
HC-O-\overset{}{\underset{\|}{C}}-R_2 + 3HOH \rightarrow \\
O \\
| \\
H_2C-O-\overset{}{\underset{\|}{C}}-R_3 \\
O
\end{array}
\quad
\begin{array}{c}
H \\
| \\
H-C-OH \\
| \\
H-C-OH \\
| \\
H-C-OH \\
| \\
H \\
\text{Glycerol}
\end{array}
+
\begin{array}{c}
HO-\overset{O}{\overset{\|}{C}}-R_1 \\
\\
HO-\overset{O}{\overset{\|}{C}}-R_2 \\
\\
HO-\overset{O}{\overset{\|}{C}}-R_3 \\
\text{Acid}
\end{array}$$

Alpha oxidation of acids:
$$4RCH_2COOH + 2HOH \rightarrow 4RCOOH + CO_2 + 3CH_4.$$

Beta oxidation of acids:
$$2RCH_2CH_2COOH + CO_2 + 2HOH \rightarrow$$
$$2RCOOH + 2CH_3COOH + CH_4,$$
$$CH_3COOH \rightarrow CH_4 + CO_2.$$

Proteins:

$$\begin{array}{c} H \\ | \\ R-C-COOH \\ | \\ NH_2 \end{array} \xrightarrow[\text{Deaminase}]{HOH} NH_3 + \begin{array}{c} H \\ | \\ R-C-COOH, \\ | \\ OH \end{array}$$

$$\begin{array}{c} H \\ | \\ R-C-COOH \\ | \\ OH \end{array} \xrightarrow{\text{Decarboxylase}} \begin{array}{c} H \\ | \\ R-C-H + CO_2, \\ | \\ OH \end{array}$$

$$2RCH_2OH + CO_2 \rightarrow 2RCOOH + CH_4,$$
$$RCOOH \rightarrow CH_4 + CO_2.$$

One hypothesis is that each molecule of methane arises from a reduction of one molecule of carbon dioxide. In other words, carbon dioxide acts as the hydrogen acceptor, while the alcohol acts as the hydrogen donor, as in the following equation:

$$\underset{\text{hydrogen donor}}{\underset{\text{ethyl alcohol}}{2C_2H_5OH}} + \underset{\text{hydrogen acceptor}}{\underset{\text{carbon dioxide}}{CO_2}} + H_2O \rightarrow \underset{\text{acetic acid}}{2CH_3COOH} + \underset{\text{methane}}{CH_4} + 2H_2O.$$

One can readily see that carbon dioxide is an important food constituent. In mixed cultures, carbon dioxide is produced by other organisms and therefore becomes more available than sulfates or nitrates. Buswell [11] describes fermentation as a chain of reactions involving the transfer of hydrogen.

The slowest reaction in the degradation process, production of methane, is therefore the rate-controlling reaction. The essential physiological characteristics of methane bacteria are: (1) they are obligate anaerobes; (2) they require carbon dioxide as a hydrogen acceptor; (3) as hydrogen donors, they use simple organic substrates, such as calcium acetate, butyrate, and ethyl and butyl alcohols; (4) their nitrogen source is ammonia; (5) they develop at a slow rate owing to low energy yields; (6) they do not form spores; (7) they are very sensitive to changes in pH.

Buswell [11] concluded that the higher the percentage of carbon atoms in the fatty-acid substrate, the higher the percentage of methane in the gas. Barker [4] established the following unique features of methane fermentation:

1) It takes place in mixed or enriched cultures and hence may be maintained continuously on a large scale.

2) It is applicable to any type of substrate except lignin and mineral oil.

3) The reaction is quantitative and converts the entire substrate to carbon dioxide and methane.

4) There is no specific temperature limitation in the range of 0 to 55°C, but once the culture has been acclimated to a certain temperature, a drop of two degrees may completely interrupt methane fermentation and render obstructive the accumulated acids.

5) The presence of inert solid matter is important, and hence addition of straw or sawdust to industrial wastes may be required.

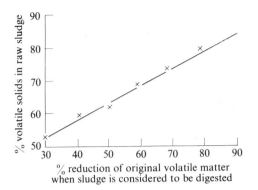

Fig. 14.1 Reduction of volatile matter in raw sludge by digestion [29].

6) If the substrate concentration is too great, volatile acids build up and inhibit the fermentation, especially when their build-up occurs faster than their subsequent conversion to methane. Keeping the volatile acid level below 3000 ppm, and closer to 2000, helps the situation, but alkali addition will not alleviate it, since it is not a pH effect. Mineral salts begin to inhibit the fermentation at 4000 ppm, and 50 ppm of nitrate nitrogen inhibit it completely.

The extent of reduction of volatile solids by digestion depends in part on the amount of volatile matter in the raw sludge. Schlenz [29] found that when volatile solids in raw sludge increased from 55 to 80 per cent, the reduction in volatile matter increased from 35 to 85 per cent. This is shown graphically in Fig. 14.1.

The usual unit-capacity requirements may be reduced, provided the operations are controlled and carried out as follows [31]: (1) tank contents must be agitated to maintain an even mixture of raw and digesting solids; (2) raw sludge must be added continuously to the digestion unit; (3) raw sludge must be concentrated or prethickened before being added to the digester. Two-stage digestion, with the first stage used primarily for active digestion and the second stage for storage and sludge consolidation, is often carried out in two separate tanks. It is usually more economical in large plants with continuous operation.

Aerobic digestion is now playing an important role in small plants. It is claimed that less-skilled operators are required; also, air is normally available in these plants, since secondary treatment of the liquid-waste fraction is becoming rather commonplace [13].

14.2 Vacuum Filtration

Vacuum filtration is a means of dewatering sludge solids which has become popular because the volume of solids for ultimate disposal is reduced and the sludge is drier than it would otherwise be, so that "handleability" is improved. Large plants are increasing their use of vacuum filtration. Some plants filter chemically precipitated and/or plain settled sludge, while others filter digested sludge. In a typical vacuum-filtration unit, a porous cylinder overlying a series of cells revolves about its axis with a peripheral speed somewhat less than one foot per minute, its lower portion passing through a trough containing the sludge to be dried. A vacuum inside the cylinder picks up a layer of sludge as the filter surface passes through the trough, and this increases the vacuum. When the cylinder has completed three-quarters of a revolution, a slight air pressure is produced on the appropriate cells, which aids the scraper, or strings, to dislodge the sludge in a thin layer. Sometimes it is necessary to add chemicals, such as lime and ferric chloride, as sludge conditioners prior to filtration. Filtering rates should be from 2 to 10 pounds of dry solids per square foot per hour. Vacuum filters are available in diameters up to about 20 feet and in many different lengths.

The quality of the filter medium (the material covering the cylinder) is important in the performance and life of the filter. In the past, woven-fabric filter media have been widely used. The physical process of solids retention on woven filters is a combination of at least three actions: (1) straining action, in which particles *larger* than the filter-medium openings cling to the filter; (2) adsorption, or attraction, to the filter of particles smaller than the openings in the filter medium; (3) filtration of particles of different sizes, which cling to already filtered, caked material. The first two actions prevail at the onset of filtration but, as the "cake" builds up, the third is responsible for the greatest amount of solids removal. Thus the problem arises that, unless the cake is removed completely and the fiber filter medium kept clean continually, the filter will clog or "blind."

Tiller and Huang [35] report that there is a paucity of theory and research on filtration through porous media. Three reasons for this deficiency are: (1) complexity of vacuum-filtration machinery, (2) difficulty of experimentally reproducing the precipitates found in filter beds, and (3) insufficient interest on the part of researchers. They also report that, although flow through filter beds is almost always viscous, no reliable theory has been developed as to the relation between permeability and porosity of the filter medium, as affected by compressive pressure.

A major step toward lengthening the life and decreasing the operational problems of vacuum-filtration systems is the use of stainless-steel, coil-spring filter media. A representative unit of this type, the Coilfilter, is shown in Fig. 14.2 [10].

14.3 Elutriation

Elutriation is a process of improving filtration by washing the sludge. It reduces the alkalinity—and therefore the lime coagulant demand—of sludge by upgrading the biochemical quality of the sludge water before chemicals are added [16]. There are three practical methods of washing sludge solids; the equipment used in all cases is relatively simple, with upward-flow tanks frequently used.

1) Single-stage elutriation, which involves one batch at a time, is a fill-and-draw procedure: sedimentation and decantation are performed in a single step.

2) Two-stage elutriation involves repeating the single-stage steps on the elutriated sludge, using fresh water on the second wash. In small plants, the same settling tank may be used for both stages.

3) In larger plants (6000 to 24,000 pounds of solids per day), a second tank, connected in series with the first, is usually employed. Such a two-tank system can also be used for countercurrent washing. With this system, the fresh water is added only to the second-stage washing, and the decanted elutriate (or top water) from this tank flows by gravity to mix with the sludge entering the first tank.

Since the degree of chemical fouling [15] resulting from digestion can be conveniently measured in terms of alkalinity, an elutriated sludge can be defined as one that has had the alkalinity of its water reduced by dilution with water of lower alkalinity, sedimentation, and decantation. Advantages of elutriation as a preliminary to sludge dewatering on vacuum filters include elimination of ammonia odors and of the need to use lime in sludge conditioning. Elutriation may also reduce the capacity requirements of secondary digesters (used for storage and additional digestion to ensure optimum filtration), and it is particularly helpful in that it permits small plants to use vacuum filters to advantage. Genter claims that elutriation reduces the ratio of sludge water to the mineralized sludge solids; thus there is a marked decrease in the chemicals required for conditioning. The savings in ferric chloride are illustrated in Fig. 14.3, which is based on Genter's data [15].

Genter [16] also discusses a method of predicting the final alkalinity of elutriated sludge by a formula. Assuming that a equals the volumes of pure water added to one volume of fouled sludge mixture, he obtains the following relationships:

$a + 1 =$ total volume of mixed sludge and clean water;

$1/(a + 1) =$ fraction of original concentration of fouling agent left if solids are allowed to settle back to a washed sludge equivalent to the original volume and the added volume of water is siphoned off;

$1/(a + 1)^2 =$ fraction of original concentration of fouling agent if this same dilution, sedimentation, and decantation technique is repeated.

Therefore, the fraction of original fouling agent left in the final sludge is $1/(a^2 + 2a + 1)$ if the second wash water is decanted for a new first wash and the two elutriation tanks are placed on countercurrent series. For example, if four volumes of pure water are used to wash a digested sludge of 3000 ppm alkalinity, the alkalinity left in the elutriated sludge after countercurrent washing in two tanks is

$$\frac{3000}{(4)^2 + (2 \times 4) + 1} = 120 \text{ ppm.}$$

14.4 Drying Beds

Sludge-drying beds remove moisture from sludge, thereby decreasing its volume and changing its physicochemical characteristics, so that sludge containing

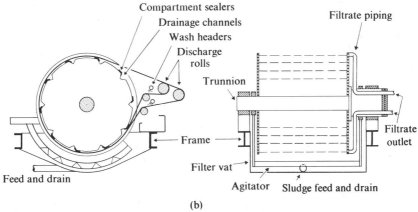

Fig. 14.2 (a) The Coilfilter, a patented machine for the vacuum filtration of sludge. This particular machine, in use since 1953 at the sewage-treatment plant at St. Charles, Illinois, has filtering media made up of two layers of alloy steel coiled springs, each spring made endless by joining its two ends with a threaded plug. These springs discharge the filter cake after each revolution of the cylinder, and are then washed before they re-enter the vat for another cycle. The material at the left which looks like a length of corduroy is actually a layer of sludge [10]. (b) Schematic drawing of the Coilfilter shown in (a). (Courtesy Komline Sanderson Co.)

25 per cent solids can be moved with a shovel or garden fork and transported in watertight containers.

Sludge filter beds are made up of 12 to 24 inches of coarse sand, well-seasoned cinders, or even washed

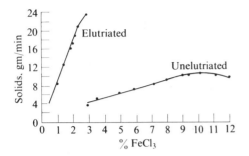

Fig. 14.3 Effect of elutriation on FeCl$_3$ required for conditioning of sludge [15].

grit from nearby grit chambers and about 12 inches of coarse gravel beneath the sand. The upper 3 inches of gravel particles are $\frac{1}{8}$ to $\frac{1}{4}$ inch in diameter. Below the gravel, the earth floor of the bed is pitched to a slight grade into open-joint tile underdrains 6 or 8 inches in diameter. These tiles may be laid from 4 to 20 feet apart on centers, depending on the porosity of the coarse gravel. Disposing of the underdrain liquor sometimes poses a problem; this should never be discharged without an analyses of its constituents and usually some form of treatment. Several smaller, rectangular beds serve the purpose better than one large filter bed. These beds may be covered with glass or plexiglass when weather conditions demand, in which case ventilation must be provided to dissipate the hot, wet air above the beds.

Generally speaking, raw settled sludge does not drain well on sand drying beds. Some form of pretreatment—digestion, elutriation and/or chemical treatment—is usually required. Well-digested sewage sludge will dewater more readily than partly digested sludge [30]. However, prolonged storage of digested sludges decreases drainability, since the gases present initially permit more drainage of moisture through the filtering medium, thus reducing the evaporation cycle. A high total-solids content in digested sludges naturally permits greater removal of dry solids per year from sludge beds.

Drying time is dependent on dosing depth, 8 inches being generally accepted as most desirable for rapid drying, and on climate. It is, naturally, short in regions of plentiful sunshine, scant rainfall, and low relative humidity, such as certain arid areas of the South where summers are long. Wind velocity also affects speed of sludge-drying on the beds. In fact, all the factors enhancing evaporation will also aid in drying sludge. Cox [12] derived the following equation for calculating the rate of evaporation of water, which may also apply to sludge drying although exact values of constants may vary from water to sludge water:

$$E = \frac{(e_a - e_d + 0.0016\,\Delta T)}{(0.564 + 0.051\,\Delta T + W/300)},$$

where

E = evaporation (inches/day)
e_a = saturated vapor pressure at air temperature
e_d = actual vapor pressure
ΔT = difference between mean temperature of the air and that of the water
W = velocity of the wind (miles/day)

Meyer's formulation is also widely used:

$$E = C(V - v)\left(1 + \frac{W}{10}\right),$$

where

E = evaporation (inches) for a given unit of time
V = saturation vapor pressure at the water temperature (inches of mercury)
v = actual vapor pressure of the air, 25 feet above ground
W = wind velocity (mph), 25 feet above ground
C = coefficient, varying with unit of time used and depth of water (varies from 0.36 to 0.50).

In addition to evaporation, the drying rate is also influenced by capillary action, which causes water to rise from the depths of the sludge to the evaporative surface.

In the case of domestic-sewage sludge, engineers estimate that approximately 20 to 25 pounds of dry solids can be loaded onto one square foot of properly designed sand-base drying bed each year. Haseltine [18] takes exception to this unit-of-loading estimate and suggests a "gross bed loading," which takes into account the number of pounds of solids applied per square foot per 30 days of actual bed use. For example, if sludge which has a density of 62.5 lb/ft^3 and contains 10 per cent solids is applied 12 inches deep and removed after 40 days, the gross bed loading is

$$\frac{62.5 \times 0.10 \times 30}{40} = 4.69 \text{ lb/ft}^2/30 \text{ days.}*$$

Haseltine also develops the following straight-line relationship between the gross bed loading (Y) and the percentage of solids in applied sludge (X) from data supplied by 14 different plants for periods of operation up to 14 years:

$$Y = 0.96X - 1.75.$$

The gross bed loading Y varied from 0 to 10 and X varied from 0 to 14. He concluded that, next to temperature, the solids content of the sludge in drying beds is the most important factor influencing bed performance. The amount of moisture to be removed from the sludge is the third most important factor.

14.5 Sludge Lagooning

Lagoons may be defined as natural or artificial earth basins used to receive sludge. Lagooning is practiced when the economics of the situation (money and land) indicate its use, since it is a relatively inexpensive method of treating waste sludges. However, there are many other factors to be considered: (1) nature and topography of the disposal area; (2) proximity of the site to populated areas; (3) meteorological conditions, especially whether prevailing winds blow toward or away from populated areas; (4) soil conditions; (5) chemical composition of sludges, with special consideration given to toxicity and odor-producing constituents; (6) proximity to surface- or ground-water supplies; (7) effect of waste materials on the porosity of the soil; (8) means of draining off the supernatant liquor to provide more space in the lagoon; (9) fencing, and other safety measures, when lagoons are deeper than five feet; (10) nuisances, such as weed growth, odors, and insect breeding.

Lagooning of wastes in limestone areas is particularly hazardous because of the channels and cavities found underground in these formations [25]. Ordinarily groundwater moves slowly, sometimes less than a foot a day, depending on the fineness of the aquiferous sand through which it percolates and the degree of saturation of the sand. In limestone country, water may travel vertically and laterally at much higher velocities, so that sludge lagooned on high ground may quickly contaminate large portions of valuable ground-water supplies. Quite often, manufacturing plants bulldoze out a sludge lagoon every year or two, the frequency depending on sludge build-up and soil conditions.

Bloodgood [5] states that at least one pound of raw sewage solids can be digested per year per 0.17 cubic foot of lagoon capacity. However, if lagoons are to be used for both digestion and dewatering, one pound of raw sludge solids requires about 0.4 cubic foot per year of lagoon capacity, provided air-dried sludge is removed as soon as it becomes ready for hauling.

14.6 The Wet Combustion Process

The Zimpro process is a relatively new treatment for sludge. It operates on the basic principles that (1) organic matter contained in an aqueous solution can be oxidized and whatever heat value it contains released and (2) oxidation at this stage is more effective than if the water were first evaporated and the residue used as fuel in a conventional boiler. Since heat is liberated by a fuel only when it is subjected to combustion in the presence of air, the Zimpro process depends on air being forced into a reactor vessel. One objective of this process is the production of the maximum number of Btu's from the organic matter in a waste effluent per pound of compressed air fed into the reactor.

Since the Zimpro process eliminates conventional filters, chemicals, sludge-digestion units, incinerators, and auxiliary equipment, it reduces space and land requirements. The end-products are steam, nitrogen, CO_2, and ash. The effluent gases from the reactor, having been "scrubbed" with water, contain no fly ash and are practically odorless.

In the treatment of sewage sludge, oxidation is brought about by continuously pumping the sludge and a proportionate amount of air (both sludge and air at elevated temperatures and pressures) into a reactor vessel. Combustion occurs as the oxygen in the compressed air combines with the organic matter in the sludge to form CO_2, N_2, and steam, while the ash remains in the residual water. The reactor, and

---

*Specific gravity of wet sludge assumed, 1.0.

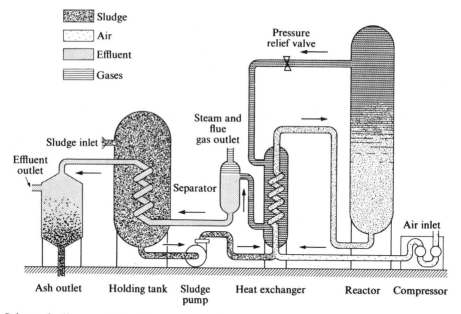

Fig. 14.4 Schematic diagram of the Zimpro process for sewage-sludge oxidation. (Courtesy Sterling Drug Co.)

the whole process system as well, is automatically maintained at a constant pressure and the products of the combustion are continuously removed from the reactor. If the concentration of volatile matter is high and the sewage sludge concentration is great enough (> 5 per cent), the steam, plus the gases (CO_2 and N_2) which are products of combustion, will contain more than enough energy to run the air compressors and pumps used in the process. The residual hot water from the reactors is utilized in heat exchangers that raise the temperature of the incoming sludge and air sufficiently to cause oxidation to begin as soon as they come together in the reactor. In this way, once the process is started, no external heat or power is required to sustain the combustion.

Equipment required for the Zimpro process includes: a compressor, an air receiver, a high-pressure sludge pump, a sludge-storage tank with agitators, heat exchangers, a reactor, a separator, and a cooler. A schematic drawing of the process is presented in Fig. 14.4. The manufacturer claims that: "Units achieve 80 to 90 per cent reduction of insoluble organic content of sewage sludge by oxidation without flame. Sludge is burned without dewatering or pretreating. The unit operates continuously at pressures of 500 to 600 psig and temperatures of 420°F. End products are substantially inorganic, inert, biologically stable ash; residual water; and odor-free gaseous products of combustion (carbon dioxide, nitrogen, and steam). The plant is designed for automatic operation with minimal maintenance. An air compressor and sludge pump are the only equipment components with moving parts. Power requirement is approximately 50 hp for a one-ton unit (dry weight). Building and land-space requirements are nominal."* Teletzke [34] describes the low-pressure Zimpro treatment, which operates in the range of 150 to 300 psi and at about 300°F. He portrays low-pressure, wet-air oxidation as an economical and flexible method of producing a sterile, drainable, and completely acceptable end-product for ultimate disposal.

New Zimpro Sludge Oxidation Units for Smaller Communities, Sterling Drug Co., Rothschild, Wisconsin.

14.7 Atomized Suspension

The atomized-suspension technique consists of atomizing the waste liquor or slurry in the top of a tower, the walls of which are maintained at an elevated temperature by hot gases circulating through a jacket, a method described by Gauvin [14, 26]. No air, or other foreign gas, is introduced into the equipment, which sharply distinguishes this technique from spray drying. The developers claim that in the immediate range of the nozzle the finely divided droplets (20 to 25 mm in diameter) quickly decelerate from the high initial velocity imparted by the atomizer to their slower terminal velocity and then become dispersed in the vapor produced by their own evaporation. The suspension thus created flows down the reactor in nearly streamline motion. Evaporation, quickly completed, is followed by drying. At the end of the drying zone, dried particles can be subjected to a sequence of chemical reactions, such as oxidation, reduction, nitration, sulfonation, and so forth, through the injection of the proper internal gaseous reactants (in the presence of a powdered catalyst, if necessary). When it leaves the reactor at the bottom, the suspension consists of a solid residue (which is recovered in cyclone collectors), large amounts of steam (which is condensed and utilized), and by-product gases (which can be further processed for recovery or piped away for disposal).

Advocates of the atomized-suspension process claim that the only outside energy required is that used for pumping of the liquid—an almost negligible amount. A striking feature of the recovery flow sheet is the complete absence of blowers or compressors, although large volumes of gases and vapors are continuously flowing through the system. Need for them is eliminated by the efficient utilization of the pressure generated in the reactor during evaporation. A typical flow sheet for Gauvin's process [14, 26] is shown in Fig. 14.5.

14.8 Drying and Incineration

A large volume of sludge can be reduced to a small volume of ash, which is free from organic matter and therefore easily disposable, by a combination of heat drying and incineration [32]. Flash drying involves drying sludge particles in suspension in a stream of hot gases, which ensures practically instantaneous

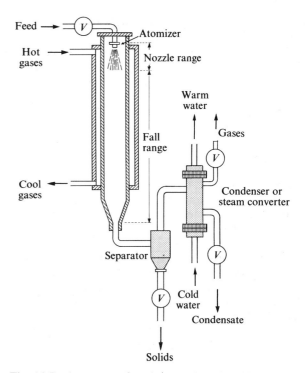

Fig. 14.5 Apparatus for the atomized-suspension technique [14, 26].

removal of moisture. When hot gases created by the drying and oxidation of the sludge itself are used directly for drying, there are no conversion losses. After the flash drying, the gas containing sludge particles usually passes to cyclone separators, where the dried sludge is separated from the moisture-carrying cooler gases.

Flash-dried sludge is utilized as fertilizer, soil conditioner, or for other valuable purposes. Unused dried sludge can be incinerated by blowing it through a duct to a burner in the combustion chamber of a furnace. The sludge blower, in addition to conveying the sludge to the furnace, also supplies the major portion of the air required for combustion. To eliminate odors, preheated gases after combustion are returned to the combusting sludge. To eliminate fly ash, the cooled gas after combustion is drawn through an ash collector by induced-draft fans, and the fly ash settles out by centrifugal action and is discharged automatically into the furnace bottom. This ash can be removed from time to time, either by shoveling or by mixing it with water and pumping it out to be used as landfill.

Whether the ultimate aim is to dry the sludge for use as a soil additive or to incinerate it to a sterile ash, it is necessary first to evaporate the free moisture from the solids, remove it in the form of a gas, and discharge it to the atmosphere. This gas is referred to as the evaporator load. Only high-temperature (1200 to 1400°F) deodorization is effective in controlling odors from sludge incinerators.

When sludge is to be incinerated, the heat released in the furnace is also of importance: the furnace volume should be ample to allow a heat release of X Btu's per cubic foot of furnace (generally held at 12,000 Btu's per cubic foot of furnace volume per hour to ensure long life of walls and furnace). The heat input is determined by multiplying the pounds of dry solids to be incinerated per hour by the gaseous products of the volatile-solids content and their heat value. The furnace volume required can therefore be computed by dividing this heat input by 12,000. Thermal efficiencies of 30 to 60 per cent can be expected from incinerators. The lower the stack temperature, the higher the thermal efficiency [20]. This relationship is shown in Fig. 14.6; in Fig. 14.7 a flow diagram is presented [20] to show heat balance for a flash-drying and incineration system.

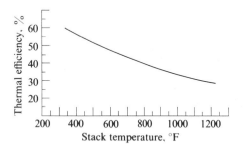

Fig. 14.6 Effect of stack temperature on thermal efficiency [20].

To calculate the rate of drying during the constant-rate period (after the temperature of the material adjusts itself to the drying conditions), either the mass-transfer or the heat-transfer equation may be used [21]:

(mass transfer) $\quad W = k'_y(H_i - H)A;$

(heat transfer) $\quad W = \dfrac{h_y(t - t_i)A}{\lambda_i},$

where

W = evaporation rate (lb/hr)
A = drying area (ft²)
h_y = heat-transfer coefficient (BTU/ft²/hr/°F)
k'_y = mass-transfer coefficient (lb/ft²/hr for a unit of humidity difference)
H_i = humidity of air at interface (lb water/lb dry air)
H = humidity of air (lb water/lb dry air)
t = temperature of air (°F)
t_i = temperature at interface (°F) and
λ_i = latent heat at temperature t_i (BTU/lb).

The heat-transfer coefficient, h_y, is estimated to be about $0.128\,G^{0.8}$ when air flows parallel to the sludge surface and about $0.37\,G^{0.37}$ when air flows perpendicular to the sludge surface (G = the mass velocity in lb/ft²/hr).

Pit incineration has been used to dispose of certain solid and semi-solid wastes. The incinerator consists of a rectangular pit lined with firebrick, to which air is supplied so as to retain particulates and to allow complete combustion. This disposal method is simple in concept and operation and is especially adaptable to situations where the waste requires batch incineration. It has been used for disposal of synthetic organics and is currently being studied for disposal of paint sludges in the automotive industry [3].

14.9 Centrifuging

Centrifugation is a method of concentrating sludge to enhance final disposal. One of the factors which made centrifugal concentration unacceptable in the earlier installations was its low efficiency—large amounts of fine particles were returned to the system with the supposedly clarified effluent. Newer installations [7], using 20-hp built-in drive motors, can handle 3000 to 4000 gallons per hour of waste sludge, containing 0.5 to 0.75 per cent solids on a dry basis. Only 11 hp is required once the centrifuge reaches operating speed (6100 rpm). The resulting sludge is concentrated to about 5 per cent solids and the effluent contains about 300 ppm solids. The centrifugal force throws the denser solid material to the wall of the centrifuge bowl, where it is discharged

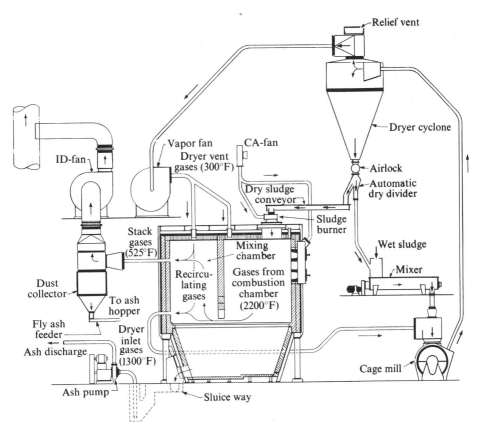

Fig. 14.7 Flow diagram for heat balance for a flash-drying and incineration system [20].

through nozzles located in the periphery. One bowl [7] is equipped with 12 nozzle openings, such that various numbers of discharge nozzles can be utilized, depending on the amount of solids in the feed liquor and the results desired. Use of the centrifuge for higher concentrations is limited by the capability of the pumps which discharge concentrated sludge from the centrifuges. The effluent from which the solids are separated travels toward the center of the centrifuge bowl through intermediate discs; as it discharges from the upper cover, it is claimed to average approximately 300 ppm solids. Centrifuged sludge is discharged from the lower cover of the centrifuge into a sump, from which it can be pumped to digesters or other final sludge-treatment units. Figure 14.8 is a schematic diagram of a centrifuge bowl. Blosser and Caron [6] expect the costs of centrifuging paper-

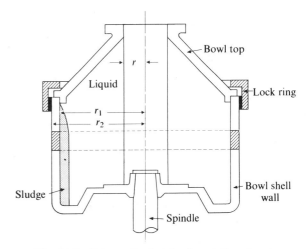

Fig. 14.8 Schematic sketch of centrifuge bowl.

mill sludges to vary from $4 to $20 per ton of dry solids, including the hauling of the cake.

Ambler [1] recently reviewed the theory of centrifugation. When a force is applied to a particle, the particle is accelerated ($F = ma$) until it reaches a velocity along the line of the force at which the resistance to its motion equals the applied force. In a settling tank, this is the force of gravity. In a centrifuge, it is the centrifugal field, $\omega^2 r$. The two differ only in direction and order of magnitude. The gravitational field is along a radius of the earth; the centrifugal field is along a radius normal to the axis of rotation and may be upward of 60,000 times that of gravity for continuous-flow centrifuges. The velocities of particle movement are generally proportional to the square root of the diameter of the particle. The effective force acting on the particle is

$$F = (m - m_1)\omega^2 r$$

and for a sphere it is

$$F = \frac{\pi}{6}(d^3)\,\Delta\rho\omega^2 r.$$

The force opposing sedimentation, according to Newton's drag law in laminar flow, is

$$F = 3\pi\mu\,dv_s.$$

At equilibrium (Stokes's law),

$$v_s = \frac{\Delta\rho\,d^2\,\omega^2 r}{18\mu}.$$

In the simplest form of a continuous centrifuge, v_s is the velocity with which the particle, if it is heavier than the fluid, approaches the bowl wall. If X is the distance the particle will travel,

$$X = v_s t - \frac{\Delta\rho\,d^2\,\omega^2 r}{18\mu}\cdot\frac{V}{Q}.$$

If X is greater than the initial distance the given particle is from the wall of the rotor, the particle will be deposited against the wall and be removed from the system. In an ideal system ($X = s/2$), half the particles of diameter d will be removed. This may be considered the cutoff point at which

$$Q = \frac{\Delta\rho\,d^2}{9\mu}\cdot\frac{V\omega^2 r}{s},$$

where Q is volume of flow per unit of time.

Since the term $\Delta\rho d^2/9\mu$ is concerned only with the parameters of the system that follow Stokes's law and the term $V\omega^2 r/s$ with the parameters of the rotor, the above equation may be written as

$$Q = 2V_g\Sigma,$$

in which

$$V_g = \frac{\Delta\rho\,d^2 g}{18\mu} \quad \text{and} \quad \Sigma = \frac{V\omega^2 r_\theta}{g s_\theta},$$

where r_θ and s_θ are the effective radius and settling distance, respectively, of the centrifuge, and Σ is an index of centrifuge size that has the dimension of (length)^2 and is the equivalent area of a settling tank theoretically capable of doing the same amount of useful work as the centrifuge.

Ambler [1] uses the above theory to formulate an index of centrifuge sizes for various centrifuge types as follows:

1) For the laboratory test-tube or bottle centrifuge,

$$\Sigma = \frac{\omega^2 V}{4.6\log[2r^2/(r_1 - r_2)]}.$$

2) For the tubular-bowl centrifuge,

$$\Sigma = \frac{\pi l \omega^2}{g}\frac{(r_2^2 - r_1^2)}{\ln[2r_2^2/(r_2^2 - r_1^2)]}.$$

3) For the disc-type centrifuge,

$$\Sigma = \frac{2\pi n \omega^2\,(r_2^3 - r_1^3)}{3gC\tan\theta},$$

where

Σ = equivalent area of the centrifuge
ω = angular velocity (rad/sec)
V = volume
r = radius from axis of rotation
r_1 = radius to inner surface
r_2 = radius to outer surface
l = light-phase discharge radius
g = gravitational constant
n = number of spaces between discs
C = concentration of solute
θ = half-included angle of the disc.

In each of these cases, Ambler bases his calculations on the behavior of a single particle under conditions of unhindered settling and on the assumption that this particle is always in equilibrium with the force field of the centrifuge under the conditions defined by Stokes's law.

14.10 Sludge Barging

Sludge barging is a means of final disposal of sludge which can be used when there is not very much space and the treatment site is adjacent to a deep body of flowing water, usually an ocean. There is little theory involved in this method of treatment. Raw, precipitated, digested, or filtered sludge solids are pumped into a waiting barge; when the barge is fully loaded, it transports the sludge to a suitable site far from shore, where it is discharged, usually by being pumped out deep under the surface of the water. The sludge should be somewhat concentrated before being loaded on the barge (to conserve space) but not so much that pumping difficulties arise.

Pumping raw or digested sludge directly into the ocean and disposing of it by the dilution technique has also been practiced in coastal cities [24, 27, 37]. This technique is ordinarily used when sludge drying is unusually expensive and the plant involved is favorably located with respect to a large body of water. These sludges should not be contaminated by grease or oil, discoloration, odors, high coliform counts, toxic materials, or be likely to create sludge banks or deposits.

San Francisco meets these requirements when disposing of its digested waste-water sludge by discharging it at a depth of 40 feet in mid-channel coincident with the ebbing of the tide, to assure rapid flushing of the sludge out of the waters of San Francisco Bay. A study of the tidal-dispersion characteristics of the receiving waters prior to using the dilution technique is of course essential.

Sylvester [33] gives the advantages of sludge pumping into the ocean or a bay as: (1) lower operating costs; (2) reduced land demands; (3) little problem with fluctuations in sludge production; (4) possibly lower initial cost and lower operating costs, which leave money available for other much needed construction and operation. He does not overlook the disadvantages to sludge pumping such as: (1) public objection; (2) discharge of poorly digested sludge; (3) long-term effects on ecology of the receiving water; (4) sludge rising to the surface at certain times; (5) need for screening to remove floating material from the sludge.

14.11 Sanitary Landfill

Sanitary landfill is used to bury garbage, refuse, and sludge in a planned and methodical manner [28]. It is a simple, effective, and inexpensive method for disposing of dry matter such as refuse, but sludge is usually too liquid for this procedure. However, vacuum-filtered or sand-bed-dried sludge can be disposed of in this manner.

The area proposed for the sanitary fill [28] should be easily accessible, yet remote from sources of water supply and recreational areas, and at the same time on land which is not too costly. The suitability of the soil and possible future use of the property are also important considerations.

For municipal refuse, the land area required is estimated at about one acre per year for 10,000 persons, when using a six-foot-deep compaction. Sanitary landfills should be located above the groundwater level and no closer than 500 feet to any sources of water supply, particularly when the soil is sandy, gravelly, or of limestone derivation. The area should be staked out for trenches and bench marks established, giving the elevation to which the finished fill is to be carried and the depth to which excavations are to be dug. Normally a trench is about 15 feet wide and about 4 feet deep. At the end of each day's dumping, the sludge should be covered and compacted by a bulldozer or tractor. Bacon [2] suggests using sludge to reclaim land as an economic method of disposal especially in marginal lands and coal strip-mining areas.

14.12 Sludge Pumping

Pumping sludge into wastebeds (excavated lagoons) or bodies of water is practiced in a number of locations in the United States. Wirts [38] lists places where sludges of various types are transported to disposal areas that include oceans and abandoned strip mines (Table 14.1). Pumping sludge away from treatment facilities is usually practiced because of space limitations, land value, or foundation conditions. Low-pressure force mains, operating at less than 200 psi, have been used to convey solids in

Table 14.1 Force mains used for sludge transportation. (After Werts [38].)

Location	Length, miles	Diameter, in.	Total head, ft.	Sludge type	Dry solids, %
Mogden, England	7	12	142	Digested	4.0–5.0
Birmingham, England	4	9 and 12	2.6 × water	Digested	8.5–10
Chicago, Ill.	17	14	210	Raw	1.0–2.0
	5	12	170	Raw	2.0–4.0
Cleveland, Ohio	13	12	391	Raw	3.0–4.0
Philadelphia, Pa.	5	8	225	Raw	3.0–4.0
Columbus, Ohio[1]	5				4.0–5.0
The Hague, Netherlands[2]	7	8		Digested	4.0–5.0
Los Angeles, Calif.[3]	7.5	24		Digested	3.73
Chicago, Ill.[4]	75–100			Raw	2.0

[1]Connecting two treatment plants.
[2]For disposal in North Sea.
[3]For disposal in Pacific Ocean.
[4]To strip-mine area.

concentrations lower than 10 per cent. The Birmingham, England, situation is of special interest to us, since the material being pumped is an "industrial" sludge from a completely combined sewerage system. In that situation, head loss—often an important factor in sludge pumping—amounted to 2.6 times that of water. The sludge varied between 8.5 and 10.25 per cent solids and contained about 60 per cent organic matter. The maintenance cost of pipelines must be compared to that of other methods of transportation, namely trucking and barging; for example, the costs of removing deposited grease, scale, and silt can be considerable. Wirts [38] gave the average (1940–1954) cost of pumping raw sludge with more than 3 per cent solids over a distance of 13 miles as $1.32 per ton or about $0.10 per ton-mile.

Miller [24] describes a 7-mile-long, 312-foot-deep ocean outfall used to dispose of digested sludge from the Hyperion treatment plant in Los Angeles into the Pacific Ocean. The 22-inch steel pipeline handles a flow of 515 mgd and is designed to withstand at least 52,000 psi. To protect it against corrosion, the pipe line has a triple-wrapped coating of coal-tar enamel, reinforced with two layers of fiber glass, and a $1\frac{1}{8}$-inch outer coat of reinforced gunite; a $\frac{1}{2}$-inch, centrifugally spun, cement-mortar lining protects the interior of the pipe. It also contains cathodic protection against corrosion.

14.13 Miscellaneous Methods

Other methods for disposing of sludge solids include sludge concentration, flotation and thickening. Biological means, aided only by temperature and time controls, are used to induce flotation of sludges [19]. The resultant solids, in concentrations of 20 per cent, do not require the addition of chemicals when they are subsequently dewatered on a vacuum filter. Optimum results with this method of concentration were found to exist at 35°C after a detention period of 120 hours [19]. However, certain types of sludge (for example, activated sludges) are not amenable to this treatment. Aside from time and temperature controls, the chief factors in the flotation method for concentration of raw sludges appear to be volatile content and pH.

In 1953 a method was developed by Torpey [36] for thickening sludge on a continuous basis without the addition of chemicals. Generally, the flow pattern permits dilute sludge—from the primary clarifiers alone or combined with secondary sludge—to be fed to the center feedwell of a thickener. A schematic drawing of one typical thickener is shown in Fig. 14.9. The solids settle, thicken in a definite "blanket" zone, and are drawn away from the bottom of the tank. The excess liquid is decanted by a peripheral weir. The thickeners also contain a mechanism with vertical pickets attached to the rake arms. The pickets are V-shaped, and their channeling action allows entrapped water (water which is caught in sludge) and gases to escape to the surface. The degree to which the sludges can be thickened depends on several factors, the chief one being the source of the sludge [8]. The nature of the sludge is also most important. Some sludges are of a gelatinous and

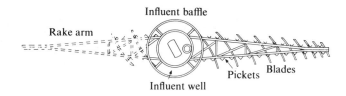

Fig. 14.9 Schematic plan of thickener mechanism, and section of tank [36].

voluminous nature which impedes thickening beyond a certain limit, regardless of detention time. Others are more granular and release entrapped water when subjected to physical action, such as the slow mechanical mixing provided by the rotating pickets and rake arms.

Composting, a method of steeping solid wastes that contain 30 to 70 per cent water in large piles and allowing microorganisms to decompose the organic fractions, has been utilized to some degree for solid wastes from industry. The process is accelerated when the piles are turned regularly by mechanical means. Mercer *et al.* [23] found that the solid wastes of apricots and clingstone peaches were amenable to this form of treatment and that aerobic conditions were maintained by an initial daily turning for 5 to 6 days followed by turning on alternate days until the process was complete.

References

1. Ambler, C. H., "Theory, centrifugation equipment," *Ind. Eng. Chem.* **53**, 430 (1961).
2. Bacon, V. W., "Sludge disposal," *Ind. Water Eng.* **4**, 27 (1967).
3. Balden, A. R., "The disposal of solid wastes," *Ind. Water Eng.* **4**, 25 (1967).
4. (a) Barker, H. A., "On the biochemistry of methane formation," *Arch. Microbiol.* **7**, 404 (1936).
   (b) Barker, H. A., "Studies on the methane producing bacteria," *Arch. Microbiol.* **7**, 720 (1936).
   (c) Barker, H. A., "The production of caproic and butyric acids by the methane fermentation of ethyl alcohol," *Arch. Microbiol.* **8**, 415 (1937).
5. Bloodgood, D. E., "Sludge lagooning," *Water Sewage Works* **93**, 344 (1946).
6. Blosser, R. O., and A. L. Caron, "Centrifugal dewatering of primary paper industry sludges," in Proceedings of 20th Industrial Waste Conference, Purdue University, May 4, 1965, p. 450.
7. Bradney, L., and R. E. Bragstad, "Concentration of activated sludge by centrifuge," *Sewage Ind. Wastes* **27**, 404 (1955).
8. Brisbin, S. G., "Sewage sludge thickening tests," *Sewage Ind. Wastes* **28**, 158 (1956).
9. Bruemmer, J. H., "Use of oxygen in sludge stabiliz-

ation," in Proceedings of 20th Industrial Waste Conference, Purdue University, May 1965, p. 544.
10. Bulletin no. 102, 5–54, Komline-Sanderson Engineering Corp., Peapack, N.J.
11. Buswell, A. M., and W. D. Hatfield, *Anaerobic Fermentations*, Bulletin no. 32, Illinois State Water Survey, Urbana, Ill. (1939).
12. Cox, G. N., *A Summary of Hydrologic Data; Bayou Duplantier Watershed, 1933–1939*, University Bulletin, Louisiana State University, Baton Rouge (1940).
13. Eckenfelder, W. W., "Studies on the oxidation kinetics," *Sewage Ind. Wastes* **28**, 983 (1956).
14. Gauvin, W. H., "The atomized suspension technique," *TAPPI*, **40**, 866 (1957).
15. Genter, A. L., "Computing coagulant requirements in sludge conditioning," *Trans. Am. Soc. Civil Eng.* **111**, 635 (1946).
16. Genter, A. L., "Conditioning and vacuum filtration of sludge," *Sewage Ind. Wastes* **28**, 829 (1956).
17. Harding, J. C., and G. E. Griffin, "Sludge disposal by wet air oxidation at a five mgd plant," *J. Water Pollution Control Federation* **37**, 1134 (1965).
18. Haseltine, T. R., "Measurement of sludge drying bed performance." *Sewage Ind. Wastes* **23**, 1065 (1951).
19. Laboon, J. F., "Experimental studies on the concentration of raw sludge," *Sewage Ind. Wastes* **24**, 423 (1952).
20. Leet, C. A., C. W. Gordon, and R. G. Tucker, *Thermal Principles of Drying and/or Incineration of Sewage Sludge*, Combustion Engineering, Inc., New York (1959).
21. McCabe, W. L., and J. C. Smith, *Unit Operations of Chemical Engineering*, McGraw-Hill Book Co., New York (1956), p. 891.
22. Malina, F. H., Jr., and H. N. Burton, "Aerobic stabilization of primary wastewater sludge." in Proceedings of 19th Industrial Waste Conference, Purdue University, 1964, p. 716.
23. Mercer, W. A., W. W. Rose, J. E. Chapman, A. Katsuyama, and F. Dwinnell, Jr., "Aerobic composting of vegetable and fruit wastes," *Compost Sci.* **3**, 3 (1962).
24. Miller, D. R., "World's deepest submarine pipeline," *Sewage Ind. Wastes* **30**, 1426 (1958).
25. Powell, S. T., "Industrial wastes," *Ind. Eng. Chem.* **46**, 95A (1954).
26. Rabinovitch, W., P. Luner, R. James, and W. H. Gauvin, "The automized suspension technique. Part III," *Pulp Paper Mag. Can.* **57**, 123 (1956).
27. Rawn, A. M., and F. R. Bowerman, "Disposal of digested sludge by dilution," *Sewage Ind. Wastes* **26**, 1309 (1954).
28. Salvato, J. A., *Environmental Sanitation*, John Wiley & Sons, New York (1958), p. 288.
29. Schlenz, H. E., "Standard practice in separate sludge digestion," *Proc. Am. Soc. Civil Engrs* **63**, 1114 (1937).
30. *Sewage Treatment Plant Design*, Manual of Engineering Practice no. 36, American Society of Civil Engineers, New York (1959), p. 265.
31. *Sewage Treatment Plant Design*, Manual of Practice no. 8, Federation of Sewage and Industrial Waste Association, Washington, D.C. (1959), p. 214.
32. *Sludge Drying and Incineration*, Bulletin no. 6791, Dorr Co., Stamford, Conn. (1941).
33. Sylvester, R. O., "Sludge disposal by dilution in Puget Sound," *J. Water Poll. Control Federation* **34**, 891 (1962).
34. Teletzke, G. H., "Low pressure wet air oxidation of sewage sludge," in Proceedings of 20th Industrial Waste Conference, Purdue University, May 4–6, 1965, p. 40.
35. Tiller, F. H., and C. J. Huang, "Theory of filtration equipment," *Ind. Eng. Chem.* **53**, 529 (1951).
36. Torpey, W. N., "Concentration of combined primary and activated sludges in separate thickening tanks," *Proc. Am. Soc. Civil Engrs* **80**, Separate no. 443 (1954).
37. West, L., "Sludge disposal experiences at Elizabeth, N.J.," *Sewage Ind. Wastes* **24**, 785 (1952).
38. Wirts, J. J., "Pipeline transportation and disposal of digested sludge," *Sewage Ind. Wastes* **28**, 121 (1956).

Part 3 | APPLICATIONS

CHAPTER 15

JOINT TREATMENT OF RAW INDUSTRIAL WASTES WITH DOMESTIC SEWAGE

INTRODUCTION

In 1951 approximately 69 per cent of all industries discharged their wastes into municipal sewers. Also, 82 per cent of the remaining 31 per cent of industries, which discharged their wastes directly into watercourses, provided no waste treatment whatsoever [7]. Good local government must give close attention to a matter so vital to the public welfare. According to Schroepfer and his committees, three choices are open to municipalities: (1) they can exclude all industrial wastes or only certain ones; (2) they can require pretreatment at the source for all industrial wastes, so that they approach the BOD and solids level of domestic sewage, or pretreatment of certain wastes because of their harmful nature; (3) they can receive all industrial wastes or all except those harmful to treatment-plant equipment.

Managers of industrial plants major have three courses of action open to them, as concerns disposal of their wastes: (1) the waste may be discharged directly to the municipal sewage plant and thence to a stream; (2) it may be sent to an industrial waste-treatment plant and after that either to a municipal plant or to a stream; (3) after a careful survey of stream flow and quality, the management may find that it can discharge waste directly into a stream. Many plants, because of their location, cannot choose where their wastes will be discharged, but, when a choice of "waste path" is possible, the answer can come only from a careful study of the current practices involved in disposal of liquid wastes. Since a study of both theories and practices is not only desirable but necessary, we shall now look into various aspects of the problem, as they concern both cities and industries.

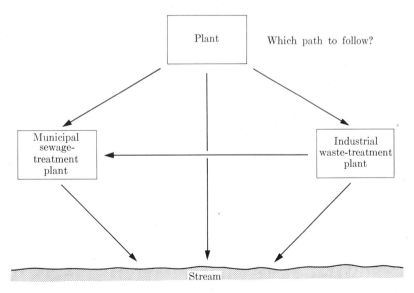

15.1 Industrial Use of Municipal Sewage Plants

It is often possible and advisable for an industry to discharge its waste directly to a municipal treatment plant, where a certain portion of the pollution can be removed [6]. A municipal sewage-treatment plant, if designed and operated properly, can handle almost any type and quantity of industrial waste [5]. Hence, one possibility that should be seriously considered is the cooperation of industry and municipalities in the joint construction and operation of a municipal waste-water treatment plant. There are many advantages to be gained from such a joint venture:

1. Responsibility is placed with one owner, while at the same time the cooperative spirit between industry and municipality is increased, particularly if division of costs is mutually satisfactory.

2. Only one chief operator is required, whose sole obligation is the management of the treatment plant; i.e., he is not encumbered by the miscellaneous duties often given to the industrial employee in charge of waste disposal, and the chances of mismanagement and neglect, which may result if industrial production men operate waste-treatment plants, are eliminated.

3. Since the operator of such a large treatment plant usually receives higher pay than separate domestic plant operators, better-trained people are available.

4. Even if identical equipment is required, construction costs are less for a single plant than for two or more. Furthermore, municipalities can apply for state and/or federal aid for plant construction, which private industry is not eligible to receive.

5. The land required for plant construction and for disposal of waste products is obtained more easily by the municipality.

6. Operating costs are lower, since more waste is treated at a lower rate per unit of volume.

7. Possible cost advantages resulting from lower municipal financing costs, federal grants, and municipal operation can be passed on to the users and may permit higher degrees of treatment at a cost to each participant no greater than the cost for separate treatment at lower removal levels.

8. Some wastes may add valuable nutrients for biological activity to counteract other industrial wastes that are nutrient-deficient. Thus, bacteria in the sewage are added to organic industrial wastes as seeding material. These microorganisms are vital to biological treatment when the necessary BOD reduction exceeds approximately 70 per cent. Similarly, acids from one industry may help neutralize alkaline wastes from another industry.

9. The treatment of all waste water generated in the community in a municipal plant or plants enables the municipality to assure a uniform level of treatment to all users of the river and even to increase the degree of treatment given to all waste water to the maximum level obtainable with technological advances.

10. Acceptance of the joint treatment project and relinquishment of individual allocations would give the municipality full control of the river's resources and permit it to use the capacity of the river to the best advantage for the public at large. The municipality has greater assurance of stream protection, since it has the opportunity for closer monitoring of effluent quality.

Among the many problems arising from combined treatment, the most important is the character of the industrial waste water reaching the disposal plant. Equalization and regulation of discharge of industrial wastes are sometimes necessary to prevent rapid change in the environmental conditions of the bacteria and other organisms which act as purifying agents, to ensure ample chemical dosage in coagulating basins, and to ensure adequate chlorination to kill harmful bacteria before the effluent is discharged to a stream.

In recent years, two factors in particular have focused attention on the subject of combined treatment for sewage and industrial wastes: the increased interest in stream-pollution abatement and the phenomenal growth of industry in the postwar years, with the subsequent increase in demand for water.

Since most sewage plants use some form of biological treatment, it is essential for satisfactory operation that extremes in industrial waste characteristics be avoided and the waste mixture be: (1) as homogeneous in composition and uniform in flow rate as possible and free from sudden dumpings (shock loads) of the more deleterious industrial wastes; (2) not highly loaded with suspended matter; (3) free of excessive acidity or alkalinity and not high in content of chemicals which precipitate on neutralization or oxidation; (4) practically free of antiseptic materials and toxic trace metals; (5) low in potential sources of high BOD, such as carbohydrates, sugar, starch, and cellulose; and (6) low in oil and grease content.

If the industrial-waste characteristics are such that the waste can be treated safely and effectively in the municipal sewage plant, there still remain two major considerations: a municipal ordinance which protects the treatment plant from any individual or industrial violation and sewer-rental charges which enable the municipality to defray the increased costs of construction and operation resulting from acceptance of the industrial wastes.

Combined municipal and industrial waste treatment is the most desirable arrangement and at the same time it is the most difficult to achieve. It is the author's contention that the difficulty is usually not a scientific one but rather one of human compatability and understanding. In numerous cases, the economics of the situation were overwhelmingly in favor of combined treatment and yet separate waste-treatment installations were finally used, primarily because of personality clashes and lack of sympathetic understanding.

What can municipalities do to assist industry in waste-treatment practices?

1. If the municipal treatment plant is new or being enlarged, it should be designed to serve the entire community, with *all* the existing and planned industry as members of a "corporation." Combined meetings, lectures, and actual visits to the treatment-plant site will enhance mutual understanding of the problems involved.

2. If the plant is already in operation, municipal officials should meet with industrial representatives and discuss the advantages and disadvantages of accepting the industrial wastes into the system. Adequate safeguards such as research, literature study, and pilot-plant experiments should precede administrative decisions, and any decision to accept the waste should be accompanied by a substantial and specifically detailed contract between the owner and user. Methods of sampling, analyses, charges, and waste characteristics should be clearly stated in this contract.

3. A municipality can purchase land, build a treatment plant for its industry, float bonds, and receive rent from industry for use of the plant and amortization of the bonds. In this way the rental becomes an operating expense which industry can deduct from income *before* taxes rather than the long-term depreciation of capital assets involved in having its own plant. This is particularly attractive to industry.

4. Most important, perhaps, is the understanding a municipality must have for its industry. All the members of the city council should be in agreement that without industry municipal survival is doubtful and its growth potential is nil. An industry located within the city limits is contributing the maximum to a city in the form of taxes and intangible benefits. Although it is common for new industry to locate outside city limits, where it can purchase sufficient land at a reasonable cost for future expansion, these industries too are an indirect but valuable addition to the community.

5. A municipality can design its treatment plant so that it will handle an industry's waste without pretreatment by the industry. An ideal arrangement is to take the industry into the business of waste disposal as a "member of the municipal corporation." Industry should pay for this service but not at the same rate as an individual householder, since a large contributor deserves concessions solely on the basis of lower unit costs for larger volumes. In addition, industry's intangible benefits to the community should be assessed and its share of the capital cost reduced proportionately.

6. Municipal controls of the influent from the industrial plant are costly and difficult to establish. Instead, it is recommended that industry control its own effluent so that the "corporation" disposal system operates efficiently. When, and if, the system ceases to function as designed, a corporation meeting should be called to decide what measures should be taken to correct the situation. In this manner, the expenses of sampling and billing, as well as the ill feelings caused by policing, are eliminated.

7. A good corporation will continually try to improve its efficiency of operation by conducting research on new methods of treatment. Research has been proven to pay off in the long haul and the lack of it has often led to plant and process obsolescence.

Industry, with its research experience, could well lead the way in this connection by supporting continued research in specific combined treatment processes.

8. The designing sanitary engineer should be selected by both the municipality and the industry for his competence and ability to work with equal ease with both groups. His fee should be on a lump-sum basis, approximating the sliding-scale percentage of estimated construction cost but not necessarily tied to this: rather than being penalized for reducing the capital costs (and hence his fees), he should be rewarded financially for economizing on construction and improving plant efficiency.

In summary, a municipality can assist its industry by encouraging mutual understanding, by embarking on a program of education, and by designing its plants to handle industrial waste. Other methods of assisting industry depend upon the formation of a "corporation treatment plant" with joint responsibility for efficient operation. Municipal ownership and reduction in charges based upon intangible as well as tangible industrial benefits should also be considered.

15.2 Municipal Ordinances

Although there are many types of municipal ordinances, all are designed to place an upper limit on the concentration of various constituents in waste. Sometimes this upper limit is zero, since any quantity of a certain pollutant would be detrimental to the plant or its component parts. In addition to their obligation to abide by municipal ordinances, many industries enter into separate contracts with the city. Generally, such contracts include: the obligation of the municipality to construct, operate, and maintain the treatment facilities and to finance the overall project by means of some type of bond; a declaration on the part of the industry as to the maximum quantity of flow, BOD, and solids; the percentage by volume of industrial waste as compared to municipal waste; the amount the industry will pay each year to cover operation and maintenance; provision for a penalty if stated limits are exceeded; and any other pertinent matters involving the joint usage of the treatment system.

The following are dangers of inadequate sewer-use control [4]: (1) explosion and fire hazards; (2) sewer clogging; (3) overloads of surface water (storm- and/or cooling-water pollution); (4) physical damage to sewers and structural damage to treatment plants; (5) interference with sewage treatment.

A comprehensive sewer ordinance [4] usually consists of the following principal parts: introduction; definition of terms; regulation requiring use of public sewers where available; regulations concerning private sewage and waste disposal where public sewers are not available; regulations and procedures regarding the construction of sewers and connections; regulations relating to quantities and character of waters and wastes admissible to public sewers; special regulations; provision for powers of inspectors; enforcement (penalty) clause; validity clause; and signatures and attest.

Since industrial wastes vary so greatly in character, only broad limits can be established in any model ordinance, and ordinances should always be based on recommendations of the consulting engineer. Most ordinances [4] provide for the control of waste substances other than sanitary sewage in the following ways:

1. They prohibit the discharge to the public sewers of flammable substances or materials that would obstruct the flow.

2. They state that industrial wastes will be admitted to the public sewers only by special permission of a stated municipal authority.

3. They ban all wastes that would damage or interfere with the operation of the sewage works, except when such wastes have been adequately pretreated, and even then their admission is to be at the discretion of a stated municipal authority.

4. They enumerate in detail, in a separate ordinance, the procedures outlined in (3).

5. They give detailed regulations to supplement the procedure in (3), stating specific limits for objectionable characteristics of industrial wastes.

A model ordinance [5] may spell out in detail the following regulations relating to quantities and character of water and wastes admissible to public sewers:

Section 1. No storm water, roof runoff, cooling water, ground-water, etc., will be allowed in the sanitary sewer.

Section 2. Storm water or other uncontaminated drainage will be discharged to sewers that are designated *combined* or *storm sewers* only.

Section 3. No person shall discharge any of the following wastes to sanitary sewers except as hereinafter provided: (a) any liquid or vapor having a temperature higher than 150°F; (b) any waste containing more than 100 ppm by weight of grease; (c) any gasoline, etc., or other flammable or explosive liquid, solid, or gas; (d) any garbage that has not been properly ground; (e) any ashes, metals, cinders, rags, mud, straw, glass, feathers, tar, plastics, wood, chicken manure, or other interfering or obstructing solids; (f) any wastes having a pH less than 5.5 or higher than 9.0, or having other corrosive effects; (g) any toxic wastes that may be a hazard to sewage plant, persons, or receiving stream; (h) any suspended solids the treatment of which at the sewage plant may involve unusual expenditures; (i) any noxious gases.

Section 4. There shall be installations of interceptors for grease, oil, and sand, when necessary.

Section 5. These installations shall be maintained by owner.

Section 6. This section establishes the conditions pertaining to the admission of any wastes having (a) a 5-day **BOD** greater than 300 ppm, (b) more than 350 ppm suspended solids, (c) any of the quantitative characteristics described in Section 3, (d) an average daily flow greater than 2 per cent of the average daily flow of the city.

Section 7. Where preliminary treatment facilities are provided for any wastes, they shall be maintained by the owner at his own expense.

Section 8. When required, the owner of any property served by a sewer carrying industrial wastes shall install a suitable manhole for observation, sampling, and measuring.

Section 9. All measurements and analyses of the characteristics of waters and wastes referred to in Sections 3 or 6 shall be determined in accordance with standard methods [1].

Section 10. No statement contained in this article shall preclude any special agreement or arrangement between the city and any industry.

15.3 Sewer-Rental Charges

Sewer-rental charges are necessary to help meet the city's budget and to ensure that industry pays a fair share of the cost of disposing of its wastes. Several methods can be used to charge for sewer service: (1) an *ad valorem* tax on property, which is the traditional method in more than 80 per cent of U.S. communities and is successful in small towns and villages; (2) special assessments, with charges set according to front footage; (3) sewer-rental charges (approximately one-sixth of municipalities having treatment plants use this method); (4) special contracts negotiated with industry; (5) combination of two or more of the above methods. In many cases a municipality charges the industry or industries solely on the basis of water consumption. Although this may not always prove equitable, it has several advantages to the municipality and to the industries. First, the billing system is simplified, omitting the need for detailed and time-consuming cost procedures. Second, the system eliminates the need for measuring flows from the industries and their strength characteristics. Thus, the municipality treats its industries just as it does its householders, rather than as a "culprit."

In considering charges, fixed expenses such as operation, maintenance, and debt retirement should all be taken into account. A portion of each of these three costs can be charged to all the users of the sewer system, and the remaining portion to property owners using the system. This is done by itemizing the cost of each component unit of the sewer system and then allocating percentages of the annual cost of each unit to the users and the rest to the property owners. Total annual charges to users and property owners are determined from the summation of the unit costs. The total share allocated to property owners may now be prorated according to individual property valuations (or sometimes front footages). The users' share necessitates additional prorating based on the following waste factors: volume, suspended solids, BOD, and (sometimes) chlorine demand. This is carried out in the computations of users' share for each unit. If the unit is designed solely upon a volume basis, such as the main sewage pumps, the entire users' share is charged to volume contributors. On the other hand, for the sludge digester, 90 per cent of the cost may be charged to contributors of suspended solids and

10 per cent to contributors of **BOD**. If the volume of sewage is based on water consumption and the supply is private (wells or river water), a meter is normally supplied by the industry for flow measurement.

When all users' charges attributed to volume, solids, and BOD are added, one obtains the total users' cost for each category. The total of the three in turn represents the users' share of the annual sewer costs, and the total of the users' and property owners' shares represents the complete annual sewer costs.

Schroepfer [7] uses the following example to illustrate a fair allocation of costs. The total annual cost of operating a sewage-disposal system in a certain town consists of:

(1) Fixed charges:
   Intercepting sewers* $ 35,000
   Treatment plant*   75,000
(2) Operating and maintenance
   costs   70,500
   Total $180,500

Table 15.1 shows the allocation of the fixed charges for the sewers and treatment plant, Table 15.2 the allocation of the operation and maintenance costs, and Table 15.3 the allocation of fixed and operational

---

*Capital investment: for intercepting sewers $700,000 and for treatment plant $1,500,000; debt retirement: 5 per cent per year (total interest and principal).

Table 15.1 Allocation of fixed charges. (After Schroepfer [7].)

Units	Total fixed charges, $	Chargeable to property owners		Chargeable to users, $	Users' share chargeable to					
					Volume		Suspended solids		BOD	
		%	$		%	$	%	$	%	$
Intercepting sewers	35,000	64	22,300	12,700	100	12,700				
Treatment plant										
Main pumping station										
Equipment	1,500	40.5	600	900	100	900				
Structures	1,250	64	800	450	100	450				
Screen and grit chambers	1,500	64	950	550	60	330	40	220		
Preliminary sedimentation tanks	4,500	40.5	1,800	2,700	85	2,300	15	400		
Trickling filters	30,000	25	7,500	22,500	10	2,250			90	20,250
Final sedimentation tanks	9,000	30	2,700	6,300	50	3,150			50	3,150
Receiving pumps	750	25	200	550					100	550
Chlorination tanks and equipment	2,000	35	700	1,300	40	520			60	780
Digester tanks and receiving filters	8,000	30	2,400	5,600			100	5,600		
Subtotal	58,500	30.3	17,650	40,850	24.2	9,900	15.2	6,220	60.6	24,730
Main control building	7,500	30.3	2,300	5,200	24.2	1,310	15.2	790	60.6	3,100
Plant water supply	2,500	30.3	800	1,700	24.2	410	15.2	260	60.6	1,030
Roads and grounds	2,500	30.3	800	1,700	24.2	410	15.2	260	60.6	1,030
Plumbing and heating	4,000	30.3	1,200	2,800	24.2	680	15.2	430	60.6	1,690
Total plant costs	75,000	30.3	22,750	52,250	24.2	12,710	15.2	7,960	60.6	31,580
Total fixed charges	110,000	41	45,050	64,950	39.2	25,410	12.2	7,960	48.6	31,580

15.3　　　　　　　　　　　　　　　　　　　　　　　　　　　　　SEWER-RENTAL CHARGES　159

Table 15.2 Allocation of operation and maintenance costs. (After Schroepfer [7]).

Unit	Total operating and maintenance cost, $	Chargeable to property owners		Chargeable to users, $	Users' share chargeable to					
					Volume		Suspended solids		BOD	
		%	$		%	$	%	$	%	$
Intercepting sewers	2,200	60	1,300	900	60	500	40	400		
Main pumping station	9,200	17	1,600	7,600	100	7,600				
Preliminary treatment	6,700	50	3,400	3,300	50	1,700	50	1,600		
Secondary treatment	13,500	15	2,000	11,500	10	1,200			90	10,300
Effluent chlorination	5,200	15	800	4,400	10	400			90	4,000
Sludge disposal	17,500	5	900	16,600			100	16,600		
General	5,000	15	800	4,200	25	1,000	43	1,800	32	1,400
Supervisory	6,200	15	900	5,300	25	1,300	43	2,300	32	1,700
Collection and billing	5,000	15	800	4,200	25	1,000	43	1,800	32	1,400
Total	70,500	17.8	12,500	58,000	25.4	14,700	42.1	24,500	32.5	18,800

charges. Figures 15.1 and 15.2 illustrate the prorating of the fixed charges of the sewers and treatment plant.

Property owners' charges should be distributed according to assessed evaluation, which in the example under discussion is taken to be $20,000,000. Therefore, $2.88 per $1000 of property valuation will be charged to property owners. Users' charges depend on flow, solids, and BOD, as stated previously. Hence, the annual flow and the quantities of each type of these waste loads must either be determined after the first year's operation or estimated prior to establishing the users' charges for the year. The third column of Table 15.4 lists the unit rates obtained from the data given in columns 1 and 2.

Table 15.3 Summary of the allocation of the fixed and operating charges. (After Schroepfer [7]).

Fixed charges	Chargeable to			
	Users		Property owners	
	%	$	%	$
Sewers	36	12,700	64.0	22,300
Treatment plant	69.7	52,250	30.3	22,750
Operation and maintenance costs	82.2	58,000	17.8	12,500
Totals		122,950		57,550
Averages	68.1		31.9	

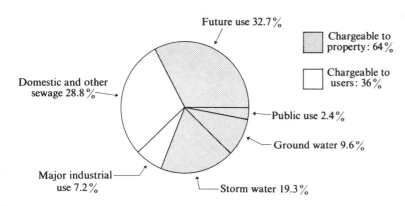

Fig. 15.1 Allocation of fixed charges on the intercepting sewers. (After Schroepfer [7].)

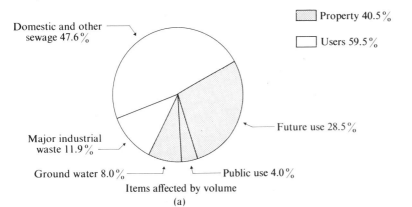

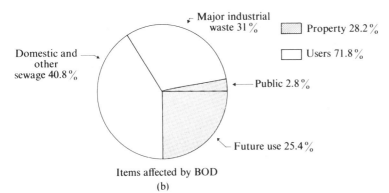

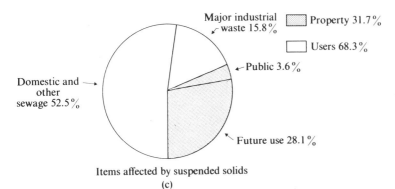

Fig. 15.2 Allocation of fixed charges on the treatment plant: (a) items affected by volume; (b) items affected by BOD; (c) items affected by suspended solids. (After Schroepfer [7].)

Table 15.4 Calculation of users' charges based on three factors. (After Schroepfer [7].)

	Annual quantity	Total, $	Unit rate, $
Volume of flow	1,370 million gal	40,000	2.93/1000 gal
Suspended solids	3,647,000 pounds	32,460	0.89/100 lb
BOD	3,847,000 pounds	50,380	1.40/100 lb

References: Introduction

1. American Public Health Association, *Standard Methods for the Examination of Water, Sewage, and Industrial Wastes*, 10th ed., New York (1955).
2. California State Water Pollution Control Board, *A Survey of Direct Utilization of Waste Waters*, Publication no. 12, Sacramento, Cal. (1955).
3. Geyer, J. C., "The effect of industrial wastes on sewage plant operation," *Sewage Works J.* **9**, 625 (1937).
4. *Municipal Sewer Ordinances*, Manual of Practice no. 3, Federation of Sewage and Industrial Wastes Association, Washington, D.C. (1957).
5. Nemerow, N. L., "Fiber losses at paper mills: Effects on streams and sewage treatment plants," *Sewage Ind. Wastes*, **23**, 880 (1951).
6. Nemerow, N. L., *Water Wastes of Industry*, Bulletin no. 5, Industrial Engineering Program, North Carolina State College, Raleigh (1956).
7. Schroepfer, G. M., "Sewer service charges," *Sewage Ind. Wastes* **23**, 1493 (1951).
8. *The Treatment of Sewage Plant Effluent for Water Reuse in Process and Boiler Feed*, Technical Reprint T-129, Graver Water Conditioning Company, New York (1954).
9. Veatch, N. T., "Industrial uses of reclaimed sewage effluents," *Sewage Works J.* **20**, 3 (1948).

CASE HISTORY OF A PROJECT FOR JOINT DISPOSAL OF UNTREATED INDUSTRIAL WASTES AND DOMESTIC SEWAGE

For the purposes of our discussion, we shall consider the case of two relatively small municipalities containing 27 small industries (mostly tanneries) which require adequate and effective treatment of their wastes. The problem presents a challenge in engineering, economics, and administration.

15.4 Existing Situation

Cayadutta Creek rises in the central part of Fulton County in New York State, flows generally south for about 14 miles through the cities of Gloversville and Johnstown, and enters the Mohawk River at Fonda (Fig. 15.3). The total catchment area covers 62 square miles above Station 6. There are no official gauging

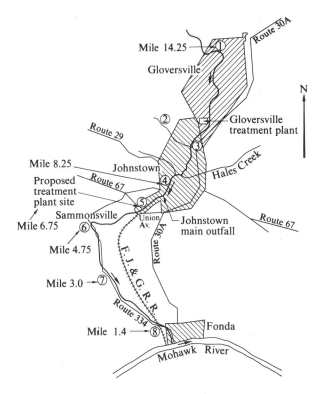

Fig. 15.3 Cayadutta Creek.

stations on this stream but approximate flow data for a comparatively short time (1898–1900) are available. This creek has been characterized by an expert state hydrologist, a member of the U.S. Geological Survey, as similar to that of Kayaderosseras Creek, which is located in Saratoga County and drains into the Hudson River basin.

Ninety-one per cent of the population of the Cayadutta Creek catchment area is concentrated in the cities of Johnstown and Gloversville and the village of Fonda. In 1952, New York State cited this creek as "one of the most grossly polluted streams in the state." From within the city of Gloversville to the junction with the Mohawk River the stream is entirely unsuited for the support of fish life, whereas formerly it was trout water throughout its entire length. It has been stated (by M. Vrooman: *March 10, 1950, Report to City of Gloversville*) that the dry weather flow of Cayadutta Creek is higher than the average for streams in the state, owing to the nature of the watershed, the sandy soil, and the larger wooded area. He also stated that "the average daily flow of the Creek at the Gloversville sewage treatment plant is 17 million gallons and the low measured dry weather flow is 4.2 million gallons." These figures were evidently obtained from separate, independent, and unofficial flow measurements. The tanning industry is an old one in American history and has a record of contributing to the damaging pollution of Cayadutta Creek. The National Tanners Association (private communication) estimates that the Fulton County area has been losing about one plant every four years. However, there has been more glove- and garment-leather demand as the population of the U.S.A. rises. They predict, therefore, that the overall demand for glove and garment leather (produced in Fulton County) will continue slowly upward, but it will be met in fewer plants with increased production.

The sewage and wastes from the cities of Gloversville and Johnstown are discharged into Cayadutta Creek. In 1960, Gloversville had a population of 21,741 while that of Johnstown was 10,390 (Fig. 15.4). Almost the entire urban population is served by public sewer systems but only half of the system is tributary to a sewage-treatment plant, which serves the people and industry of Gloversville. It consists of a bar screen, grit chamber, two antiquated Dortmund-type primary settling basins, a fixed-nozzle trickling filter, one final Dortmund-type settling basin, and some sludge-drying beds. The plant was built in the early 1900s and is incapable of handling more than half the waste water at the present flow rates.

The tanning industry retained a New York City consulting firm to represent their interests in this problem. The two cities retained a local consulting engineering firm, which in turn retained the present author to advise them on study procedures and solutions to the pollution problem in Cayadutta Creek.

15.5 Stream Survey

A stream survey is an essential part of any well-conceived waste-treatment study. Ideally, a survey designed to study the oxygen-sag curve should be carried out during extremely hot weather, extremely low stream flow, and typical high organic matter loading. It is seldom possible to conduct a stream

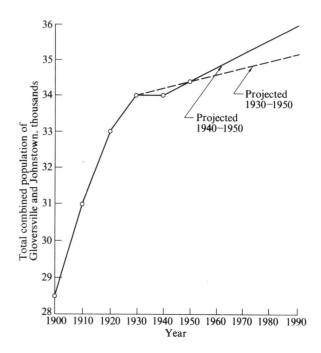

Figure 15.4

survey under all of these "ideal" conditions. In this study we were particularly fortunate to collect stream samples during extremely low flow conditions—comparable to those which may be expected to occur for a seven-day period only once in ten years—while, at the same time, the municipal and industrial pollution loads were considered to be above average. Although stream temperatures were not high (11° to 14°C), these values are never very high, owing to the relatively cold mountain water diluting the wastes. For example, during the state survey of 1951, samples collected on August 22 and 23 at Stations 5 and 6 showed temperatures of only 15° to 19°C.

During October 1964, the creek was visited and examined at various locations and dissolved-oxygen values were determined in order to locate the sag curve points. After an initial appraisal and a trial survey, the creek was sampled at the eight stations shown in Fig. 15.3. After the first day samples were collected only from Stations 1, 5, and 6, on six days at different times during each day. Composite samples were analyzed for dissolved oxygen, BOD, and temperature and, in addition, the creek flow was measured on each sampling date at Station 6; these results are shown in Table 15.5. Flow times are shown in Table 15.6, a summary of the BOD and flow data in Table 15.7, the BOD curves for Stations 5 and 6 in Fig. 15.5, flow data from a similar, gauged creek in Table 15.8, the probability of the minimum flow data occurring in Table 15.9 and Fig. 15.6, and a multiple-regression technique analysis of the stream data in Tables 15.10 and 15.11.

Cayadutta Creek analysis. Using the multiple regression method described in Chapter 4, we take the following three equations. When solved simultaneously, these will yield the "best" equation, which relates the dissolved-oxygen sag to the BOD, flow, and temperature at the bottom of the sag (Station 6).

$$b_1 \Sigma X_1^2 + b_2 \Sigma X_1 X_2 + b_3 \Sigma X_1 X_3 = \Sigma X_1 Y, \quad (1)$$

$$b_1 \Sigma X_1 X_2 + b_2 \Sigma X_2^2 + b_3 \Sigma X_2 X_3 = \Sigma X_2 Y, \quad (2)$$

$$b_1 \Sigma X_1 X_3 + b_2 \Sigma X_2 X_3 + b_3 \Sigma X_3^2 = \Sigma X_3 Y. \quad (3)$$

From Table 15.11 we can substitute numerical data in order to find b_1, b_2, and b_3. From Eq. (1) we obtain

$$364.78 b_1 + 4.58 b_2 + 125.62 b_3 = -43.78. \quad (4)$$

Dividing Eq. (4) by 364.78 yields

$$b_1 + 0.01256 b_2 + 0.3435 b_3 = -0.1200. \quad (5)$$

Multiplying Eq. (4) by 0.01255 gives

$$4.58 b_1 + 0.05748 b_2 + 1.5728 b_2 = -0.5494. \quad (6)$$

We next apply Eq. (2) and obtain

$$4.58 b_1 + 10.68 b_2 + 0.14 b_3 = 3.02. \quad (7)$$

Subtracting Eq. (6) from (7) yields

$$10.62252 b_2 - 1.4328 b_3 = 3.5694. \quad (8)$$

Dividing Eq. (8) by 10.62252, we obtain

$$b_2 - 0.1349 b_3 = 0.3360. \quad (9)$$

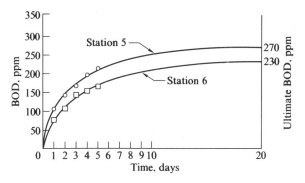

Fig. 15.5 Laboratory BOD's for Stations 5 and 6 at 20°C. Each point represents an average of seven samples collected from the creek on seven different days at different times of day, all during a drought flow period (October 8–18, 1964).

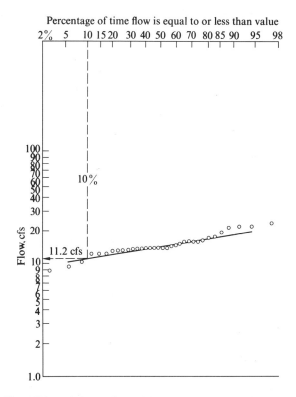

Fig. 15.6 Minimum flow of Cayadutta Creek and expected recurrence of this level.

Table 15.5 Cayadutta Creek analyses during October 1964.

Station and milage	Date	Time	DO, ppm	Flow, cfs	BOD, ppm					Water temp., °C
					Day 1	Day 2	Day 3	Day 4	Day 5	
1. Bleeker St. Bridge, Gloversville (clean H_2O site) (14.25)	10/8		10.12						8.0	8
	10/12		9						8.5	13
	10/13		10.0						9.5	12
	10/14		10.5						10.0	10
	10/15		9						8.5	11
	10/17		9.5						9.3	12
	10/18		10.0						8.9	11
4. Main St., Johnstown (8.25)	10/8		1.0						310	12.5
	10/12	1:30	1.9						240	15
	10/13	10:15	0.5						290	13
	10/14	9:00	5.1						340	10
	10/15	4:00	2.7						280	12
	10/17	11:05	6.2						240	12
	10/18	1:30	6.0						30	12
5. Harding property below Johnstown (6.75)	10/8		2.75		140	230	260	290	340	11.5
	10/12	2:00	2.8		110	180	220	260	290	15
	10/13	10:25	5.0		100	130	130	140	150	14
	10/14	9:30	3.1		120	190	170	180	200	14
	10/15	4:30	3.0		100	140	160	210	220	13
	10/17	11:25	5.4		100	70	130	150	160	14
	10/18	2:05	5.5		70	70	100	140	140	15
6. Sammonsville Bridge (4.75)	10/8		2.2	59.4	130	200	270	300	330	13
	10/12	2:25	1.8	51	140	170	250	230	240	14
	10/13	10:45	1.1	44	40	70	60	80	140	12
	10/14	9:45	4.6	39	110	150	140	150	180	11
	10/15	4:50		39	40	110	150	160	160	12
	10/17	11:55	4.8	12.2	40	30	100	100	70	12
	10/18	2:40	4.8	12.2	40	30	50	50	40	13
	10/31			34						
	11/1			42						
7. Rt. 334, adjacent to Peresse Rd., Berryville Cross (3.0)	10/8		1.25						190	11
	10/12	2:45	0						200	14
	10/13	11:10	4.1						140	12
	10/14	10:05	5.0						170	11
	10/15	5:15	3.9						80	10
	10/17	12:15	1.8						70	11
	10/18	3:10	4.0						50	11
8. Rt. 334, 1 mile north of Fonda next to Cannarella house (1.4)	10/8		5.1						180	10.5
	10/12	3:05	3.5						90	13
	10/13	11:30	5.1						140	11
	10/14	10:20	5.1						160	11
	10/15	4:30	5.0						80	12
	10/17	12:45	3.6						100	11
	10/18	3:30	4.5						40	

Table 15.6 Time of flows from Station 5 downstream taken about one week before stream-sampling program in October 1964.

Site	Distance between points, mil	Time	Fall, feet
Start at Harding farm (Station 5)			
To old power dam	1	45 min	70
To bridge at Sammonsville (Station 6)	1	2 hr	20
To bridge at Fonda Ave.	1.75	2 hr	80
To railroad bridge	1.50	2 hr 40 min	55
Begin slack water of Mohawk River	1	2 hr 30 min	50
Total	6.25	9 hr 55 min	275

Table 15.7 Summary of 7-day sampling of Cayadutta Creek during dry period from 10/8/64 to 10/18/64.

Station	Reading	Average of 7 samples	Range
1	DO, ppm	9.73	9–10.5
	Temp., °C	11	8–13
	5-day BOD, at 20°C, ppm	8.96	8–10
5	DO, ppm	3.93	2.75–5.5
	Temp., °C	13.8	11.5–15
	5-day BOD, at 20°C, ppm	214	140–340
	L	270 (projected factor 1.26)	
	1-day	106	70–140
	2-day	144	70–230
	3-day	167	100–260
	4-day	196	140–290
6	DO, ppm	3.17	1.1–4.8
	Temp., °C	12.4	11–14
	5-day BOD, at 20°C, ppm	166	40–330
	L	230 (projected factor 1.39)	
	1-day	77	40–140
	2-day	109	30–200
	3-day	146	50–270
	4-day	153	50–300
	*Flow, cfs†	36.7	12.2–59.4

*Time of travel between Stations 5 and 6 was 2 hr 45 min (0.115 days).
†Flow for 10/12/64 computed as arithmetic average of flow on 10/8/64. This flow represents approximate value of 7-day consecutive low flow likely to occur in Cayadutta Creek below Johnstown once in ten years: $\frac{62}{90} \times 20 = 13.8$ cfs for entire Cayadutta Creek (slightly less for below Johnstown). The 20 cgs in the equation is the minimum 7-day flow for Kayaderosseras Creek (*New York State Upper Hudson River Drainage Basin Survey Series Report no. 2*, p. 244).

Multiplying Eq. (4) by -0.3435, we get

$$-125.32b_1 - 1.573b_2 - 43.047b_2 = +15.038. \quad (10)$$

Next we multiply Eq. (8) by $+0.1443$:

$$1.533b_2 - 0.20675b_3 = 0.51506. \quad (11)$$

Finally, we apply Eq. (3) and obtain

$$125.32b_1 + 0.04b_2 + 51.83b_3 = -20.55. \quad (12)$$

Adding Eqs. (10), (11), and (12), we have

$$8.5763b_3 = -4.99694, \quad (13)$$

$$b_3 = -0.5826. \quad (14)$$

From Eq. (9),

$$b_2 - (-0.07859) = 0.3360,$$
$$b_2 = +0.2574. \quad (15)$$

From Eq. (5),

$$b_1 + (0.00323) + (-0.20012) = -(0.1200),$$
$$b_1 + 0.00323 - 0.20012 = -0.1200,$$
$$b_1 = 0.0769. \quad (16)$$

As a check, we substitute in Eq. (12):

$$125.32\,(0.0769) + 0.04\,(0.2574) + 51.83\,(-0.5826)$$
$$= -20.55$$
$$9.6471 + 0.010296 - 30.1962 = -20.55$$
$$-20.5488 = -20.55$$

$$\bar{Y} = a + b_1\bar{X}_1 + b_2\bar{X}_2 + b_3\bar{X}_3$$

$$a = \bar{Y} - (b_1\bar{X}_1 + b_2\bar{X}_2 + b_2\bar{X}_3)$$
$$a = 6.56 - [(0.0769 \times 9.35) + (0.2574 \times 12.4) + (-0.5826 \times 3.92)]$$
$$a = 6.56 - [0.7190 + 3.1918 - 2.2838]$$
$$a = 6.56 - 1.6270$$
$$a = 4.9330.$$

Therefore, the Cayadutta Creek equation is

$$Y = a + b_1X_1 + b_2X_2 + b_3X_3$$

or

$$Y = 4.9330 + 0.0769X_1 + 0.2574X_2 - 0.5826X_3,$$

where $X_1 = $ 5-day BOD (1000/ppm) 20° C, at Station 6
$X_2 = $ temperature (°C)
$X_3 = $ flow (100/cfs)
$Y = $ DO drop from Station 1 to Station 6 (ppm).

To verify this stream equation, we can substitute our actual observed stream values for X_1, X_2, and X_3 and obtain a calculated Y value, which can then be compared with the observed value for accuracy:

Date (1964)	Observed Y (ppm)	Calculated Y (ppm)
10/8	7.9	7.53
10/12	7.2	7.72
10/13	8.9	7.25
10/14	5.9	6.70
10/15	6.1	7.01
10/17	4.7	3.37
10/18	5.2	5.42

During the October 1964 study the lowest flow was 12.2 cfs, the highest temperature 14°C, and the DO sag allowed was between 9.0 and 2.0 (i.e. 7.0) ppm. To find the BOD load at sag, we calculate as follows:

$$Y = 4.9330 + 0.0769X_1 + 0.2574X_2 - 0.5826X_3$$

$$7.0 = 4.9330 + 0.0769X_1 + 0.2574\,(14) - 0.5826\left(\frac{100}{12.2}\right)$$

$$7.0 = 4.9330 + 0.0769X_1 + 3.6036 - 4.7773$$

$$3.2407 = 0.0769X_1$$

$$X_1 = 42.1417 = \frac{1000}{\text{ppm}}.$$

$$\text{ppm allowed} = \frac{1000}{42.1417} = 23.73 \text{ ppm}$$

The average BOD at sag was 166 ppm during the entire 7-day survey. Therefore, the BOD reduction required in the stream at Station 6 was

$$\frac{166 - 23.73}{166} \times 100 = 85.70\%.$$

Although the Streeter-Phelps method yielded values of k_1 and k_2 which gave a Fair's f of about 35,[*] the results are not dependable because of the variability of wastes from one moment to the next as well as the multiple entrances of wastes into the stream. The only reliable procedure for evaluating the oxygen-sag characteristics is to collect many stream

---

[*] $f = k_2/k_1 = $ reaeration rate/deoxygenation rate.

15.5 STREAM SURVEY

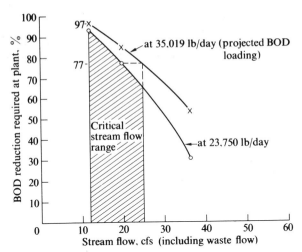

Fig. 15.7 BOD reduction required at Station 5, at 12.4°C and 2 ppm DO remaining and based on a waste discharge of 23,750 lb/day.

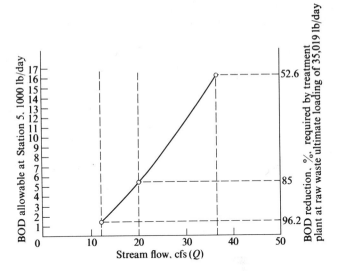

Fig. 15.8 Special design curve for computing treatment plant requirements at 12.4°C and 2 ppm DO at Station 5.

samples under these critical conditions and statistically correlate the data in order to obtain a stream equation. This method is commonly referred to as the Churchill multiple-regression technique. The stream equation represents the line which best fits the data for the conditions under which the samples were collected. Projection of the line beyond this range of conditions is not recommended, but extrapolations to different conditions within the range of existing data can be made with a reasonable degree of certainty. The stream equation for Cayadutta Creek, developed by extensive analysis, can be used to compute the BOD reductions necessary to maintain a certain minimum dissolved-oxygen level at a given temperature. These calculations are given below and in Figs. 15.7 and 15.8.

To find the waste-treatment requirements, we use the stream equation developed during the low-flow* critical period for Cayadutta Creek during the October 1964 study. We obtain

$$Y = 4.9330 + 0.0769 X_1 + 0.2574 X_2 - 0.5826 X_3,$$

when $X_1 = 1000/\text{ppm BOD}$
$X_2 = \text{temperature} = 12.4$ (°C)
$X_3 = 100/\text{cfs} = \dfrac{100}{12}$ (see Table 15.10) $= 8.33$
$Y = $ DO sag from Station 1 to Station 6
$= 9.73 - 2.00 = 7.73$

$$7.73 = 4.9330 + 0.0769 X_1 + 0.2574 (12.4) - 0.5826 (8.33)$$
$$7.73 + 4.85 - 4.9330 - 3.200 = 0.0769 X_1$$
$$12.58 - 8.133 = \frac{4.447}{0.0769} = X_1 = 57.75$$

$$\text{ppm} = \frac{1000}{57.75} = 17.3 \text{ ppm at sag}$$

The BOD values for the seven days of the survey, at Station 5, were 340, 290, 150, 200, 220, 160, and 140 ppm, giving an average of 214 ppm. At Station 6, the values were 330, 240, 140, 180, 160, 70 and 40, with an average of 166 ppm. The percentage decrease in BOD between Stations 5 and 6 was

$$\frac{214 - 116}{214} = \frac{48}{214} = 22.5\%.$$

The BOD (in pounds) being discharged at Station 5 on the seven days of the survey was a total of 142,500 lb (see Table 15.14) and an average of 23,750 lb/day.

Taking $23,750 \times 77.5\% = 18,400$ lb/day as the BOD left at Station 6, with no treatment at 12 cfs, 12.4°C, and an allowable DO deficit at the sag point of 8.65 ppm (10.65−2.0), we obtain an allowable 17.3 ppm BOD at the sag or

$$\frac{17.3 \times 12}{1.54} \times 8.34 = 1122 \text{ lb/day}.$$

Table 15.8 Minimum flow data of measured creek compared with that of Cayadutta Creek.

Year	Minimum daily flow of Kayaderosseras Creek*, cfs	Equivalent minimum daily flow† of Cayadutta Creek (Station 6), cfs
1927	13	8.9
1928	18	12.4
1929	24	16.5
1930	19	13.1
1931	19	13.1
1932	20	13.8
1933	19	13.1
1934	20	13.8
1935	34	23.4
1936	20	13.8
1937	21	14.5
1938	20	13.8
1939	20	13.8
1940	23	15.9
1941	15	10.3
1942	23	15.9
1943	29	19.3
1944	21	14.5
1945	26	17.9
1946	20	13.8
1947	22	15.2
1948	18	12.4
1949	14	9.6
1950	23	15.9
1951	32	22
1952	31	21.4
1953	20	13.8
1954	23	15.9
1955	20	13.8
1956	32	22
1957	19	13.1
1958	20	13.8
1959	18	12.4
1960	25	17.2

*Hydrologically similar to Cayadutta Creek.

†Calculated by dividing the drainage area of Cayadutta Creek (62 mile²) by that of Kayaderosseras Creek (90 mile²) and multiplying by the known rate of flow for the latter, e.g.,

$$\frac{62 \times 13}{90} = 8.9.$$

Table 15.9 A normal probability distribution analysis of data (1927–1960) from Table 15.8.

Flow, cfs	Magnitude (M)	Plotting position*
8.9	1	0.0286
9.6	2	0.0572
10.3	3	0.0858
12.4	4	0.1143
12.4	5	0.1430
12.4	6	0.1715
13.1	7	0.2000
13.1	8	0.2290
13.1	9	0.2570
13.1	10	0.2860
13.8	11	0.3140
13.8	12	0.333
13.8	13	0.371
13.8	14	0.400
13.8	15	0.428
13.8	16	0.458
13.8	17	0.486
13.8	18	0.515
13.8	19	0.544
14.5	20	0.571
14.5	21	0.600
15.2	22	0.629
15.9	23	0.658
15.9	24	0.685
15.9	25	0.715
15.9	26	0.744
16.5	27	0.770
17.2	28	0.800
17.9	29	0.829
19.3	30	0.858
21.4	31	0.887
22.0	32	0.916
22.0	33	0.945
23.4	34	0.974

*Calculated from the formula $M/(N + 1)$, where M = magnitude in decreasing order of drought severity and N = number of values.

Since $23,750 - 0.225 = 18,400$ lb/day, the BOD reduction required is

$$\frac{18,400 - 1122}{18,400} \times 100 = \frac{17,278}{18,400} = 93.8\%.$$

At 12.4°C and the same DO sag (7.73 ppm), but at the increased stream flow of 20 cfs, we obtain an allowable BOD at the sag of

$Y = 4.9330 + 0.0769\, X_1 + 0.2574\, X_2 - 0.5826\, X_3$

$7.73 - 4.9330 - 3.2000 = 0.0769 X_1 - 0.5826\left(\dfrac{100}{20}\right)$

$\dfrac{7.73 + 2.4130 - 8.1330}{0.0769} = X_1 = \dfrac{2.0100}{0.0769} = 26.05$

$\text{ppm} = \dfrac{1000}{26.05} = 38.4 \text{ ppm}$

at 38.4 ppm $\times\, 8.34 \times \dfrac{20}{1.54} = 4155$ lb BOD at Station 6

BOD reduction required $= \dfrac{18,400 - 4155}{18,400} \times 100$

$= \dfrac{14,245}{18,400} \times 100 = 77.5\%$

At the same temperature (12.4°C) and the same DO sag (7.73 ppm), but at the average stream flow at the sag of 36.7 cfs (October 1964 survey), we obtain an allowable BOD at the sag of

$Y = 4.9330 + 0.0769\, X_1 + 0.2574\, X_2 - 0.5826\, X_3$
$7.73 = 4.9330 \times 0.0769\, (X_1) + 0.2574\, (12.4°C)$
$\qquad - 0.5826 \left(\dfrac{100}{36.7}\right)$

$7.730 + 1.590 - 4.9330 - 3.2000 = 0.0769\, X_1$

$\dfrac{9.3200 - 8.1330}{0.0769} = \dfrac{1.1890}{0.0769} = 15.45 = X_1$

$\text{ppm} = \dfrac{1000}{15.45} = 64.8$

at 64.8 ppm $\times\, 8.34 \times \dfrac{36.7}{1.54} = 12,850$ lb BOD at Station 6

BOD reduction required $= \dfrac{18,400 - 12,850}{18,400} \times 100$

$= \dfrac{5,550}{18,400} = 30.2\%$

Figures 15.7 and 15.8 indicate that to maintain 2 ppm of dissolved oxygen (a preselected safe value for this class of stream) at loadings of 23,750 pounds of BOD per day at the bottom of the sag at a temperature of 12.4°C, BOD reductions of 65 to 94 per cent at 23,750 pounds per day loading and 77 to 97 per cent at 35,019 pounds per day would be required for critical stream flows of 12 to 25 cfs.

15.6 Composite Waste Sampling

Waste samples were collected hourly for a 24-hour period from three sources, the Johnstown 30-inch sewer (main), the Johnstown 8-inch sewer (Tynville), and the Gloversville sewage treatment plant, on November 17 and December 3, 1964. Similar samples were collected for a 24-hour period on January 21, 1965, except that the 8-inch Johnstown sewer was eliminated as being relatively insignificant. Weirs were installed in the Johnstown lines to record the total flows from Johnstown as well as from Gloversville. Samples were collected and composited according to the rate of flow at the hour of sampling. A summary of the proportionate pollutional loads and volumes for these three days is shown in Table 15.12. Additional 24-hour composites from each line were collected according to flow and analyzed on February 18, March 30, April 21, and May 6, 1965 (Table 15.13).

Table 15.10 Summary of data required from Cayadutta Creek analyses October 1964 for Churchill method of analysis.

Date	DO, ppm		Drop in DO, ppm (Y)	BOD at sag, ppm	Temp. at sag, °C	Flow at sag, cfs
	Station 1	Station 6				
10/8	10.1	2.2	7.9	330	13	59.4
10/12	9.0	1.8	7.2	240	14	51
10/13	10.0	1.1	8.9	140	12	44
10/14	10.5	4.6	5.9	180	11	39
10/15	9.0	2.9	6.1	160	12	39
10/17	9.5	4.8	4.7	70	12	12.2
10/18	10.0	4.8	5.2	40	13	12.2

Table 15.11 Churchill analysis applied to Cayadutta Creek data.

Date	Dissolved oxygen At Station 1, ppm	Dissolved oxygen At Station 6, ppm	Drop in DO, ppm (Y)	BOD at sag, 1000/ppm (X_1)	Temp. at sag, °C (X_2)	Flow at sag, 100/cfs (X_3)	Y^2
10/8/64	10.1	2.2	7.9	3.03	13	1.68	62.41
10/12/64	9.0	1.8	7.2	4.17	14	1.96	51.84
10/13/64	10.0	1.1	8.9	7.14	12	2.27	79.21
10/14/64	10.5	4.6	5.9	5.56	11	2.56	34.81
10/15/64	9.0	2.9	6.1	6.25	12	2.56	37.21
10/17/64	9.5	4.8	4.7	14.29	12	8.20	22.09
10/18/64	10.0	4.8	5.2	25.00	13	8.20	27.04
Totals			45.9	65.44	87	27.43	314.61
Means			$\bar{Y}$	$\bar{X}_1$	$\bar{X}_2$	$\bar{X}_3$	
			6.56	9.35	12.4	3.92	$n\bar{Y}^2$
Corrected items*							301.21
Corrected totals							13.40

*n = number of samples (7).

Table 15.12 Summary of 24-hour sampling results.

Source	Date	BOD load lb/day	BOD load % in peak period*	BOD load % of daily total	Flow mgd	Flow Ratio of peak period to daily avg.	Flow % of daily total	Suspended solids lb/day	Suspended solids % in peak period	Suspended solids % of daily total
Gloversville	11/17/64	13,350	45.2	67.5	3.80	1.38	66	27,550	80.8	87.5
	12/3/64	13,350	50.2	56.3	3.67	1.34	50	11,100	61.6	53
	1/21/65	23,400	41.1	63.1	3.96	1.35	60.5	21,600	25.8	63
Johnstown Main	11/17/64	6,400	44.8	32.2	1.92	1.55	33.4	4,100	57.7	12.5
	12/3/64	15,000	27.5	42	3.60	1.46	49	18,400	42.0	45
	12/1/65	13,600	43.2	36.9	2.58	1.33	39.5	12,800	37.0	37
Tynville	11/17/64	102.4	25.4	0.53	0.0449	1.24	0.6	32.9	56	
	12/3/64	223	25.2	1.4	0.0809	0.925	1	213	17.9	2.0
Totals	11/17/64	19,852			5.7649			31,683		
	12/3/64	28,573			7.3509			29,713		
	1/21/65	37,000			6.54			34,400		

*The peak period was taken to be 6.00 a.m. to 12 noon.

15.7 Composite Waste Analyses

The hourly data reveal several significant findings.

a) Slugs, which can be defined for this situation as instantaneous discharge of high volumes of waste, concentrated acid or alkali, or BOD, are apparently not a major problem. The flow increases by about 100 per cent of the daily average, for a period of about 12 hours during the daytime. The pH becomes alka-

Table 15.11 (continued)

YX_1	YX_2	YX_3	X_1^2	X_1X_2	X_1X_3	X_2^2	X_2X_3	X_3^2
23.94	102.7	13.27	9.18	39.39	5.09	169	21.84	2.82
30.02	100.8	14.12	17.39	58.38	8.17	196	27.44	3.84
63.55	106.8	20.20	50.98	85.68	16.21	144	27.24	5.15
32.80	64.9	15.10	30.91	61.16	14.23	121	28.16	6.55
38.13	73.2	15.62	39.06	75.00	16.00	144	30.72	6.55
67.16	56.4	38.54	204.20	171.48	117.48	144	98.40	67.24
130.00	67.6	42.64	625.00	325.00	205.00	169	106.60	67.24
385.60	572.4	159.49	976.72	816.00	381.18	1087	340.40	159.39

$n\overline{YX}_1$	$n\overline{YX}_2$	$n\overline{YX}_3$	$n\overline{X}_1^2$	$n\overline{X}_1\overline{X}_2$	$n\overline{X}_1\overline{X}_3$	$n\overline{X}_2^2$	$n\overline{X}_2\overline{X}_3$	$n\overline{X}_3^2$
429.38	569.38	180.04	611.94	811.51	256.56	1076.32	340.26	107.56
−43.78	3.02	−20.55	364.78	4.58	125.62	10.68	0.14	51.83

line (8 to 10) during the same period but returns to normal (7 to 8) during the twelve night hours. The BOD varies considerably during both day and night but is generally high and is confined to a range of 300 to 700 ppm from 6 a.m. until midnight. There is little pattern of discharge of BOD and no apparent practical gain as far as BOD is concerned from separate equalization basins. Since the flow and pH are largely higher during the entire daytime period, equalization to level out these factors would require very large basins. The cost of such units and the potential danger of septicity seem to the author to far outweigh the benefits derived from levelling the flow and pH. In this instance, it seems that the great numbers of varied tanneries themselves contribute to equalization of waste simply by their diversity.

b) The total BOD loads and flows given in Tables 15.12 and 15.13 can be examined more easily by referring to Table 15.14. The total flow measured averages 6.724 million gallons per day and contains an average of 23,442 pounds of 5-day, 20°C BOD and approximately 20,650 pounds of suspended solids. These values do not include any flows or loads not connected to the Gloversville sewage treatment plant or the 30-inch sewer outfall in Johnstown. The 8-inch sewer outfall in Johnstown, although measured and sampled at the beginning, contains less than 1 per cent of the total volume or BOD load and can therefore be considered insignificant in these surveys.

c) The maximum variations in flow and load from day to day were found to be 18 per cent from the average flow and 22 per cent from the average BOD. These variations are considered well within normal values and tend to substantiate the use of the average daily values given under the previous section in designing waste-treatment facilities.

Industrial production records during the sampling days (Tables 15.15 and 15.16) demonstrated that all major industries connected to the two major sewer systems were in operation at almost full capacity during these days. This provides some measure of assurance in using the average flow and BOD values obtained during this period. Table 15.17 shows industry's percentage of the total measured flow on these days. This reveals an industrial waste problem of about 50 per cent by volume when industry is operating near its rated capacity.

Table 15.13 Composite analyses (24-hr) of Gloversville and Johnstown waste water.

Characteristic	Feb. 18–19, 1965 Gloversville	Feb. 18–19, 1965 Johnstown*	March 30, 1965 Gloversville	March 30, 1965 Johnstown 32	April 21, 1965 Gloversville	April 21, 1965 Johnstown	May 6, 1965 Gloversville	May 6, 1965 Johnstown
Flow, mgd								
24-hr average	3.41	3.05	3.05	4.88	3.63	2.84	4.3	2.07
6 a.m. to 12 noon average			4.73	6.33	4.29	3.00		
6 a.m. to 2 p.m. average								
pH	9.4	8.3					5.9	2.92
Total solids, ppm	3130	2450	2430	1970	2542	1840	3120	2962
Suspended solids, ppm	258	81	265	145	418	213	475	322
Volatile suspended solids, ppm			196	96	265	135	305	250
BOD (5-day, 20°C), ppm	405	385	300–435	285–330	371–435	371–386	485–520	540–585
			(367)†	(307)	(403)	(378)	(502)	(562)
BOD, lb/day	11,500	9800						
Settleable solids, ml/liter			20.0	5.5	13.0	4.5	14.0	16.0
On supernatant								
Suspended solids, ppm			97	70	128	86	172	135
Volatile solids, ppm			78	52	92	60	120	110
BOD, ppm			95–180	225–355	266–281	326–386	210–300	375–405
			(285–330)					
8-hr readings								
Suspended solids, ppm							690	423
Volatile solids, ppm							420	310
BOD, ppm							405–495	405–435
							(450)	(420)
Total ash, ppm	2250	1570						
Total volatile, ppm	880	880						
Suspended ash, ppm	140	48						
Suspended volatile, ppm	118	33						
Suspended volatile, %	46	41						
Analysis of settled 2-hr sludge								
Total solids, %	1.54	0.88						
Total ash, %	34.4	37.5						
Total organics, %	63.6	62.5						

*30-inch sewer.
†Average values are given in parentheses.

Table 15.14 Summary of total loads for treatment.

Date	Total flow, mgd	Total BOD, lb/day	Total suspended solids, lb/day
11/17/64	5.765	19,852	31,683
12/3/64	7.351	28,573	29,713
1/21/65*	6.540	37,000	34,400
2/18/65	6.460	21,300	9,405
3/30/65	7.930	21,925	12,700
4/21/65	6.470	21,150	17,700
5/6/65	6.370	27,850	22,700
Average	6.724	23,442	20,650

*Since an unusually large percentage of deerskin was tanned, this day was not considered typical of even maximum normal operation and therefore it was excluded from the average.

15.8 Laboratory Pilot-Plant Studies

To form more definite conclusions on the proper units to be included in the waste-treatment plant, certain small-scale laboratory studies were necessary, including sludge digestion and activated-sludge treatment.

Sludge digestion. A mixture of primary and secondary settled sludge was collected from the settling basins at the Gloversville treatment plant. A pilot digester, consisting of a glass container and a gas-collecting system maintained at 37°C, was set up in a private laboratory in Johnstown. The raw sludge sample selected was analyzed for organic matter at the start of the "batch" digestion period and again after 50 days of digestion, and gas volume measured almost daily (see Table 15. 18).

Although this was a batch-type experiment, over the 50-day period 9.07 cubic feet of gas were produced per pound of volatile matter destroyed. Greater amounts of gas may be expected from a continuous digestion operation maintained at optimum environmental conditions. In this experiment, more gas would have evolved after an increased digestion period, but the rate of gas production did slow down considerably after 50 days. Normal gas production for sewage sludge is about 15 cubic feet per pound of organic matter destroyed. Digestion experiments on a continuous basis and over a longer period would be needed to find whether an accumulated toxic effect exists. However, Vrooman and Ehle [16] reported successful digestion of this waste sludge.

Activated-sludge treatment. The apparatus used in the study consisted of an aeration tank with two mixers and three separate air-diffuser tubes fitted with porous stones. Air flow (cubic feet of air/hour) was measured by a previously calibrated rotameter. The tank was 23.5 in. long, 8.5 in. wide, and filled to a depth that gave an aeration volume of 6 gal.

Since settling was expected to be an integral part of a biological treatment plant of this type, various mixtures of settled tannery waste (1:1 mixture of beamhouse and tanyard wastes) and settled domestic sewage were added to the aeration tank in a semibatch procedure to simulate continuous operation as closely as possible. The standard average aeration period of 6 hours was used and the waste mixture was added in three increments of 2 gallons each at 2-hour intervals. The tank contents (6 gallons of a mixture of tannery wastes and domestic sewage) were first settled for a 1-hour period; then 2 gallons of the supernatant were siphoned off and 2 gallons of the waste mixture added; the tank contents were aerated for 2 hours and settled for 15 minutes, 2 gallons of supernatant were withdrawn, and 2 more gallons of waste mixture added. This procedure resulted in the addition of 6 gallons of waste mixture in a period of 4 hours for a total aeration time of 8 hours and an average aeration period of 6 hours as shown below:

Time (hr)	Volume added (gal)	Aeration time (hr)
0	2	8
2	2	6
4	2	4
		Average: 6

The results obtained are summarized in Table 15.19 and shown graphically in Fig. 15.9. Each loading represents about one week's aeration data with samples being taken for analysis several times during this week of adaptation and acclimation.

Laboratory experiments verified that 65 to 75 per cent of this waste would degrade biologically even when loaded at the high rate of 95 to 115 pounds of BOD per 1000 cubic feet of aerator (see Fig. 15.9). Higher BOD reductions (75 to 85 per cent) were obtained with lower BOD loadings (60 to 82 pounds per 1000 ft^3) and increased dilution of the tannery waste with domestic sewage.

These studies showed that the activated-sludge

Table 15.15 Industrial production during sampling days.

Company	November 17, 1964		December 3, 1964		January 21, 1965	
	Water used, gpd	% production*	Water used, gpd	% production	Water used, gpd	% production
Wood and Hyde Leather	200,000	100	225,000	100	225,000	100
Filmer Leather	41,310	50	41,310	50	41,300	50
Twin City Leather	120,000	50	120,000	100	120,000	100
Wilson Tanning	49,920	80	54,337	80	38,000	50
Leavitt-Berner Tanning	151,700	100	151,700	100	151,700	100
F. Rulison & Sons	76,300	100	76,300	100	76,300	100
Peerless Tanning	24,000	$33\frac{1}{3}$	48,000	$66\frac{2}{3}$	48,000	$66\frac{2}{3}$
Karg Bros.	266,000	68	254,000	65	386,000	90–95
Decca Records†	10,807	100	10,807	100	10,807	100
U.S. Rabbitt Tanning Co.†	10,000	100	7,000	75	7,000	75
Gloversville Continental Mill†	200,000	66	190,000	100	190,000	100
Independent Leather	134,000	60	107,000	40	161,000	60
Liberty Dressing	78,950	70–75			120,803	60
G. Levor		85		80		
Framglo Tanners (1)	500,000	100	500,000	100	500,000	100
Framglo Tanners (2)		80		80		80
Rebel Dye†	26,250	10	30,500	10	24,500	9
Lee Dyeing† (Johnstown)	175,000	40	1,000	0	1,000	0
Adirondack Finishing†	450,000	80	450,000	80	500,000	90
Crown Finishing (Maranco Leather)	42,352	100	42,000	100	42,352	100
Simco Leather	61,300	100	61,000	100	61,000	100
Johnstown Tanning	60,000	70	35,000	60	10,000	40
Napatan	21,072	100	21,000	80	21,000	100
Ellithorp Tanning	105,000	100	110,000	100	110,000	100
Johnstown Knitting†			100,000	45	100,000	75
Gloversville Leather						
Riss Tanning						
Industrial total, gpd	2,803,961		2,735,954		2,845,662	
Total flow, mgd	5.7649		7.3509		6.54	
Total BOD, lb/day	19,852		28,573		37,000	
Industrial portion of total water flow, %	48.7		37.2		43.4	

*Percentage of plant's total productive capacity.
†Figures are based upon yearly consumption (average figure).

process or a modification of it could be utilized successfully in the overall treatment of the combined tannery and sewage wastes. Larger prototype field experiments would disclose whether these results can be projected directly to full-scale operation.

The present Gloversville treatment plant has experienced much difficulty due to its overloaded condition. However, from the best records available, when all the flow units of the plant were operating, about 58 to 60 per cent of the BOD was removed. This reduction in BOD also shows that biological degradation by trickling filtration is possible with this waste under full-scale field conditions. The exact degree of this oxidation could be determined more easily in a properly designed and operated field pilot plant.

Table 15.16 Water consumption related to production percentage.*

Industry	Water consumption, gal		Production percentage				
			Beamhouse		Tanning		
	From meter reading	From other sources	Type of skin	Rated potential	Compared 1/21/65	Operation potential	Type
Leather tanneries							
Wood and Hyde Leather	150,000	75,000	Burn – sheep	100	100	100	Combination
Filmer Leather	90,000	Pond (in future)	Horse, cow, jacks, deer	100	75	80	Combination
Twin City Leather	10,028		Sheep	100	100	100	Combination
Wilson Tanning	25,215		Sheep, goat, deer	0	0	$33\frac{1}{3}$	Combination
Leavitt-Berner Tanning (not contributing to sewage treatment plant)	140,000		Sheep, goat, deer	100	100	100	Combination
F. Rulison & Sons	60,150		Horse, cow	80	80	80	Combination
Peerless Tanning	21,766					75	Chrome
Karg Bros.	400,000		Pig, deer	100	100	100	Chrome
Independent Leather	230,000					100	Combination
Liberty Dressing	85,582		Goat, calf			45	Combination
G. Levor	Drinking water only	600,000	Pig, calf, goat	80		80	Combination
Framglo Tanners (1)	Drinking water only	400,000	Sheep, deer			100	
Framglo Tanners (2)	Drinking water only	300,000		80		None	
Crown Finishing (Maranco Leather)	26,465						
Simco Leather	45,030					100	Chrome, some comb. & veg.
Johnstown Tanning (1)	11,000					10	
(2)	0						
Ellithorp Tanning	248,000	From Levor 152,000				90	Chr. & comb.
Gloversville Leather	0					25	Chr. & comb.
Riss Tanning	27,970					$33\frac{1}{3}$	Chr. & comb.
Nonleather industries							
Rebel Dye	61,000 (8,145 ft³)			20			
Adirondack Finishing	474,310			80			
Lee Dyeing (Johnstown)	8,800 (1,175 ft³)			0			
Gloversville Continental Mill	208,000 gal			$66\frac{2}{3}$		Anticipate 330,000 gal due to dyeing technique but not necessarily increased production	
Johnstown Knitting	150,200 (20,178 ft³)			75			
Diane Knitting	10,590 ft³			75			
Decca Records	17,745 (2,366 ft³)			75			
Mohawk Cabinet				80			
U.S. Rabbitt Tanning Co.	1,000			0		Water running in tubs to keep them soaked. Should be disregarded	
Total industrial flow, gal	4,019,262						
Total flow, gal	6,460,000						
Total BOD	21,300 lb/day						
Industrial flow (% of total)	62.5%						

*Data are for the 24-hr period from 6 a.m. on February 18 to 6 a.m. on February 19, 1965.

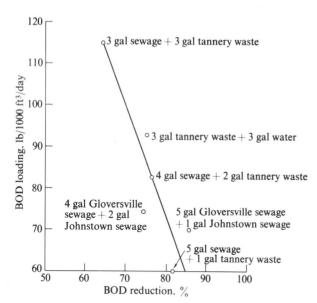

Fig. 15.9 Activated-sludge treatment: BOD reduction related to BOD loading.

Table 15.17 Industrial waste flow.

Date	Total flow, mgd	Industrial flow (estimated by survey), mgd	Industrial portion of total flow, %
11/17/64	5.765	2.804	48.7
12/3/64	7.351	2.736	37.2
1/21/65	6.540	2.846	43.4
2/18/65	6.460	4.019	62.5
Average			ca.48

15.9 Literature Survey

A study was also made of previous research work or reported practice dealing with combined treatment of domestic sewage and tannery waste by the activated-sludge process. The reports of Chase and Kahn [2], Braunschweig [1], Jansky [6], Thebaraj et al. [15], Snook [14], Hubbel [5], Pauschardt and Furkert [12], Kubelka [8, 9], Fales [3], Kalibina [7], Furkert [4], and Mausner [10] provide some evidence that the activated-sludge treatment process is feasible for tannery–sewage waste mixtures. This review of previous work tends to substantiate the biological pilot studies described above. Most of this reported work, however, has been on a research or pilot-plant basis or of a more sewage-diluted waste. There is a definite lack of reviews of full-scale biological treatment used on tannery and sewage waste mixed 50:50 (by volume).

Braunschweig [1] disproves the notion that chromium in the newer tannery processes interferes with aerobic biological treatment. Jansky's work [6] is typical of that of the more recent supporters of activated sludge as a treatment method. Thebaraj et al. [15] found that all aerobic biological systems were effective and that the choice depended upon several economic and practical considerations. Fales [3] much earlier was of the same opinion as Thebaraj but pointed out the higher operating costs of activated sludge compared to trickling filtration. Furkert [4] also verifies our digestion studies and noted that except for the high H_2S content of the digester gas the composition of the gas is normal.

15.10 Conclusions from Study

The following specific conclusions and recommendations were made as a result of this study:

1. The stream survey was instrumental in providing evidence that secondary treatment of the combined industrial and sanitary wastes of the area is required and that 65 to 94 per cent BOD reduction will be needed depending upon the dilution available in the stream. Use of the curves plotted in Figs. 15.6 and 15.7 would allow for a more precise selection of BOD reduction required for specific critical stream flows.

2. The existing dry-weather flow to be treated averages 6.724 million gallons per day with peaks of two to three times this rate; about 50 per cent of this flow originates from the industries of the area.

3. The combined area waste contains a daily average of 23,442 pounds of 5-day, 20°C BOD and 20,650 pounds of suspended solids. These loadings are affected considerably by the type of skin tanned, deerskin being an especially significant contributor of high BOD and solids loads.

4. Laboratory pilot studies demonstrated that the conventional activated-sludge treatment process is capable of reducing the BOD of the combined waste from 65 to 85 per cent (depending primarily upon the organic loading) at loadings ranging from 60 to 115 pounds of BOD per 1000 cubic feet of aerator capacity.

15.10

Table 15.18 Sludge digestion (laboratory study).*

Accumulated gas produced, cc	Days of digestion at 37° C
250	1
290	2
460	3
640	4
730	5
850	6
940	7
990	8
1040	9
1140	10
1140	11
1150	12
1190	13
1240	14
1290	15
1340	16
1380	17
1420	18
1480	19
1540	20
1600	21
1650	22
1700	23
1820	24
1900	25
1940	26
1980	27
1980	28
2010	29
2020	30
2030	33
2040	36
2050	40
2060	41
2080	42
2090	43
2110	44
2150	47
2170	49
2200	50

*The results were analyzed as follows:
organic matter (raw sludge)	= 5.4533 gm
organic matter (after 50 days)	= 1.5542 gm
loss of organic matter	= 3.8991 gm (71.2%)
gas produced:	
total	= 2200 cc
per gram of organic matter destroyed	= 567 cc
per gram of volatile matter added	= 403 cc

CONCLUSIONS FROM STUDY

5. A digestion batch experiment yielded about 9 cubic feet of gas per pound of volatile matter destroyed and effected a 71 per cent reduction in organic matter.

6. A literature study confirmed the findings of the laboratory results—that the combined wastes of this type are amenable to biological oxidation.

7. Because of the unique nature of the volume and characteristics of the tannery–sewage waste mixture, as well as the size and cost of the project, field prototype studies should precede full-scale plant construction.

8. There should be additional laboratory research on development of improved methods of aerobic biological treatment of tannery wastes to allow for greater BOD reduction at higher BOD loadings.

The decision reached as a result of this study was that the cost was too high, the risk too great, and previous reported experience too slight for full-scale biological treatment to be recommended at the time. A prototype in the field—preferably at the site of the Gloversville treatment plant—was to be built and operated for about 6 months to obtain detailed data for the final design and to obtain greater certainty that the earlier findings were valid. This prototype should contain both trickling-filtration and activated-sludge units (as well as provision for its modification). It should also allow for experimentation with series and parallel operation of the units and both diffused and mechanical aeration. Some sludge-digestion studies should be carried out over the entire period.

A schematic drawing of this field prototype is shown in Fig. 15.10. It consists of two sets of screens in series ($\frac{1}{2}$-inch openings followed by $\frac{1}{4}$-inch openings), pump, primary settling, trickling filter, aeration, and final settling. The plant began operation in early August 1965 and sampling was begun on August 16. Table 15.20 gives data for the first 7 weeks of operation. Table 15.21 shows how the prototype operating results influenced the final design parameters.

To convert them from the metric system:

$2200 \times 0.000353 = 0.0777$ ft³ of gas produced (total)

$\frac{3.8991}{454} = 0.00858$ lb of volatile (organic) matter destroyed

$\frac{0.00777}{0.00858} = 9.07$ ft³ of gas per pound of organic matter destroyed

15.11 Overall Planning Study Conclusions

The conclusions reached through data collection, pilot and prototype plant studies, engineering evaluations, and reviews of design and operational experiences in major municipal sewage-treatment plants treating large amounts of tannery waste may be summarized as follows.

1. *Degree of treatment.* Primary treatment by settling, followed by secondary treatment through biological processes, is required to meet New York State standards for plant effluent quality that may be accepted by the Cayadutta Creek under conditions of minimum dissolved oxygen content (at times of low flow and high temperature). The efficiency of treatment units and processes must be high, with an overall plant removal of approximately 85 per cent of the incoming BOD.

2. *Pretreatment at mills.*
   a) Tanneries should remove fleshings, hair, hide pieces, and trimmings to make discharges transportable in gravity sewers. This can be accomplished by means found most efficient and economical, including primary settling tanks and/or mechanical screening. Animal greases plus petroleum solvents should be removed at the tanneries.

   b) The glue factory should remove settleable solids to make discharges transportable in gravity sewers. This can be accomplished by means found most efficient and economical, including primary settling tanks and/or mechanical screening.

3. *Sewage-treatment plant: processes (general).* The liquid wastes treatment will include (a) pretreatment by mechanical screening, grit removal, preaeration of grease for removal in primary treatment, and pretreatment for pH control and chemical coagulation (in future); (b) primary treatment by settling; and (c) secondary treatment by trickling filters (high-rate roughing filters), then through activated sludge, followed by settling, with provisions for chemical for more complete solids removal in the future.

Table 15.19 Activated-sludge pilot laboratory studies.

Waste treated		BOD loading, lb/1000 ft³	Suspended solids under aeration, ppm	Air rate, ft³/lb BOD removed	BOD		
Origin	Quantity, gal				Influent, ppm	Effluent, ppm	Reduction, %
Sewage + Tannery waste mixture	5 1	60	2330	2450	239	44	81.6
Sewage + Tannery waste mixture	4 2	82.8	2221	1900	331	78	76.5
Sewage + Tannery waste mixture	3 3	114.7	2768	1735	459	165	64.1
Sewage wastes Gloversville Johnstown	 5 1	70	3386	2000	280	39	86
Sewage wastes Gloversville Johnstown no. 1 Johnstown no. 2	 4.015 2.112 0.044	73.8	2508	2116	295	68	76.9
Tannery waste mixture + Tap water	3 3	93.0	2646	1070	374	91	75.6

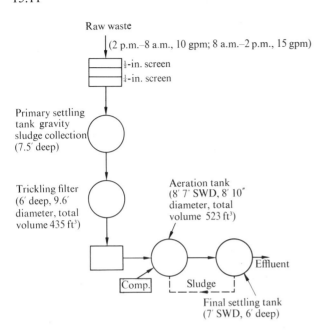

Fig. 15.10 Field prototype of the Gloversville–Johnstown joint treatment plant.

precipitation for more complete solids removal in the future, with discharge to the Cayadutta Creek where further treatment is by dilution and the oxygenation capacity of the stream.

Sludge treatment and disposal will include high-rate digestion with sludge-gas utilization followed by (a) dewatering by lagoons and disposal by approved landfill methods; (b) dewatering by vacuum filters and disposal by approved landfill methods; or (c) dewatering by vacuum filters, incineration (multiple hearth) and disposal of ash by landfill.

4. *Sewage treatment plant: processes (recommendations).* The following treatment units and processes are specifically recommended and should be included in preliminary planning.

a) Three mechanically cleaned bar-rack screens to remove large debris from the flow

b) Two circular grit-removal units designed on surface overflow rate to remove grit and sand prior to primary settling (separation and washing of settled grit and organic matter by two hydrocyclone classifying devices). Prior to final design, consideration to be given to utilization of an aerated grit-removal unit

c) Disposal of screenings and grit in sanitary landfill

d) Grease removal by skimming in the primary settling tanks; grease flotation facilitated by aeration following or incorporated with grit-removal unit and immediately preceding the primary settling tanks

e) Possible future chemical application in the aeration structure for pH control and introduction of chemicals to aid precipitation of wastes, for short periods, at times of exceptionally low flows in Cayadutta Creek

f) Possible future addition of coagulating chemicals in the flow to the secondary clarifiers for "polishing" effluent and/or in the discharge from the secondary clarifiers for control of algal nutrients, if found necessary (structure provided for addition of coagulating chemicals in inflow)

g) Six rectangular primary settling tanks with mechanical sludge collectors and scum skimmers

h) Biological secondary treatment in two stages by two high-rate (roughing) filters, with stone or plastic media and rotating arm distributors, and activated-sludge treatment (in multiple units) in two sections, with mechanical aeration units directly powered by electric motors

i) Two circular secondary settling tanks with sludge- and scum-collection mechanisms, sludge collectors to be of the "vacuum cleaner" type

j) High-rate digestion provided through a primary digester followed by a secondary digester; floating covers on both digesters with gas-collection and holder facilities; gas utilization for heating of sludge and buildings; gas recirculation mixing provided in both digesters, with possible operation of either digester as primary

k) Dewatering of digested sludge and disposal of sludge cake by approved landfill methods

Final recommendations and determinations on sludge dewatering and disposal must consider the net annual costs, reflecting capital costs and operation, the physical problems involved in handling sludge as amounts increase over the years, and the future utilization of the land considered for landfill. Certain of these evaluations can be made from the engineering and cost points of view. However, the Cities of Gloversville and Johnstown and the Town of Johnstown, as well as the New York State Department of Health, must also consider future land use and future

Table 15.20 Prototype operating data.

Date	Raw waste			Primary effluent			Trickling-filter effluent			Final effluent			Under aeration	
	BOD, ppm	pH	SS,* ppm	BOD, ppm	pH	SS, ppm	BOD, ppm	pH	SS, ppm	BOD, ppm	pH	SS, ppm	SS, ppm	pH
8/16/65	655	7.6	603	448	7.3	248	353	7.3	251	93	7.2	139	1430	7.2
8/17/65	468	8.4	528	373	8.3	256	241	7.6	202	81	7.2	44	1603	7.3
8/18/65	468	9.1	578	380	9.0	170	230	8.5	262	73	7.4	75		
8/19/65	563	8.0	369	390	8.0	211	268	7.6	187	85	7.2	78	1850	7.3
8/20/65	493	8.5		408	8.7	281	256	8.3	166	73	7.4	84	2631	7.6
8/23/65	443	7.0	382	433	7.4	145	316	7.5	243	72	7.5	43	2313	7.4
8/24/65	555	8.1	393	425	8.1	184	330	7.8	250	90	7.4	89	2066	7.5
8/25/65	370	7.9	392	313	7.8	169	398	7.8	394	79	7.4	197	2487	7.5
8/26/65	408	8.3	454	360	7.8	243	231	7.5	168	102	7.3	127	2788	7.2
8/27/65	273	8.8	337	308	8.5	135	215	8.1	155	19	7.6	53		
8/30/65	298	7.9	347	210	8.1	111	124	7.5	116	25	7.4	47	1848	7.6
8/31/65	250	7.9	379	230	7.9	230	163	7.8	214	24	7.5	67	1934	7.4
9/1/65	423	7.8	460	260	7.9	292	256	7.8	260	54	7.5	160	854	7.6
9/2/65	483	8.6	550	408	8.4	350	435	8.4	360	96	7.5	166	2243	7.8
9/3/65	455	7.9	431	315	7.9	311	235	8.0	205	141	7.5	216	2775	7.6
9/9/65	563	8.1	486	418	8.2	82	275	7.9	87	76	7.4	264	2430	7.9
9/10/65	563	7.9	262	413	7.4	110	290	7.4	137	51	7.1	218	2500	7.4
9/13/65	563	8.4	329	453	8.1	154	290	8.2	142	112	7.4	180	2219	7.6
9/14/65	628	8.7	392	538	8.6	240	351	7.6	210	162	7.4	366	1899	7.4
9/15/65	658	7.6	356	568	7.6	257	349	7.5	178	69	7.0	22	2653	7.2
9/16/65	623	9.0	235	593	8.5	108	368	8.5	105	62	7.5	19	3025	7.6
9/17/65	635	7.6	284	460	7.6	60	348	7.5	96	75	7.1	44	3226	7.5
9/21/65	720	7.9	425	425	7.8	199	215	7.6	178	135	7.4	130	2300	7.5
9/22/65	530	7.2	435	450	7.0	167	204	7.0	166	55	7.0	338	2892	6.9
9/23/65	490	8.0	349	393	8.0	173	183	7.6	224	33	7.3	51	2790	7.5
9/24/65	730	8.7	288	523	8.5	192	170	8.3	115	111	7.5	140	2490	7.8
9/27/65	543	8.5	306	383	8.4	247	233	7.8	240	117	7.4	93	2326	7.6
9/28/65	480	8.1	403	420	8.0	197	289	7.8	193	127	7.6	124	2361	7.8
9/29/65	516		315	463		177	258		180	126		141	2664	
9/30/65	650	8.1	307	490	7.6	145	344	7.5	195	192	7.4	157	2096	7.4
10/1/65	435	7.8	312	363	7.6	162	280	7.6	229	158	7.2	241	1495	7.4

*SS = suspended solids.

disposal of both sewage-treatment-plant wastes and municipal refuse. As a guide in the financial comparisons, the following is an estimate of the approximate net annual costs (capital plus operational) for the three possible methods.

A. Dewatering by lagoons and disposal by approved landfill methods $20,000

B. Dewatering by vacuum filters and disposal by approved landfill methods $50,000

C. Dewatering by vacuum filters, incineration (multiple-hearth), and disposal of ash by landfill $65,000

Dewatering at Fond du Lac, Wisconsin, was formerly achieved by vacuum filters. To reduce high de-

15.11 OVERALL PLANNING STUDY CONCLUSIONS

Table 15.21 Prototype operating results and design parameters.

Unit and effect	Prototype results	Design parameters	Comment
Primary settling tank Suspended solids removal, % Surface settling rate 5-day BOD removal	49(75)* 390(330) 24(39)	60 800 30	Prototype tank construction and inherent limitations in small tanks resulted in lower settling efficiencies. Better results are expected in full scale tanks with scum- and sludge-removal facilities and improved hydraulic characteristics. Additional settling-tank efficiency could be obtained by using flocculating agents if needed.
Roughing filter, with loading of 150 lb of $BOD_5/1000$ ft^3	33% BOD_5 removal	30% BOD_5 removal	Performance of roughing filter established by test.
Aeration tank Process loading of 0.26 lb BOD/lb MLSS† (8-hr peak) Process loading of 0.4 lb BOD/lb MLSS (24-hr average)	81% BOD_5 removal 77% BOD_5 removal	81% BOD_5 removal 77% BOD_5 removal	Performance of activated sludge system established by test.
Aeration tank (oxygen requirements, lb/day)	0.7 lb BOD_{5R} + 0.02 lb MLSS	1 lb O_2/lb BOD_5 removed/day	Aerator capacity designed for mean peak 8-hr BOD loading with 25% present safety factor. Oxygen transfer ratio α to be determined in laboratory prior to final specifications on aerators.
Secondary settling tanks, surface settling rate	760 gal/ft^2/day	600 gal/ft^2/day (at peak 8-hr rate of 13.1 mgd)	Selection of design overflow rates not on basis of pilot plant results. Inclusion of skimming devices on secondary settling tanks due to experience with pilot plant.
Flotation (thickening of waste-activated sludge)	Waste sludge of less than 0.5%. Solids thickened to 5% or greater at loadings greater than 2 lb/ft^2/hr	Design loading: 2 lb/ft^2/hr	Review of prototype data indicates that the design loading is suitable and that this loading should be achieved without the use of chemical conditioning.
Sludge digestion tank	Digestion studies did not develop digestion rate curves.	Displacement time Primary, 25 days Secondary, 25 days	Digestion studies were not conclusive.

*The data in parentheses give the results achieved by using polymer.

†MLSS = mixed liquor suspended solids.

watering costs, vacuum filtration was abandoned and replaced by evaporation and drainage in lagoons. Dewatered sludge removed from the lagoons is disposed of in landfill. This technique has been in operation since 1962. Sewage-treatment-plant loadings at Fond du Lac are only slightly less than those anticipated in our Gloversville–Johnstown design (6.4 million gallons per day including 3 million gallons per

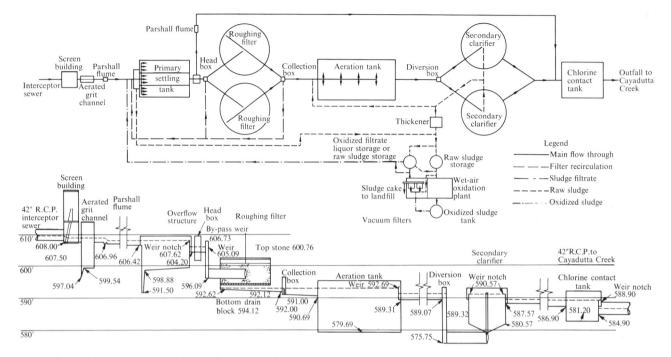

Fig. 15.11 Line diagram and hydraulic profile of the Gloversville–Johnstown joint waste-water treatment plant. (Courtesy Morrell Vrooman Engineers.)

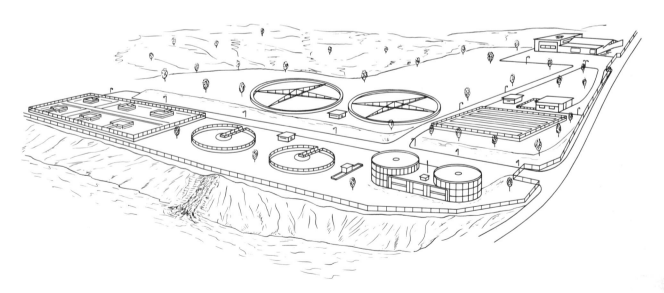

Fig. 15.12 General view of the Gloversville–Johnstown joint treatment plant. (Courtesy Morrell Vrooman Engineers.)

day of tannery wastes). Fond du Lac suspended solids are 173,000 pounds per week (compared to Gloversville–Johnstown 195,578 pounds per week).

Based on the above operational experience and consideration of the low annual costs compared to dewatering by vacuum filtration, dewatering by lagooning is recommended. In accordance with the requirements of the New York State Department of Health, the alternative to dewatering of the digested sludge by lagooning would be dewatering by vacuum filtration. For both methods, ultimate disposal would be by landfill. Both primary and secondary treatment units are shown schematically in Fig. 15.11 and in a general view in Fig. 15.12.

15.12 Solids Handling

The estimated quantity of raw solids to be handled at the plant is 170,000 pounds per week, of which approximately 70 per cent are of industrial origin, primarily tannery wastes.

A review of existing secondary waste-water treatment plants in the United States and Canada handling a large percentage of solids disclosed four plants that were treating tannery effluents in combination with municipal waste-waters. All four plants utilize digestion. Two dispose of the digested solids in liquid form on wastelands using tank trucks, one lagoons the digested solids, and one uses vacuum filteration, drying, and landfill. All of the plants are able to digest the solids effectively.

The problems with digestion of tannery–municipal solids have not been primarily chemical or biological but physical. Hair and scum have caused serious problems in the digesters themselves and in digestion-tank appurtenances. Extensive pretreatment, including fine screens, have been necessary at some plants to reduce such problems. The screens have in some cases introduced another problem, blinding.

The cities of Gloversville and Johnstown are surrounded by a rural area providing land for landfill of the final residue from the sludge-handling system. Both cities operate refuse landfill operations and own large areas of land designated for future landfill use. Therefore, landfill of the dewatered solids from the waste-water treatment plant could be accomplished.

A review of construction and operational costs indicated that digestion and dewatering in lagoons would provide the most economical solution to conditioning the solids. This solution was discussed extensively with the regulatory agencies and with the Gloversville–Johnstown Joint Sewer Board, who have responsibility for administering the project, but was eventually eliminated.

There was considerable interest in the wet-air oxidation system being used to condition and to destroy organic solids. This system was viewed as capable of treating industrial solids without possible upset by the changing chemistry of the leather industry. The Zimpro Division of the Sterling Drug Company had considerable experience in handling similar solids at South Milwaukee and in Kempen, Germany. The wet-air process offered the added advantage of producing a solid that was readily filterable. The filterability of either raw or digested solids has been a potentially troublesome and expensive feature of the operation.

The sludge-handling process finally selected was a low-pressure (300 psi), wet-air oxidation plant, vacuum filters for dewatering, and ultimate disposal of solids in landfill. The wet-air plant as designed will reduce the nonsoluble organic solids by 40 to 50 per cent. The high BOD filterate will be pumped to a holding tank and discharged to the head of the plant during low BOD load periods. This plant will have the capacity to handle the weekly solids loading in 5 days on a 16-hour schedule or in $3\frac{1}{2}$ days on a 24-hour schedule. It is basically one unit with several key items of equipment duplicated. The two vacuum filters will dewater cake from the oxidized-sludge holding tank. This tank will be equipped with overflow weirs and a sludge collector. The dense sludge pulled from the tank will be filtered during the 8-hour day shift, 5 days per week. This solution to the sludge-handling problem is not conventional but it fits the unique problem of these particular communities in their location and with their industries.

15.13 Final Design of the Gloversville–Johnstown Joint Sewage Treatment Plant

1. *Flow*

24-hr mean	9.50 mgd
8-hr mean peak	13.12 mgd
1-hr mean peak	16.22 mgd
1-hr maximum peak (to be used in hydraulic design of conduits)	19.3 mgd

Maximum hydraulic capacity through primary	30 mgd	Estimated removals Suspended solids 5-day BOD	60% 30%

2. *Process loading to sewage treatment plant*
   - BOD weekday 24-hr mean — 35,019 lb/day
   - BOD weekday 8-hr mean — 53,351 lb/day
   - Suspended-solids load — 195,578 lb/week

3. *Mechanically cleaned bar screens*
   - Number of units — 3
   - Width of screen channel — 3 ft
   - Bar spacing — 1-in. clear opening
   - Velocity through screen at
     - 5.0 mgd — 1.2 ft/sec
     - 9.5 mgd — 1.6 ft/sec
     - 30 mgd — 2.7 ft/sec

4. *Grit-removal units*
   - Number of units — 2
   - Type — Aerated
   - Dimensions — 7 ft wide × 30 ft long × 7 ft deep
   - Particle size removed — 100% of 0.2 mm at 20 mgd
   - Grit-cleaning devices (cyclone with screw classifier) — 2 units
   - Grit disposal — Landfill

5. *Primary settling tanks*
   - Number of tanks — 6
   - Flows and surface settling rates
     - 8-hr peak (basic design) — 13.12 mgd; 800 gal/ft²/day
     - 24-hr — 9.50 mgd; 580 gal/ft²/day
     - Maximum peak hour — 19.3 mgd; 1180 gal/ft²/day
   - Total surface area required — 16,400 ft²
   - Surface area of each tank — 2736 ft²
   - Tank dimensions — 152 ft long × 18 ft wide × 8 ft deep
   - Displacement time at
     - 13.12 mgd — 1.8 hr
     - 9.50 mgd — 2.5 hr

6. *High-rate (roughing) filters*
   - Number of units — 2
   - Process loading:
     - Peak 8-hr BOD — 150 lb/1000 ft³
     - 24-hr average BOD — 98 lb/1000 ft³
   - Diameter of filters — 165 ft
   - Depth of filters — 6 ft
   - Hydraulic loading
     - 8-hr peak — 315 gal/ft²/day
     - 24-hr average — 227 gal/ft²/day
   - Volume of filter media — 250,000 ft³
   - Filter media — 4-in. stone or plastic
   - Recirculation pumps — 3–3500 gpm variable speed
   - Removal, 5-day BOD — 30% at 150 lb/1000 ft³

7. *Aeration tank and equipment*
   - Number of tanks — 1 with 2 compartments
   - Volume division of each tank — 1/4 and 3/4
   - Tank dimensions — 260 ft long × 130 ft wide × 13 ft deep
   - Total tank volume — 439,000 ft³
   - Process loading:
     - 24-hr average (39 lb/100 ft³) — 0.25 lb BOD/lb MLSS at MLSS of 2500 ppm
     - 8-hr peak (60 lb/1000 ft³) — 0.39 lb BOD/lb MLSS at MLSS of 2500 ppm
   - Displacement time including 33% recirculation
     - 24-hr flow, 9.50 mgd — 6.3 hr
     - 8-hr flow, 13.12 mgd — 4.5 hr
   - Aeration equipment
     - Type — Mechanical
     - Number of aerators — 8
     - Connected horsepower — 800 hp
     - Aerator oxygenation capacity (each) — 310 lb/hr

15.14 ESTIMATED COSTS AND FINANCING

8. *Secondary settling tanks*
   - Number of tanks: 2
   - Diameter: 120 ft
   - Side water depth: 10 ft
   - Surface settling rate
     - 13.12 mgd flow: 600 gal/day/ft^2
     - 9.50 mgd flow: 434 gal/day/ft^2

9. *Chlorination equipment and contact tank*
   - Number of tanks: 1 with 2 compartments
   - Dimensions: 100 ft long × 60 ft wide (30 ft each) × 7 ft deep
   - Detention time at
     - 30 mgd: 15 min
     - 9.5 mgd: 47 min
     - 5.0 mgd: 90 min
   - Chlorinators
     - Number: 2
     - Rating: 4000 ppd
   - Evaporators
     - Number: 2
     - Rating: 4000 ppd
   - Residual analyzer
     - Number: 1

10. *Secondary return sludge pumps*
    - Number of pumps: 3
    - Type: Variable speed
    - Maximum capacity each: 2500 gpm

11. *Waste sludge pumps*
    - Number of pumps: 3
    - Type: Variable speed
    - Maximum capacity each: 200 gpm

12. *Sludge thickener*
    - Number of units: 2
    - Type of thickener: Flotation
    - Loading: 2 lb/ft^2/hr
    - Operation: 100 hr/week
    - Total surface area required: 264 ft^2
    - Total surface area provided: 300 ft^2

13. *Wet-air oxidation unit*
    - Number of units: 1
    - Capacity: 25 tons/day
    - Operating pressure: 300 psig
    - Insoluble organic matter reduction: 50%
    - Operating volume and schedules
      - 170,000 lb/week: 65% volatile content; 5% solids
      - 24 hr/day continuous operation: 3½ days/week
      - 16 hr/day operation: 5 days/week
    - Oxidized-sludge storage: 120,000 gal for 1 day
    - Duplicate items: Boiler, High-pressure pump

14. *Raw-sludge holding tanks*
    - Number of tanks: 2
    - Total holding capacity: 7 days
    - Tank proportions
      - Side water depth: 20 ft
      - Diameter: 42 ft

15. *Vacuum filtration*
    - Number of units: 2
    - Filter area: 400 ft^2 each
    - Design filter rate
      - Oxidized sludge: 5 lb/ft^2/hr
      - Raw sludge: 5 lb/ft^2/hr
    - Operating volume and schedules
      - Oxidized sludge, 115,000 lb/week: 29 hr filter time/week
      - Raw sludge, 170,000 lb/week: 43 hr filter time/week

16. *Ultimate disposal of oxidized filter cake*
    Sludge to be landfilled in city refuse areas or other selected areas. Because of the character of oxidized-sludge cake, sludge need not be covered.

15.14 Estimated costs and financing

The estimated costs of the plant units described above are as follows:

	Amount
Site development	$300,000
Screen building and grit tanks	158,000
Primary settling tank	323,000

Roughing filters (two)	437,000	Administration building	228,000
Aeration tank	645,000	Electrical contract	250,000
Secondary settling tanks	310,000	Subtotal	$4,386,000
Chlorine contact tank	67,000	Contingency (5%)	219,300
Recirculation building	203,000	Total	$4,605,300
Overflow and Parshall flume structures	10,000		
Sludge building, including thickeners, vacuum filters, and wet-air plant	1,056,000		
Sludge and oxidized liquor holding tanks	77,000		
Yard piping and conduits	279,000		
Waterline to plant site and meter pit	32,000		
Fencing	11,000		

The project costs and the Federal and State grants are estimated as follows.

Interceptor sewers	$1,227,000
Waste-water treatment plant	4,600,000
Subtotal	5,827,000

Table 15.22 Inconsistencies between theory of design and actual practice in design.

Situation	Theoretical solution	Actual practice	Reason for violating theoretical solution
Flow BOD	Measured as 5 to 7 mgd Measured as 20,000–25,000 lb/day	Designed for 9.5 mgd Designed for 35,019 lb/day	Addition of a large glue-manufacturing plant and another tannery; consideration of future loads
Sludge handling	Digestion plus lagooning was least costly and proved acceptable in the laboratory	Zimpro plus vacuum filtration plus landfill	Lack of State approval because of health hazards of lagooning digested sludge; lack of confidence in efficiency of sludge digestion
Charging for services	Incentive plan based upon unit costs for a pound of BOD and suspended solids and a gallon of waste	Percentage of water bill	Ease of charging; elimination of need to sample, police, and analyze industrial wastes; strong representation on joint Sewer Board of tannery industry
Equalization	Theory would indicate that it is needed because of great fluctuations in instantaneous flow and character	In practice, not required because the great number of tanneries and length of travel in sewers provide equalization	Great number of tanneries; increase in cost of construction and operation of equalization basin
Biological treatment	Theory and laboratory results indicate that activated sludge is an excellent method of reducing high BOD	Practice shows it more suitable to use a combination of roughing filters and activated-sludge treatment in series	Activated sludge computed to be about twice the cost*

*The costs are compared as follows:

	Roughing filter	Activated sludge
Fixed charge	$0.55/lb	$0.39/lb
Operating costs	$0.03/lb	$0.72/lb
Total costs	$0.58/lb	$1.11/lb

All other costs and contingencies	1,212,000
Total	7,039,000
Less Federal and State grants	3,981,000
Net cost to community	3,058,000

It is estimated that the plant will have twenty full-time employees and that the annual operational and maintenance costs will be $262,000. The local communities will receive a reimbursement from the State of New York to the amount of one-third of this cost, leaving a local net cost of $175,000.

The estimated total annual costs for the project and their distribution between the cities are as follows:

Annual debt service (30 years at $4\frac{1}{2}\%$)	$220,000
Annual net operation and maintenance costs	175,000
Total	$395,000

Distribution

City of Gloversville (55%)	$217,250
City of Johnstown (45%)	177,750
Total	$395,000

The agreement between the cities of Gloversville and Johnstown calls for a 55/45 per cent split of capital costs and of operation and maintenance for three years. The division of operation and maintenance charges will be reviewed every three years, to reflect the results of samples collected and analyzed.

The final method of allocating costs to the users has not been fully established at this time, although certain principles have been established. The two cities would like to keep the rates in the cities the same, based on the volume of water used, probably with surcharges for industrial users. It is the aim of the Sewer Board and the cities to avoid a rate structure dependent upon repeated and critical sampling of industrial users. The average home-owner in the cities will pay for the service based upon his water usage and present estimates put the average annual cost at less than $20 per year per home.

15.15 Application of the Plan in Practice

There are many lessons which this case history serves to teach us. Several of the theoretical solutions to this problem had to be abandoned because of the situation existing in the social, economic, and governmental world of today. Table 15.22 describes five such inconsistencies along with the reasons for deviating from the theory.

References: Case History

1. Braunschweig, T. D., "Studies on tannery sewage," *J. Am. Leather Chemists' Assoc.*, **60**, 125 (1965).
2. Chase, E. S., and P. Kahn, "Activated sludge filters for tannery waste treatment," *Wastes Eng.*, **26**, 167 (1955).
3. Fales, A. L., Discussion of paper by W. Howalt, "Studies of tannery waste disposal," *Trans. Am. Soc. Civil Eng.*, 1394 (1928); *Hide and Leather*, **75**, 48 (1928).
4. Furkert, H., "Mechanical clarification and biological purification of Elmshorn sewage, a major portion of which is tannery wastes," *Tech. Gemeindebl.* **39**, 285 (1936) and **40**, 11 (1937); *Chem. Abstr.*, **31**, 5076 (1937).
5. Hubbel, G. E., *Water Works Sewerage*, **82**, 331 (1935).
6. Jansky, K., "Tannery waste water disposal," *Kozarstvi*, **11**, 327 and 355 (1961); *J. Am. Leather Chemists' Assoc.*, **57**, 281 (1962).
7. Kalibina, M. M., "The application of the biological method of judging the efficiency of a purifying plant. Observations on the growth of organisms in an activated sludge tank...", *Chem. Abstr.*, **25**, 3422 (1931).
8. Kubelka, V. *Veda Vyzkum Prumyslu Kozedelnem*, **1**, 113 (1952).
9. Kubelka, V., "Principles of a final biological purification of tannery effluents," *Veda Vyzkum Prumyslu Kozedelnem*, **1**, 113 (1956).
10. Mausner, L., "Tannery waste water," *Gerber*, **41**, 1519 (1938).
11. Nemerow, N. L., and H. S. Sumitomo, *Pollution Index for Benefit Analysis*, to be published.
12. Pauschardt, H., and H. Furkert, *Stadtereinigung*, 411 and 427 (1936).
13. Research Report no. 10, Civil Engineering Department, Syracuse University, January 1969.
14. Snock, A., *Collegium*, **703**, 612 (1928).
15. Thebaraj, G. J., S. M. Bose, and Y. Nayudamma, "Comparative studies on the treatment of tannery effluents by trickling filter, activated sludge and oxidation pond systems," *Central Leather Research Institute, Madras, India*, **13**, 411 (1962).
16. Vrooman, M. and V. Ehle, *Sewage Ind. Wastes*, **22**, 94 (1950).

CHAPTER 16

JOINT TREATMENT OF PARTIALLY TREATED INDUSTRIAL WASTES AND DOMESTIC SEWAGE

Most industrial wastes contain only a few harmful constituents and the removal of these leaves the remaining wastes amenable to treatment along with domestic sewage. The well-trained waste engineer will recognize the situations which call for this approach, whereas the less-knowledgeable engineer will demand complete separation of all industrial wastes from city sewers and treatment plants. There is a fine line of distinction between a wholly untreatable waste and one containing only certain components which are untreatable when the waste is combined with domestic sewage. It is the objective of this chapter to define this difference more clearly. Again, it should be stressed that the concern of a municipality for its industries and, in return, that of industry for the effectiveness of the municipal treatment facilities are significant factors in the success of joint treatment.

AN EXAMPLE OF COMBINED TREATMENT FOLLOWING PRETREATMENT

A certain highly industrialized city had for 25 years been handling all the wastes from within its environs in its secondary-type treatment plant. This disposal plant was now overloaded and a decision had to be made concerning the treatment of industrial wastes. The author was employed by the municipality to advise its industries on treatment. Besides the overloaded conditions which reduced plant efficiency, other operating difficulties were being caused by grease which clogs the fixed nozzles or filters, feathers which clog the distributing devices and build up in the digester, toxic chemicals which inhibit bacterial action in the digester, build-up of grease and feathers which reduces or stops digestion, and excess floating solids which go over the effluent weir into the receiving stream. The plant produced inconsistent removal efficiencies and was at least partially inoperative about 25 per cent of the time owing to the character and quantity of industrial wastes.

The existing plant contained a bar screen, grit chamber, circular rim-driven clarifier, fixed-nozzle trickling filter, secondary circular clarifier, separate sludge digester, and sand drying beds. The average daily flow was 3.88 million gallons per day, the suspended-solids load 9276 pounds per day, and the BOD load 11,533 pounds per day; 44.1 per cent of the average flow, 47.5 per cent of the average BOD, and 41.2 per cent of the suspended-solids load occurred during the nine daylight hours between 8 a.m. and 5 p.m.

The industries and their contributing pollution loads were as follows:

1. A dyeing and finishing mill for synthetic textiles contributed 33 per cent of the treatment-plant flow and 80 per cent of the total industrial flow, 25 per cent of the total plant BOD and 62 per cent of the total industrial BOD, but only 4 per cent of the total suspended solids.
2. A laundry contributed 6.2 per cent of the total plant BOD and suspended solids. This waste included rags and lint, which lead to stoppage of sewer lines, but it comprised less than 1 per cent of the plant flow.
3. A rendering plant contributed 4.5 per cent of the total plant BOD. There was a great deal of grease and odor in this waste, and 1.91 per cent of the flow and suspended solids came from it.
4. A poultry-plant waste consisted of blood, feathers, and paunch manure and contributed 1 per cent of the flow, 1.4 per cent of the BOD, and 1.7 per cent of the suspended solids of the total combined wastes.
5. A slaughterhouse and meat-packing plant con-

tributed 2.8 per cent of the BOD and 1.3 per cent of the suspended solids, but only 0.28 per cent of the flow.
6. An electrical-plating company contributed 2.2 per cent of the flow, 0.69 per cent of the BOD, and 0.87 per cent of the suspended solids. High chromium concentrations were found in several samples of these wastes.
7. The domestic customers contributed 84 per cent of the plant's suspended solids, 60.4 per cent of the flow, and 57 per cent of the BOD.

Overloaded plant capacity is a serious problem in many municipalities in this country. Precedent plays an important part in the solution to such a problem; people who have been disposing of their wastes in a certain way for a number of years will tend to resist any proposed change in overall methods. The initial method of plant financing will also tend to set a pattern for subsequent financing. Therefore, continued treatment in the same general system as in the past, but with improved efficiency, will be welcomed by all parties concerned. The task of designing waste-treatment facilities is difficult enough in itself, without attempting to convince the public that its approach has been, and continues to be, erroneous.

With these principles in mind, the author felt that the problem under discussion could best be solved by continued use of combined treatment, provided all other factors favored it. Since both the public and the private corporations concerned favored this approach and since it was more economical, the only major deterrent to its acceptance was the presence of toxic wastes and wastes which interfered with the proper operation of the plant, although only the chromium waste from the plating plant appeared to present any difficulty. Since this waste was a small percentage of the total flow, the author believed that a satisfactory arrangement could be made to eliminate it.

Three steps were now necessary to complete the technical solution to the problem: (1) to ascertain the capacity of the various existing treatment-plant units, (2) to reduce the incoming waste load to a minimum by proper pretreatment of industrial wastes at each individual factory, and (3) to reevaluate the present plant and suggest the additions required to handle the future waste load effectively.

16.1 Ascertaining Present Plant Capacity

Table 16.1 presents the load-ratings for each treatment unit in the present plant. A rated capacity of 100 per cent means that the unit is currently loaded to its maximum for optimum removal efficiency. A rating greater than 100 per cent means that the unit is overloaded by the percentage above 100 and a percentage less than 100 indicates that it is not being used to capacity. Since a minimum of 85 per cent efficiency was required to protect the receiving stream, it is clear that only the grit chambers (detritors) were adequate to handle even the present pollutional load.

Table 16.1 Capacity of present plant units.

Units	Rated capacity (average)*	Percentage of rated capacity being used
Detritors (two)	8.00 mgd	48.5
Primary settling basin	3.18 mgd	122
Secondary settling basin	2.00 mgd	194
Trickling filters (loading after primary settling, 9240 lb/day)	2210 lb/day (for 90% efficiency)	418†
	5600 lb/day (for 85% efficiency)	165†
	11,300 lb/day (for 80% efficiency)	82†
Digester	65,000 ft³	518
Sludge-drying beds	One loading per month of 9 in. of sludge (825 ft³ of sludge per day)	202
	2260 lb dry solids per day	246

*Based on normal accepted design loadings for the various units.
†Assuming 20 per cent reduction in primary settling basin.

16.2 Reducing the Incoming Load

Each industrial plant was visited and helped to assess the load it was discharging to the treatment plant. (At this point in any project, the engineer can suggest the design and construction of pretreatment methods discussed more fully in Chapters 6 through 10.) Methods of reducing or eliminating the undesirable wastes were proposed; all the industries were receptive to suggestions about changes in processes or disposal practices.

The textile industry, contributor of the largest volume, agreed to carry out investigations on the possibility of water reuse, but, after several months of laboratory study, no practical method of rendering the water reusable was found. Little change, therefore, could be expected in the load from the finishing mill. On the other hand, the hosiery-mill section of the textile operation made substitutions in soaps, detergents, and so forth, that considerably lowered the BOD issuing from this mill.

The rendering plant agreed to install new baffles in its existing grease tank trap and, in addition, to construct a second grease tank trap and new pipeline, which would eliminate the escape of grease into the sewer. This grease had been contributing significantly to the clogging of the fixed nozzles on the trickling filter at the treatment plant. The slaughterhouse and packing plant eliminated all blood from their discharge by separating the killing area completely from the rest of the plant. In this way, the blood wastes, which are used for fertilizers, glues, and animal feeds, were drained into large drums in the basement for separate disposal. All grease and floor drainings went to the grease tank before discharge to the city sewer, and orders were given for the grease tank to be opened and cleaned periodically. All paunch manure was "screw conveyed" out of the basement of the plant into tank trucks, which carried it to the country for sale as soil fertilizer. In this way, all solid manure was eliminated from the sewer.

The poultry plant presented another facet of the same problem. Feathers were clogging filter nozzles at the treatment plant and resisted settling in the sedimentation tank, because of their large surface area relative to density. Some feathers, however, after becoming thoroughly wetted, did settle and finally ended up in the digester, where they interfered with that process because they did not decompose. The poultry plant installed a series of three fine screens, which are cleaned daily, to remove feathers and other suspended matter which formerly escaped from the plant into the effluent sewer. Four cement gutters covered with perforated steel plates lead into the screens, with Wade-type drains (see any plumber's supply manual) placed at each end of the gutters as primary aids in keeping feather discharge to a minimum.

The laundry installed a metal frame with a heavy wire-screen cloth attached and a series of baffle plates in the catch basin located at the rear of the building. The cloth is changed as often as necessary and eliminates much of the suspended solids leaving the plant. Additional screens were installed within the plant to catch larger objects such as bits of cloth and towels.

The metal-plating plant installed automatic pH recorders on the plating-room effluent line. Ammonium hydroxide (NH_4OH) was added to stabilize the pH value at about 7.0, thus preventing excess acidity from entering the city sewer and ensuring elimination of slug acid discharges. In addition, a study was made of the plating-bath operations and precautions were taken against overflows of the metal baths (especially those containing expensive cyanides and chromium).

A small chemical company made minor discharge-line changes to eliminate storm water from the sewer. (The treatment of uncontaminated storm water in waste-treatment plants is usually an uneconomical procedure and can result in lowered plant efficiency.) They also agreed not to discharge any toxic chemical to the sewer unless adequately diluted. Two separators were used to facilitate the removal of these foreign constituents from the process waste waters. One was designed to remove all oily ester-type material originating from floor washings; the other is a two-stage system that is used for separation of all process water.

Frequent discussions with individual industries improved their attitudes toward pretreatment. Overall waste-treatment objectives were pointed out in these meetings and tours of the city waste-treatment plant greatly aided the industries to understand the problems involved.

16.3 Reevaluation of Present Plant and Suggestions for Additions

The third phase in the technical solution to this problem involved evaluating the final load reaching the city treatment plant after all industries had installed pretreatment devices. The wastes from each contributing industry were analyzed several times, over prolonged periods, to ascertain the pollution loads they contained. In addition, the total load reaching the city treatment plant was measured (Tables 16.2 and 16.3).

Table 16.2 Total loading of treatment plant.

Characteristic	Before pretreatment*	After pretreatment†
Flow, mgd	3.88	3.36
BOD, lb/day	11,533	10,130
Suspended solids, lb/day	9,276	8,129

*Average of 6 sampling days during June to September 1954.
†Average of 20 sampling days during June to September 1955.

The reader will notice, in Table 16.3, that certain plants' waste loads were increased after pretreatment. Although in theory this increase should not take place, it often occurs in practice, because of the variability of industrial operations. This set of figures, therefore, points out the need for numerous samplings over extended periods to ascertain precise loadings. The engineer should not be discouraged by a few results which may be somewhat misleading if interpreted directly. Moreover, a major objective in this particular case was the removal of nuisance wastes rather than reduction of loading. It may not necessarily follow that removal and reduction will occur simultaneously.

These new pollution-load measurements indicated a decrease of 1403 pounds of BOD per day since the installation of the pretreatment devices. At the average rate of $150 per pound of BOD for capital expenses of treatment, this represented a considerable saving. Also, this measured load included 322 pounds of BOD and 118 pounds of suspended solids from the packing plant, which would be removed when the packing plant's renovations were completed. The industries had eliminated many of the nuisances which existed before. The removal of feathers, blood, toxic chemicals, rags, and inflammable materials from the city waste waters resulted in increased efficiency of the existing treatment plant: even without further additions, the plant was able consistently to remove about 80 per cent of the BOD, 75 per cent of the suspended solids, and 66 per cent of the color of the total waste water, whereas previously it functioned irregularly and was only about 50 per cent efficient when it was operating. These results were obtained even though the plant was still obviously overloaded.

Additional plant capacity was advocated—in the form of increased clarifier volume, digester volume, filter-stone area, and final sludge-handling equipment—to achieve maximum removal of pollution and to protect the best usage of the receiving waters. These additions were designed and constructed; once they were in operation, efficiencies of the operations rose to over 90 per cent. The new sections of the plant were built at a relatively low cost and operated at maximum efficiency, due in large part to the excellent cooperation of the industries and the city.

This example illustrates the importance of individual consultations with industry and the need for

Table 16.3 Results of pretreatment practiced by industries.

Plant	Flow, gpm		BOD, lb/day		Suspended solids, lb/day	
	Before*	After*	Before	After	Before	After
Synthetic dyeing and finishing mills	857	838	3048	2736	376	509
Laundry	50	100	719	222	576	433
Rendering plant	51	86	514	1995†	179	1323†
Poultry plant	28	80	162	187	159	122
Slaughterhouse and packing plant‡	6.95		322		118	
Plating plant	59	69	80	47	81	25

*Before and after pretreatment.
†Increase due to insufficient sampling before pretreatment rather than decreased efficiency.
‡Improvements not completed at time of final survey.

rendering small-factory wastes compatible with domestic sewage, as a prelude to combined treatment. Although alternatives such as separate treatment were not pursued, this solution was considered the best for all concerned because the precedent of joint treatment had been established in this community for a long time. However, the effort and time involved for the waste engineer who handles a problem this way should not be underestimated; unfortunately, because such an expenditure of time and energy is often not financially rewarding to the engineer, this thoroughness of approach is rarely attained in actual practice.

CHAPTER 17

DISCHARGE OF COMPLETELY TREATED WASTES TO MUNICIPAL SEWER SYSTEMS

In many situations, an industry may find it advisable to discharge its wastes into the city sewer system even after complete treatment. It is, of course, an advantage to an industry to have someone else assume the responsibility for final handling and disposal of its residual liquid wastes, because there is always the possibility of a mishap at the plant which would make the wastes from the industry's treatment plant unacceptable for direct discharge to the receiving stream. The municipal treatment plant thus serves as an added protective device for the stream. It may also be more convenient for an industry to discharge by gravity into a nearby municipal sewer than to pump treated waste a long distance to a suitable location on a river. The attitude of the municipality and the cost of its sewer-service charges play an important part in an industry's decision to utilize the municipal sewer and treatment plant. Acceptance of industry as an integral part of a city's life and equitable charges for the use of city sewers encourage industry to make use of the city's sanitary facilities. Let us emphasize, once again, that the means of disposal of industrial wastes must be determined on an individual basis: each case is unique and often a single factor, such as location or precedent, can influence an industry's selection of the final stage of its waste treatment.

AN EXAMPLE OF THE DISCHARGE OF A COMPLETELY TREATED WASTE INTO A MUNICIPAL SEWER SYSTEM

For our discussion, we shall consider an industry which is erecting a new plant near its existing one, to manufacture electrical and mechanical business machines. The site for the new plant was selected for its convenience, setting and surroundings, and availability; but, as is often the case, at the time the site was selected little consideration was given to the waste-disposal problem. Though this oversight has been a common occurrence in the past, fortunately more and more companies are according waste treatment its proper importance when they select future plant locations, since they are beginning to realize that it can be a costly experience when an industry does not look into all aspects of waste disposal prior to selecting a site for a new plant.

The plant is located in a small town and is the only major industry. Production in the new plant will be essentially similar to that of the old, on an expanded basis. The existing plant manufactures about 1500 machines per day and discharges all its wastes untreated, through two holding lagoons, into a small creek. The new plant, however, will be served by the municipal sewer system and the municipal primary-treatment plant. Effluent from this plant discharges by gravity to the only flowing stream in the area. The municipality has agreed in advance to construct the necessary additional sewer lines to the industrial site and the industry has agreed to pay for amortizing the bonds covering the sewer construction. The entire waste can flow by gravity from the industry to the municipal treatment plant. In fact, the topography of the area is such that all creeks, streams, and even underground waters drain toward the city. The city obtains its water supply from underground wells located near the city limits, on a direct line between the industrial-plant site and the municipal sewage-treatment plant. Should the industry decide not to use the municipal sewer facilities, it would have to pump its entire waste output about four miles, against a 200- to 300-foot elevation rise, into the receiving stream having adequate dilution.

The business-machine corporation needed to determine the quantity and quality of the wastes being produced at the existing plant, so that it could estimate the wastes which would result from future production at the new plant and then determine the degree and type of treatment to be given those wastes before they entered the city waste-water treatment system.

17.1 The Sampling Program

The first step was to sample existing wastes. All the wastes from the existing plant were discharged into two pipelines. Both of these were weired and boxed for sampling and flow measurement; photographs of these weirs and boxes are shown in Figs. 17.1 and 17.2. The wastes from Pipeline 1 (Fig. 17.1) originated primarily in the plating room and those from Pipeline 2 (Fig. 17.2) came from the pickling room. Slug discharges and changes in production were included in the samples. Each line was sampled for a 24-hour period on each day of the week except Sunday. The days selected, however, were in different weeks over a two-month period in 1958, as follows: Monday, November 17; Tuesday, September 23; Wednesday, October 8; Thursday, November 6; Friday, October 24; Saturday, November 15. (See Table 17.4 for figures for the production of machines per plating line on these sampling days.)

Samples were collected from each weir box every half-hour during the 24-hour period, in volumes according to the rate of waste discharge at the time. Flows were recorded and samples composited each half-hour over the entire period of sampling, as illustrated graphically in Figs. 17.3 through 17.6. Batch dumpings of 55-gallon drums from the pickling and blacking rooms were also recorded during the sampling period (September 29, 1958, to November 18, 1958); the results are presented in Table 17.1.

17.2 Analyses of Wastes

The industry decided to construct its own laboratory and make its own analyses under qualified supervision; hence a chemical laboratory was designed and equipped by the author. All chemical analyses were carried out by qualified chemists employed by the industry. Metals were analyzed in accordance with the American Public Health Association's manual, *Standard Methods for the Examination of Water, Sewage, and Industrial Wastes*. The provisional method for heavy metals was used and a modified method for preliminary treatment of samples, to induce ionization of metals, was utilized.

The 24-hour composite samples from each production line were analyzed for pH, copper, chromium,

Fig. 17.1 Pipeline 1 (plating-room wastes), showing weirs and boxes set up for sampling and measuring.

Fig. 17.2 Pipeline 2 (pickling-room wastes).

zinc, cyanide, iron, and nickel. These analytical results are presented in Table 17.2. In addition, three 55-gallon barrels of waste from the pickling and blacking rooms were sampled and analyzed for cyanide only. These results are shown in Table 17.3.

17.3 Plant-Production Study

In order to ascertain the level of production during the sampling days, a detailed production study was made. With this information, one can establish the relation between production and waste discharge; it is then relatively easy to predict future waste volumes and pollution loads resulting from increased plant production. Present normal production at the plant ranges from 1300 to 1700 machine units per day. The equivalent number of machines produced per plating line is presented in Table 17.4.

17.4 Suggested In-Plant Changes to Reduce Waste

An in-plant production survey was carried out to determine means of reducing metal impurities in waste effluents. This brought out the fact that the plant

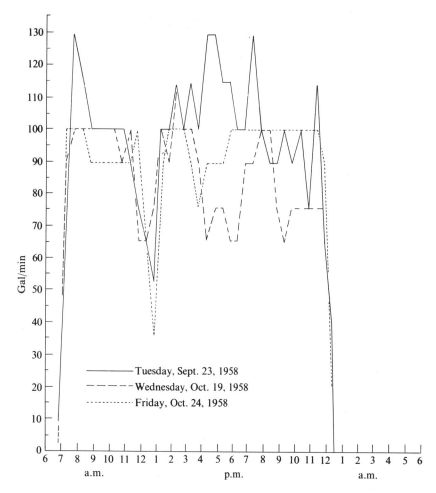

Fig. 17.3 Waste-water flow, Weir 1 (plating room).

operates five plating lines (as shown in Table 17.4), each of which warrants some alteration from the standpoint of decreasing the amount of waste it contributes. Some of the more pertinent suggestions made by the author for changes are outlined in the following paragraphs.

1. Bright nickel cycle (see Fig. 17.7)
a) Add a high-pressure fog spray of water in the last 2-foot section of the nickel-plating line, to reduce the "drag-out" which results when a metal part is removed from a plating bath. Fog sprays utilize a fine spray of water under pressure to rinse the acids and metallic ions off the work (machine parts). A relatively small quantity of water is consumed by the spray, with the same net effect as a stream or bath of water.
b) Utilize a static rinse as make-up for nickel plate. Static (rather than continuous) rinses reduce the quantity of metal and acids lost in waste. As the static-rinse bath builds up in concentration of rinsed-off metals, it becomes increasingly less capable of rinsing the work. The rinse water can then be concentrated somewhat and returned to the plating bath as make-up water.
c) Use high-pressure jet sprays in the cold-water rinse and recirculate the cold-water rinse to a holding tank, from which it is bled as often as possible to the static-rinse tank as make-up water. There must be an overflow in the holding tank, leading to

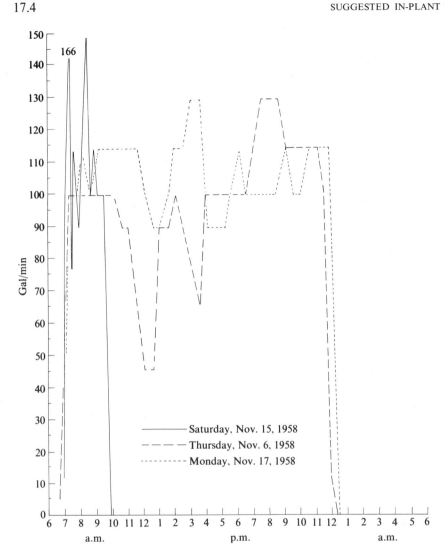

Fig. 17.4 Waste-water flow, Weir 1 (plating room).

the sewer. By this practice, waste waters are used to the maximum before final loss to the sewer.

d) Replace the cyanide dip with an acid dip. Cyanides are not only detrimental to biological treatment of waste water, but also hazardous and even lethal to humans, in the form of hydrogen cyanide gas. Even in small concentrations, cyanide compounds can be toxic to fish. One purpose of the cyanide dip is to brighten the work and to remove the last traces of contaminating matter; acid dips have proved to be adequate substitutes in many cases, as well as being more economical.

2. Zinc cycle (see Fig. 17.8)

a) Add a high-pressure fog spray of water in the last 2-foot section of the zinc-plating line, to reduce "drag-out."

b) Install a static rinse to replace the cold-water rinse and reuse static-rinse water as make-up for the zinc-plating bath.

c) Use small, high-pressure sprays in cold-water rinses and overflow each in the direction of the zinc-plating tank. (This utilizes the principle of counter-current flow.) If the flow of rinse water is small enough, it may serve eventually as make-up water

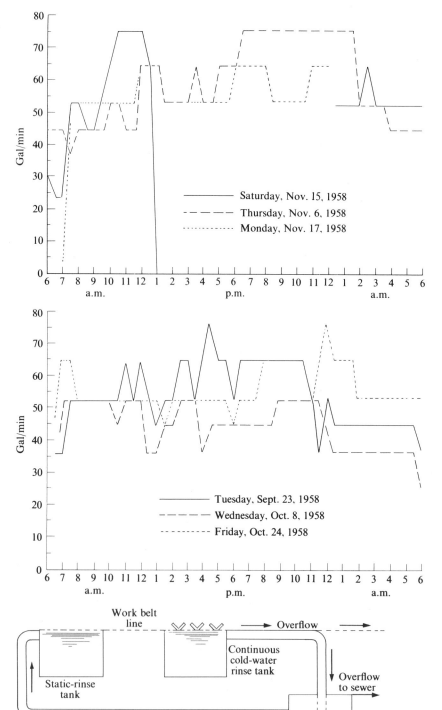

Fig. 17.5 Waste-water flow, Weir 2 (pickling room).

Fig. 17.6 Waste-water flow, Weir 2 (pickling room).

Fig. 17.7 Schematic diagram of nickel-cycle process in the business-machine plant (side view).

Table 17.1 Additional batch waste discharge.

Nickel strip from pickle room (Ni plus CN)		Cyanide from pickle room* (CN plus Fe)		Cyanide from blacking room (case hardening) (CN)	
N†	Date	N	Date	N	Date
1	9/29	1	10/13	6	10/7
1	9/30	1	10/18	6	10/21
2	10/3	2	10/24	7	11/13
1	10/6	1	10/31		
1	10/7	1	11/3		
1	10/8	1	11/7		
2	10/9	2	11/8		
2	10/13	1	11/13		
3	10/17	2	11/15		
1	10/20	1	11/15		
1	10/21	1	11/18		
2	10/24				
1	10/27				
2	10/29				
2	11/5				
1	11/6				
2	11/7				
1	11/15				

*Usually dumped Saturday.
†Number of 55-gallon barrels.

for the plating tank. An overflow pipe on the cold-water-rinse tank closest to the plating tank can serve as entrance to the sewer, if there is no room for an overflow in the zinc-plating tank.

3. Chrome cycle

a) Add a high-pressure fog spray of water just before racks move to the static rinse, to reduce "drag-out" from chrome-plating bath.
b) Return rinse water to the chromate plating tank, instead of discharging it to the sewer twice a week.

4. Combination copper and nickel cycle

a) Remove nickel part of cycle to new plant, where there will be a more modern automatic nickel cycle.
b) Design a new copper cycle so that no "drag-out" occurs across floors, where drippings are subsequently washed into the sewer. This is purely a matter of clean housekeeping and is often overlooked by management.
c) Install a high-pressure fog spray of water near the end of the copper-cyanide plating tank to reduce "drag-out" prior to rinse.
d) Install a static rinse to follow the plating bath, instead of the present flowing rinse.
e) Use static rinse as make-up water for the plating bath.

5. Barrel nickel cycle

a) Install a static rinse following the acid nickel-plating tank to replace one of the two continuously flowing rinses in use at present.
b) Reuse static-rinse-tank contents as make-up for the acid nickel-plating tank.

6. Blacking room

Replace the molten-cyanide case-hardening tank with a carburonitriding electrical furnace. This will eliminate pollution due to cyanide from the case-hardening process.

7. Nickel stripping

Reduce nickel stripping to a minimum or eliminate it completely. This is a source of a great percentage of cyanide in the waste (see Tables 17.1 and 17.3).

17.5 City Waste-Water Treatment Plant

The city possesses a primary-treatment plant with separate sludge-digestion facilities. The disposal plant contains the following units: 30-inch sewer-influent line; 1-inch bar screen (mechanically cleaned, 30-minute time switch); two auxiliary hand-cleaned bar screens (not used at present); screenings grinder (used three times a day, grindings returned to flow); two grit chambers (each closed off once a month for cleaning); Parshall flume for flow measurement; two horizontal settling basins (operated in parallel); chlorinator (750 lb/day capacity, prechlorination used in summer); two fixed-cover digesters (contents circulated 35 minutes per day, sludge withdrawn every two weeks); gas holder; gas boiler plus auxiliary oil boiler used only in winter (when, and if, digester gas fails); four vertical-type waste-water pumps (capacity: three 4-mgd, one 2-mgd at 100 per cent efficiency*); two

---

*Normal operating efficiency, 60 to 80 per cent.

Table 17.2 Analytical results of CN sampling* of two waste lines: metal content† and pH.

Date	Weir no. 1 (plating); average volume, 100 gal/min							Weir no. 2 (pickling); average volume, 50 gal/min						
	Cu	Cr	Zn	CN	Fe	Ni	pH	Cu	Cr	Zn	CN	Fe	Ni	pH
Mon., 11/17/58	3.2	12.4	15	17.9	1.1	5.5	7.6	0.8	0.28	<1	1.8	0.6	6.0	8.2
Tues., 9/23/58	2.8	32	7.5	11.4	2.5	6.3	9.4	1.6	0	<1	2.3	5.0	6.0	8.6
Wed., 10/8/58	3.8	22	7.0	20.8	7.6	5.0	9.8	0.8	0.2	<1	2.3	3.2	7.5	8.0
Thurs., 11/6/58	5.2	19.6	20	26.8	2.8	8.5	8.0	0.4	0.1	<1	3.1	2.6	13	8.6
Fri., 11/14/58	4.2	20	13.5	30.5	2.8	7.5	6.5	0.4	0	<1	1.3	4.8	11	7.8
Sat., 11/15/58	7.8	5.6	215	14.3	4.2	35	2.6	2.2	0.24	1.7	3.7	1.1	32	3.0
Average	4.5	18.6	46.3	20.3	3.5	11.3	7.3	1.0	0.12	<1.0	2.4	2.9	12.6	7.3
Range	2.8–7.8	5.6–32	7.0–215	11.4–30.5	1.1–7.6	5.0–35	2.6–9.8	0.4–2.2	0–0.28	1–1.7	1.3–3.7	0.6–5.0	6.0–32	3.0–8.6

*As 24-hour composites.
†In parts per million.

Table 17.3 Analyses of cyanide (in ppm) in pickling- and blacking-room barrels.

Sample number	Nickel strip	Blacking room	Pickling room
1	41,340	38,220	53,040
2	19,980	38,940	34,320
3		61,360	

sludge pumps (duplicate units pump every morning to digester). The effluent of the plant discharges to a Class B (best usage for bathing or recreation) stream.

Tables 17.5, 17.6, 17.7, and 17.8 present recent operating data for this plant. Careful consideration should be given to the sludge digesters, which appear to be amply designed for present loading conditions, providing a total volume of about 51,000 cubic feet. Active digestion should require about 17,500 cubic feet, leaving 33,500 cubic feet for sludge storage—about 53 days of storage for the digested sludge. This may be of some concern to the operator during the long winter months at this plant location, when sludge cannot be pumped to open drying beds. The presence of toxic metals in the waste water would further aggravate this situation by retarding active digestion.

17.6 Toxic Limits for Metals

Literature on the subject reveals that even small concentrations of metals can deter sludge digestion. Table 17.9 shows that generally no more than 1 ppm of Cu, CN, or Cr and 2 to 5 ppm of Zn or Ni should be allowed in the sewage-plant influent. Dilution of plating wastes with municipal sewage often dictates the degree of pretreatment, if any, required to meet these limits.

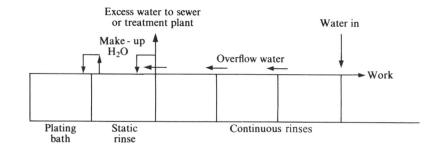

Fig. 17.8 Zinc-cycle process tanks in the business-machine plant (view from above).

17.7 Treatment of Industrial Wastes

Because of the difficulty in predicting the effects of these wastes on the municipal treatment plant, it was decided to recommend pretreatment by the industry in two stages. The first stage was to be installed immediately and the second one only if it is shown to be required.

Stage 1. Drag-out and excessive metal contamination in the wastes should be reduced as recommended in Section 17.4. All chromium, cyanide, and acid or alkaline industrial wastes should be segregated and piped separately from the plant to the industrial waste-treatment plant. A treatment unit of the flotation type should be constructed to remove oils and greases as well as solid matter. This tank would also serve as an equalizing device. The "float" from the flotator-equalizer can be burned rather than delivered to the

Table 17.4 Number of machine units per plating line on waste-sampling days.

Date	Nickel	Chrome	Zinc	Barrel nickel	Copper
Mon., 11/17/58	2400	940	1560	1650	740
Tues., 9/23/58	1990	1720	1080	1870	1340
Wed., 10/8/58		No production data available			
Thurs., 11/6/58	2290	1280	1670	1700	1500
Fri., 10/24/58	2330	1380	1640	1800	3840
Sat., 11/15/58			No production		

Table 17.5 Waste-water influent at city disposal plant.*

| Month | 1956 | | 1957 | | 1958 | | BOD, ppm | |
	Average flow, mgd	Suspended solids, ppm	Average flow, mgd	Suspended solids, ppm	Average flow, mgd	Suspended solids, ppm	1957†	1958‡
January	5.25	138	5.52	66	5.20	95	24	25
February	5.19	135	5.39	78	5.06	87	29	108
March	8.47	70	6.06	82	7.31	45	30	77
April	8.67	53	6.36	72	8.30	75	32	328
May	6.39	81	5.55	95	7.12	124	39	132
June	5.33	80	4.98	89	6.47	59	44	88
July	0.71	94	4.69	112	5.23	98	42	80
August	4.07	129	4.26	99	4.57	99	64	142
September	3.86	128	3.94	174	4.44	90	44	131
October	4.17	139	3.59	135			71	
November	3.84	123	3.36	138			61	
December	4.83	92	4.58	150			74	
Average	5.40	105	4.86	108			46	

*Contains a considerable quantity of storm water.
†Samples collected more than once per month; values appear low owing to storm water diluting sewage.
‡Samples collected once per month; values appear low owing to storm water diluting sewage.

Table 17.6 Efficiency of sedimentation units at city disposal plant (1957).

Sanitary characteristic	Maximum	Minimum
Flow, mgd	6.78	3.28
Raw suspended solids, ppm	356	52
Final suspended solids, ppm	126	24
Removal of suspended solids, %	75	35
Raw BOD, ppm	73.5	24.5
Final BOD, ppm	58.5	14.7
Removal of BOD, % (average)	63	20

Table 17.7 Analysis of sludge solids at city disposal plant (1958).

	Raw sludge		Digested sludge	
Month	Total solids, %	Volatile matter, %	Total solids, %	Volatile matter, %
January	3.85	85	7.9	56
February	5.0	84	7.7	56
March	3.5	94	8.5	60
April	3.5	83	9.5	59
May	4.6	85	8.0	59
June	3.8	80	10.1	54.5
July	4.5	83	8.0	59
August	4.45	76	8.8	55
September	4.6	81	9.5	58

Table 17.8 Gas produced in digester at city disposal plant.

Year	Gas production total, ft^3
1948	6,253,883
1949	6,475,236
1950	6,464,516
1951	6,350,273
1952	5,699,839
1953	6,220,937
1954	5,994,279
1955	5,964,300
1956	5,463,108
1957	5,562,321

concentration tank. Oils and greases can be either skimmed and burned or concentrated in other units.

Sludge from Stage 1 should be pumped to a concentration tank, from which it can be removed to dump areas by septic-tank cleaning trucks. The supernatant from the concentration tank should be returned to the beginning of the treatment plant's cycle.

Stage 2. If the effluent from the industry's treatment plant does not meet the standards set forth by the city or by the author in Section 17.6, the wastes should be further treated by the means outlined in the schematic drawing in Fig. 17.9. The detailed design of all treatment units should be prepared at the outset, and both stages submitted to the city and the state prior to construction of a new plant, with the agreement that the units in Stage 2 will be constructed only after the sampling and analysis program is completed and only if these analyses show Stage 2 treatment to be necessary. It should be stipulated that no wastes which exceed the concentrations listed in Table 17.9, when they reach the sewage treatment plant (Sampling Point 2) will be discharged into the city sewage system.

Sampling and analysis program. For a complete sampling and analysis program, the following samples should be collected:

Sampling Point 1. Daily composite of effluent from industrial plant, drawn from sampling manhole at industrial waste-treatment plant.

Sampling Point 2. Daily composite of sewage influent to city treatment plant to check degree of dilution available.

Sampling Point 3. Daily composite of raw sludge before it gets to digester at city treatment plant.

Sampling Point 4. Weekly composite of sludge in digester at city treatment plant to check the digester efficiency.

When these samples have been collected, analyses should be made for cyanide, zinc, chromium, and nickel. Volume of flow at Points 1, 2, and 3 should be recorded daily. A total organic-solids determination should be made daily on the composite raw sludge collected at Point 3.

Table 17.10, taken from Jones [4], indicates goals of a plating-waste treatment designed to protect fish life in a receiving stream.

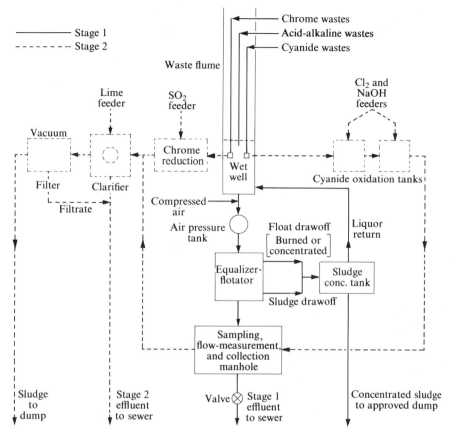

Fig. 17.9 Waste-treatment sequence, with Stage 1 units indicated by solid lines and Stage 2 units by dashed lines.

Table 17.9 Toxic limit for metals in raw sewage subject to sludge digestion.*

Metal	Reference								
	1	8	5†	3	10	2	7	6	9
Chromium	5.0	5.0	0.05			1.0		1.5	
Cyanide	2.0	1.0	0	0.1	1–1.6				
Copper	1.0	1.0	0.30	0.2		1.0	0.7		
Iron	5.0								
Zinc		5.0	0.3	0.3					> 5.0
Nickel			2.0						

*Concentrations given are in ppm. References are listed at end of this chapter.
†For streams and sewers.

Table 17.10 Lethal limits for metals as salts. (After Jones [10].)*

Salt	Fish tested	Lethal concentration, ppm		Exposure time, hr
Aluminum nitrate	Stickleback	0.1	Al	144
Aluminum potassium sulfate (alum)	Goldfish	100		12–96
Barium chloride	Goldfish	5,000		12–17
Barium chloride	Salmon	158		?
Barium nitrate	Stickleback	500	Ba	180
Beryllium sulfate	Fathead minnow	0.2	Be	96
Beryllium sulfate	Bluegill	1.3	Be	96
Cadmium chloride	Goldfish	0.017		9–18
Cadmium chloride	Fathead minnow	0.9		96
Cadmium (salt?)	Rainbow trout	3	Cd	168
Cadmium nitrate	Stickleback	0.3	Cd	190
Calcium nitrate	Goldfish	6,061		43–48
Calcium nitrate	Stickleback	1,000	Ca	192
Cobalt chloride	Goldfish	10		168
Cobalt (salt?)	Rainbow trout	30	Co	168
Cobalt nitrate	Stickleback	15	Co	160
Copper nitrate	Salmon	0.18		?
Copper nitrate	Stickleback	0.02	Cu	192
Copper nitrate	Rainbow trout	0.08	Cu	20
Copper sulfate	Stickleback	0.03	Cu	160
Copper sulfate	Fathead minnow	0.05	Cu	96
Copper sulfate	Bluegill	0.2	Cu	96
Copper sulfate	Minnow	1.0	Cu	80
Copper sulfate	Brown trout	1.0	Cu	80
Cupric chloride	Goldfish	0.019		3–7
Lead chloride	Fathead minnow	2.4	Pb	96
Lead nitrate	Minnow	0.33	Pb	?
Lead nitrate	Stickleback	0.33	Pb	?
Lead nitrate	Brown trout	0.33	Pb	?
Lead nitrate	Stickleback	0.1	Pb	336
Lead nitrate	Goldfish	10		1–2
Lead nitrate	Rainbow trout	1	Pb	100
Magnesium nitrate	Stickleback	400	Mg	120
Manganese nitrate	Stickleback	50	Mn	160
Manganese (salt?)	Rainbow trout	75	Mn	168
Mercuric chloride	Rainbow trout	0.01	Hg	204
Mercuric chloride	Rainbow trout	0.15	Hg	168
Mercuric chloride	Rainbow trout	1.0	Hg	600
Nickel chloride	Goldfish	10		200
Nickel chloride	Fathead minnow	4	Ni	96
Nickel nitrate	Stickleback	1	Ni	156
Nickel (salt?)	Rainbow trout	30	Ni	168
Potassium chloride	Goldfish	74.6		5–15
Potassium chloride	Straw-colored minnow	373		12–29
Potassium nitrate	Stickleback	70	K	154

Table 17.10 *(continued)*

Salt	Fish tested	Lethal concentration, ppm		Exposure time, hr
Silver nitrate	Stickleback	0.004	Ag	180
Sodium chloride	Goldfish	10,000		240
Sodium chloride	Plains killifish	16,000		96
Sodium chloride	Green sunfish	10,713		96
Sodium chloride	Gambusia	10,670		96
Sodium chloride	Red shiner	9,513		96
Sodium chloride	Fathead minnow	8,718		96
Sodium chloride	Black bullhead	7,994		96
Sodium nitrate	Stickleback	600	Na	180
Sodium nitrate	Goldfish	1,282		14
Sodium sulfate	Goldfish	100		96
Strontium chloride	Goldfish	15,384		17–31
Strontium nitrate	Stickleback	1,500	Sr	164
Titanium sulfate	Fathead minnow	8.2	Ti	96
Uranyl sulfate	Fathead minnow	2.8	U	96
Vanadyl sulfate	Fathead minnow	4.8	V	96
Vanadyl sulfate	Bluegill	6		96
Zinc sulfate	Stickleback	0.3	Zn	204
Zinc sulfate	Goldfish	100		120
Zinc sulfate	Rainbow trout	0.5		64

*In this table the concentration values are all lowest at which definite toxic action is indicated by research. It must not be assumed that lower concentrations are harmless. Most of the data are for temperatures between 15° and 23° C. Exposure times have been approximated in some cases.

References

1. Coburn, S. E., "Limits for toxic wastes in sewage treatment," *Sewage Works J.* **21**, 522 (1949).
2. Connecticut State Water Commission, 8th Report (1938–1940), Hartford, Conn.
3. Dodge, B. F., and W. Zabban, "How a small electroplater can treat cyanide plating waste solutions with hypochlorite," *Plating* **42**, 71 (1955).
4. Jones, E., *Fish and River Pollution*, Butterworth, Washington, D.C. (1964), p. 74.
5. Kittrell, F. W., "Metal plating wastes in municipal sewerage systems," in Proceedings of 5th Southern Municipal and Industrial Waste Conference, 1956, at Chapel Hill, N.C., p. 216.
6. Pagano, J. F., R. Teweles, and A. M. Buswell, "The effect of chromium on the methane fermentation of acetic acid," *Sewage Ind. Wastes* **22**, 336 (1950).
7. Rudgal, H. T., "Bottle experiments as guide in operation of digesters receiving copper–sludge mixtures." *Sewage Works J.* **13**, 1248 (1941).
8. Rudolfs, W. (Chairman Research Committee), "Review of literature on toxic materials affecting sewage treatment processes, streams, and BOD determinations," *Sewage Ind. Wastes* **22**, 1157 (1950).
9. Sierp, F., and H. Ziegler, "Influence of zinc upon the purification of waste water," *Chem. Abstr.* **44**, 2152 (1950).
10. Whitlock, E. A., "The significance and treatment of cyanides in industrial waste waters," *Water Sanit. Engr.* **4**, 249 (1953).

CHAPTER 18

DISCHARGE OF RAW WASTES TO STREAMS

One of the legitimate uses of a watercourse is as a channel to receive and carry away wastes from humans and industry. There are still numerous instances of receiving waters that are capable of assimilating considerable quantities of untreated wastes and of sustaining this load without any noticeable detriment; however, their number is decreasing. Although, in these cases, it is not ordinarily necessary for a community to require industrial-waste treatment, indiscriminate discharge of untreated wastes to a stream (regardless of the dilution afforded by the stream or the uses to which it is put) should not be allowed. Discharge of untreated wastes (although almost an academic problem in 1968) should be permitted only after a detailed survey, by competent and certified sanitary engineers, of the existing condition and future uses of the receiving stream.

As our population grows and industry expands to meet the needs of the people, fewer opportunities will exist for the discharge of untreated wastes to rivers, since there will be more competition for water and more use of streams. The need for (and extent of) waste treatment will increase, probably more rapidly than the population.

It is psychologically and physically easier for an industry to install waste-treatment measures initially than to begin abatement practices after many years of operating without them. With this in mind, it is imperative for the plant engineer to set forth a program of future pollution-abatement steps, regardless of the company's present waste-treatment requirements.

An industry is usually liable legally for any wastes originating on its property and reaching public watercourses and the public shows little mercy for an industry which provides no treatment of waste waters which may be a nuisance. An industry, therefore, should carefully consider the possibility of lawsuits and adverse public opinion before it decides on direct discharge of raw wastes to a nearby stream.

AN EXAMPLE OF THE DISCHARGE OF AN UNTREATED WASTE DIRECTLY TO A STREAM

A large pickle-processing company is located in a small town. Both factory and town discharge wastes into a nearby stream, in which there is an annual problem: springtime fish deaths. The town uses an inadequately designed and poorly operated Imhoff tank for the treatment of its wastes; the pickle company's wastes receive no treatment. The problem was to determine adequate remedial measures to be taken by both the town and the pickle company. Of major concern in the study carried out by the author was the handling of the wastes from the pickle factory.

18.1 The Pickle-making Process and Its Wastes

Cucumbers from one to four inches long are picked in the "green season" (May to July), shipped to the pickle company, and stored in 5000-gallon open wooden vats. During the off-season or out-of-use periods, these vats are filled with water to which is added 50 pounds of lime, to "sweeten" the vat and keep the wood joints swollen tight. During the green season, to make way for the cucumbers, the vat is emptied and the lime water discharged to the sewer—the first type of waste discharged in the spring.

The vats are then filled with a 15 per cent brine solution, the cucumbers are added, and a wooden head is clamped down, to keep them submerged in the brine. Active fermentation, accompanied by gasification and foaming, begins about a week after the cucumbers go into the vats. For a week or so the fermentation is at its peak; thereafter it slowly ceases. The cucumbers are left in the brine vats for about three months, after which they are sorted according to size and briefly returned to the vats. The spent brine, discharged to the sewer, is the second source of waste.

From the brine vats, the sorted cucumbers go to wooden tanks, with a solution of turmeric and alum added for coloring, swelling, and firming purposes. Air is also pumped into the tank to keep the pickles in constant motion. This process continues for about eight hours, after which the turmeric and alum water is discharged to the sewer and becomes the third major source of waste water.

Up to this point all cucumbers, whether they are intended for sweet or sour pickles, have been receiving the same treatment. Now their pathways part.

Sweet pickles, if they are to be cut, are transferred to a tank fitted with a rotating cutter and screen catch. The pickles are sliced by the rotating cutter and washed by a jet of water at the same time. The washings, seeds, and some small pieces of pickle go through the screen and into the sewer, creating the fourth source of waste and the one which contains the greatest quantity of suspended organic matter. The cut and midget pickles are then put in vats containing vinegar and sugar. Fortunately, this solution is usually not wasted after the pickles have been removed.

Sour pickles are also washed and sliced, with the loss of some seeds and pieces in the wash water, and are put into vats containing vinegar, herbs, and a very small amount of sugar. Again, there is normally no waste from this vat, as the liquid is used over and over again.

Both the sweet and the sour pickles are then packed in jars, with a syrup or vinegar solution as required; in the case of dill pickles, brine and a dill solution are added. The sealed jars are washed with water, which becomes the fifth major source of waste, containing traces of sugar, salt, vinegar, and dill. The jars are delivered to a pasteurizer, where boiling water is flushed over them for a certain contact period. This hot water is wasted and, although it is not highly contaminated, its considerable volume and high temperature can be harmful to the receiving waters. After this sterilization step, the jars are labeled, packed, and shipped to consumers all over the world.

To summarize, there are five major wastes from the transformation of cucumbers into pickles, any one of which could result in the gross pollution of receiving waters. These wastes are lime, brine, alum and turmeric, pickle pieces, and syrup (vinegar and sugar).

The brine, alum, and turmeric wastes are usually discharged every day throughout the year. The irregular addition of the alkaline lime water or the acid vinegar-and-sugar wastes determines the resulting pH and BOD of the wastes from the pickle factory.

18.2 General Stream and Waste Survey

A waste and stream survey of the existing pollution conditions was made on August 23 and 30 and September 6, 1952. With lower stream flows and higher stream temperatures, oxygen conditions are generally more critical during this time of year. However, fishes do not normally spawn at this time. The survey consisted of recording all the wastes discharged by the pickle factory, analysis (on one of the days) of the town Imhoff-tank effluent, and the condition of the receiving stream from the point of discharge of the pickle waste to a point 12 miles downstream on the river. The following sample points were chosen to establish the extent of pollution downstream from the pickle factory: (1) flume box and V-notch weir installation in the discharge line of the pickle factory (see Fig. 18.1); (2a) branch just above junction with (2b) small stream forming beginning of river just

Fig. 18.1 The author measuring volume of waste flow and pH standard in the effluent from the pickle factory.

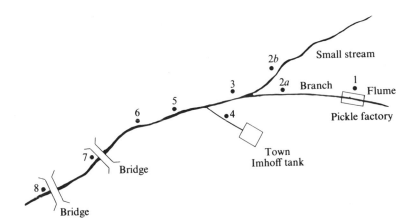

Fig. 18.2 Map of stream which received waste from the pickle factory and the town, showing locations chosen by the survey team as sampling points.

above junction with branch; (3) river just above town Imhoff-tank outfall; (4) town Imhoff-tank outfall; (5) river just below town Imhoff-tank outfall; (6) river 2 miles below Imhoff-tank outfall; (7) river 8 miles below Imhoff-tank outfall; and (8) river 12 miles below Imhoff-tank outfall. The sampling points are shown on the map in Fig. 18.2.

The approximate average volumes and rates of flow of the river are presented in Tables 18.1, 18.2, and 18.3, to give the reader some idea of the sluggishness of the river. The results of the analyses of composite pickle-factory waste, Imhoff-tank effluent, and receiving waters at the various sampling points for the three survey days are presented in Tables 18.2, 18.3, and 18.4. The volumes and types of pickle-factory wastes for the three days are given in Table 18.5. A summary of the relative pollution loads contributed by the town and the pickle factory is shown in Table 18.6.

18.3 Evaluation of Survey Results with Regard to Fish Killings

This preliminary survey was intended only to find the cause and extent of the pollution which resulted in springtime fish killings. The lowest flow in the river, observed 8 miles below the town, was 12.7 cfs; 12 miles below the town it was 29.4 cfs. The U.S. Geological Survey supplied data showing that the average flow in the river 12 miles below the town was 55.3 cfs but the minimum June flow was only 2.16 cfs. This variation between average and minimum flow is considerable and worth noting carefully. The June flow is of primary importance as far as preservation of fish life is concerned. The lowest flow observed in the river during our survey (12.65 cfs) was approximately six times the minimum June flow on record. This means that—all conditions except river flow remaining constant—the pollution concentration observed

Table 18.1 Average rate of flow in river.

Station number	Average velocity		Distance from		Time of travel from	
	ft/sec	miles/hr	Town Imhoff tank, miles	Upstream station, miles	Town Imhoff tank, hr	Station just upstream, hr
4	2.5	1.7	0	2	0.0	3.4
7	0.3	0.205	8	6	9.8	6.4
8	0.29	0.198	12	4	29.8	20.0

Table 18.2 Chemical and sanitary characteristics of wastes and receiving stream on August 23.

Characteristic	Station number*							
	1	2a	2b	3	4	5	7	8
Flow								
cfs	0.716				0.462	6.38	36.2	40.5
24-hr, total gal†	96,775				300,000			
Dissolved oxygen, ppm	4.0	1.0	6.0	1.4	0	1.0	0	5.0
Chlorides, as ppm NaCl	3,000			5300		3500	600	0
pH	5.4	4.6	6.2	4.6	6.6	5.9	6.2	6.5
Color	Yellow	Slight yellow	Slight brown	Slight yellow	Gray	Milky	Slight tannin	Slight tannin
Odor	Pickle	Pickle	None	Pickle	Stale sewage	Slight pickle	None	None
Temperature, °C	24.8			23	23	24	23.5	23.5
BOD (20°C, 5-day), ppm	780				49			
Suspended solids, ppm	96				148			
Total solids, ppm	8,670				350			
Dissolved solids, % of total	99				58			

*No figures for Station 6 on this day.   †Estimated.

Table 18.3 Chemical and sanitary characteristics of wastes and receiving stream on August 30.

Characteristic	Station number				
	1	4	6	7	8
Flow					
cfs	0.818	0.462		Lower than on August 23	
24-hr total, gal*	68,040	300,000			
Dissolved oxygen, ppm	0		0	0	4.7
Chloride, as ppm NaCl	2,832		708	325	295
pH	4.0		5.2	6.5	6.2
Color	Yellow		Gray-brown	Slight tannin	Tannin
Odor	Pickle		None	None	None
Temperature, °C	24		27.5	25	24
BOD (20°C, 5-day), ppm	2,040				
Suspended solids, ppm	224				
Total solids, ppm	24,274				
Dissolved solids, % of total solids	99				
Total volatile matter, % of total solids	4.6				
Total ash, %	95.4				
Suspended volatile matter, % of total suspended solids	59				
Suspended ash, % of total suspended solids	41				

*Estimated.

Table 18.4 Chemical and sanitary characteristics of wastes and receiving stream on September 6.

Characteristic	Station number			
	1	4	7	8
Flow				
cfs	0.840	0.462	12.7	29.4
24-hr total, gal*	90,600	300,000		
Dissolved oxygen, ppm	2.0	0	2.7	0
pH	6.3	6.4	5.8	5.1
Color	Yellow-brown	Gray-brown	Slight tannin	Slight tannin
Odor	Pickle	None	None	None
Temperature, °C	24	24	22.5	22.5
BOD (20°C, 5-day), ppm	390	99		
Suspended solids, ppm	124	162		
Total solids, ppm	10,268	408		
Dissolved solids, % of total	99	60.4		

*Estimated.

during our survey might be six times as great at other times. The extremely low flow in the spring plays a major role in the fish deaths.

There are numerous reasons why great numbers of fish are killed in watercourses during the spring. Fish are known to be susceptible to low or high pH or to abrupt changes in pH. They are also susceptible to high salt concentrations, to excessively high temperatures or rapid changes in temperature, to toxic chemicals, to fish diseases, and to lack of dissolved oxygen, as well as to many other conditions. The purpose of our survey was to ascertain which factor was most likely to be causing the fish deaths and what wastes were responsible.

pH. Ellis [4] states that fish and the common aquatic organisms prefer pH values of 6.5 to 8.4 and that pH values below 5.0 or above 9.0 are definitely detrimental, even lethal. Changes in these values within a few hours, coupled with a slight increase in temperature or lowering of the dissolved oxygen, can be fatal to various warm-water fish and trout. Stiemke and Eckenfelder [6] found that the fish reacted differently according either to their stage of development or to the season. In spring they were killed at pH values higher than those producing death in winter.

Although the pH of the pickle-plant effluent varied from 3.9 to 8.8, that of the receiving waters below the town's septic-tank outfall did not go below 5.1. The pH of the stream was considerably above this value (but not over 7) at most of the stations during the three days of the survey and it did not change mat-

Table 18.5 Wastes discharged per day by pickle factory.*

Date	Turmeric and alum,† gal (pH 3.2)	Brine,† gal (pH 3.2)	Lime, gal (pH 9.9)	Vinegar and sugar, gal (pH < 1.0)	Cooling water, gal	Washer water	
						Jar washer, gal	Cut pickle washer, gal
8/23/52	3720	4000	45,000	2055	0	0	0
8/30/52	4557	4000	0	1170	0	0	0
9/6/52	8025	0	30,000	0	0	0	1,850
Average normal daily discharge‡	7500	1000	5,000	1600	27,800	9600	12,000
Maximum	8500	6000	50,000	2300			
Minimum	4700	0	0	700			

*Estimated from tank dumpings.
†Discharged almost every day in the year.
‡As indicated by plant supervisor.

18.3 EVALUATION OF SURVEY RESULTS

Table 18.6 Relative pollution loads contributed by town sewage and pickle-factory wastes.

Characteristic	Pollution source	
	Town	Factory
Average volume, gal/day	300,000*	85,142†
BOD		
ppm	74‡	1,070
lb/day	185	760
% of total	19.6	80.4
Suspended solids		
ppm	155	148
lb/day	388	105
% of total	78.8	21.2

*Based on water consumption.
†Based on actual flow measurements.
‡After treatment.

erially from one day to the next. These results tend to eliminate pH in itself as the determining factor in the fish killings. However, during June and minimum flows, it might be a contributing factor, owing to the greatly reduced amount of water available for dilution.

Salt concentration. Ellis [4] also states that changes in salinity, as reflected by the specific conductance of a stream, can be critical to fish. Eldridge [3] claims that fresh-water fish and other aquatic life can stand only a limited concentration of the various minerals contained in brine. Concentrations of 5000 to 10,000 ppm do not permanently affect most types of fish, unless these concentrations are maintained for longer than 24 hours. When the mineral content is continuously high, concentrations of 500 to 1000 ppm will eventually result in the death of the fish. The probable order of toxicity of the various chemicals found in brine is as follows: $KCl > K_2SO_4 > MgCl_2$ and $CaCl_2 > NaCl$. Brines cause wildfowl to seek fresher water, and vegetation is soon destroyed in localities and streams into which there is a continuous discharge of brine. Anderson [1] asserts that all salts are toxic in concentrations high enough to exert unfavorable osmotic pressure. Some salts, namely sodium acetate, bromide, chloride, formate, and nitrate, have approximately the same threshold concentrations in terms of molarity. Anderson found that 3780 ppm of NaCl, added to Lake Erie water, was the threshold concentration required to exert this unfavorable osmotic pressure.

Although the salt concentration was 3000 ppm in the pickle-plant effluent, it was considerably less in the receiving water. Eight miles below the town on the river, only 600 ppm were found on one day and 325 ppm on another. From these results, it appears that the salt concentration in the river was not high enough (at least on the days of our survey) to cause fish to die. However, samples would have to be taken during low-flow periods to determine accurately the chloride concentration at the time when the fish were dying.

Temperature. Dimick and Merryfield [2] concluded that at the highest temperature recorded in their experiments, 89°F (32°C), no trout, salmon, or whitefish were found and that temperatures of this magnitude would probably be lethal to these fish. On the other hand, largemouth bass, pumpkin-seed sunfish, bluegill sunfish, white crappie, bullhead catfish, and carp did not appear to be distressed.

Although the sampling days were not the hottest days of the summer, they were sufficiently hot for the survey team to be able to ascertain the effect of weather on the temperature of the water. Since the pickle waste never reached 25°C and the river water below the Imhoff-tank outfall usually had a lower temperature than the pickle waste, temperature alone was eliminated as the factor mainly responsible for the fish killings.

Toxic chemicals. A review, with the proper authorities, of the pickle-factory processes and the sanitary constituents of the sewage revealed no unusual chemicals which might be toxic if they were in low concentration and did not cause a pH change of the receiving water.

Fish diseases. No dead fish were in evidence during the survey. A conference with interested fishermen along the river failed to disclose any signs, such as spots or other markings on the dead fish, which would indicate fish diseases brought about by natural causes. Fishermen also stated that they had not noticed any fish with red or otherwise irritated gills. This further confirmed the fact that there were no toxic chemicals or irritating constituents in the water.

Lack of dissolved oxygen. Dimick and Merryfield [2] further state that the breathing rate of fish is, in general, increased in water of low oxygen concentration, and under such conditions the toxicity to the fish of poisonous materials may be greatly intensified. They found no live salmon or trout at any time during August and September (months during which these fish are usually most in evidence) in water having less than 5.2 ppm of dissolved oxygen. They therefore suggest that 5 ppm of dissolved oxygen is the lowest rate satisfactory to fish. Southgate [5] decided that no general statement can be made, even from the mass of data now available, on the minimum concentration of dissolved oxygen required to support fish life. It is known, for example, that the oxygen requirements of fish vary with temperature, with the species and age of the fish, with the concentration of other substances dissolved in the water, and with other factors. Most ichthyologists, however, seem to believe that 4 ppm of dissolved oxygen must be present if the environment is to be habitable by fish. Some species can survive on as little as 2 ppm, but it would be difficult to find an expert who would be willing to swear to this.

Our survey of this river showed, by a process of elimination and also by observation, that there was a lack of dissolved oxygen in the river, from the pickle factory to 12 miles downstream. This proved that the organic loading on the receiving water was too high. The bacteria naturally present in the river water and those added in the sewage from the Imhoff-tank effluent needed food to survive and multiply. This food was being supplied in the pickle-factory wastes and in the sewage itself. But the bacteria, in consuming this food, utilized the oxygen dissolved in the river water to oxidize (digest) the food, thus removing the oxygen from the water at a rate faster than it could be supplied from natural sources.

However, on two of the three days the river had recovered sufficiently, 12 miles from the town, for ample dissolved oxygen to be once again available at that point. The lack of oxygen would be even more acute in June, when the water would be hotter and not able to dissolve as much oxygen and when the bacteria metabolism would operate at a faster rate. It is easy to visualize the fish coming up the river to spawn in June, meeting the combined pollution from the town and the pickle factory, and literally suffocating from the lack of oxygen.

The potential of the pickle waste to use up oxygen in the river is indicated by the BOD test. The results of the survey showed that the three-day composite pickle waste exerted a BOD of 1070 ppm in an average of 85,142 gallons per day, while the town-sewage effluent had a BOD of 74 ppm in 300,000 gallons per day. As a percentage, this means that the pickle waste exerted 80.4 per cent of the oxygen demand on the receiving waters, while the town sewage exerted the remaining 19.6 per cent. There were many other indices of pollution, but the BOD appeared to be the one which applied to this case, since the dearth of dissolved oxygen was the main consideration.

18.4 Preliminary Conclusions and Suggestions

The first phase of the pollution investigation, therefore, revealed the following facts. It was apparent that the fish deaths were being caused, at least in part, by too small an amount of dissolved oxygen. The concentration of chlorides and the low pH could also be contributing to the deaths, in low-flow periods. These facts should be verified in the spring of the following year, if such pollution were permitted to continue.

The cause of the oxygen shortage was in part the town sewage and in part the wastes being discharged from the pickle factory. More precisely, the town's sewage accounted for one-fifth of the lack of oxygen, while the pickle factory was responsible for four-fifths. The chlorides and acids came mainly from the pickle factory. The organic-matter pollution, causing lack of dissolved oxygen, existed for at least 12 miles down the river from the town. On some days the lack of oxygen might extend beyond that point.

Our recommendations, based on these facts, were: (1) the town should clean out the Imhoff tank, repair the mechanisms for diverting flow from one side to the other, calculate the capacity and loading on the treatment device, and carefully study its efficiency. If the present system is overloaded, it should be discarded and a more modern and efficient method of treatment, with qualified supervision, employed to treat the sewage. (2) The pickle factory should analyze its operation and housekeeping practices and try to decrease the discharge of polluting material into the river. A complete study of the BOD values of each of its constituent wastes should be made at the same time and some type of waste treatment installed.

After the town and the pickle factory had concluded the investigations outlined above, consideration should be given to storing the brine from the factory during the low-flow period the following June. Observation of stream conditions during that period would determine definitely the cause of the fish deaths. When eliminating a problem, it is always best to know exactly what is causing it before one spends money on projects which may not, in the long run, prove effective.

18.5 Findings of the Pickle-Factory Survey and Detailed Recommendations

The special pickle-factory survey was made in October 1952 and was originally conducted to recommend screening the plant waste to eliminate solids from the river, as one step toward restoring normal fish life in the river during the following spring. Many samples of waste were collected and analyzed for suspended solids, pH, and BOD. Representative analyses are presented in Table 18.7.

The information presented in Table 18.7 shows that all the wastes are acid to some degree. These acids can kill fish and for that reason alone should not be allowed to enter the sewer unneutralized. It was also clear that the high concentration of solids came from the alum and turmeric vat wastes and from the drainings from the relish machine and box. Control of these two wastes alone would greatly reduce solids in the plant waste entering the river. From an oxygen-depletion standpoint all the wastes, except the cut-pickle washing-cylinder waste, are very pollutional. The syrup wastes contained an unusually high oxygen demand and should never be allowed to enter the river without adequate treatment. The 5-day 20°C BOD of this waste was greater than 18,000 ppm (or more than 90 times as pollutional as normal domestic sewage). The volumes of each of these wastes determine the oxygen demand on the stream.

After observing plant practices, the author decided that there were many places where solid wastes could be kept out of the sewers by slight modifications in equipment, methods, and operations. These changes should be made before considering capital expenditure for waste treatment.

The recommended changes in plant practice to reduce discharge of solids into the sewers were:

1. A permanent screen should be installed over the holes in the drain plugs in the syrup tanks to aid in the retention of small pickles and peelings.
2. Plug boxes, which allow the release of spent

Table 18.7 Analysis of wastes from pickle factory.

Number	Source	Suspended solids, ppm	pH	BOD, ppm
1	Alum and turmeric waste from whole pickles	1024	4.1	420
2	Drainings from slicing machine and box	7980	3.2	2,500
3	Syrup waste from cucumbers and onions*	868	2.5	>18,000
4	Cut-pickle brine-wash waste directly from vat before discharge	122	4.3	2,880
5	Cut-pickle washing-cylinder waste	254	5.8	160
6	Cut-pickle brine-wash waste (first wash)	647	4.0	1,770
7	Alum and turmeric waste from chips	1185	3.2	1,890

*Both are pickled in a similar way.

alum and turmeric, should be repaired and made longer for the alum and turmeric and first-wash vats.
3. The vats should not be filled too full with the alum and turmeric water and the air should be controlled, to stop pickles jumping into the plug boxes.
4. All plug boxes should have four sides and covers which fit tightly.
5. Pickles fallen on the floor should be swept into a pile, picked up, and put in garbage barrels, not flushed or swept down the sewer or into cracks around the vats.
6. More care should be taken when loading vats, especially with cut pickles, to ensure that the pickles go into the vats and not onto the floor.
7. A metal tray or trough should be fitted under the slotted wooden collection box for shredded relish in the relish machine, to keep the valuable (and highly pollutional) relish solids and drainings out of the sewer.
8. A new box with a fine-mesh screen should be placed under the drain in the rotary-cut pickle washer. The screen can be cleaned once or twice a day and the recovered solids reused or removed to the garbage.
9. More care should be exercised in removing (with the nets) the cut pickles from the turmeric and alum vats into the slotted boxes.
10. All holes in floors, especially the ones specifically used for drains (as under the relish shredder), should be covered with screening to retain solids.
11. The fine stringy matter deposited underneath the pickle classifier (which now goes through a large open drain hole in the floor and thence to the sewer) should be reclaimed for canning. The hole should be covered and replaced by a handhole or a smaller drain hole covered with screen.
12. The drain holes near the cut-pickle washer should be covered with screening, since many seeds and pickles go right through the floor to the ditch leading to the sewer.
13. Screens should be installed on the grating of the floor of Assembly Room 2, where jars are filled and packed for shipping. The floor drainings, mainly from broken jars and jar washings, are presently flushed through the grating into the sewer flume lines.

These recommendations for specific modifications exemplify the thinking behind this approach to waste treatment.

When most of these improvements have been made, the plant personnel should be lectured on the meaning of stream pollution. Such a lecture will not only help to ensure the effectiveness of the plant improvements but will also make every employee pollution-conscious. This can benefit the town as well as the plant. After all this is accomplished, screenings on a large scale and other waste-treatment practices should be investigated.

Now let us briefly summarize what was achieved during this first period of study: the process waste waters were tested for pH, suspended solids, and 5-day BOD; it was found that the wastes were all acid, that the alum and turmeric wastes and the relish drainings and "losings" were very high in suspended-solids concentration, and that all process waste waters were high in BOD, except the cut-pickle washer waste. The syrup wastes, the strongest, were more than 90 times as strong as normal domestic sewage in BOD.

The author suggested many solids-recovery practices to reduce suspended-solids pollution of the river and proposed an educational program stressing the effects of stream pollution on the individual. He recommended that the remedial and educational programs should be carried out before any other waste-treatment measures were undertaken.

18.6 Effects of Changes on Factory and Stream

The author made a detailed survey the following spring of the progress made in the recommended elimination of solids in the pickle-factory waste. It was found that recommendations 2, 3, 4, 5, 7, 9, 10, 11, and 12 (Section 18.5) were being followed satisfactorily enough. Further suggestions were made to improve conditions outlined in items 1, 6, 8, and 13. Item 1 was to be remedied in the near future. More care should be exercised in item 6 to return the supposedly empty boxes of cut pickles to the cart right side up. In item 8, the screen under the washer should be upright at all times, to prevent pieces of pickle from going down the drain under the washer. Also, a removable wooden cover over the spray chamber would save a considerable quantity of water. The screens suggested in item 13 had not been installed. However, after watching the wash-up period, the author was still convinced that they were needed to retain the many pieces of pickle and relish being flushed through the grating and into the river.

In addition, two changes were suggested to reduce the volume of the highly pollutional vinegar-sugar syrup waste: a pump with a suction nozzle should be used to retain more of the syrup from the vats; and the excess syrup should be clarified with activated carbon and reused, thus not only reducing a strong waste but also saving money.

The plant manager himself suggested the possibility of using an extracted turmeric, without the starch base. It is hoped that this will aid in reducing the organic content of the turmeric acid.

The plant employees were given a 20-minute lecture by the author on the effects of wastes on rivers, as part of the company's educational program to reduce wastes. A procedure was also outlined for discharging brine slowly and at the proper times during the spring months.

A visual field survey was also carried out, by the author and several interested officials, from the plant to 15 miles below the town. No signs of fish deaths or indications of pickle wastes were observed in the river. Continued dissolved oxygen measurements at various stations were not made by the author, since funds were not available for this phase.

Since it was the opinion of the fishermen who used the river that pollution would "tell" on the red-bellied perch coming up to spawn within the next three weeks, the following pollution-abatement program was outlined for the pickle factory:

1. Make vigorous efforts, with existing methods, to prevent pieces of pickle from getting into the watercourse.
2. Investigate methods for reuse of all vinegar–sugar syrup or at least try to prevent its discharge during the summer months.
3. Withhold the discharge of brine as long as possible, then let it slowly siphon into the river, preferably during a rainfall, with adequate dilution from the lime water and other plant wastes.
4. Follow the further remedial measures suggested to prevent pieces of pickle from getting into the river.
5. Keep the public informed about the steps which are being taken to reduce stream pollution.

To the author's knowledge, no further fish deaths resulted from the discharge of pickle wastes into the river. In this case, in-plant improvements and changes in operation were sufficient to prevent fish killings in the receiving stream during the early and mid-1950s. There was little doubt, however, that some form of waste treatment would eventually be required as industrial production and municipal growth increased. In the meantime, essentially untreated wastes were being discharged without detriment to the best usage of the river.

References

1. Anderson, B. G., "The toxicity thresholds of various sodium salts determined by the use of daphnia magna," *Sewage Works J.* **18**, 82 (1946).
2. Dimick, R. E., and F. Merryfield, "The fishes of the Willamette River system in relation to pollution," *Sewage Works J.* **19**, 958 (1947).
3. Eldridge, E. F., *Industrial Waste Treatment Practice*, McGraw-Hill Book Co., New York (1942), p. 335.
4. Ellis, M. M., "Industrial wastes and fish life (abstract)," *Sewage Works J.* **18**, 764 (1946).
5. Southgate, B. A., *Treatment and Disposal of Industrial Waste Waters*, H. M. Stationery Office, London (1948), p. 25.
6. Stiemke, R. E., and W. W. Eckenfelder, *Effects of pH, Acids and Alkalis on Fishes*, Engineering Research Bulletin no. 33, North Carolina State College, Raleigh (January 1947).

CHAPTER 19

DISCHARGE OF PARTIALLY TREATED INDUSTRIAL WASTES DIRECTLY TO STREAMS

Large industries located outside city limits often have water requirements so great that they must develop their own sources of water and likewise they must dispose of their own wastes. This is one price an industry must pay to obtain sufficient space and escape municipal taxes. There are scattered instances, of course, of cities extending their sewer lines to accept the wastes of a nearby industry; but usually a plant draws its process water from a nearby river or well and discharges its wastes to the same stream, after a careful analysis of the uses of the stream and its condition. Wastes from these large plants contain so much pollution that some treatment is required before discharge into the stream. Since treatment of large volumes of waste water is expensive, an industry should investigate many alternative methods of protecting the receiving waters. Although such studies require time-consuming survey, analysis, and evaluation, an industry cannot bypass these steps without spending an excessive amount of money for waste treatment. In short, the more the industry knows about its wastes, generally the lower the cost of the actual treatment.

AN EXAMPLE OF DISCHARGING WASTE TO A STREAM AFTER PARTIAL TREATMENT

A large textile mill is located on a river, in a small mill town with a small Imhoff tank providing sewage-treatment facilities. The stream receiving the mill waste has been classified C (having fish survival for its best usage). Two dams downstream from the mill cause impoundment of the stream and subsequent nuisance conditions. The mill takes its water supply from the stream just above the plant, where it maintains a reservoir containing about one billion gallons of water when full. The state has directed the industry to maintain the stream condition as classified, so it has little choice except to treat its wastes before discharging them directly to the stream. Since extensive treatment of such large quantities of waste as there are in this instance can be economic suicide in many cases, a great deal of planning, research, and analysis must be carried out in order to minimize costs and yet attain the best possible efficiency. The state has granted the mill a six-month period of grace, to set up a pilot plant to determine the feasibility of biological treatment and aeration as means of reducing the BOD of the mill waste.

With the above problems in mind, the textile mill asked the author to supervise the investigation and to analyze, interpret, and evaluate the results of the six-month pilot-plant study. In order to make positive recommendations to management, the author also made in-plant and river studies and held conferences with management to impress upon them the following facts.

Waste treatment must be considered an integral part of production. In manufacturing, the objective is to produce, with the least expense, an article which will satisfy the consumer. Similarly, in waste disposal, the objective is to reduce, with the least expense, the pollutional nature of the effluent, in order to maintain the standards of the receiving river. Waste treatment which costs more than it should or does more than is required is contrary to the laws of good business; similarly, inadequate waste treatment reflects poor business judgment, since public hostility breeds sales resistance and a bad corporate image. A plan for waste treatment in this instance must, therefore, include the minimum of aeration, neutralization, and retention which will still accomplish adequate reduction of the pollutional load on the river.

In addition to these three pollution-reducing

devices, certain microorganisms can be developed during aeration that will assist in decomposing the organic matter in the waste. But these bacteria require the proper environment, two important features of which are air and pH, and the maintenance of sufficient dissolved air and a near-neutral pH costs money. For instance, the amount of detention time required determines the size of the holding tanks.

19.1 Procedure

The pilot plant for the textile-mill waste consisted of a holding basin, an aeration tank, and a final settling basin (Figs. 19.1 to 19.4). It took six months of operation to decide whether these devices, could—at a reasonable cost—maintain the "fishing" classification of the river. The six-month study was divided into four separate investigations: (1) efficiency of the pilot plant; (2) characteristics of the receiving river; (3) substitution of soluble for insoluble sizing, to reduce the strength of the waste; (4) volumes and loads of individual waste lines. The pilot plant was operated continuously over the entire period. Rates of flow of air and waste were varied and BOD was used as the criterion of treatment efficiency. The results are presented in Table 19.1.

The river was sampled during a two-week period in mid-October and again for one day in November. The river flow was purposely maintained at about 15 mgd, the present minimum flow, by means of overflow control at the reservoir. The results are presented in Table 19.2. The oxygen-sag curve and BOD profile are plotted in Fig. 19.5. Monthly averages of the flow of the river for the last five years are given in Table 19.3.

Experiments in starch substitution were carried out by the author: carboxymethyl cellulose (CMC) and Penfer gum 300, two substitutes for starch, were compared with pearl starch, the sizing most commonly

Fig. 19.1 Waste-treatment engineer observing high pH of incoming waste.

Fig. 19.2 Side view of pilot plant built to test methods of treating textile-mill wastes.

used. The results are shown in Tables 19.4 and 19.5. Samples taken from the waste pipeline were collected and analyzed when the mill was using starch only and when it was using starch-substitute products. These results are presented in Table 19.6.

A 24-hour survey of all four waste lines and of the combined effluent was made on a normal operating day, August 22; the flows and BOD's of each are given in Table 19.7. There are four main wastes from this plant: dye, starch, kier (or bleach), and desize (Table 19.7). The desize waste, the main offender, contributes 60.8 per cent of the BOD load and 40.5 per cent of the flow. Since the other three wastes contribute smaller but significant flows and BOD loads, segregation does not appear feasible, especially because of the high BOD reduction required. The total flow measured by means of a weir box was 6.09 mgd on the test day. The total BOD load was between 32,000 and 35,550 pounds per day. The waste was highly colored, warm, and had a composite pH of about 11 or higher. It definitely needed to be treated for the stream standards for the river (Table 19.8) to be maintained. The quality standards for this class of river water are described in Table 19.8.

19.2 River Studies

The oxygen curve of the river below the mill showed that the bottom of the sag occurred at the first dam (Station 3) (see Fig. 19.5). Below this point the river showed some recovery of oxygen. The plotted results illustrate that the critical stretch of the river was from the mill outfall to the first dam. This represents a distance of only about two miles and an average flow time of about 450 minutes ($7\frac{1}{2}$ hours). The BOD in this two-mile stretch diminished precipitately, indicating a rapid utilization of oxygen. The BOD of the river, determined from an 8-day average (Fig. 19.5), was 360 ppm just below the mill effluent and only 219 ppm at the dam. This means that a reduction of 141 ppm of BOD (61 per cent) was taking place in this critical stretch. If the *initial* river BOD could be reduced to 141 ppm and the organic matter reduced at the same rapid rate, theoretically there would be none left at the first dam. However, since organic matter

Fig. 19.3 Operator observing anemometer, which registers total air being fed to the wastes in aeration tank.

remaining after-treatment probably will oxidize more slowly, some would remain at the first dam.

Deoxygenation rates of 0.3 to 0.6 per day can be attributed to the decomposition of the organic matter between the mill outfall and the first dam. Although reaeration rates cannot be accurately determined in this stretch because of the absence of any oxygen at Station 3, extremely high values are evident from calculations using certain assumptions. The use of these approximate K_1- and reasonable K_2-values resulted in BOD reductions of 60 to 70 per cent, which were necessary to maintain the required 2 ppm of dissolved oxygen. This is expressed graphically in Fig. 19.5. A line parallel to the BOD curve was drawn so as to give zero BOD at a point just above the first dam. A BOD of about 120 ppm (rather than 360 ppm) is required at Station 2, which constitutes a BOD reduction of 67 per cent. Thus it can be shown, both numerically and graphically, that 60 to 70 per cent BOD reduction must be achieved to maintain 2 ppm of dissolved

Fig. 19.4 Aeration basin, viewed from above. Final settling basins are at the top of the picture.

oxygen at Station 3. The addition of clean, highly oxygenated dilution water during low flows can be construed as equivalent to a certain BOD reduction. For example, at normal deoxygenation rates, 1 pound of dissolved oxygen will be utilized during the first day by about 5 pounds of 5-day BOD. As stated before, the river studies were made at the present minimum flow of 15 mgd; any additional flow, carrying clean water, would naturally increase the total oxygen assets of the river. The following computations clarify, by the use of actual figures, the value of clean-water storage upstream and the resulting increased minimum flow available below the mill.

The effect of river dilution on BOD reduction is as follows. Keeping in mind the fact that there are 8.34 pounds of water in a gallon, with present river conditions (storage upstream, one billion gallons; summer minimum discharge, 15 mgd), we find

Table 19.1 Summary of six months of operation of pilot plant.

Treatment or characteristic of waste	Waste-flow rate, gpm						
	2	2	2	3	4	4	6
Holding-basin detention, hr	27	10	10	6	5.25	5.25	3
Aeration-tank detention, hr	12	12	12	8	6	6	4
Final settling basin detention, hr	8	8	8	5.5	4	4	2.75
Air rate, cfm	90	90	60	60	60	60	86–95
Air, ft^3/lb BOD	5400	5400	3600	2650	2400	2065	2130
Number of 4-hr composites collected	68	16	18	19	22	16	30
BOD of raw waste at 20°C, ppm	834	840	798	842	750	872	822
BOD of primary settled waste							
ppm	374	710	711	701	700	755	730
% reduction	56.5	15.5	10.9	15.5	6.7	13.4	11.2
BOD of aerated waste							
ppm	210	196	232	333	346	569	419
% reduction	74.8	76.9	71.0	60.4	53.9	34.8	49.0
BOD of final settled waste							
ppm	198	193	194	330	350	583	402
% reduction	76.3	77.1	75.9	60.9	53.8	33.1	51.0
CO$_2$ used per day* prior to aeration, lb							
Minimum	0	0	3	12	27	0	43
Average	2	10	17	48	47	0	65
Maximum	10	62	52	121	76	0	94

*To reduce pH to about 9.0

15 mgd × 8.34 lb/gal × 8 ppm DO upstream
$= \sim 1000$ lb O$_2$/day

available in the river. At normal deoxygenation rates of 0.1 per day, this oxygen, exclusive of reaeration, will take care of a pollutional load containing about 3300 pounds of 5-day 20°C BOD (assuming only about 30 per cent of the 5-day BOD will be satisfied in the first day of critical flow time).

Changing the river conditions by increased dilution (storage upstream, 1.5 billion gallons; summer minimum discharge, 22.5 mgd), we have

22.5 mgd × 8.34 lb/gal × 8 ppm DO upstream
$= \sim 1500$ lb O$_2$/day

available in the river. At normal deoxygenation rates of 0.1 per day, this oxygen, exclusive of reaeration, will take care of a pollution load containing about 5000 pounds of BOD. Therefore, instead of a 70 per cent reduction (2300/3300) being required to meet stream standards, only 2300/5000 (about 45 per cent) will be necessary. In other words, this dilution results in approximately 24 per cent BOD reduction.

19.3 Pilot-Plant Results

The pilot plant, consisting of a combination of holding, aeration, and final-settling basins, showed BOD removals of 33 to 77 per cent. A total of 189 four-hour composite samples were collected and analyzed during the 6-month period. The percentage of BOD removal depended on the detention time of the mill waste in the holding basin, the quantity of aeration applied to the waste, and the degree of pH reduction (Table 19.1). Holding the waste for 10 hours and aerating it for 12 hours at 60 cfm resulted in an overall BOD reduction of 71 per cent. This would be an acceptable method of treatment without any other disposal devices. However, the cost of installing and maintaining equipment to produce such great amounts of air and CO$_2$ would

Table 19.2 Results of sampling of river receiving textile-mill wastes.

Sampling days*	Station number and location	Flow, mgd	Dissolved oxygen		Temperature, °C	BOD (20°C, 5-day), ppm	pH	Time of flow to next station, min	Accumulated time, min
			ppm	% saturation					
October 11–14†	1. Water plant	15.4	8.9	89	16	0.75	6.8	0	0
	2. Just below waste entry	15.4	3.7	42	22	400	10.0	75	75
	3. First dam	15.4	0	0	23	180	7.4	375	450
	4. Bridge	15.4	0.3	3.2	18.5	160	7.3	90	540
	5. Second dam	15.4	1.9	19	16	142	7.4	720	1260
	6. Bridge	15.4	0.3	3	15	122	7.4	75	1335
	7. Bridge	15.4	1.6	15.7	15	100	7.4	130	1465
October 18–21‡	1. Water plant	16.0	8.9	84	13	0.50	7.1	0	0
	2. Just below waste entry	16.0	3.8	41	20	320	10.3	75	75
	3. First dam	16.0	0.1§	0	19	258	7.9	375	450
	4. Bridge	16.0	0.0	0	17	202	7.6	90	540
	5. Second dam	16.0	2.4	23	13.5	171	7.4	720	1260
	6. Bridge	16.0	0.8	7.4	12	131	7.4	75	1335
	7. Bridge	16.0	3.6	33.3	12	145	7.5	130	1465
November 16	1. Just below waste entry	14	8.0	94	24	280	10.7	0	0
	2. Just below waste entry	14	4.6	52.3	22	240	10.4	30	30
	3. First dam	14	0.0	0	20.5	200	8.0	375	405
	4. Just below dam	14	0.0	0	18	190	7.2	60	465

*All figures are averages of four consecutive days of stream sampling.
†With chlorination of sewage from Imhoff tank below Station 1.
‡No chlorination of sewage from Imhoff tank below Station 1.
§One sample showed 0.4 ppm; three showed 0.0 ppm.

be prohibitive. The mill would do well to investigate the possibility of using boiler-flue gas as a source of CO_2 and giving less aeration time to the wastes. It was found that little or no CO_2 was needed prior to aeration when the waste was held for a 24-hour period. One experiment, using a 5-hour holding period and 6 hours of aeration without any neutralization with CO_2, gave about 35 per cent BOD reduction (Table 19.1). Several alternative plans for disposal of the wastes, which would achieve 60 to 70 per cent BOD reduction, were suggested, based on the pilot-plant results.

1) Lagoon 27 hr + substitution of soluble sizing on 40% of grey goods

BOD reduction: (56.5%) + (14.5%)* = 71%

2) Lagoon 27 hr + aeration 4 hr

BOD reduction: (56.5%) + (15%) = 71.5%

3) Holding basin 3 hr + aeration 4 hr + substitution of soluble sizing on 40% of grey goods

---

*Estimated from soluble-sizing reduction.

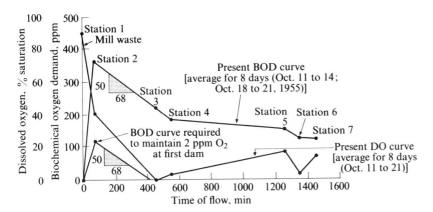

Fig. 19.5 Dissolved-oxygen and BOD profile of the river discussed in this chapter.

BOD reduction: (13.5%) + (36.5%) + (14.5%)
= 64.5%

4) Holding basin 10 hr + aeration 12 hr

BOD reduction: (10.9%) + (60.1%) = 71%

5) Holding basin 6 hr + aeration 8 hr + substitution of soluble sizing on 40% of grey goods

BOD reduction: (15.5%) + (44.9%) + (14.5%)
= 74.9%

6) Segregation* of kiering and desizing wastes 3 and 4 + holding lagoon 27 hr + aeration 4–12 hr

BOD reduction: (56.5 × 61%) + (23.5 × 61%)
= ~50%

7) Increase upstream storage reservoir to 1.5 billion gallons, which means a minimum flow of 22.5 mgd during a 3-month summer drought period + build a rock-lined trench leading to biological-oxidation basins (27-hr detention)

BOD equivalent reduction: (24%) + (56.5%)
= 80.5%

After a careful analysis of these methods, we see that there is great merit in using plan 7. Since there is an adequate amount of clean water available upstream (Table 19.3), it could be stored to supplement the river in the low-flow months, at a reasonable cost and with little or no future operating expense involved. Not only would this plan provide dilution of the river during the low-flow months, but detaining the waste for 27 hours in the biological-oxidation basins would result in a BOD reduction of 56.5 per cent. Adding this to the equivalent BOD reduction of 24 per cent provided by dilution, one obtains, for the total treatment, the equivalent of about 80 per cent reduction in BOD.

*Segregation would reduce the volume of waste to be treated from about 6.2 mgd to 3.4 mgd, but aeration of the remaining wastes for up to 12 hours would result in an overall BOD reduction of only 0.80 × 61 per cent = about 50 per cent of total. Because of the necessity of obtaining 70 per cent reduction, segregation would therefore not be practical, unless some treatment were also given to both dye and starch wastes.

Table 19.3 Volume of flow of river receiving textile-mill wastes.

Month	Average daily flow, mgd				
	1954	1953	1952	1951	1950
January	167.2	73.4	54.7	44.5	73.8
February	55.0	98.5	70.0	52.9	62.5
March	71.9	80.0	151.1	76.4	70.6
April	77.1	48.4	93.8	62.7	68.6
May	56.0	69.6	56.0	37.3	49.3
June	37.7	67.5	38.8	48.2	68.0
July	26.0	30.9	30.5	25.2	63.1
August	18.6	26.5	40.8	28.3	35.9
September	10.6	23.8	27.7	34.2	48.4
October	8.3*	24.7	26.5	22.0	47.7
November	11.6*	27.7	32.0	36.8	38.1
December	23.3*	70.1	41.4	145.1	81.0
Daily average for year	46.9	53.4	55.3	51.1	58.9

*Estimated.

Table 19.4 Results of experiments in starch substitution.

Sizing compound	Method of preparation	Laboratory BOD studies			
		BOD (5-day, 20°C)		BOD (10-day, 20°C)	
		ppm	% red.	ppm	% red.
Pearl starch (100%)	Heated at 160°F for 4 hr with Rhozyme; final solution 0.1%	800		874	
Pearl starch (65%) and carboxymethyl cellulose (35%)	Same as above	336	58.3	525	40
Penfer gum 300 (100%)	Same as above	369	53.9	511	41.5
Pearl starch (65%) and carboxymethyl cellulose (35%)	Heated at 205°F for 20 min, no enzyme added; final solution 0.1%	283	64.6	265	69.7
Penfer gum 300 (100%)	Same as above	321	60.0	318	63.7

Storage of clean water was found to be economically feasible, as shown in Table 19.3, since storage for any 3-month period from January to July during the 5-year period would have yielded more water than required to maintain flows of greater than 22.5 mgd in the stream. During a 5-year period, no monthly average flow fell below 26 mgd between January and July.

19.4 Substitution of Soluble Sizing

A mixture of 65 per cent pearl starch and 35 per cent carboxymethyl cellulose (CMC) resulted in a 5-day 20°C BOD which was 64.6 per cent lower than that obtained by using pure pearl starch (Table 19.4). In addition, no enzymes were needed, since the sizing is easily washed off the cloth with warm water. Penfer gum of the 300 series showed a 60 per cent 5-day BOD reduction. Even after 10 days, the BOD values were 63.7 to 69.7 per cent lower than those obtained with pure starch. The approximate 5-, 10-, and 20-day 20°C BOD values were verified in additional experiments (Table 19.5). There did not appear to be any lag in the oxidation of the soluble sizes. Field studies (Table 19.6) substantiated the BOD results of the laboratory; however, these involved only one day of actual testing. There is no question in the mind of the author as to the advantages of using either a mixture of starch

Table 19.5 Additional results of starch-substitution experiments with 65% pearl starch and 35% carboxymethyl cellulose in a 0.1% solution.

BOD* at 20°C, ppm

5-day	10-day	15-day	20-day
386	483	600	450†

*Average of four bottles (2 dilutions, duplicate samples) each day.
†Insufficient samples within BOD range.

Table 19.6 Tests for BOD outlet of No. 4 waste line (desize).

Date	Flow, mgd	BOD		Sizing compound
		ppm	lb/day	
Aug. 22–23	2.49	1040	21,600	Starch (100%)
Dec. 9	~2.49*	813	~16,900	Pure starch (60%); CMC, starch, and Penfer gum (40%)

*No exact flow was determined.

Table 19.7 Results of measurements of waste flow and BOD load for 24 hours (Aug. 22–23).

Waste line		Flow		BOD		
Number	Source	mgd	% of total	ppm	% of total	lb/day
1	Dye	1.735	28.2	320	13.02	4,635
2	Starch	0.950	15.45	440	9.81	3,485
3	Kier (bleach)	0.971	15.80	720	16.40	5,830
4	Desize	2.490	40.50	1040	60.80	21,600
Total (sum of 4 lines) as measured individually		6.146				35,550
Totals of effluent as measured		6.09		630		32,000

and CMC or Penfer gum of the 300 series. A definite BOD reduction is to be obtained by using them. However, the higher cost of the soluble sizes and an increased percentage of rejects have acted as deterrents to the changeover. If these objections could be overcome, less pollutional waste would be discharged.

The following is a list of recommendations.

1. Provide storage for 1.5 billion gallons of water upstream. This should provide a 3-month low-flow minimum of 22.5 mgd during summer operating periods.
2. Transport the entire waste of the mill through a shallow, rock-lined trench to a biological-oxidation basin, for a minimum of 24 hours of storage. The basin should be no deeper than 4 feet and constructed so as to minimize short-circuiting. The plant complied with both recommendations. Figure 19.6 shows the shallow rock-lined trench carrying the waste to the biological-oxidation basin and Fig. 19.7 shows the approach to the first oxidation basin, before it was filled.
3. Assign a full-time qualified engineer (preferably with water- and waste-treatment experience) to the position of waste-treatment supervisor. This man should keep *daily* records of the pollutional characteristics of both the raw waste and the oxidation-basin effluent, as well as the river conditions. These data should be kept on file at the mill and copies forwarded to the water-pollution control board each month.

Table 19.8 Standards of quality for fishing (C) classification.*

Items	Specifications
Floating solids, oils, settleable solids, sludge deposits	None [permitted] which are readily visible and attributable to sewage, industrial ... or other wastes, and which measurably increase the amounts of these constituents in receiving waters, after opportunity for reasonable dilution and mixture [of the waters] with the wastes discharged thereto
pH	May range between 6.0 and 8.5, except that swamp waters may range between 5.0 and 8.5
Dissolved oxygen	Not less than 2 ppm
Toxic wastes, deleterious substances, colored or other wastes, heated liquids.	[Not permitted if,] alone or in combination with other substances or wastes, [they are] in sufficient amounts or at such temperatures as to be injurious to fish survival or impair the waters for any other best usage, as determined by the Water Pollution Control Authority for the specific waters which are assigned to this class

*In accord with state regulations, these waters are suitable for fish survival, industrial and agricultural uses, and other uses requiring water of lower quality.

Fig. 19.6 Textile-mill waste traveling in a shallow, rock-lined trench to the biological-oxidation basin, where it will be held for a minimum of 24 hours.

4. Continue to investigate means of overcoming the drawbacks to the use of soluble sizes. The extra BOD reduction obtained by the use of these sizes may be needed in the near future, for additional protection to the stream.
5. Continue to operate the pilot plant, as time and personnel permit, to obtain additional information on minimum air and CO_2 requirements for secondary treatment.
6. Attempt to maintain continuous flow through the first dam at all times.

The first three recommendations should be carried out immediately.

Without the use of carbon dioxide to acidulate the wastes, BOD reductions were less. When the raw, unneutralized finishing-mill waste was held for only 5 hours and aerated for only 6 hours, an average of 28.5 per cent BOD reduction occurred.* That almost one-third of the BOD was oxidized at a pH of between 11 and 12 was quite a revelation. Some questions existed about the nature of the oxidation: mainly, whether it was chemical or biological. When the flow through the pilot plant was reduced (Table 19.9) to a holding time of 10 hours and an aeration period of 12 hours, a BOD reduction of 38.8 per cent occurred, despite the fact that the initial pH averaged 11.3. In earlier experiments at the same rate of flow (Table 19.1), 17 pounds of CO_2 per day were required to give a BOD reduction of 71 per cent. To install equipment to supply CO_2 in adequate amounts for this mill would cost approximately $150,000. In addition, it would cost about $275 per day for power and fuel to generate the CO_2. This is a considerable expense, even for a textile mill as large as this one. However, in many cases a mill can eliminate the high neutralization cost by providing sufficient detention time and sacrificing some efficiency; also, as mentioned previously, the use of flue gas for neutralization is often an inexpensive way to achieve extra BOD reduction.

---

*Average of Tables 19.1 and 19.9.

Fig. 19.7 Biological-oxidation basin built to hold textile wastes for 24 hours. Note shallowness: it is not over four feet in depth. Smokestacks of textile mill can be seen in background.

Table 19.9 Summary of additional results of pilot-plant study.

Treatment	Waste-flow rate	
	4 gpm	2 gpm
Holding-basin detention, hr	5.25	10
Aeration-basin detention, hr	6	12
Air rate		
cfm	60	60
ft³/lb BOD	2370	4650
Number of 4-hr composites	34	48
BOD of raw waste, ppm	760	765
BOD reduction (5-day, 20°C)		
Primary settling tank, %	6.7	8.3
Aeration tank, %	22.1	38.8
Final settling tank, %	19.1	37.5
CO_2 used, lb/day	0	0

Table 19.10 Environmental conditions during aeration.*

Number of 4-hr composites averaged	48
Temperature, °F	
Maximum	70
Minimum	55
Mean	64
pH level	
Raw	11.3
Primary settled	11.1
Aerated	10.0
Final settled	9.8
Plate counts in aeration basin, total bacteria/ml	
Maximum	14,000,000
Minimum	10,800
Mean	5,351,853

*Waste flow, 2 gpm; no pH adjustment.

The pH dropped only slightly, from 11.3 to 11.1, in the holding basin, but dropped more rapidly to 10.0 in the aerator (Table 19.10). The reaction of CO_2 in the air with caustic alkalinity in the waste, producing carbonates, and the CO_2 given off by bacterial action on organic matter, could have been responsible for the lowered pH. Total bacterial plate counts (similar to those made in water analyses) showed that an average of somewhat over 5 million microorganisms per milliliter (as reported by the mill chemist) were living in the aeration basin. The numbers varied from as high as 14 million to as low as 10,000. This concentration of bacteria and a **BOD** reduction of about 40 per cent signify that bacteria can survive, and apparently metabolize organic matter, at a pH as high as 11.3. The possibility of utilizing bacteria for oxidizing dissolved organic matter at elevated pH values should not be overlooked.

On the basis of these findings, a full-scale treatment plant was constructed, complying with the major recommendations of the author.

CHAPTER 20

DISCHARGE OF COMPLETELY TREATED WASTES TO STREAMS

Complete treatment of wastes prior to direct discharge to a receiving stream is gradually receiving more and more consideration. The amount of dilution water in streams is not increasing and, on the other hand, pollution loads unfortunately *are* increasing. With the population explosion and industrial expansion, we can expect more extensive waste-treatment requirements. At present, complete treatment is required only in special instances and in the case of the large, wet industries—for example, textiles, pulp and paper, steel, and chemicals.

There is some doubt as to what is meant by the expression "complete treatment." It is generally conceded that complete treatment refers to secondary treatment; that is, the removal of about 85 to 90 per cent of the BOD by a combination of physical, biological, and/or chemical means. According to this definition, one is removing only two polluting constituents: suspended solids and dissolved organic matter (including colloidal solids). Does this definition, then, imply that the removal of *any* two forms of pollution—such as color and suspended matter, oils and alkalinity (high pH), or acids and organic matter—also constitutes complete treatment? The author doubts that this is the original meaning of the term; and in these days when "complete treatment" is insufficient and certainly not complete in some cases, a reevaluation of our terminology is in order. For example, an industry may have little or no dissolved organic matter in its waste and yet be required to remove two or more other forms of pollution. In the author's mind, this also constitutes complete treatment, as the term is currently defined. The expression "complete treatment" will hardly be satisfactory, with its present definition, when the public begins to accept and include "tertiary treatment" in its thoughts on the subject. Tertiary treatment presently provides for the removal of three or more forms of contamination: suspended solids, dissolved organic solids, and dissolved inorganic solids. True complete treatment would remove refractory solids as well.

An industry requiring complete treatment for its waste usually discharges a large volume of waste and is located outside, and some distance from, a municipality, on a stream requiring the maintenance of high standards of water quality. This author prefers to consider "complete treatment" as that which renders waste waters reusable for industrial and (in some cases) municipal water supplies. This normally will mean a fairly complete removal of all suspended, dissolved, and colloidal solids, including both inorganic and organic fractions. Since this is, at present, rarely practiced, we are forced to accept as a definition of "complete treatment" the removal merely of a major portion of the suspended solids and dissolved organic matter.

AN EXAMPLE OF COMPLETE WASTE TREATMENT BY A FIRM PRIOR TO DIRECT DISCHARGE INTO THE RECEIVING STREAM

20.1 The Problem

Townsends, Inc., an integrated poultry operation, consists of a hatchery, feedmill, soybean mill, and poultry-processing plant located about two miles east of Millsboro, Delaware. It is owned privately by the Townsend family, and the raising and processing of chickens is their main business. The waste problem is at the poultry-dressing plant. This plant, built in 1957, is located about 50 yards from Swan Creek, a tributary of the tidal Indian River. The relative locations of the plant, town, and receiving waters are shown in Fig. 20.1. Of special significance is the location of the Millsboro extended-aeration sewage-treatment plant which discharges into the Indian River about 3 miles above the confluence with Swan Creek. The poultry-plant waste from Townsends is discharged after screening and ineffective flotation treatment into Swan Creek about 1 mile upstream of the confluence with the Indian River. The proximity of these waste discharges to the shellfish area only $2\frac{1}{2}$ miles below Swan Creek is a major concern to the regulatory authorities. Although the main portion of the town of Millsboro (Fig. 20.1) is served by the extended-aeration sewage-treatment plant followed by chlorination, many of the homes along Route 24 are individually served with septic tanks and well-water supplies located in relatively sandy soils. The underground disposal in sandy areas of sewage or wastes may represent some danger to these water supplies. During each of the last three years many areas of the Indian River Bay have had to be closed periodically during the summer for cleaning because of bacterial contamination. Coliform standards have been set at 70/100 ml for shellfish and at 1000/100 ml for swimming.

In 1956, the Delaware Water Pollution Commission concluded in their *Indian River Drainage Basin Survey* that:

1. A portion of the fresh-water flow within the Indian River watershed originates from swampy and marshlike areas which have a decided effect upon the chemical and physical composition of the runoff waters. These waters are generally high in iron and color, low in turbidity, suspended solids, and dissolved oxygen, and acid in pH.
2. Average dry-weather flow in this basin area is approximately 0.25 cfs per square mile.
3. Small tributaries predominate in the drainage basin. The only surface supply location with sufficient volume for either domestic or industrial use is at Millsboro dam.
4. Studies made by the Delaware State Water Pollution Commission when the former owner of the plant was in operation clearly indicated that dry cleaning of manure solids, coagulated blood, and feather removal followed by satisfactory removal of settleable solids with heavy disinfection will effectively and satisfactorily protect state waters downstream from this plant.

Millsboro (the closest and most significant municipal co-polluter in this case) is located on the Indian River, 13 miles from the ocean and is one of the prin-

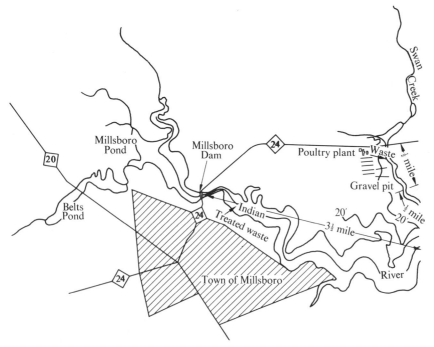

$2\frac{1}{2}$ miles to shell-fishing area (sheltered coves)

$7\frac{1}{2}$ miles to Indian River Bay

10 miles to Indian River Inlet (Atlantic Ocean) with Lewes, Delaware, on the north and Bethany Beach on the south

Figure 20.1

cipal towns of the Indian River County of Delaware. It is a distributing point for carloads of poultry feed and coal for the broiler chicken industry. The Commission's historical survey revealed that as late as 1956 the pool at the base of Millsboro Dam was still noted for its herring run in April and May; in good seasons as many as a million have been taken in a few weeks. At times crabbers brought thousands of soft-shell crabs to Millsboro for shipment alive in boxes filled with wet grass. The alternate opening and closing of Indian River Inlet prior to 1938 nearly ruined the industry, though a few soft-shell crabs were still shipped from there during the periods of transformation. The new inlet revived the market for crabs, fish, oysters, and clams taken in Indian River. The main body of the Indian River from Millsboro to the ocean is tidal, with an elevation of less than 10 feet at Millsboro. This, the flattest stretch of the area, yields a slope of only 0.7 foot per mile.

In 1956 about one-third (18,600 acres) of the Rehoboth and Indian River Bays, which receive the poultry waste, was utilized for oyster cultivation at an annual "take" of $800,000. At the same time an additional $237,500 revenue resulted from the growing and harvesting of clams in these bays. Each year as much as $250,000 is spent in the shore-line areas between Lewes and Bethany Beach for fishing tackle, bait, and other small items associated with the sport. Boat rentals have been estimated to bring $96,000 per year. The Indian River is a vital link to the tremendous menhaden fishing industry in the mid-Atlantic states. In 1953 the U.S. Fisheries Statistics Report stated that Lewes, Delaware, was the nation's leading fishing port poundwise with landings of about 363 million pounds, consisting almost entirely of menhaden. This catch had a reported value of $4,117,000. Duck hunting is also estimated to contribute about $25,000 per season and muskrat trapping about $15,000. Despite the value of the shell-fishing, fishing and hunting industries, the 1956 Delaware Report stated that "there is little doubt that bathing and swimming is a primary interest in this drainage basin area." They were referring to the areas of Rehoboth Beach, Lewes, Rehoboth Bay, Dewey Beach, Indian River Bay, and Bethany Beach.

The foregoing information led the Delaware Water Pollution Commission in 1956 to conclude that "this entire basin must, of necessity, be classified as an unusually clean water area which has as its major interests bathing, swimming, boating, sports fishing, commercial fishing, shellfish, wildlife, recreation, and seasonal real estate."

20.2 Stream Studies

The Delaware State Water Pollution Commission conducted many studies of the Indian River and Inlet Bay areas during 1952–55. Figure 20.2 shows the drainage basin and the location of the sampling points (described in Table 20.1).

The State of Delaware Water Pollution Commission investigated the quality of the Indian River (Fig. 20.1), which extends from the Millsboro Dam to the vicinity of Oak Orchard, on July 1 and 28, 1953. The results are shown in Tables 20.2 to 20.3. Fresh-water flow in the Indian River Basin was determined during two periods, May 1, 4, and 5, 1953, and May 5, 1955. These results are shown in Table 20.4. One may note that in Tables 20.2 and 20.3 samplings were taken as near to high tide as possible. Thus the increased volume of dilution water from the bay might tend to minimize the effects of pollution. It may also be noted that the water temperatures were high—a fact which is not considered abnormal since the Indian River Bay is broad and quite shallow. The high dissolved oxygen (although the upper reaches near the Millsboro Dam are relatively low) may indicate that little or no pollution is present.

The reader should recognize the scarcity of meaningful analytical data on the sanitary characteristics of the receiving stream. The evidence for pollution comes from instances of fish deaths rather than direct stream analytical measurements. However, coliform bacteria counts have been run on many samples of the Indian River at the sampling points shown in Fig. 20.3. Typical data on coliforms at some of these points collected as late as 1964 are shown in Table 20.5. These data indicate that considerable attention should be given to the bacteriological quality of the receiving water, since these are primarily recreational and fishing waters.

20.3 State Decision

The stream data illustrate a lack of positive evidence on the effect of organic loading, especially from the poultry plant. They do show bacterial contamination

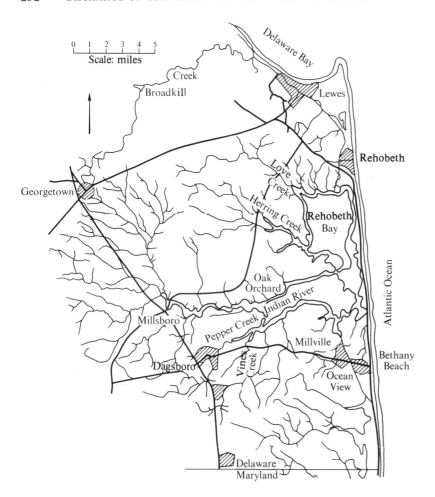

Fig. 20.2 Delaware Water Pollution Commission Survey of the Indian River drainage basin.

in recreational, fishing, and shellfish-producing waters. The author had to decide whether to recommend a complete stream survey to determine the exact degree of treatment required for the poultry wastes. He decided against this survey for the following important reasons: (1) it would be costly and time-consuming with no apparent financial support available from the poultry processer or the state; (2) the state commission had already decided that a high degree of treatment was required (and only this would be approved) in order to protect the valuable resources of the receiving waters downstream.

20.4 Poultry-Waste Characteristics

During the normal 8- to 11-hour working day at the poultry plant 9000 to 10,000 chickens weighing $3\frac{3}{4}$ pounds each are processed every hour. The processes and their associated wastes are summarized in Fig. 20.4. A separate septic-tank sewage-disposal system serves the 225 plant employees. The chickens are not force-fed (a procedure of fattening before killing to produce more weight) at the plant and dry removal is practiced. Although the killing room is separated from the rest of the processing operation and blood

Table 20.1 Indian River basin sampling stations.

Description	Miles from Indian River inlet	Description	Miles from Indian River inlet
Assawoman Canal, Ocean View	5.50		
White Creek tributary, Ocean View	6.20	Iron Branch at railroad	13.52
White Creek, Millville	6.34	Vines Creek, Frankford	13.70
Indian River	6.59	Indian River at Millsboro	14.10
Indian River	7.16	Iron Branch near Millsboro	14.40
Lewes–Rehoboth Canal jetty	7.60	Vines Creek, below feather-rendering effluent	14.65
Indian River	7.73		
Indian River, mouth of Island Creek	8.30	Vines Creek tributary receiving feather effluent	14.70
Blackwater Creek	8.46	South tributary of Iron Branch	14.80
Indian River	9.01	South tributary of Iron Branch	14.82
Stokely Cannery, Rehoboth Beach	9.40	Lewes–Rehoboth Canal Bridge on Route 18	14.86
Vines Creek	9.43		
Lewes–Rehoboth Bridge, Route 41	9.45	Vines Creek near Frankford	14.90
Love Creek	10.00	Vines Creek	14.95
Unity Branch in Fairmont	10.16	Shoals Branch at Betts Pond	15.20
Indian River	10.20	Famys Branch near Millsboro	15.50
Chapel Branch in Angola	10.24	Betts Pond at Route 113	15.70
Indian River	11.00	Vines Creek near Millsboro	15.90
Vines Creek	11.70	Roosevelt Inlet	16.59
Indian River	11.74	Cow Bridge Branch below Morris Millpond	17.62
Indian River	11.80		
Pepper Creek	11.82	Stockley Branch, Hospital for the Mentally Retarded	17.90
Love Creek	11.95		
Swan Creek near Millsboro	12.10	Wood Branch near Morris Millpond	18.75
Indian River	12.57	Deep Branch near Morris Millpond	18.85
Iron Branch near Millsboro	12.58	Cow Bridge near Morris Millpond	19.60
Vines Creek, Frankford	13.30	Wood Branch near Georgetown	20.07
Indian River	13.31	Wood Branch near Georgetown	21.40
Pepper Creek	13.45	Wood Branch, Georgetown	22.20
Pepper Creek	13.50		

Table 20.2 Quality of water in the Indian River from Millsboro to Oak Orchard on July 1, 1953.*

Sampling station, miles from inlet	Time	Temperature, °C	D.O., ppm	D.O. saturation, %	Salinity as NaCl, ppm	Depth, ft
Millsboro Dam (14.10)	1:00 p.m.	27.8	4.65	60.7	3,500	
13.31		27.5	5.5	71.0	3,200	
12.57		30.5	7.8	105.7	2,400	
11.74		29.5	7.3	97.3	2,500	
11.00†	1:30 p.m.	30.5	7.35	101.8	4,500	5
10.20		30.0	7.5	105.5	6,800	
9.01		30.0	7.4	105.5	8,050	
8.30	2:00 p.m.	30.0	8.3	131.0	16,300	
8.55		31.0	8.6	137.0	16,450	2
11.80		31.5	8.65	139.5	16,450	1.5
7.73		30.0	8.7	136.6	16,700	

*High tide (from the U.S. Geological Survey Table) was at 2 p.m. at Indian River inlet. The water throughout the stretch being studied was quiescent before 1.30 p.m. and choppy thereafter.
†High water.

Table 20.3 Quality of water in the Indian River from Millsboro to Oak Orchard on July 28, 1953 (high tide).

Sampling station, miles from inlet	Time (p.m.)	Temperature, °C	Air temperature, °C	D.O., ppm	D.O. saturation, %	Salinity as NaCl, ppm
Millsboro Dam (14.10)	2:00	17.5	25.0	3.5	17.1	7,100
13.31	2:05	28.0		5.0	68.6	8,200
12.57	2:10	30.0		11.8	170.0	9,000
11.74	2:20	30.0		8.3	123.0	11,800
11.00	2:28	30.0		7.9	118.0	12,200
10.20	2:35	30.0		7.7	118.0	14,300
9.01	2:45	29.5		7.7	120.0	18,200
8.30	2:55	30.0		7.6	124.0	20,000
7.73	3:04	28.5		8.3	134.0	21,400
7.16	3:12	28.0	26.0	8.5	138.0	23,000
6.59	3:22	27.5		8.5	139.0	28,000

is scooped out of the killing-floor area and disposed of with the screenings, the film which collects on the walls is washed into the sewer at the end of each working day. The feathers, which constitute about 14 per cent of the raw chicken weight, are sold for rendering for about $16 per ton; the offal, making up 16 per cent of the weight of the chicken, is sold for about $21 per ton. The processing waste-water is screened through four Sweco vibrating screens which are cleaned daily with alkali (1 pound/day) to keep them clean of feathers.

The Delaware Water Pollution Commission carried out composite analysis of the poultry-plant effluent on July 21 through 24, 1964. The results are shown in Table 20.6. The BOD averaged about 630 ppm with total nitrogen (mostly organic) of about 60 ppm and a slightly alkaline pH (7–8). Suspended solids were about 200 to 600 ppm, mostly organic.

Although quite accurate water-flow records are kept at the poultry plant and indicate a consumption of about 800 gallons per minute (500,000 gallons per day or an average of 5 gallons per bird), for additional information the plant effluent was weired and measured every half-hour during a typical operating day on January 27, 1965 (Table 20.7). Some reduction in process wash water may be achieved by closer control and using higher pressure nozzles. The flow rate, however, must be approved by the Department of Agriculture, which supervises cleanliness within poultry-processing plants.

Table 20.4 Fresh-water volume within Indian River basin.

Station (miles from inlet)	Flow, mgd	
	May 1, 4, and 5, 1953	May 5, 1955
11.95	10.25	4.04
10.24	7.2	2.80
10.16	13.0	4.03
12.10	6.20	3.58
14.10	111.1	39.40
14.40	3.54	1.55
14.80	3.02	1.60
11.82	1.59	1.63
13.45*		3.60
11.70	3.68	
13.70	0.36	
Total	159.37	60.60

*Upstream from 11.82 and not added into total.

20.5 The Solution

Since the Indian River and its receiving bays were already contaminated in the 1950s before the poultry plant began operations in 1957 and since the poultry-plant waste was also found to be highly pollutional

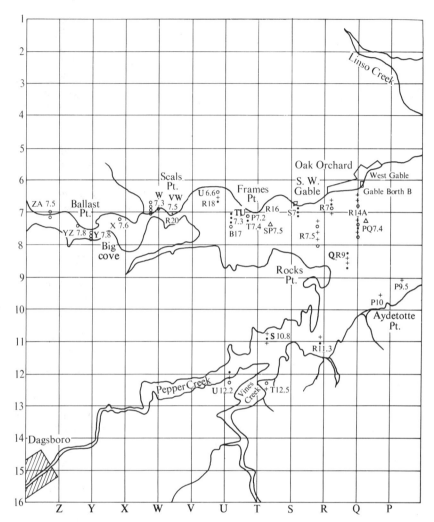

Figure 20.3

(both from analysis and stream observations), adequate treatment of the waste was necessary. From Tables 20.6 and 20.7 the total BOD load was computed to be 2550 pounds—equivalent to a population of 12,750 persons. The major question was what type of treatment should be used to protect the best uses of the stream. Obviously, the major concern is bacterial contamination, so that chlorination of the poultry-plant waste would be a minimum requirement. Chlorination in the presence of 2550 pounds of BOD and the other suspended and floating matter normally found in poultry-plant effluents would be difficult and costly. Organic matter reacts rapidly with chlorine and the chlorine necessary would be expected to cost well over $250 per day. Therefore, more economical means for removing a major portion of the organic matter prior to chlorination were demanded.

A two-stage, oxidation-pond treatment system was chosen to perform the task because of the low construction and operation costs compared with other biological treatment systems. The first stage consists of a baffled, high-rate, deep pond to allow sedimentation of heavy solids, flotation of grease or feathers which escape preliminary treatment by the screens, and bacterial degradation of the organic matter. This pond is 595 feet long, 109 feet wide, and 8 feet deep;

Table 20.5 Selected data on coliforms at stations in Indian River.

Date	Sampling point*	Coliform count, MPN/100 ml	Salinity, ppm
8/14/61	U6.6	790	14,500
7/25/62	QR9	430	23,600
7/29/63	VW7.3	430	
7/25/62	W7.3	4,600	17,700
8/8/62	W7.3	2,400	
8/21/62	W7.3	11,000	18,600
8/8/62	W7.3	2,400	
7/15/64	T7.2	430	
10/13/64	ST7.5	430	
	ST7.5		
7/24/62	R7	430	21,600
6/21/61	S7	1,600	
7/24/62	R7	930	22,500
7/15/64	R7.5	430	

*See Fig. 20.3.

30 over-and-under baffles on 15-foot centers cover the middle 435 feet. The second stage is a shallow photosynthetic pond designed to remove more organic matter and convert inorganic phosphates and ammonia nitrogen to an algal mass. It is 635 feet long, of nonuniform width, and about 2 feet deep, covering an area of about 212,000 square feet. The effluent from this two-stage treatment is chlorinated before discharge into Swan Creek. Detailed drawings of this plant are shown in Fig. 20.5. The area chosen for this two-stage treatment plant was predominantly sandy.

Four tests in the area confirmed that the soil was about 99.5 per cent inorganic matter (stable at 900°C) and only about 5 to 6 per cent moisture.

A rough cost estimate of $83,738 was given to Townsends by the author on May 25, 1965 (Table 20.8) and some minor revisions were made in the original plans on July 22, 1965. On August 25, 1965, George and Lynch Construction Company signed a contract with Townsends, Inc., for construction of the treatment plant at a cost of $90,000 and a construction period of about 45 days. Some photographs of the treatment plant during construction are shown in Fig. 20.6. The plant was officially inaugurated on March 14, 1966, although operation actually began about January 1, 1966.

20.6 Results

Some details of the design and operation of the facility were presented in *Poultry Meat* (August 1966). In a letter dated January 30, 1967, Mr. Donald J. Snyder, manager of the Dressed Poultry Division of Townsends, stated, "The system has been working very well and has never given us any trouble... and we are happy to have anyone inspect the system if they should care to."

Samples were collected and analyzed by the Delaware Water Pollution Commission on April 7, June 29, and December 6, 1966, and March 23, 1967. These results are shown in Table 20.9.

From these four samples it was apparent that the treatment facilities were operating satisfactorily at a BOD loading of about 1390 lb/day. The loading on

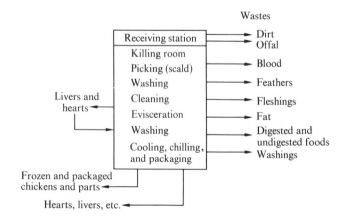

Fig. 20.4 Flow sheet of poultry-processing plant.

Table 20.6 Sanitary characteristics of the poultry-plant effluent as reported by the Delaware State Water Pollution Commission.

Characteristic*	Date (July 1964)					
	21	21†	22	22	23	24
Sample no.	638	640	642	672	674	676
5-day BOD††	425	1200	395	800	500	457
Chloride, ppm	64	74	37	55	45	24
COD††	1710	3690	3250	1590	2700	2780
Total N	62.4	91.8	57.4	60.2	57.7	59.3
Organic N	56.3	80.2	54.3	54.3	53.2	55.7
NH_3N	6.1	11.6	3.1	5.9	4.5	3.6
NO_2-N	0.076	0.018	0.09			
Acidity				28	5	20
Total alkaline (as $CACO_3$)				48	73	39
pH			7.6	6.9	8.1	7.4
Total suspended solids			360	606	254	204
Suspended volatile solids			360	584	244	180
Suspended ash			0	22	10	24
Total solids			801			
Total volatile solids			482			
Total ash			319			

*All results are given in milligrams/liter unless otherwise indicated.
†Plant washdown during sample collection on this day led to unusual results.
††An additional plant effluent sample was composited and analyzed on 3/17/65 and found to contain 418 ppm of BOD and 880 ppm of COD.

the first basin is

$$\frac{1390}{(595 \times 109)/43{,}560} = 1390/1.49 = 935 \text{ lb BOD/acre.}$$

At a daily waste-flow rate of about 530,000 gallons the detention time in this first basin is 7.35 days:

$$\frac{595 \text{ ft} \times 109 \text{ ft} \times 8 \text{ ft} \times 7.5}{11 \text{ hrs/day} \times 800 \text{ gal/min} \times 60 \text{ min/hr}}$$

This unusually high loading resulted in a BOD reduction of about

$$\frac{313 - 87}{313} \times 100 = 72.5 \text{ per cent.}$$

The second basin handled a BOD loading of 385 lb/day or

$$\frac{385}{212{,}000/43{,}560} = 385/4.87 = 79 \text{ lb/acre}$$

and effected an addditional BOD reduction of

$$\frac{87 - 79}{87} \times 100 = 9.2 \text{ per cent}$$

when the algae are not removed from the effluent and

$$\frac{87 - 26}{87} \times 100 = 70 \text{ per cent}$$

when the algae are filtered out of the final effluent. Detention time in the second basin averages about 6 days. No attempt is made to remove algae from the final effluent but the effluent is withdrawn slightly below the surface.

The overall BOD reduction obtained during the first year of operation (based upon only four samples) was about

$$\frac{313 - 26}{313} \times 100 = 92 \text{ per cent}$$

Table 20.7 Poultry-plant effluent flow on January 27, 1965.*

Time	Flow, gpm	Time	Flow, gpm
6:00 a.m.	645	12:30 p.m.	645
6:30	1190	1:00	645
7:00	800	1:30	645
7:30	1020	2:00	645
8:00	800	2:30	645
8:30	645	3:00	525
9:00	1190	3:30	380
9:30	800	4:00	352
10:00	1020	4:30	380
10:30	1020	5:00	408
11:00	645	5:30	380
12 noon	645	6:00	408
		6:30	380

*Average rate was 674 gpm or 40,400 gal and chickens were processed on this day at the rate of 8570 per hour, so that $\frac{40{,}440}{8{,}570} = 4.7$ gal/chicken were used.

when the algae were filtered from the final effluent and

$$\frac{313 - 79}{313} \times 100 = 75 \text{ per cent}$$

when the algae cells were left in the final effluent. Although scum removal in the first basin was frequently required in 1966, no operating difficulties or nuisance resulted from an overall plant BOD loading of

$$1390/(1.49 + 4.87) = 1390/6.36$$
$$= 219 \text{ lb BOD/acre/day}.$$

The preliminary results point out some other interesting phenomena, for example, that the expected rise in pH in the second pond was coupled with a corresponding reduction in phosphates and coliform bacteria. Total coliform counts in the chlorinated effluent approximate 10/100 ml and apparently meet current shellfish standards. Although it is too early

Table 20.8 Rough cost estimate of waste-treatment system for poultry-processing plant.

Item*	Cost per item, $
Pumps (2)	3,708
Installation and delivery of pipeline	500
Pipeline at $7.50/ft for 955.5 ft	7,180
Chlorinator (duplicate of existing one)	1,500
Excavation for two basins at $0.50/yd^3	30,000
Cement of soil cement at $4.50/barrel	15,800
Wood at $200/mbf	10,000
Poured concrete at $25/yd^3	2,500
Steel at $0.20/lb	10,000
Concrete block at $0.50/unit	1,100
C.I. pipe and fittings	750
Chlorination shack	200
Flagstone	500
Total	83,738

*These figures do not include some additional items which should be considered by the company such as

Fencing, especially for the no. 1 basin;
Ditching around basins to prevent groundwater intrusion;
Seeding of the birms to prevent erosion;
Landscaping to improve the aesthetic appearance of the system.

to formulate any firm and final conclusions, one can observe that elevated BOD loadings were handled in a properly designed two-stage, oxidation-pond treatment plant system and produced satisfactory operating results.

Continued sampling of the treatment plant facilities on March 29, June 14, and July 26, 1967, yielded the results shown in Table 20.10. Excellent BOD reduction continues, in the range of 85 to 90 per cent. In addition, coliform bacteria counts are less than 10/100 ml, which is acceptable for discharge into water used primarily for shellfish cultivation.

This example shows how a large poultry plant discharging about half a million gallons of waste per

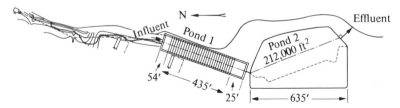

Fig. 20.5 Diagram of the two-stage, oxidation-pond treatment system designed for the poultry-processing plant.

Table 20.9

Characteristic	Influent pond no. 1			Effluent pond no. 1				Effluent pond no. 2			
	4/7/66	6/29/66	3/23/67	4/7/66	6/29/66	3/23/67	4/7/66	8/29/66	12/6/66	3/23/67
Physical										
Color, units			380			175				220
Turbidity, units			110			60				55
Dissolved oxygen, mg/liter		3.84			0			5.18	2.5	
Temperature, °C		21.5			24			25	7.0	
Minerals										
pH	6.4	6.4	6.9	6.8	6.7	6.8	7.4	8.7	7.0	7.1
Acidity (CaCO$_3$), ppm	31	26	42	41	55	50	27	0	52	36
Alkalinity (CaCO$_3$), ppm	38	32	16.4	135	143	155	138	8	144	153
Hardness (CaCO$_3$), ppm			65			69				68
Chloride (Cl), ppm		80	82		69	84		73		83
Nitrogen balance (mg/liter as N)										
Total Kjeldahl N	98.3	54.3	42.5	15	31.7	33.3	17	22.9	30	28.0
Organic N	89.9	45.4	28	4.2	8.7	5.3	9.0	17.4	4.5	8.0
NH$_3$-N	8.4	8.9	14.5	10.8	23	28.0	8.0	5.5	25.5	20.0
NO$_2$-N	0.47	2.44	0.013	<0.02	0	0	<0.02	0	0	0
NO$_3$-N	4.3	2.98	0.4	<2	0.1	0.24	0.38	0.4	0.5	
Waste analyses										
BOD, mg/liter	300	380	260	70	86	105	97(u)* 24(f)	83(u) 27(f)	65	70
COD, mg/liter	600	560	370	185	150	150	196	270(u) 190(f)	130	120
Total PO$_4$, mg/liter	8.6	0.74	10	11.4	2.8	9.9	7.4	1.6	9.7	9.2
Ortho PO$_4$, mg/liter					1.9			1.5		
Methylene blue alkyl benzene sulfonate		16								
Solids balance										
Settleable solids, mg/liter	1.2	2.0	0.6	<0.1	<0.1	<0.1	0.2			<0.1
Total suspended solids, mg/liter		338	148		70	50		218	52	52
Volatile suspended solids, mg/liter		326	140		64	50		208	52	52
Total solids, mg/liter		592	470		318	332		56.9	29.7	323
Total volatile solids, mg/liter		362	238		125	118		33.5	207	110
Bacteriological analysis										
Total coliform/100 ml	2.5 × 10^6	6 × 10^6		0.2 × 10^6	9 × 10^4		5.4 × 10^4	3 × 10^3		
Fecal coliform/100 ml	6 × 10^4	1.2 × 10^6		5 × 10^4	1 × 10^4		6 × 10^3	1 × 10^3		
Fecal streptococci/100 ml	1.3 × 10^6	2.0 × 10^6		3.2 × 10^5	4.5 × 10^4		6.3 × 10^4	<1 × 10^3		

*Unfiltered = u; filtered = f.

Fig. 20.6 Two views of oxidation basin no. 1: (a) down length, showing scum collection area in foreground and baffles in background; (b) on diagonal across basin in baffled area.

Table 20.10 Continued analyses of treatment plant.

Characteristic	3/29/67			6/14/67		10/31/67		7/26/67	
	Influent	Effluent basin no. 1	Effluent basin no. 2	Influent	Final chlorinated effluent	Influent	Final chlorinated effluent	Effluent	Final chlorinated effluent
pH	6.4	6.7	7.1	6.4	8.3	6.0	6.9	7.2	7.1
Acidity ($CaCO_3$), ppm	29	66	35	30	5	36	34	32	35
Alkalinity ($CaCO_3$), ppm	35	152	134	40	121	29	110	135	135
Hardness ($CaCO_3$), ppm	49	72	73	130	180	93	86		
Chloride (Cl), mg/liter	98	84	84	149	124	160	135		
Total Kjeldahl N	49.8	32	30.5			87.2	30.8	30.5	
Organic nitrogen	49.8	5.0	10			74.0	7.3	106	
NH_3–N	0	27	20.5	23.6	21.2	13.2	23.5	19.9	
NO_2–N	0.042	0	0	1.1	0.41	0.39	0.10	0.120	
NO_3–N	6.8	0.43	0.14	50.0	6.40	4.3	1.3	0.05	
BOD, ppm	340	100	55	365	39	470		38	30
COD, ppm	420	125	90	280	160	560	110		
Total PO_4	10	17	14	13.5	12.2	25	13	1.3	
Settleable solids, ml/liter				2.5		1.0	<0.1	0.3	<0.1
Total suspended solids, ppm	292	68	72	220	110	274	80	54	68
Total solids, ppm	727	443	350	776	479	820	459	523	514
Color	395	180	195	115	115	86	79	40	95
Turbidity	162	62	54	21	36	115	40	16	40
Temperature, °C								27	27
Coliform bacteria, 100 ml							<10		<10

day solved its pollution problem in a satisfactory manner. It was forced by circumstances to provide the equivalent of secondary treatment but did so at a cost of less than $100,000 capital expense. Adequate screening, followed by two-stage oxidation utilizing over-and-under contact baffles, and final chlorination gave 85 to 95 per cent BOD reduction. The cooperative spirit exhibited by both the plant and the regulatory authority, combined with some engineering innovations in design, resulted in success.

CHAPTER 21

SITE SELECTION

Selecting the best site for a new plant is one of the most difficult problems faced by any industrial firm [5]. To avoid spending an unnecessarily large sum of money and to forestall the possibility of spending days fraught with indecision and needless headaches, a company should make a careful site analysis before making other decisions concerning plant operation. Site selection is an especially difficult problem for the chemical and "varied-product" industries. There are many pitfalls encountered and the effects of choosing a wrong site can be catastrophic. Too often a company selects a site without sufficient consideration of water: not only water supply but water for dilution in waste disposal. The lack of either of these water resources can prevent the successful operation of any type of plant, on any type of location.

Many other factors, of course, are important in selecting a site: amount and type of labor available; union activity; attitude in the area toward industry; transportation facilities; electric power; state taxes; fuel (oil, gas, coal); climate; supply of raw materials; distance from market; waste-treatment requirements; disaster hazards; production hazards; prevailing winds; soil conditions; appearance of the area; recreation facilities; educational facilities.

Some of these factors are tangible and their values readily assessed; others, such as the psychological attitude of the community, are difficult to evaluate. The author cannot give a comprehensive discussion and evaluation of all the factors in this chapter, since to do so would involve writing an entire textbook. Readers of this book are naturally most concerned with the influence of clean water supplies and dilution waters for waste disposal on site selection. However, so that the industrial waste engineer may retain a perspective on the overall problem of plant location, a few comments on the other factors will be included.

21.1 Evaluation on Cost-of-Product Basis

A workable method for evaluating sites for new plants, given by Hoyer [5], shows the effect of many factors on the ultimate cost of operation per 100 pounds of product and presents factual data for those items that constitute major costs (Table 21.1). To look at the table and judge the sites on a cost-of-product basis only, one would think that site 1 was the obvious choice. But this assumption might not always be correct. It is interesting to note, for example, that the maximum difference in the cost of water at the eight sites is only $0.02 per 100 pounds of product, so that, from a cost standpoint, water does not appear to be an important factor. However, a more detailed study might reveal that the *quality* of the water supply varies with each site and this would have a direct effect on quality and cost of production. If the product being manufactured is such that waste disposal will constitute a major factor in the cost of the product, then waste treatment should be added to the breakdown in Table 21.1.

Another method of presenting the same comparative values is shown in Fig. 21.1, which gives a summary of a plant-location survey [9]. Here various costs, on an annual basis, are shown in bar-chart form. Such data, in this consolidated and summarized form, can be very helpful to an industry trying to decide on a site for a new plant.

21.2 Tangible and Intangible Factors

The importance of labor supply cannot be overemphasized. On account of the scarcity of labor in many cities in the northern United States, companies have begun to relocate in the South; this trend has been especially evident since 1954. The type of labor available is equally important, i.e., unskilled, semiskilled, or skilled, as required for the manufacture of the particular product. Often an industry finds it

Table 21.1 Plant-location study based on cost per 100 pounds of product. (After Hoyer [5].)

Site number*	Cost per 100 lb of product, $							Difference in cost from Site 1				
								$1000/yr		Total, $/100 lb	Cost of steam, $	Taxes, $
	Steam	Power	Water	Labor	Total freight	Taxes	Total	Before federal taxes	After taxes			
1	2.15	0.79	0.75	5.03	3.31	1.16	13.19					
2	3.70	0.72	0.73	5.03	3.19	1.97	15.34	215	133	2.15	1.55	0.81
3	3.77	0.49	0.73	5.03	3.32	2.26	15.60	241	150	2.41	1.62	1.10
4	3.52	0.87	0.75	5.83	3.78	1.68	16.43	324	200	3.24	1.37	0.52
5	4.42	0.89	0.75	4.27	3.11	3.06	16.50	331	205	3.31	2.27	1.90
6	4.64	0.79	0.73	4.27	3.27	3.06	16.76	357	220	3.57	2.49	1.90
7	3.94	0.89	0.73	4.70	3.58	3.07	16.91	372	230	3.72	1.79	1.91
8	4.64	0.79	0.73	4.38	3.34	3.61	17.49	430	265	4.30	2.49	2.45
Maximum difference	2.49	0.40	0.02	1.56	0.67	2.45	4.30					

*At Sites 1, 5, 7, and 8 no water transportation was available; at Sites 2, 3, 4, and 6, water transportation was possible.

possible to collaborate with a college or university in the area in training industrial personnel. In general, the potential employees should be intelligent, easily trained, cooperative, and sympathetic to industrial objectives. A knowledge of union activities can best be acquired by investigating past records of labor difficulties and by ascertaining the number of unions in the area, their numerical strength, and their attitude toward management.

Intangible factors include attitude of people in the area toward industry, appearance of the area, condition of the air, transportation, and recreation. The attitude of people in the area can best be judged from conversation and correspondence with city officials regarding location in their town. A favorable appearance involves clean and well-repaired streets, homes, business districts, parks, and schools, and, in addition, evidence that new construction is going on in the area. The condition of the air is of increasing importance to new industry, since the nuisances and potential health hazards arising from polluted air are manifold.

An industry should evaluate the ease of access to the town by road, rail, and air, as well as distance to major markets. Recreational facilities to be taken into account include playgrounds, swimming pools, golf courses, tennis courts, public parks and beaches, libraries, and cultural centers.

Wood [17] states that the major reasons given for plant relocation have been labor, markets, transportation, raw materials, power, and fuel. Another reason given by industries for changes from former locations is a continuing policy of decentralization, which brings better control of operations, social and economic benefits to employees, and lower costs.

Smith-Corona Marchant, Inc., when closing its Syracuse plant and moving one of its operations to Cortland, New York, said [12], "In reaching a decision to move, the executive committee was influenced primarily by the necessity of achieving the economies of operation which would result from the consolidation, and by the larger number of employees located in the Cortland area as compared with the Syracuse area." It is interesting to note that this move took place even though complete waste-treatment facilities will be required at the new site, whereas there were no major facilities required at the Syracuse plant.

The Thiokol Chemical Corporation decided to locate a new engine plant at one of seven southern sites [8]. The choice was said to depend on acceptance by the community, the labor situation, the transportation problem, and the necessity for a deep-water port. The Lee-Schoeller Paper Company announced

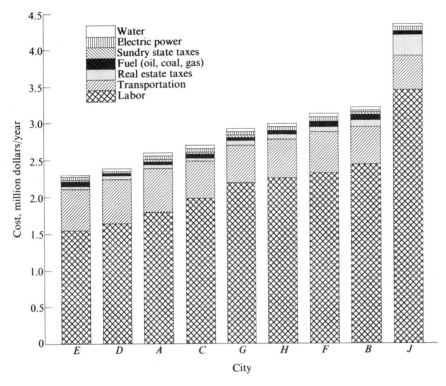

Fig. 21.1 Summary of plant-location survey. (After Neuhoff[9].)

recently that it would build a new plant for the manufacture of photographic paper at Pulaski, New York. State Commissioner McHugh was quoted [13] as saying, "It is interesting that the choice of site was determined by the high-quality water supply, in addition to the excellent transportation facilities by rail, and the new Interstate Highway 81."

Brennan [14], associate editor of *Plant Management and Engineering*, recently stated: "Company officials who plan construction of a new plant emphasize that while the conveniences offered by a superhighway site certainly add to its attraction, equally important facets of the site-selection process are just as much studied: space available, utilities provided, labor force on hand, and potential for future expansion."

Safford [11] finds that, in most cases, an engineer's first steps toward site location consist of estimating the total pollution load which will be discharged into the stream from the proposed plant at both its initial and its ultimate size and then establishing several tentative locations.

Toal [15] says that the perfect plant site would consist of a large acreage of level, flood-free land, with excellent bearing surface, bounded on one side by a railroad, on another by a stream carrying an unlimited quantity of virtually distilled water, on the third side by a high-tension power line, and on the fourth by a major highway, affording ready access and advertising value. The site should be located outside (but relatively close to) a substantial community, which contains all necessary community facilities and cultural advantages. A single owner, willing to sell at a reasonable price, finishes off this picture-book site. Toal noted that he has never found such a utopian site.

21.3 The Importance of Long-Term Planning

There is an important relation between city planning and an industry's decision on where to locate a new plant [17]. The advantages of one location over another consists of slight differences in such factors as plant sites, local laws and taxes, utilities, amusements, transportation, and so forth. If a city looks ahead, it can plan to allocate certain areas to industrial development, complete with railroad sidings and access highways, so that there can be factory sites of various sizes and the individual plants do not have to construct long spur tracks, pave roads, or put in water, gas, sewage, or electric power lines to connect with the nearest existing system. The cost of land should be reasonable, so that it does not restrict the size of the site. Large areas should be set aside, to allow room not only for parking space and space around hazardous areas but also for future expansion. The exact prediction of future expansion is difficult, of course, and usually falls short of eventual need. However, wise city planning can usually go a long way toward eliminating the majority of these difficulties.

Most industries which produce large amounts of waste choose sites outside city limits, primarily to avoid city taxes but at the same time to get a big piece of land without paying too much for the site. It does not necessarily follow that the city's income will be reduced, in these cases, by an amount corresponding to what the industries would have paid in taxes if these plants had been located within the city limits. A city gets its income indirectly, since wages paid by the industry to its employees are spent in the city and this spending supports the business of the city; also, workers who own homes in the city pay real-estate taxes. If the industry moved out of the area entirely, these benefits would disappear. It is, therefore, sound economics for the city to encourage industry to locate wherever there is sufficient room for expansion, whether it is inside or outside city limits; it is in the city's interests to have the industry expand at this site rather than move at some later date to another city where expansion is possible. A great many industrial plants are moving for this very reason (lack of expansion space) although they often do so under the guise of some other reason.

Lewis [16] believes that a community that is well planned and administered, as a result of good city and regional planning, may find industries coming to it primarily because of these characteristics. He clasfies industry [7] into urban and suburban types. In his urban grouping he puts small-scale industries with a seasonal labor force and therefore a high turnover in employees, which use little or no water, and therefore contribute little to the waste problem. His suburban group consists of the heavy-metals and chemical industries, refineries, textile mills, and so forth, i.e., industries requiring large quantities of water and producing significant waste-disposal problems. He also mentions a third group, industries capable of employing a sufficient number and variety of workers to enable them to go into new areas and establish communities of their own. He classes in this group large-scale munitions manufacturers, mining companies, and factories for the assembly and testing of airplanes and names steel cities such as Gary, Indiana, and rubber cities such as Akron, Ohio, as examples.

21.4 Waste Disposal as a Critical Factor

The disposal of waste from industrial processes has been neglected in the past, but such neglect is no longer permissible. There are a few isolated sites where it is still legal for an industry to discharge offensive wastes, but it is much better to face the disposal problem when a plant is being built than to evade it in the hope that somebody else will do something about it later; the problem may be much more difficult later on.

Pollution-control authorities are generally stricter in their specifications for waste-disposal facilities for new plants than for old ones. The reasons for this are sound. The construction of new plants is proceeding at such a rapid rate that they will outnumber older plants in the next few decades, and a new plant can better afford to install extensive waste-treatment measures than an old plant which should install them but does not have money for the purpose set aside in its budget. Also, a new plant can be designed to incorporate the waste-treatment system in its production scheme, with the proper space and piping facilities included from the start. This state of affairs is usually impossible, or difficult to the point of being a severe handicap, in older plants.

Not only pollution-control authorities, but also the public, apply pressure on industries concerning this matter. For example, fear of public outcry forces oil refineries to be constantly aware of the waste-disposal problem when selecting a site. A rather caustic editorial printed in the *New York Times* on June 13, 1962, asked, "How many more bitter and losing battles will they have to fight before corporation managers learn to look for industrial sites that will not invade or endanger areas that have been dedicated to wildlife conservation? The Shell Oil Company should have foreseen the buzz-saw it ran into by trying to establish a refinery in the midst of the famous waterfowl marshes of the upper Delaware Bay.... Shell would be wise to look elsewhere for a site." In a letter to the editor the following day, Eugene H. Harlow commented that "it is true that oil spills are a disaster to nature's children, since petroleum derivatives float about endlessly, killing many forms of life and driving away others, until they deposit their sticky blackness on some possibly distant shore. It is up to the oil industry to prevent this scourge." He pleads for simply eliminating oil spills rather than turning the refinery away from the chosen site.

Goudy [4] points out that many industries have to relocate because of trouble with waste disposal at their old locations. He also mentions that industries have a habit of minimizing their waste-disposal problems and so making it difficult for consulting engineers to advise them as to proper location. Since a great deal of money is spent by both a municipality and an industry, after the industry has located on a certain site, in an attempt to solve waste problems, Goudy makes a plea for consideration of these matters while the location of the industry is being planned. He gives many examples of incompatible solutions of industrial and municipal waste problems in California. In summary, he states that the question of waste disposal should be given far more attention than it has been in the past. It would greatly assist industries considering the problem of relocation to deal with a coordinating committee, which would have at hand all the information the industrial managements need to arrive at a final decision where to locate. In some metropolitan areas, this committee might consist of representatives from public and private utilities, as well as heads of city and county departments supplying services to industry.

Wood [17], however, points out the disadvantage of utilizing local coordinating committees. Competition—among localities and among industries interested in almost every industrial location—makes it necessary to keep the project confidential until a decision has been reached. In such cases, a consulting engineer can function as a coordinator, because he can assemble information impartially without disclosing the identity of his client.

Safford [11] lists municipal and industrial growth (hence greater pollutional loads, but at the same time shorter work weeks, more extensive use of recreational facilities, and higher standards of living) as the major factors in the modern emphasis on waste treatment as a factor in plant location. Another aspect of this problem which cannot be expressed in dollars is the fact that unless the waste-disposal problem is solved satisfactorily it will require undue attention by top mill executives, demanding time and energy which should be spent on their regular duties. Safford emphasizes the importance of the size and topography of the site, since, in so many cases, the most feasible method to dispose of waste is by storage in shallow basins which may serve as both equalization tanks and oxidation lagoons or by storage in reservoirs which provide a substantial amount of impoundment, with release according to the ability of the nearby stream to assimilate pollution.

Another consideration is that changes in manufacturing process or upgrading of regulatory standards may require more complete waste treatment than management currently envisions. Ample acreage is therefore doubly a necessity.

Several engineers [16] were presented with a hypothetical plant-location problem, for which they presented the solution with respect to industrial-waste discharge step by step. In their panel discussion, the background of the plant was given first; then the interests of the company's management, the state pollution-control authorities, and the industrial-sewage specialist were discussed. The consulting engineer described the sort of facilities he would design for the waste-treatment plant; a hypothetical contractor built the plant; and finally the operator's part was outlined. The reader would do well to refer to this original paper [16] to become familiar with the details involved in an industrial waste-treatment plant from inception to operation.

21.5 Water Supply as a Critical Factor

Generally speaking, today's water-quality requirements are so exacting that few natural waters are good enough to be used without treatment.

Barlow [1] states that water as such can be

considered always available, if the word "available" is used in its connotation of existence, since it is always possible to obtain water, at a cost. In industrial-site selection, however, "availability" not only means existence and obtainability, it also means reliability. A water supply for an industrial plant is virtually useless unless it is reliable. This adjective is just as applicable when one is considering waste disposal, for a certain degree of dilution is necessary to effect certain kinds of treatment. Barlow believes that, even though the costs of developing a water supply represent a considerable amount of money for capital investment and operating expenditure, water is actually very cheap. In fact, it is too cheap, and this is the reason why water supply is sometimes lightly regarded and becomes important to industry only when it is depleted. Water supply is seldom *the* deciding factor in site selection, although it is always an important consideration; sometimes an industry decides that one site is more desirable than others on the basis of water availability. A site should have both ground and surface waters available and it should be so located as to minimize the effect of the industry's water usage on other users in the area. Site selection from a water-availability standpoint must take into consideration all the complications which could develop at a source, and extreme care must be exercised to evaluate the supply closely. This frequently requires the expenditure of fairly large amounts of money and time, but the dividends of a proper choice offset the costs of obtaining the advance information on which the choice is based.

Toal [15] is of the opinion that, next to its labor supply, water is the most important resource of the South; yet there is no doubt that good water sites are a limited resource in the South as elsewhere. Toal knows of no responsible industry which does not recognize the fact that, if it uses large quantities of water, it has a responsibility in the matter of stream pollution. He recommends a course of action which will neither keep industry out of a specific area of the country nor so pollute our streams as to interfere unreasonably with other uses of water and diminish the number of usable industrial water sites.

He observes that the last two du Pont plants to be constructed in the Carolinas utilize ground water, although both are located alongside large rivers. In these cases, their engineer wanted the rivers for insurance; but certainly the du Pont company engineer had a preference for ground water, for which nature does much or all of the filtering necessary and exerts a certain amount of control over temperature.

Goudy [4] brings out a point that the author has been trying to emphasize throughout this book, namely, a company will often go through the preliminary stages of site selection and will have made a final decision to locate its new plant at a given place when, for the first time, engineers for the industry learn that the quality of water is not satisfactory for their purposes; or that they are unable to dispose of liquid waste without polluting the underground water supply; or that the statement of the cost of water, as indicated by the meter-rate schedule, did not disclose the fact that there was an additional district tax of 50 cents per $1000 valuation; or that adequate protection from floods could be obtained only at exorbitant cost. The cause of such tardy awakenings is improper coordination of engineering information from the various departments involved in the problem.

Inexpensive water is essential for such industries as woolen mills and pulp and paper mills, which use vast quantities for cooling, processing, and fire protection. Goudy [4] suggested as long ago as 1947 that, when water is expensive at a chosen site, reclaimed waste water should be considered for plants having large water requirements. He cites examples of sewage effluent being reused for industrial purposes: the Bethlehem Steel Company at Baltimore uses treated city sewage for cooling water, as does the Barnsdall Oil Company, of Corpus Christi, Texas, and many others. These examples illustrate the fact that reclamation is not only practical but also economical, in cases where the original cost of water is excessive or where there is a water shortage.

Safford [11] mentions as particularly troublesome sites below large impoundments, where the low-level outlets may discharge water with scant levels of dissolved oxygen. He also says that a factor of growing importance in the Southeast, affecting the quantity of flow available for assimilation, is the practice of irrigation. Owning land on both sides of the receiving stream is very advantageous: first, it prevents claims arising from diverse ownership and, second, it permits a diffuser pipe to be laid across the stream to the opposite shore, which increases dispersion of waste waters.

Peterson [10] announced that "presently, at Elmira, New York, a vast food processing plant—35 acres under one roof—is under construction by the Great Atlantic and Pacific Tea Company. One of the major factors involved in the location of this plant ... was a plentiful supply of pure ground water, which is a prime necessity to the food processing industry.... While we may not consider water in regard to research and development activities, at least one important industrial research laboratory was located in our State because of the presence of large quantities of good quality water available for cooling purposes."

21.6 Site Selection for Atomic-energy Plants

Gorman [3] was faced with the unusual but contemporary problem of selecting a site for an atomic-energy plant. Site selection is one of the most important decisions made by the management of such a plant, because the site profoundly affects (1) the layout and design of costly structures and facilities, (2) the pattern of future expansion, (3) the day-to-day operations, and (4) in case of unforeseen incidents, the safety of employees or persons and property in the vicinity of the plant. These are important factors in company policy, finance, and public relations. Thus, in this new and rapidly expanding industry, perhaps more than in any other, decisions as to the site focus on the character and quantities of wastes which will, or may, be released, particularly in the case of nuclear reactors and chemical processing plants where levels of radioactivity in the product and the wastes are high. In general, this also holds true for fuel-processing and fabricating plants, research laboratories, and other places where materials with lower levels of radioactivity are used.

During World War II the government selected sites for its atomic-energy plants in more or less isolated areas, partly for reasons of security, but also because of the availability of power and water at the sites. The safety record of on-site personnel has been good and there has been no exposure of the public to environmental hazards. However, these power plants have quickly become obsolescent, because technological advances have been so rapid.

It seems likely that those who organize and finance companies devoted to the peaceful uses of atomic energy will want to locate their plants at places which are strategic in relation to the market for their products and services, i.e., reasonably near populated areas—just the reverse of the situation when sites were being sought for the government's wartime atomic plants. One of the first considerations will be the possible exposure of the citizens of nearby communities to hazards (real or imaginary) which a plant using nuclear energy may bring into the area. The situation presents a delicate problem in public relations. While public officials usually welcome new industries to their areas, they may, in the case of a "hazardous" plant, be concerned as to the future impact of this new industry on the health and safety of their people and on the environmental assets of their communities. This situation arose after the Shippingport, Pennsylvania, reactor was put into operation. Although no environmental contamination had been observed from this operation, the public became alarmed by publicity scares and misguided informants. Once the public's anxiety feelings have been aroused, it takes a great deal of time and effort on the part of responsible persons to allay them.

References

1. Barlow, A. C., "Site selection," in Conference on Industrial Water Conservation, Continued Education Series no. 83, University of Michigan School of Public Health, Ann Arbor (1959), pp. 49–56.
2. Bower, B. T., *The Location Decision of Industry and Its Relationship to Water*, Report no. 13 to the Western Agricultural Economics Research Council, San Francisco (1964).
3. Gorman, A. E., "Waste disposal as related to site selection," Paper read at Nuclear Engineering and Science Congress, Cleveland, Ohio, December 12–16, 1955, Preprint 3.
4. Goudy, R. F., Discussion of article by Wood, *Trans. Am. Soc. Civil Engrs*. **112**, 589 (1947).
5. Hoyer, C. O., "Industrial development in the South," *Trans. Am. Soc. Civil Engrs* **120**, 411 (1955).
6. Lewis, H. M., Discussion of article by Wood, *Trans. Am. Soc. Civil Engrs*. **112**, 589 (1947).
7. *Major Economic Factors in Metropolitan Growth and Arrangement, Regional Survey of New York and Its Environs*, Vol. I (1927), pp. 19–30 and 104–107.

8. *Miami* (Fla.) *Herald*, July 15, 1961.
9. Neuhoff, M. C., *Techniques of Plant Location*, Studies in Business Policy no. 61, National Industrial Conference Board, New York (1953).
10. Peterson, R. (First Deputy Director of the Department of Commerce, New York State), Speech given in Syracuse, N.Y., on April 15, 1965.
11. Safford, T. H., "Industrial waste treatment as a factor in the location of wet process industries," in Procedings of 7th Ontario Industrial Waste Conference, Honey Harbour, Ontario, June 1960.
12. *Syracuse* (N.Y.) *Herald Journal*, May 4, 1960.
13. *Syracuse* (N.Y.) *Herald Journal*, August 27, 1961.
14. *Syracuse* (N.Y.) *Herald Journal*, November 19, 1961.
15. Toal, F. C., "Industrial expansion and water use," in Proceedings of Fourth Southern Municipal and Industrial Waste Conference, Duke University, March 31, 1955, pp. 29–32.
16. Watson, K. S., *et al.*, "Some factors in the location of a new chemical plant: A panel discussion," *Sewage Ind. Wastes*, **28**, 1247 (1956).
17. Wood, C. P., "Factors controlling the location of various types of industry," *Trans. Am. Soc. Civil Engrs.* **112**, 577 (1947).

Part 4 | MAJOR INDUSTRIAL WASTES

12 × 9

INTRODUCTION

The purpose of Part 4 is to provide the reader with a fairly complete list of references to the majority of publications concerning industrial wastes. The origin, character, and methods of treatment of the major types of industrial waste are described. It is not the purpose of this section, however, to provide the reader with details of each and every waste or method of treating it, but rather to give a condensation of existing information and guide to the literature. The author believes that solving waste-treatment problems should demand not an ability to memorize details but rather the application of scientific principles and judgment to the solution of practical problems. Although the reader may have to search the literature for the detailed information he needs, he should be capable of applying this information once he has absorbed the contents of this text.

The following table provides the reader with a brief summary of the major liquid wastes, their origin, characteristics, and current methods of treatment, which are described in the last six chapters of this text. It should be useful as a quick reference, but in no way can it be considered complete for any specific industry.

The author has divided industrial wastes roughly into five major classifications—apparel, food and drugs, materials, chemicals, and energy—and has devoted a chapter to each category, with the exception of the energy industry, which merits two chapters because the wastes from energy industries such as steam power plants and coal processing are one thing and those from atomic-energy plants are quite another. Nuclear wastes present such unique problems that the author felt they merited a chapter to themselves.

Summary of Industrial Waste: Its Origin, Character, and Treatment

Industries producing wastes	Origin of major wastes	Major characteristics	Major treatment and disposal methods
Apparel [Chapter 22]			
Textiles	Cooking of fibers; desizing of fabric	Highly alkaline, colored, high BOD and temperature, high suspended solids	Neutralization, chemical precipitation, biological treatment, aeration and/or trickling filtration
Leather goods	Unhairing, soaking, deliming, and bating of hides	High total solids, hardness, salt, sulfides, chromium, pH, precipitated lime, and BOD	Equalization, sedimentation, and biological treatment
Laundry trades	Washing of fabrics	High turbidity, alkalinity, and organic solids	Screening, chemical precipitation, flotation, and adsorption
Food and Drugs [Chapter 23]			
Canned goods	Trimming, culling, juicing, and blanching of fruits and vegetables	High in suspended solids, colloidal and dissolved organic matter	Screening, lagooning, soil absorption or spray irrigation
Dairy products	Dilutions of whole milk, separated milk, buttermilk, and whey	High in dissolved organic matter, mainly protein, fat, and lactose	Biological treatment, aeration, trickling filtration, activated sludge
Brewed and distilled beverages	Steeping and pressing of grain; residue from distillation of alcohol; condensate from stillage evaporation	High in dissolved organic solids, containing nitrogen and fermented starches or their products	Recovery, concentration by centrifugation and evaporation, trickling filtration; use in feeds; digestion of slops

Summary of Industrial Waste *(continued)*

Industries producing wastes	Origin of major wastes	Major characteristics	Major treatment and disposal methods
Meat and poultry products	Stockyards; slaughtering of animals; rendering of bones and fats; residues in condensates; grease and wash water; picking of chickens	High in dissolved and suspended organic matter, blood, other proteins, and fats	Screening, settling and/or flotation, trickling filtration
Animal feedlots	Excreta from animals	High in organic suspended solids and BOD	Land disposal and anaerobic lagoons
Beet sugar	Transfer, screening, and juicing waters; drainings from lime sludge; condensates after evaporator; juice and extracted sugar	High in dissolved and suspended organic matter, containing sugar and protein	Reuse of wastes, coagulation, and lagooning
Pharmaceutical products	Mycelium, spent filtrate, and wash waters	High in suspended and dissolved organic matter, including vitamins	Evaporation and drying; feeds
Yeast	Residue from yeast filtration	High in solids (mainly organic) and BOD	Anaerobic digestion, trickling filtration
Pickles	Lime water; brine, alum and turmeric, syrup, seeds and pieces of cucumber	Variable pH, high suspended solids, color, and organic matter	Good housekeeping, screening, equalization
Coffee	Pulping and fermenting of coffee bean	High BOD and suspended solids	Screening, settling, and trickling filtration
Fish	Rejects from centrifuge; pressed fish; evaporator and other wash water wastes	Very high BOD, total organic solids, and odor	Evaporation of total waste; barge remainder to sea
Rice	Soaking, cooking, and washing of rice	High BOD, total and suspended solids (mainly starch)	Lime coagulation, digestion
Soft drinks	Bottle washing; floor and equipment cleaning; syrup-storage-tank drains	High pH, suspended solids, and BOD	Screening, plus discharge to municipal sewer
Bakeries	Washing and greasing of pans; floor washings	High BOD, grease, floor washings, sugars, flour, detergents	Amenable to biological oxidation
Water production	Filter backwash; lime-soda sludge; brine; alum sludge	Minerals and suspended solids	Direct discharge to streams or indirectly through holding lagoons

Materials [Chapter 24]

Pulp and paper	Cooking, refining, washing of fibers, screening of paper pulp	High or low pH, color, high suspended, colloidal, and dissolved solids, inorganic fillers	Settling, lagooning, biological treatment, aeration, recovery of by-products
Photographic products	Spent solutions of developer and fixer	Alkaline, containing various organic and inorganic reducing agents	Recovery of silver; discharge of wastes into municipal sewer
Steel	Coking of coal, washing of blast-furnace flue gases, and	Low pH, acids, cyanogen, phenol, ore, coke, limestone,	Neutralization, recovery and reuse, chemical coagulation

Summary of Industrial Waste *(continued)*

Industries producing wastes	Origin of major wastes	Major characteristics	Major treatment and disposal methods
	pickling of steel	alkali, oils, mill scale, and fine suspended solids	
Metal-plated products	Stripping of oxides, cleaning and plating of metals	Acid, metals, toxic, low volume, mainly mineral matter	Alkaline chlorination of cyanide; reduction and precipitation of chromium; lime precipitation of other metals
Iron-foundry products	Wasting of used sand by hydraulic discharge	High suspended solids, mainly sand; some clay and coal	Selective screening, drying of reclaimed sand
Oil fields and refineries	Drilling muds, salt, oil, and some natural gas; acid sludges and miscellaneous oils from refining	High dissolved salts from field; high BOD, odor, phenol, and sulfur compounds from refinery	Diversion, recovery, injection of salts; acidification and burning of alkaline sludges
Fuel oil use	Spills from fuel-tank filling waste; auto crankcase oils	High in emulsified and dissolved oils	Leak and spill prevention, flotation
Rubber	Washing of latex, coagulated rubber, exuded impurities from crude rubber	High BOD and odor, high suspended solids, variable pH, high chlorides	Aeration, chlorination, sulfonation, biological treatment
Glass	Polishing and cleaning of glass	Red color, alkaline nonsettleable suspended solids	Calcium-chloride precipitation
Naval stores	Washing of stumps, drop solution, solvent recovery, and oil-recovery water	Acid, high BOD	By-product recovery, equalization, recirculation and reuse, trickling filtration
Glue manufacturing	Lime wash, acid washes, extraction of nonspecific proteins	High COD, BOD, pH, chromium, periodic strong mineral acids	Amenable to aerobic biological treatment, flotation, chemical precipitation
Wood preserving	Steam condensates	High in COD, BOD, solids, phenols	Chemical coagulation; oxidation pond and other aerobic biological treatment
Candle manufacturing	Wax spills, stearic acid condensates	Organic (fatty) acids	Anaerobic digestion
Plywood manufacturing	Glue washings	High BOD, pH, phenols, potential toxicity	Settling ponds, incineration

Chemicals [Chapter 25]

Acids	Dilute wash waters; many varied dilute acids	Low pH, low organic content	Upflow or straight neutralization, burning when some organic matter is present
Detergents	Washing and purifying soaps and detergents	High in BOD and saponified soaps	Flotation and skimming, precipitation with $CaCl_2$
Cornstarch	Evaporator condensate or bottoms when not reused or recovered, syrup from final washes, wastes from "bottling-up" process	High BOD and dissolved organic matter; mainly starch and related material	Equalization, biological filtration, anaerobic digestion

Summary of Industrial Waste *(continued)*

Industries producing wastes	Origin of major wastes	Major characteristics	Major treatment and disposal methods
Explosives	Washing TNT and guncotton for purification, washing and pickling of cartridges	TNT, colored, acid, odorous, and contains organic acids and alcohol from powder and cotton, metals, acid, oils, and soaps	Flotation, chemical precipitation, biological treatment, aeration, chlorination of TNT, neutralization, adsorption
Pesticides	Washing and purification products such as 2,4-D and DDT	High organic matter, benzene-ring structure, toxic to bacteria and fish, acid	Dilution, storage, activated-carbon adsorption, alkaline chlorination
Phosphate and phosphorus	Washing, screening, floating rock, condenser bleedoff from phosphate reduction plant	Clays, slimes and tall oils, low pH, high suspended solids, phosphorus, silica and fluoride	Lagooning, mechanical clarification, coagulation and settling of refined waste
Formaldehyde	Residues from manufacturing synthetic resins and from dyeing synthetic fibers	Normally high BOD and HCHO, toxic to bacteria in high concentrations	Trickling filtration, adsorption on activated charcoal
Plastics and resins	Unit operations from polymer preparation and use; spills and equipment washdowns	Acids, caustic, dissolved organic matter such as phenols, formaldehyde, etc.	Discharge to municipal sewer, reuse, controlled-discharge

Energy [Chapter 26]

Steam power	Cooling water, boiler blowdown, coal drainage	Hot, high volume, high inorganic and dissolved solids	Cooling by aeration, storage of ashes, neutralization of excess acid wastes
Coal processing	Cleaning and classification of coal, leaching of sulfur strata with water	High suspended solids, mainly coal; low pH, high H_2SO_4 and $FeSO_4$	Settling, froth flotation, drainage control, and sealing of mines

Nuclear power and radioactive materials [Chapter 27]

	Processing ores; laundering of contaminated clothes; research-lab wastes; processing of fuel; power-plant cooling waters	Radioactive elements, can be very acid and "hot"	Concentration and containing, or dilution and dispersion

CHAPTER 22

THE APPAREL INDUSTRIES

The apparel industry may be subdivided into three classifications: textiles, leather goods, and laundry trades. Each of these is concerned with wearing apparel—shirts, suits, shoes, work clothes, and so forth.

TEXTILE WASTES

Textile mill operations consist of weaving, dyeing, printing, and finishing. Many processes involve several steps, each contributing a particular type of waste, e.g. sizing of the fibers, kiering (alkaline cooking at elevated temperature), desizing the woven cloth, bleaching, mercerizing, dyeing, and printing. Textile wastes are generally colored, highly alkaline, high in BOD and suspended solids, and high in temperature. Wastes from synthetic-fiber manufacture resemble chemical-manufacturing wastes and their treatment depends on the chemical process employed in the fiber manufacture. Equalization and holding are generally preliminary steps to the treatment of those wastes, because of their variable composition. Additional methods are chemical precipitation, trickling filtration, and, more recently, biological treatment and aeration. The textile industry has long been one of the largest of water users and polluters and there has been little success in developing low-cost treatment methods, which the industry urgently needs to lessen the pollution loads it discharges to streams.

22.1 Origin and Characteristics of Textile Wastes

According to a recent and authoritative publication on textile wastes [73], the sources of polluting compounds are the natural impurities extracted from the fiber and the processing chemicals which are removed from the cloth and discharged as waste. It is necessary for the industrial waste engineer to have a working knowledge of the various processes which produce the wastes and which vary with the particular material. The materials can be subdivided into three groups: cotton, wool, and synthetic fibers.

Cotton. Raw cotton is carded, spun, spooled and warped, slashed (filled with starch), drawn, and woven or knitted into cloth before being sent to the finishing mill. No water-borne pollution originates in this sequence of operations, since they are all mechanical processes, except slashing. In slashing, the warp thread is sized with starch to give it the tensile strength and smoothness necessary for subsequent weaving. The starches used for sizing are cellulose derivations. The sized cloth, referred to as "grey goods," contains 8 to 15 per cent slashing compound, which must be removed in the finishing operation. The grey goods are desized to allow further wet processing, kiered to remove natural impurities, bleached to render them white, mercerized to give the fabric luster, strength, and dye affinity, printed or dyed, and finally filled or sized again, to make them more resistant to wear and smoother to the touch. In addition, some goods are waterproofed, with aluminum acetate or formate mixed with gelatin and a dispersed wax. Each of these processes may involve many steps and may be carried out simultaneously with different machines in different parts of the mill. Figure 22.1 based on a 1967 survey [90] summarizes the operations involved in cotton textile finishing.

Masselli and Burford [68] found that the major wastes and their respective BOD loads resulting from cotton finishing are as shown in Tables 22.1 and 22.2. Brown's findings [13] indicate that starch waste constitutes about 16 per cent of the total volume of the waste produced, 53 per cent of the BOD, 36 per cent of the total solids, and 6 per cent of the alkalinity. Caustic waste constitutes about 19 per cent of the total volume, 37 per cent of the BOD, 43 per cent of the total solids, and 60 per cent of the total alkalinity. General waste is com-

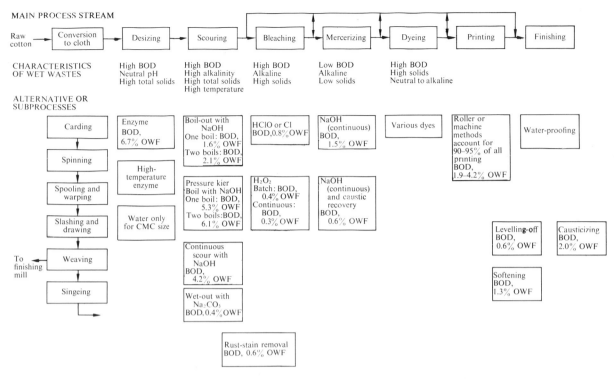

Fig. 22.1 Cotton-textile finishing process flow chart. (Taken from a chart prepared for the F.W.P.C.A.)

Table 22.1 Pollutional loads contributed by various textile processes. (After Masselli and Burford [68].)

Department	Process	lb BOD/1000 lb cloth*	% of total
Desizing		53	35
Scouring	Either {pressure kier, first scour	53	16
	{pressure kier, second scour	8	1
	Or continuous scour	42	15
	Average	47	
	Subtotal (scouring)		32
Dyeing		0.5–32	15–30
Printing	Color-shop wastes	12	7
	Wash after printing, with soap	17–30	17–30
	Wash after printing, with detergent	7	7
	Subtotal (printing)		15–35
Bleaching	Hypochlorite bleach	8	3
	Peroxide bleach	3	1
Mercerizing		6	1
Total		125–250	

*Approximately 800 to 1000 lb of impurities are discharged in the waste per 1000 lb of cotton processed.

Table 22.2 BOD contributed in the dye process. (After Masselli and Burford [68].)

Process	lb BOD/1000 lb cloth
Vat dye, continuous	18
Vat dye, jig	32
Naphthol, jig	14
Direct, jig	0.5
Sulfur, jig	31

posed of wastes from all other processes (washing, bleaching, dyeing, and finishing) and constitutes 65 per cent of the total volume, 10 per cent of the BOD, 21 per cent of the total solids, and 34 per cent of the total alkalinity.

Wool. Wool wastes originate from scouring, dyeing, oiling, fulling, carbonizing, and washing processes. Practically all the natural and acquired impurities in wool are removed by scouring it in hot detergent–alkali solutions. Because of the high pollutional content of these scouring wastes, some wool is scoured with organic solvents; the grease-laden solvent is then recovered by distillation, leaving behind a recoverable wool grease. During the dyeing process, the hot dye solution is generally circulated by pumps through the wool, which is packed in a removable metal basket suspended in a kettle. In oiling, the carding oil is usually mixed with water and sprayed on the wool. This oil, usually olive oil or a lard–mineral oil mixture, varies from 1 to 11 per cent of the wool by weight. Oiling increases the cohesion of the fibers and aids in the spinning, but all of the oil has to be washed out of the cloth later in the finishing process.

Fulling is the process whereby the loosely woven wool from the loom is shrunk into a tight, closely woven cloth. In most plants, soap mixed with soda ash and a sequestering agent is used in the fulling process; excess fulling solution must be squeezed and/or washed out of the fabric. Carbonizing is a process utilizing hot concentrated acids to convert vegetable matter in the wool into loose, charred particles, which are mechanically crushed and then shaken out of the cloth in a machine called a "duster." In addition, piece dyeing, bleaching, and return fulling may take place, but these operations usually involve a small percentage of the total cloth processed.

A summary of alternative processes for finishing wool is given in Fig. 22.2, based on a recent government survey [1967, 90].

The actual wool-fiber content in "grease wool," as taken from the sheep's back, averages only 40 per cent, the remaining 60 per cent being composed of natural impurities such as sand, grease, suint (dried sheep perspiration), and burrs. Consequently, when $2\frac{1}{2}$ pounds of grease wool are scoured, only 1 pound of scoured wool is obtained; in other words, for every 1000 pounds of wool scoured and produced, 1500 pounds of impurities are discharged to waste. In addition, 300 to 600 pounds of process chemicals are discharged. In terms of BOD, 200 to 250 pounds of BOD are discharged per 1000 pounds of scoured wool produced.

Wool scouring and finishing mills produce a composite effluent having a pH of 9 to 10.5 and containing approximately 900 ppm BOD, 3000 ppm total solids, 600 ppm total alkalinity, 4 ppm total chromium, and 100 ppm suspended solids. The waste is brown in color and mainly colloidal in nature. The major BOD source is the wool grease and suint removed in the scouring and the soap used in fulling and washing. Approximately 70,000 gallons of water are used to process each 1000 pounds of wool, but this waste is only slightly affected by plain sedimentation. Most woolen mills in the U.S. are dyeing and finishing mills, which purchase scoured wool. In such mills the contribution of wool waste to the BOD in the plant waste is negligible. About 24 per cent of the BOD in a woolen mill's waste originates from the dye process, 75 per cent from the wash after fulling, and only 1 per cent from neutralizing after carbonizing. Tables 22.3, 22.4, and 22.5 detail the waste characteristics.

Synthetic fibers. Synthetic fibers are essentially composed of pure chemical compounds and have no natural impurities. Because of this, only light scouring and bleaching are necessary to prepare the cloth for dyeing. Processing of the fibers and cloth is readily done on the conventional machinery used for cotton and wool. At present, the major synthetic fibers are rayon, acetate, nylon, Orlon, and Dacron. Rayon is chiefly composed of regenerated cellulose; acetate is a cellulose–acetate fiber; "nylon" is the generic term for any long-chain synthetic polymeric amide; Orlon is a trade name for synthetic, orientable fibers from

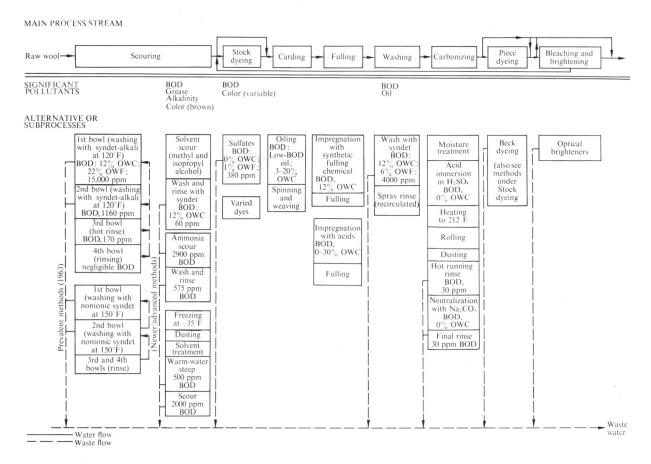

Fig. 22.2 Wool-textile production process flow chart. (Taken from a chart prepared for the F.W.P.C.A.)

polymers containing a preponderance of acrylic units, the newest of which are the acrylonitriles and ethyl or methyl acrylate; Dacron is a polyester fiber manufactured from ethylene glycol and terephthalic acid. All the pollution from treatment of these fibers originates in the various scouring and dyeing chemicals used to process them. (See Figs. 22.3 and 22.4 based on a U.S. Department of the Interior survey [1967, 90].) The volumes and BOD's of such compounds are presented in Tables 22.6, 22.7, and 22.8.

22.2 Treatment of Textile Wastes

Masselli et al. [69] emphasize certain preliminary practices to reduce the quantity and strength of textile wastes: good housekeeping, closer process control, process-chemical substitution, and recovery. They admit that even the best housekeeping practices will reduce the BOD load by only 5 to 10 per cent; however, closer control of cotton kiering, sizing, and the amount of chemicals used in the various other processes may reduce pollution loads up to a maximum of 30 per cent. The author also agrees with the statement of Masselli and Burford that no treatment plant should be planned until serious consideration has been given to pollution reduction through chemical substitution. These authors list cotton- and woolen-mill processes where substitution would be effective.

Table 22.3 Inventory of process chemicals and BOD in woolen-mill wastes. (After Masselli et al. [69].)

Process chemical	Chemical composition	Use	% OWF used* Scouring and carding	% OWF used* Finishing	% OWF used* Total	Concentration in effluent, ppm	BOD* % OWC	BOD* % OWF
Soap	Fatty-acid soap	Scouring, fulling	2.1	5.5	7.6	152	155	11.7
Soda ash	Na_2CO_3	Scouring, fulling	14.2	2.8	17.0	340	0	0
Quadrafos	$Na_6P_4O_{13}$	Washing		0	0.5	10	0	0
Pine oil	Pine oil	Washing	0.5	0	0.5	10	0	0
Paragon 500	?		0.5	0	0.5	10	108	0.5
Proxol T	Mineral oil, plus nonionic emulsifier	Carding	0.5	0	0.5	10	20	0.1
Acetic acid, 84%	CH_3COOH	Dyeing		1.2	1.2	24	62	0.7
Olive sub C3	Oil	Spinning	0.4	0	0.4	8		
Sulfuric acid	H_2SO_4	Carbonizing, dyeing	0	0.2	0.2	4	0	0
Chrome mordant	$Na_2Cr_2O_7 + (NH_4)_2SO_4$	Dyeing	0	0.4	0.4	8	0	0
Chrome	$Na_2Cr_3O_7$	Dyeing	0	0.6	0.6	12	0	0
Glauber salt	Na_2SO_4	Dyeing	0	0.4	0.4	8	0	0
Monochlorobenzene	C_6H_5Cl	Dyeing	0	0.2	0.2	4	3	0
Nopco 1656	Soluble fatty ester	Spinning	0	0.2	0.2	4	12	0
Iversol	Blend of soaps, solvents, and detergents	Washing	0	1.6	1.6	32	60	1.0
Rinsol	Detergent	Washing	0	2.9	2.9	58	72	2.1
Supertex E	Fatty-acid soaps, solvent cresylic acid	Washing	0	0.2	0.2	4	25	0.1
Wool finish B	High carbohydrates and enzymes	Finish	0	2.3	2.3	46	57	1.3
Added impurities (subtotal)			18.7	18.5	37.2	748		17.5
Natural impurities (grease, suint, dirt)			150.0	0	150.0	3000	16.7	25.0
Grand total			168.7	18.5	187.2	3748		42.5

*Per cent OWC is BOD inherent in chemical, based on its weight; per cent OWF is BOD due to the chemical, based on weight of wool.

Cotton mills. (1) Substitution of low-BOD synthetic detergents (syndets) (0 to 20 per cent BOD*) for soap (140 to 155 per cent BOD). The maximum reduction is approximately 35 per cent, in plants where considerable soaping is done. (However, a disadvantage to substituting syndets for soap is their persistence in streams and underground water supplies.) (2) Substitution of steam ranges for oxidation of dyes, in place of dichromate–acetic acid baths (5 to 15 per cent reduction). (3) Use of less caustic in kiering (10 to 20 per cent BOD reduction, 10 to 30 per cent caustic reduction). (4) Use of low-BOD dispersing, emulsifying, leveling, etc., agents in place of high-BOD agents (5 to 15 per cent reduction). (5) Substitution of low-BOD sizes (carboxymethyl celluloses, 3 per cent;

*Pounds of BOD per pound of chemical, that is, 0 to 0.02 lb BOD/lb detergent.

Table 22.4 Comparison of pollution potential of woolen processes. (After Masselli et al. [69].)

Process	Scour and finish mill, BOD			Finish mill, BOD		
	% OWF*	% of total†	% reduction‡	% OWF*	% of total†	% reduction§
Method I						
Scour with soap	25.0	55.4				
Stock dye with acetic acid	4.9	10.9		4.9	24.4	
Card with 100% BOD oil						
Full with soap	15.0	33.3		15.0	74.6	
Wash with soap						
Neutralize after carbonizing	0.2	0.4		0.2	1.0	
Total	45.1			20.1		
Method II						
Scour with 12% BOD detergent	22.1	74.6				
Stock dye with ammonium sulfate	0.9	3.0		0.9	12.0	
Card with 20% BOD oil						
Full with 12% BOD detergent	6.4	21.6		6.4	85.3	
Wash with 12% BOD detergent						
Neutralize after carbonizing	0.2	0.7		0.2	2.7	
Total	29.6		34	7.5		63
Method III						
Solvent scour, recover grease only	0	0				
Wash out suint salts and dirt with 12% BOD detergent	10.0	58.9				
Stock dye with ammonium sulfate	0.9	5.3		0.9	12.9	
Card with 3% BOD oil						
Full with 12% BOD detergent	5.9	34.7		5.9	84.3	
Wash with 12% BOD detergent						
Neutralize after carbonizing	0.2	1.2		0.2	2.9	
Total	17.0		62	7.0		65
Method IV						
Scour with methyl and isopropyl alcohols; recover grease and suint	0	0				
Wash out dirt with detergent	1.0	19.6				
Stock dye with ammonium sulfate	0.9	17.6		0.9	22.0	
Card with 3% BOD oil						
Full with 12% BOD detergent	3.0	58.8		3.0	73.2	
Wash with 12% BOD detergent						
Neutralize after carbonizing	0.2	3.9		0.2	4.9	
Total	5.1		89	4.1		80

*Based on oven-dry wool.
†Based on the total of that particular method.
‡Based on the total of Method I (45.1% OWF).
§Based on the total of Method I (20.1% OWF).

Table 22.5 Analysis of waste from woolen mill. (After Masselli et al. [69].)

Method	pH	Alkalinity* CO$_3^=$, ppm	Alkalinity* HCO$_3^-$, ppm	Solids Total, ppm	Solids Fixed, ppm	Solids Volatile, ppm	BOD, ppm
Grease scour, 1st bowl, soap–alkali	9.7	4870	7340	64,448	19,133	45,315	21,300
Grease scour, 1st bowl, detergent–Na$_2$SO$_4$	8.0	0	6442	60,593	19,889	40,012	15,400
Grease scour, 2nd bowl, soap–alkali	10.4	9153	2214	25,624	15,131	10,493	4,780
Grease scour, 2nd bowl, detergent–Na$_2$SO$_4$	8.3	16	463	6,368	2,086	4,478	1,160
Grease scour, 3rd bowl, soap–alkali	9.7	355	154	1,129	555	574	255
Grease scour, 3rd bowl, detergent–Na$_2$SO$_4$	7.3	0	75	1,609	525	1,083	170
Stock dyeing, acetic acid	7.3	18	803	3,855	2,248	1,266	2,182
Stock dyeing, ammonium sulfate	6.7	0	194	8,315	3,782	4,533	379
Wash after fulling, 1st soap, soap used for fulling	10.0	2117	584	19,267	4,771	14,489	11,455
Wash after fulling, 1st soap, detergent used for fulling	9.7	380	60	4,830	977	3,853	4,000
Neutralization following carbonizing, 1st running rinse	2.2	0	0	2,241	193	1,048	28
Neutralization following carbonizing, 1st soda–ash bath	8.5	517	2788	9,781	9,559	222	28
Optical wool bleaching in dye kettles	6.0	0	281	908	376	532	390

*No free hydroxide present in any waste samples.

hydroxymethyl cellulose, 3 per cent; polyacrylic acid, 1 per cent; polyvinyl alcohol, 1 per cent) for the high-BOD starch (50 to 70 per cent) now widely used. This could theoretically reduce the total BOD from a cotton mill by 40 to 90 per cent. (6) Replacement of acetic acid in dyeing with an inorganic salt, such as ammonium sulfate or chloride (0 per cent BOD).

Woolen mills. (1) Replacement of soap used in scouring by low-BOD detergents (maximum BOD reduction possible is 5 per cent). (2) Replacement of Na$_2$CO$_3$ with detergent–Na$_2$SO$_4$ mixture, to reduce high alkalinity of waste. (3) Replacement of carding oils (100 per cent BOD) with mineral oils with non-ionic emulsifiers (20 per cent BOD); BOD reductions of approximately 10 per cent will be effected in scouring-and-finishing mills and 25 per cent in finishing mills. (4) Replacement of the soap used for fulling and wash after fulling with low-BOD detergents; 15 to 30 per cent BOD reductions may be obtained; H$_2$SO$_4$ may also be substituted for soap for fulling. (5) Use of ammonium sulfate instead of acetic acid will reduce the total plant BOD 5 to 10 per cent in a scouring-and-finishing mill.

It should be noted that, if a so-called "no-BOD" compound is substituted for a "high-BOD" compound, other difficulties may arise. For example, when soluble sizes are used to replace starch, the sizes may persist for many miles in a river. In due time they may find their way into water supplies, interfere with treatment, and because of their resistance to treatment may be found in a domestic water supply. This should not cause plants to eliminate substitution of these sizes as a preliminary step in waste treatment, but rather cause them to examine water uses and treatment methods carefully. In some cases, where effluent water does not find its way into the domestic water supply or where adequate water-treatment methods are used, substitution of soluble sizes in the textile industry may prove invaluable. Recovery of certain materials should be considered by all mills, since some 200,000 tons of cotton impurities—such as waxes, pectins, and alcohols—are now being dumped into sewers each year [69]. Caustic soda and

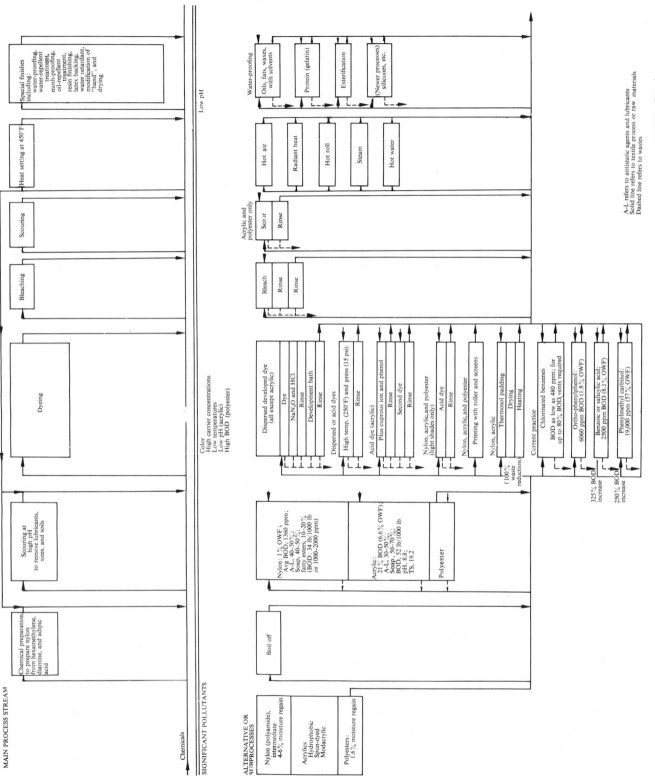

Fig. 22.3. Noncellulose synthetic textile finishing process flow chart. (Taken from a chart prepared for the F.W.P.C.A.)

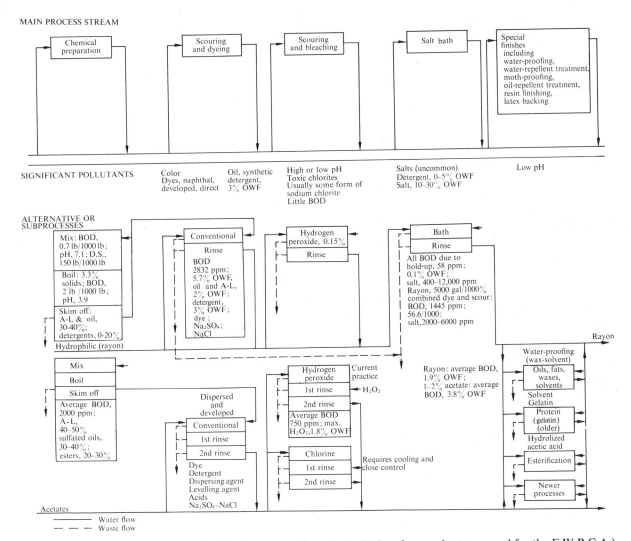

Fig. 22.4 Cellulose synthetic-textile finishing process flow chart. (Taken from a chart prepared for the F.W.P.C.A.)

slashing starch are examples of recoverable chemicals. Many mills are already recovering caustic soda, but mainly so that the remaining waste can be treated more effectively by biological means. About half the caustic soda is found in the contaminated kiering liquors, and the other half comes from mercerizing. Dialysis [84] and evaporation have been used to purify and recover this caustic soda. Since the kier liquors are relatively dilute solutions of caustic soda, which also contain many other colloidal and suspended impurities such as gums, pectins, and hemicelluloses, it has not been found practical to dialyze or evaporate the waste in order to recover caustic soda.

Masselli [68] calculates that 400,000 tons of glucose might be recovered annually, solely from the starch–desize wastes from our textile mills. This figure was based on an average addition of 10 to 15 per cent

Table 22.6 BOD loads and concentration from processing of various fibers. (After Masselli et al. [69].)*

Fiber	gal/1000 lb	BOD, % OWF	Average BOD, ppm
Rayon	5,000	5.0–8.0	1200–1800
Acetate	9,000	4.0–6.0	500–800
Nylon	15,000	3.5–5.5	300–500
Orlon	25,000	10.0–15.0	500–700
Dacron			
o-phenylphenol	12,000	15.0–25.0	1500–2500
Monochlorobenzene	12,000	3.0–5.0	300–500
Benzoic acid	12,000	60.0–80.0	6000–8000
Salicylic acid	12,000	50.0–70.0	5000–7000
Phenylmethyl carbinol	12,000	40.0–60.0	4000–6000
Cotton	70,000	12.5–25.0	220–600
Wool	70,000	40.0–60.0	700–1200

*Based on actual plant surveys, with the exception of data for Dacron.

starch during the sizing process. Recovery of glucose would not only be of economic benefit to the industry but would also reduce the BOD load to be treated by 45 to 94 per cent. Steam produced from the evaporation of desize wastes could be used in the mill.

Lanolin in wool grease from woolen-mill wastes has often been recovered through solvent extraction. A cleaning-solution solvent, such as carbon tetrachloride or benzene, is generally used. A potential supply of 50,000 to 100,000 tons of wool grease exists in our woolen mills, and BOD reductions of 20 to 30 per cent are effected by grease recovery.

Suint can also be recovered by a new alcohol-extraction process and sold for detergent manufacture or potassium salts. Recovery of suint results in an additional 20 to 30 per cent BOD reduction and 20,000 to 40,000 tons of suint can be produced in this manner in the U.S. per year.

Soap is also a valuable product recoverable from woolen-mill wastes. Although little soap recovery is practiced in this country, it could result in 30 to 70 per cent reductions in the BOD of wool-scouring and finishing-mill wastes and the recovered fat might be rendered or used as a fuel source.

It therefore appears that recovery, with its inherent large BOD reductions, is an important step in any waste-treatment plan. However, if recovery, chemical substitution, process control, and good housekeeping practices are not sufficient to eliminate pollution, additional waste-treatment methods must be utilized.

Table 22.7 BOD contribution of process chemicals used in finishing of synthetic fibers. (After Masselli et al. [69].)

Fiber	Process chemical*	BOD†		
		% OWC	% OWF	% total
Acetate	Scour and dye			
	Antistat-lubricant		1.5	44
	2% sulfonated oil	52	1.0	31
	1% syndet‡	5	0.1	2
	2% aliphatic ester	41	0.8	24
	2% softener	0	0	0
	Total		3.4	
Nylon	Scour			
	Antistat-lubricant		1.5	29
	1% soaps	150	1.5	29
	1% fatty esters	55	0.6	11
	Dye			
	2–4% sulfonated oils	56	1.7	32
	Total		5.2	

Table 22.7 (continued)

Fiber	Process chemical*	BOD† % OWC	BOD† % OWF	BOD† % total
Dacron	Scour			
	Antistat-lubricant		1.5	9
	1% nonionic syndet	5	0.1	0
	Dye			
	4% acetic acid (84%)	58	2.3	13
	10% o-phenylphenol	138	13.8	78
	Other possible carriers			
	6% p-phenylphenol			
	40% benzoic acid	165	(66.0)	(94)
	40% salicylic acid	141	(56.4)	(94)
	30% phenylmethyl carbinol	150	(45.0)	(92)
	Total		17.7	
Rayon	Scour and dye			
	Antistat-lubricant		1.5	50
	3% syndet	14	0.4	14
	2% soluble oil	53	1.1	36
	10–30% common salt	0	0	0
	Total		3.0	
Orlon	Scour			
	Antistat-lubricant		1.5	12
	2% soaps	150	3.0	24
	0.5% syndet	0	0	0
	3.0% formic acid	20	0.6	5
	Second dye			
	1% wetting agent	14	0.1	1
	3% phenolic compounds	200	6.0	48
	3% copper sulfate			
	2% hydroxy ammonium sulfate	4	0.1	1
	Scour			
	2% syndet	0	0	0
	1% pine oil	108	1.1	8
	Total		12.4	

*Per cent figure before process chemical indicates amount used (OWF).
†Per cent OWC is BOD inherent in chemical, based on its own weight; per cent OWF is the BOD due to the chemical based on weight of the wool; per cent of total is the chemical's contribution to the total BOD load.
‡Syndet indicates synthetic detergent.

22.3 Final Waste Treatment

Generally, chemical coagulation and biological treatment are the chief methods used for final removal of excess BOD, though both processes have their limitations, as pointed out in Chapters 11 and 13. Alum, ferrous sulfate, ferric sulfate, or ferric chloride are used as coagulants, in conjunction with lime or sulfuric acid for pH control. Calcium chloride has also been found effective in such procedures as coagulating

Table 22.8 Estimated pollution load from the processing of various fibers. (After Masselli et al. [69].)*

Fiber	Natural impurities	Sizes, oils, antistats	Scouring	Dyes, emulsifiers, carriers, etc.	Special finishes, waterproof, etc.	Total
Cotton	3–5	0.5–10.0	0.5–6.0	0.2–8.0	0.2–8.0	4.4–37.0
Greasy wool	20.0†–30.0	0.2–9.0	1.5–15.0‡	0.5–10.0	0.2–8.0	21.9†–72.0
Scoured wool	1.0–2.0	0.2–9.0	1.0–15.0‡	0.5–10.0	0.2–8.0	2.9–44.0
Rayon	0	0.5–6.0	0.5–5.0	0.2–5.0	0.2–8.0	1.4–24.0
Acetate	0	0.5–6.0	0.5–5.0	0.2–5.0	0.2–8.0	1.4–24.0
Orlon	0	0.5–6.0	0.5–5.0	0.5–10.0	0.2–8.0	1.7–29.0
Nylon	0	0.5–6.0	0.5–5.0	0.2–5.0	0.2–8.0	1.4–24.0
Dacron	0	0.5–6.0	0.5–5.0	3.0–60.0	0.2–8.0	4.2–78.0

*These estimates are based on surveys and processing methods described in the literature. All results give the per cent OWF.
†If grease and suint are removed by solvent extraction, this load may be reduced to approximately 2 per cent.
‡High values include soap used for fulling also.

wool-scouring wastes. Each coagulant, when used with a specific waste, has its own optimum isoelectric point (pH for maximum coagulation) which must be determined experimentally. Some wastes are more easily coagulated with one chemical than another; others cannot be coagulated with any known economical coagulant. Masselli [69] presents data (Table 22.9) to assist the waste engineer in estimating the BOD reduction possible by chemical coagulation of a particular waste. Figure 22.5 shows schematically the major treatments used today for cotton-finishing wastes.

The author [77–79] gives five ways that textile-dye wastes must be treated before discharge to a stream: (1) equalization, (2) neutralization, (3) proportioning, (4) color removal, and (5) reduction of organic oxygen-demanding matter. It was found that $Al_2(SO_4)_3 \cdot 18H_2O$ (alum) completely removed the apparent color from a sewage–dye-waste mixture and also reduced the BOD by 63 per cent. The dosage of alum required was 200 ppm at the existing pH of 8.3 and 140 ppm at a pH of 7.0.

Chamberlain [23] reported that, instead of coagulating dye wastes chemically, he used chlorine, in the form of chlorinated copperas, to oxidize or bleach many dyes and to remove BOD from sulfur dyes. Chlorine can also be added at the same time

Table 22.9 BOD reductions through chemical coagulation of certain wastes. (After Masselli et al. [69].)

Chemical waste	BOD reduction, %
Soap*	90
Phenol	0
Glucose	0
Starch	57
Gelatin*	65
Glue*	33
Emulsified mineral oil	80
Sulfonated castor oil	82
Sulfonated vegetable oil	44
Coconut oil*	92
o-phenylphenol	0
Salicylic acid	17
Benzoic acid	8
Acetic acid	8
Oxalic acid	86
Sodium acetate	0
Alum–wax emulsion	85

*The following wastes required more than 3 lb of alum as coagulant per 1000 gal: soap, 5 lb; gelatin, 18 lb; glue, 5 lb; coconut oil, 10 lb.

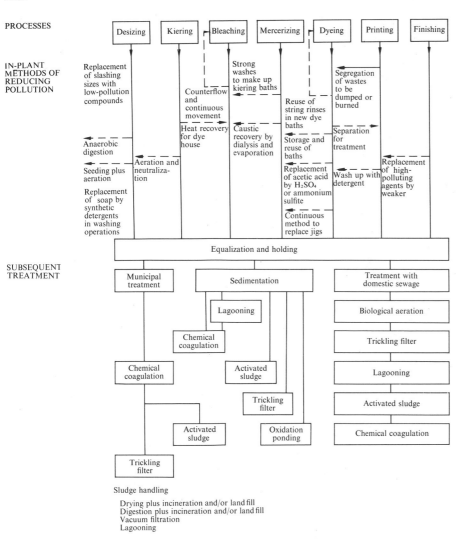

Fig. 22.5 Cotton-textile finishing waste-treatment flow chart. (Taken from a chart prepared for the F.W.P.C.A.)

as the chemicals, to aid in coagulation and color removal, as well as in the final stage of waste processing. The chlorine requirements are normally 100 to 250 ppm. The action of chlorine is primarily one of oxidizing organic dyes to colorless end-products.

Of the biological means of treating textile wastes, trickling filtration, activated sludge, and dispersed-growth aeration have been most successful. Trickling filtration is generally desirable from the standpoint of flexibility, lower operating costs, and capability of handling shock loads of wastes. Activated-sludge treatment gives greater BOD reduction, but entails large units to provide the long detention periods (12 to 48 hours) usually needed and also requires highly qualified supervision. Dispersed-growth aeration generally gives somewhat lower BOD reduction than activated-sludge treatment, but does away with the sludge problem; it also takes a minimum of operation and maintenance.

The initial pH of the waste is a controlling factor in the efficiency of any biological treatment. Optimum BOD reductions are obtained when the pH is between

7 and 9, but some BOD reduction takes place when the pH is between 9 and 11, the extent in this range depending on the character of the waste and the equalization it receives prior to aeration. Little or no BOD reduction occurs when the pH exceeds 11.5. Since methods of controlling the pH can be expensive, the cost factor may limit the acceptability of biological treatment. The pH is normally lowered by the addition of acid (H_2SO_4), compressed gas (CO_2), or flue gas. The first two methods are quite effective but relatively expensive. However, flue gases, which usually contain 12 to 14 per cent carbon dioxide, can be used to lower the pH of caustic solutions and, once the capital expense of a pipeline, scrubber, and blower has been overcome, the operating costs are low and the operation is practical. (The principles of neutralization with flue gas were discussed in Chapter 8.) Neutralization is not only feasible, by means of flue gas, but necessary in the biological treatment of alkaline textile wastes. The Beaches [4]

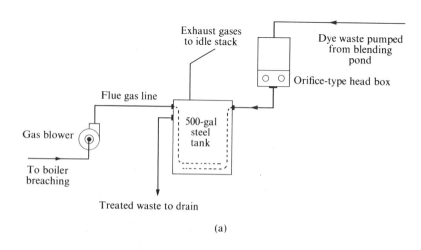

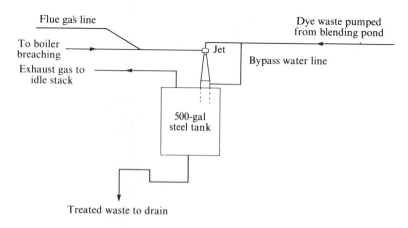

Fig. 22.6 Gas pilot plants: (a) blower flue; (b) jet flue. (After Beach and Beach [4].)

showed that certain dye wastes had a pH ranging from 2 to 11 and contained up to 30 ppm sulfides. Various methods of passing flue gas into the wastes were studied and they found that, by using a commercial fume scrubber operating on the aspirator principle, the pH could be reduced from 9.0 to 6.1 and 98 per cent of the H_2S eliminated. Their pilot plants are diagrammed in Figs. 22.6 and 22.7. Other workers have also found flue-gas treatment feasible [14, 27, 51, 52, 54, 75, 108].

Dispersed-growth aeration is gaining acceptance as a method of treating highly dissolved organic textile wastes. As pointed out in Chapter 13, laboratory results for dispersed-growth aeration of cotton-kiering liquors, enzyme-desizing wastes, and starch-rinse wastes have indicated satisfactory treatment. Pilot-plant results of this treatment on finishing-mill wastes were so good [48] that a full-scale treatment plant, based on these results, was designed.

A combination of trickling filters and activated sludge was used successfully in treating a 40 to 60 per cent mixture of textile-finishing-mill waste and domestic sewage [40]. Another laboratory research project [7] illustrated that, at a high loading of 2.73 pounds of BOD per cubic yard of stone and at a relatively high pH of 10.5, a 58 per cent reduction was obtained by trickling filtration. These results indicate that it may be feasible to treat biologically a highly alkaline sewage–waste mixture without prior neutralization. Such a procedure might mean lower efficiency of treatment, but this disadvantage would be offset by considerable savings in chemical costs to the municipalities and industries involved. Also, the reduction in pH of the filter effluent (from 10.5 to 9.1)

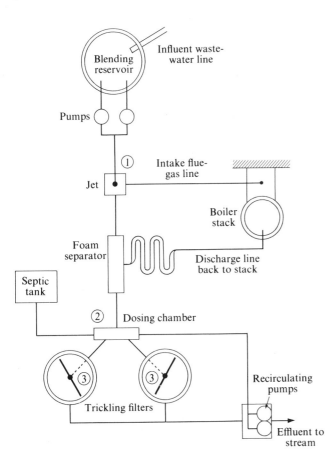

Fig. 22.7 Diagrammatic sketch of a waste-water treatment plant. At (1), the pH is adjusted and the sulfides are removed; at (2), dye waste is blended with settled sewage; and at (3), the combined waste is treated on trickling filters to lower the BOD. (After Beach and Beach [4].)

will produce a higher degree of efficiency in any subsequent biological treatment unit. Therefore, the trickling filter can be used to advantage as a "roughing" or preliminary biological treatment. The 42.5 per cent color removal obtained by this process is a desirable secondary result. A complete activated-sludge system [63] was used in one case to treat the waste from a small bleachery, which had an influent pH of 11.4 and a BOD of 950 ppm.

These examples of how wastes from textile mills can be successfully handled should do much to encourage engineers and textile-plant managers to keep trying to improve their methods of waste disposal.

References: Textile Wastes

1. Allen, L. A., and G. E. Eden, "Effects of continuous aeration on bacterial oxidation of organic matter," *J. Hyg.* **44**, 508 (1946).
2. Baity, H. G., "Textile wastes treatment," *Am. Dyestuff Reptr.* **27**, 544 (1938).
3. Bassett, H. P., "Super-filtration by dialysis," *Chem. Met. Eng.* **45**, 254 (1938).
4. Beach, C. J., and M. G. Beach, "Treatment of alkaline dye wastes with flue gas," in Proceedings of 5th Southern Municipal and Industrial Waste Conference, April 1956, p. 162.
5. Beedham, C. C., "Some experiments on the treatment of a sewage containing wool-scouring refuse," *Surveyor* **79**, 335 (1931).
6. Besselievre, E. B., *Industrial Waste Treatment*, 1st ed., McGraw-Hill Book Co., New York (1952), p. 110.
7. Sanitary Engineering Division, Research Report no. 1, "Biological treatment of highly alkaline textile mill waste-sewage mixture," *Proc. Am. Soc. Civil Engrs.* **81**, Paper no. 750, 1955.
8. Bogren, G. G., "Treatment of cotton finishing waste liquors," *Ind. Eng. Chem.* **42**, 619 (1950).
9. Bogren, G. G., "Waste treatment in cotton finishing plants," *Am. Dyestuff Reptr.* **39**, 669 (1950).
10. Bogren, G. G., "Treatment of cotton finishing wastes at the Sayles Finishing Plants, Inc.," *Sewage Ind. Wastes* **24**, 994 (1952).
11. Brown, J. L., Jr., "Organization of waste control program in the textile industry," in Proceedings of 2nd Southern Municipal and Industrial Waste Conference, March 1953, p. 128.
12. Brown, J. L., Jr., "Bleaching and dyehouse waste studies," *Am. Dyestuff Reptr.* **44**, 379 (1955).
13. Brown, J. L., Jr., "Combined treatment, textile waste and domestic sewage," in Proceedings of 6th Southern Municipal and Industrial Waste Conference, April 1957, p. 179.
14. Brown, K. M., and S. Kurtis, "Solved difficult waste disposal problem," *Petrol. Refiner* **29**, 111 (1950).
15. Burford, M. G., H. F. Berger, and J. W. Masselli, *Textile Wastes, A Review*, New England Interstate Water Pollution Control Commission (1950).
16. Burford, M. G., H. F. Berger, and J. W. Masselli, *A Survey of Three Textile Mills in Connecticut*, Hall Laboratory of Chemistry, Wesleyan University, Middletown, Conn. (1951).
17. Burford, M. G., and J. W. Masselli, *An Analytical and Inventory Survey of Industrial Wastes from a Connecticut Cotton Mill*, Hall Laboratory of Chemistry, Wesleyan University, Middletown, Conn. (1953).
18. Burford, M. G., J. W. Masselli, W. J. Snow, H. Campbell, and F. J. Deluise, *Industrial Waste Surveys of Two New England Cotton Finishing Mills*, New England Interstate Water Pollution Control Commission (1953).
19. Buswell, A. M., and H. F. Mueller, "Treatment of wool wastes," in Proceedings of 11th Industrial Waste Conference, Purdue University, May 1956, p. 160.
20. Campbell, M. S., "Disposal and recovery of textile wastes," *Textile Res. J.* **3**, 490 (1933) and **4**, 29 (1933).
21. Catlett, G. F., *Textile Waste as Related to Stream Pollution*, North Carolina Stream Sanitation and Conservation Committee (1948).
22. Cawley, W. A., and C. C. Wells, "Lagoon system for chemical cellulose waste," *Ind. Wastes* **4**, 37 (1959).
23. Chamberlain, N. S., "Application of chlorine and treatment of textile wastes," in Proceedings of 3rd Southern Municipal and Industrial Waste Conference, March 1954, p. 176.
24. Chrisco, H. F., M. White, and H. G. Baity, "The effect of precipitants on textile waste liquors," *Sewage Works J.* **5**, 674 (1933).
25. Coburn, S. E., "Treatment of cotton printing and finishing wastes," *Ind. Eng. Chem.* **42**, 621 (1950).
26. U.S. Public Health Service, *Cotton Wastes*, Ohio River Survey, Supplement D (1943), p. 1106.
27. Curtis, H. A., and R. L. Copson, "Treating alkaline factory waste liquors, such as kier liquor from treat-

ing cotton with caustic soda," U.S. Patent no. 1802806, 28 April 1931.
28. Dickerson, B. W., "A solution to the cotton sizing problem: soluble sizes," paper presented at 4th Southern Municipal and Industrial Waste Conference, April 1955.
29. Dickerson, B. W., "A solution to the cotton sizing waste problem," *Ind. Wastes* **1**, 910 (1955).
30. *Dyeing and Finishing Dacron*, E. I. duPont de Nemours Co., Wilmington, Del. (1956).
31. Eldridge, E. F., *Industrial Waste Treatment Practice*, McGraw-Hill Book Co., New York (1952), p. 236.
32. Eynon, D. J., "Operation of Cerini dialysers for recovery of caustic soda solutions containing hemicellulose," *J. Soc. Chem. Ind.* **52**, 173T (1933).
33. Foute, K. E., "Two industrial waste problems at New Haven, Conn.," *Sewage Ind. Wastes* **24**, 1305 (1952).
34. Frisk, P. W., "Industrial waste problems in the rayon industry," *Rayon Syn. Textiles* **30**, 52 (1949).
35. Geyer, J. C., and W. A. Perry, *Textile Waste Treatment and Recovery*, Textile Foundation, Inc., Washington, D.C. (1936).
36. Goode, W. B., "Waste treatment cost need not be excessive," *Textile World* **100**, 100 (1950).
37. Gurnham, C. F., *Principles of Industrial Waste Treatment*, John Wiley & Sons, New York (1955), p. 26.
38. Gurnham, C. F., "Textile wastes abatement policies and practices," in Proceedings of 4th Southern Municipal and Industrial Waste Conference, March–April 1955.
39. Hart, W. B., "Cotton bleaching and dyeing wastes, a specific solution and a general prescription," *Ind. Eng. Chem.* **49**, 81A (1957).
40. Hazen, R., "Pilot plant studies on treatment of textile wastes and municipal sewage," in Proceedings of 6th Southern Municipal and Industrial Waste Conference, April 1957, p. 161.
41. Helmers, E. M., J. D. France, A. E. Greenberg, and C. N. Sawyer, "Nutrition requirements in the biological stabilization of industrial wastes," *Sewage Ind. Wastes* **23**, 884 (1951).
42. Helmers, R. N., J. D. France, A. E. Greenberg, and C. N. Sawyer, "Nutritional requirements in the biological stabilization of industrial wastes," in Proceedings of 6th Southern Municipal and Industrial Waste Conference, February 1951, p. 376.
43. Hess, R. W., "Organic chemicals manufactured," *Ind. Eng. Chem.* **44**, 494 (1952).
44. Hoover, C. R., "Treatment of liquid wastes from the textile industry," *Ind. Eng. Chem.* **31**, 1353 (1939).
45. Horton, R. K., and H. G. Baity, "The disposal of textile wastes with domestic sewage," *Textile Res. J.* **10**, 2 (1938).
46. Howard, E. F., G. W. Gleeson, and F. Merryfield, "The pollutional character of flax retting wastes," *Sewage Works J.* **6**, 597 (1934).
47. Hughes, J. W., "Industrial waste treatment at a viscose-rayon factory," *Surveyor* **110**, 781 (1951).
48. Hutto, G., and S. Williams, "Pilot plant studies of processing wastes of cotton textiles," in Proceedings of 9th Southern Municipal and Industrial Waste Conference, April 1960.
49. Ingols, R. S., "Textile waste problems," *Sewage Ind. Wastes* **30**, 1273 (1958).
50. Jabobs, H. L., "Rayon waste recovery and treatment," *Sewage Ind. Wastes* **25**, 296 (1953).
51. Jung, H., "Purifying water," French Patent no. 767586, 20 July 1934.
52. Jung, H., "Chemical treatment of sewage," *Gesundh. Ing.* **69**, 305 (1948).
53. Kehren, M., and H. Denks, "Treatment of dye plant waste with iron," *Fachorgan Textilveredlung* **8**, 1 (1953).
54. King, J. C., "A solution to highly alkaline textile dye wastes—flue gas treatment," in Proceedings of 4th Southern Municipal and Industrial Waste Conference, April 1955.
55. Kirk, R. E., and D. F. Othmer, *Encyclopedia of Chemical Technology*, Vol. 5, Interscience Publishers, New York (1950).
56. Lee, J. A., "Caustic soda recovery in the rayon industry," *Chem. Met. Eng.* **42**, 482 (1935).
57. Lovett, L. E., "Application of osmosis to recovery of caustic soda solutions containing hemicellulose in rayon industry," *Trans. Electrochem. Soc.* **73**, 163 (1938).
58. Ludwig, R. G., and H. F. Ludwig, "Neutralization of acid felt processing waste," *Sewage Ind. Wastes* **24**, 1248 (1952).
59. McCarthy, J. A., "The textile industry and stream pollution," *Am. Dyestuff Reptr.* **39**, 732 (1950).
60. McCarthy, J. A., "What do you know about textile wastes?" in Proceedings of 1st Southern Municipal and Industrial Waste Conference, March 1952, p. 91.
61. McCarthy, J. A., "Use of calcium chloride in the

treatment of industrial wastes," *Sewage Ind. Wastes* **24**, 473 (1952).
62. McCarthy, J. A., "Reducing textile waste problems by using starch substitutes," *Sewage Ind. Wastes* **28**, 334 (1956).
63. McKinney, R. E., J. M. Symonds, W. G. Shifrin, and M. Vezina, "Design and operation of a complete mixing activated sludge system," *Sewage Ind. Wastes* **30**, 287 (1958).
64. Margulies, P. H., "Versatility of hydrogen peroxide," *Textile Inds.* **118**, 111 (1954).
65. Marsh, J. T., *An Introduction to Textile Finishing*, John Wiley & Sons, New York (1951).
66. Masselli, J. W., and M. G. Burford, *Pollution Sources in Wool Scouring and Finishing Mills and Their Reduction through Process and Process Chemical Changes*, New England Interstate Water Pollution Control Commission (1954).
67. Masselli, J. W., and M. G. Burford, "Pollution reduction in cotton finishing wastes through process chemical changes," *Sewage Ind. Wastes* **26**, 1109 (1954).
68. Masselli, J. W., and M. G. Burford, "Pollution reduction program for the textile industry," *Sewage Ind. Wastes* **28**, 1273 (1956).
69. Masselli, J. W., N. W. Masselli, and M. G. Burford, *A Simplification of Textile Waste Survey and Treatment*, New England Interstate Water Pollution Control Commission (1959).
70. Mauersberger, H. R., *Mathew's Textile Fibers*, John Wiley & Sons, New York (1948).
71. Mauersberger, H. R., *American Handbook of Synthetic Textiles*, Textile Book Publishing, New York (1952).
72. Merrill, G. R., A. R. Macommac, and H. R. Mauersberger, *American Cotton Handbook*, 2nd ed., New York (1949).
73. Miles, H. J., and R. Porges, "Treatment of sulfur dye wastes by activated sludge process," *Sewage Works J.* **10**, 322 (1938).
74. Miles, H. J., and R. Porges, "Textile waste studies in North Carolina," *Am. Dyestuff Reptr.* **27**, 736 (1938).
75. Murdock, H. R., "Stream pollution alleviated—processing sulfur dye wastes," *Ind. Eng. Chem.* **43**, 77A (1951).
76. Neas, G. M., "Treatment of viscose rayon wastes," in Proceedings of 14th Industrial Waste Conference, Purdue University, May 1959, p. 450.
77. Nemerow, N. L., "Textile dye wastes," in Proceedings of 1st Southern Municipal and Industrial Waste Conference, March 1952, p. 165.
78. Nemerow, N. L., "Textile dye wastes," in Proceedings of 7th Industrial Waste Conference, Purdue University, May 1952, p. 282.
79. Nemerow, N. L., "Textile dye wastes," *Chem. Age* **66**, 887 (1952).
80. Nemerow, N. L., "Oxidation of cotton kier wastes," *Sewage Ind. Wastes* **25**, 1060 (1953).
81. Nemerow, N. L., "Oxidation of enzyme desize and starch rinse textile wastes," *Sewage Ind. Wastes* **26**, 1231 (1954).
82. Nemerow, N. L., "Holding and aeration of cotton mill finishing wastes," in Proceedings of 5th Southern Municipal and Industrial Waste Conference, April 1956, p. 149.
83. Nemerow, N. L., "Color and methods for color removal," in Proceedings of 11th Industrial Waste Conference, Purdue University, May 1956, p. 584.
84. Nemerow, N. L., and W. R. Steele, "Dialysis of caustic textile wastes," in Proceedings of 10th Industrial Waste Conference, Purdue University, May 1955, p. 74.
85. New England Conference on Industrial Wastes, "Waste disposal presents dilemma for New England industry," *Chem. Eng. News* **28**, 2342 (1950).
86. Palmer, C. W., "Wool scouring wastes," *Trans. Am. Inst. Chem. Eng.* **12**, 113 (1919).
87. Philips, R. W., "Textiles and the chemical industry," *Ind. Eng. Chem.* **27**, 437 (1955).
88. Porges, R., R. K. Horton, and H. G. Baity, "Textile wastes and studies of pH control," *Sewage Works J.* **11**, 828 (1939).
89. Porges. R., R. K. Horton, and H. B. Gotaas, "Chemical precipitation of sulfur dye wastes on a pilot plant scale," *Sewage Works J.* **13**, 308 (1941).
90. Procopi, J., and G. E. Shaffer, Jr., "Solution of some industrial water and waste treatment problems," *Ind. Wastes* **1**, 963 (1955).
91. Rhode Island Section, Subcommittee on Stream Pollution, American Association of Textile Chemists and Colorists, "The BOD of textile chemicals," *Am. Dyestuff Reptr.* **44**, 385 (1955).
92. Roberts, C. B., and H. T. Farrar, "The treatment of gaseous and liquid effluents attendant in producing viscose cellulose film," *Ind. Wastes* **1**, 282 (1956).
93. Roetman, E. T., "Stream pollution control at Front Royal, Va., rayon plant," *Southern Power Ind.* **62**, 86 (1944).

94. Roetman, E. T., "Viscose rayon manufacturing wastes and their treatment," *Water Works Sewerage* **91**, 265 (1944) and **91**, 295 (1944).
95. Rudolfs, W., "Industrial waste developments," *Sewage Works J.* **9**, 998 (1937).
96. Ryder, L. W., "The design and construction of the treatment plant for wool scouring and dyeing wastes at manufacturing plant, Glasgow, Va.," *J. Boston Soc. Civil Engrs.* **37**, 183 (1950).
97. Sadow, R. D., "The treatment of zefran fiber wastes," in Proceedings of 15th Industrial Waste Conference, Purdue University, May 1960, p. 359.
98. Snell, F. D., "Treatment of rayon boil-off waste," *Ind. Eng. Chem.* **29**, 1438 (1937).
99. Snyder, D. W., "Pollution control in the textile industry by process change," in Proceedings of 2nd Southern Municipal and Industrial Waste Conference, March 1953, p. 136.
100. Snyder, D. W., "Cotton slashing with synthetic compounds as a means toward pollution abatement," *Am. Dyestuff Reptr.* **44**, 382 (1955).
101. Snyder, D. W., "Pollution abatement resulting from the practical use of synthetic compounds in cotton slashing," in Proceedings of 5th Southern Municipal and Industrial Waste Conference, April 1956, p. 157.
102. Souther, R. H., "Research in textile waste problems," in Proceedings of 1st Southern Municipal and Industrial Waste Conference, March 1952, p. 102.
103. Souther, R. H., "In-plant process control for the reduction of wastes," *Am. Dyestuff Reptr.* **42**, 656 (1953).
104. Souther, R. H., "A tool that makes dyeing easier," *Textile Inds.* **117**, 124 (1953).
105. Souther, R. H., and T. A. Alspaugh, "Biological treatment of mixtures of highly alkaline textile mill waste and sewage," *Am. Dyestuff Reptr.* **44**, 390 (1955).
106. Speel, H. C., *Textile Chemicals and Auxiliaries*, Reinhold Publishing Corp., New York (1952), p. 24.
107. Stafford, W., and H. J. Northrup, "The BOD of textile chemicals," *Am. Dyestuff Reptr.* **44**, 355 (1955).
108. Steel, W. R., "Application of flue gas to the disposal of caustic textile wastes," in Proceedings of 3rd Southern Municipal and Industrial Waste Conference, March 1954, p. 190.
109. Steel, W. R., and J. V. McMahon, "Disposal of caustic liquor from textile industry," U.S. Patent no. 2632732, 24 March 1953.
110. Stream Sanitation Committee of Piedmont Section, American Association of Textile Chemists and Colorists, "Bibliography on textile wastes," *Am. Dyestuff Reptr.* **44**, 168 (1955).
111. "Symposium on waste disposal problems of southern textile mills," *Am. Dyestuff Reptr.* **44**, 379 (1955).
112. Taylor, E. F., G. C. Cross, C. E. Jones, and R. F. Rocheleau, "Biochemical oxidation of wastes from the new plant for manufacturing Orlon at Waynesboro, Va.," in Proceedings of 15th Industrial Waste Conference, Purdue University, May 1960, p. 508.
113. Taylor, W. H., "Alcohol extraction of wool wastes," *Sanitalk* **2**, 25 (1954).
114. New England Interstate Water Pollution Control Commission, *Textile Wastes—A Review 1936–1950*, Hall Laboratory of Chemistry, Wesleyan University, Middletown, Conn. (1950).
115. North Carolina State Stream Sanitation and Conservation Committee, *Textile Wastes as Related to Stream Pollution* (1948), p. 189.
116. Thornton, H. A., and J. R. Morse, "Adsorbents in waste water treatment, dye absorption and recovery studies," *Sewage Ind. Wastes* **23**, 497 (1951).
117. "Treatment of alkaline sulfur dye wastes with flue gas," Research Report no. 8, *J. Sanit. Eng. Div.* **82**, 1078 (1956).
118. Turnbull, S. G., Jr., "Waste problems associated with the dyeing and finishing of synthetic fibers," in Proceedings of 5th Southern Municipal and Industrial Waste Conference, April 1956, p. 170.
119. "Viscose rayon manufacturing wastes and their treatment," *Water Works Sewerage* **91**, 265 (1944) and **91**, 295 (1944).
120. Vollrath, H. B., "Applying dialysis to colloid-crystalloid separations," *Chem. Met. Eng.* **43**, 303 (1936).

Suggested Additional Reading

The following references represent most of the recent publications in textile wastes since 1962.

1. Alspaugh, T. A., "Interrelationships among water resources pollution control and growth of the textile industry in the southeast," in Proceedings of 13th Southern Municipal and Industrial Waste Conference, April 1964, p. 178.
2. Alspaugh, T. A., "Tannery, textile and wool scouring wastes," *Lit. Rev., J. Water Pollution Control Federation*, April 1965.

3. Alspaugh, T. A., "Textile and tannery wastes," *Lit. Rev., J. Water Pollution Control Federation*, June 1967.
4. }
5. } U.S. Public Health Service, *An Industrial Waste Guide to the Cotton Textile Industry*, Washington, D.C. (1959).
6. Anderson, C.A., and G. F. Wood, "Lanolin née wool grease," *J. Am. Oil Chem. Soc.* **40**, 301 (1963).
7. Anderson, C.A., "Wool grease and suint," *J. Textile Inst. Abstr.* **54**, 225 (1963).
8. Agranonik, E. Z., and Ya. Z. Sorokin, "Purification of waste waters from rayon fibre production," *Chem. Abstr.* **58**, 2265 (1963).
9. Agranonik, E. Z., "Purification of waste waters from rayon production with reagents in a fluidized layer," *Chem. Abstr.* **58**, 5370 (1963).
10. Agrononik. E. Z., "Removal of zinc and other impurities from the collection sewage of viscose fibre factories by lining in a fluidized layer," *Chem. Abstr.* **61**, 1433 (1964).
11. "Treatment of wool scouring effluent," *Textile Technicians' Dig.* **20**, 5580 (1963).
12. "Treatment of wool scouring effluent and the recovery of wool grease," *Dyer* **130**, 23 (1963); *Abstr. Am. Dyestuff Reptr.* **25**, 949 (1963); *Textile Technicians' Dig.* **20**, 4884 (1963).
13. "Treating wool and yarn scouring liquor," *Chem. Abstr.* **59**, 308 (1963).
14. "Waste water in the textile industry," *J. Textile Inst. Abstr.* **54**, A140 (1963).
15. Antonie, M., "Waste waters from cold water steeping of hemp and their purification," *Chem. Abstr.* **58**, 340 (1963).
16. Barboi, V. M., *et al.*, "The use of weak-acid cation exchangers for the removal of zinc from waste waters from viscose fibre plant," *J. Textile Inst. Abstr.* **54**, A548 (1963).
17. Barochina, I. Ya., *et al.*, "Complex purification of waste waters from zinc and ventilation air from hydrogen sulfide in the manufacture of synthetic fibres," *Chem. Abstr.* **59**, 6124 (1963).
18. Bashwed, A. M., "Anaerobic digestion treats cotton mill de-size wastes," *Wastes Eng.* **33**, 402 (1962).
19. Bovy, R., "Waste waters from wool washing," *Berner Intern. Vortragstag Pro Aqua* **108**, 1961.
20. Brawn, P., "Effects of trade effluents on activated sludge treatment and observations upon residual impurities in effluent," *Chem. Abstr.* **59**, 11092 (1963).
21. Brown, J. C., Jr., "Waste treatment experiences reported," *Southern Power Ind.* **79**, 18 (1961).
22. Buswell, A. M., *et al.*, "Anaerobic digestion treats cotton mill de-size wastes," *Wastes Eng.* **33**, 402 (1962).
23. Caldwell, D. H., "Sewage oxidation ponds—performance operation and design," *Sewage Works J.* **18**, 433 (1946).
24. Carter, L., "Bioassay of trade waste," *Textile Technicians' Dig.* **20**, 880 (1963).
25. Chesner, L., and F. W. Roberts, "Some effluent studies with bleaching liquors," *Textile Technicians' Dig.* **20**, 3869 (1963).
26. Culver, R. H, "Pilot plant testing for domestic and industrial waste treatment," *J. Water Pollution Control Federation* **35**, 1455 (1963).
27. Dean, B. T., "Nylon waste treatment," *J. Water Pollution Control Federation* **33**, 864 (1961).
28. Downing, D. T., "Solvent fractionation of wool wax acids," *Australian J. Appl. Sci.* **14**, 50 (1963); *J. Textile Inst. Abstr.* **54**, A315 (1963).
29. *Dyeing and Finishing Fabrics Containing Dacron Staple Combined with Other Fibers*, E. I. Dupont de Nemours & Co., Wilmington, Del. (1967).
30. *Dyeing and Finishing Filament Yarn and Fabric of Dacron*, E. I. Dupont de Nemours & Co., Wilmington, Del. (1966).
31. Evers, D., "Trade effluent control in the carpet industry," *Textile Inst. Ind.* **3**, 237 (1965).
32. Franklin, J. S., J. F. Calville, and E. Bowes, "Treatment of wool scour effluent with calcium chloride using rotary vacuum filtration," *Chem Abstr.* **61**, 2818 (1964).
33. Goode, S. T., "Condensed report of the wool grease industry," *J. Am. Oil Chem. Soc.* **40**, 4 (1963).
34. Griffe, J., "Textile industrial effluent waters," *Textile Technicians' Dig.* **20**, 2275 (1963).
35. Grishina, E. E., "Anaerobic fermentation of wool wash waters," *Chem Abstr.* **58**, 13599 (1963).
36. Harwood, J., "Problems of effluent disposal," *Textile Technicians' Dig.* **20**, 1585 (1963); *Textile Manuf.* **89**, wash waters," *Chem Abstr.* **58**, 13599 (1963).
37. "How to treat textile wastes," *Wastes Eng.* **32**, 188 (1961).
38. Huddleston, R. L., "Biodegradeable detergents for the textile industry," *Am. Dyestuff Reptr.* **55**, 42 (1966).
39. Iida, H., and J. Kuwabara, "Fundamental studies of decoloration of dye waste," *Chem. Abstr.* **58**, 6564 (1963).

40. Ioffe, A. Z., and E. Z. Agranonik, "Purification of waste water from the production of viscose containing zinc and carbon disulfide by a method of thermolysis," *Chem. Abstr.* **61**, 434 (1964).
41. "Is waste treatment a business expense or capital investment?" *Wastes Eng.* **6**, 240 (1961).
42. Ivanov, B. I., N. F. Sharanova, N. A. Kuzmina, and L. N. Karazeeva, "Purification of waste waters from the production of vinyl acetate and related polymers," *Chem. Abstr.* **61**, 433 (1964).
43. Jones, E. L., et al., "Aerobic treatment of textile mill wastes," *J. Water Pollution Control Federation* **34**, 495 (1962).
44. Jones, L. L., "Textile waste treatment at Canton cotton mills," *Am. Dyestuff Reptr.* **54**, 61 (1965).
45. Kashiwaya, M., "Activated sludge treatment of textile and dyeing mill wastes," *J. Water Pollution Control Federation* **36**, 291 (1964).
46. Kehren, M., "Comments on solving the problem of the disposal of textile effluents," *J. Textile Inst. Abstr.* **54**, A268 (1963).
47. Kehren, M., "Effluent water problems in the textile finishing industry," *J. Textile Inst. Abstr.* **54**, A268, (1963); *Textile Technicians' Dig.* **20**, 2276 (1963).
48. Kehren, M., "Purification of waste waters from the textile finishing industry. IV," *J. Textile Inst. Abstr.* **54**, A188 (1963); *Textile Technicians' Dig.* **20**, 1584 (1963).
49. Kehren, M., "Purification of effluent waters from the textile finishing industry, V," *J. Textile Inst. Abstr.* **54**, A268 (1963); *Textile Technicians' Dig.* **20**, 2277 (1963).
50. Kehren, M., "Purification of waste waters from the textile finishing industry. I and III," *Textile Technicians' Dig.* **20**, 262 (1963) and **20**, 881 (1963).
51. Kehren, M., "Special hints on setting up disposal plants of textile wastes," *J. Textile Inst. Abstr.* **54**, A140 (1963).
52. Kehren, M., "Use of Niers process for textile waste purification," *J. Textile Inst. Abstr.* **54**, A140 (1963).
53. Kehren, M., "Waste disposal in the textile industry," *Textil-Rundschau* **16**, 372 (1961).
54. Klein, L., "Stream pollution and effluent treatment, with special reference to textile and paper mill effluents," *Chem. Ind.* **21**, 866 (1964).
55. Koganovskii, O. M., and Ya. M. Zagrai, "Use of cation exchange resin in a bed for purification of industrial waste waters from zinc ions," *Chem. Abstr.* **59**, 1371 (1963).
56. Kuchinskii, M. E., "Dehydration of sediments formed during clarification of sewage from viscose fiber production," *Chem. Abstr.* **61**, 433 (1964).
57. Kulakov, E. A., "Biochemical purification of sewage from viscose fibre fractures," *Chem. Abstr.* **61**, 432 (1964).
58. Little, A. H., "Effluent from textile works," *Textile Manuf.* **89**, 1586 (1963).
59. Little, A. H., "Textile works effluent treatment." *Textile Inst. Ind.* **1**, 9 (1963); *J. Textile Inst. Abstr.* **54**, A402 (1963).
60. Matsnow, A. I., "Purification of sewage from viscose fiber production by flotation," *Chem. Abstr.* **61**, 432 (1964).
61. Matuskov, Ya. E., "Ion exchanger for purification of waste waters of rayon production," *Chem. Abstr.* **58**, 2263 (1963).
62. Modrzejewski, B., "Automatic neutralization of sewage in the synthetic fiber works in Lodz," *Przemysl Chem.* **40**, 226 (1961).
63. Mokrzycki, J., et al., "Purification of strongly alkaline waste water," *Prace Inst. Wlokiennictwa* **9**, 19 (1959).
64. Mongait, I. L., E. A. Kulakov, E. B. Ishkhanova, and N. V. Vandyck, "Biological purification of waste waters from synthetic fibre manufacturing, industrial purification of general industrial waste," *Chem. Abstr.* **59**, 6124 (1963).
65. Mongait, I. L., and G. I. Fishman, "Precipitation methods for zinc removal from viscose fibre wastes," *Chem. Abstr.* **59**, 3640 (1963).
66. Mongait, I. L., and G. I. Fishman, "Waste purification in viscose fibre plants," *Chem. Abstr.* **59**, 1372 (1963).
67. National Stream Sanitation Committee of the American Association of Textile Chemists and Colorists in Cooperation with the National Technical Task Committee on Industrial Wastes, *Cotton Textile Industry*, U.S. Public Health Service, Bureau of State Service Division of Water Supply and Pollution Control (1959).
68. Nosek, J., M. Ratkousky, and O. Ruzicka, "Clarification of textile waste water through the use of wet transfer of slag and ash from steam boilers," *Chem. Abstr.* **61**, 4056 (1964).
69. Nosek, J., "Waste water purification in the textile industry," *Chem. Abstr.* **61**, 1600 (1964).
70. Nowacki, J., "The problem of sewage water of the

textile industry," *Przemysl Wlokienniczy* **10**, 462 (1956).
71. Oakum, D. A., et al., "Textile and wool scouring wastes," *Lit. Rev., J. Water Pollution Control Federation*, June 1963.
72. Pinault, R. W., "Waste disposal systems—how they shape up today," *Textile Worker* **114**, 100 (1964).
73. Podrezova, A. S., "Removal of zinc salts from waste waters," *Chem. Abstr.* **58**, 12290 (1963).
74. Popp, P., "Methods for the purification of waste waters from cord rayon production," *J. Textile Inst. Abstr.* **54**, A268 (1963); *Textile Technicians' Dig.* **20**, 2274 (1963).
75. Rhyn, H., "Turofloc process and pH control for waste purification," *J. Textile Inst. Abstr.* **54**, A140 (1963).
76. Riemer, H., "Flotation principle for use in textile waste treatment," *J. Textile Inst. Abstr.* **54**, A140 (1963).
77. Sadow, R. D., "Acrylonitrile and zinc wastes treatment," *Ind. Water Wastes* **6**, 66 (1961).
78. Smith, A. L., et al., "Finding the best way to treat wastes from a blanket mill," *Wastes Eng.* **32**, 230 (1961).
79. Smith, A. L., and J. C. Grey, "The evaluation of a waste treatment scheme at Chatham Manufacturing Co., Elkin, North Carolina," in Proceedings of 9th Southern Municipal and Industrial Waste Conference, 1960, p. 105.
80. Smith, A. L., "Waste disposal by textile plants," *J. Water Pollution Control Federation*, **37**, 1607 (1965).
81. Smith, L. J., "Plant for treatment of textile effluents," *Textile Technicians' Dig.* **20**, 1587 (1963); *Textile Manuf.* **89**, 34 (1963).
82. Sontheimer, H., "New technical viewpoints on textile waste treatment," *J. Textile Inst. Abstr.* **54**, A140 (1963).
83. Starobinets, G. L., V. M. Akulovich, and A. I. Pakrovskaya, "Combined cation-anion exchanger method for removing zinc from sewage from viscose fibre factories," *Chem. Abstr.* **61**, 434 (1964).
84. Stempkovska, L. A., "Effluent methods for purifying industrial sewage," *Chem. Abstr.* **58**, 1233 (1963).
85. Stempkovska, L. A., "Use of absorbents and precipitating agents for purifying industrial waste waters," *Chem. Abstr.* **58**, 12239 (1963).
86. Stack, V. T., et al., "Biological treatment of textile waste," in Proceedings of 16th Industrial Waste Conference, Purdue University, 1961.
87. Tanaka, M., M. Dasai, Y. Sonoda and H. Ono, "Biological treatment of wool spinnery waste," *Chem. Abstr.* **61**, 6771 (1964).
88. Taylor, E. F., et al., "Biochemical oxidation of wastes from the new plant for manufacturing Orlon at Waymenboro, Virginia," in Proceedings of 15th Industrial Waste Conference, Purdue University, 1960, p. 508.
89. Taylor, E. F., et al., "Orlon manufacturing wastes treatment," *J. Water Pollution Control Federation* **33**, 1076 (1961).
90. U.S. Department of the Interior, "The cost of clean water," Industrial Waste Profile no. 4, *Textile Mill Products*, Washington, D.C. (1967).
91. Van Achter, R., F. Edeline, L. deLaey, and L. Creyten, "Bleachery waste water study," *Chem. Abstr.* **61**, 4055 (1964).
92. Vasil'kova, L. P., "Biochemical purification of waste water from the production of vinyl acetate and related polymers," *Chem. Abstr.* **61**, 434 (1964).
93. Wilke, H., "Waste and refuse disposal," *Textile Technicians' Dig.* **20**, 6635 (1963).
94. Williams, S. W., Jr., et al., "Treatment of textile mill wastes in aerated lagoons," in Proceedings of 16th Industrial Waste Conference, Purdue University, 1961.
95. Willis, C.A., "Developing patterns for efficient water utilization of textile dyeing and finishing industries," in Proceedings of 14th Southern Water Resources and Pollution Control Conference, 1965, p. 100.
96. Wilson, I. S., "Concentration effects in the biological oxidation of trade waste," *Textile Technicians' Dig.* **20**, 5582 (1963).
97. Zagrai, Ya. M., L. A. Kulsku, and A. M. Loganovsku, "The use of the bubbling layer of the cation exchangers for the removal of zinc from waste waters," *J. Textile Inst. Abstr.* **54**, A548 (1963).

TANNERY WASTES

Tannery wastes originate from the beamhouse and the tanyard. In the beamhouse, curing, fleshing, washing, soaking, dehairing, lime splitting, bating, pickling, and degreasing operations are carried out. In the tanyard, the final leather is prepared by several processes. These include vegetable or chrome tanning, shaving, and finishing. The finishing operation includes bleaching, stuffing and fat-liquoring, and coloring. The discharge from a tannery averages 8000 to 12,000 gallons of waste per 1000 pounds of wet, salted hide processed. The waste averages 8000 ppm total solids, 1500 ppm volatile (organic) solids, 1000 ppm protein, 300 ppm NaCl, 1600 ppm total hardness, 1000 ppm sulfide, 40 ppm chromium, 60 ppm ammonium nitrogen, and 1000 ppm BOD. It has a pH of between 11 and 12 and normally produces a 5 to 10 per cent sludge concentration because of the lime and sodium sulfide contents. The generally accepted procedure for waste treatment is equalization, sedimentation, trickling filtration, or activated-sludge treatment. The latter two biological treatments generally reduce the BOD by 85 to 95 per cent and the sulfide by 100 per cent.

22.4 Origin and Characteristics of Tannery Wastes

Tanning is the act of converting animal skins into leather. For a detailed discussion of the composition of animal skin, the reader is referred to the work of Masselli et al. [32]. The dry matter of the skin is almost entirely protein, of which 85 per cent is collagen. The skin also contains minor amounts of lipids, albumins, globulin, and carbohydrates. The preliminary processes prepare the hide protein (mainly collagen) so that all undesirable impurities are removed, leaving the collagen in a receptive condition to absorb the tannin or chromium used in tanning.

Curing involves dehydration of the hide by drying it with salt or air in order to stop proteolytic enzyme degradation. Fleshing removes the areolar (fatty) tissues from the skin by mechanical means. Washing and soaking remove the dirt, salts, blood, manure, and nonfibrous proteins and restore the moisture lost during preservation and storage. Unhairing is accomplished by the use of lime, with or without sodium sulfide; this makes the skins more attractive and more amenable to the removal of trace protein impurities.

Lime splitting separates the skin into two layers: one is the more valuable grain layer; the other, the lower or flesh side, is called the "split." Bating prepares the hide for tanning by reducing the pH, reducing the swelling, peptizing the fibers, and removing the protein-degradation products. Bating is generally accomplished with ammonium salts and a mixture of commercially prepared enzymes (predominantly trypsin and chymotrypsin). The bating bath renders the grain silky, slippery, smoother, and more porous, increases its width, and diminishes its wrinkles. Pickling generally precedes chrome tanning and involves treatment of the skin with salt and acid to prevent precipitation of the chromium salts on the skin fibers. Degreasing removes natural grease, thus preventing formation of metallic soaps and allowing the skin to be more evenly penetrated by tanning liquors.

Chrome tanning is used primarily for light leather, while vegetable tanning is still preferred for most heavy-leather products. The process of chrome tanning is of shorter duration and produces a more resistant leather. Vegetable tanning produces leathers which are fuller, plumper, more easily tooled and embossed, and less affected by body perspiration or changes in humidity. In the United States mainly quebracho and an extract of chestnut wood are used. Bleaching with dilute Na_2CO_3, followed by H_2SO_4, gives the leather a lighter and more uniform color before dyeing. The process of incorporating oils and greases into the tanned skins is called stuffing and fat-liquoring, and makes the hides soft, pliable, and resistant to tearing. Dyeing to produce the final colored leather product is usually done with basic dyestuffs.

A recent study by the Department of the Interior [1967, 170] revealed changes in the processing of leather, which are shown in a detailed flow chart (Fig. 22.8).

An equalized tannery waste, including rinses, is high in total solids (6000 to 8000 ppm) of which about half (3000 ppm) is NaCl. It contains about 900 ppm BOD, 1600 ppm total hardness, 120 ppm sulfide, 1000 ppm protein, and 30 to 70 ppm chromium. Of importance to the industrial-waste engineer are the high BOD, hardness, sulfide, chromium, and sludge content. About one gallon of this waste is produced for each pound of hide received by the tannery. Masselli

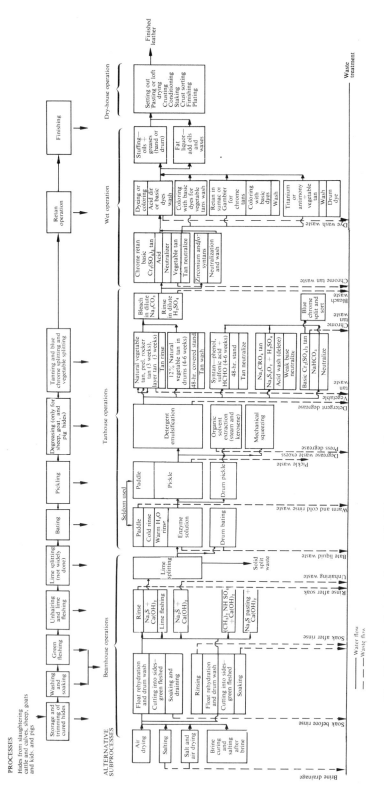

Fig. 22.8 Leather tanning and finishing process flow chart. (Taken from a chart prepared for the F.W.P.C.A.)

et al. [32] also state that for every 1000 pounds of wet, salted hide, there is a BOD load of 76 pounds, 52 per cent of which is discharged in the unhair waste, 20 per cent in the soaks, and 13 per cent in the delime and bate wastes.

Wide fluctuations in the nature of tannery wastes, due to intermittent dump discharges, make these wastes difficult to treat, especially in combination with municipal sewage. Protein and other material extracted from the hides are estimated to produce 50 to 70 per cent of the BOD load and process chemicals are estimated to produce 30 to 50 per cent.

The average composition and waste contribution of major tannery wastes are presented in Tables 22.10 and 22.11, after the findings of Masselli *et al.* [32]. In Table 22.12 the annual consumption of process chemicals for a cattle-skin tannery is presented, showing, among other things, that the total of process chemicals (including curing salt) used per year is 14,080,000 pounds—approximately 61 per cent of the weight of hides. About 71 per cent of this total, or about 440 pounds of chemicals for each 1000 pounds of hide received by the tannery, is discharged to waste. Salt constitutes 57 per cent of the chemicals in this waste.

22.5 Treatment of Tannery Wastes

In 1952 there were 443 tanneries in the United States, approximately 60 per cent of them located in the North-east [2]. (The number had decreased to about 250 by 1967.) Treatment of tannery wastes is limited in most cases to equalization and sedimentation, although some chemical coagulation and sludge digestion are also practiced. Equalization is necessary to minimize the wide fluctuations in the composition of the waste caused by intermittent dump discharges of strong liquors, and sedimentation is necessary because of the large volumes of sludge (5 to 10 per cent) present in the waste. If secondary treatment is necessary, trickling filters and activated-sludge systems can be used (when the pH has been reduced to 9.0 and the hardness to 200 ppm) to produce an 85 to 95 per cent BOD reduction. The evolution of H_2S from the waste, as a result of contact with another acid waste, should be prevented. When discharging tannery waste to a municipal treatment plant, provisions must be made to remove hair and fleshings (usually by screening) and to avoid deposition of scale in the sewer line. A schematic description of tannery waste treatment is given in Fig. 22.9.

Table 22.10 Cattlehide tannery survey, average composition and contribution of strong liquors. (After Masselli *et al.* [32].)

Process	Volume		BOD				Sodium chloride, ppm	Total hardness, ppm	Protein, ppm	Total solids, ppm	Volatile solids, ppm
	gal/day	% of total	ppm	lb/day	% of total	lb/1000 lb hide					
Soaks	73,100	42	2,200	1310	20	15*	20,000	670	1,900	30,000	3,600
Unhair	27,200	16	15,500	3510	52	40*	18,000	25,000	22,700	78,000	18,000
Relime	27,200	16	650	147	2	2*	3,500	25,000		20,300	2,500
Delime and bate	17,600	10	6,000	880	13	10†	10	4,100	4,300	15,000	8,800
Pickle	9,800	6	2,900	237	4	3†	47,000	2,400		79,000	7,200
Chrome tan	8,500	5	6,500‡	425	6	8†	26,000	1,800		93,000	13,000
Color and fat-liquor											
First dump	5,100	3	2,000	85	1	3§				16,000	8,000
Second dump	5,100	3	2,200	93	1	3§	250	2,600		9,500	4,900
Total	173,600			6687							

*Based on wet, salted hide.
†Based on fleshed, split hide, after relime.
‡Estimated at 50 per cent concentration of volatile solids.
§Based on chrome-tanned leather.

Table 22.11 Pigskin tannery survey, average composition and contribution of strong liquors. (After Masselli et al. [32].)

Process	Volume		BOD			lb/1000 lb hide	Sodium chloride, ppm	Total hardness, ppm	Protein, ppm	Total solids, ppm	Volatile solids, ppm
	gal/day	% of total	ppm	lb/day	% of total						
Soaks	3,000	19	2,400	60	8	17*	35,000			28,000	2,300
Unhair	4,000	26	14,000	467	61	70†	5,700	38,000	18,400	55,000	12,900
Delime and bate	4,000	26	4,400	147	19	23†	640	4,200	1,600	14,000	7,400
Pickle	700	5	4,200	25	3	9†	80,000			98,000	12,000
Degrease											
Kerosene layer	340	2		(1210)‡		435§					
Brine layer	800	5	2,600	17	2	7	100,000			110,000	2,300
Vegetable tan	30		24,000	6	2§	1				93,000	25,000
Chrome tan	600	4	2,300††	12	2	5§	51,000			80,000	4,600
Color and fat-liquor											
First dump	1,000	6	490	4	1	1**	410			3,950	890
Second dump	1,000	6	3,950	33	4	8**	135			3,980	3,030
Total	15,470			771							

*Based on wet, salted hide.
†Based on fleshed hide after soaking (30 per cent of flesh removed).
‡Calculated from kerosene (53 per cent BOD) only; not included in total.
§Based on pickled hide.
**Based on leather.
††Estimated at 50 per cent concentration of volatile solids.

There have been three comprehensive studies of the treatment of tannery wastes. The first was made by the Leather Chemists' Association on Stream Pollution [3]. It found that all 32 plants reporting used primary sedimentation, but only five employed secondary treatment. Similarly, spent tans were lagooned and released at periods of high stream flow by only five of the plants. Although the U.S. Public Health Service [19] recommends chemical precipitation as one of the treatments to be used on combined tannery wastes, the survey [42] showed that only one plant used chemical precipitation or applied its effluent to a filter bed. The Tannery Waste Disposal Committee of Pennsylvania [58] found that mixing of wastes and sedimentation removed 85 per cent of the suspended solids and 40 per cent of the BOD. This committee recommended equalization and settling for at least eight hours, then trickling filtration and sedimentation. As an intermediate step it recommended the addition of coagulants to the filter effluent. In a more recent study, Haseltine [18] found combining tannery wastes with domestic sewage feasible. The tannery waste is equalized separately for three to four days; it is then mixed with twice its volume of sewage, aerated, and settled. The sludge is concentrated, dewatered on vacuum filters, and incinerated.

Masselli et al. [32] noted that treatment of tannery wastes is limited in most cases to equalization and sedimentation. Many tanneries discharge without treatment directly into coastal waters or to a municipal sewage-treatment plant. They recommend that, when treatment is required, the strong liquors should be segregated and the dilute rinses discharged directly without treatment. The wastes should be screened to prevent damage to, or clogging of, pumps and pipelines. Holding basins or storage tanks should be used to provide equalization for at least one day's flow of waste. When sedimentation is called for, it is necessary to provide mechanical sludge removal followed by centrifugation or vacuum filtration. The dewatered sludge can be finally disposed of by sand or cinder-bed filtration or by lagooning.

Some important information pertaining to the re-

Table 22.12 Annual consumption of process chemicals in cattlehide tannery. (After Masselli et al. [32].)

Process chemical	lb used	% BOD	lb BOD	ppm in waste
Sodium chloride	1,368,000	0	0	684
Sodium chloride (used in curing 19% OWH)	4,408,000	0	0	2200
Lime	2,470,000	0	0	1235
Sodium sulfide (62% Na_2S, 25% S^-)	981,000	40	392,000	490
Sulfuric acid	350,000	0	0	175
Soda ash	161,000	0	0	80
Oropon [95% $(NH_4)_2SO_4$]	144,000	5	7,200	72
Calcium formate	88,000	12	11,000	44
Lactic acid (30%)	77,000	32	25,000	38
Sodium formate	56,000	2	1,100	28
Sterizol	42,000			20
Ammonium chloride	20,000	0	0	10
Chemicals absorbed by hide*				
Tanolin R (16% chromium)	1,670,000	0	0	626†
Tamol L	729,000	0	0	36
D-1 oil	37,000	83	3,100	2
Other oils (total of 12)	650,000	80	52,000	33
Quebracho	146,000	5	700	7
Soyarich flour	100,000			5
Tanbark H	88,000	11	1,000	4
Titanium dioxide	88,000	0	0	4
Ade 11 tan	38,000			2
Gambade	156,000	4	600	8
Maratan B	136,000			7
Methocel	20,000	6	120	1
Orotan TV	30,000	5	150	2
Semisol glue	37,000			2
Upper tan	28,000			1
Totals	14,080,000		494,300	5343

*Absorption estimated at 90%, only 10% discharged to waste. Pounds BOD and ppm in waste are based on 10% of the pounds used.
†Based on 75% discharged to waste.

sults of primary treatment of tannery wastes, including settling, lagooning, chemical precipitation, and sludge digestion, is provided in references 7, 9, 11, 15, 20, 21, 23, 25, 30, 39, 43, 46, 47, 48, 64, 65, and 68. Results of secondary treatment of tannery wastes are given in references 1, 5, 26, 34, 40, 49, 51, and 56.

Under proper conditions, the activated-sludge process developed on a pilot-plant scale by Chase and Kahn [1955, 27] was able to render tannery wastes entirely suitable for discharge into streams. An effluent was produced which was practically clear, free from objectionable odor, and perfectly stable. During a period of over four months, between March and July 1916, the following reductions were obtained: albuminoid ammonia, dissolved, 80 per cent, suspended, 69 per cent; free ammonia nitrogen, 65 per cent; total suspended solids, 78 per cent; fats (by ether extraction), 79 per cent.

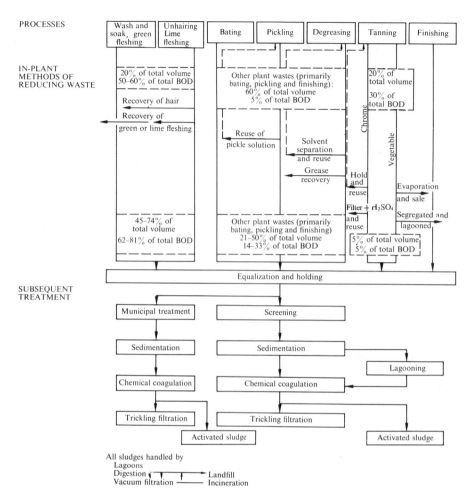

Fig. 22.9 Leather tanning and finishing waste-treatment flow chart. (Taken from a chart prepared for the F.W.P.C.A.)

Owing to the variability of the wastes it was found necessary to equalize their flow and quality, in order to preserve the bacterial activity of the sludge and to make the wastes more amenable to treatment. Milk of lime was added periodically to avoid an acid effluent from the equalizing tank and to maintain the alkalinity level of the wastes above the minimum required for good clarification.

Return activated sludge (sludge which is returned to the aeration tank from the final settling tank after activated-sludge treatment), as high as 50 per cent of the influent, was found to be quite economical when compared with the relative quantities of air required per gallon of wastes treated by aeration of activated sludge. Within certain limits it was found that the rate of air supply and the period of aeration could be considered inversely proportional, with economic advantages favoring longer periods of aeration.

The pilot-plant operation indicated that, with the most economical use of air, the volume of sludge to be disposed of would probably not exceed 10,000 gallons per million gallons of wastes treated. It could be dried on sludge beds, if applied in shallow doses not over 10 inches deep, because the activated-sludge process produced no objectionable odors and had no fly problems (such as occurred with trickling filters); furthermore, aeration renders the sludge more stable and less likely to create malodors when drying.

Gates and Lin [1966, 60] studied in two pilot plants the possibility of using an anaerobic lagoon with an oxygen cover for treating tannery wastes. They obtained 88.5 per cent reduction in BOD at organic loadings of about 1000 pounds of BOD acre/day. The final effluent, although colored, was clear and odorless. Most of the reduction occurred in the anaerobic section—which included sludge digestion—the aerobic lagoon being utilized primarily for odor control.

Emerson and Nemerow recently reported and will soon publish results of laboratory and pilot-plant experiments on biological treatment of equalized combined beamhouse and tanhouse wastes. The treatment without neutralization by high-solids (15,000 ppm suspended solids) aeration, at loadings of 140 pounds of BOD per 1000 ft^3 with a detention time of 12 hours, yielded 90 per cent BOD reduction.

References: Tannery Wastes

1. Alsop, E. C., "Purification of liquid tannery waste by forced oxidation," *J. Am. Leather Chemists' Assoc.* **7**, 72 (1912).
2. Ball, W. J., "Operation of inadequate facilities at Ballston Spa, N.Y.," *Sewage Ind. Wastes* **25**, 1345 (1953).
3. Clark, H. W., "Carbonation of industrial wastes to prevent clogging," *Water Works Sewerage* **79**, 52 (1932).
4. *Dimethylamine Sulfate*, Leather Chemicals Department, Rohm & Haas Co., Philadelphia, Pa. (1955).
5. Eddy, H. P., and A. L. Fales, "The activated-sludge process in treatment of tannery wastes," *Ind. Eng. Chem.* **8**, 548 (1916).
6. Eich, J. F., "Tannery wastes disposal by spray irrigation," *Ind. Wastes* **1**, 271 (1956).
7. Eldridge, E. F., "Report on sanitary engineering projects," *Mich. State Univ. Eng. Exp. Sta. Bull.* **67**, 32 (1936), **83**, 3 (1939), and **87** (1939).
8. Fair, G. M., and J. C. Geyer, *Water Supply and Waste-Water Disposal*, John Wiley & Sons, New York (1954), p. 879.
9. Fales, A. L., "Treatment of industrial wastes from paper mills and tannery on Neponset river," *Ind. Eng. Chem.* **21**, 216 (1929).
10. Foster, W., "Discussion on trade waste (tannery) purification plant," *J. Proc. Inst. Sewage Purif.* Part 2, 1950, p. 98.
11. Foster, W., "Chrome tannery treatment plant, description," *Sewage Ind. Wastes*, **24**, 927 (1952).
12. Gurnham, C. F., *Principles of Industrial Waste Treatment*, John Wiley & Sons, New York (1955), p. 6.
13. Hamor, W. A., "Tannery wastes in sewage," *Ind. Eng. Chem.* **4**, 382 (1912).
14. Harnley, J. W., "Liquid industrial wastes—symposium," *Ind. Eng. Chem.* **44**, 520 (1952).
15. Harnley, J. W., R. F. Wagner, and H. G. Swope, "Treatment at Griess-Pfleger Tannery, Waukegan, Ill.," *Sewage Works J.* **12**, 771 (1940).
16. Hartman, B. J., "Combined treatment, tannery wastes and domestic sewage, Fond du Lac, Wis.," *Sewage Ind. Wastes* **25**, 1419 (1953).
17. Haseltine, T. R., "Central primary plant will handle sewage from three municipalities," *Wastes Eng.* **28**, 176 (1957).
18. Haseltine, T. R., "Combined treatment, tannery wastes and sewage, Williamsport, Pa.," *Sewage Ind. Wastes* **30**, 65 (1958).
19. Hammon, H. B., "Purification of tannery wastes," *U.S. Public Health Serv. Bull.* **100**, part III, 1919; *J. Am. Leather Chemists' Assoc.* **39**, 385 (1944).
20. Howalt, W., and E. S. Cavett, "Studies on tannery waste disposal," *Proc. Am. Soc. Civil Engrs.* **53**, 1675 (1927); *Trans. Am. Soc. Civil Engrs.* **92**, 1351 (1928).
21. Hubbell, G. E., "Tannery waste disposal at Rockford, Mich.," *Water Works Sewerage* **82**, 331 (1935).
22. Hubbell, G. E., "Waste treatment plant, Wolverine Shoe and Tanning Corp., Rockford, Mich.," in Proceedings of 10th Industrial Waste Conference, Purdue University, May 1955.
23. U.S. Public Health Service, *Industrial Waste Guide*, Ohio River Pollution Survey, Supplement D (1943).
24. Ingols, R. D., "The toxicity of chromium," in Proceedings of 8th Industrial Waste Conference, Purdue University Engineering Extension Series, Bulletin no. 23, 1953, p. 86.

25. Künzel-Mehner, A., "Treatment with ferric chloride," *Sewage Works J.* **17**, 412 (1945).
26. Loveland, F. A., *J. Am. Leather Chemists' Assoc.* **7**, 474 (1912).
27. McCarthy, J. A., and B. L. Rosenthal, "Biological treatment of tannery wastes," *Public Works Mag.* **88**, 82 (1957).
28. McLaughlin, G. D., and E. R. Theis, *The Chemistry of Leather Manufacture*, Reinhold Publishing Corp., New York (1945), p. 47.
29. McKee, J. E., and T. R. Camp, "Tanning wastes—some special problems," *Sewage Ind. Wastes* **22**, 803 (1950).
30. Maskey, D. F., "Study of tannery waste disposal," *J. Am. Leather Chemists' Assoc.* **36**, 121 (1941).
31. Masselli, J. W., and M. G. Burford, "Pollution reduction program for the textile industry," *Sewage Ind. Wastes* **28**, 1283 (1956).
32. Masselli, J. W., N. W. Masselli, and M. G. Burford, *Tannery Wastes*, New England Interstate Water Pollution Control Commission (1958).
33. Milligan, F. B., "Tannery waste treatment in Pennsylvania," *Am. City* **53**, 50 (1938).
34. Mohlman, F. W., "Treatment of packing-house, tannery, and corn-products wastes," *Ind. Eng. Chem.* **18**, 1076 (1926).
35. O'Flaherty, F., *Encyclopedia of Chemical Technology*, Vol. 8, R. E. Kirk and D. F. Othmer (eds.), Interscience Publishers, New York (1952), p. 219.
36. O'Flaherty, F., W. T. Roddy, and R. M. Lollar, "Preparation for tannage," in *The Chemistry and Technology of Leather*, Vol. 1, Reinhold Publishing Corp., New York (1956).
37. Parker, R. R., "Spray irrigation for disposal of tannery wastes," in Proceedings of 6th Ontario Industrial Waste Conference, June 1959, p. 3.
38. Pennsylvania Sewage and Industrial Waste Association, "Tannery waste and pickle liquor," *Ind. Wastes* **3**, 18 (1958).
39. Porter, W., "Operating problems from tannery wastes, Ballston Spa, N.Y.," *Sewage Works J.* **21**, 738 (1949).
40. Power, R. M., *Sanitalk* **5**, 19 (1957).
41. Department of Scientific and Industrial Research, *Report of the Water Pollution Research Board, 1932*, H.M.S.O., London (1933), p. 31.
42. Reuning, H. T., "Report of stream pollution committee," *J. Am. Leather Chemists' Assoc.* **38**, 292 (1943) and **39**, 378 (1944).
43. Reuning, H. T., "Tanning wastes," *J. Am. Leather Chemists' Assoc.* **42**, 573 (1947).
44. Reuning, H. T., "Disposal of tannery wastes," *Sewage Works J.* **20**, 525 (1948).
45. Reuning, H. T., "Tannery waste treatment," *Sewage Works Eng.* **20**, 133 (1949).
46. Reuning, H. T., and R. F. Coltart, "An effective tannery waste treatment plant," *Public Works Mag.* **78**, 21 (1947).
47. Riffenburg, H. B., and W. W. Allison, "Treatment of tannery wastes with flue gas and lime," *Ind. Eng. Chem.* **33**, 801 (1941).
48. Rosenthal, B. L., "Treatment of tannery waste and sewage mixture on trickling filters," *Sanitalk* **5**, 21 (1957).
49. Rosenthal, B. L., "Treatment of tannery wastes by activated sludge," *Sanitalk*, **6**, 7 (1957).
50. Rudolfs, W., *Industrial Wastes*, Reinhold Publishing Corp., New York (1953), Chapter 8.
51. Sarber, R. W., "Tannery waste disposal," *J. Am. Leather Chemists' Assoc.* **36**, 463 (1941).
52. Siebert, C. H., *Digest of Industrial Waste Treatment*, Pennsylvania Department of Health, Harrisburg Pa., (1940).
53. Smith, W. R., "Discussion on combined treatment," *Sewage Ind. Wastes* **22**, 101 (1950).
54. Snow, B. F., "South Essex, Mass., sewage system," *Sewage Works J.* **4**, 851 (1932).
55. Southgate, B. A., *Treatment and Disposal of Industrial Waste Waters*, H.M.S.O., London (1948).
56. Sutherland, R., "Tanning industry," *Ind. Eng. Chem.* **39**, 628 (1947).
57. U.S. Public Health Service, *Tannery Wastes*, Ohio River Survey, Supplement D (1943), p. 1218.
58. "Tannery Waste Disposal Committee of Pennsylvania, a report," *J. Am. Leather Chemists' Assoc.* **26**, 70 (1931).
59. Taylor, W. H., "Treatment of wool scouring wastes," *Sanitalk*, **1**, 13 (1953).
60. *Technical Manual and Yearbook of American Association of Textile Chemists and Colorists*, Howes Publishing Co., New York.
61. *Treatment of Tannery Wastes*, Pennsylvania Department of Health (1930).
62. Van der Leeden, R., "By-product recovery," *Sewage Works J.* **8**, 350 (1936).

63. Veitch, F. P., *J. Am. Leather Chemists' Assoc.* **8**, 10 (1913).
64. Vrooman, M., and V. Ehle, "Digestion of combined tannery and sewage sludge," *Sewage Works J.* **22**, 94 (1950).
65. Warrick, L. F., and E. J. Beatty, "Treatment with domestic sewage," *Sewage Works J.* **8**, 122 (1936).
66. Wilson, J. A., *The Chemistry of Leather Manufacture*, American Chemical Society Monograph, The Chemical Catalog Co., New York (1923).
67. Wilson, J. A., *Modern Practice in Leather Manufacturing*, Reinhold Publishing Corp., New York (1941), p. 272.
68. Wimmer, A., "The sewage from the tannery city Backnang," *Sewage Works J.* **9**, 529 (1937).

Suggested Additional Reading

The following recent references have been published since 1962 and include some older publications not referred to above.

1. "A condensed report by the Tannery Waste Disposal Committee of Pennsylvania to the Sanitary Water Board on the treatment of tannery wastes," *J. Am. Leather Chemists' Assoc.* **26**, 70 (1931).
2. Abramovich, I. A., "Separation of chromium from tannery effluents," *Chem. Abst.* **56**, 7068 (1962).
3. Appelius, W., "The purification of waste waters from tanneries," Report of the Freiburg Tanning School for 1911–12, *Ledertech. Rundschau* **4**, 113 (1912); *J. Am. Leather Chemists' Assoc.* **7**, 342 (1912).
4. Asendorf, E., "A new method of purification of waste water in the tannery," *Chem. Abstr.* **60**, 13008 (1964).
5. Bechard, E., "Depollution of tannery waste waters," *Rev. Tech. Ind. Cuir* **57**, 228 (1965); *Chem. Abstr.* **63**, 12864 (1965).
6. Bechard, E., "Depollution of tannery waste waters," *Chem. Abstr.* **65**, 16670 (1966) and **65**, 11971 (1966).
7. Bechard, E., "Purification of tannery waste waters," *Chem. Abstr.* **62**, 2739 (1965).
8. Bechard, E., "Treatment of tannery waste waters," *Chem. Abstr.* **63**, 8022 (1965).
9. Bianucci, G., and G. De Stefani, "Tannery wastes," *Effluent Water Treat. J.* **5**, 407 (1965); *Chem. Abstr.* **64**, 17245 (1966).
10. Berg, K., "Design basis for the treatment of sewage from factories of the leather industries," *Wasser Boden* **8**, 340 (1956).
11. Besse, J., "Purification of tannery effluents by argillaceous colloids," *Cuir Tech.* **18**, 244 (1929); *J. Am. Leather Chemists' Assoc.* **25**, 244 (1930).
12. Besse, J., "Purification of tannery effluents by colloidal clay," *Chem. Abstr.* **23**, 5608 (1929).
13. Besse, J., "Tannery effluents," *Cuir Tech.* **16**, 96 (1927); *J. Am. Leather Chemists' Assoc.* **22**, 296 (1927).
14. Besselievre, E. B., "Treatment of tannery waste to prevent stream pollution," *J. Am. Leather Chemists' Assoc.* **17**, 605 (1922).
15. Bianucci, G., and G. Stefani, "Tannery wastes. I and II," *Effluent Water Treat. J.* **3**, 18 (1963); *Chem Abstr.* **59**, 308 (1963).
16. Bianucci, G., and G. De Stefani, "Treatment and disposal of tannery waste," *Aqua Ind.* **4**, 35 (1962); *Chem. Abstr.* **64**, 13904 (1966).
17. Billingham, A., "Winslow Brothers and Smith Company," *JIEC*, **18**, 1357 (1926); *J. Am. Leather Chemists' Assoc.* **22**, 108 (1927).
18. Bolde, A., and B. Rosenthal, "High lime tannery wastes cause sewer incrustation," *Wastes Eng.* **31**, 150 (1960).
19. Braunschweig, T. D., "Tannery sewage," *J. Am. Leather Chemists' Assoc.* **60**, 125 (1965); *Chem. Abstr.* **62**, 12891 (1965).
20. Bravo, G. A., "Examination and treatment of water for tannery operations," *Assoc. Ital. Chim.* **17**, 94 (1939); *J. Am. Leather Chemists' Assoc.* **36**, 99 (1941).
21. Bruhne, A., "Report on the construction and effect of an experimental plant for the treatment of tannery waste in Waldhol," *Limnol. Schriftenreich* **4**, 97 (1957).
22. Bycichin, I. A., and C. Holamek, "Purification of water polluted with tannery waste," *Tech. Hlidkakozeluzska* **24**, 33 (1949); *J. Am. Leather Chemists' Assoc.* **44**, 619 (1949).
23. Ceamis, M., "Noisousness and purification of tannery waste waters," *Ind. Usoara* **2**, 208 (1955).
24. Ceamis, M., "Purification of tannery waste, mechanical, mechanical-chemical, or separate purification," *Ind. Usoara* **13**, 149 (1966); *Chem. Abstr.* **65**, 16670 (1966).
25. Ceamis, M., "Waste waters from tanneries—a study of their characteristics," *Rev. Tech. Ind. Cuir* **56**, 36 (1964); *Chem. Abstr.* **60**, 14248 (1964).

26. Cerny, A., "Waste waters from tanneries and leather works," *Wasser, Aswasser, Luft. Boden* **7**, 221 (1958); *Chem. Abstr.* **57**, 12271 (1962).
27. Chase, E. S., and P. Kahn, "Activated sludge and filters for tannery waste treatment," *Wastes Eng.* **26**, 167 (1955); *J. Am. Leather Chemists' Assoc.* **50**, 366 (1955).
28. Cille, G. G., *Treatment of Tannery Effluents*, South African Council for Scientific and Industrial Research, Pretoria, p. 619.
29. "Clarification of effluents in a modern tannery," *Cuir Tech.* **26**, 103 (1937); *J. Am. Leather Chemists' Assoc.* **33**, 34 (1938).
30. Collins, F. L., "Tannery waste disposal at Bolivar, Tennessee," *J. Am. Leather Chemists' Assoc.* **46**, 176 (1951).
31. "Compressed air solves tannery waste problem," *Water Waste Treat. J.* **8**, 139 (1960).
32. Dawson, F. M., and H. W. Ruf, "Developments and trends in industrial waste disposal," *Munic. Sanit.* **25**, 1937; *J. Am. Leather Chemists' Assoc.* **39**, 407 (1944).
33. Depner, and Germans, "Views on the article 'Serial experiments on the clarification of waste water from leather factories in Worms'," *Leder* **6**, 133 (1955); *J. Am. Leather Chemists' Assoc.* **50**, 5261 (1955).
34. Desmurs, G., "Purification of tannery effluents," *Cuir Tech.* **23**, 78 (1934); *J. Am. Leather Chemists' Assoc.* **30**, 35 (1935).
35. Domanski, J., "Chemical purification of tannery wastes," *Gaz, Woda Tech. Sanit.* **36**, 344 (1962); *Chem. Abstr.* **60**, 7786 (1964).
36. Domanski, J., "Thickening and dewatering of sludge precipitated from tanning industry sewage on sludge beds," *Gaz, Woda Tech. Sanit.* **39**, 362 (1964); *Chem. Abstr.* **62**, 15897 (1965).
37. Domanski, J., "Sedimentation of suspension in coagulation of sewage from tanning industry," *Gaz, Woda Tech. Sanit.* **38**, 279 (1964); *Chem. Abstr.* **62**, 15897 (1965).
38. Dorr, E. S., "The Miles-acid process on tannery waste," *Public Works* **49**, 403 (1920); *J. Am. Leather Chemists' Assoc.* **16**, 162 (1921).
39. Duyck, M, "Tannery effluents," *Halle aux Cuirs* **18**, 1913; *J. Am. Leather Chemists' Assoc.* **8**, 168 (1913).
40. Dyukov, A. I., "The work of horizontal sedimentation tanks and their design," *Chem. Abstr.* **25**, 3750 (1931).
41. Eagle, R. H., "Treatment of tannery sewage at the National Calfskin Co., Peabody, Mass," *J. Am. Leather Chemists' Assoc.* **14**, 577 (1919).
42. Eddy, H. P., and A. L. Fales, "The activated sludge process in treatment of tannery wastes," *J. Am. Leather Chemists' Assoc.* **11**, 441 (1916).
43. "Effluent disposal plant," *Leather Trades Rev.* **1261**, 298 (1957); *J. Am. Leather Chemists' Assoc.* **53**, 241 (1958).
44. Eitner, W., "Purification of tannery waste waters by the biological process," *Gerber* **32**, 199 (1906); *J. Am. Leather Chemists' Assoc.* **2**, 58 (1907).
45. Eick, J. F., "Tannery waste disposal by spray irrigation," *Ind. Wastes* **1**, 271 (1956).
46. Eldridge, E. F., "A fill-and-draw sedimentation plant for tannery wastes," *Mich. State Univ. Eng. Exp. Sta. Bull.* **83**, 3 (1939).
47. Eldridge, E. F., "Tannery waste sedimentation studies," *Mich. State Univ. Eng. Exp. Sta. Bull.* **11**, 5 (1936); *J. Am. Leather Chemists' Assoc.* **39**, 403 (1944).
48. Esten, P. A., "Bibliography of tannery waste treatment," *J. Am. Leather Chemists' Assoc.* **6**, 464 (1911).
49. Eye, J. D., and S. P. Graef, "Literature survey on tannery effluents," *J. Am. Leather Chemists' Assoc.* **62**, 194 (1967); *Chem. Abstr.* **66**, 98277 (1967).
50. Eye, J. D., "The treatment and disposal of tannery wastes," in *Chemistry and Technology of Leather*, Reinhold Publishing Corp., New York (1962).
51. Eye, J. D., "Waste management in the tanning industry," *Leather Shoes* **151**, 41 (1966).
52. Eye, J. D., "Waste treatment in the tanning industry," in *Chemistry and Technology of Leather*, Reinhold Publishing Corp., New York (1962).
53. Fales, A. L., "Industrial wastes disposal," *Proc. Am. Soc. Civil Engrs.* 1928; *Through Hide Leather* **75**, 48 (1928); *J. Am. Leather Chemists' Assoc.* **24**, 98 (1929).
54. Fassina, M. L., "Purification of tannery effluents," *Cuir Tech.* **27**, 22 (1938); *J. Am. Leather Chemists' Assoc.* **33**, 380 (1938).
55. Fassina, L., "Purification of waste waters. II: tanning industry," *Chim. Ind.* **38**, 847 (1937); *Chem. Abstr.* **32**, 3532 (1938).
56. Feikes, L., and E. Roth, "The development and operation of a purification plant for a tannery," *Leder* **8**, 114 (1957); *J. Am. Leather Chemists' Assoc.* **53**, 301 (1958).
57. Fontenelli, L. J., and W. Rudolfs, "Effect of industrial

wastes on the operation of a sewage treatment plant," *Sewage Works J.* **17**, 692 (1945).
58. Frendrup, W., "Common purification of tannery waste waters and domestic waste waters," *Leder* **17**, 79 (1966).
59. Furkert, H., "Mechanical clarification and biological purification of Elmshom sewage, a major portion of which is tannery waste," *Tech. Gemeindebl.* **39**, 285 (1944) and **40**, 36 (1944); *J. Am. Leather Chemists' Assoc.* **39**, 407 (1944).
60. Gates, W. E., and S. Lin, "Pilot plant studies on the anaerobic treatment of tannery effluents," *J. Am. Leather Chemists' Assoc.* **61**, 10 (1966).
61. Gale, S., and N. L. Nemerow, "Activated sludge treatment of tannery waste–domestic sewage mixtures," unpublished research paper.
62. Genin, G., "The treatment of waste tannery waters," *Halle aux Cuirs* **58**, 100 (1931); *J. Am. Leather Chemists' Assoc.* **27**, 361 (1932).
63. Guerree, H., "Purification of tannery waste waters," *Bull. Assoc. Franc. Chimistes Ind. Cuir Doc. Sci. Tech. Ind. Cuir.* **26**, 95 (1964); *Chem Abstr.* **61**, 2818 (1964).
64. Harnley, J. W., "Wastes from tanneries," *Ind. Eng. Chem.* **44**, 520 (1952).
65. Harnley, J. W., and D. C. Benrud, "Treatment and recovery of chrome wastes," *J. Am. Leather Chemists' Assoc.* **46**, 169 (1951).
66. Harwood, J. H., "The uses and limitations of aluminum sulphate in the chemical coagulation of trade effluents," *J. Inst. Sewage Purif.* Part 1 (1954), p. 26.
67. Haseltine, T. T., "Tannery waste treatment with sewage at Williamsport, Penn.," *Sewage Ind. Wastes* **30**, 65 (1958).
68. Heizig, H. M., and J. Brower, "Practices of industrial waste disposition at Milwaukee," *Sewage Works J.* **4**, 680 (1932).
69. Hepler, J. M., "Tannery sewage disposal," *Hide Leather* **71**, 40 (1926); *J. Am. Leather Chemists' Assoc.* **21**, 320 (1926).
70. Hommon, H. B., "Purification of tannery wastes (at Luray, Salenca, Elkton Tanneries)," *J. Am. Leather Chemists' Assoc.* **12**, 307 (1917).
71. Hommon, H. B., "Studies on the treatment and disposal of industrial wastes, purification of tannery wastes," *U.S. Public Health Serv. Bull.* **100**, 1919; *J. Am. Leather Chemists' Assoc.* **39**, 385 (1944).
72. "Industrial waste treatment processes and plant design," *Mich. State Univ. Eng. Exp. Sta. Bull.* **82**, (1938); *J. Am. Leather Chemists' Assoc.* **39**, 409 (1944).
73. Ivanov, G. I., "Anaerobic purification of tannery waste waters," *Kozh. Obuvn. Prom.* **4**, 30 (1962); *Chem. Abstr.* **58**, 6566 (1963).
74. Jackson, D. D., and A. M. Buswell, "The sterilization of tannery wastes," *J. Am. Leather Chemists' Assoc.* **12**, 56 (1917).
75. Jansky, K., "Sludge processing in tannery waste waters," *Věda Vyzkum Prumyslu Kozědělném* **7**, 91 (1963); *Chem. Abstr.* **60**, 14248 (1964).
76. Jansky, K., P. Safarik, and J. Suec, "The influence of tannery wastes on the receiving stream," *Chem. Abstr.* **54**, 1777 (1960).
77. Jansky, K., "Tannery wastes studies," *Kožářství* **11**, 327 (1961); *Chem. Abstr.* **58**, 6566 (1963).
78. Jansky, K., "Tannery waste water disposal," *Kožářství* **11**, 327 (1961); *J. Am. Leather Chemists' Assoc.* **57**, 281 (1962).
79. Januszkiewicz, T., "Water economy and characteristics of waste water from vegetable tanneries of pigskin," *Przeglad Skorzany* (1959), p. 106; *Chem. Abstr.* **57**, 13563 (1962).
80. Jettmar, J., "Tannery effluents and their treatment," *Collegium* **513**, 5 (1913); *J. Am. Leather Chemists' Assoc.* **8**, 229 (1913).
81. Kahler, H. L., and J. K. Brown, "Treating tannery waste liquor," U.S. Patent no. 3184407; *Chem. Abstr.* **63**, 4009 (1965).
82. Kalibina, M. M., "The application of the biological method in judging the efficiency of a purifying plant," *Trans. Inst. Structural Res.* **3**; Water Preservation Committee Publication no. 9, p. 68; Department of Scientific and Industrial Research, *Surv. Current Lit.* **3**, 404; *Chem. Abstr.* **25**, 3422 (1931).
83. Kohler, R., "The influence of sewage from tanneries and leather factories on the waters, on biological purification and on sludge digestion," *Leder Kurier.*
84. Kubelka, V., "Principles of a final biological purification of tannery effluents," *J. Am. Leather Chemists' Assoc.* **52**, 637 (1957).
85. Kubelka, V., "Purification of waste waters from the tanning industries," *Bull. Inst. Polit. Iasi* **7**, 145 (1961); *Chem. Abstr.* **58**, 11094 (1963).
86. Kubelka, V., "Recent advances in the treatment of industrial wastes, with special consideration of the tanning industry," *Voda* **31**, 159 (1958); *Chem. Abstr.* **50**, 11046g (1958).

87. Ludvick, J., and S. Siska, "Chemical oxygen demand in tannery waste waters. II: Some problems of optimum oxidant consumption in the $KMnO_4$–4–hour test," *Kozařství* **13**, 263 (1963); *Chem. Abstr.* **60**, 10377 (1964).
88. Ludvik, J., "Oxygen demand of tannery wastes. I: Reaction conditions in permanganate number—four hour determination," *Kozařství* **13**, 187 (1963); *Chem. Abstr.* **60**, 3865 (1964).
89. Magda, M., and J. Grabowska, "Modernization of purification technology of a tannery sewage," *Gaz, Woda Tech. Sanit.* **36**, 245 (1962); *Chem. Abstr.* **60**, 7786 (1964).
90. Malek, M., and J. Ludvik, "Some problems of mechanical treatment of tannery waste. I: A description of plant removal of insolubles, and sedimentation," *Věda Vyzkum Průmyslu Koželdělném* **7**, 63 (1963); *Chem. Abstr.* **60**, 11743 (1964).
91. Maskey, D. A., "A study of tannery waste disposal," *J. Am. Leather Chemists' Assoc.* **36**, 121 (1941).
92. Masselli, J. W., *Tannery Waste Pollution Sources and Methods of Treatment*, New England Interstate Water Pollution Control Commission, Boston, Mass. (1958), p. 41.
93. Mausner, L. "Tannery waste water," *Gerber* **41**, 1519 (1938); *J. Am. Leather Chemists' Assoc.* **33**, 488 (1938).
94. Mehner, A., "Treatment of tannery waste with ferric chloride," *Sewage Works J.* **17**, 412 (1945).
95. Merkel, W., "The purification of tannery waste liquors," *Leder* **3**, 9 (1952); *J. Am. Leather Chemists' Assoc.* **47**, 498 (1952).
96. Meunier, L., "The purification of tannery effluents," *Collegium* **507**, 268 (1912); *Through Leather Trades Rev.*, 16 October 1912; *J. Am. Leather Chemists' Assoc.* **7**, 687 (1912).
97. Morrison, J. A. S., "The treatment of tannery effluents," *J. Am. Leather Chemists' Assoc.* **6**, 326 (1911).
98. Moore, E. W., "Wastes from the tanning, fat processing and laundry soap industries," in *Industrial Wastes, Their Disposal and Treatment*, Reinhold Publishing Corp., New York (1953); American Chemical Society Monograph no. 118, 1953, p. 141.
99. Munteanu, A., and L. Weiner, "Purification of waste water resulting from the leather industries," *Hidrotehnica* **8**, 48 (1963); *Chem. Abstr.* **62**, 321 (1965).
100. Murphy, H. S., J. M. Hepler, and E. A. Eldridge, "Water pollution control—tannery wastes," Michigan State Department of Health Bulletin, May 1927; *J. Am. Leather Chemists' Assoc.* **39**, 388 (1944).
101. Najda, J., "Coagulation of chrome alum tanning sewages," *Gaz, Woda Tech. Sanit.* **40**, 14 (1966); *Chem. Abstr.* **64**, 10914 (1966).
102. Nylander, R. A., *Tannery Waste Clarification by Chemical Coagulation*, M.S. Thesis, University of New Hampshire (1963).
103. Oberlander, T. F., "A study in the treatment of tannery wastes," *J. Am. Leather Chemists' Assoc.* **21**, 393 (1932).
104. Orlita, A., and I. Orlitova, "Influence of tannery chemicals on *bacillus megatherium*," *Kozařství* **12**, 25 (1962); *Chem. Abstr.* **57**, 10953 (1962).
105. Orlita, A. and J. Suec, "Tannery effluents from the biological point of view. I: Beam house waste. II: Wastes from vegetable and chrome tannage," *Chem. Abstr.* **54**, 12437 (1960).
106. Parker, J. S., and J. S. Watkins, "Combination plant treats sewage and tannery wastes at Middlesborough, Ky.," *Water Works Sewerage* **88**, 109 (1941).
107. Paszto, P., "Pilot plant model experiments on treatment of tannery sewage," *Beszamolo Vizgazdalkodasi Tud. Kut. Int. Munkajarol* (1960), p. 188; *Chem. Abstr.* **61**, 2818 (1964).
108. Peck, C. L., "The profitable recovery of proteins from tannery waste waters," *J. Am. Leather Chemists' Assoc.* **13**, 417 (1918).
109. Peck, C. L., "Treatment of tannery sewage," *J. Am. Leather Chemists' Assoc.* **12**, 422 (1917).
110. & 111. Perkowski, S., "Mechanical and chemical purification of tannery waste waters," *Chem. Abstr.* **53**, 11724a (1959); *J. Am. Leather Chemists' Assoc.* **54**, 659 (1959).
112. Plsko, E., "Spectrochemical determination of chromium in tannery waste waters by using copper foil electrode," *Chem. Anal.* **7**, 239 (1962); *Chem. Abstr.* **57**, 12271 (1962).
113. "Purification of tannery effluents," *J. Appl. Chem.* **5**, 505 (1955).
114. Redlick, H. H., "The problem of tannery waste disposal," *J. Am. Leather Chemists' Assoc.* **48**, 422 (1953).
115. "Report on sanitary engineering projects (1938)," *Mich. State Univ. Eng. Exp. Sta. Bull.* **83**, (1938); *J. Am. Leather Chemists' Assoc.* **39**, 410 (1944).
116. Reuning, H. T., "Tannery waste treatment," *J. Am. Leather Chemists' Assoc.* **38**, 392 (1943).
117. Rolants, E., "Purification of tannery effluents,"

Marché Cuirs; *J. Am. Leather Chemists' Assoc.* **9**, 195 (1914).
118. Rosenthal, B. L., "Treatment of tannery waste by activated sludge," *Leather Mfr.* **75**, 26 (1958).
119. Rosenthal, B. L., "Treatment of tannery waste sewage mixture on trickling filters," *Leather Mfr.* **74**, 20 (1957).
120. Roth, A., "Disposal of tannery wastes," *J. Am. Leather Chemists' Assoc.* **9**, 512 (1914).
121. Roubaudi, D., "Tannery waste waters," *Bull. Assoc. Franc. Chimistes Ind. Cuir Doc. Sci. Tech. Ind. Cuir* **28**, 246 (1966); *Chem. Abstr.* **65**, 16670 (1966).
122. Sawyer, C. N., "Some new concepts concerning tannery wastes and sewers," *J. Water Pollution Control Federation* **37**, 722 (1965).
123. Schoatzle, T. C., and A. W. Blohm, "Efficiency of a tannery waste treatment works," Maryland State Department of Health Bulletin, April 1928; *J. Am. Leather Chemists' Assoc.* **30**, 392 (1944).
124. Schlichting, "Serial experiments on clarification of waste waters from leather factories in Worms," *Leder* **5**, 214 (1954); *Public Health Eng. Abstr.* **35**, 20 (1955).
125. Scholz, H. G., "An automatic process and an apparatus for the purification of the waste waters of the tannery," *Rev. Tech. Ind. Cuir* **55**, 368 (1963); *Chem. Abstr.* **60**, 9008 (1964).
126. Scholz, H. G., "Neutralization and depoisoning of tanning waste waters in an electrically driven apparatus," *Prakt. Chem.* **14**, 445 (1963); *Chem. Abstr.* **60**, 9008 (1964).
127. Scholz, H., "Purification of waste waters from European tanneries," *Leder Kurier* **4**, 159 (1960); *Chem. Abstr.* **55**, 6740 (1961).
128. Scholtz, H., "Treatment and disposal of tannery waste waters," *Leder* **4**, 121 (1953); *J. Am. Leather Chemists' Assoc.* **49**, 71 (1954).
129. Scholz, I. H., "Waste water disposal and treatment in the European leather industry," *Chem. Abstr.* **53**, 10608 (1959).
130. Schweining, H. L., "Industrial wastes effects at South San Francisco, California, sewage treatment plant," *Sewage Ind. Wastes* **29**, 1377 (1957); *J. Am. Leather Chemists' Assoc.* **53**, 650 (1958).
131. Seltzer, J. M., "Sewage disposal," *J. Am. Leather Chemists' Assoc.* **10**, 370 (1915).
132. Shankin, V. F., "The purification of waste waters from tanneries," *Chem. Abstr.* **53**, 375 (1959); *J. Am. Leather Chemists' Assoc.* **54**, 415 (1959).
133. Sheuchenko, M. A., and R. S. Kas 'Yanchuk, "Absorption of tanning substances from water and their stability of destruction oxidation," *Ukr. Khim. Zh.* **30**, 1103 (1964); *Chem. Abstr.* **62**, 5057 (1965).
134. Shuttleworth, G., "The problem of tannery effluent disposal in South Africa," *J. Proc. Inst. Sewage Purif.* Part 3 (1965), p. 244; *Chem. Abstr.* **63**, 17677 (1965).
135. Smith, E. H., "Chemical engineering in the heavy leather industry," *Trans Am. Inst. Chem. Engrs.* **33**, 162 (1937); *Chem. Abstr.* **31**, 6917 (1937).
136. Smoot, C. C., "Sewage disposal and use of tannery wastes," *J. Am. Leather Chemists' Assoc.* **9**, 523 (1914).
137. Snoek, A., "The purification of the effluents of the town of Elmshorn, with special reference to the tannery wastes," *Collegium* **703**, 612 (1928); *J. Am. Leather Chemists' Assoc.* **24**, 520 (1929).
138. Southgate, B. A., "Notes on treatment of industrial wastes," *Water Sanit. Eng.* **3**, 88 (1952).
139. Schankin, V. F., "Treatment of waste waters from leather factories," *Light Ind.* **18**, 44 (1958).
140. Southgate, B. A., "Treatment of waste water from tanneries," *J. Sci. Food Agr.* **5**, 1159 (1954).
141. Southgate, B. A., "Waste disposal in Britain," *Ind. Eng. Chem.* **44**, 524 (1952); *J. Am. Leather Chemists' Assoc.* **47**, 492 (1952).
142. Sproul, O. J., P. F. Atkins, Jr., and F. E. Woodard, "Investigations on physical and chemical treatment methods for cattleskin tannery wastes," *J. Water Pollution Control Federation* **38**, 508 (1966).
143. Stankey, S. A., "Effects of chromium wastes on activated sludge," *Sewage Ind. Wastes* **31**, 496 (1959).
144. Strell, M., "Purification of tannery wastes," *Städtereinigung* **27**, 357 (1935); *J. Am. Leather Chemists' Assoc.* **33**, 271 (1938).
145. *Studies on Tannery Sewage*, Loewangalt & Co., (1964).
146. Tanaka, M., A. Kasahara, and H. Ond, "Treatment of industrial waste by activated sludge. V: Treatment of tannery wastes," *Kogyo Gijutsuin, Hakko Kenkyusho Kenkyu Hokoku* **24**, 239 (1963); *Chem. Abstr.* **64**, 17242 (1966).
147. "Tannery effluent disposal," *Chem. Age* **67**, 570 (1952).
148. "Tannery effluent purification," *Engineer* **210**, 444 (1960).
149. Ohio River Survey, *Tannery Wastes*, Supplement D (1953), p. 1218.

150. New England Interstate Water Pollution Control Commission, *Tannery Wastes, Pollution Sources and Methods of Treatment,* Prepared by Wesleyan University (1958).
151. Michigan Stream Control Commission, "Tannery wastes," Second Biennial Report, 1933–34; *J. Am. Leather Chemists' Assoc.* **39**, 401 (1944).
152. "Tanneries waste treatment—difficult but not impossible," *Wastes Eng.* **30**, 194 (1959).
153. Tarcsay, F., "Problems connected with tannery effluents," *Bor-Cipotech.* **11**, 11 (1961); *Chem. Abstr.* **55**, 11715 (1961).
154. Taylor, W. H., "Disposal of tannery wastes, Hartnett Tanning Co.," *Public Works Mag.* **84**, pamphlet no. 339, November 1953.
155. Thabaraj, G. J., S. M. Bose, and Y. Nayudamma, "Comparative studies on the treatment of tannery effluence by trickling filter, activated sludge, and oxidation pond systems," Centre Leather Research Institute, Madras, Bulletin no. 8, 1960, p. 411; *Chem. Abstr.* **57**, 10953 (1962).
156. Thabaraj, G. J., S. M. Bose, and Y. Nayudamma, "Effect of pretanning operation on removal of globular proteins and collagenous constituents from hides and skins," *Leather Sci.* **10**, 109 (1963).
157. Thabaraj, G. J., S. M. Bose, and Y. Nayudamma, "Utilization of tannery effluents for agricultural purposes," *Environ. Health (India)* **6**, 18 (1964).
158. "The purification of tannery effluents," *Leather World* **4**, 144 (1912); *J. Am. Leather Chemists' Assoc.* **1**, 243 (1912).
159. Toyoda, H., T. Yarisawa, A. Futami, and M. Kikkawa, "Research in treatment of waste water from tanneries. III: The effect of chemical treatment," *Nippon Hikaku Gijutsu Kyokaishi* **8**, 79 (1963); *Chem. Zentr.* **20**, 1903 (1964); *Chem. Abstr.* **62**, 3795 (1965).
160. Toyoda, H., T. Yarisawa, A. Futami, and M. Kikkawa, "Research in the treatment of waste water from tanneries. IV: Federal treatment," *Nippon Hikaku Gijutsu Kyokaishi* **8**, 123 (1963); *Chem. Abstr.* **62**, 6252 (1965).
161. Toyoda, H., T. Yarisawa, A. Futami, and M. Kikkawa, "Studies on treatment of tannery wastes. III: Effects of chemical treatments. IV: Treatment with ferric chloride," *J. Japanese Assoc. Leather Tech.* **3**, 79 (1963); *J. Am. Leather Chemists' Assoc.* **58**, 446 and 704 (1963).
162. Toyoda, H., T. Yarisawa, A. Futami, and M. Kikkawa, "Treatment of tannery waste. I and VII," *Tokyo Kogyo Shikensho Hokoku* **59**, 246 (1964); *Chem. Abstr.* **62**, 14321 (1965).
163. Veitch, E. P., "The purification of tannery effluents and the recovery of by-products therefrom," *J. Am. Leather Chemists' Assoc.* **10**, 126 (1915).
164. Villa, L., "Waste waters in the tanning and towing industry and their purification," *Bull. Assoc. Franc. Chimistes Ind. Cuir Doc. Sci. Tech. Ind. Cuir* **26**, 263 (1964); *Chem. Abstr.* **62**, 11524 (1965).
165. Vryburg, R., "The influence of chrome compounds on activities of a sewage treatment plant," *Sewage Ind. Wastes* **25**, 240 (1953); *J. Am. Leather Chemists' Assoc.* **48**, 301 (1953).
166. Ward, E. W., "Treatment of wastes at Hartnett Tannery," *Water Works Sewerage* **101**, 42 (1954).
167. Warrick, L. F., and E. J. Beatty, "The treatment of industrial wastes in connection with domestic sewage," *Sewage Works J.* **8**, 122 (1936).
168. "Waste water disposal and treatment in the European leather industry," *Chem. Abstr.* **53**, 10608e (1960); *J. Am. Leather Chemists' Assoc.* **54**, 539 (1959).
169. Wims, F. J., "Treatment of chrome tanning wastes for acceptance by an activated sludge plant," in Proceedings of 18th Industrial Waste Conference, Purdue University Engineering Extension Series, Bulletin no. 115, 1964, p. 534; *Chem. Abstr.* **62**, 1439 (1965).
170. U.S. Department of the Interior, *The Leather Industry—The Cost of Water,* Profile Report no. 6 (1967).

LAUNDRY WASTES

The laundry industry is a service—not a manufacturing—industry and therefore no specific subdivisions exist. According to a statement issued by the American Laundry Institute in January 1961, professional laundries have become the nation's largest personal-service industry, with an annual sales volume of more than $1,600,000,000. The industry processes more than 5 billion pounds of laundry per week, including more than 50 million men's shirts. Laundry wastes originate from the use of soap, soda, and detergents in removing grease, dirt, and starch from soiled clothing. The waste has a high turbidity and alkalinity and a readily putrescible organic content with a BOD of 400 to 1000 ppm. The usual method of treatment is chemical precipitation, after adjustment of pH by dilution or chemical addition. If secondary treatment is required, laundry wastes may be oxidized readily on trickling filters. The activated-sludge process is sometimes used, but it is not as satisfactory as the trickling filter. The spread of laundromats in non-sewered areas is creating especially critical problems.

22.6 Origin and Characteristics of Laundry Wastes

Wastes originate from the washing of clothes, which are usually placed in a double cylinder with water, soap, and other washing agents. Rotation of the inner perforated cylinder (the outer cylinder is stationary) produces the agitation necessary to free or dissolve the impurities (dirt) from the fabrics. A detailed discussion of commercial laundry methods is given by Smith [65]. The amount of alkali used (and hence the washing formulas) varies with the type and amount of soil content; the present tendency is to use alkalis, such as sesqui- or orthosilicate, which have low buffering values.

The U.S. Public Health Service [39] estimates laundry water consumption, and hence waste production, as four gallons per pound of clothes. Boyer [9] describes the highly putrescible character of laundry wastes as strongly alkaline, exceedingly turbid, highly colored, and containing large quantities of soap, soda ash, grease, dirt, dyes, and scourings from cloth; its BOD will be on average twice that of domestic sewage and at times it will be five times as great. Rudolfs [58] presents analyses (Table 22.13)

Table 22.13 Comparison of the sanitary characteristics of commercial and domestic laundry wastes.* (From Rudolfs [58].)

Analysis	Commercial	Domestic
pH	10.3	8.1
Total alkalinity, ppm	511	678
Total solids, ppm	2114	3314
Volatile solids, ppm	1538	2515
BOD, 5-day, ppm	1860	3813
Oxygen consumed, ppm	868	1045
Grease, ppm	554	1406

*"Commercial" presumably refers to large-scale operations such as linen service for hotels and restaurants, while "domestic" refers to laundries processing home apparel [author's note].

of both commercial and domestic laundry wastes. The U.S. Public Health Service in its survey [39] indicated that most laundry waste waters have compositions within the limits shown in Table 22.14. (The reader will note some discrepancies between the pollution loads given in Tables 22.13 and 22.14.)

Table 22.14 Typical laundry waste water composition [39].

Analysis	Ranges of values
pH	9.0–9.3
Alkalinity above pH 7.0, as Na_2CO_3, ppm	60–250
Total solids, ppm	800–1200
BOD, 5-day, ppm	400–450

Eckenfelder and Barnhart [18] studied the treatment of wastes from laundromats and small laundry operations. Most installations contain between 25 and 35 machines, each using 25 to 30 gallons of water per washing cycle. Of this water, 22 gallons were hot (140°F) and 8 gallons cold, so that the average temperature of the waste discharged was 100°F. An average waste-water volume of 50,000 gallons per week per installation can be expected. Approximately 100 pounds of commercial detergent are used per week. The characteristics of the composite waste are summarized in Table 22.15.

Table 22.15 Analysis of typical laundromat effluent (24-hr composites). (After Eckenfelder and Barnhart [18].)

Characteristic	Range of values
Turbidity*	208–300
COD, ppm	344–445
Detergent, as ABS†, ppm	50–90
pH	7.0–8.1
Suspended solids, ppm	140–163

*Based on an arbitrary scale setting, with pure water equal to zero.
†Alkyl benzene sulfonate.

22.7 Treatment of Laundry Wastes

In 1944 Gehm [28] came to the following conclusions concerning treatment of laundry wastes.

1. To remove about 75 per cent of oxygen-consuming solids and grease, laundry wastes can be treated most economically by acidification with H_2SO_4, CO_2, or SO_2, followed by coagulation with alum or ferric sulfate. Coagulation by other salts and lime may be effective in some cases, but is usually too costly.

2. Laundry waste can be effectively treated by trickling filtration or by the activated-sludge process, with long aeration periods.

3. Sludge obtained can be dried directly on sand beds, digested anaerobically, or filter-pressed. As a final recovery, soap or the dried sludge can be reclaimed.

4. After chemical coagulation, laundry wastes can be further purified by biological filtration or activated-sludge treatment.

5. Domestic sewage can contain laundry waste up to 20 per cent of its volume when being treated by the activated-sludge process and any amount of laundry waste when being treated by biological filtration.

Both Eliassen and Schulhoff [23] and the Florida State Board of Health [27] demonstrated that flotation produced better results than sedimentation. This is undoubtedly due to the relatively large percentage of emulsified grease in these wastes.

Eckenfelder and Barnhart [18] concluded that through a combination of physical adsorption, with seven parts of carbon to one part of detergent, and chemical coagulation, with 100 grains per gallon of alum, it is possible to remove almost all of the anionic synthetic detergent in wastes from laundromat operation. They also found that settling for a period of four hours will result in a sludge containing one to two per cent solids.

Andres et al. [20] present three methods of treatment for laundromat wastes prior to their discharge into ground water: 1) Separmatic treatment, with pressure diatomaceous-earth filters; 2) Lansing treatment—a flotation process, using pH reduction, followed by air flocculation and floating of coagulated sludge; 3) activated carbon, alum, and soda ash coagulation as proposed above [18]. Although the three systems, used either in combination or separately, may remove 85 to 95 per cent of suspended solids, BOD, and synthetic detergents, none appears to effect a significant reduction in dissolved solids.

The U.S. Public Health Service [39] summarized the most recent information available in 1956, stating that appreciable waste reduction could be accomplished by avoiding overuse of washing agents and by controlling washer loads. Most commercial laundry wastes are discharged directly to municipal sewage systems and treated together with domestic sewage. Considerable experimental work on the separate treatment of laundry wastes with trickling filters has been reported in the literature. In practice, some laundry wastes are being treated by chemical flocculation and sedimentation, with further purification obtained by lagooning and sand filtration.

Sigworth [63] concludes that the aesthetically objectionable foam and flavor from syndets (synthetic detergents) in potable water supplies can be controlled at the treatment plant by carbon dosages in the range of one to two parts (per million) of carbon for each part of syndet formulation. He states that "of all presently known water-purification processes, activated carbon is the only tool which will assure complete success." Table 22.16 gives the dosages of activated carbon required to remove the taste and foam characteristics of five typical detergents.

Table 22.16 Foam and flavor studies on synthetic detergents. (After Sigworth [63].)

Synthetic detergent	Amount producing characteristic		Carbon dosages necessary to treat 25 ppm concentration of detergent	
	Noticeable foam, ppm	Unpalatable water, ppm	To correct foam, ppm	To correct flavor, ppm
A	10.0	17.5	34	43
B	2.0	10.0	44	20
C	500.0	20.0	0	50
D	1.0	8.0	25	50
E	5.0	15.0	30	40

Current treatment methods include the following systems: (1) natural methods of disposal—into deep wells, fissures in cavernous limestone strata, and artificial lagoons—although there are serious objections to each of these methods; (2) trickling filtration; (3) the activated-sludge process; (4) chemical precipitation; (5) sand filter and high-rate trickling filter; (6) acidification with H_2SO_4, CO_2, or SO_2, followed by coagulation with alum and ferric sulfate; (7) the activated-carbon method and diatomaceous-earth filter. Spade [1962, 14] discusses the basic features of this last method and the properties and characteristics of diatomaceous earth. Since the treatment does not use coagulation and settling basins, it makes possible a great saving in space, and also includes completely automatic filtration. However, its main disadvantage is the inability to handle large flows; the maximum economical range at present seems to be 40,000 gal/day. If the ability to handle greater flows can be achieved, consideration may be given to developing diatomaceous filters as polish units for the final effluent of sewage-treatment plants. Typical analyses for the Maric diatomaceous-earth filter and the Bruner clear stabilizing-filter systems are given in Table 22.17.

Table 22.17 Comparison of the Bruner and Maric filter systems. (After Spade [1962, 14].)

Filter system	Waste characteristic*				
	BOD	D.O.	Detergents, as ABS	Total solids	Suspended solids
Bruner					
Raw	162	1.6	38	563	100
Final	28	5.2	1.1	1020	2.0
Maric					
Raw	132	4.7	35	340	220
Final	12	3.8	3.0	16	12

*All results are given as parts per million.

References: Laundry Wastes

1. American Association of Textile Chemists and Colorists—Stream Pollution Abatement Committee, "Learn to live with laundry wastes, most of it goes into your sewers," *Wastes Eng.* **28**, 189 (1957).
2. Ardern, E., and W. T. Lockett, "Pretreatment of occasional abnormal sewage, as adjunct to activated sludge process," *Surveyor* **89**, 499 (1936).
3. Besselievre, E. B., *Industrial Waste Treatment*, McGraw-Hill Book Co., New York (1952), p. 84.
4. Bloodgood, D. E., "Tenth Purdue Conference highlights industrial waste problems," *Ind. Wastes* **1**, 33 (1955).
5. Bogan, R. H., and C. N. Sawyer, "Biochemical degradation of synthetic detergents, preliminary studies," *J. Water Pollution Control Federation* **26**, 1069 (1954).

6. Bogan, R. H., and C. N. Sawyer, "Biochemical degradation of synthetic detergents, studies on the relation between chemical structure and biochemical oxidation," *J. Water Pollution Control Federation* **27**, 917 (1955).
7. Bogan, R. H., and C. N. Sawyer, "Biochemical degradation of synthetic detergents; relationship between biological degradation and froth persistence," *J. Water Pollution Control Federation* **28**, 637 (1956).
8. Bogan, R. H., and C. N. Sawyer, "The biochemical oxidation of synthetic detergents," in Proceedings of 14th Industrial Waste Conference, Purdue University, May 1959, p.231.
9. Boyer, J. A., "The treatment of laundry wastes," *Texas Agr. Exp. St. Bull.* **42**, 1 October 1933.
10. Calvert, C. K., and E. H. Parks, "The population equivalent of certain industrial wastes," *Sewage Works J.* **6**, 1159 (1934).
11. Campenni, L. G., "Synthetic detergents in ground water. I," *Water Sewage Works* **108**, 188 (1961).
12. "Commercial laundering industry," *U.S. Public Health Serv. Bull.* **509**, (1956).
13. Daniels, F. E., "Treatment of laundry wastes," *Public Works Mag.* **54**, 190 (1923).
14. Degens, D. N., Jr., H. Van der Zee, and J. D. Kommer, "Influence of anionic detergents on the diffused-air activated-sludge process," *J. Water Pollution Control Federation* **27**, 10 (1955).
15. "Detergents are degrading in sewage treatment plants," *Wastes Eng.* **30**, 36 (1959).
16. Ohio River Valley Water Sanitation Commission Committee Report, "Detergents in sewage and surface water," *Ind. Wastes* **1**, 212 (1956).
17. Dobbins, W. E., "Treatment of radioactive laundry wastes," *Public Works* **88**, 85 (1957).
18. Eckenfelder, W. W., and E. Barnhart, *Removal of Synthetic Detergents from Laundry and Laundromat Wastes*, New York State Water Pollution Control Board, Research Report no. 5, March 1960.
19. "Effect of detergents on sewage and water treatment," *Chem. Eng. News* **31**, 1072 (1953).
20. New York State Water Pollution Control Board, *Effect of Synthetic Detergents on the Ground Waters of Long Island*, Research Report no. 6, by C. W. Lauman, Inc., and Suffolk County Health Department, June 1960.
21. Eldridge, E. F., "Laundry wastes," *Mich. State Univ. Eng. Exp. Sta. Bull.* **82**, (1938).
22. Eldridge, E. F., *Industrial Waste Treatment and Practice*, McGraw-Hill Book Co., New York (1942), p. 879.
23. Eliassen, R., and B. Schulhoff, "Laundry waste treatment by flotation," *Water Works Sewerage* **90**, 418 (1943).
24. Fair, G. M., and J. C. Geyer, *Water Supply and Waste Water Disposal*, John Wiley & Sons, New York (1954).
25. Finch, J., "Synthetic detergents in sewage," *Water Sewage Works* **103**, 482 (1956).
26. Finch, J., "Synthetic detergents," *Water Sewage Works* **105**, 979 (1958).
27. Florida State Board of Health, "Experimental pilot plant studies treatment wastes," *Wastes Eng.* **24**, 512 (1953).
28. Gehm, H. W., "Volume, characteristics, and disposal of laundry wastes," *Sewage Works J.* **16**, 571 (1944).
29. Gibbs, F. S., "The removal of fatty acids and soaps from soap-manufacturing waste waters," in Proceedings of 5th Industrial Waste Conference, Purdue University, November 1949.
30. Gloyna, E. F., "Radioactive contaminated laundry waste and its treatment," *Sewage Ind. Wastes* **26**, 869 (1954).
31. Grune, W. N., "Waste treatment at a quartermaster laundry," *Ind. Wastes* **3**, 112 (1958).
32. Gurnham, C. F., *Principles of Industrial Waste Treatment*, John Wiley & Sons, New York (1955), p. 6.
33. Hernandez, J. W., and D. E. Bloodgood, "Effects of ABS on anaerobic sludge digestion," *J. Water Pollution Control Federation* **32**, 1261 (1960).
34. Holden, J. T., and J. N. Fowler, *The Technology of Washing*, British Launderers' Research Association, London (1935).
35. Hood, J. W., Proceedings of New Jersey Sewage Works Association, 1941, p. 31.
36. Howells, D. H., and C. N. Sawyer, "Effects of synthetic detergents on chemical coagulation of water," *Water Sewage Works* **103**, 71 (1956).
37. Hurley, J., "Some experimental work on the effects of synthetic detergents on sewage treatment," Paper read at Public Works and Municipal Services Congress, 7 November 1952.
38. Hurwitz, E., R. E. Beaudoin, T. Lothiam, and M. Sniegowski, "Assimilation of ABS by an activated-sludge treatment plant—waterway system," *J. Water Pollution Control Federation* **32**, 1111 (1960).
39. "Industrial waste guide to the commercial laundering

industry," *U.S. Public Health Serv. Bull.* **509** (1956).

40. Keefer, C. E., "Detergents in sewage," *Water Sewage Works* **99**, 89 (1952).
41. Kessler, L. H., and J. T. Norgaard, "Sewage treatment at army camps," *Sewage Works J.* **14**, 757 (1942).
42. Key, A., *Progress toward the Solution of the Synthetic Detergent Problem*, Institute of Sewage Purification (1960).
43. Kline, H. S., "Characteristics of laundry wastes," in Ohio Conference on Sewage Treatment, 10th Annual Report, 1936, p. 67.
44. Lumb, C., *Experiments on the Effects of Certain Synthetic Detergents on Biological Oxidation of Sewage*, Institute of Sewage Purification, (1953).
45. Lynch, W. O., and C. N. Sawyer, "Effects of detergents on oxygen transfer in bubble aeration," *J. Water Pollution Control Federation* **32**, 25 (1960).
46. McCarthy, J., "Study of laundry waste treatment," *Public Works Mag.* **73**, 13 (1942).
47. McGauhey, D. H., and S. A. Klein, "Removal of ABS by sewage treatment," *Sewage Ind. Wastes* **31**, 877 (1959).
48. McKinney, R. F., and J. M. Symons, "Synthetic detergents," *Water Sewage Works* **105**, 425 (1958).
49. McKinney, R. F., and J. M. Symons, "Bacterial degradation of ABS, fundamental biochemistry," *Sewage Ind. Wastes* **31**, 549 (1959).
50. Malaney, G. W., W. D. Sheets, and J. Ayres, "Effects of anionic surface active agents on waste-water treatment units," *J. Water Pollution Control Federation* **32**, 1161 (1960).
51. Manganelli, R. M., "Effects of synthetic detergents on activated sludge," in Proceedings of 11th Industrial Waste Conference, Purdue University, May 1956, p. 611.
52. Mills, E. V., J. T. Calvert, and G. H. Cooper, *J. Proc. Inst. Sewage Purif.* (1947), p. 7.
53. Newell, C. W., *et al.*, "Plutonium removal from laundry wastes, laboratory studies," *Sewage Ind. Wastes* **23**, 1464 (1951).
54. Newell, J. R., C. W. Christenson, *et al.*, "Laboratory studies of removal of plutonium from laundry wastes," *Ind. Eng. Chem.* **43**, 1516 (1951).
55. Porterhouse, W., *J. Proc. Inst. Sewage Purif.* Part 1, 1939, p. 56.
56. Riker, I. R., "Effect of laundry wastes on Imhoff tanks and trickling filter," in Proceedings of New Jersey Sewage Works Association, Trenton, N. J., 1927; *Public Works* **58**, 337 (1927).
57. Rudolfs, W., and L. L. Setter, "Laundry wastes," *New Jersey Agr. Exp. Sta. Bull.* **610**, 1936.
58. Rudolfs, W., *Industrial Wastes*, Reinhold Publishing Corp., New York (1953), p. 471.
59. Ryckman, D. W., and C. N. Sawyer, "Chemical structure and biological oxidizability of surfactants," in Proceedings of 12th Industrial Waste Conference, Purdue University, May 1957, p. 270.
60. Sakers, L. E., and F. M. Zimmerman, "Treatment of laundry wastes," *Sewage Works J.* **1**, 79 (1928).
61. Sawyer, C. N., "Effects of synthetic detergents on sewage treatment processes," *Sewage Ind. Wastes* **30**, 757 (1958).
62. Sheets, W. D., and G. W. Malaney, "Chemical oxygen demand values of syndets, surfactants, and builders," in Proceedings of 11th Industrial Waste Conference, Purdue University, May 1956, p. 185.
63. Sigworth, E. A., "Synthetic detergents and their correction with activated carbon," *J. North Carolina Section Am. Water Works Assoc.*; in Proceedings of 40th Meeting of Water Pollution Control Association, 1960, p. 45.
64. Singleton, P., U.S. Patent no. 2196480 (1940).
65. Smith, R. B., *Washroom Methods and Practice in the Power Room Laundries*, Moore-Robbins Publishing Co., New York (1948).
66. Snell, F. D., and J. M. Fain, "Chemical treatment of trade wastes; laundry wastes," *Ind. Eng. Chem.* **34**, 970 (1942).
67. Steel, E. W., *Water Supply and Sewerage*, 3rd ed., McGraw-Hill Book Co., New York (1953).
68. National Lime Association, "The use of lime in industrial waste treatment," Trade Waste Bulletin no. 1, Washington, D. C., April 1948.
69. Todd, A. R., "Water purification upset seriously by detergents," *Water Sewage Works* **101**, 80 (1954).
70. Waddams, A. L., "Synthetic detergents and sewage processing," *J. Proc. Inst. Sewage Purif.*, part 1, 1950, p. 32.
71. Weaver, P. J., "Household detergents in water and sewage," in Proceedings of 7th Ontario Industrial Waste Conference, June 1960, p. 71.
72. Weaver, P. J., "Review of detergent research program," *J. Water Pollution Control Federation* **32**, 288 (1960).
73. Wise, R. S., "Work of the Connecticut State Water Commission," *Proc. Inst. Chem. Engrs.* **27**, 101 (1931).

74. Wollner, H. J., V. M. Kumin, and P. A. Kahm, "Clarification by flotation, reuse, laundry waste water," *Sewage Ind. Wastes* **26**, 509 (1954).

Suggested Additional Reading

The following recent material has been published since 1962.
1. Acharya, C. V., "Certain industrial wastes," *Environ. Health* **30**, 30 (1966).
2. Andres, B. D. and J. M. Flynn, *Report on Launderette Waste Treatment Processes*, Suffolk County, Long Island, New York (1963).
3. Bready, J. W., and W. M. Bready, "The clarification and reclamation of industrial waste water from laundries," U.S. Patent no. 3192155 (1965); *Chem. Abstr.* **64**, 3199 (1966).
4. Davis, D. W., J. M. Baloga, and F. B. Hutto, "The removal of ABS from laundry effluent by diatomite filtration," *Am. Chemists' Soc. Div. Water Wastes Chem.* Preprints nos. 59–61 (1963); *Chem. Abstr.* **61**, 11737 (1964).
5. "Detergents removed from laundry waste," *Public Works Mag.* **93**, 152 (1962).
6. "Detergents removed from laundry wastes by filtration," *Wastes Eng.* **33**, 372 (1962).
7. Eckenfelder, W. W., "Removal of ABS and phosphate from laundry waste waters," Purdue University, Engineering Extension Series, Bulletin no. 117, 1964, p. 467.
8. Eckenfelder, W. W., "Synthetic detergent removal from laundry wastes," *Water Sewage Works* **108**, 347 (1961).
9. Flynn, J. M., and B. Andres, "Launderette waste treatment processes," *J. Water Pollution Control Federation* **35**, 783 (1963).
10. Flynn, J. M., "Launderette waste treatment system," *Water Sewage Works* **110**, 83 (1963); *Chem. Abstr.* **60**, 14245 (1964).
11. Heidler, K., and E. Motlora, "Determination of anion active detergents in washing and waste water from washing of raw wool," *Textil* **18**, 16 (1963); *Chem. Abstr.* **61**, 1598 (1964).
12. Pitter, P., and J. Chudoba, "Surface active agents in waste waters. Treatment of laundry wastes by activated sludge process," *Sb. Vysoke Skoly Chem. Technol. Praze* **10**, 31 (1966); *Chem. Abstr.* **66**, 13216 (1967).
13. Pitter, P., and J. Chudoba, "Surface-active agents in waste waters and laundry waste treatment by coagulation," *Sb. Vysoke Skoly Chem. Technol. Praze* **9**, 123 (1966); *Chem. Abstr.* **66**, 9235 (1967).
14. Spade, J. F., "Treatment methods for laundry wastes," *Water Sewage Works* **109**, 110 (1962).
15. Swami, A. K., "Treatment methods in laundry wastes," *Indian Ind. Wastes J.* **5**, 140 (1965).
16. "The international laundry, dry cleaning and allied trades exhibition," *Chem. Ind.* **32**, 1367 (1966).
17. "Treating 10,000 gallons of laundry wastes per week," *Wastes Eng.* **32**, 209 (1961).
18. Delaey, J., R. van Achter, L. Creyten, and F. Edetine, "Laundry waste waters. I: Wastewater analysis. II: Oxygen conversion," *Tribune Centre Belg. Etude Doc. Eaux* **17**, 116 (1964); *Chem. Abstr.* **63**, 5365 (1965).
19. Wagg, R. E., "Disinfection of textiles in laundering and dry cleaning," *Chem. Ind.* **44**, 1830 (1965).
20. Winker, R., and K. Langecker, "Decontamination of laundry sewage," *Kernenergie* **8**, 637 (1965); *Chem. Abstr.* **64**, 19179 (1966).

CHAPTER 23

FOOD-PROCESSING INDUSTRIES

23.1 Introduction

Food-processing industries are those whose main concern is the production of edible goods for human or animal consumption. Processing plants included in this group are: (1) canneries, (2) dairies, (3) breweries and distilleries, (4) meat-packing and -rendering plants (including poultry plants and animal feedlots), (5) beet-sugar refineries, (6) pharmaceutical plants, (7) yeast plants, and (8) miscellaneous plants, producing such foods as pickles, coffee, fish, rice, soft drinks, bakeries, and water production. The production processes usually consist of the following steps: cleaning the raw material, removal of inedible portions, preparation of the foodstuff, and packaging. The wastes to be considered are: spoiled raw material or spoiled manufactured products; rinsing or washing waters; condensing or cooling waters; transporting waters; process waters; floor- and equipment-cleaning liquids; product drainage; overflow from tanks or vats; and unusable portions of the product.

The characteristics of food-processing wastes exhibit extreme variation. The BOD may be as low as 100 ppm or as high as 100,000 ppm. Suspended solids, almost completely absent from some wastes, are found in others in concentrations as high as 120,000 ppm. The waste may be highly alkaline (pH 11.0) or highly acidic (pH 3.5). Mineral nutrients (nitrogen and phosphorus) may be absent or may be present in excess of the (BOD/N) or (BOD/P) ratio necessary to promote good environmental conditions for biological treatment. Similarly, the volume of wastes may be almost negligible in some industries, but reach one or more million gallons per day in others.

Food-processing wastes usually contain organic matter (in the dissolved or colloidal state) in varying degrees of concentration, so that biological forms of waste treatment are indicated. Since these wastes differ from domestic sewage in general characteristics and in particular by their higher concentrations of organic matter, pretreatment is required to produce an equivalent effluent. In addition, one or more of the following adjustments are frequently necessary to provide the proper environmental conditions for the microorganisms upon which biological treatment depends: continuous feeding, temperature control, pH adjustment, mixing, supplementary nutrients, and microorganism population adaptation.

Among the aerobic or anaerobic biological treatments available, the major and more effective methods make use of activated sludge, biological filtration, anaerobic digestion, oxidation ponds, lagoons, and spray irrigation. The loadings of the biological units must be carried out with care, since many of the wastes contain high concentrations of organic matter. Quite frequently long periods of aeration or high-rate two-stage biofiltration is required to produce an acceptable effluent. The type of treatment selected will depend on the following aspects: degree of treatment required, nature of the organic waste, concentration of organic matter, variation in waste flow, volume of waste, and capital and operating costs.

CANNERY WASTES

The canning industry is one of the most important to the people of the United States because (1) its total annual retail value today (1967) exceeds $5 billion for canned food and $3 billion for frozen food and (2) it utilizes great quantities of water—in 1964 the "pack" of 944 million equivalent cases of canned and frozen fruits and vegetables required 76 billion gallons of water. The industry is extremely diversified; about 200 plants in the United States can or freeze literally dozens of different raw products. Cannery wastes are classified according to the product being processed, its season of growth, and its geographic location. Since the harvesting and processing periods of the three main groups of products—vegetables, fruits, and citrus fruits—are short, many canneries are designed to process more than one product. Wastes from these plants are primarily organic and result from trimming, juicing, blanching, and pasteurizing of raw materials, the cleaning of processing equipment, and the cooling of the finished product. The four most common and effective methods of treatment are: discharge to municipal treatment plant, lagooning with the addition of chemical stabilizers, soil absorption or spray irrigation, and anaerobic digestion.

23.2 Origin of Cannery Wastes

Figure 23.1 gives a fairly detailed schematic illustration of the canning and freezing processes for both fruits and vegetables. Peas, beets, carrots, corn, squash, pumpkins, and beans are among the vegetables which produce strong wastes when processed for canning. Since the preparation for processing differs with each vegetable, the methods used should be studied individually, but there is little other difference in cannery procedures, and hence the origin of all vegetable wastes is similar. The process waste usually consists of: wash water; solids from sorting, peeling, and coring operations; spillage from filling and sealing machines; and wash water from cleaning floors, tables, walls, belts, and so forth.

Among fruits, the processing of peaches, tomatoes, cherries, apples, pears, and grapes presents the most common problems of waste discharges. Waste flows may originate from lye peeling, spray washing, sorting, grading, slicing and canning, exhausting of condensate, cooling of cans, and plant cleanup. Other wastes originate from specific operations not necessarily common to all fruit processing.

The three main citrus fruits—oranges, lemons, and grapefruit—are usually processed in one plant, which produces canned citrus juices, juice concentrates, citrus oils, dried meal, molasses, and other by-products. Liquid wastes from citrus-fruit processing include cooling waters, pectin wastes, pulp-press liquors, processing-plant wastes, and floor washings. Citrus-cannery waste is a mixture of the peel, rag, and seed of the fruit, surplus juice from the washing operations, and blemished fruits.

Table 23.1 Volume and characteristics of cannery wastes. (From Sanborn [134].)

Product	Volume per case, gal	5-day BOD, ppm	Suspended solids, ppm
Asparagus	70	100	30
Beans, green or wax	26–44	160–600	60–85
Beans, lima	50–257	189–450	422
Beans, baked	35	925–1440	225
Beets	27–65	1580–5480	720–2188
Carrots	23	520–3030	1830
Corn, cream style	24	623	302
Corn, whole-kernel	25–70	1123–6025	300–4000
Peas	14–56	380–4700	272–400
Mushrooms	6600*	76–390	50–242
Potatoes, sweet	3500*	295	610
Potatoes, white	†	200–2900	990–1180
Pumpkin	20–42	2850–6875	785–3500
Sauerkraut	3	6300	630
Spinach	160	280–730	90–580
Apples, sauce	†	1685–3453	
Apricots	57–80	200–1020	260
Tomatoes, whole	3–15	570–4000	190–2000
Tomatoes, juice	38–100	178–3880	170–1168

*Per ton.
†Not given.

23.3 Characteristics of Cannery Wastes

The volume and characteristics of waste waters vary considerably from one plant to the next, and within the same plant from day to day. Data presented by Sanborn [134] illustrate the variability of the wastes after screening (Table 23.1). Eckenfelder [50] gives additional data on the characteristics of cannery wastes (Table 23.2).

Citrus-cannery waste is a slick, slimy, nonuniform mass, with a moisture content of about 83 per cent. A complete breakdown of the wastes of a cannery processing 700 tons per day of oranges, lemons, and grapefruit and producing 0.7 mgd of waste containing 6 tons of BOD is given by Ludwig et al. [104] (Table 23.3).

Wakefield [171] presents data on Florida citrus-plant wastes, which we reproduce in Table 23.4. Sanborn [134] also shows variation in the BOD of cannery wastes within the same plant canning different fruits and vegetables (Table 23.5).

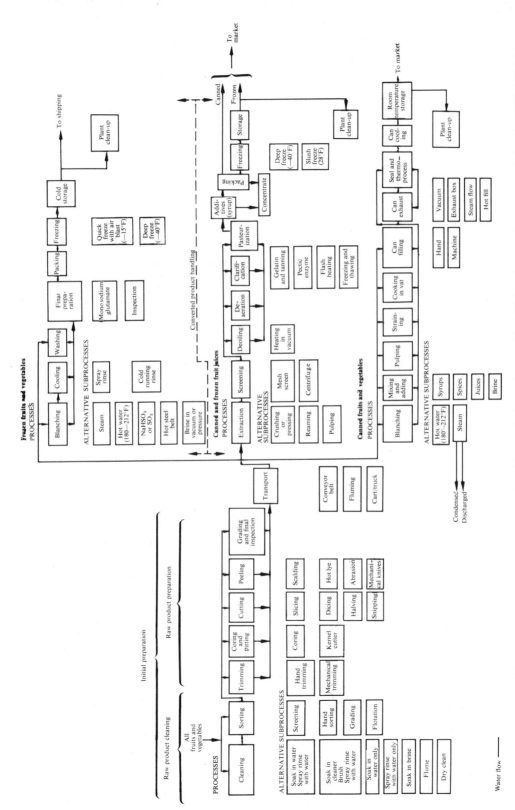

Fig. 23.1 Process flow chart for preparation of canned and frozen fruits and vegetables. (Taken from a chart prepared for the F.W.P.C.A.)

Table 23.2 Cannery-waste characteristics. (From Eckenfelder [50].)

Product	Flow, gal/case	BOD, ppm	Suspended solids, ppm
Tomatoes	4.5–78.0	616–1870	550–925
Corn	30–116	885–2936	530–2325
Green beans	104.5	93	291
Green beans and corn	99.5	270	264
Mixed vegetables	12.2	750	593
Pears	32.4–42.5	238–468	340–637
Peaches	37.5	1070	250
Apples	26.8	1600	300
Cherries	16.0	800	185

Table 23.3 Composition of citrus wastes. (From Ludwig et al. [104].)

Type of waste	Waste flow, gpd	BOD, ppm	Suspended solids, ppm
Cooling water	285,000	100	765
Pectin wastes	225,000	2720	1790
Pulp-press liquor	120,000	9850	780
Processing-plant waste	40,000	3230	3400
Floor washings	30,000	970	685
Composite wastes		2100	7200

23.4 Treatment of Cannery Wastes

Canning is a highly seasonal business, and hence the treatment of cannery wastes presents unique problems. Orlob et al. [1965, 171] present a detailed description of their six categories of cannery-waste treatment: (1) in-plant modifications, (2) preliminary treatment, (3) chemical treatment, (4) biological treatment, (5) land disposal, and (6) municipal treatment. The selection of the type of treatment most suitable for a particular plant must be guided by two sets of considerations: one comprising the standard aspects of volume and character of wastes and treatment required, the other taking into account the unique conditions of number and duration of packing periods. Cannery wastes are most efficiently treated by screening, chemical precipitation, lagooning, and spray irrigation. Digestion and biological filtration are also used, but to a lesser extent.

Screening is a preliminary step, designed to remove large solids prior to the final treatment or discharge of the waste to a receiving stream or municipal waste-water system. Only slight reductions in BOD are realized by screening. Mechanically operated screens (mesh size varying from 12 to 30) of either the rotating or vibrating type are used. Typical screening loads are about 40 to 50 pounds per 1000 gallons of waste water. Vibrating screens will produce solids having a moisture content between 70 and 95 per cent, depending on the product. The waste retained on the screens is disposed of in various ways: it can be spread on the ground, used for sanitary fill, dried and burned, or used to supplement animal feed.

Chemical precipitation, in conjunction with other treatment methods, is used to adjust the pH and to reduce the concentration of solids in the wastes. It has been quite effective for treating apple, tomato, and cherry wastes. Ferric salts or aluminate and lime have produced 40 to 50 per cent BOD reductions. Dosage rates amount to about 5 to 10 pounds of lime, plus 1 to 8 pounds of ferrous sulfate or alum, for each 1000 gallons of screened waste. Chemical precipitation produces about 10 to 15 per cent by volume of sludge, which will normally dry on sand beds in about a week without producing an odor.

Treatment in lagoons involves biological action (both aerobic and anaerobic), sedimentation, soil absorption, evaporation, and dilution. Some engineers advocate lagooning as the only practical and economical treatment of cannery wastes when adequate land is available. Lagoons in which aerobic conditions are not maintained give off unpleasant odors and provide a breeding ground for mosquitoes and other insects. To eliminate odors, $NaNO_3$ is used, at a dosage equal to 20 per cent of the applied oxygen demand (normally 20 to 200 pounds per 1000 no. 2 cases). However, using $NaNO_3$-treated lagoons for complete treatment may be impractical from the standpoint of cost because of the large volumes of wastes involved. Also, with strong wastes such as pea wastes, odors may still persist even with the $NaNO_3$ treatment. Surface sprays have been used to reduce the fly and other insect nuisance and in some cases to combat the odors arising from these lagoons. Seepage

Table 23.4 Citrus wastes. (From Wakefield [171].)

Plant or process	Flow, gpd	BOD, ppm	lb BOD/ 1000 cases	Suspended solids, ppm
Juice	158,610–813,200	182	12.7–43.1	25–85
Sectioning of grapefruit	211,700–420,260	873–945	384–887	124–140
Concentration of orange juice (average of four plants)	2,396,500	82	57.1*	27

*Per 1000 gal of concentrate.

Table 23.5 Cannery waste variation. (After Rudolfs [134].)

Product	Origin of waste	BOD, ppm
Peas	Pea washer	3,700
	Blancher overflow	13,815
	Blancher dump	34,490
	Ensilage-stack liquor	35,000–78,000
Corn	Corn washer	2,800
	Whole-kernel washer	7,000
	Ensilage-stack liquor	22,000–33,000
Kidney beans	Soak water	10,500
	Blancher	3,600
Cherries, sour	Pitter drippage	38,000–55,000
	Pit flume water	950–3,330
Grapefruit sections	Fruit-wash water	20–110
	Peeling table	38,080
	Sectioning tables	2,480
	Exhaust-box overflow	1,000
	Floor-wash water	4,000
	Peel-bin drippage	50,000

must also be considered, especially when lagoons are located near underground sources of water.

Spray irrigation is another economical and unobjectionable method that can be used whenever the cannery waste is nonpathogenic and nontoxic to plants. Its use is primarily limited by the capacity of the spray field to absorb the waste water. High BOD reductions may be expected as the waste percolates through the vegetation and soil. Some spray-irrigation performances are given in Table 23.6. Use of ridge-and-furrow irrigation or absorption beds is limited to soils of relatively high water-absorbing capacity. Permanent pasture grasses are apparently able to handle a heavier organic load that alfalfa can. Wastes should be screened before spraying, although comminution alone has been used successfully in conjunction with spray irrigation.

Oxygen-demanding materials in cannery wastes can be removed by biological oxidation. When the operation is limited by seasonal conditions, it is difficult to justify capital investment for bio-oxidation facilities. However, in many instances cannery wastes can be combined with domestic sewage, and then bio-oxidation processes provide a practical and economic solution. High-rate trickling filters have reduced the BOD of pea, green-bean, and tomato wastes by as much as 97 per cent, and cider, apple, cherry, tomato, and citrus wastes have also been successfully treated. BOD loadings to filters having a removal rate of 90 per cent range from 0.5 to 2.0 pounds per cubic yard per day. Activated-sludge treatment has also been used to produce a clear, odorless citrus-waste effluent with at least 90 per cent BOD reduction. Mixed cannery wastes and pea and carrot wastes have been handled by conventional activated-sludge plants effecting 91 to 95 per cent BOD reduction on wastes whose raw BOD ranged from 1350 to 1500 ppm. At a BOD loading of 1.7 to 2.5 pounds of BOD per day per pound of sludge, detention times vary from three to five hours.

In a study of canning wastes in which the author was involved [1965, 172], Orlob sought to provide

Table 23.6 Spray irrigation performance. (After Eckenfelder [50].)

Product	Pump rate, gpm	Total area sprayed, acres	Rate of application, gpm/acre	Average application, in./day	Average loading	
					lb BOD/ acre/day	lb suspended solids/ acre/day
Tomatoes	1000	5.63	178	2.96	413	364
	500	6.4	86	0.70	155	139
Corn	350	2.28	153.5	3.35	864	500
Asparagus and beans	253	0.9	282	3.5	22.5	356
Tomatoes, corn, and lima beans	430	9.18	43.8	0.375	40.5	14.7
Lima beans	430	6.65	65	0.375	65	46
Cherries	216	2.24	96.5	3.61	807	654

an insight (not heretofore evident in the literature) into the interrelationship of the many technical and economic considerations which govern decisions between alternative methods of coping with the cannery-waste treatment problem. This revealing study (1) characterized the physical, chemical, and biochemical in-plant waste streams and composite flows resulting from processing of peaches and tomatoes; (2) evaluated the technical and economic feasibility of in-plant separation and/or treatment of cannery-waste flows; and (3) developed engineering-economic systems for cannery-waste treatment and/or disposal.

References: Cannery Wastes

1. Adams, S. L., "Utilization of cannery fruit waste by continuous fermentation," *Wash. State Inst. Technol. Div. Ind. Res. Bull.* **207** (1950).
2. Adams, S. L., "Fruit waste, cannery, utilization by continuous fermentation," *Sewage Ind. Wastes* **26**, 927 (1954).
3. Arnold, P. T. D., R. B. Becker, and W. N. Neal, "The feeding value and nutritive properties of citrus by-products," *Florida Univ. Agr. Exp. Sta. Bull.* **354** (1941).
4. Atkins, C. D., E. Wiederhold, and E. L. Moore, "Vitamin C content of processing residues from Florida citrus fruits," *Fruit Products J.* **24**, 260 (1945).
5. Baker, C. M., "Pea canning waste disposal," *Canning Age* **6**, 895 (1925).
6. Baker, C. M., L. F. Warrick, and J. P. Smith, *Treatment of Pea Canning Wastes*, Wisconsin State Board of Health, Madison, Wis. (1926).
7. Beidler, J. W., "Apple, cherry and tomato waste treatment," in Proceedings of 7th Industrial Waste Conference, Purdue University, May 1952, p. 156.
8. Benson, H. K., "Industrial utilization of cannery waste," in Proceedings of Conference on Industrial Wastes, Civil Engineering Department, University of Washington, 28 April 1949.
9. Billings, C. H., "Citrus waste treatment experiments in Texas," *Sewage Ind. Wastes* **13**, 847 (1941).
10. Black, H. H., "Treating corn cannery wastes," *Canning Age* **23**, 325 (1942).
11. Black, H. H., "Treating corn cannery waste," *Sewage Ind. Wastes* **14**, 928 (1942).
12. Bolton, P., "Cannery waste disposal by field irrigation," *Food Packer* **28**, 42 (1947).
13. Bolton, P., "The disposal of canning waste by irrigation," in Proceedings of 3rd Industrial Waste Conference, Purdue University, May 1947, p. 273.
14. Bordenca, C., and R. K. Allison, "*l*-Carvone and *d*-limonene," *Sewage Ind. Wastes* **24**, 684 (1952).
15. Brown, H. D., H. N. Hall, and W. D. Sheets, "Can-

nery waste disposal by irrigation," *Ind. Wastes* **1**, 204 (1956).
16. Buswell, A. M., "Sodium nitrate reactions in stabilizing organic wastes," *Sewage Ind. Wastes* **19**, 628 (1947).
17. Canham, R. A., "Anaerobic treatment of food canning waste," in Proceedings of 5th Industrial Waste Conference, Purdue University, November 1949.
18. Canham, R. A., and D. E. Bloodgood, "Anaerobic digestion of tomato and pumpkin wastes," *Sewage Ind. Wastes* **22**, 1095 (1950).
19. Canham, R. A., "Anaerobic treatment of food canning wastes," *Sewage Ind. Wastes* **23**, 695 (1951).
20. Canham, R. A., "Some problems encountered in spray irrigation of canning waste," in Proceedings of 10th Industrial Waste Conference, Purdue University, May 1955.
21. Canham, R. A., "Waste handling problems common to the Canadian and U.S. canning industries," in Proceedings of 4th Ontario Industrial Waste Conference, June 1957, p. 113.
22. Canham, R. A., "Comminuted solids inclusion with spray-irrigated canning waste," *Sewage Ind. Wastes* **30**, 1028 (1958).
23. U.S. Public Health Service, "Cannery wastes," *Ohio River Survey*, Supplement D (1943), p. 1047.
24. Carpenter, W. T., "Sodium nitrate used to control nuisance," *Water Works Sewerage* **79**, 175 (1932).
25. Colker, D. A., and R. K. Eskew, "Processing vegetable wastes for high-protein, high-vitamin leaf meals," Eastern Regional Research Laboratory, Bureau of Agricultural and Industrial Chemistry, Circular no. *AIC*-78, 1945.
26. Crist, M. L., "Industrial wastes: sewage = 4 : 1," *Sewage Works Eng. Munic. Sanit.* **18**, 207 (1947).
27. "Dehydrating canning waste," *Food Packer* **28**, 40 (1947).
28. De Martini, F. E., W. A. Moore, and G. E. Terhoeven, *Food Dehydration Wastes*, U.S. Public Health Service, Supplement 191 (1946).
29. Dennis, J. M., "Spray irrigation," *Sewage Ind. Wastes* **25**, 591 (1953).
30. Dickinson, D., "Purification of wastes from fruit and vegetable canneries," *Surveyor* **105**, 1001 (1946).
31. Dickinson, D., "Fruit and vegetable wastes, character and disposal," *Sewage Ind. Wastes* **17**, 1040 (1945) and **19**, 533 (1947).
32. Dickinson, D., "Performance of recirculating plant for purification in biological filters," *Sewage Ind. Wastes* **21**, 757 (1949).
33. Dougherty, R. J., "The treatment of food processing wastes," *Sewage Ind. Wastes* **24**, 925 (1952).
34. Dougherty, M. H., and R. R. McNary, "Activated sludge treatment," *Sewage Ind. Wastes* **27**, 821 (1955).
35. Dougherty, M. H., and R. R. McNary, "Activated citrus sludge, vitamin and feed potential," *Sewage Ind. Wastes* **30**, 1151 (1958).
36. Dougherty, M. H., and R. R. McNary, "Activated sludge, temperature effects," *Sewage Ind. Wastes* **30**, 1263 (1958).
37. Dougherty, M. H., R. W. Wolford, and R. R. McNary, "Activated sludge treatment—laboratory study," *Sewage Ind. Wastes* **27**, 821 (1955).
38. Dougherty, M. H., R. W. Wolford, and R. R. McNary, "Citrus waste water treatment of activated sludge," *Sewage Ind. Wastes* **27**, 821 (1955).
39. Dougherty, R. J., "The treatment of food processing wastes," in Proceedings of 6th Industrial Waste Conference, Purdue University, February 1951, p. 479.
40. Drake, J. A., "Strength of waste from frozen food processing," in Proceedings of 5th Industrial Waste Conference, Purdue University, November 1949.
41. Drake, J. A., and F. K. Bieri, "Disposal of liquid waste by the irrigation method at vegetable canning plants in Minnesota, 1948–1950," in Proceedings of 6th Industrial Waste Conference, Purdue University, February 1951.
42. Drake, J. A., and F. K. Bieri, "Strength of wastes from frozen food processing," *Sewage Ind. Wastes* **23**, 693 (1951).
43. Drake, J. A., and F. K. Bieri, "Vegetable canning wastes, disposal by irrigation in Minnesota," *Sewage Ind. Wastes* **24**, 803 (1952).
44. Duncan, J. G., "Total oxidation as applied to cannery waste," in Proceedings of 5th Ontario Industrial Waste Conference, May 1958, p. 28.
45. Dunstan, G. H., and J. V. Lunsford, "Irrigation disposal field studies," *Sewage Ind. Wastes* **27**, 827 (1955).
46. Dunstan, G. H., G. V. Leete, and J. V. Lunsford, "Blancher waste disposal by trickling filtration," *Ind. Wastes* **1**, 60 (1955).
47. Dunstan, G. H., and J. V. Lunsford, "Cannery waste disposal by irrigation," *Sewage Ind. Wastes* **27**, 827 (1955).
48. East, C. A., "Modern sewage treatment plant at

Vernon, British Columbia," *Eng. Contract Record* **53**, 11 (1940).
49. Eckenfelder, W. W., "Pilot plant investigations of biological sludge treatment of cannery and related waste," in Proceedings of 7th Industrial Waste Conference, Purdue University, May 1952, p. 181.
50. Eckenfelder, W. W., et al., "Study of fruit and vegetable processing waste disposal methods in the eastern region," Final Report, New York State Canners' Association, September 1958.
51. Eckenfelder, W. W., and E. R. Grich, "Plant scale studies on biological oxidation of cannery waste," in Proceedings of 10th Industrial Waste Conference, Purdue University, May 1955, p. 549.
52. Eldridge, E. F., "Experience with special cannery wastes—beets, tomatoes and squash," *Mich. State Coll. Eng. Exp. Sta. Bull.* **78**, 3 (1938).
53. Eldridge, E. F., "The treatment of red beet, tomato, and squash cannery wastes," *Mich. State Coll. Eng. Exp. Sta. Bull.* **83**, 15 (1939).
54. Eldridge, E. F., "Treatment of red beet, tomato, and squash waste," *Sewage Ind. Wastes* **10**, 914 (1938) and **11**, 712 (1939).
55. Eldridge, E. F., "Symposium on industrial wastes—canning industry," *Ind. Eng. Chem.* **39**, 619 (1947).
56. Eldridge, E. F., "Truck and vegetable waste disposal practices review," *Sewage Ind. Wastes* **20**, 773 (1948).
57. Erickson, F. K., "Treatment versus utilization of cannery wastes," in Proceedings of Conference on Industrial Wastes, Civil Engineering Department, University of Washington, 28 April 1949.
58. Everts, W. S., "Disposal of wastes from fruit and vegetable canneries," *Sewage Ind. Wastes* **16**, 944 (1944).
59. Feustal, I. C., and J. H. Thompson, "Pear canning waste may be valuable for yeast culture," *Western Canner Packer* **38**, 60 (1946).
60. Foley, M. B., R. J. Mauterer, and G. J. Dustin, "Aerobic oxidation of corn wet milling processes," in Proceedings of 10th Industrial Waste Conference, Purdue University, May 1955, p. 173.
61. Foley, M. B., R. J. Mauterer, and G. J. Dustin, "Aerobic oxidation of corn wet milling waste," in Proceedings of 10th Industrial Waste Conference, Purdue University, May 1955.
62. Froehlich, C. W., "Fruit and vegetable waste disposal to public sewers at Fullerton, Calif.," *Sewage Ind. Wastes* **16**, 940 (1944).
63. "Fruit and vegetable processing wastes, availability of selected bibliography," *Sewage Ind. Wastes* **22**, 1191 (1950).
64. "Fruit wastes used for soluble food casings," *Sewage Ind. Wastes* **19**, 957 (1947).
65. Furphy, H. G., and J. B. Ley, "Country town sewage; installation at Shepparton, Vic., treating canning factory waste," *Commonwealth Engr.* **25**, 413 (1938).
66. Gray, H. F., and H. F. Ludwig, "Characteristics and treatment of potato dehydration wastes," *Sewage Works J.* **15**, 71 (1943).
67. Greenfield, R. E., "Safe handling of hexane in soybean processing," in Proceedings of 6th Industrial Waste Conference, Purdue University, February 1951.
68. Gregory, T. R., and J. H. Kimball, "Cannery wastes at Palo Alto," *Sewage Works J.* **9**, 607 (1937).
69. Hall, H. W., "Disposal of citrus wastes," *Civil Eng.* **14**, 15 (1944).
70. Halvorson, H. O., D. W. Johnson, and H. Tsuchiya, "Treatment of corn canning wastes," *Canner* **90** (no.7), 12 and (no.8), 12 (1940).
71. Harwood, H. J., "Fats—a source of chemicals," *Sewage Ind. Wastes* **24**, 1552 (1952).
72. Hatfield, R., E. R. Strong, F. Hernsohm, H. Powell, and T. G. Stone, "Corn products, waste treatment, trickling pilot-plant studies," *Sewage Ind. Wastes* **28**, 1240 (1956).
73. Hert, O. H., "Tomato canning plant wastes," *Food Packer* **28**, 40 (1947).
74. Hert, O. H., "Tomato and pumpkin waste," in Proceedings of 3rd Industrial Waste Conference, Purdue University, May 1947, p. 244.
75. Hert, O. H., "Research on aerobic treatment of canning plant waste," in Proceedings of 4th Industrial Waste Conference, Purdue University, September 1948.
76. Hert, O. H., "Research in anaerobic treatment of canning plant wastes," *Sewage Ind. Wastes* **22**, 1095 (1950).
77. Heid, J. L., "Drying citrus canning wastes and disposing of effluents," *Food Inds.* **17**, 1479 (1945).
78. Heider, W., "Indiana cannery waste problem and disposal practice," *Sewage Ind. Wastes* **18**, 170 (1946).
79. Holmes, J. A., and C. J. Fink, "Sodium aluminate as a coagulant in chemical treatment of cannery waste waters," *Ind. Eng. Chem.* **21**, 150 (1929).
80. Hommon, H. B., *The Purification of Tomato Cannery*

Factory Waste, U.S. Public Health Service, Bulletin no. 118, 1921.
81. Hoppe, T. C., "The disposal of waste from evaporated salt production," in Proceedings of 11th Industrial Waste Conference, Purdue University, May 1956.
82. "How to treat wastes," *Food Inds.* **9**, 430 (1937).
83. Hyde, C. G., and G. L. Sullivan, "Treatment of solid and liquid wastes from the processing of fresh fruits and vegetables in the San Jose area," *Sewage Ind. Wastes* **22**, 581 (1950).
84. Hyde, C. G., and G. L. Sullivan, "Industrial waste survey, San Jose, Calif.," *Sewage Ind. Wastes* **23**, 1060 (1951).
85. Ingols, R. S., "Treating food processing waste," *Ind. Wastes* **3**, 95 (1958).
86. Ingols, R. S., "Citrus waste by-product recovery, Florida," *Sewage Ind. Wastes* **17**, 320 (1945).
87. Ingols, R. S., "Review of older methods for treating food processing wastes," *Ind. Wastes* **1**, 288 (1956).
88. *The Investigation on the Disposal of Canning Factory Wastes at Washington, Ill.*, Illinois Water Survey, Series no. 11 (1913).
89. "Ion exchange plant recovers sugar from fruit waste," *Food Inds.* **18**, 1846 (1946).
90. Jones, E. E., "Waste waters from the manufacture of cider, and their treatment in percolating filters," *Sewage Ind. Wastes* **22**, 1097 (1950).
91. Jones, B. R., "Studies of pigmented non-sulfur purple bacteria in relation to cannery waste lagoon odors," *Sewage Ind. Wastes* **28**, 883 (1956).
92. Jones, E. E., "Disposal of waste waters from the preparation of vegetables for drying," *J. Soc. Chem. Ind.* **64**, 80 (1945).
93. Kennedy, C. C., "Improvement in sewage treatment at Stockton, Calif., as effected by cannery operations," *Sewage Works J.* **9**, 271 (1937).
94. Kester, E. B., and G. R. Van Atta, "Minor oil-producing crops of the United States," *Oil Soap* **19**, 119 (1942).
95. Kimball, J. H., and H. L. May, "Development in cannery waste studies at Palo Alto, Calif.," *Sewage Works J.* **13**, 731 (1941).
96. Kimberly, A. E., *Progress Report on Cannery Waste Treatment Studies*, Ohio State Department of Health, Columbus (1927).
97. Kimberly, A. E., "Vegetable wastes, Wisconsin experiments," *Sewage Ind. Wastes* **3**, 553 (1931), **7**, 144 (1935) and **11**, 339 (1939).
98. Lackey, J. B., "Citrus waste treatment, experimental and applied," in Proceedings of 11th Industrial Waste Conference, Purdue University, May 1956, p. 179.
99. Lackey, J. B., W. T. Calaway, and G. B. Morgan, "Biological purification of citrus waste," *Sewage Ind. Wastes* **28**, 538 (1956).
100. "Lagooning with sodium nitrate, forum discussion," *Sewage Ind. Wastes* **25**, 720 (1953).
101. Lane, L. C., "Disposal of liquid and solid wastes by means of spray irrigation in the canning and dairy industries," in Proceedings of 10th Industrial Waste Conference, Purdue University, May 1955.
102. Leete, G. V., G. H. Dunstan, and J. V. Lunsford, "Blancher waste disposal by trickling filtration," *Ind. Wastes* **1**, 60 (1955).
103. Logan, R. P., "Scum removal by vacuator at Palo Alto, Calif.," *Sewage Ind. Wastes* **21**, 799 (1949).
104. Ludwig, H. F., G. W. Ludwig, and J. A. Finley, "Citrus by-product waste at Ontario, Calif.," *Sewage Ind. Wastes* **23**, 1254 (1951).
105. Lunsford, J. V., "Effect of cannery waste removal on stream conditions," *Sewage Ind. Wastes* **29**, 428 (1957).
106. Lutz, H. C., "Cannery waste treatment in Pennsylvania," Paper presented at Pennsylvania Sewage Works Association, June 1939.
107. McKinney, R. E., "The use of biological waste treatment systems for stabilization of industrial waste," in Proceedings of 11th Industrial Waste Conference, Purdue University, May 1956.
108. McKinney, R. E., L. Poliakoff, and R. G. Weichlein, "Citrus waste treatment studies," *Water Sewage Works* **101**, 123 (1954).
109. McNary, R. R., M. H. Dougherty, and R. W. Wolford, "Determination of COD," *Sewage Ind. Wastes* **29**, 894 (1957).
110. McNary, R. R., "Citrus cannery waste disposal," *Sewage Works J.* **21**, 944 (1949).
111. McNary, R. R., R. W. Wolford, and M. H. Dougherty, "Experimental treatment of citrus waste water," in Proceedings of 8th Industrial Waste Conference, Purdue University, May 1953.
112. McNary, R. R., R. W. Wolford, and M. H. Dougherty, "Activated sludge pilot plant studies," *Sewage Ind. Wastes* **28**, 894 (1956).
113. McNary, R. R., R. W. Wolford, and H. D. Marshall, "Pilot plant treatment of citrus waste water by activated sludge," *Sewage Ind. Wastes* **28**, 894 (1956).
114. Marshall, E. A., "Beet canning wastes, effects on

sewage treatment at Geneva, N.Y.," *Sewage Ind. Wastes* **19**, 266 (1947).
115. Miller, P. E., "Cannery wastes," in Proceedings of 3rd Industrial Waste Conference, Purdue University, May 1947, p. 263.
116. Monson, H., "Development of vegetable cannery waste disposal by land irrigation," in Proceedings of 8th Industrial Waste Conference, Purdue University, May 1953.
117. Monson, H., "Cannery waste disposal by spray irrigation—after ten years," in Proceedings of 13th Industrial Waste Conference, Purdue University, May 1958.
118. Morris, R. H., "Vegetable wastes, their availability and utilization," in Proceedings of 2nd Industrial Waste Conference, Purdue University, January 1946, p. 54.
119. Morris, R. H., "Vegetable wastes, by-product utilization," *Sewage Ind. Wastes* **19**, 1103 (1947).
120. Morris, R. H., D. A. Colker, and M. F. Chernoff, "Vegetable wastes, availability and utilization," Bureau of Agricultural and Industrial Engineering, Eastern Regional Research Laboratory, Circular no. *AIC*-51, 1944.
121. Nelson, F. G., "Pretreatment of cannery wastes," *Sewage Ind. Wastes* **20**, 530 (1948).
122. Nolte, A. J., H. W. von Loesecke, and G. N. Pulley, "Feed yeast and industrial alcohol from citrus-waste press juice," *Ind. Eng. Chem.* **34**, 670 (1942).
123. Norgaard, J. T., R. Hicks, and D. A. Reinsch, "Treatment of combined sewage and fruit canning wastes," *J. Water Pollution Control Federation* **32**, 1088 (1960).
124. O'Connell, W. J., "California vegetable cannery waste disposal practices," *Sewage Ind. Wastes* **29**, 268 (1957).
125. Olivier, G. E., and C. H. Dunstan, "Anaerobic digestion of pea-blancher waste," *Sewage Ind. Wastes* **27**, 1171 (1955).
126. Pence, I. V., "Vegetable wastes, effects on sewage treatment at Peru, Ind.," *Sewage Ind. Wastes* **18**, 585 (1946).
127. Poole, B. A., "Report of the 1946 situation with reference to wastes from Indiana food-processing industries," in Proceedings of 3rd Industrial Waste Conference, Purdue University, May 1947.
128. Porges, N., "Waste loading, potato-chip plants," *Sewage Ind. Wastes* **24**, 1001 (1952).
129. Porges, N., "Waste from the processing of corn chips," *J. Water Pollution Control Federation* **32**, 182 (1960).
130. Ohio Department of Health, *Progress Report on Canning Waste Treatment Studies*, undated, studies conducted in 1926–7.
131. Pulley, G. N., E. L. Moore, and C. D. Atkins, "Grapefruit-cannery wastes yield crude citrus pectin," *Food Inds.* **16**, 285 (1944).
132. Ricks, W. H., "Disposal of fruit canning waste by spray irrigation," in Proceedings of 7th Industrial Waste Conference, Purdue University, May 1952.
133. Riedesel, P. W., and W. R. Lawson, "Freezing as a factor in the stabilization of corn-cannery wastes," *Sewage Works J.* **17**, 952 (1945).
134. Rudolfs, W., *Industrial Waste Treatment*, Reinhold Publishing Corp., New York (1953).
135. Runyan, M. W., "Columbia river dilution simplifies treatment of sewage and cannery wastes," *Wastes Eng.* **28**, 20 (1957).
136. Ryan, W. A., "Effect of cannery wastes on operation of sewage treatment plants," discussion by C. J. Bernhardt, *Sewage Works J.* **12**, 99 (1940).
137. Ryan, W. A., and C. J. Bernhardt, "Effects on sewage treatment," a discussion, *Sewage Ind. Wastes* **12**, 105 (1940).
138. Ryan, W. A., "Experience with sodium nitrate treatment of cannery wastes," *Sewage Works J.* **17**, 1227 (1945).
139. Ryan, W. A., "Industrial waste lagoons," *Sewage Ind. Wastes* **22**, 71 (1950).
140. Sanborn, N. H., "Nitrate treatment of cannery wastes," *Canner* **92**, 12 (1941).
141. Sanborn, N. H., "Treatment of vegetable cannery wastes," *Ind. Eng. Chem.* **34**, 911 (1942).
142. Sanborn, N. H., "Vegetable waste treatment," *Sewage Ind. Wastes* **14**, 1366 (1942) and **15**, 155 (1943).
143. Sanborn, N. H., "Food canning waste utilization," in Proceedings of 1st Industrial Waste Conference, Purdue University, November 1944, p. 89.
144. Sanborn, N. H., "Toxicity to fish of chemicals used in canning," *Sewage Ind. Wastes* **17**, 1296 (1945).
145. Sanborn, N. H., "Utilization of canning wastes," *Sewage Ind. Wastes* **18**, 763 (1946).
146. Sanborn, N. H., "Disposal of food processing wastes by spray irrigation," *Sewage Ind. Wastes* **25**, 1034 (1953).
147. Sessler, R. E., "Waste water use, soap and edible oil plant," *Sewage Ind. Wastes* **27**, 1178 (1955).

148. "Sewage treatment meets amazing growth problem," *Public Works Mag.* **84**, 59 (1953).
149. Southgate, B. A., "Food wastes contain high concentrate of organic matter; notes on industrial waste," *Water Sanit. Engr.*, July 1952.
150. "Spray irrigation, Yakima, Wash.," *Sewage Ind. Wastes* **30**, 1199 (1958).
151. Stander, G. J., "Corn products wastes, anaerobic digestion and filtration; maize, starch waste," *Sewage Ind. Wastes* **29**, 1093 (1957).
152. Stiemke, R. E., "Disposal of wastes from small abattoirs," in Proceedings of 4th Industrial Waste Conference, Purdue University, September 1948, p. 178.
153. Stilz, W. P., "Vibrating screens in industrial waste water treatment," *Ind. Wastes* **1**, 68 (1955).
154. Wisconsin Board of Health, *Stream Pollution in Wisconsin*, Madison (1927).
155. Templeton, C. W., "Cost studies of dehydrating tomato wastes," *Food Packer* **28**, 53 (1947).
156. Templeton, C. W., "Dehydration of tomato wastes," in Proceedings of 3rd Industrial Waste Conference, Purdue University, May 1947.
157. Templeton, C. W., "Cannery waste treatment and disposal," *Sewage Ind. Wastes* **23**, 1540 (1951).
158. Tomhave, A. E., and E. Hoffman, "A preliminary investigation on the use of certain dried vegetable wastes as poultry feeds," *Univ. Delaware Agr. Exp. Sta. Bull.* **247** (1944).
159. New York State Board of Health, *The Treatment of Canning Wastes* (1930).
160. Texas Department of Health, Bureau of Sanitary Engineering, *Treatment of Wastes from Citrus Juice Canning Plants* (1940).
161. Uhlmann, P. A., "Three years' treatment of sewage at Celina, Ohio," *Public Works Mag.* **74**, 13 (1943).
162. Uhlmann, P. A., "Charges to industries for treating their wastes in a municipal plant," *Public Works Mag.* **75**, 23 (1944).
163. Vasam, O. R., and W. L. Shaw, *A Selected Bibliography of Periodical Literature on Fruit and Vegetable Processing Waste*, U.S. Department of Agriculture, Bureau of Agricultural and Industrial Chemistry, Washington, D.C.
164. Veldhuis, M. K., and W. D. Gordon, "Experiments on production of feed yeast from citrus press liquor," *Citrus Ind.* **29**, 7 (1948).
165. Vincent, D. B., "Processing citrus cannery waste," in Proceedings of 6th Industrial Waste Conference, Purdue University, February 1951, p. 123.
166. Vincent, D. B., "Processing citrus cannery waste," *Sewage Ind. Wastes* **24**, 805 (1952).
167. Von Loesecke, H. W., "Citrus fruit industry," *Ind. Eng. Chem.* **44**, 476 (1952).
168. Von Loesecke, H. W., "Citrus fruits industry," *Sewage Ind. Wastes* **24**, 1443 (1952).
169. Von Loesecke, H. W., G. N. Pulley, A. J. Nolte, and H. E. Gorseline, "Experimental treatment of citrus canning effluent in Florida," *Sewage Works J.* **13**, 115 (1941).
170. Van Pollen, E. M., and G. H. McIntosh, "Corn products manufacture," *Ind. Eng. Chem.* **44**, 483 (1952).
171. Wakefield, J. W., "Semitropical industrial waste problems," in Proceedings of 7th Industrial Waste Conference, Purdue University, May 1952, p. 495.
172. Wakefield, J. W., "Waste disposal, research review," *Sewage Ind. Wastes* **26**, 1189 (1954).
173. Wakefield, J. W., B. F. O'Neal, and S. F. Kelso, "Citrus waste disposal practices and research," in Proceedings of 9th Industrial Waste Conference, Purdue University, May 1954.
174. Walters, G. L., "Fruit and vegetable waste disposal to public sewers at Fullerton, Calif.," *Sewage Ind. Wastes* **15**, 771 (1943).
175. Warrick, L. F. (presiding), "Symposium on food-canning waste," in Proceedings of 1st Industrial Waste Conference, Purdue University, November 1944, p. 156.
176. Warrick, L. F., and E. J. Beatty, "The treatment of industrial wastes in connection with domestic sewage," *Sewage Works J.* **8**, 122 (1936).
177. Warrick, L. F., F. J. McKee, E. H. Wirth, and N. H. Sanborn, *Methods of Treating Cannery Waste*, Bulletin no. 28-*L*, National Canners' Association, 1939.
178. Warrick, L. F., T. F. Wisniewski, and N. H. Sanborn, *Cannery Waste Disposal Lagoons*, Bulletin no. 29-*L*, National Canners' Association, 1945.
179. Warrick, L. F., T. F. Wisniewski, and N. H. Sanborn, "Vegetable waste lagoon studies," *Sewage Ind. Wastes* **17**, 1044 (1945).
180. Michigan Department of Health, *Water Pollution Control*, Cannery Wastes Bulletin (April 1927).
181. Weber, G. L., "Plastics from citrus wastes," *Pacific Plastics Mag.* **1**, 44 (1943).
182. Webster, R. A., "Vegetable processing wastes, pilot plant studies," *Sewage Ind. Wastes* **25**, 1432 (1953).

183. Wells, W. N., P. W. Rohrbaugh, and G. A. Doty, "Wastes, total and reducing sugars, determination," *Sewage Ind. Wastes* **24**, 212 (1952).
184. White, J. F., *Waste disposal and Fremont Lake.* Bulletin no. 111, Michigan State College (1952).
185. Wiegand, E. H., "Dollars from factory wastes," *Canner* **98**, 12 (1944) and **98**, 20 (1944).
186. Williamson, G., "Spray irrigation for disposal of food processing waste," in Proceedings of 6th Ontario Industrial Waste Conference, June 1949, p. 17.
187. Wisniewski, T. F., and D. T. Sharow, "Odor control at cannery lagoons," in Proceedings of 7th Industrial Waste Conference, Purdue University, May 1952.
188. Wolman, A., and G. L. Hall, Maryland Department of Health, Bureau of Sanitary Engineering, Annual Report, 1938, p. 216.
189. "Workshop on wastes from food processing," in Proceedings of 2nd Industrial Waste Conference, Purdue University, January 1946.
190. Young, H. H., "Lagoons, nitrate treatment," *Sewage Ind. Wastes* **22**, 1380 (1950).
191. Young, H. H., "Treatment of cherry waste waters," *Public Works Mag.* **74**, 25 (1943).
192. Young, H. H., "Nitrate treatment of lagoons," in Proceedings of 4th Industrial Waste Conference, Purdue University, September 1948, p. 208.

Suggested Additional Reading

The following are recent publications and important older references to make this section complete.

1. "A review of the literature of 1963 on waste water and water pollution control," *J. Water Pollution Control Federation* **36**, 654 (1964).
2. "A review of the literature of 1965 on waste water and water pollution control," *J. Water Pollution Control Federation* **38**, 869 (1966).
3. Adams, S. L., "Industrial alcohol from cannery wastes," *Food Packer* **30**, 26 (1949).
4. Aerojet—General Corporation, "California waste management study, a report to the State of California Department of Public Health," Report no. 3056 (final), Contract no. 347, August 1965.
5. Agar, C. C., "Cannery, milk and food products
6. wastes, effect and treatment (forum on industrial waste problems, 1949)," *Sewage Ind. Wastes* **22**, 1046 (1950).
7. Alikonis, J. J., and J. V. Ziemba, "Waste treatment," *Food Eng.* **39**, 84 (1967).
8. Ambrose, T. W., and C. O. Reiser, "Wastes from potato plants," *Ind. Eng. Chem.* **46**, 1331 (1954).
9. Ammerman, G. R., and N. W. Destrosier, "Odor studies with canning wastes," *Food Technol.* **3**, 253 (1954).
10. "Boosts waste plant performance," *Food Eng.* **32**, 80 (1960).
11. Ammerman, G., and N. DesRosier, "Aeration curbs waste disposal odors," *Food Eng.* **30**, 115 (1958).
12. Anderson, D. R., W. D. Bishop, and H. L. Ludwig, "Percolation of citrus wastes through soil," Purdue University Engineering Extension Series, Bulletin no. 121, 1966, p. 892; *Chem. Abstr.* **67**, 57117 (1967).
13. "Air flotation can economize your waste-water treatment," *Food Eng.* **27**, 107 (1955).
14. "Asparagus waste fed to microbes," *Food Ind.* **16**, 713 (1944).
15. "Cannery waste disposal by lagooning," *Public Works Mag.* **74**, 25 (1943).
16. "Don't discard cannery wastes," *Food Packer* **25**, 30 (1944).
17. "Food yeast," *Food Ind.* **15**, 123 (1943).
18. "How to treat wastes," *Food Ind.* **9**, 430 (1937).
19. "Ion exchange plant recovers sugar from fruit wastes," *Food Ind.* **18**, 1846 (1946).
20. "Vibrating screen solves pollution problem," *Canner* **118**, 24 (1954).
21. "Leamington treats municipal industrial wastes," *Water Pollution Control* **104**, 6 (1966).
22. "Grappling with that problem of citrus waste odors," *Food Eng.* **26**, 133 (1954).
23. "How packers and borough solve Pennsylvania waste disposal problems," *Canner-Packer*, May 1959, p. 27.
24. "Strides in waste disposal," *Food Eng.* **35**, 85 (1963).
25. "Treats waste with steam injection," *Food Eng.* **40**, 207 (1968).
26. "Aeration curbs waste disposal odors," *Food Eng.* **30**, 115 (1958).
27. "How to handle high BOD fruit processing wastes," *Wastes Eng.* **34**, 128 (1963).
28. Allen, L. H., et al., "Microbiological problems in manufacture of sugar from beets," *J. Soc. Chem. Ind.* **67**, 450 (1948).
29. Allison, C., "Some rights and wrongs of cannery waste treatment," *Water Works Sewerage* **95**, 227 (1948).

30. Askew, M. W., "Plastics in waste treatment. II," *Process Biochem.* **2**, 31 (1967); *Chem. Abstr.* **66**, 58697m (1967).
31. Atkins, P. F., and O. J. Sproul, "Feasibility of biological treatment of potato processing wastes," in Proceedings of 19th Industrial Waste Conference, Purdue University, May 1964, p. 303.
32. Batina, R. T., and J. V. Ziemba, "3-Pronged approach to handling wastes," *Food Eng.* **40**, 74 (1968).
33. Beechmans, I., and P. Beaujeau, "The study of treatment of industrial wastes (in the sugar industry)," *Centre Belge Etude Doc. Eaux Bull. Trimestr.* **21**, 147 (1953); *Chem. Abstr.* **48**, 10363 (1954).
34. Belick, F. M., "Canning wastes complicate treatment in San Jose—Santa Clara water pollution control plant," *Civil Eng.* **36**, 49 (1966).
35. Bell, J. W., "Spray irrigation and research of canning plant wastes," *Can. Food Ind.* **32**, 31 (1961); *Water Pollution Abstr.* **35**, 737 (1962).
36. Bell, J. W., "Spray irrigation for poultry and cannery wastes," *Public Works Mag.* **9**, 111 (1955).
37. Black, H. H., and G. McDermott, "Industrial waste guide," *Sewage Works J.* **24**, 181 (1952).
38. Black, H. H., "Industrial waste treatment as varied as industry," *Civil Eng.* **25**, 688 (1955).
39. Bloodgood, D. E., "Twenty years of industrial waste treatment," Purdue University Engineering Extension Series, Bulletin no. 118, 1965, p. 182.
40. Bower, B. T., "Developing patterns for efficient water utilization: the canning industry," in Proceedings of 14th Southern Municipal and Industrial Waste Conference, 1965, p. 78.
41. Bohstedt, G., "Comparative feed value—sweet corn, pea vine, field corn silage," *Canning Trade* **75**, 7 (1953).
42. Bondi, H. S., "Food wastes made into profitable product," *Food Eng.* **27**, 84 (1955).
43. Bradakis, H. L., "A joint municipal-industry spray irrigation project," *Ind. Water Wastes* **6**, 117 (1961).
44. Braun, R., "Utilization of organic industrial wastes by composting," *Compost Sci.* **3**, 34 (1962).
45. Brooke, D. L., and G. L. Capei, "An economic method of alternative methods of cull tomato disposal in Dade County, Florida," Agriculture Economic Report no. 59-2, Florida Experimental Station, September 1958.
46. Brown, A. H., W. D. Ramage, and H. S. Owens, "Progress in processing pear canning wastes," *Food Packer* **31**, 30 (1950).
47. Brown, A. H., W. D. Ramage, and H. S. Owens, "Pomace and molasses from pear wastes," *Food Packer* **31**, 50 (1950).
48. Brown, H. D., "Cannery waste disposal research in Ohio, 1953," *Canning Trade* **76**, 10 (1953).
49. Brown, H. D., "Disposal of cannery wastes in Ohio," *Canning Trade* **78**, 24 (1956).
50. Brown, H. D., et al., "Disposal of cannery wastes by irrigation. II," *Food Packer* **36**, 41 (1955).
51. Burbank, N. C., Jr., and J. S. Kumagai, "A study of a pineapple cannery waste," in Proceedings of 20th Industrial Waste Conference, Purdue University Engineering Extension Series, Bulletin no. 118, 1966, p. 365; *J. Water Pollution Control Federation* **39**, 867 (1967).
52. Burch, J. E., E. S. Lipinsky, and J. H. Litchfield, "Technical and economic factors in the utilization of waste products," *Food Technol.* **17**, 1266 (1963).
53. California State Water Quality Control Board, *Cannery Waste Treatment Utilization and Disposal. A Literature Review*, Sacramento, Calif. (1965).
54. Canham, R. A., "Spray irrigation for disposal and crop growth," *Ind. Wastes* **2**, 57 (1957).
55. Canham, R. A., "The program of the National Canners Association on waste disposal," *Food Technol.* **11**, 670 (1957).
56. Canham, R. A., "Anaerobic treatment of food canning wastes," *Cannery Trade* **72**, 41 (1950).
57. Canham, R. A., "Current trends in handling canning waste," in Proceedings of 5th Annual Water Symposium, 21 February 1956; *Louisiana State Univ. Eng. Exp. Sta. Bull.* **55**, 112 (1956).
58. Canham, R. A., "Food canning waste disposal," *Penn. Packer* **21**, 15 (1955).
59. Carl, C. E., "Waste treatment by stabilization ponds," *J. Milk Food Technol.* **24**, 147 (1961).
60. Cessna, J. O., "Modesto's pollution explosion," *Water Works Wastes Eng.* **1**, 50 (1964); *J. Water Pollution Control Federation* **37**, 736 (1965).
61. Chakrabarty, R. M., "Cane sugar wastes and their disposal," *Environ. Health* **6**, 265 (1964).
62. Chipperfield, P. N. J., "Performance of plastic filter media in industrial and domestic waste treatment," *J. Water Pollution Control Federation* **39**, 1860 (1967).
63. Chong, G., and W. Y. Cruess, "Obtaining juice

from waste pear peels and cores," *Canner* **109**, 14 (1949).
64. Ciabattari, E. J., and R. J. Nogaj, "Aeration apparatus for industrial wastes," Yeomans Bros. Co., 12 September 1967; U.S. Patent no. 3341450, *Chem. Abstr.* **68**, 42985 (1968).
65. Clevenger, M. A., "California canners tackle waste problem," *Canning Trade* **73**, 7 (1949); *Western Canner Packer* **41**, 15 (1949); *Canner* **109**, 22 (1949).
66. Cyr, R., "Treatment of waste waters from food industry," *Ingenieur* **42**, 25 (1956); *Water Pollution Abstr.* **33**, 65 (1960).
67. Cooley, A. M., E. D. Wahl, and G. O. Fossum, "Characteristics and amounts of potato wastes from various process streams," in Proceedings of 19th Industrial Waste Conference, Purdue University, 1964, p. 379.
68. & 69. Czirfusz, M., "Food industrial waste water purification by hydrocyclone," *Elelm. Ipar* **16**, 197 (1962).
70. Davis, N. E., "Control of canning waste oxidation ponds," *Public Works Mag.* **90**, 133 (1959); *Water Pollution Abstr.* **33**, 102 (1960).
71. "Destroys fluidized waste without pollution," *Food Eng.* **39**, 204 (1967).
72. Dickinson, D., "Food production effluents—case for composting," *Compost Sci.* **4**, 14 (1963).
73. Dickinson, D., "Purification of effluents from food manufacture," *Effluent Water Treat. J.* **3**, 373 (1963).
74. Dickinson, D., "The purification of cannery waste waters in biological filters," *J. Sci. Food Agric.* **5**, 94 (1954); *Ind. Agr. Aliment.* **70**, 301 (1953).
75. Dickinson, D., "Performance of recirculating plant for the purification of cannery wastes or biological filters," *Surveyor* **108**, 89 (1949); *Food* **18**, 104 (1949); *J. Proc. Inst. Sewage Purif.*, part 1, 1959, p. 54.
76. Dickinson, D., "The origin, treatment and disposal of effluents in the food canning and food freezing industries," in Proceedings of 2nd Symposium on the Treatment of Waste Waters, Newcastle-upon-Tyne, September 1959, Pergamon, Oxford (1960), p. 385.
77. Dickson, G., "A multipond flow-through lagoon system for treatment of cannery wastes," in Proceedings of 12th Pacific Northwest Industrial Waste Conference, University of Washington, 1965; *J. Water Pollution Control Federation* **38**, 870 (1966).
78. Dickson, G., "The application of stabilization basins to vegetable cannery wastes," in Proceedings of 18th Industrial Waste Conference, Purdue University Engineering Extension Series, Bulletin no. 115, 1963, p. 95.
79. Dickson, G., "Vegetable cannery liquid waste treatment by the 'ever-full' lagoon system," in Proceedings of 11th Ontario Industrial Waste Conference, 1964; *Chem. Abstr.* **66**, 793932 (1967).
80. Dietrich, K. R., "Improved chemical treatment of waste waters from canneries," *Ind. Obst- Gemüseverwert.* **52**, 6 (1967); *Chem. Abstr.* **66**, 793499 (1967).
81. Dietz, M. R., and R. C. Frodey, "Cannery waste disposal by Gerber Products," *Compost Sci.* **1**, 22 (1960).
82. Dougherty, M. H., "Activated sludge treatment of citrus wastes," *J. Water Pollution Control Federation* **36**, 72 (1964).
83. Donnerhack, W., "Sewage from the fruit and vegetable processing industry and possibilities of treatment," *Wasserwirtsch.-Wassertech.* **17**, 117 (1967); *Chem. Abstr.* **67**, 57116 (1967).
84. Drehwing, F. J., and N. L. Nemerow, "Aeration of pumpkin cannery wastes," in Proceedings of 18th Industrial Wastes Conference, Purdue University, April 1963, p. 102.
85. Dunstan, G. H., and L. L. Smith, "Experimental operation of industrial waste stabilization ponds," *Public Works Mag.* **91**, 93 (1960).
86. Dutton, C. S., and C. P. Fisher, "The use of aerohydraulic guns in the biological treatment of organic wastes," Purdue University Engineering Extension Series, Bulletin no. 121, 1966.
87. Eckenfelder, W. W., Jr., *Industrial Water Pollution Control*, McGraw-Hill Book Co., New York (1966).
88. Eckenfelder, W. W., and D. J. O'Connor, *Treatment of Organic Wastes in Aerated Lagoons*, Manhattan College Civil Engineering Department, Research Report no. 7, New York State Water Pollution Control Board (1960).
89. Edeline, F., et al., "Recycling of sludge waters of sugar refineries," *Centre Belge Etude Doc. Eaux Bull. Trimestr.* **107**, 292 (1959).
90. Edeline, F., G. Lambert, W. Binet, and H. Fattecceoni, "Waste water from canneries. Comparison of two processes: aspersion and oxidation ditch," *Tribune Centre Belge Etude Doc. Eaux* **20**, 335 (1967); *Chem. Abstr.* **68**, 42958 (1968).
91. Edwards, P. W., R. K. Eskew, A. Hersch, Jr., N. C. Aceto, and C. J. Redfield, "Recovery of tomato processing wastes," *Food Technol.* **6**, 383 (1952).

92. Egan, J. T., and R. W. Brouillette, "Applications of PVC as a trickling filter media," in Proceedings of 9th Ontario Industrial Waste Conference, June 1962, p. 1.
93. Eldridge, E. F., "Experiences with special cannery wastes," *Sewage Works J.* **10**, 914 (1938).
94. Eldridge, E. F., "The role of chemicals in industrial waste treatment—cannery wastes," *Water Works Sewerage* **89**, 342 (1942).
95. Engelbrecht, R. S., B. B. Ewing, and R. C. Hoover, "Soybean and mixed-feed plant processing wastes," *J. Water Pollution Control Federation* **36**, 434 (1964).
96. "Experiments on treatment of cannery waste," *Sewage Works J.* **11**, 339 (1939).
97. Felton, G. E., "Use of ion exchange in by-product recovery from pineapple waste," *Food Technol.* **3**, 40 (1949).
98. Fisk, W. W., "Food processing waste disposal," *Water Sewage Works* **3**, 417 (1964).
99. Flay, R. B., "Pea cannery wastes, periodate oxidation for control," *Sewage Ind. Wastes* **25**, 953 (1953).
100. Forbes, W. J., "Handling of pea straw wastes," in Proceedings of 4th Ontario Industrial Wastes Conference, 1957.
101. Fossum, G. O., A. M. Colley, and E. D. Wahl, "Stabilization ponds receiving potato wastes with domestic sewage," in Proceedings of 10th Industrial Waste Conference, Purdue University, p. 96.
102. Francis, R. L., "Characteristics of potato flake processing wastes," *J. Water Pollution Control Federation* **34**, 291 (1962).
103. Francis, R. L., "Summary of potato products wastes studies," North Dakota Water and Sewage Works Conference, Official Bulletin no. 30, July 1962.
104. U.S. Department of Health, Education and Welfare, *Fruit Processing Industry—An Industrial Waste Guide*, U.S. Public Health Service Publication no. 952, Washington, D.C. (1962).
105. Grant, S. L., "Sewage–apple wastes plant (Winchester, Va.)," *Sewage Ind. Wastes* **21**, 210 (1950).
106. Green, F. G., *Cannery Waste Disposal—a Selected Annotated Bibliography*, National Research Council, Ottawa, (1957); *Batelle Tech. Rev.* **6**, 609a (1957).
107. Grau, P., "Pilot plant studies of cannery waste treatment," *Sb. Vysoke Skoly Chem. Technol. Praze* **5**, 59 (1962).
108. Grau, P., "Pilot plant studies on the treatment of waste waters from the canning of fruit and vegetables," *Sb. Vysoke Skoly Chem. Technol. Praze* **5**, 59 (1962); *J. Water Pollution Control Federation* **39**, 737 (1965).
109. Hamlin, G. H., "Super-rate biological filter," *Ind. Water Wastes* **7**, 148 (1962).
110. Hart, S. A., and P. H. McGauhey, "Wastes management in food producing and processing industries," *Oregon State Coll. Eng. Exp. Sta. Circ.* **29**, 92 (1963).
111. Heider, R. W., "Cannery waste treatment," *Sewage Works J.* **18**, 588 (1946).
112. Helmers, E. N., et al., "Nutritional requirement in the biological stabilization of industrial wastes treatment with domestic sewage," *Sewage Ind. Wastes* **23**, 834 (1951).
113. Hemens, J., et al., "Full-scale anaerobic digestion of effluents from production of maize-starch," *Water Waste Treat. J.* **9**, 16 (1962).
114. Hess, R. W., "Discussion on lagooning with sodium nitrate," *Sewage Ind. Wastes* **25**, 720 (1953).
115. Heukelekian, H., "Disposal of waste waters from the preparation of vegetables for drying," *J. Soc. Chem. Ind.* **63**, 3 (1945).
116. Hicks, W. M., "Disposal of fruit canning wastes by spray irrigation," in Proceedings of 7th Industrial Wastes Conference, Purdue University, 1952, p. 130.
117. Hindin, E., and G. H. Dustan, "Anaerobic digestion of potato processing wastes," *J. Water Pollution Control Federation* **35**, 486 (1963).
118. Hindin, E., and G. H. Dustan, "Effect of potato chip wastes on digestion," *Water Sewage Works* **108**, 432 (1961).
119. Ingols, R. S., H. W. Hodgen, and H. H. Miller, "Canning plant waste problem solved by simplified procedure," *Water Sewage Works* **101**, 227 (1954).
120. Jackson, F. A., "Spray irrigation uses cannery wastes," *Western Canner Packer* **45**, 14 (1957).
121. Jenks, H. N., "Aeration ponds handle cannery wastes," *Sewage Ind. Wastes Eng.* **21**, 62 (1950).
122. Jones, E. E., "Waste waters from the manufacture of cider and their treatment, in percolating filters," *J. Proc. Inst. Sewage Purif.*, Part 2, 1939, p. 212.
123. Jones, E. E., "Waste waters from manufacture of cider," *J. Proc. Inst. Sewage Purif.*, Part 2, 1949, p. 212.
124. Jones, F. O., "A storage lagoon for wastes from canning company plants," *Public Works Mag.* **75**, 21 (1944).
125. Judell, T. L., "Effluent treatment in the food proces-

sing industries," *Food Technol. Australia* **17**, 446 (1965); *J. Water Pollution Control Federation* **38**, 870 (1966).

126. Kefford, J. F., "Composting trials with pear canning wastes," *J. Water Pollution Control Federation* **38**, 870 (1966).

127. Kefford, J. E., "Composting trials with pear canning wastes," *J. Water Pollution Control Federation* **39**, 867 (1967).

128. Kefford, J. E., "Composting trials with pear canning wastes," *Compost Sci.* **6**, 24 (1966).

129. Kennedy, J. G., "The modern factory problems of effluent disposal," *Food Manuf.* **40**, 68 (1965); *J. Water Pollution Control Federation* **38**, 870 (1966).

130. Kestoren, J. W., and R. Hendrickson, "Florida turns waste disposal problem into profitable by-products industry," *Canner/Packer* **133**, 34 (1964); *J. Water Pollution Control Federation* **37**, 737 (1965).

131. Kimberly, A. E., "Status of pea-canning waste treatment," *Water Works Sewerage* **78**, 126 (1931).

132. Kingston, T. M. S., "Chatham, Ontario's approach to joint municipal—industrial waste treatment," *J. Water Pollution Control Federation* **35**, 544 (1963).

133. Kirsch, J. T., "Lagoon and viner stack odor control," *Canning Trade* **75**, 7 (1953).

134. Kubelka, V., and V. Koran, "Waste waters from beet sugar mills," *Chem. Abstr.* **49**, 7275 (1955).

135. Kumagai, J. S., and N. C. Burbank, Jr., "Treatment by activated sludge studied as aid in canning waste disposal," *Food. Eng.* **37**, 101 (1965).

136. Lefebure, P. H., "The problem of purification of vegetable canning waste waters," *Centre Belge Etude Doc. Eaux Bull. Trimestr.* **15**, 37 (1952); *Water Pollution Abstr.* **26**, 80 (1953).

137. Lefebure, P. H., and D. Dickinson, "Waste waters of canning industries," *Centre Belge Etude Doc. Eaux Bull. Trimestr.* **88**, 114 (1958).

138. LeVine, R. Y., "Lime treatment of tomato waste," *Public Works Mag.* **79**, 28 (1948).

139. Levine, M., and G. H. Nelson, "Experiments on the purification of beet sugar wastes by stream flow aeration," *Sewage Works J.* **1**, 40 (1929).

140. Ludwig. R. G., and R. V. Stone, "Disposal effects of citrus by-products wastes," *Water Sewage Works* **110**, R-202 (1963).

141. Luley, H. G., "Spray irrigation of vegetable and fruit processing wastes," *J. Water Pollution Control Federation* **35**, 1252 (1963).

142. Marsh, R. J., "Pilot plant produces commercial by-product from canner waste," *Canner* **113**, 12 (1951).

143. Mather, J. R., "The disposal of industrial effluent by woods irrigation," in Proceedings of 8th Industrial Waste Conference, Purdue University, 1953, p. 439.

144. McIndoe, W. C., "Western wastes as materials for alcohol production," *Chem. Met. Eng.* **48**, 111 (1944).

145. McIntosh, G. H., and G. C. McGeorge, "Year-round lagoon operation," *Food Process.* **25**, 82 (1964); *J. Water Pollution Control Federation* **37**, 735 (1965).

146. McNary, R. R., "The treatment of citrus processing waste water," in Proceedings of 17th Industrial Waste Conference, Purdue University, 1962, p. 834.

147. McNary, R. R., and R. W. Wolford, "Citrus processing wastes," *J. Southern Research* **4**, 31 (1952); *Chem. Abstr.* **47**, 791 (1953).

148. Mercer, W. A., and G. K. York, "Bacteriological studies of water reuse systems in canneries," *Canner* **117**, 11 and 30 (1953).

149. Mercer, W. A., "Aeration-digestion sedimentation treatment of canning waste water," in Proceedings of 5th Pacific Northwest Industrial Waste Conference, 1958, p. 39.

150. Mercer, W. A., J. A. Chapman, F. Duinnell, Jr., W. W. Rose, and A. Katsuyama, "Aerobic composting of vegetable and fruit wastes," *Compost Sci.* **3**, (1962).

151. Mercer, W. A., and W. W. Rose, "Current research on canning waste problems," in Proceedings of 12th Pacific Northwest Industrial Waste Conference, University of Washington, 1965.

152. Mercer, W. A., "Physical characteristics of recirculated waters as related to their sanitary condition," *Food Technol.* **18**, 335 (1964); *J. Water Pollution Control Federation* **37**, 737 (1965).

153. Mercer, W. A., "Water supply and waste disposal problems of the food industries," in Proceedings of 11th Pacific Northwest Industrial Waste Conference, 1963.

154. Mercer, W. A., "Water supply and waste disposal problems of food industries," *Oregon State Coll. Eng. Exp. Sta. Circ.* **29**, 913 (1964).

155. Miller, P. R., "Spray irrigation at Morgan Packing Co.," in Proceedings of 8th Industrial Waste Conference, Purdue University, 1953, p. 284.

156. Miller, L. B., "Cannery waste—an example of complete treatment and disposal," *Public Works Mag.* **78**, 19 (1947).

157. Molloy, D. J., "Instant wastes treatment," *Eng. Index*, 1964, p. 913.
158. Monson, H., "Cannery waste disposal by spray irrigation," *Compost Serv.* **1**, 44 (1960).
159. Monson, H. G., "Vegetable cannery waste disposal," *Canner* **116**, 12 (1953).
160. Monson, H., "Successful disposal of cannery waste materials," *Food Can.* **16**, 42 (1956).
161. National Canner's Research Foundation Group, "Aerobic composting of vegetable and fruit wastes," *Compost Sci.* **3**, 9 (1963).
162. Neil, J. H., "The use and construction of oxidation ponds for industrial waste treatment," in Proceedings of 7th Ontario Industrial Waste Conference, 1960.
163. Nelson, L. E., "Cannery waste disposal by spray irrigation," *Wastes Eng.* **23**, 398 (1952).
164. Nemerow, N. L., and F. J. Drehwing, "Aeration of pumpkin cannery wastes," in Proceedings of 18th Industrial Waste Conference, Purdue University, April 1963, p. 102.
165. Neubert, A. M., *et al.*, "Recovery of sugars from pear canning waste," *J. Agr. Food Chem.* **2**, 30 (1954).
166. Neuhaus, G. M., "Waste water problems of the food industry with reference to the proposed water law, Fette v. Seifen 59, 106 (1957)," *Water Pollution Abstr.* **34**, 96 (1961).
167. Norgaard, J. T., "Significant characteristics of sewage and industrial wastes," *Sewage Ind. Wastes* **22**, 1024 (1950).
168. O'Connell, W. J., and K. A. Fitch, "Waste disposal," *Food Inds.* **22**, 71 (1950).
169. Olson, O. O., W. Van Heuvelen, and J. W. Vennes, "Aeration of potato waste," in Proceedings of 19th Industrial Waste Conference, Purdue University, 1964, p. 180.
170. Olson, O. O., and J. W. Vennes, "Mechanical aeration of potato processing and domestic sewage," North Dakota Water and Sewage Works Conference, Official Bulletin no. 30, May 1963.
171. Orlob, G. T., F. G. Agardy, R. C. Cooper, N. L. Nemerow, and R. G. Spicker, *Cannery Waste Treatment, Utilization and Disposal—A Literature Review*, California State Water Quality Control Board and Water Resources Engineers, Inc. (1965).
172. Orlob, G. T., "Cannery waste treatment, utilization and disposal—a literature review," prepared for California State Water Quality Control Board, Std. Agreement nos. 12–39 (September 1965); *J. Water Pollution Control Federation* **38**, 870 (1966).
173. "Ninth Ontario Industrial Waste Conference," *Ind. Water Wastes* **7**, 146 (1962).
174. Pailthorp, R., "After only 10 years of primary treatment—high-rate filters," *Wastes Eng.* **34**, 282 (1963).
175. Popper, K., *et al.*, "Recycles process brine," *Food Eng.* **39**, 78 (1967).
176. Parker, C. D., "Food cannery waste treatment by lagoons and ditches at Shepparton, Victoria, Australia," Purdue University Engineering Extension Series, Bulletin no. 121, 1966, p. 284.
177. Parker, R. R., "Spray irrigation for industrial wastes disposal," *Munic. Utilities* **103**, 28 (1965); *Eng. Index*, 1966, p. 1346.
178. "Plastic waste treatment successful in treating cannery wastes," *Water Sewage Works* **114**, 86A (1967).
179. Popper, K., W. M. Camviand, G. G. Watters, R. J. Bouthelet, and F. P. Boyle, "Recycled process brine prevents pollution," *Food Eng.* **39**, 78 (1967).
180. "Potato waste treated by dissolved air floatation," *Water Pollution Control* **104**, 61 (1966).
181. Powell, H. W., "Spray irrigation for the disposal of cannery wastes," *Munic. Utilities* **92**, 30 (1954).
182. Redfield, C. S., and R. E. Esken, "Apply essence recovery costs," *Glass Packer* **32**, 33 (1953).
183. Rehm, L. F., and F. N. van Kirk, "San Jose-Santa Clara Kraus Process Plant copes with variable canning load," *Water Works Waste Eng.* **1**, 32 (1964).
184. Reiser, C. O., "Torula yeast from potato wastes," *J. Agr. Food Chem.* **2**, 70 (1954).
185. Reyes, E. and C. C. Campillo, "Aminoacid content in food yeast grown in pineapple canning industry waste," *Rev. Latinoam. Microbiol.* **4**, 113 (1961).
186. Rhoads, A. T., "Cannery waste disposal," *Proc. Florida State Hort. Soc.* **77**, 284 (1964).
187. Rigg, K., "Cannery waste—a gardener's viewpoint," *Compost Sci.* **3**, 43 (1962).
188. Roberts, W. M., "Water and waste problems and the development of the food industry," in Proceedings of 13th Southern Municipal Waste Conference, 1964, p. 163.
189. Rosa, J., "Review of the qualitative composition and possible use of fruit and vegetable wastes," *Prace Inst. Lab. Badawczych Przemyslu Spozywecgo* **16**, 58 (1966); *Chem. Abstr.* **66**, 1080 (1967).
190. Rose, W. R., J. Chapman, and W. A. Mercer, "Composting fruit waste solids," *Oregon State Coll. Eng.*

Exp. Sta. Circ. **29**, 32 (1963); *Eng. Index*, 1964, p. 913.
191. Rose, W. W., et al., "Composting fruit waste solids," in Proceedings of 11th Pacific Northwest Industrial Waste Conference, 1963, p. 18.
192. Rudolfs, W., and H. Heukelekian, "Food wastes," *Food Eng.* **24**, 104 (1952).
193. Rudolph, V. J., and R. E. Dils, "Irrigating trees with cannery waste water," *Mich. State Univ. Agr. Exp. Sta. Quart. Bull.* **37**, 407 (1955).
194. Rupp, V. R., P. DeLeon, and C. W. Templeton, "Workshop on wastes from food processing," *Sewage Works J.* **19**, 1107 (1947).
195. Ryan, W. A., "Sodium nitrate in waste treatment," *Water Works Sewerage* **92**, 227 (1945).
196. Sak, J. G., "Plastic biological oxidation media for industrial waste treatment needs," in Proceedings of 11th Ontario Industrial Waste Conference, 1967, p. 187.
197. Sanborn, N. H., "Cannery waste treatment," *Sewage Works J.* **14**, 1366 (1942).
198. Sanborn, N. H., "Cannery wastes bring dollars," *Food Packer* **26**, 35 (1945).
199. Sanborn, N. H., "Disposal of cannery wastes," *Food Inds.* **18**, 521 (1946).
200. Sanborn, N. H., "Effective sanitation of pea viner stations," *Canner* **113**, 16 (1951).
201. Sanborn, N. H., "Food canning wastes," *Water Works Sewerage* **92**, 71 (1945).
202. Sanborn, N. H., "Nitrate kills odor in waste," *Food Inds.* **13**, 57 (1941).
203. Sanborn, N. H., "Sanitation of pea viner stations," *Canner* **110**, 10 (1950).
204. Sanborn, N. H., "Spray irrigation as a means of cannery waste disposal," *Canning Trade* **74**, 17 (1952).
205. Sanborn, N. H., "Treatment of cannery waste," *Sewage Works Eng.* **20**, 112 and 199 (1949).
206. Sanborn, N. H., "Waste disposal methods for food industry," *Chem. Eng. News* **24**, 1059 (1946).
207. Sawyer, F. G., "Cannery alcohol," *Ind. Eng. Chem.* **40**, 16A (1948).
208. Schraufnagel, F. H., "Ridge and furrow irrigation," *J. Water Pollution Control Federation* **34**, 1117 (1962).
209. Schraufnagel, F. H., "Waste disposal by ridge and furrow irrigation," Wisconsin Commission on Water Pollution, Report no. WP-108; *Wisconsin Univ. Eng. Exp. Sta. Res. Rept.* **20** (1959); *Public Health Rept.* **74**, 133 (1959).
210. Seabrook, B. L., "This woodland spray system disposes of a billion gallons of waste water annually," *Food Eng.* **29**, 112 (1957).
211. Sen, B. P., and T. R. Bhasharan, "Anaerobic digest of liquid molasses distilling waste," *J. Water Pollution Control Federation* **34**, 1015 (1962).
212. Severance, H. D., "Control of cannery odors at Monterey," *Sewage Works J.* **3**, 152 (1932).
213. Shaw, R., "Recovery of edible and industrial products from effluent streams of potato processing plants," in Proceedings of 21st Industrial Waste Conference, Purdue University, 1966, p. 13.
214. Sherman, W. C., and C. J. Koehn, "Beta-carotene from sweet potatoes," *Ind. Eng. Chem.* **40**, 1445 (1948).
215. & 216. Shikaze, K. H., "Aerated lagoons for the treatment of cannery wastes," in Proceedings of 12th Ontario Industrial Waste Conference, June 1965, p. 17; *Chem. Abstr.* **66**, 79399A (1967).
217. Siebert, C. L., and C. Allison, "Some rights and wrongs of cannery waste treatment," *Water Works Sewerage* **95**, 227 (1948).
218. Skrinde, R. T., and G. H. Dunstan, "Aerobic treatment of pea processing wastes," *Intern. J. Air Water Pollution* **5**, 339 (1963).
219. Smith, A. K., A. M. Nash, A. C. Eldridge, and W. J. Wolf, "Recovery of soybean whey protein with edible gums and detergents," *J. Agr. Food Chem.* **10**, 302 (1962).
220. Stubbs, J. J., W. M. Noble, and J. C. Lewis, "Fruit juices yield food yeast," *Food Inds.* **16**, 694 (1944).
221. Swamson, E., and J. V. Ziemba, "Seek more profits from plant wastes," *Food Eng.* **39**, 110 (1967).
222. Taggart, R. S., "Dry waste disposal," *Food Eng.* **39**, 98 (1967).
223. Tanaka, M., A. Yasuhiro, and H. Ono, "Treatment of industrial wastes by activated sludge. XII. Treatment of confectionery wastes," *Kogyo Gijutsuin, Hakko Kenkyusho Kenkyu Hokoku* **30**, 35 (1966); *Chem. Abstr.* **68**, 32997 (1968).
224. Toth, T. S., "Some experiences with spray irrigation," in Proceedings of 4th Ontario Industrial Waste Conference, 1957, p. 4.
225. Treadway, R. H., "Utilization of white potatoes," *Am. Potato J.* **25**, 300 (1948).
226. Troemper, A. P., "By-product recovery features Illinois wastes progress," *Wastes Eng.* **22**, 543 (1951).
227. Townshend, A. R., and R. G. Barrens, "Cannery waste treatment in Ontario," in Proceedings of 9th

Ontario Industrial Waste Conference, June 1962, p. 25.
228. Uber, W. J., and H. G. Rogers, "Dairy and food-processing wastes predominate in Minnesota," *Wastes Eng.* **22**, 538 (1951).
229. U.S. Department of the Interior, Federal Water Pollution Control Administration, "The cost of clean water," in Industrial Waste Profile no. 6, *Canned and Frozen Fruits and Vegetables*; Federal Water Pollution Control Association Publication no. 1 (Water Pollution), September 1967.
230. Waicollen, C., "Study of waste water and cider factories and apple distilleries," *Chim. Ind.* **26**, 30 (1931).
231. Wail, M. B., R. C. Kelley, and J. J. Willaman, "Carotene concentrates from vegetable leaf wastes," *Ind. Eng. Chem.* **36**, 1057 (1944).
232. Walters, G. L., "Problems of disposal of peach and tomato waste," *Sewage Works J.* **9**, 607 (1937).
233. Walters, G. L., "Problems of disposal of peach and tomato waste," *Sewage Works J.* **15**, 3 (1943).
234. Warrick, L. F., "Cannery waste treatment in Wisconsin," *Water Works Sewerage* **81**, 346 (1934).
235. Warrick, L. F., "Experiments on treatment of canning wastes," *Public Works Mag.* **70**, 10 (1939).
236. Warrick, L. F., H. W. Ruf, and J. M. Holderby, "Digestion of solids from pea and corn canning waste," *Water Works Sewerage* **83**, 425 (1936).
237. Warrick, L. F., "The disposal of food plant wastes," *Food Inds.* **10**, 20 (1938).
238. Warrick, L. F., and T. F. Wisniewski, "Use of sodium nitrate in treating canning wastes," *Public Works Mag.* **76**, 48 (1945).
239. Weaver, E. A., et al., *Aerobic Microbiological Treatment of Potato Starch Factory Wastes*, U.S. Department of Agriculture, Bureau of Agricultural and Industrial Chemistry, A.I.C. 350 (1953).
240. Wisniewski, T. E., "Food-processing waste treatment in Wisconsin," in Proceedings of 4th Industrial Waste Conference, Purdue University, 1948, p. 203.
241. Wisniewski, T. E., "How to handle a variety of wastewater problems," *Natl. Engr.* **58**, 17 (1954); *Chem. Abstr.* **48**, 11688 (1954).
242. Wisniewski, T. F., "Irrigation disposal of industrial wastes," *Public Works Mag.* **92**, 96 (1961).
243. Wisniewski, T. F., and O. J. Muegge, "Wisconsin's stream pollution control program," *Wastes Eng.* **22**, 540 (1951).
244. VanLuven, A. L., "Some considerations of food industry wastes problems in Quebec," in Proceedings of 10th Ontario Industrial Wastes Conference, June 1963, p. 91.
245. VanLuven, A. L., "Food industry waste problems in Quebec, Canada," *Ind. Water Wastes* **8**, 22 (1963); *Eng. Index*, 1964, p. 913.
246. VanLuven, A. L., "Table 1—relation between waste and unit processes," based on a speech made at 10th Ontario Industrial Wastes Conference, June 1963.
247. "Pilot-plant studies on treatment of fruit and vegetable canning wastes," *Sb. Vysoke Skoly Chem. Technol. Praze, Oddil Fak. Technol. Paliv Vody* **5**, 59 (1962).
248. Vennes, J. W., and E. G. Olmstead, "Experimental lagooning of potato flake wastes," North Dakota Water and Sewage Works Conference, Official Bulletin no. 30, May 1963.
249. Vennes, John W., and Edwin G. Olmstead, "Stabilization of potato wastes," North Dakota Water and Sewage Works Conference, Official Bulletin no. 29, October 1961, p. 10.
250. Zickefoose, C. S., "Pea wastes and city sewage treated in parallel plants," *Wastes Eng.* **34**, 242 (1963).
251. Wahl, E. D., A. M. Cooley, and G. O. Fossum, "Digestion of potato waste substances—laboratory conditions," in Proceedings of 19th Industrial Waste Conference, Purdue University, May 1964.
252. Zimmermann, F. J., "Sewage sludge treatment by wet air oxidation," in Proceedings of 12th Industrial Waste Conference, Purdue University, 1958, p. 409.

DAIRY WASTES

In 1963, 16 million cows in the United States produced approximately 127 billion pounds of milk. This was distributed among five related industries as follows: 52 billion pounds of fluid milk, including 1.5 billion pounds of cottage cheese; 34 billion pounds of butter; 14.4 billion pounds of cheese; 11.9 billion pounds of ice-cream and frozen desserts; and 10.8 billion pounds of condensed and powdered milk. The plants handling or processing milk and milk products may be roughly classified as follows: receiving stations, bottling plants, cheese factories, creameries, condenseries, and dry-milk and ice-cream plants.

The receiving station serves as a collection point for raw milk from the farmers. Here milk-cans are emptied into a weighing vat, the milk is sampled, and it is then loaded into tank cars or trucks for shipment to processing plants. Waste-producing operations are washing and sterilizing of cans, vats, tanks, cooling equipment, and floors. At the bottling plant, the raw milk is dumped, sampled, weighed, clarified, filtered, preheated, pasteurized, cooled, and poured into glass or paper containers. Waste-producing operations here include washing of bottles, cases, cans, processing equipment, and floors. Figure 23.2 illustrates this process.

A typical milk-receiving station* now receives about 25 per cent of its milk in conventional 10-gallon milk-cans and the other 75 per cent in the new 2500-gallon pickup tank cars. About $2\frac{1}{2}$ gallons of rinse water are required per 10-gallon can in the total cleaning process. When cans are processed, the rinse water leaves the station in a steady discharge. For cleaning the pickup tanks, an automatic prerinse of about 200 gallons per truck is followed by a rinse of about 500 gallons, which is reused for 5 to 10 trucks before being discharged. An additional 500 gallons of cold-water rinse per truck concludes the tank cleaning. Thus, a similar volume of rinse water per gallon of milk received is used for the old-type cans and the new tank cars.

A cheese factory receives whole milk, cream, or separated milk, which is weighed, preheated, filtered, pasteurized, and cooled. A 1 to 3 per cent lactic-acid bacterial culture is sometimes used to attain the proper pH. It then is placed in cheese vats to which rennet, acid, or other souring agents are added. This causes a separation of casein in the form of a curd. The whey is then withdrawn, and the cheese washed with water. Other ingredients, such as cream, may be added (depending on the final product desired) and the cheese is shaped and packaged for sale (see Fig. 23.3).

A creamery processes whole milk, sour cream, and/or sweet cream into butter and other products (Fig. 23.4). When whole milk is received, it is centrifuged to separate the cream from the milk; the cream is churned into butter, while the separated milk may be processed into other dairy products, either for human consumption or for animal feed.

In the condensery, whole milk or other dairy products are evaporated to obtain a concentrated product. Unsweetened milk is the most important product and is produced by preheating whole milk and then evaporating and homogenizing it. Cans of evaporated milk are filled and sealed by machine, after which they are sterilized. Sweetened condensed milk is made in much the same manner, except that sugar is added. Other products include condensed nonfat milk, whey, and buttermilk.

A dry-milk plant uses atmospheric drying, vacuum drying, or spray drying to produce powdered whole or nonfat skimmed milk. The ice-cream plant, employing various formulas, combines milk and milk products with cream in certain proportions with flavorings, sugar (or other sweetening agent), and a stabilizer. The resulting mixture is homogenized, pasteurized, and cooled, after which fruits, nuts, or candies may be added, and the entire mixture frozen.

23.5 Origin and Characteristics of Dairy Wastes

Dairy wastes are made up, for the most part, of: various dilutions of whole milk, separated milk, buttermilk, and whey from accidental or intentional spills; drippings allowed to escape into the waste through inefficient design and operation of process equipment; washes containing alkaline or other chemicals used to remove milk and milk products, as well as partially caramelized materials, from cans, bottles, tanks, vats, utensils, pipes, pumps, hot wells, evaporating coils, churns, and floors; and process washes of butter, cheese, casein, and other products.

---

*Private communication to author.

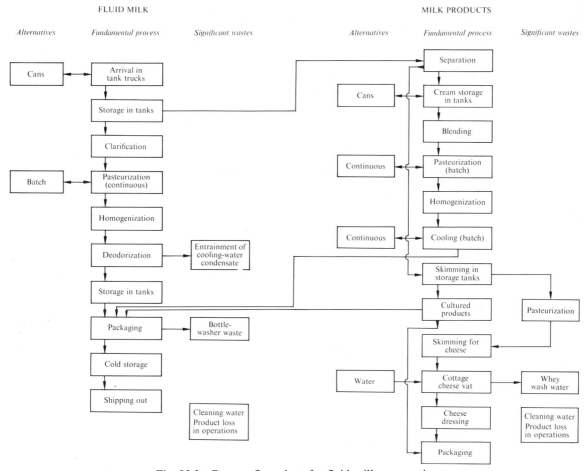

Fig. 23.2 Process flow chart for fluid-milk preparation.

Although dairy plants are found in most communities, there is considerable variation in the size of plants and in the type of product they manufacture. Table 23.7 shows the average composition of milk constituents, as reported by three different investigators. Dairy wastes are largely neutral or slightly alkaline, but have a tendency to become acid quite rapidly, because of the fermentation of milk sugar to lactic acid. Lactose in milk wastes may be converted to lactic acid when streams become devoid of oxygen, and the resulting lowered pH may cause precipitation of casein. Cheese-plant waste is decidedly acid, because of the presence of whey. Milk wastes contain very little suspended material (except the fine curd found in cheese waste) and their pollution effects are almost entirely due to the oxygen demand which they impose on the receiving stream. Heavy black sludge and strong butyric-acid odors, caused by decomposing casein, characterize milk-waste pollution. Roughly, 100 pounds of whole milk will result in about 10 pounds of BOD. The average composition of milk, milk by-product, and cheese waste is given in Table 23.8.

23.6 Treatment of Dairy Wastes

Milk-plant wastes are generally high in dissolved organic matter, contain about 1000 ppm BOD, and

Because there is a wide variation in the flow rate and strength of milk wastes, holding and equalization are desirable to provide a uniform waste for treatment. Aeration is desirable, either as a means of treatment in itself or as pretreatment before biological processes. Aeration for one day often results in 50 per cent BOD reduction and eliminates odors during conversion of the lactose to lactic acid. Higher BOD reductions (50 to 80 per cent) have been obtained by aerating milk waste in the presence of some seed material, and following this with a period of settling. High-rate recirculating filters are quite commonly used for treating milk wastes. Some two-stage filters

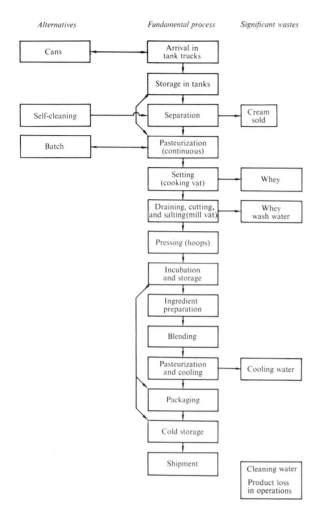

Fig. 23.3 Process flow chart for the preparation of cheese (natural and processed).

are nearly neutral in pH. Since these wastes are mainly composed of soluble organic materials, they tend, if stored, to ferment and become anaerobic and odorous. Therefore they respond ideally to treatment by biological methods. Aerobic processes are most suitable, but the final selection of a treatment method hinges on the location and size of the plant. The six conventional methods generally used which are most effective are: (1) aeration, (2) trickling filtration, (3) activated sludge, (4) irrigation, (5) lagooning, and (6) anaerobic digestion.

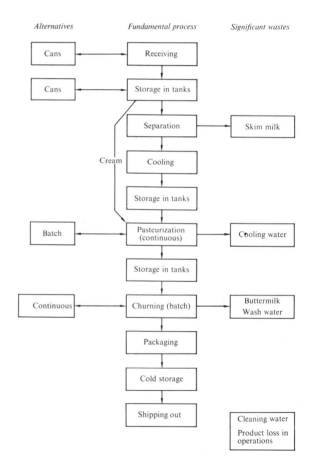

Fig. 23.4 Process flow chart for creamery butter.

23.6   TREATMENT OF DAIRY WASTES

Table 23.7 Composition of whole milk.*

Investigator	Water	Fat	Protein			Lactose	Ash
			Casein	Albumin	Total		
Roberts [101]	87.25	3.80			3.50	4.80	0.65
Eldrige [20]	87.30	3.60			3.80	4.50	0.80
Van Slyke	87.10	3.90	2.5	0.7	3.2	5.10	0.70

*Results are given as percentages.

Table 23.8 Average composition of milk, milk by-products, and cheese wastes.

Characteristics	Whole milk, ppm	Skim milk, ppm	Butter-milk, ppm	Whey, ppm	Process wastes, ppm	Separated whey, ppm
Total solids	125,000	82,300	77,500	72,000	4516	54,772
Organic solids	117,000	74,500	68,800	64,000	2698	49,612
Ash solids	8,000	7,800	8,700	8,000	1818	5,160
Fat	36,000	1,000	5,000	4,000		
Soluble solids					3956	54,656
Suspended solids					560	116
Milk sugar	45,000	46,000	43,000	44,000		
Protein (casein)	38,000	39,000	36,000	8,000		
Total organic nitrogen					73.2	1,300
Free ammonia					6.0	31
Na					807	648
Ca					112.5	350
Hg					25	78
K					116	1,000
P					59	450
BOD, 5-day	102,500	73,000	64,000	32,000	1890	30,100
Oxygen consumed	36,750	32,200	28,600	25,900		

yield greater than 90 per cent BOD reduction, while single-stage filters loaded at the rate of about one pound of BOD per cubic yard (1610 pounds of BOD per acre · foot) yield about 75 to 80 per cent BOD reduction.

The activated-sludge process has proved a successful method for the complete treatment of milk wastes. This system employs aeration to cause the accumulation of an adapted sludge. The flora and fauna in the active sludge, when supplied with sufficient air, oxidize the dissolved organic solids in the waste. Excess sludge is settled out and subsequently returned to the aeration units. There is some indication that the treatment can be carried out without wasting any sludge, although this requires an aeration period sufficient to "burn up" most of the excess sludge. Properly designed plants which provide ample air for handling the raw waste plus returned sludge are not easily upset, nor is the control procedure difficult. Operation costs, however, may be higher than those involved in trickling filtration. Both continuous-flow and batch-operated activated-sludge treatment plants have removed 90 to 97 per cent of the BOD. Generally, a tank volume of about 80 to 100 gallons is needed for each pound of BOD in the waste under treatment. Since milk wastes average about 1000 ppm BOD, the volume of returned sludge is usually 6 to 7 times the raw-milk flow.

References: Dairy Wastes

1. Agar, C. C., "Cannery, milk, and food products waste problems forum," *Sewage Ind. Wastes* **22**, 776 (1950).
2. Backmeyer, D. P., "Milk wastes in sewage sludge digestion tanks," *Sewage Ind. Wastes* **23**, 700 (1951); in Proceedings of 5th Industrial Waste Conference, Purdue University, November 1949, p. 411.
3. Backmeyer, D. P., "Condensed milk disposal by digestion," *Sewage Works J.* **21**, 1076 (1949).
4. Backmeyer, D. P., "Digestion of cheese whey, Marion, Ind.," *Sewage Ind. Wastes* **20**, 1115 (1948).
5. Backmeyer, D. P., "The effect of whey upon the operation of an activated sludge plant," in Proceedings of 3rd Industrial Waste Conference, Purdue University, May 1947.
6. Barrett, N. W., "British treatment practice," *Sewage Ind. Wastes* **11**, 560 (1939).
7. Bloodgood, D. E., "Milk waste disposal," *Sewage Works J.* **20**, 695 (1948).
8. Bloodgood, D. E., and A. J. Steffen, "Disposal of milk wastes, review," *Sewage Ind. Wastes* **20**, 707 (1948).
9. Bragstad, R. E., "Waste saving by improvements in milk plant equipment," in Proceedings of 1st Industrial Waste Conference, Purdue University, November 1944, p. 22.
10. Breska, G. J., P. R. Erikson, G. A. Rohlich, L. E. Engelbert, and N. Porges, "Objectives and procedures for a study of spray irrigation of dairy wastes," in Proceedings of 14th Industrial Waste Conference, Purdue University, May 1959.
11. Campbell, W., "The Dairy Industry Committee exhibit on waste saving," *Sewage Ind. Wastes* **23**, 700 (1955); in Proceedings of 5th Industrial Waste Conference, Purdue University, November 1949, p. 407.
12. Dairy Industry Committee, "Dairy waste prevention and treatment," *Milk Plant Monthly* **45**, 11 (1956).
13. Davis, D., "British treatment practice," *Sewage Ind. Wastes* **11**, 150 (1939).
14. Davy, P. S., "Pretreatment of milk wastes reduces treatment-plant load," *Public Works Mag.* **83**, 56 (1952).
15. Dean, W. A., "The Dairy Industry Committee's task committee on waste disposal," in Proceedings of 3rd Industrial Waste Conference, Purdue University, May 1947, p. 282.
16. De Martini, F. E., W. A. Moore, and G. E. Moore, *Food Dehydration Wastes*, U.S. Public Health Service, Supplement 191 (1946).
17. Eldridge, E. F., "Dairy waste disposal at two plants," *Sewage Works J.* **21**, 68 (1949).
18. Eldridge, E. F., "Waste prevention in the dairy industry," in Proceedings of Conference on Industrial Waste, Civil Engineering Department, University of Washington, 28 April 1949.
19. Eldridge, E. F., "Research needed in milk waste disposal," in Proceedings of 3rd Industrial Waste Conference, Purdue University, May 1947, p. 316.
20. Eldridge, E. F., *Industrial Waste Treatment Practice*, 1st ed., McGraw-Hill Book Co., New York (1942).
21. Eldridge, E. F., "Oxidized sludge and regenerative digestion, Zeeland, Mich.," *Sewage Ind. Wastes* **14**, 748 (1942).
22. Eldridge, E. F., "Bio-filtration," *Sewage Ind. Waste* **13**, 105 (1941).
23. Eldridge, E. F., "High-rate filter treatment, Perrington, Mich.," *Sewage Ind. Wastes* **12**, 658 (1940).
24. Eldridge, E. F., "Guggenheim process studies," *Sewage Ind. Wastes* **11**, 710 (1939).
25. Eldridge, E. F., "A function study of milk waste treatment processes," *Mich. State Univ. Eng. Exp. Sta. Bull.* **77** (1937).
26. Eldridge, E. F., "Trickling filter treatment, Sheridan, Mich.," *Sewage Ind. Wastes* **9**, 534 (1937).
27. Eldridge, E. F., and W. E. Zimmer, "Treatment with sewage, Mason, Mich.," *Sewage Ind. Wastes* **3**, 199 (1931).
28. Eldridge, E. F., "Treatment experiments, Michigan," *Sewage Ind. Wastes* **2**, 130 (1930), **4**, 225 (1932), and **5**, 194 (1933).
29. Forney, C., and R. R. Kountz, "Activated sludge

total oxidation metabolism," in *Proceedings of 13th Industrial Waste Conference, Purdue University, May 1958*, p. 313; *Sewage Ind. Wastes* **31**, 819 (1959).

30. Galligan, W. E., and M. Levine, "Purification of creamery waste on filters at two Iowa creameries," *Iowa State Univ. Eng. Exp. Sta. Bull.* **115**, 44 (1934).
31. Galligan, W. E., and M. Levine, "Trickling filter treatment, Slater and Glidden, Iowa," *Sewage Ind. Wastes* **6**, 1015 (1934).
32. Gaummitz, E. W., "Methods of instigating research in dairy waste under Research and Marketing Act," *Sewage Ind. Wastes* **23**, 701 (1951); in *Proceedings of 5th Industrial Waste Conference, Purdue University, November 1949*, p. 421.
33. Gloyna, E. F., "Milk waste treatment on an experimental trickling filter," *Water Sewage Works* **97**, 473 (1950).
34. Granstrom, M. L., "Techniques of research as applied to rendering plant wastes," in *Proceedings of 1st Southern Municipal and Industrial Waste Conference, March 1952*, p. 158.
35. Gurnham, C. F., "Latest developments in the treatment of dairy waste," *Ind. Wastes* **1**, 227 (1956).
36. Halvorson, H. O., and R. L. Smith, "Trickling filter treatment, Waconia, Minn.," *Sewage Ind. Wastes* **7**, 776 (1935).
37. Harding, H. G., "Dairy industry," *Sewage Ind. Wastes* **24**, 1443 (1952); "Symposium on liquid wastes," *Ind. Eng. Chem.* **44**, 487 (1952).
38. Harding, H. G., "Dairy industry," *Ind. Eng. Chem.* **44**, 487 (1952).
39. Harding, H. C., and H. A. Trebler, "Fundamentals in the control and treatment of dairy waste," *Sewage Ind. Wastes* **27**, 1369 (1955).
40. Hasfurther, W. A., and C. W. Klassen, "Aeration, milk wastes, plant description and data," *Sewage Ind. Wastes* **23**, 701 (1951); in *Proceedings of 5th Industrial Waste Conference, Purdue University, November 1949*, p. 424.
41. Hatch, B. F., and J. H. Bass, "Comparison of trickling filter, Guggenheim, Ohio, and activated sludge treatment," *Sewage Ind. Wastes* **13**, 171 (1941).
42. Hauer, G. E., "Successful treatment of dairy waste by aeration," *Sewage Ind. Wastes* **24**, 1271 (1952).
43. Hindin, E., "Determination of lactose in dairy wastes," *Sewage Ind. Wastes* **25**, 188 (1953).
44. Hoover, S. R., "Biochemical oxidation of dairy wastes, review," *Sewage Ind. Wastes* **25**, 201 (1953).
45. Hoover, S. R., *et al.*, "Aeration as partial treatment of dairy waste," in *Proceedings of 6th Industrial Waste Conference, Purdue University, February 1951*, p. 313.
46. Hoover, S. R., J. B. Pepinsky, L. Jasewicz, and N. Porges, "Aeration as partial treatment," *Sewage Ind. Wastes* **24**, 812 (1952).
47. Hoover, S. R., and N. Porges, "Dairy waste assimilation by activated sludge," *Sewage Ind. Wastes* **24**, 306 (1952).
48. Hoover, S. R., L. Jasewicz, and N. Porges, "Biochemical oxidation of dairy waste," in *Proceedings of 9th Industrial Waste Conference, Purdue University, May 1954*.
49. Hoover, S. R., L. Jasewicz, and N. Porges, "Biochemical oxidation of dairy wastes, endogenous respiration and stability of aerated dairy waste sludge," *Sewage Ind. Wastes* **24**, 1144 (1952).
50. Hoover, S. R., L. Jasewicz, and N. Porges, "Endogenous respiration and stability of aerated dairy waste," in *Proceedings of 7th Industrial Waste Conference, Purdue University, May 1952*, p. 541.
51. Hoover, S. R., L. Jasewicz, J. B. Pepinsky, and N. Porges, "Activated sludge assimilation," *Sewage Ind. Wastes* **23**, 167 (1951).
52. Horton, J. P., and H. A. Trebler, "Recent developments in design of small milk waste disposal plant," *Sewage Ind. Wastes* **25**, 941 (1953).
53. Horton, J. P., and H. A. Trebler, "Recent developments in the design of small milk waste disposal plants," in *Proceedings of 8th Industrial Waste Conference, Purdue University, May 1953*.
54. "Imhoff tank operation—milk wastes," *Sewage Ind. Wastes* **7**, 330 (1935).
55. *Industrial Waste Guide to the Milk Processing Industry*, U.S. Public Health Service Bulletin no. 289, revised 1959.
56. U.S. Public Health Service, *Industrial Waste Guide to the Milk Processing Industry*, (1953).
57. Jasewicz, L., and N. Porges, "Aeration of wastes, nitrogen supplement and sludge oxidation," *Sewage Ind. Wastes* **30**, 555 (1958).
58. Jasewicz, L., and N. Porges, "Biochemical oxidation of dairy wastes," *Sewage Ind. Wastes* **28**, 1130 (1956).
59. Jasewicz, L., N. Porges, and S. R. Hoover, "Borax as a preservative of dairy wastes for BOD test," in *Proceedings of 8th Industrial Waste Conference, Purdue University, May 1953*.

60. Jenkins, S. H., "The treatment of waste waters from dairies," Public Works, Roads, and Transport Congress, Paper no. 22, 1937.
61. Jenkins, S. H., "British treatment studies," *Sewage Ind. Wastes* **9**, 702 (1937).
62. Jens and Heinsmann, "Dairy waste treatment in fish ponds," *Dairy Sci. Abstr.* **21**, 3044 (1959).
63. Johnson, W. S., "An aeration plant for milk waste disposal," *Sewage Ind. Wastes* **22**, 1093 (1950); in Proceedings of 4th Industrial Waste Conference, Purdue University, September 1948, p. 54.
64. Kessener, H., and W. Rudolfs, "Activated sludge treatment," *Sewage Ind. Wastes* **6**, 318 (1934).
65. Kimberly, A. E., "Treatment and disposal practice, Ohio," *Sewage Ind. Wastes* **2**, 303 (1930).
66. Klein, A. H. V., "Dairy wastes," *Sewage Ind. Wastes* **30**, 1393 (1958).
67. Kountz, R. R., "Aerobic digestion of dairy wastes," in Proceedings of 5th Southern Municipal and Industrial Waste Conference, April 1956, p. 238.
68. Kountz, R. R., "Dairy waste treatment pilot plant," in Proceedings of 8th Industrial Waste Conference, Purdue University, May 1953.
69. Lane, L. C., "Disposal of liquid and solid wastes by means of spray irrigation in the canning and dairy industries," in Proceedings of 10th Industrial Waste Conference, Purdue University, May 1955.
70. Lawton, G. W., G. Breska, L. E. Engelbert, G. A. Rohlich, and N. Porges, "Spray irrigation of dairy wastes," *Sewage Ind. Wastes* **31**, 923 (1959).
71. Levine, M., "Trickling filter studies," *Sewage Ind. Wastes* **7**, 316 (1935).
72. Levine, M., "Biological treatment," *Sewage Ind. Wastes* **4**, 322 (1932).
73. Levine, M., "Septic tank treatment," *Sewage Ind. Wastes* **2**, 303 (1930).
74. Levine, M., G. W. Burke, and J. H. Watkins, "Lath filter experiments," *Sewage Ind. Wastes* **2**, 469 (1930).
75. Levine, M., G. H. Nelson, and H. E. Goresline, "Trickling filter studies," *Sewage Ind. Wastes* **8**, 691 (1936).
76. Levine, M., and L. Soppeland, "Proteolysis by bacteria from creamery waste," *Iowa State Univ. Eng. Exp. Sta. Bull.* **82** (1926).
77. Levine, M., L. Soppeland and G. W. Borke, "Aeration studies on creamery waste purification," *Iowa State Univ. Eng. Exp. Sta. Bull.* **88** (1923).
78. McKee, F. J., "Dairy waste disposal by spray irrigation," *Sewage Ind. Wastes* **29**, 157 (1957).
79. McKee, F. J., "Spray irrigation for the disposal of milk wastes," in Proceedings of 3rd Ontario Industrial Waste Conference, June 1956, p. 17.
80. McKee, F. J., "Spray irrigation of dairy waste," in Proceedings of 10th Industrial Waste Conference, Purdue University, May 1955.
81. McKee, F. J., "New items in milk waste saving and treatment," in Proceedings of 8th Industrial Waste Conference, Purdue University, May 1953.
82. McKee, F. J., "Milk waste treatment by aeration," *Sewage Ind. Wastes* **22**, 1041 (1950).
83. Maloney, T. E., H. F. Ludwig, J. A. Harmon, and L. McClintock, "Effects of whey wastes on stabilization ponds," *J. Water Pollution Control Federation* **32**, 1283 (1960).
84. Manus, L. J., "Should dairy waste be converted into by-products or discharged to sewers?" *Wastes Eng.* **28**, 342 (1957).
85. "Milk processing industry waste guide," *Sewage Ind. Wastes* **26**, 299 (1954).
86. U.S. Public Health Service, *Milk Wastes*, Ohio River Survey, Supplement D (1943), p. 1162.
87. Minotto, J., "Chlorination and land disposal, Phoenix, Ariz.," *Sewage Ind. Wastes* **2**, 393 (1930).
88. Montagna, S. D., "Activated sludge treatment, Somerset, Pa.," *Sewage Ind. Wastes* **12**, 108 (1940).
89. Morgan, P. E., and E. R. Baumann, "Trickling filters successfully treat milk waste," *J. Sanit. Eng. Div. Am. Soc. Civil Engrs.* **83**, 1336 (1957).
90. Murdock, H. R., "Industrial wastes," *Ind. Eng. Chem.* **44**, 99 (1952).
91. Naack, K., "A hygienic problem of agricultural utilization of dairy wastes," *Sewage Ind. Wastes* **26**, 1510 (1954).
92. Neill, D. G., "Waste treatment facilities of the Belle Center Creamery and Cheese Company," *Sewage Ind. Wastes* **22**, 1093 (1950); in Proceedings of 4th Industrial Waste Conference, Purdue University, September 1948, p. 45.
93. Nelson, G. H., M. Levine, and H. E. Goresline, "Effects of nature of filling material and dosing cycle on purification of creamery waste," *Iowa State Univ. Eng. Exp. Sta. Bull.* **124** (1935).
94. Oeming, L. F., "Trickling filter treatment, Ovid, Mich.," *Sewage Ind. Wastes* **20**, 512 (1948).
95. Porges, N., "Bio-oxidative treatment of dairy wastes," in Proceedings of 3rd Ontario Industrial

Waste Conference, June 1956, p. 55.
96. Porges, N., and S. R. Hoover, "Treatment of dairy wastes by aeration, methods of study," *Sewage Ind. Wastes* **23**, 694 (1951); in Proceedings of 5th Industrial Waste Conference, Purdue University, November 1949, p. 213.
97. Porges, N., and L. Jasewicz, "Aeration of whey wastes, a COD and solids balance," *Sewage Ind. Wastes* **31**, 443 (1959).
98. Porges, N., J. B. Pepinsky, N. C. Hendler, and S. R. Hoover, "Biochemical oxidation of dairy wastes, corporation study of yeasts," *Sewage Ind. Wastes* **22**, 888 (1950).
99. Porges, N., J. B. Pepinsky, N. C. Hendler, and S. R. Hoover, "Biochemical oxidation of dairy wastes, methods of study," *Sewage Ind. Wastes* **22**, 318 (1950).
100. Porges, N., J. B. Pepinsky, L. Jasewicz, and S. R. Hoover, "Biochemical oxidation of dairy wastes, failure of $NaNO_3$ as a source of oxygen," *Sewage Ind. Wastes* **24**, 874 (1952).
101. Roberts, C. R., "Progress in milk waste disposal," *Sewage Works J.* **8**, 489 (1936).
102. Rower, "Treatment of dairy waste by aeration," *Sanitalk* **3**, 2 (1955).
103. Ruf, H., and L. F. Warrick, "Trickling filter studies," *Sewage Ind. Wastes* **6**, 580 (1934).
104. Rugaber, J. W., "Milk losses," in Proceedings of 7th Industrial Waste Conference, Purdue University, May 1952.
105. Rugaber, J. W., "Operation of a milk waste treatment plant employing a trickling filter," *Sewage Ind. Wastes* **23**, 1425 (1951).
106. Saal, H., "New whey disposal system," *Milk Plant Monthly* **21**, 32 (1959).
107. Sanborn, N. H., "Disposal of food processing wastes by spray irrigation," *Sewage Ind. Wastes* **25**, 1034 (1953).
108. Sanders, M. D., "Industrial sanitary engineering in dairy plant management," *Sewage Ind. Wastes* **22**, 1093 (1950); in Proceedings of 4th Industrial Waste Conference, Purdue University, September 1948, p. 63.
109. Sanders, M. D., "What can be done with whey?" in Proceedings of 3rd Industrial Waste Conference, Purdue University, May 1947.
110. Schoepfer, G., and N. Ziemke, "Development of the anaerobic contact process," *Sewage Ind. Wastes* **31**, 164 and 697 (1959).
111. Schanfama, J. H. A., "Purification and utilization of dairy wastes by overhead irrigation of meadows and arable lands," *Sewage Ind. Wastes* **26**, 926 (1954).
112. Schaufnagel, F. H., "Dairy waste disposal by ridge and furrow irrigation," in Proceedings of 12th Industrial Waste Conference, Purdue University, May 1957.
113. Schulze, K. L., "New developments in dairy waste treatment," in Proceedings of 9th Industrial Waste Conference, Purdue University, May 1954.
114. Schulze, K. L., "Experience with a new type of dairy waste treatment," in Proceedings of 3rd Annual Dairy Engineering Conference, Michigan State University, March 1955.
115. Schwarzkopf, V., "Reduction of milk waste in dairy plants," *Sewage Ind. Wastes* **24**, 1489 (1952).
116. Siebert, C. L., "Pennsylvania milk waste treatment methods," *Sewage Ind. Wastes* **29**, 1038 (1957).
117. Silvester, D., "Some experiences in the disposal of milk wastes," *J. Soc. Dairy Technol.* **12**, 228 (1959).
118. Southgate, B. A., "Treatment and disposal of waste waters from dairies," *Sewage Ind. Wastes* **27**, 1112 (1955); *Dairy Sci. Abstr.* **16**, 428 (1954).
119. Southgate, B. A., "Waste disposal in Britain," *Ind. Eng. Chem.* **44**, 524 (1952).
120. Southgate, B. A., "British research on treatment methods," *Sewage Ind. Wastes* **14**, 917 (1942) and **20**, 951 (1948).
121. Spaulding, A. R., "The anaerobic digestion rates of milk wastes," *Sewage Ind. Wastes* **22**, 1092 (1950).
122. Steel, E. W., and P. J. A. Zeller, "Texas treatment, lath filters and chemical treatment," *Sewage Ind. Wastes* **2**, 305 (1930).
123. Steffen, A. J., "Some recent developments in dairy waste research," in Proceedings of 12th Industrial Waste Conference, Purdue University, May 1957.
124. Sutermeister, E., *et al., Casein and Its Industrial Applications*, 2nd ed., Reinhold Publishing Corp., New York (1959).
125. "Symposium on milk wastes," in Proceedings of 1st Industrial Waste Conference, Purdue University, November 1944.
126. Thayer, P. M., "Milk waste treatment by activated sludge," *Water Sewage Works* **100**, 34 (1953).
127. Thayer, P. M., "Activated sludge plant, new design, milk wastes," *Sewage Ind. Wastes* **23**, 1537 (1951) and **24**, 507 (1952); in Proceedings of 6th Industrial

Waste Conference, Purdue University, February 1951, p. 171.
128. Thayer, P. M., "Design features of activated sludge plants for treatment of milk waste," in Proceedings of 7th Industrial Waste Conference, Purdue University, May 1952, p. 509.
129. Thayer, P. M., "Design for activated sludge plant to treat milk waste," in Proceedings of 6th Industrial Waste Conference, Purdue University, February 1951.
130. "Treatment plant, Beltsville, Md., research center," *Sewage Ind. Wastes* **8**, 157 (1936).
131. Trebler, H. A., "Waste prevention," *Sewage Ind. Wastes* **18**, 759 (1946).
132. Trebler, H. A., "Experience and questions concerning aeration methods as applied to dairy waste," in Proceedings of 10th Industrial Waste Conference, Purdue University, May 1955.
133. Trebler, H. A., R. P. Ernsberger, and C. T. Roland, "Waste prevention and treatment," *Sewage Ind. Wastes* **10**, 868 (1938).
134. Trebler, H. A., and H. G. Harding, "Fundamentals of the control and treatment of dairy waste," *Sewage Ind. Wastes* **27**, 1369 (1955).
135. Trebler, H. A., and H. G. Harding, "Two-stage, high-rate trickling filters for dairy waste treatment," *Sewage Ind. Wastes* **22**, 1093 (1950); in Proceedings of 4th Industrial Waste Conference, Purdue University, September 1948, p. 67.
136. Trebler, H. A., and H. G. Harding, "Milk waste problem, review," *Sewage Ind. Wastes* **20**, 594 (1948).
137. Walker, C. L., "Treatment and disposal," *Sewage Ind. Wastes* **2**, 123 (1930).
138. Walzholz, Pester, Lembke, and Schmidt, "Dairy waste treatment with aquapura ferrobion trickling filter process, and aero-accelerator," *Dairy Sci. Abstr.* **21**, 1304 (1959).
139. Warrick, L. F., and E. J. Beatty, "Treatment with domestic sewage," *Sewage Ind. Wastes* **8**, 122 (1936).
140. Wasserman, A. E., W. J. Hopkins, and N. Porges, "Rapid conversion of whey to yeast," in Proceedings of 15th International Dairy Conference, London, 1959, part 2, p. 1241.
141. Wasserman, A. E., W. J. Hopkins, and N. Porges, "Whey utilization—growth conditions for *saccharomyces fragilis*," *Sewage Ind. Wastes* **30**, 913 (1958).
142. Wasserman, A. E., and N. Porges, "Whey utilization: summary of laboratory investigations in yeast propagation," in Proceedings of 14th Industrial Waste Conference, Purdue University, May 1959.
143. Watson, C. W., Jr., "Practical aspects of dairy waste treatment," in Proceedings of 15th Industrial Waste Conference, Purdue University, May 1960, p. 81.
144. Watson, C. W., Jr., "In-plant control of dairy wastes," *Sewage Ind. Wastes* **27**, 55 (1955).
145. Wheatland, A. B., "Treatment of waste waters from milk-products factories," in *Waste Treatment*, Pergamon Press, London (1959).
146. Wisely, W. H., "Treatment of sewage containing milk and brewery wastes," *Sewage Ind. Wastes* **8**, 624 (1936).
147. Wittmer, E. F., "Trickling filtration, alternating two-stage, Marysville, Ohio," *Sewage Ind. Wastes* **20**, 846 (1948).
148. Woolings, N. D., "Milk wastes," *Can, Milk, Ice Cream J.* **34**, 27 (1953).
149. Woolings, N. D., "Treatment of milk wastes," *Munic. Utilities* **90**, 25 (1952).
150. "Workshop on milk wastes," in Proceedings of 2nd Industrial Waste Conference, Purdue University, January 1946.
151. Wright, E., "Waste control, Boston Metropolitan District Commission," *Sewage Ind. Wastes* **3**, 287 (1931).
152. Zack, S. I., "Trickling filter performance on wastes at two milk processing plants," *Sewage Ind. Wastes* **28**, 1009 (1956).
153. Zack, S. I., "Treatment and disposal of milk wastes," *Sewage Ind. Wastes* **25**, 177 (1953).

Suggested Additional Reading

The following recent references have been published since 1962.
1. Adamse, A. D., *Bacteriological Studies on Dairy Waste Activated Sludge*, D. Agric. Thesis, Agricultural University, Wageninge, Netherlands (1966).
2. Anders, H., "Water and waste waters in dairy plants," *Molkerei-Ztg.* **15**, 419 (1961).
3. Allen, L. A., "Treatment and disposal of effluents with special reference to the dairy industry. I: Pollution and legal requirements," *Dairy Ind.* **29**, 90 (1964); *Chem. Abstr.* **60**, 15582 (1964).
4. Ammon, F. V., "Milk trade waste waters and their treatment," *Ber. Abwassertech. Verein* **11**, 236 (1960); *Water Pollution Abstr.* **34**, 2115 (1961).
5. Anderson, M. E., and H. A. Morris, "This problem

of waste disposal. I: Spray irrigation systems. II: Lagoon and trickling filter disposal systems. III: Municipal waste disposal systems," *Mfd. Milk Prod. J.* **57**, 8 (1966); *Public Health Eng. Abstr.* **47**, 250 (1967); *J. Water Pollution Control Federation* **39**, 868 (1967).

6. "Dairy waste disposal," *Svensk. Mejeritidu* **52**, 492 (1960).
7. "Spray drying adapted for high acid whey," *Chem. Eng. News* **39**, 66 (1961).
8. "Water and waste problems of the dairy industry," *Molkerei-Ztg.* **15**, 905 (1961).
9. "Whey disposal by heat coagulation," *Ind. Wastes* **5**, 95 (1960).
10. Baltjes, J., "Irrigation with dairy waste water," *Misset's Zuivel* **70**, 348 (1964); *Dairy Sci. Abstr.* **25**, 1883 (1964).
11. Bergman, T., A. Magnusson, A. Berglof, and K. Joost, "Influence of a standstill on the quantity and composition of dairy waste water," in Proceedings of 17th International Dairy Congress, Munich, 1966, p. 753; *Dairy Sci. Abstr.* **28**, 3448 (1966); *J. Water Pollution Control Federation* **39**, 868 (1967).
12. Bloodgood, D. E., "Twenty years of industrial waste treatment," Purdue University Engineering Extension Series, Bulletin no. 118, 1965, p. 182.
13. Carl, C. E., "Waste treatment by stabilization ponds," *J. Milk Food Technol.* **24**, 147 (1961); *Public Health Eng. Abstr.* **41**, 35 (1961).
14. Clews, J., "Wastes from milk processing," *Compost Sci.* **2**, 42 (1961).
15. Cojocaru, R., "Fat separators in the inner sewage of food industry units," *Ind. Aliment.* **15**, 121 (1964); *Chem. Abstr.* **61**, 10436 (1964).
16. Colvin, C. H., "Liquid manure disposal systems on dairy farms," New York State Association Milk and Food Sanitarians, 40th Annual Report, 1966, p. 171; *J. Water Pollution Control Federation* **39**, 868 (1967).
17. Dendy, P., "Rural wastes," *Water Waste Treat. J.* **8**, 390 (1961); *Public Health Eng. Abstr.* **41**, 34 (1961).
18. Dieterich, K. R., "Waste water problems in the dairy industry," *Molkerei-Ztg.* **15**, 263 (1961).
19. Fairall, J. M., "Biological treatment of certain organic industrial wastes," in Proceedings of 10th Ontario Industrial Waste Conference, June 1963, p. 113.
20. Farrall, A. W., *Engineering for Dairy and Food Products*, John Wiley & Sons, New York (1963).
21. Fritz, A., "Determination of water pollution by dairy effluent," *Milchwissenschaft* **15**, (1960); *Dairy Sci. Abstr.* **23**, 272 (1961).
22. Froebrich, G., "Biological sewage-treatment plant of the cheese plant of the Kraft Company in Schwabmuenchen," *Ber. Abwassertech.-Verein* **11**, 247 (1960); *Chem. Abstr.* **60**, 3857 (1964).
23. Frobrich, G., "Operation report on the biological treatment plant at the cheese factory of the firm of Kraft in Schwabmunchen," *Ber. Abwassertech.-Verein* **11**, 247 (1960); *Water Pollution Abstr.* **34**, 388 (1961).
24. Fuerhoff, W., "Experiment on the biological treatment of dairy wastes," *Vom Wasser* **28**, 430 (1964); *J. Water Pollution Control Federation* **37**, 740 (1965).
25. Gillar, J., and P. Marvan, "First biological studies of a single-stage fermentation method for the purification of dairy effluent," *Prumysl Potravin* **17**, 328 (1966); *Dairy Sci. Abstr.* **28**, 3086 (1966).
26. Gillar, J., J. Salplachta, and Halvka, "Single stage fermentation of waste waters at dairies having new purification plants," *Prumysl Potravin* **18**, 341 (1967).
27. Gockel, H., "Waste water and sludge clarification," *Vom Wasser* **28**, 94 (1961); *Chem. Abstr.* **60**, 15581 (1964).
28. Gotz, A., "The treatment of dairy waste waters in an oxidation channel," *Ber. Abwassertech.-Verein* **11**, 254 (1960); *Water Pollution Abstr.* **34**, 388 (1961).
29. Grewis, O. E., and C. A. Burkett, "Two-thousand town treats twenty-thousand waste," *Water Wastes Eng.* **3**, 54 (1966); *J. Water Pollution Control Federation* **39**, 868 (1967).
30. Gromski, L., "Technical and economic aspects of using activated sludge in purification of small amounts of sewage," *Gaz, Woda Tech. Sanit.* **39**, 150 (1965); *Chem. Abstr.* **63**, 1286 (1965); *J. Water Pollution Control Federation* **38**, 872 (1966).
31. Groves, F. W., and T. F. Graf, "An economic analysis of whey utilization and disposal in Wisconsin," University of Wisconsin, College of Agriculture, *Agr. Econ.* (1965), p. 44; *J. Water Pollution Control Federation* **38**, 872 (1966).
32. Guillaume, F., "Treatment of industrial wastes by means of the oxidation ditch," in Proceedings of 12th Ontario Industrial Waste Conference, June 1965, p. 59; *Chem. Abstr.* **66**, 7452 (1967).
33. Hanrohen, F. P., and B. H. Webb, "Spray drying cottage cheese whey," *J. Dairy Sci.* **44**, 1171 (1961).
34. Harding, H. G., "Milk waste treatment," in Proceed-

ings of 8th Ontario Industrial Waste Conference, June 1961, p. 67.
35. Hlavka, M., "Effect of detergents on the treatment of sewage waters of dairies," *Prumysl Potravin* **17**, 558 (1966); *Chem. Abstr.* **66**, 1360 (1967).
36. Ingram, W. T., "Trickling filter treatment of whey wastes," *J. Water Pollution Control Federation* **33**, 844 (1961).
37. Johansson, S., and A. B. Buhrgard, "Separation of butter wash water," in Proceedings of 17th International Dairy Congress, Munich, 1966, p. 757; *Dairy Sci. Abstr.* **28**, 3447 (1966); *J. Water Pollution Control Federation* **39**, 868 (1967).
38. Kiss, P., and J. Timar, "Problems of the purification of dairy effluent," *Tejipar* **12**, 6 (1963); *Dairy Sci. Abstr.* **27**, 723 (1965); *J. Water Pollution Control Federation* **38**, 872 (1966).
39. Kosman, J. L., "New York State control of milk plant waste disposal," in New York State Association of Milk and Food Sanitarians, 39th Annual Report, 1965, p. 115; *J. Water Pollution Control Federation* **38**, 872 (1966).
40. Kurian, B. A., "An aerobic biological treatment of dairy wastes processes," *Indian Dairy Sci.* **50**, 39 (1965).
41. Laurie, K. W., and K. E. Flynn, "Effluent disposal trials at Leongatha South," *Dairy Eng. Dig.* **9**, 35 (1962); *Dairy Sci. Abstr.* **24**, 2482 (1962).
42. Lebedeva, K., and A. Mironov, "New whey product," *Mul. Prom.* **22**, 15 (1961); *Dairy Sci Abstr.* **23**, 587 (1961).
43. Leighton, F. R., "The challenge to and the changing role of chemists in the dairy industry," *Australian J. Dairy Technol.* **19**, 43 (1964); *J. Water Pollution Control Federation* **37**, 740 (1965).
44. "Lightning aerator solves waste treatment problem at milk processing plant," *Water Sewage Works* **114**, 143 (1967).
45. Loehr, R. C., and J. A. Rut, "Anaerobic lagoon treatment of milking-parlor wastes," *J. Water Pollution Control Federation* **40**, 83 (1968).
46. Magnusson, F., "Disposal and purification of dairy effluents in the Netherlands and Great Britain (Mejeriavloppsvattnets Avledning och rening i Holland och Storbritannien)," *Svensk. Mejeritidu* **52**, 455 (1960); *Dairy Sci Abstr.* **23**, 993 (1961).
47. Maloney, T. E., et al., "Effect of whey wastes on stabilization ponds," *J. Water Pollution Control Federation* **32**, 1283 (1960).
48. McKee, F. J., "Dairy waste disposal," *Mfd. Milk Prod. J.* **56**, 10 (1965); *Public Health Eng. Abstr.* **46**, 140 (1966); *J. Water Pollution Control Federation* **38**, 872 (1966).
49. Meinecke, K., et al., "Purification of dairy sewage in the oxidation trench of Holzthaleben," *Wasserwirtsch.-Wassertech.* **11**, 336 (1961).
50. "Milk wastes cause odors in ponds," *Wastes Eng.* **V–34**, 25 (1963).
51. Mishukov, B. G., "Performance of a low-pressure aeration tank for biochemical clarification of dairy waste waters," *Sanit. Teckhn.*, 1964, p. 26; *Chem. Abstr.* **62**, 12889 (1965); *J. Water Pollution Control Federation* **38**, 872 (1966).
52. Mishukov, B. G., "Purification of wastes from milk processing enterprises," *Sb. Tr. Leningr. Inzh.-Stroct. Inst.* **47**, 72 (1964); *Chem. Abstr.* **63**, 2737 (1965); *J. Water Pollution Control Federation* **38**, 872 (1966).
53. Mitchell, W. D., and N. G. Cassidy, "Utilization of casein wastes for pasture irrigation," in Proceedings of 17th International Dairy Congress, Munich, 1966, p. 745; *Dairy Sci. Abstr.* **28**, 3453 (1966); *J. Water Pollution Control Federation* **39**, 868 (1967).
54. Muers, M. M., "Dairy water conservation and effluent disposal," *J. Soc. Dairy Technol.* **18**, 3 (1965); *Dairy Sci. Abstr.* **27**, 1089 (1965); *J. Water Pollution Control Federation* **38**, 874 (1966).
55. Nehrkorn, A., and H. Reploh, "Experiences obtained with oxidation trenches in the purification of dairy effluent," *Gesundh. Ingr.* **87**, 143 (1966); *Dairy Sci. Abstr.* **29**, 98 (1967).
56. Nehrkorn, A., and H. Reploh, "Experience with oxidation trenches, a contribution to dairy effluent hygiene," *Arch. Hyg. Bakteriol.* **150**, 237 (1966); *Dairy Sci. Abstr.* **29**, 97 (1967); *J. Water Pollution Control Federation* **39**, 868 (1967).
57. Neil, J. H., "The use and construction of oxidation ponds for industrial waste treatment," in Proceedings of 7th Ontario Industrial Waste Conference, June 1960, p. 84.
58. Norman, J. D., and A. Busch, "Application of research information to design and operation of industrial waste facilities," in Proceedings of 13th Ontario Industrial Waste Conference, June 1966, p. 157.
59. Palmer, L. M., "What's new in manure disposal," *Eng. Index*, 1964, p. 913.

60. Parker, R. R., "Spray irrigation for industrial wastes," *Can. Munic. Utilities* **103**, 28 (1965); *Eng. Index*, 1966, p. 1346.
61. Petrov, P. I., and V. I. Krechetov, "Reagents conditions in electroflotation clarification of waste waters for a buttermaking plant," in Proceedings of 2nd Scientific and Technical Conference at the Kishinev Technical Institute, 1966, p. 295; *Chem. Abstr.* **67**, 6358 (1967).
62. Reynolds, D. J., "Methods of estimating the 'strength' of dairy effluents," in Proceedings of 17th International Dairy Congress, Munich, 1966, p. 775; *Dairy Sci. Abstr.* **28**, 3454 (1966); *J. Water Pollution Control Federation* **39**, 868 (1967).
63. Rinn, M., "Detoxication and neutralization of corrosive effluent in dairies, in particular of residues from butyrometers," *Molkerei-Ztg.* **86**, 107 (1965); *Dairy Sci. Abstr.* **27**, 1061 (1965); *J. Water Pollution Control Federation* **38**, 872 (1966).
64. Rook, J. A. F., and J. E. Storry, "Energy and nutrition and milk secretion in the dairy cow," *Chem. Inds.*, 24 October 1964, p. 1778.
65. Salplachtec, J., C. M. Hlanka, M. Svoboda, and D. Stelcova, "The one-step fermentation of waste waters as viewed from dairying practices," *Inform. Publ. Strediske Tech. Inform. Petravinar.* **224**, 89 (1962); *Chem. Abstr.* **60**, 6609 (1964).
66. Salvato, J. A., and F. O. Bogedain, "The disposal of liquid wastes and dairy farms, receiving stations and small plants," in New York State Association of Milk and Food Sanitarians, 40th Annual Report, 1966, p. 174; *J. Water Pollution Control Federation* **39**, 870 (1967).
67. Schropp, N., "Waste water problems in cheese plants," *Molkerei-Ztg.* **18**, 349 (1964); *J. Water Pollution Control Federation* **37**, 740 (1965).
68. Schropp, W., and K. F. Vogt, "Utilization of dairy waste in agriculture," in Proceedings of 17th International Dairy Congress, Munich, 1966, p. 735; *Dairy Sci. Abstr.* **28**, 3452 (1966); *J. Water Pollution Control Federation* **39**, 868 (1967).
69. Scott, R. H., "Land disposal of industrial wastes," *Oregon State Coll. Eng. Exp. Sta. Circ.* **29**, 261 (1963); *Eng. Index*, 1964, p. 912.
70. Schulz-Falkenhain, H., "Standards for costing effluent disposal," in Proceedings of 17th International Dairy Congress, Munich, 1966, p. 763; *Dairy Sci. Abstr.* **28**, 3445 (1966); *J. Water Pollution Control Federation* **39**, 868 (1967).
71. Sharratt, J., "Whey, its effect on soil and plant growth," University of Wisconsin, University Microfilms, Order no. 61–316 1; *Dissertation Abstr.* **22**, 691 (1961).
72. Shifrin, S. M., and V. N. Burtsev, "Clarification of sewage from cheese processing plants," Sanit. Tekh., Dokl. Nauch. 23rd. Konf. Leningrad. Inzh. Stroit Inst., Leningrad, (1965), p. 69; *Chem. Abstr.* **66**, 2107 (1967).
73. Shifrin, S. M., and Ya. A. Burtsev, "Clarification of sewage from the Leningrad Dairy Products Plant on high-load biological filters," Sanit. Tekh., Dokl. Nauch. 23rd. Konf. Leningrad. Inzh. Stroit. Inst., Leningrad, (1965), p. 80; *Chem. Abstr.* **66**, 9485 (1967).
74. Shifrin, S. M., and V. Burtsev, "Composition of the waste water from a cheese plant," *Molochn. Prom.* **27**, 12 (1965); *Chem. Abstr.* **63**, 17575 (1965); *J. Water Pollution Control Federation* **38**, 872 (1966).
75. Shifrin, S. M., and V. P. Burtsev, "Treatment of effluents from cheese making plants," *Vodosnabzh. Sanit. Tekhn.*, 1965, p. 30; *Chem. Abstr.* **64**, 9419 (1966).
76. Siddiqi, R. H., R. S. Engelbrecht, and R. E. Speece, "The role of enzymes in the contact stabilization process," in *Advances in Water Pollution Research*, Proceedings of 3rd International Conference on Water Pollution Research, Washington, D.C., Vol. 2, 1967, p. 353; *J. Water Pollution Control Federation* **39**, 868 (1967).
77. Sigworth, E. A., "Potentiality of active carbon in treatment of industrial wastes," *Eng. Index*, 1964, p. 912; Proceedings of 10th Ontario Industrial Waste Conference, June 1963, p. 177.
78. Sladka, A., "The biocenosis of a trickling filter," *Vodni Hospodarstvi* **15**, 263 (1965); *Chem. Abstr.* **63**, 17675 (1965); *J. Water Pollution Control Federation* **38**, 872 (1966).
79. Svoboda, M., *et al.*, "Final purification of dairy effluent in high trickling filter," *Sb. Praci Vyzk. Ust. Mlek.* 1965, p. 130; *Dairy Sci. Abstr.* **27**, 3054 (1965).
80. Svoboda, M., J. Salplachta, M. Hlavka, and D. Stelcova, "Further experiments with single stage fermentation of dairy effluent," *Prumysl Potravin* V-**14**, 193 (1963); *Dairy Sci. Abstr.* **25**, 2232 (1963).
81. Svoboda, M., *et al.*, "Purification of dairy waste by means of biological tower filter plant," in Proceedings of 17th International Dairy Congress, Munich, 1966,

p. 723; *Dairy Sci. Abstr.* **28**, 3450 (1966); *J. Water Pollution Control Federation* **38**, 868 (1967).
82. Svoboda, M., et al., "Purification of dairy waste waters by means of lagoons," in Proceedings of 17th International Dairy Congress, Munich, 1966, p. 715; *Dairy Sci. Abstr.* **28**, 3451 (1966); *J. Water Pollution Control Federation* **39**, 868 (1967).
83. Svoboda, M., Gillar, Jr., J. Salplachta, G. Hlavka, D. Stelclova, and P. Marvan, "Secondary stage treatment of dairy waste water on deep trickling filter," *Vodni Hospodarstvi* **14**, 219 (1964); *Chem. Abstr.* **61**, 13034 (1964).
84. Svoboda, M., et al., "Single stage fermentation of dairy waste water," *Chem. Abstr.* **55**, 23882 (1961).
85. Svoboda, M., "Utilization of dairy effluent in agriculture," *Dairy Sci. Abstr.* **23**, 426 (1961).
86. Standeven, W. E., "Milk plant waste disposal," in New York State Association of Milk and Food Sanitarians, 39th Annual Report, 1965, p. 111; *J. Water Pollution Control Federation* **38**, 872 (1966).
87. Strouse, D. C., "Waste stabilization ponds in Iowa—a survey," *Water Works Waste Eng.* **1**, 52 (1964).
88. Thom, R., "The determination of the pollution load in dairy waste water with the dichromate method of Eldridge," in Proceedings of 17th International Dairy Congress, Munich, 1966, p. 781; *J. Water Pollution Control Federation* **39**, 868 (1967).
89. Thom, R., et al., "Purification of dairy wastes by the activated sludge method," *Acta Microbiol. Polon.* **8**, 175 (1959); *Dairy Sci. Abstr.* **23**, 166 (1961).
90. Thom, R., "Treatment of dairy waste waters by the activated sludge method with large bubble aeration," in Proceedings of 17th International Dairy Congress, Munich, 1966, p. 709; *Dairy Sci. Abstr.* **28**, 3449 (1966); *J. Water Pollution Control Federation* **39**, 868 (1967).
91. Tookos, I., and L. Tiefenbach, "Analysis of dairy waste waters," *Elelm. Ipar* **18**, 171 (1964); *Dairy Sci. Abstr.* **27**, 1060 (1965); *J. Water Pollution Control Federation* **38**, 872 (1966).
92. Tookos, I., "Experiments on the purification of dairy effluent," *Wasserwirtsch.-Wassertech.* **16**, 133 (1966); *Chem. Abstr.* **65**, 11969 (1966).
93. Tookos, I., "Model experiments for the purification of dairy effluents by aeration," *Elelm. Ipar* **19**, 367 (1965); *Dairy Sci. Abstr.* **28**, 1125 (1966); *J. Water Pollution Control Federation* **39**, 868 (1967).
94. "Treating milk wastes in deep trickling filters," *Wastes Eng.* **33**, 28 (1962).
95. "Treatment of dairy effluent by the activated sludge method," *Dairy Sci. Abstr.* **25**, 2231 (1963).
96. Van Luven, A. L., "Food industry waste problems in Quebec, Canada," *Eng. Index*, 1964, p. 913; *Ind. Water Wastes V* **8**, 22 (1963).
97. Van Luven, A. L., "Some considerations of food industry wastes problems in Quebec," *Eng. Index*, 1964, p. 913; Proceedings of Ontario Water Resources Commission 10th Industrial Waste Conference, June 1963, p. 91.
98. Viehl, K., "The waste water problem in dairy plants," *Molkerei-Kaserei-Ztg.* **12**, 369 (1961).
99. Vrignaud, Y., "Purification of effluent from cheese factories. The Fromageries Bel process," *Tech. Lait* 1964, p. 40; *Public Health Eng. Abstr.* **45**, 1556 (1965); *J. Water Pollution Control Federation* **38**, 872 (1966).
100. Walzholz, G., and A. Pester, "Engineering problems in the collection and evaluation of dairy effluents," *Dairy Sci. Abstr.* **23**, 115 (1961).
101. Walzholz, G., "Treatment of dairy sewage," *Fette Seifen Anstrichmittel* **63**, 276 (1961); *J. Sci. Food Agr.* **12**, ii (1961).
102. Watson, C. W., Jr., "New developments in dairy waste treatment," *Ind. Water Wastes* **6**, 55 (1961).
103. Wheatland, A. B., "The treatment of effluents from the milk industry," *Chem. Ind.* **16**, 1547 (1967).
104. Wieselberger, F., "Treatment of dairy waste by alternating double filtration," *Ber. Abwassertech.-Verein* **11**, 264 (1960); *Water Pollution Abstr.* **34**, 2117 (1961).
105. Wisniewski, T. F., "What's being done in Wisconsin about dairy wastes disposal," *Mfd. Milk Prod. J.* **56**, 6 (1965); *Dairy Sci. Abstr.* **27**, 2758 (1965); *J. Water Pollution Control Federation* **38**, 872 (1966).
106. Yoney, Z., "By-products of Turkish milk industries and possibilities of their utilization," *Ankara Univ. Ziraat Fak. Yayinlari*, 1962, p. 193; *Chem. Abstr.* **60**, 6135 (1964).
107. Zabierzewski, C., "Technical progress in dairy effluent treatment," *Przeglad Mlecz.* **11** (no. 2), 27 and (no. 4), 15 (1963); *Dairy Sci. Abstr.* **27**, 2758 (1965); *J. Water Pollution Control Federation* **38**, 872 (1966).
108. U.S. Department of the Interior, *The Dairies*, Profile Report no. 9, Vol. III, Washington, D.C. (1967).

BREWERY, DISTILLERY, AND PHARMACEUTICAL WASTES

The fermentation industries include breweries and distilleries, manufacturers of alcohol and certain organic chemicals, and some parts of the pharmaceutical industry, such as producers of antibiotics. Fermentation has been defined as the decomposition of complex organic substances into material of simpler composition, under the influence of nitrogenous organic substances called ferments. The transformation of grape juice into wine, the manufacture of alcohol from molasses, and the use of yeast in dough to make bread are familiar examples of fermentation.

Two main types of raw material are used for producing alcohol or alcoholic products: starchy materials, such as barley, oats, rye, wheat, corn, rice, and potatoes, and materials containing sugars, such as blackstrap and high-sugar molasses, fruits, and sugar beet. The process of converting these raw materials to alcohol varies somewhat, depending on the particular raw material and the desired alcoholic product. For instance, flavor is of prime importance in the manufacture of beer, and this concern accordingly influences the process used. Manufacturers of distilled products, on the other hand, are more concerned with alcohol yield. The process of brewing and distilling consists of: (1) conversion of malt to a finely divided state in a malt mill; (2) preparation of the mash by mixing malt with hot water and, in some cases, with raw grain; (3) transformation of starches to sugar by the action of hops; (4) draining and washing the "sweet" water from the mash to fermentation tanks; (5) fermentation of sugars to alcohol by yeasts; (6) cooling, skimming, and clarification of the fermented liquor; and (7) locking in casks (if used for beer); storing in vats (if used for alcohol).

Two types of yeast are used for fermentation: "bottom yeast," used in beer manufacture, and "top yeast," used in top-fermented beers and whiskey mash. The latter type is also used almost exclusively in the manufacture of commercial compressed (bakers') yeast. The production of yeast, whether it be intended for fermentation purposes or for direct sale for baking purposes, is accompanied by the formation of by-product wastes of a highly pollutional character.

The growth of biological and pharmaceutical plants was greatly accelerated during and after World War II. Manufacture of new products, particularly antibiotics, has greatly increased the waste-disposal problem of this industry.

23.7 Origin of Brewery, Distillery, and Pharmaceutical Wastes

The brewing of beer has two stages, malting of barley and brewing the beer from this malt. Both these operations are carried on in the same plant. The malting process consists of the following steps: (1) grain is removed from storage and screened; (2) screened grain is placed in a tank and steeped with water to bleach out the color; (3) grain is then allowed to germinate, while air and water are introduced to stimulate growth of enzymes to be used for inoculum; (4) the grain malt is removed after five to eight days of aeration to the dryer, where it is dried for about four days, to a predetermined moisture content; (5) the finished malt, after the sprouts have been screened out, is stored and aged in large elevators.

The malting process produces two major wastes: those arising from the steep tank after grain has been removed, and those remaining in the germinating drum after the green malt has been removed. In the actual brewing process, considerable water is required, most of which is used for cooling purposes. Brewery wastes are composed mainly of liquor pressed from the wet grain, liquor from yeast recovery, and wash water from the various departments. After the distillation of the alcohol process, a residue remains which is referred to as "distillery slops," "beer slops," or "still bottoms."

There are several sources of wastes in a distillery. Of major concern are the "dealcoholized" still residue and evaporator condensate, when the stillage is evaporated. Minor wastes include redistillation residue and equipment washes. In the manufacture of compressed yeast seed, yeast is planted in a nutrient solution and allowed to grow under aerobic conditions until maximum cell multiplication is attained. The yeast is then separated from the spent nutrient solution, compressed, and finally packaged. The yeast-plant effluent consists of: (1) filter residues resulting from the preparation of the nutrient solutions, (2) spent nutrients, (3) wash waters, (4) filter-press effluents, and (5) cooling and condenser waters.

Pharmaceutical wastes originate principally from the spent liquors from the fermentation processes, with the addition of the floor washings and laboratory wastes. Liquid wastes from pharmaceutical plants producing antibiotics and biologicals can be classed as: (1) strong fermentation beers, (2) inorganic solids, such as diatomaceous earth, which are utilized as a precoat or an aid to the filtration process, (3) washing of floors and equipment, (4) chemical waste, and (5) barometric condenser water from evaporation.

23.8 Characteristics of Brewery, Distillery, and Pharmaceutical Wastes

Malt wastes were shown [110] to contain an average of 72 ppm of suspended solids and 390 ppm of BOD, when 6996 bushels of barley were processed per day. The volume of wastes averaged 75 gallons per bushel, or 524,700 gallons per day. The solids were mainly organic and high in nitrogen, indicating considerable protein material. A major portion of the solids were in solution, as indicated by the low suspended-solids content.

A summary of individual wastes from a typical bourbon distillery is given in Table 23.9. Table 23.10 shows the composition of some distillery "slops" [28]; all the slops have an acid reaction and contain from 2 to 7 per cent total solids. Beer slops from the fermentation of rye contained 3.3 per cent total solids, of which 91 per cent was volatile. Other data on the solids concentration and BOD content of fermentation wastes are given in Table 23.11.

Yeast wastes consist primarily of the spent nutrient (although only 20 per cent of the wastes by volume, they account for 75 to 80 per cent of the total BOD). They are brown, have the typical odor of yeast, and are highly hygroscopic. The solids are almost entirely dissolved and colloidal; the suspended-solids content is seldom above 200 ppm. The composition of this waste, according to Trubnick and Rudolfs [127], is given in Table 23.12. The average characteristics of a fermentation process-plant waste are shown in Table 23.13.

Wastes from pharmaceutical plants producing penicillin and similar antibiotics are strong (high

Table 23.9 Summary of distillery waste. (From Ruf et al. [110].)

Source of waste	Volume, gal	Suspended solids		pH	BOD, ppm	Population equivalent
		Total, ppm	Volatile, ppm			
Cooker condensate	54,000	7	6	7.2	5.4	14
Cooker wash water	2,750				1,370	185
Redistillation residues	1,000			7.7	1,700	84
Floor and equipment wash						227
Evaporator condensate	29,490	35	30	4.5	375	540
Thick slop to feed house	35,235	50,000	48,000	4.3	20,000	35,300

Table 23.10 Composition of distillery slops. (From Buswell [28].)

Item	Spirit type	Bourbon type	Molasses	Apple brandy
pH	4.1	4.2	4.5	3.8
Total solids, ppm	47,345	37,388	71,053	18,866
Suspended solids, ppm	24,800	17,900	40	50
BOD, ppm	34,100	26,000	28,700	21,000
Total volatile solids, ppm	43,300	34,226	55,608	16,948

Table 23.11 Solids and BOD of fermentation wastes. (From Boruff [15].)

Fermentation waste	Solids, %	BOD, ppm
Brewery-press liquor	3	10–25,000
Yeast plant	1–3	7–14,000
Industrial alcohol	5	22,000
Distillery slops	4.5–6	15–20,000

BOD) and generally should not be treated with domestic sewage, unless the extra load is considered in the design and operation of the treatment plant. Some wash waters range as high as 14,000 ppm BOD, and average values of combined wastes are 2500 to 5000 ppm. Brown [23] reports five main pharmaceutical wastes and their characteristics as follows: (1) strong fermentation beers: small in volume but having 4000 to 8000 ppm BOD; (2) inorganic solids: waste slurry with little BOD; (3) washings of floor and equipment: large percentage of total volume and BOD from 600 to 2500 ppm; (4) chemical waste: solution of solvents which exert a substantial BOD when diluted with other wastes; (5) barometric-condenser water: resulting from solids and volatile gases being mixed with condenser water, causing 60 to 120 ppm BOD.

Wastes from the production of fine chemicals and antibiotics, including vitamins B_1, B_2, and B_{12}, streptomycin, lysine, sulfaquinazoline, nicarbazin, and glycamide, possess the following pollutional characteristics [1962, 28].

Characteristic	General antibiotic wastes	Specific antibiotic wastes	
		Terramycin	Penicillin
BOD, ppm	1500–1900	20,000	8,000–13,000
Suspended solids, ppm	500–1000	10	
pH	1–11	9.3	2–4

Table 23.12 Composition of spent nutrient of yeast plant. (From Trubnick and Rudolfs [127].)

Characteristic	Concentration or value
Total solids, ppm	10,000–20,000
Suspended solids, ppm	50–200
Volatile solids, ppm	7,000–15,000
Total nitrogen, ppm	800–900
Organic nitrogen, ppm	500–700
Total carbon, ppm	3,800–5,500
Organic carbon, ppm	3,700–5,500
BOD, ppm	2,000–15,000
Sulfate, as ppm SO_4	2,000–2,500
Phosphate, as ppm P_2O_5	20–140
pH	4.5–6.5

Table 23.13

Characteristic	Concentration or value
BOD, ppm	4,500
pH	6–7
Total solids, ppm	10,000
Settleable suspended solids, ml/liter	25

23.9 Treatment of Brewery, Distillery, and Pharmaceutical Wastes

The principal pollutional load from a distillery is stillage, the residual grain mash from distillation columns. As much of this as possible is recovered by the industry as a by-product for manufacturing animal feed or for conversion to chemical products. Without such recovery, the population equivalent of distillery wastes, based on BOD, would be about 50,000 for each 1000 bushels of grain mash. Screening the dried grains reduces the population equivalent to 30,000. Complete stillage recovery makes possible a population equivalent of only about 2500 and a large-volume, but weak, waste. Dried yeast and dried spent grains are two valuable by-products recovered from these fermentation-waste slops, and much research is being devoted to recovering additional valuable materials from them. If additional BOD removal is required after recovery processes have been carried out, the residual waste can be treated effectively by trickling filtration or anaerobic digestion. Centrifuging has also been used to concentrate distillery slops. Trickling filtration has effected BOD removals of 60 to 98 per cent [34, 66, 95, 101], while anaerobic digestion removes 60 to 90 per cent of the BOD [29, 39, 66, 94, 122]. Trubnick and Rudolfs [127] investigated the treatment of compressed-yeast wastes, with the results shown in Table 23.14.

Table 23.14 Treatment efficienes for fermentation wastes. (From Trubnick and Rudolfs [127].)

Treatment process	Average BOD reduction, %
Electrodialysis	28
Chemical treatment	10
Anaerobic digestion	83
Activated sludge	30
Trickling filter	72

Anaerobic digestion and controlled aeration have both been used to reduce the BOD of pharmaceutical wastes by approximately 80 per cent. Effluents from such treatments can be further processed in sand filters, to produce an effluent with about 35 ppm of BOD. An interesting biofiltration plant [95], consisting of two aerators, two clarifiers, and two high-rate filters, has given greater than 90 per cent BOD reduction and 65 per cent suspended-solids removal. The initial waste contains 5700 ppm BOD in a flow of 90,000 gallons per day. Milcher [1962, 41] and Winar [1967, 50] recommended deep well injection while McKinney [1962, 40], Dazai [1966 27], and Howe [1962, 34] report success with activated-sludge treatment of antibiotic wastes. At times the only possible treatment of antibiotic wastes is by evaporation and incineration. Because bulk rubbish is usually handled in addition to process wastes at a pharmaceutical plant, incineration is readily carried out. Residues from penicillin and other antibiotics can be dried and used in stock food. A vacuum-dried mycelium from the manufacture of penicillin can be digested to produce methane; at the same time this reduces the organic matter by about 55 per cent. The traces of antibiotics and mystery factors in spent mycelium permit sale of these waste as animal-growth-promoting agents.

References: Brewery, Distillery, and Pharmaceutical Wastes

1. Acklin, O., K. Bram, and M. Zurcher, "Waste treatment plant, Switzerland," *Sewage Ind. Wastes*, 6, 1027 (1934).
2. Ambrose, T. W., "Wastes from potato starch plants," *Ind. Eng. Chem.* 46, 1331 (1954).
3. Arbogast, C. H. P., "Incineration of pharmaceutical wastes," in Proceedings of 4th Industrial Waste Conference, Purdue University, September 1948, p. 255.
4. Arbogast, C. H. P., "Incineration of wastes, large pharmaceutical establishments," *Sewage Ind. Wastes* 23, 120 (1951).
5. Barker, W. G., R. H. Otto, D. Schwarz, and G. F. Shipton, "Pharmaceutical waste disposal studies," in Proceedings of 13th Industrial Waste Conference, Purdue University, May 1958.
6. Barker, W. G., R. H. Otto, D. Schwarz, and B. C. Tjarksen, "Pharmaceutical waste disposal studies," in Proceedings of 15th Industrial Waste Conference, Purdue Unversity, May 1960, p. 58.
7. Beeson, W. M., "Feeding of fermentation wastes to livestock," in Proceedings of 3rd Industrial Waste Conference, Purdue University, May 1947.
8. Beeson, W. M., "Livestock feed from fermentative wastes," *Sewage Ind. Wastes* 22, 969 (1950).
9. Blohm, A. W., "Effect on sewage treatment, Westminster, Md.," *Sewage Ind. Wastes* 9, 638 (1937).
10. Bloodgood, D. E., "Status of waste production and disposal," *Sewage Ind. Wastes* 19, 607 (1947).
11. Bohonos, N., "What do we have left after a fermentation process?" in Proceedings of 3rd Industrial Waste Conference, Purdue University, May 1947.
12. Bohonos, N., "Fermentative wastes, characteristics," *Sewage Ind. Wastes* 22, 1254 (1950).
13. Bollen, W. B., "Douglas-fir ethanol spillage, waste disposal studies," *Sewage Ind. Wastes* 22, 962 (1950).
14. Bonacci, L. N., and W. Rudolfs, "Electrodialysis treatment," *Sewage Ind. Wastes* 14, 1281 (1942).
15. Boruff, C. S., "Waste problems in the fermentation industry," *Ind. Eng. Chem.* 31, 1335 (1939).
16. Boruff, C. S., "Fermentative waste problems, review," *Sewage Ind. Wastes* 12, 398 (1940).
17. Boruff, C. S., "Waste treatment, status," *Sewage Ind. Wastes* 24, 1443 (1952).
18. Boruff, C. S., "Recovery and uses of grain distillery

stillage," in Proceedings of 2nd Ontario Industrial Waste Conference, June 1955, p. 87.
19. Boruff, C. S., and R. K. Blaine, "Distillery feeds and waste," *Sewage Ind. Wastes* **25**, 1179 (1953).
20. Boruff, C. S., and L. P. Weiner, "Feed by-products from grain alcohol and whiskey stillage," in Proceedings of 1st Industrial Waste Conference, Purdue University, November 1944.
21. U.S. Public Health Service, "Brewery wastes," *Ohio River Survey*, Supplement D (1943), p. 1039.
22. Brown, E. M., "Brandy and molasses wastes," *Sewage Ind. Wastes* **16**, 949 (1944).
23. Brown, J. M., "Treatment of pharmaceutical wastes," *Sewage Ind. Wastes* **23**, 1017 (1951).
24. Brown, J. M., and J. G. Niedercorn, "Antibiotic waste treatment," *Sewage Ind. Wastes* **24**, 1442 (1952).
25. Bueltman, C. G., "Bio-oxidation of brewery wastes," in Proceedings of 14th Industrial Waste Conference, Purdue University, May 1959.
26. Bushee, R. J., "Pilot plant experiments, Gulf Brewery Company, Houston, Texas," *Sewage Works J.* **11**, 295 (1939).
27. Buswell, A. M., "Beer slop waste treatment," *Sewage Ind. Wastes* **7**, 773 (1935).
28. Buswell, A. M., "Treatment of 'beer slop' and similar wastes," *Water Works Sewerage* **82**, 135 (1935).
29. Buswell, A. M., "Anaerobic fermentation plants," in Proceedings of 5th Industrial Waste Conference, Purdue University, November 1949, p. 168.
30. Buswell, A. M., and H. F. Mueller, "Methane fermentation mechanism," *Sewage Ind. Wastes* **24**, 1445 (1952).
31. Carson, C. T., "Treatment of distillery wastes," *Water Sewage Works* **98**, 312 (1951).
32. "Combined treatment in municipal plant, Golden, Colo.," *Sewage Ind. Wastes* **25**, 1014 (1953).
33. Cushman, J. R., and J. R. Hayes, "Pilot plant studies of pharmaceutical waste at Upjohn Company," in Proceedings of 11th Industrial Waste Conference, Purdue University, May 1956.
34. Davidson, A. B., "Designing a distillery waste disposal plant," in Proceedings of 5th Industrial Waste Conference, Purdue University, November 1949, p. 159.
35. Davidson, A. B., "The treatment of distillery waste," *Sewage Ind. Wastes* **22**, 654 (1950).
36. Davidson, A. B., "Waste disposal plant, design," *Sewage Ind. Wastes* **23**, 695 (1951).
37. Davidson, A. B., and J. F. Banks, "Anaerobic treatment of distillery waste pilot plant studies," in Proceedings of 4th Industrial Waste Conference, Purdue University, September 1948, p. 94.
38. Davidson, A. B., and J. F. Banks, "Distillery waste, anaerobic treatment, pilot plant studies," *Sewage Ind. Wastes* **22**, 1094 (1950).
39. Davidson, A. B., and H. B. Brown, "Rapid anaerobic digestion studies," in Proceedings of 7th Industrial Waste Conference, Purdue University, May 1952, p. 142.
40. U.S. Public Health Service, "Distillery waste," *Ohio River Survey*, Supplement D (1943), p. 1130.
41. Eckenfelder, W., "Bio-oxidation of brewery wastes," in Proceedings of 6th Ontario Industrial Waste Conference, June 1959, p. 195.
42. Eden, G. E., and G. F. Lowden, "The treatment of waste waters from the manufacture of penicillin," *Mfg. Chemist* **23**, 144 (1952).
43. Edmondson, K. H., "Disposal of antibiotic waste spent beers by triple effect evaporation," in Proceedings of 8th Industrial Waste Conference, Purdue University, May 1953, p. 46.
44. Eldridge, E. F., *Industrial Waste Treatment Practice*, McGraw-Hill Book Co., New York (1942).
45. Feightner, H. C., "Spent grain utilization," *Sewage Ind. Wastes* **22**, 968 (1950).
46. Feightner, H. C., "Comments at the opening of the Fermentation Process Symposium," in Proceedings of 3rd Industrial Waste Conference, Purdue University, May 1947.
47. Gabaccia, A. J., "Composting waste sludge from pharmaceutical manufacturing," *Sewage Ind. Wastes* **31**, 1175 (1959).
48. Gallager, A., L. J. Keefe, S. A. Mayer, and W. D. Hanlon, "Waste disposal, Bristol Laboratories," *Sewage Ind. Wastes* **26**, 1355 (1954).
49. Gehm, H. W., "Aerobic treatment of waste high in BOD concentration," in Proceedings of 8th Industrial Waste Conference, Purdue University, May 1953.
50. Greely, S. A., and W. D. Hatfield, "The sewage disposal works of Decatur, Ill.," *Trans. Am. Soc. Civil Engrs.* **94**, 544 (1930).
51. Greenleaf, J. W., "Perkins, Ill., waste treatment plant," *Sewage Ind. Wastes* **14**, 250 (1942).
52. Gurnham, C. F., *Principles of Industrial Waste Treatment*, John Wiley & Sons, New York (1955).

53. Gurnham, C. F., "Waste solvent incineration successful at Upjohn Co., Kalamazoo, Mich.," *Ind. Wastes* **2**, 29 (1957).
54. Hale, F., "Source of BOD in brewery waste," *Sewage Ind. Wastes* **25**, 1187 (1953).
55. Hall, R. L., "Hops and spent grain press liquor," in Proceedings of 3rd Industrial Waste Conference, Purdue University, May 1947, p. 138.
56. Hall, R. L., "By-product value of brewery and distillery wastes," *Sewage Ind. Wastes* **22**, 1254 (1950).
57. Halperin, Z., "Tartrate recovery," *Sewage Ind. Wastes* **18**, 169 (1946).
58. Hatfield, W. D., "Operation of the pre-aeration plant at Decatur, Ill.," *Sewage Works J.* **3**, 62 (1931).
59. Hatfield, W. D., "Cornstarch processes," in *Industrial Wastes, Their Disposal and Treatment*, ed. W. Rudolfs, Reinhold Publishing Corp., New York (1953).
60. Haupt, H., "By-product recovery," *Sewage Ind. Wastes* **8**, 350 (1936).
61. Helmers, E. N., J. D. Frame, A. E. Greenberg, and C. N. Sawyer, "Biological stabilization of waste, nutrient requirements," *Sewage Ind. Wastes* **24**, 496 (1952).
62. Hilgart, A. A., "Penicillin and streptomycin wastes, treatment plant, design and operation," *Sewage Ind. Wastes* **22**, 207 (1950).
63. Hodgson, H. J. N., and J. Johnston, "Waste treatment studies, Genelg, South Australia," *Sewage Ind. Wastes* **12**, 321 (1940).
64. Hoover, C. R., and F. K. Burr, "Distillery wastes—chemical and filtration studies," *Ind. Eng. Chem.* **28**, 38 (1936).
65. Horne, W. R., and U.S. Rinaca, "Treatment of pharmaceutical wastes," in Proceedings of 15th Industrial Waste Conference, Purdue University, May 1960, p. 235.
66. Howe, R. H. L., and S. M. Paradiso, "Miracle drug waste and plain sewage treatment by modified activated sludge and biofiltration units," *Wastes Eng.* **27**, 210 (1956).
67. Hurwitz, E., K. H. Edmondson, and S. M. Clarke, "Pharmaceutical waste treatment," *Water Sewage Works* **99**, 202 (1952).
68. "Incineration of wastes," *Sewage Ind. Wastes* **20**, 776 (1948).
69. Jackson, C. J., "Whiskey and industrial alcohol distillery wastes," Paper read at Institute of Sewage Purification Annual Conference, June 1956.
70. Klassen, C. W., and A. P. Troemper. "Wastes from a whiskey distillery," in Proceedings of 3rd Industrial Waste Conference, Purdue University, May 1947, p. 153.
71. Klassen, C. W., and A. P. Troemper, "Waste sources, characteristics," *Sewage Ind. Wastes* **22**, 1254 (1950).
72. Knoedler, E. L., and S. H. Babcock, "Character and disposal of wastes," *Sewage Ind. Wastes* **19**, 950 (1947).
73. Lin, P. W., "Yeast growing on brewery waste," in Proceedings of 5th Industrial Waste Conference, Purdue University, November 1949, p. 181.
74. Lin, P. W., "Yeast, growth on brewery waste," *Sewage Ind. Wastes* **23**, 695 (1951).
75. Liontas, J. A., "Mixed waste treatment, high-rate filters," *Sewage Ind. Wastes* **26**, 310 (1954).
76. Lundberg, "Methane fermentation of penicillin mycelium," *Arkiv. Kemi* **5**, 1 (1952).
77. Mann, U. T., "Penicillin waste, effect on activated sludge plants," *Sewage Ind. Wastes* **23**, 1457 (1951).
78. Mauriello, C. G., "Biological and pharmaceutical wastes, treatment problem," *Sewage Ind. Wastes* **30**, 1397 (1958).
79. Mohlman, F. W., "Brewery wastes," *Mod. Brewer* **21**, 35 (1939).
80. Mohlman, F. W., "Brewery waste survey, Chicago," *Sewage Ind. Wastes* **11**, 721 (1939).
81. Mohlman, F. W., "Utilization and disposal of industrial wastes," in Proceedings of 1st Industrial Waste Conference, Purdue University, November 1944, p. 43.
82. Mohlman, F. W., "Status of waste production and disposal," *Sewage Ind. Wastes* **19**, 473 (1947).
83. Mohlman, F. W., and A. J. Beck, "Disposal of industrial wastes," *Ind. Eng. Chem.* **21**, 205 (1929).
84. Morgan, E. H., and A. J. Beck, "Effects of activated sludge, Chicago," *Sewage Ind. Wastes* **1**, 46 (1928).
85. Muss, D. L., "Penicillin waste treatment methods," *Sewage Ind. Wastes* **23**, 486 (1951).
86. Newton, D., "Brewery wastes BOD: sewage = 15:1—best solution is series filters," *Wastes Eng.* **27**, 500 (1956).
87. Niles, C. F., Jr., "Composting spent hops," in Proceedings of 4th Industrial Waste Conference, Purdue University, September 1948.
88. Niles, C. F., Jr., "Spent hops, composting for fertilizer," *Sewage Ind. Wastes* **22**, 1094 (1950).

89. Nissen, B. H., "Brewery waste utilization," in Proceedings of 2nd Industrial Waste Conference, Purdue University, January 1946, p. 82.
90. Nissen, B. H., "By-product recovery," *Sewage Ind. Wastes* **19**, 1104 (1947).
91. Painter, H. A., "Treatment of malt whiskey distillery wastes by anaerobic digestion," *Brewers' Guardian*, August 1960.
92. Pattee, E. C., "Research in the disposal of distillery wastes," in Proceedings of 4th Industrial Waste Conference, Purdue University, September 1948, p. 122.
93. Pattee, S. C., "Waste disposal research," *Sewage Ind. Wastes* **22**, 1094 (1950).
94. Pearson, E. A., D. F. Feuerstein, and B. Onodera, "Treatment and utilization of winery wastes," in Proceedings of 10th Industrial Waste Conference, Purdue University, May 1955, p. 34.
95. Pitts, H. W., "Treatment problems," *Sewage Ind. Wastes* **27**, 970 (1955).
96. Pulfrey, A. L., R. W. Kerr, and H. R. Reintjes, "Wet milling of corn," *Ind. Eng. Chem.* **32**, 1483 (1940).
97. Renard, M., "Spray irrigation waste treatment, Montierchaume Distillery," *Sewage Ind. Wastes* **24**, 1551 (1952).
98. Ridgeway, J. W., et al., "Spray drying of distillers' solubles," in Proceedings of 3rd Industrial Waste Conference, Purdue University, May 1947, p. 128.
99. Ridgeway, J. W., W. V. Baldyge, and M. Scarba, "Distillers' solubles, spray drying," *Sewage Ind. Wastes* **22**, 968 (1950).
100. Rimers, F. E., U. S. Rinaca, and L. E. Poese, "Fine chemical waste, pilot plant trickling filter studies," *Sewage Ind. Wastes* **25**, 51 (1953).
101. Roberts, N., and J. B. Hardwick, "Pilot-plant studies on distillery waste," in Proceedings of 6th Industrial Waste Conference, Purdue University, February 1951, p. 80.
102. Roberts, N., and J. B. Hardwick, "Distillery, residual waste load, high-rate trickling filter pilot plant studies," *Sewage Ind. Wastes* **24**, 804 (1952).
103. Rudolfs, W., "By-products and treatment of brewery and yeast waste," in Proceedings of 1st Industrial Waste Conference, Purdue University, November 1944.
104. Rudolfs, W., "By-product recovery and waste treatment," *Sewage Ind. Wastes* **18**, 762 (1946).
105. Rudolfs, W., and E. H. Trubnick, "Character of wastes," *Sewage Ind. Wastes* **20**, 1084 (1948).
106. Rudolfs, W., and E. H. Trubnick, "Biological treatment, compressed yeast waste treatment," *Sewage Works J.* **21**, 109 (1949).
107. Rudolfs, W., and E. H. Trubnick, "Pilot plant digestion studies, compressed yeast waste treatment," *Sewage Works J.* **21**, 295 (1949).
108. Rudolfs, W., and E. H. Trubnick, "Trickling filter studies, compressed yeast waste treatment," *Sewage Works J.* **21**, 491 (1949).
109. Rudolfs, W., and E. H. Trubnick, "Treatment of compressed yeast wastes," *Ind. Eng. Chem.* **42**, 612 (1950).
110. Ruf, H. W., L. F. Warrick, and M. S. Nichols, "Malt house waste treatment studies in Wisconsin," *Sewage Ind. Wastes* **17**, 564 (1935).
111. Schmidt, T. L., "Utilization of brewery wastes," in Proceedings of Conference on Industrial Wastes, Civil Engineering Department, University of Washington, 28 April 1949.
112. Schneider, R., "Brewery waste disposal, Azusa, Calif.," *Sewage Ind. Wastes* **22**, 1307 (1950).
113. Schuck, C., "Yeast as a human food," in Proceedings of 4th Industrial Waste Conference, Purdue University, September 1948.
114. Shaw, P. A., "California, Mokelumne river pollution by winery wastes," *Sewage Ind. Wastes* **9**, 599 (1937).
115. Singruen, E., "Uses for yeasts," in Proceedings of 5th Industrial Waste Conference, Purdue University, November 1949.
116. Sjostrom, O. A., "Treatment of waste water from a starch and glucose factory," *Ind. Eng. Chem.* **3**, 100 (1911).
117. Snook, W. F. A., "Treatment of wastes," *Sewage Ind. Wastes* **5**, 750 (1933).
118. Stecher, E., "Effect on municipal sewage, Munich, Germany," *Sewage Ind. Wastes* **11**, 720 (1939).
119. *The Story of Corn and Its Products*, Corn Industries Research Foundation, Inc.
120. "Symposium on food canning wastes," in Proceedings of 1st Industrial Waste Conference, Purdue University, November 1944.
121. Tarzwell, C. M., "The use of bio-assay in relation to the disposal of toxic wastes," in Proceedings of 3rd Ontario Industrial Waste Conference, June 1956, p. 117.
122. Tatlock, M. W., "Treatment of yeast products wastes," in Proceedings of 3rd Industrial Waste

Conference, Purdue University, May 1947, p. 111.
123. Tidswell, M. A., "Experiment in purification of sewage containing a large proportion of brewery waste at Burton-upon-Trent," Paper read at Institute of Sewage Purification, Midland Branch (England), 28 October, 1959.
124. Tolman, S. L., "Operation of the waste treatment plant of the Gulf Brewery Company, Houston, Texas," Ohio Conference on Sewage Treatment, 14th Annual Report.
125. Tolman, S. L., "Treatment plant, Gulf Brewery Company, Houston, Texas," *Sewage Ind. Wastes* **13**, 1014 (1941).
126. Tompkins, L. B., "Two-stage filtration, Upjohn Company," *Sewage Ind. Wastes* **29**, 1161 (1957).
127. Trubnick, E. H., and W. Rudolfs, "Treatment of compressed yeast effluents," in Proceedings of 4th Industrial Waste Conference, Purdue University, September 1948, p. 109.
128. "Upjohn Company Report—antifoamant solves frothing problem," *Ind. Wastes* **2**, 61 (1957).
129. Van Patten, E. N., and G. H. McIntosh, "Liquid industrial wastes, corn products manufacture," *Ind. Eng. Chem.* **44**, 483 (1952).
130. Vaughn, R. H., and G. L. Marsh, "Disposal of California winery wastes," *Ind. Eng. Chem.* **45**, 2686 (1953).
131. Vogler, J. F., J. M. Brown, and G. E. Griffin, "Chemical and antibiotic waste treatment," *Sewage Ind. Wastes* **24**, 485 (1952).
132. Wagner, T. B., "Disposal of starch factory wastes," *Ind. Eng. Chem.* **3**, 99 (1911).
133. Wallach, A., and A. Wolman, "Chemical treatment experiments," *Sewage Ind. Wastes* **14**, 382 (1942).
134. Warcollien, C., "Apple distillery, character and disposal of wastes," *Sewage Ind. Wastes* **4**, 410 (1932).
135. Ward, A. R., "Effect on sewage treatment, Stockport, England," *Sewage Ind. Wastes* **8**, 346 (1936).
136. Wheeler, M., "Beer slop, feed recovery plant," *Sewage Ind. Wastes* **23**, 695 (1951).
137. Wisely, W. H., "Treatment with sewage, Highland, Ill.," *Sewage Ind. Wastes* **8**, 624 (1936).

Suggested Additional Reading

The following references have been published since 1962.
1. Alekonis, J. J., and J. V. Ziembo, "Waste treatment," *J. Food Eng.* **39**, 89 (1967).
2. Bhaskaran, T. R., "Utilization of materials derived from treatment of wastes from molasses distilleries," in Proceedings of 2nd International Conference on Water Pollution Control Research, Paper no. 5, August 1966.
3. Blaine, R. K., and J. M. Van Lanen, "Application of waste-to-product ratios in the fermentation industries," *Biotechnol. Bioeng.* **4**, 129 (1962).
4. Bueltman, C. G., "Biooxidation of brewery wastes," Purdue University Engineering Extension Series, Bulletin no. 14, 1959, p. 656.
5. Chipperfield, P. N. J., "Performance of plastic filter media in industrial and domestic waste treatment," *J. Water Pollution Control Federation* **39**, 1860 (1967).
6. Dieterick, K. R., *Waste Utilization and Waste Water Treatment in the Biochemical Industries*, Alfred Hutching, Heidelburg (1960).
7. Garvie, W. R., and D. E. Waite, "Aqua ammonia used to neutralize acid effluent," *Water Pollution Control* **103**, 35 (1965).
8. "High speed bottle washing," *Engineering* **204**, 334 (1967).
9. Knoebel, I. G., "Belleville lick sludge bulking," *Wastes Eng.* **34**, 396 (1963).
10. Leary, R. D., and L. A. Ernest, "Industrial and domestic wastewater control in the Milwaukee Metropolitan District," *J. Water Pollution Control Federation* **39** 1223 (1967).
11. Martin, R. K., "Malt distillery effluent plant," *Brit. Chem. Engr.* **8**, 736 (1963).
12. Nakagawa, M., C. Tatsumi, and E. Sinno, "Production of fat by microorganisms, VI: Composition of fats produced by rhodotorula from various carbon sources," *Hakko Kyokaishi* **19**, 623 (1961); *Chem. Abstr.* **59**, 13317g (1963).
13. "New know-how prompts brewers to diversity," *Chem. Eng.*, 28 August 1967, p. 60.
14. O'Rourke, J. T., and H. D. Tomlinson, "Effects of brewery wastes on treatment," *Ind. Wastes* **7**, 119 (1962).
15. O'Rourke, J. T., and H. D. Tomlinson, "Extreme variations in brewery waste characteristics and their effect on treatment," in Proceedings of 17th Industrial Waste Conference, Purdue University, May 1962, p. 254.
16. Schaake, W., "The water requirements and waste water output in the manufacture of industrial pro-

ducts," *Wasserwirtsch.-Wassertech.* **10**, 291 (1960); *Chem. Abstr.* **55**, 5824 (1961).
17. Sen, P. B., and T. R. Bhaskaran, "Anaerobic digestion of liquid molasses distillery wastes," *J. Water Pollution Control Federation* **34**, 1015 (1962).
18. Sproul, O. J., and D. W. Ryckman, "Significant physiological characteristics of organic pollutants," *J. Water Pollution Control Federation* **35**, 1136 (1963).
19. Tidswell, M. A., "Experimental purification of sewage containing a large proportion of brewery wastes at Burton-upon-Trent," *J. Proc. Inst. Sewage Purif.* Part 2 (1960), p. 139.
20. Van Lewen, A. L., "Some considerations of food industry waste problems in Quebec," in Proceedings of 10th Ontario Industrial Waste Conference, Ontario Water Resources Commission, June 1963, p. 91.

Specific pharmaceutical and antibiotic wastes

21. Asendorf, E., "Purification of industrial wastes," *Chemiker-Ztg.* **9**, 573 (1965); *Chem. Abstr.* **65**, 15064g (1966).
22. Barker, W. G., et al., "Pharmaceutical waste disposal studies," in Proceedings of 15th Industrial Waste Conference, Purdue University Engineering Extension Series, Bulletin no. 106, 1961, p. 5.
23. Blaine, R. K., and J. M. Van Lanen, "Application of waste-to-product ratios in the fermentation industries," *Biotechnol. Bioeng.* **4**, 129 (1962).
24. Burgess, J. V., "Treatment of chemical plant effluents," *Effluent Water Treat. J.* **7**, 533 (1967); *Chem. Abstr.* **68**, 244366 (1968).
25. Colovos and Tinklenberg, "Land disposal of pharmaceutical manufacturing wastes," *Biotechnol. Bioeng.* **4**, 153 (1962).
26. Dazai, M., Y. Yoshida, M. Ogawa, Y. Sagano, and H. Ono, "Treatment of industrial wastes by activated sludge, VII: Treatment of antibiotic wastes, report 1," *Kogyo Gijutsuin Hakko Kenkyusho Kenkyu Hokoku* **27**, 69 (1965); *Chem. Abstr.* **64**, 17243 (1966).
27. Dazai, M., T. Higashikara, M. Ogawa, and H. Ono, "Treatment of industrial wastes by activated sludge. VIII: Treatment of antibiotic wastes, report 2," *Kogyo Gijutsuin Hakko Kenkyusko Kenkyu Hokoko* **27**, 77 (1965); *Chem. Abstr.* **64**, 17243 (1966).
28. Eckenfelder, W. W., Jr., and E. L. Barnhart, "Biological treatment of pharmaceutical wastes," *Biotechnol. Bioeng.* **4**, 171 (1962).
29. Genetelli, E. J., and H. Heukelekian, "Rational approach to design for complex chemical waste," in Proceedings of 5th Texas Water Pollution Assoc. Industrial Water and Waste Conference, 1965, p. 372; *Eng. Index*, 1966, p. 1342.
30. Herion, R. W., Jr., H. O. Loughhead, and A. L. Vellutao, "Pharmaceutical wastes," *Ind. Water Wastes* **8**, 28 (1962).
31. Herion, R. W., Jr., and H. O. Loughhead, "Two treatment installations for pharmaceutical wastes," Purdue University Engineering Extension Series, Bulletin no. 115, 1963, p. 218.
32. Horne, W. R., and V. S. Rinaca, "Treatment of pharmaceutical wastes," in Proceedings of 15th Industrial Waste Conference, Purdue University, Engineering Extension Series, Bulletin no. 106, 1961, p. 235.
33. Howe, R. H., "Biological degradation of wastes containing certain toxic chemical compounds," in Proceedings of 16th Industrial Waste Conference, Purdue University, May 1961, p. 262.
34. Howe, R. H. L., "Complete biological treatment of antibiotic production wastes," *Biotechnol. Bioeng.* **4**, 161 (1962).
35. Howe, R. H. L., "Handling wastes from the billion dollar pharmaceuticals industry," *Wastes Eng.* **31**, 752 (1960).
36. Kempe, L., "Introductory remarks on Symposium on fermentation waste disposal," *Biotechnol. Bioeng.* **4**, 127 (1962).
37. Khan, R. K., "Use of algae for treatment of synthetic drug wastes," *Environ. Health* **4**, 193 (1962).
38. Ludzack and J. W. Mandia, "Behavior of 3-amino-1,2,4-triazole in surface water and sewage treatment," in Proceedings of 16th Industrial Waste Conference, Purdue University, May 1961, p. 540.
39. Madera, V., V. Solin, and M. Spacek, "Wastes from pharmaceutical production," *Sb. Vysoke Skoly Chem. Technol. Praze Oddil Fak. Technol. Paliv Vody* **6**, 57 (1962); *Chem. Abstr.* **63**, 2738 (1965).
40. McKinney, R. E., "Complete mixing activated sludge treatment of antibiotic wastes," *Biotechnol. Bioeng.* **4**, 181 (1962).
41. Melcher, R. R., "Pharmaceutical waste disposal by soil injection," *Biotechnol. Bioeng.* **4**, 147 (1962).
42. Molof, A. H., and N. Zalechs, "Parameter of disposal of waste from pharmaceutical industry," *Ann. N. Y. Acad. Sci.* **130**, 851 (1965).

43. Molof, A. H., "Pharmaceutical waste treatment by the trickling filter, activated sludge and compost process," *Biotechnol. Bioeng.* **4**, 197 (1962).
44. Otto, Barker, Schwarz, and Tjarksen, "Laboratory testing of pharmaceutical wastes for biological control," *Biotechnol. Bioeng.* **4**, 139 (1962).
45. Patil, D. M., T. K. Srinivasan, and G. K. Seth, "Treatment and disposal of synthetic drug wastes," *Environ. Health* **4**, 96 (1962).
46. Purice, V., "Methane production fermentation of slurries in residual waters, effected in the waste-recovery station of the antibiotics plant—Sasi (Romania)," *Rev. Chim.* **16**, 84 (1965); *Chem. Abstr.* **63**, 2737 (1965).
47. Sharonova, N. F., A. A. Lifshits, and A. M. Vlasova, "The composition and methods of purification of waste waters formed during the manufacture of pharmaceutical preparations," *Tr. Vses. Nauchn.-Issled. Inst. Pererabotki Ispol'z. Topliva* **12**, 253 (1963); *Chem. Abstr.* **61**, 1599 (1964).
48. Stickelberger, D., "Composting of industrial wastes, especially from the pharmaceutical industry," *Chemiker-Ztg.* **90**, 642 (1966); *Chem. Abstr.* **66**, 55996 (1967).
49. Walker, J. F., and R. W. Herion, "Design of pharmaceutical wastes plant was based on laboratory studies," *Wastes Eng.* **33**, 610 (1962).
50. Winar, R. M., "The disposal of waste water underground," *Ind. Water Eng.* **4**, 21 (1967).

MEAT PACKING, RENDERING, AND POULTRY-PLANT WASTES

The meat industry has three main sources of waste: stockyards, slaughterhouses, and packinghouses. Animals are kept in the *stockyards* until they are killed. The killing, dressing, and some by-product processing are carried out in the *slaughterhouse*, or abattoir. To obtain the finished product—namely, the fresh carcass, plus a few fresh meat by-products such as hearts, livers, and tongues—the following operations are performed in the slaughterhouse. The animals are stuck and bled on the killing floor (cattle being stunned prior to sticking). Carcasses are trimmed, washed, and hung in cooling rooms. Livers, hearts, kidneys, tongues, brains, etc., are sent to the cooling rooms to be chilled before being marketed. Hides, skins, and pelts are removed from the cattle, calves, sheep, and pigs, and are salted and placed in piles until they are shipped to tanners or wool-processing plants. Viscera are removed and, together with head and feet bones, are sent to the rendering plant; other bones are shipped to glue factories. Many slaughterhouses are equipped to render their own inedible offal into tallow, grease, and tankage; other independent rendering plants convert inedible poultry, fish, and animal offal and waste products into animal feed and grease. This is accomplished by cooking the inedibles at a high temperature for several hours. The cooked material is pressed to remove the grease and the pressings are ground for feed. Thus, slaughterhouse wastes are produced on the killing floor, in the process of dressing the carcass, in rendering operations, in the hide cellar, and in the cooling room.

Although *packinghouses* also perform some of the operations carried out in abattoirs, they are primarily concerned with the production of salable products. Carcasses are trimmed, cleaned, and cooled as in the slaughterhouse, but the packinghouse processes some of the meat further by cooking, curing, smoking, and pickling. Packinghouse operations also include the manufacture of sausages, canning of meat, rendering of edible fats into lard and edible tallow, cleaning of casings, drying of hog's hair, and some rendering of inedible fats into grease and inedible tallow. In addition, the packinghouse is equipped to process, to varying degrees, the by-products of the slaughterhouses. Blood is usually collected, coagulated, dried, and finally formed into edible and inedible products. Tanning, wool palling, and the manufacture of glues, soaps, and fertilizers are usually carried out in separate plants. Packinghouse wastes, therefore, issue from various operations on the killing floor, during carcass dressing, rendering, bag-hair removal and processing, casing, and cleaning, in the making of tripe, during the manufacture of by-products such as glue, soap, and fertilizer, and from laundering of linens and uniforms of the packinghouse workers.

ORIGIN AND CHARACTERISTICS OF MEAT-PACKING WASTeS   339

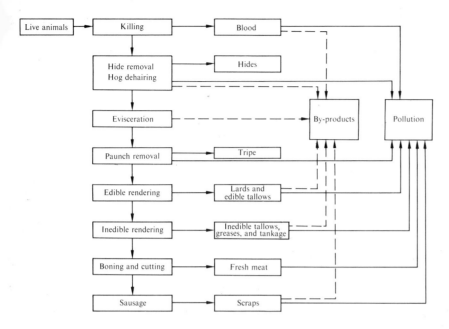

Fig. 23.5 Fundamental processes in the meat-packing industry.

Figure 23.5 is a schematic illustration of the operation of meat-processing plants.

Poultry processing differs from the processing of other meats enough to warrant a separate discussion. The operations of the poultry industry consist, in general, of the following steps: (1) the processor furnishes baby chicks and feed to the grower; (2) the grower, after about six weeks, sends the broilers to the processor; and (3) the processor prepares and markets the broilers. Figure 23.6 depicts the steps involved in processing poultry.

Broilers are delivered to the processing plants and are hung live by their feet on a moving chain which delivers them to the killing table, where their throats are slit. The blood usually flows into a trough and thence to drums for storage. (In some plants, the blood is sprayed with a coagulant and the residue shoveled into drums, while the supernatant goes to the sewers.) As the endless chain continues through the plant, the birds are mechanically plucked, washed, cleaned, washed again, and finally removed from the chain. At certain locations along the chain, drums are placed for storing feathers, heads, feet, offal, and the scrap-waste products, since it is important not to allow any of this waste to be flushed into the sewer. The dressed bird is then cut, frozen, or

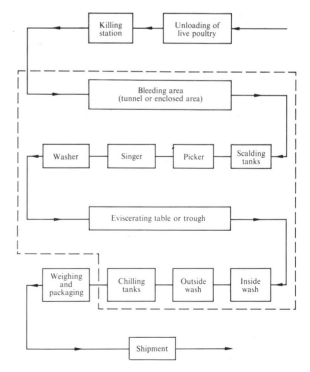

Fig. 23.6 Major steps in poultry processing.

refrigerated, depending on the operation of the day. Rendering plants in the area offer a ready market for offal, feet, heads, scraps, and even blood, which they convert into tankage for feed additives or process as pet food or fertilizer. In spite of all the apparently profitable methods for disposal of these poultry wastes, a large percentage unfortunately ends up in the sewers.

23.10 Origin and Characteristics of Meat-Packing Wastes

Stockyard wastes contain excreta, both liquid and solid. The amount and strength of the wastes vary widely, depending on the presence or absence of cattle beams (horns), how thoroughly or how often manure is removed, frequency of washing, and so forth. An analysis of wastes from one Chicago stockyard is given in Table 23.15. Another study of these same wastes showed a volume of 623,000 gallons per day for a 27-acre section of the yard and an average BOD of 100 ppm (population equivalent of 3100) [35].

Table 23.15 Stockyard waste [35].

Characteristics	Concentration, ppm
Total suspended solids	173
Volatile suspended solids	132
Organic nitrogen	11
Ammonia nitrogen	8
BOD	64

In a recent [1967, 49] survey, five processes were singled out for analysis, because of their potential impact on the waste load of the meat-packing industry. Some are actually by-product processes, but in an industry well known for "utilizing all parts but the 'squeal,'" by-product recovery has essentially become part of the process.

1. *Blood recovery* is an "all or nothing" situation: either it is recovered or it escapes to the sewer. Recovery means a 42 per cent reduction in the gross waste load of a meat-packing plant. In 1966 over 95 per cent of the industry, on a live-weight basis, was recovering blood. Blood is a rich source of protein and, hence, for all but the very small plant it is economically rational to recover it. The very small plant, which does not render (inedible products), which does not produce tankage, and which is not located in an area where it can sell raw blood to others, will probably dump blood in the sewer.

2. *Paunch material* is a source of pollution problems if it is dumped into the sewer, as the total-solids concentration becomes so large that it interferes with the efficient workings of traditional waste-treatment methods.

3. *Edible rendering* can be highly pollutional, depending upon the method. The worst offender is wet rendering without evaporating tank water. This process is the oldest type and is not being adopted by new plants. If tank water is evaporated waste loads are cut in half. Newer methods of rendering, including dry rendering and low-temperature rendering, cut waste loads by 60 per cent.

4. *Inedible rendering* similarly is either a dry or a wet process. Wet rendering must be followed by evaporation of tank water in order to cut waste loads in half. Both forms of dry rendering, batch and continuous, will produce waste loads 60 per cent lower than those from a wet rendering system without evaporation of tank water. Continuous dry rendering is the latest method in terms of technology.

5. *Cleanup* by water from high-pressure hoses has been and continues to be the general practice in the meat-packing industry. Pollution loads could be substantially reduced by the use of dry cleanup prior to the wet cleanup. This could also mean greater recovery of scraps for inedible rendering, instead of hosing them into the sewer. Another effect of dry cleanup is the reduction of waste-water volume. Data have indicated a high, direct correlation between water use per 1000 pounds (live weight) killed and waste load. Decrease in waste-water volume seems to be accompanied by utilization of dry cleanup and, hence, lower waste load per unit of product as well as lower waste-water volumes per unit of product.

Slaughterhouse processes are centered about the killing floor. The wastes produced here have a deep reddish-brown color, high BOD, and contain a considerable amount of suspended material. Blood, being highly nitrogenous, decomposes readily. In addition, the wastes contain varying amounts of manure, hair, and dirt. Analysis of several samples of killing-floor waste taken from an average slaughter-

Table 23.16 Packinghouse wastes.

Source	Suspended solids, ppm	Organic nitrogen, ppm	BOD, ppm	pH
Killing floor	220	134	825	6.6
Blood and tank water	3,690	5400	32,000	9.0
Scalding tub	8,360	1290	4,600	9.0
Meat cutting	610	33	520	7.4
Gut washer	15,120	643	13,200	6.0
Sausage department	560	136	800	7.3
Lard department	180	84	180	7.3
By-products	1,380	186	2,200	6.7

house showed an average BOD of 2000 ppm and a total nitrogen content of about 500 ppm at a flow of 5000 gallons per day.

The content of individual process wastes from a packinghouse is shown in Table 23.16. Table 23.17 presents the characteristics for combined slaughterhouse and packinghouse wastes.

Meat-plant wastes are similar to domestic sewage in regard to their composition and effects on receiving bodies of water. However, the total organic contents of these wastes are considerably higher than those of domestic sewages. On the other hand, the danger from pathogenic organisms in packing- and slaughterhouse wastes is slight when compared with domestic sewage. In the absence of adequate dilution, the principal deleterious effects of meat-plant wastes are oxygen depletion, sludge deposits, discoloration, and general nuisance conditions.

Rendering-plant wastes depend on the process used. If the "wet" process is used, a liquor waste containing high concentrations of organic matter, especially nitrogenous compounds, may result. Some plants evaporate the liquor and mix the residue with the product of the rendering process. Although "dry" rendering does not result in a liquor, there is a small amount of drainage and press liquor, most of which is returned to the rendering vat. Wash waters from the rendering plant may add considerable pollution material to the waste.

The total liquid waste from the poultry-dressing process contains varying amounts of blood, feathers, fleshings, fats, washings from evisceration, digested and undigested foods, manure, and dirt. The manure from receiving and feeding stations and blood from the killing and sticking operations contribute the largest amount of pollution from the process. The composition of poultry-plant wastes is given in Table 23.18. Some recent data on poultry-plant characteristics before and after various treatments are given by Vasuki and Sabis [1967, 45] (Table 23.19).

Table 23.17 Slaughterhouse and packinghouse wastes.

Type of kill	Volume per animal, gal	Suspended solids, ppm	Organic nitrogen, ppm	BOD, ppm	Population equivalent per animal
Mixed	359	929	324	2240	40.2
Cattle	395	820	154	996	19.6
Hogs	143	717	122	1045	7.5
Mixed	996	457	113	635	30.7
Cattle	2189	467		448	49.2
Hogs	552	633		1030	28.6

Table 23.18 Composition of poultry-plant wastes.

Characteristic	Concentration
Volume	3.26 gal/bird
Total solids	26.6 lb/1000 birds
Suspended solids	15.3 lb/1000 birds
Settleable solids	9.4 lb/1000 birds
Grease	1.3 lb/1000 birds
BOD, 5-day	30.0 lb/1000 birds

23.11 Treatment of Meat-Packing Wastes

In-plant recovery practices, screening, flotation, and biological treatment are the major methods used to treat meat-plant wastes. Some practices aimed at reducing the quantity or strength of wastes from meat plants are profitable, at least in larger plants. Among them are recovery of blood and grease, and utilization of tank waters and tankage-press liquors. Grease recovery is usually accomplished by means of baffled basins or traps on waste lines. Practically all plants recover the major portion of the blood from the killing operation. While separate disposal of paunch manure is seldom profitable, it is nonetheless desirable; even in plants where this is done, a certain amount of the liquid waste reaches the sewers. The extent to which waste loads from packinghouses are reduced by plant practices depends to some extent on whether or not such methods cost less than handling the waste load in a municipal disposal system. The latter cost, in turn, usually depends on the degree of difficulty the municipality encounters in removing the pollution at its treatment plant.

The most common methods used for treatment of meat-plant wastes are fine screening, sedimentation, chemical precipitation, trickling filters, and activated sludge. Screening by rotary wire-mesh screen removes materials such as hair, flesh, paunch, floating solids. Removals of 9 per cent of the suspended solids on a 20-mesh screen and 19 per cent on a 30-mesh screen have been reported. Sedimentation in Imhoff tanks is also satisfactory, being capable of removing 63 per cent of the suspended solids and 35 per cent of the BOD, with a 1- to 3-hour detention period. Trickling filters, at rates varying from 0.6 to 1.0 mgad, can give 81 to 90 per cent removal, with no accompanying nuisances. Activated-sludge treatment produces a satisfactory effluent after about 9 hours of aeration, at a rate of 3.5 cubic feet of air per gallon of waste. Experiences with double filtration in Mason City, Iowa, and Wells Forge, North Dakota, showed overall BOD reductions in excess of 95 per cent.

Few meat-plant wastes are chemically coagulated, because of the high costs. However, one plant uses $FeCl_3$ to reduce the BOD from 1448 to 188 ppm and the suspended solids from 2975 to 167 ppm. Operating costs were at one time reported to be $68 per million gallons. However, the effluent was sold for irrigation, grease was recovered, and sludge was utilized, thus reducing the net cost to about $25 per million gallons.

Chlorine and alum, if used in sufficient quantities, appreciably reduce the BOD and color of rendering-plant wastes and make possible improved clarification. Once again the chemical cost is high, but the BOD of raw wastes ranging from 1500 to 3800 ppm can be reduced to 400 to 600 ppm.

A trickling filter operating in conjunction with an air-flotation unit produced satisfactory results (61 per cent BOD reduction) at a loading of 2.6 pounds of BOD per cubic yard.

Poultry-plant wastes should and do respond readily to biological treatment; if troublesome materials such as feathers, feet, heads, and so forth, are removed beforehand, satisfactory biological treatment is attainable. Treatment facilities for one poultry-dressing plant's wastes include stationary screens in pits, septic tanks, and lagoons; an overall removal of 93 percent of the BOD was reported to result from the use of these measures. The author has designed a two-stage oxidation-basin treatment system which provides 85 to 90 per cent BOD reduction.

Table 23.19 Characteristics of poultry-plant waste before and after treatment.

Characteristic	Plant 1 Influent	Plant 1 Effluent	Plant 2 Influent*	Plant 2 Effluent	Plant 3 Influent	Plant 3 Effluent	Plant 4 Influent	Plant 4 Effluent	Plant 5 Influent†	Plant 5 Effluent	Plant 6 Influent‡	Plant 6 Effluent
BOD	30.6§	28.8	37.0	11.6	38.5		17.7	5.8	28.9		25.5	5.0
Organic nitrogen		2.6	6.1	2.3			1.8	1.3	1.9		2.7	0.8
Ammonia nitrogen	0.4	0.3	0.7	2.2			0.2	0.3	0.2		1.2	1.3
Nitrite nitrogen		0.003	0.010	0.001		Treated in municipal plant	0.006	0.004	0.006	Secondary treatment recently started	0.06	0.006
Nitrate nitrogen		0.7	0.4	0.1					0.2		0.9	0.1
Suspended solids	23.1	22.0	29.9	17.2	24.4		15.1	17.2	12.6		15.2	7.3
Volatile suspended solids	21.7	20.4	29.3	15.9	23.4		14.5	15.1	11.9		14.6	7.1
Total solids	41.5	41.6	72.4	48.4			36.0	34.3	29.8		42.7	33.1
Volatile total solids	33.0	28.4	52.9	29.3			20.6	18.6	22.0		21.6	13.8
pH	6.4	6.4	7.3	7.1	7.3		7.5	7.5	6.4		6.5	7.5
Kill rate (birds/day)	52,000		75,000		80,000		43,500		76,800		80,000	

* After settling.
† After screening.
‡ After flotation.
§ Results are given in pounds per 1000 birds, except for the pH.

References: Meat-Packing, Rendering, and Poultry-Plant Wastes

1. Aikins, G. A., "Meat packing waste disposal," *Ind. Wastes* **3**, 55 (1958).
2. "Anaerobic digestion, forum discussion," *Sewage Ind. Wastes* **25**, 723 (1953).
3. Beard, A. H., Jr., "Design of a grease recovery plant for a meat packer," in Proceedings of 4th Industrial Waste Conference, Purdue University, May 1947.
4. Beard, A. H., "Grease recovery plant for a meat packer, design," *Sewage Ind. Wastes* **22**, 1095 (1950).
5. Bolton, J. M., "Wastes from poultry processing plants," in Proceedings of 13th Industrial Waste Conference, Purdue University, May 1958.
6. Bradney, L., W. Nelson, and R. E. Bragstad, "Meat packing waste treatment," *Sewage Ind. Wastes* **22**, 807 (1950).
7. Bradney, L., W. Nelson, and R. E. Bragstad, "Waste treatment, Sioux Falls, S.D.," *Sewage Ind. Wastes* **22**, 807 (1950).
8. Bragstad, R. E., and L. Bradney, "Treatment with sewage, Sioux Falls, S.D.," *Sewage Ind. Wastes* **9**, 959 (1937).
9. Brévot, G., "France, disposal practice," *Sewage Ind. Wastes* **20**, 595 (1948).
10. Eldridge, E. F., "Textile mill waste treatment and recirculating filter for slaughterhouse wastes," *Mich. State Univ. Eng. Exp. Sta. Bull.* **96** (1942).
11. Carrick, C. W., "Use of industrial wastes in poultry feeding," in Proceedings of 6th Industrial Waste Conference, Purdue University, February 1951.
12. Eldridge, E. F., "Chemical treatment studies," *Sewage Ind. Wastes* **7**, 768 (1935).
13. Eldridge, E. F., *Industrial Waste Treatment Practice*, McGraw-Hill Book Co., New York (1942).
14. Eldridge, E. F., "High-rate filter treatment, Owosso, Mich.," *Sewage Ind. Wastes* **15**, 979 (1943).
15. Eldridge, E. F., "The meat packing plant waste disposal problem," *Mich. State Univ. Eng. Exp. Sta. Bull.* **105** (1946).
16. Eldridge, E. F., "Treatment methods, review," *Sewage Ind. Wastes* **20**, 181 (1948).
17. Farrell, L. A., "The why and how of treating rendering plant wastes," *Water Sewage Works* **100**, 172 (1953).
18. Foote, K. E., "Pork processing waste, effect on treatment plant," *Sewage Ind. Wastes* **24**, 1305 (1952).
19. Fridy, C. D., "Combining efficiency with economy in industrial waste treatment," *Consulting Engr.* **4**, 51 (1954).
20. Fullen, W. J., "Packing plant waste, anaerobic digestion," *Sewage Ind. Wastes* **25**, 576 (1953).
21. Fullen, W. P., and E. N. Anderson, "Superchlorination, Austin, Minn.," *Sewage Ind. Wastes* **3**, 761 (1931).
22. Gold, D. D., "Summary of treatment methods for slaughterhouse and packinghouse wastes," *Tenn. Univ. Eng. Exp. Sta. Bull.* **17** (1953).
23. Granstrom, M. L., "Rendering plant waste treatment studies," *Sewage Ind. Wastes* **23**, 1012 (1951).
24. Green, H. R., "Agreement between city and packers, Cedar Rapids, Iowa," *Sewage Ind. Wastes* **9**, 950 (1937).
25. Halvorson, H. O., "An innovation in meat-packing waste treatment," *Sewage Ind. Wastes* **3**, 334 (1931).
26. Halvorson, H. O., "Operating and economic factors involved in the study of packinghouse wastes," *Sewage Ind. Wastes* **25**, 170 (1953).
27. Hansen, P., and K. V. Hill, "Treatment with sewage, Austin, Minn.," *Sewage Ind. Wastes* **11**, 1045 (1939).
28. Heukelekian, H., H. E. Orford, and J. L. Cherry, "Chicken packinghouse waste characteristics," *Sewage Ind. Wastes* **22**, 521 (1950).
29. Hicks, R., "Waste treatment, Auckland, New Zealand," *Sewage Ind. Wastes* **28**, 594 (1956).
30. Hill, K. V., "Treatment practices, results," *Sewage Ind. Wastes* **17**, 292 (1945).
31. Hill, K. V., "Designing a combined treatment works for municipal sewage and packinghouse wastes at Austin, Minn.," in Proceedings of 13th Industrial Waste Conference, Purdue University, May 1958.
32. Hirlinger, K., and C. E. Gross, "Packinghouse waste, rapid analysis," *Sewage Ind. Wastes* **25**, 958 (1953).
33. Hirlinger, K., and C. E. Gross, "Packinghouse waste and trickling filter efficiency following air flotation," *Sewage Ind. Wastes* **29**, 165 (1957).
34. Howson, L. R., "Packinghouse waste treatment," *Water Works Sewerage* **87**, 217 (1940).
35. U.S. Public Health Service, *Industrial Waste Guide, Ohio River Survey*, Supplement D, (1943).
36. U.S. Public Health Service, *Industrial Waste Guide to the Meat Industry* (1954).
37. Jones, H. E., "British problem waste characteristics," *Sewage Ind. Wastes* **20**, 947 (1948).
38. Klassen, C. W., and W. H. Hasfurther, "Treatment of waste from small packinghouses," *Sewage Works Eng.* **20**, 136 (1949).
39. Knechtges, O. J., F. M. Dawson, and M. S. Nichols,

"Digestion with sewage sludge, experimental," *Sewage Ind. Wastes* **7**, 3 (1935).

40. Kountz, R. R., "Treatment of waste from small slaughterhouses," in Proceedings of 9th Industrial Waste Conference, Purdue University, May 1954.
41. Kountz, R. R., "Treatment of waste from slaughterhouses and poultry dressing," in Proceedings of 2nd Ontario Industrial Waste Conference, June 1955, p. 34.
42. Levine, M., "Treatment experiments," *Sewage Ind. Wastes* **7**, 316 (1935).
43. Levine, M., F. G. Nelson, and E. Dye, "Purification of packinghouse wastes," *Iowa Univ. Studies Eng. Bull.* 1937, p. 130.
44. Levine, M., H. N. Jenks, and F. G. Nelson, "Activated sludge and stream flow aeration experiments," *Sewage Ind. Wastes* **1**, 425 (1929).
45. Levine, M., F. G. Nelson, and E. Dye, "Treatment experiments, Mason City, Iowa," *Sewage Ind. Wastes* **9**, 530 (1937).
46. Lloyd, R., and G. C. Ware, "Anaerobic digestion of waste water from slaughterhouses," *Food Manuf.* **31**, 511 (1956).
47. Mahlie, W. S., "Treatment with sewage, Fort Worth, Texas," *Sewage Ind. Wastes* **15**, 521 (1943).
48. U.S. Public Health Service, "Meat wastes," *Ohio River Survey*, Supplement D (1943), p. 1145.
49. Miller, P. E., "Poultry wastes," in Proceedings of 6th Industrial Waste Conference, Purdue University, February 1951, p. 176.
50. Miller, P. E., "Poultry dressing wastes, sources and characteristics," *Sewage Ind. Wastes* **24**, 807 (1952).
51. Miller, P. E., "Spray irrigation at Morgan Packing Company," in Proceedings of 8th Industrial Waste Conference, Purdue University, May 1953.
52. Milling, M. A., and B. A. Poole, "Activated sludge treatment, Muncie, Ind.," *Sewage Ind. Wastes* **10**, 738 (1938).
53. Mohlman, F. W., "Treatment methods, review," *Sewage Ind. Wastes* **20**, 774 (1948).
54. Mohlman, F. W., and K. V. Hill, "Packinghouses," *Ind. Eng. Chem.* **44**, 498 (1952).
55. Mohlman, F. W., and K. V. Hill, "General review," *Sewage Ind. Wastes* **24**, 1444 (1952).
56. Mortenson, E. N., "Grease recovery in meat packing industry," in Proceedings of 1st Industrial Waste Conference, Purdue University, November 1944, p. 178.
57. Mortenson, E. N., "Grease recovery," *Sewage Ind. Wastes* **19**, 1101 (1947).
58. Nemerow, N. L., "South Buffalo creek sewage and industrial wastes," Report to the City of Greensboro, N.C., 1955.
59. Nichols, M. S., and J. C. Mackin, "Trickling filter experiments, Madison, Wis.," *Sewage Ind. Wastes* **2**, 435 (1930).
60. Porges, R., "Wastes from poultry dressing establishments," *Sewage Ind. Wastes* **22**, 531 (1950).
61. Putnam, E. G., "Waste prevention in the meat packing industry," in Proceedings of the Conference on Industrial Waste, April 1949, p. 43.
62. Roberts, J. M., "Combined treatment of poultry and domestic waste," *Sewage Ind. Wastes* **30**, 1186 (1958).
63. Rowntree, J. B., "General discussion, American practice," *Sewage Ind. Wastes* **24**, 1549 (1952).
64. Rudolfs, W., and V. Del Guercio, "Slaughterhouse waste digestion," *Water Sewage Works* **100**, 60 (1953).
65. Rupp, V. R., "Paunch manure, composting," *Sewage Ind. Wastes* **24**, 922 (1952).
66. Sanders, M. D., "Meat packing waste, chemical precipitation," *Sewage Works J.* **21**, 373 (1949).
67. Sanderson, W. W., "Studies of the character and treatment of wastes from duck farms," in Proceedings of 8th Industrial Waste Conference, Purdue University, May 1953.
68. Schroepfer, G. J., "New developments in packinghouse waste treatment," in Proceedings of 8th Industrial Waste Conference, Purdue University, May 1953.
69. Schroepfer, G. J., W. J. Fullen, A. S. Johnson, N. R. Ziemke, and J. J. Anderson, "The anaerobic process as applied to packinghouse waste," *Sewage Ind. Wastes* **27**, 460 (1955).
70. Singleton, K. B., "The investigation into the disposal of blood by anaerobic digestion," *Sewage Ind. Wastes* **29**, 1174 (1957).
71. "Slaughterhouse and packinghouse waste, bibliography bulletin," *Sewage Ind. Wastes* **24**, 176 (1952).
72. Sollo, F. W., "Pond treatment of meat packing plant wastes," in Proceedings of 15th Industrial Waste Conference, Purdue University, May 1960, p. 386.
73. Staub, C. P., "Statistical evaluation of packinghouse waste data," in Proceedings of 8th Industrial Waste Conference, Purdue University, May 1953.
74. Steffen, A. J., "What to do about paunch waste," in Proceedings of 3rd Industrial Waste Conference, Purdue University, May 1947.
75. Steffen, A. J., "Digestion, anaerobic, full-scale facility," *Sewage Ind. Wastes* **27**, 1364 (1955).

76. Steffen, A. J., "New developments in the treatment of meat processing wastes," in Proceedings of 5th Southern Municipal and Industrial Waste Conference, April 1956, p. 245.
77. Steffen, A. J., "The new and the old in slaughterhouse waste treatment processes," *Wastes Eng.* **28**, 401 (1957).
78. Stiemke, R. E., "Abattoir waste treatment," in Proceedings of Conference of Inter-American Association of Sanitary Engineers, March 1950, p. 307.
79. Stiemke, R. E., "Small abattoirs, disposal of wastes," *Sewage Ind. Wastes* **22**, 1380 (1950).
80. Stiemke, R. E., "Abattoir waste treatment, pilot plant studies," *Sewage Ind. Wastes* **23**, 1063 (1951).
81. Stilz, W. P., "Vibrating screens in industrial waste water treatment," *Ind. Wastes* **1**, 68 (1955).
82. "Symposium on meat packing plant wastes," in Proceedings of 1st Industrial Waste Conference, Purdue University, November 1944.
83. "Treatment plant, Beltsville Research Center, Md.," *Sewage Ind. Wastes* **8**, 157 (1936).
84. "Treatment with sewage, Cedar Rapids, Iowa," *Sewage Ind. Wastes* **8**, 352 (1936).
85. Uhlman, P. A., "Activated sludge treatment of rendering wastes," *Sewage Works Eng. Munic. Sanit.* **20**, 330 (1949).
86. Uhlman, P. A., "Packinghouse wastes treated by activated sludge," *Ind. Wastes* **1**, 72 (1955).
87. Van Kleeck, L. W., "How to treat meat wastes by filtration with sewage," *Wastes Eng.* **28**, 76 (1956).
88. Van der Leeden, R., "By-product recovery," *Sewage Ind. Wastes* **8**, 350 (1936).
89. Van Luven, A. L., "Treatment of packinghouse wastes," in Proceedings of 6th Ontario Industrial Waste Conference, June 1959, p. 141.
90. Watson, K. S., "Industrial waste treatment and recovery in West Virginia," in Proceedings of 4th Industrial Waste Conference, Purdue University, September 1948, p. 20.
91. West Virginia Water Commission, "Slaughterhouse waste treatment guide," *Sewage Ind. Wastes* **20**, 362 (1948).
92. Wolf, H. W., and W. T. Woodring, "Poultry dressing waste, small plants," *Sewage Ind. Wastes* **25**, 1429 (1953).
93. Wymore, A. H., "Design and operation of a waste treatment plant for a small packing plant," in Proceedings of 6th Industrial Waste Conference, Purdue University, February 1951, p. 413.
94. Wymore, A. H., "Small packing plant, waste treatment, design and operation," *Sewage Ind. Wastes* **24**, 923 (1952).

Suggested Additional Reading

The following references have been published since 1962.
1. Anderson, J. S., and A. J. Koplovsky, "Oxidation pond studies on evisceration wastes from poultry establishments," in Proceedings of 16th Industrial Waste Conference, Purdue University, May 1961.
2. Bryan, E. H., "Two stage biological waste treatment," *Ind. Water Wastes* **8**, 31 (1963).
3. Chesnova, L. M., and Trudov, "A 2-stage anaerobic fermentation process," *Sb. Tr. Leningr. Inst. Inzh. Stroct.* **47**, 93 (1964).
4. Clark, C. E., "Hog waste disposal by lagooning," *J. Sanit. Eng. Div. Am. Soc. Civil Engrs.* **91**, 27 (1965); *Chem. Abstr.* **64**, 13903 (1966).
5. Coerver, J. F., "Industry's idea clinic," *J. Water Pollution Control Federation* **36**, 944 (1964).
6. Coerver, J. F., "Anaerobic and aerobic ponds for packinghouse waste treatment in Louisiana," in Proceedings of 19th Industrial Waste Conference, Purdue University Engineering Extension Series, Bulletin no. 117, 1965, p. 200.
7. Cojacaru, R., "Fat separators in the internal sewer systems of food industry units," *Ind. Aliment.* **15**, 121 (1964).
8. "Converts waste-fat nuisance into solid income," *Food Process.* **23**, 41 (1962).
9. Dazai, M., M. Ogawa, S. Nomura, and H. Ono, "Treatment of industrial wastes by activated sludge. IX: Treatment of slaughterhouse wastes," *Kogyo Gijutsuin, Hakko Kenkyusho Kenkyu Hokoku* **27**, 85 (1965).
10. Deyl, Z., "Tower-like filters, the biological unit for slaughterhouse wastes," *Sb. Vysoke Skoly Zemedel. Brne* **8**, 113 (1965); *Chem Abstr.* **64**, 17245 (1966).
11. Deyl, Z., "Mechanism of destruction of hemaglobin in high rate percolating filters," *Vodni Hospodarstvi* **10**, 41 (1960).
12. Earle, R. L., "Aerobic digestion of meat wastes," in Proceedings of 7th Meat Industry Research Conference, Hamilton, New Zealand, 1965, p. 48.
13. Eby, H. J., *Disposal of Poultry Manure and Other*

Waste, U.S. Department of Agriculture, Agriculture Research Service Bulletin. A.R.S. 42.
14. Emdo, R., "Chemical precipitation of slaughterhouse wastes," *Hokkaido-ritsu Eisei Kenkyushoho* **12**, 178 (1961); *Chem. Abstr.* **62**, 158976 (1965).
15. Gates, C. D., "Treatment of Long Island duck farm wastes," *J. Water Pollution Control Federation* **25**, 1596 (1963).
16. "Georgia sewage plant for poultry wastes," *Water Waste Treat. J.* **9**, 132 (1962).
17. Hart, S. A., "Digestion tests of livestock wastes," *J. Water Pollution Control Federation* **35**, 748 (1963).
18. Hart, S. A., and M. E. Turner, "Lagoons for livestock manure," *J. Water Pollution Control Federation* **37**, 1578 (1965).
19. Hicks, R., "Self-purification of wastes from meat industries," in Proceedings of 7th Meat Industry Research Conference, Hamilton, New Zealand, 1965, p. 39.
20. Horasawa, I., and T. Suzuki, "Correlation between biochemical oxygen demand and chemical oxygen demand for slaughterhouse waste," *Suido Kenkyusho Hokoku* **2**, 47 (1963); *Chem. Abstr.* **60**, 1558 (1964).
21. Jackson, R., L. Bradney, and R. E. Bragstad, "Short-term aeration solves activated sludge expansion problems at Sioux Falls," *J. Water Pollution Control Federation* **37**, 255 (1965).
22. Joncour, J., "Apparatus for straining of effluents from abbatoirs," French Patent no. 1436052, 22 April 1966.
23. Lohr, R. C., "Packinghouse waste treatment," *Kansas Univ. Eng. Agr. Bull.* **50** (1962).
24. MacKenzie, D. S., "How meat packing company should plan to meet water and waste disposal problems," *Natl. Prov.* **149**, 103 (1963).
25. Macon, J. A., and D. N. Cote, "Study of meat packing wastes in North Carolina. I: Introduction and plant effluent studies," Carolina State College, Industrial Extension Service, August 1961.
26. Melzer, O., "Purification of slaughterhouse wastes in tower filters," *Sb. Vysoke Skoly. Chem.-Technol. Praze Oddil Fak. Technol. Paliv Vody* **8**, 135 (1965).
27. Nusbaum, I., and L. Burtman, "The determination of floatable matter in waste discharges," *J. Water Pollution Control Federation* **37**, 577 (1965).
28. Pesenson, I. B., "Removal of fat from waste waters by settling," *Ref. Zh. Khim.*, abs. no. 171343 (1961).
29. Polzin, P. G., "Effluent problems in meat processing factories," *Fleischwirtschaft* **44**, 1127 (1964).
30. Porges, R., "Industrial waste stabilization ponds in the United States," *J. Water Pollution Control Federation* **35**, 456 (1963).
31. Porges, R., and E. J. Struzeski, Jr., "Wastes from the poultry processing industry," Technical Report no. W62, Robert A. Taft Sanitary Engineering Center, U.S. Public Health Service, Cincinnati, Ohio (1962).
32. Rollag, D. A., and J. N. Dornbush, "Design and performance evaluation of an anaerobic stabilization pond system for meat processing wastes," *J. Water Pollution Control Federation* **38**, 1805 (1966).
33. Schaffer, R. B., "Polyelectrolytes in industrial waste treatment," in Proceedings of 18th Industrial Waste Conference, Purdue University, 1963, p. 447.
34. Schraufnagel, F. H., "Ridge-and-furrow irrigation for industrial waste disposal," *J. Water Pollution Control Federation* **34**, 1117 (1962).
35. Steffen, A. J., and R. M. Leman, "Degasifying digester liquor," *Wastes Eng.* **32**, 400 (1961).
36. Steffen, A. J., "Operating experiences in anaerobic treatment of packinghouse wastes," in Proceedings of Research Conference, Research Advisory Council, American Meat Institute Foundation, University of Chicago, no. 64 (1961), p. 81.
37. Steffen, A. J., "Operation of full-scale anaerobic contact treatment plant for meat packing wastes," in Proceedings of 16th Industrial Waste Conference, Purdue University, May 1961.
38. Steffen, A., and M. Bedker, "Separation of solids in the anaerobic contact process," *Public Works Mag.* **91**, 100 (1960).
39. Steffen, A. J., "Stabilization ponds for meat packing wastes," *J. Water Pollution Control Federation* **35**, 440 (1963).
40. "Strides in waste disposal," *Food Eng.* **35**, 85 (1963).
41. "Swift waste treatment by ponding low in cost," *Natl. Prov.* **143**, 93 (1960).
42. Taiganides, E. P., "Characteristics and treatment of wastes from a confinement hog production unit," *Dissertation Abstr.* **24**, 213 (1963).
43. Teletzke, G. H., "Chicken for the barbecue—wastes for aerobic digestion," *Wastes Eng.* **32**, 135 (1961).
44. "Unique Wilson waste plant cuts BOD to low level," *Natl. Prov.* **143**, 26 (1960); U.S. Department of the Interior, *Meat Products*, Industrial Profiles, Report no. 8, 1967.
45. Vasuki, N. C., and W. R. Sabis, "In-plant control of poultry waste discharges," Paper presented at 1st

Mid-Atlantic Industrial Waste Conference, University of Delaware, November 1967, p. 143.
46. Wagemann, H., "Suggestions for the drainage of slaughterhouse waste water into public sewers," *Fleischwirtschaft* **15**, 425 (1963).
47. Whiten, G., "Treating combined sewage and poultry wastes," *Public Works Mag.* **92**, 98 (1961).
48. Willoughby, E., and V. D. Patton, "Design of a modern meat packing waste treatment plant," *J. Water Pollution Control Federation* **40**, 132 (1968).

23.12 Feedlot Wastes

Large-scale livestock operations have removed animals from pasturage and now handle large numbers in small confinement areas (feedlots), where feed and water are brought to the livestock. Poultry, cattle, and swine are the major animals involved. Loehr [8] gives the average characteristics of these wastes as obtained from Hart and Turner [6] (Table 23.20). It is reported that, on a BOD basis, one dairy cow is the equivalent of 20 to 25 persons, one beef animal, 18 to 20 persons, one hog, 2 or 3 persons, and that 10 to 15 chickens are the equivalent of one person. Cassell [2] reports that in 1963 the American poultry industry produced approximately 63 billion eggs for consumption, from a hen population of about 365 million birds. It is also estimated that about 5 million tons of organic material was produced by the laying-hen industry. The farmer has traditionally disposed of chicken manure by spreading on land. This is no longer desirable because the attendant odors and flies offend nearby residents and vacationers. Chicken

Table 23.20 Average characteristics of the surface liquid-manure lagoons. (From Loehr [8].)

Characteristic	Poultry manure					Dairy manure			Swine manure		
Loading, lb volatile solids/ft³	0.001	0.004	0.006	0.010	0.016	0.004	0.008	0.01	0.005	0.01	0.014
pH	7.6	7.0	7.3	7.1	7.3	7.4	7.0	7.0	6.9	6.7	6.3
Total solids, mg/liter	600	1600	5500	4000	8000	2200	7000	7000	1500	2200	3000
Volatile solids, % of TS	55	45	45	50	60	70	75	75	60	60	65
BOD$_5$, mg/liter	40	100	300	300	1500	500	500	1200	500	500	1000
Alkalinity, mg/liter	400	1300	5500	4800	7500	1300	2500	3500	700	2000	3000

Table 23.21 Effluent quality of field livestock waste lagoons.

Parameter	Type of livestock		
	Swine	Swine	Poultry
Loading rate,* lb volatile solids/day/1000 ft³	0.36–3.9	10–15	4–11
pH	6.7–8.0		6.8–7.9
Total solids, mg/liter		1000–1500	
Volatile solids, mg/liter	850–2330	400–800	
Volatile acids, mg/liter	72–528		
Alkalinity, mg/liter	1120–2220		
BOD$_5$, mg/liter			320–1350
COD, mg/liter	940–3850	500–1300	590–2550
Total nitrogen, mg/liter		150–350	113–290

*Note: lb/1000 ft³ × 16 = g/m³.

Table 23.22 Production of wastes by livestock in the United States. (From Wadleigh [1968, 9].)

Livestock	U.S. animal population (1965), millions	Solid wastes, gal per capita/day	Total production of solid waste, million tons/yr	Liquid wastes, gal per capita/day	Total production of liquid wastes, million tons/yr
Cattle	107	23,600	1004.0	9000	390.0
Horses	3	16,100	17.5	3600	4.4
Hogs	53	2,700	57.3	1600	33.9
Sheep	26	1,130	11.8	680	7.1
Chickens	375	182	27.4		
Turkeys	104	448	19.0		
Ducks	11	336	1.6		
Total			1138.6		435.4

manures vary from 10 to 15 per cent total solids, of which about 75 per cent is organic matter with a BOD of about 30,000 ppm. Cassell [2] reports a wet weight of 0.1 to 0.4 pound per hen, a wet volume of 6 to 8 cubic inches per hen, and about 0.01 to 0.015 pounds of BOD per hen.

Cassell found chicken manure conditioned with anionic-cationic polyelectrolyte mixtures can be effectively dewatered and the volume reduced by as much as 65 per cent. In laboratory experiments, he found that the following conditions were necessary to digest chicken manure anaerobically:

pH	7.5
Volatile acids	1500 mg/liter
Alkalinity	10–12,000 mg/liter
NH_3–N	1500 mg/liter
Detention time	20 days
Loading	0.088 lb volatile solids/ft^3/day
Temperature	35°C
Na^+	0.018 molar

Loehr found that treatment and disposal of animal wastes are complicated by the nature of the wastes, the volume to be handled, the lack of interest of the livestock producer in waste treatment, and the proximity of the suburban population. In the past, most of the wastes have been recycled through the soil environment with a minor release to the nearby waters. He found that anaerobic treatment in lagoons offers a possible approach for handling the tremendous quantities of manure that originate from confinement feedlot operations, but admits that this is not the complete answer to the problem. In combination with units to treat the effluent from the lagoons, anaerobic lagoons may be a useful *part* of the process for treating livestock and feedlot wastes that have a high solids content.

Clark [3] and Dornbush and Anderson [4] give data on the character of effluent from two-cell anaerobic lagoons treating hog and poultry feedlot wastes (Table 23.21). In a recent publication, Wadleigh [1968, 9] provides information on the amount of solid and liquid wastes produced by livestock in the United States (Tables 23.22 and 23.23). Liquid wastes are

Table 23.23 Population equivalent of the fecal production by animals, in terms of biochemical oxygen demand (BOD). (From Wadleigh [1968, 9].

Biotype	Fecal, gal per capita	Relative BOD per unit of waste	Population equivalent
Man	150	1.0	1.0
Horse	16,100	0.105	11.3
Cow	23,600	0.105	16.4
Sheep	1,130	0.325	2.45
Hog	2,700	0.105	1.90
Hen	182	0.115	0.14

reported to amount to over 400 million tons per year.

The New York State Health Department has reported on the treatment of duck-farm wastes from Suffolk County, representing 6,250,000 ducks (60 to 70 per cent of the nation's total). The waste waters from these farms contain two major objectionable impurities, manure and waste grain. Each of the 21 farms with treatment facilities has an aerated lagoon with 5-day detention, two or three sedimentation tanks in parallel with 12-hour detention (only one is used at a time while the others dewater and dry), and a chlorine contact chamber.

References: Animal Feedlot Wastes

1. Adams, J. L., "Hydraulic manure systems," in Proceedings of 2nd National Symposium on Poultry Industry Waste Management, University of Nebraska, 1964.
2. Cassell, E. A., "Studies on chicken manure disposal. I," New York State Health Department, Laboratory Studies, Research Report no. 12, 1966.
3. Clark, C. E., "Hog waste by lagooning," *J. Sanit. Eng. Div. Am. Soc. Civil Engrs.* **91**, 27 (1965).
4. Dornbush, J. N., and J. R. Anderson, "Lagooning of livestock wastes," in Proceedings of 19th Industrial Waste Conference, Purdue University Engineering Extension Series, Bulletin no. 117, 1964, p. 317.
5. Hart, S. A., "Digestion of livestock wastes," *J. Water Pollution Control Federation* **35**, 748 (1963).
6. Hart, S. A., and M. E. Turner, "Lagoons for livestock manure," *J. Water Pollution Control Federation* **37**, 1578 (1965).
7. Johnson, C. A., "Liquid handling processes for poultry manure utilization," in Proceedings of 2nd National Symposium on Poultry Industry Waste Management, University of Nebraska, 1964.
8. Loehr, R. C., "Effluent quality from anaerobic lagoons treating feedlot wastes," *J. Water Pollution Control Federation* **39**, 384 (1967).
9. Wadleigh, C. H., *Wastes in Relation to Agriculture and Forestry*, U.S. Department of Agriculture, Miscellaneous Publication no. 1065, March 1968.

BEET-SUGAR WASTES

The process of extracting sugar from beets is essentially the same in all factories in the United States. The sugar "campaign" usually starts in the fall and lasts from 60 to 100 days, operating 24 hours a day during this "on" season. The majority of factories operate only what is known as the "straight house," in which sugar is extracted to the point at which a heavy molasses is obtained. Some factories also operate Steffen houses, in which powdered lime is added to the beet molasses to precipitate the sugar as calcium sucrate [18]. A typical beet-sugar plant conducts the following operations [50]: (1) beets are weighed, unloaded, screened, and washed; (2) beets go to slicers and diffusers, and the resulting pulp is sent to pulp presses and then to storage, from which raw juice is withdrawn; (3) product (raw juice) is put through first and second carbonation tanks, where lime and CO_2 are added; (4) in the sulfidizer, SO_2 is added; (5) product is filtered and the cake is removed; (6) residue is evaporated; (7) the evaporated liquor is centrifuged to draw off sugar; (8) sugar is dried and stored. Southgate [55] presents a typical flow sheet of these processes (see Fig. 23.7).

23.13 Origin and Characteristics of Beet-sugar Wastes

There are five sources of waste water at a beet-sugar plant which employs the Steffen process: (1) the flume (transport) water, which is used to wash the beets and to transport them from stockpiles in the factory to the site where they are to be processed; (2) the process waste water, consisting of (a) the battery-wash water, from the operation of flushing the exhausted (desugared) cossettes (sliced beets) from the diffusion battery cells, and (b) the pulp-press water, from the partial dewatering of the exhausted pulp; (3) the lime-cake or lime-slurry residue from the carbonation process; (4) the condensate from the multiple-effect evaporators and vacuum pans used to concentrate the sugar solution; (5) the Steffen waste, resulting from the extraction of sugar from the straight-house molasses by the Steffen process.

Flume wash waters vary considerably in content of

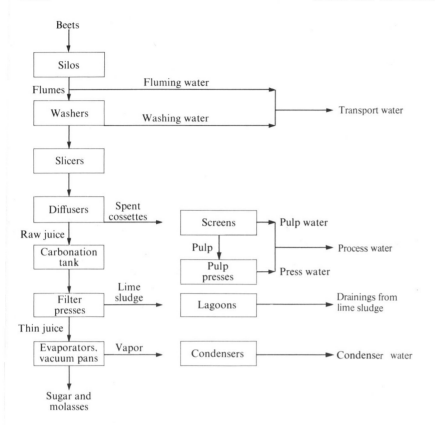

Fig. 23.7 Flow sheet showing discharges of waste waters from a beet-sugar factory. (After Southgate [55].)

soil, stones, beet leaves, roots, and dissolved solids. Pulp-press water is high in organic material and suspended solids, as is the lime-cake slurry. Condenser water may contain organic matter entrained with the vapors from the last effect. The flume water represents about 72 per cent of the total beet-sugar waste volume; the BOD of this water, however, is comparatively low, about on a level with domestic sewage. Data collected by Eldridge [18] give the composition of beet-sugar factory wastes (Table 23.24); similar data by Pearson and Sawyer [41] are presented in Table 22.25. More recently, Rogers and Smith [50] found the general analysis to be as presented in Table 23.26.

The Steffen process consists of diluting the molasses to a specific concentration and treating it with enough powdered lime to precipitate calcium sucrate, which can be removed from the liquor by filtration, and the sugar recovered by treating with carbon dioxide.

Table 23.24 Beet-sugar factory wastes. (After Eldridge [18].)

Characteristic	Flume water	Process water	Lime-cake drainage	Steffen waste
Volume, gal/ton beets	2200	660	75	120
BOD, ppm	200	1230	1420	10,000
Total solids, ppm	1580	2220	3310	43,600
Suspended solids, ppm	800	1100	450	700
Dissolved solids, ppm	780	1120	2850	42,900

23.14 Treatment of Beet-sugar Wastes

Elimination of all unnecessary water usage and reuse of waste waters should precede any actual treatment in waste-treatment plant is uneconomical. Some ferable, because large volumes are discharged during a relatively short period of the year, so that investment in waste-treatment plant is uneconomical. Some

Table 23.25 Beet-sugar wastes. (After Pearson and Sawyer [41].)

Characteristic	Flume water	Process waste water
Suspended solids, ppm	400	1300
Volatile solids, %	35	75
Total solids, ppm		3800
BOD, ppm	200	1600
COD, ppm	175	1500
Protein-N, ppm	10	65
NH_3-N, ppm	3	15
Sucrose, ppm	100	1500
Volume, gal/ton beets	2000–3000	325

Table 23.26 Beet-sugar wastes. (After Rogers and Smith [50].)

Characteristic	Amount or value
BOD, ppm	445
Total solids, ppm	6470
Suspended solids, ppm	4920
pH	7.9
Alkalinity, ppm	250

factories have had satisfactory experiences with the reusing of flume water. Battery wash water and pulp-press water may be successfully eliminated by the use of somewhat more costly and less efficient diffusers, known as rack continuous diffusers. The lime-cake drainage may be eliminated by "dry" transportation from the pressure filters.

Any actual treatment of beet-sugar wastes is accomplished mainly by the use of lagoons. With lagooning and land spraying, BOD reductions of 50 to 67 per cent have been obtained [27, 44, 50]. Coagulation has also been used [18, 41] and may be followed by sedimentation and/or biological filtration, if conditions require a higher degree of purification.

Hopkins *et al.* [27] found that, if total beet-sugar wastes were discharged uniformly across the upper end of 5-acre shallow lagoons, with a detention time of about one day, virtually all suspended solids, 55 per cent of the concentration of BOD, and 63 per cent of the pounds of BOD were removed. This procedure also reduced the alkalinity by 69 per cent, completely eliminated nitrate nitrogen, and reduced ammonia nitrogen by 94.3 per cent. Coliform-type bacteria increased, but phosphates were unchanged. Water loss was 3.27 acre feet per day, of which 0.18 acre foot was due to evaporation and 3.09 acre-feet due to infiltration.

Five methods have been receiving attention recently:

1. Absorption by sawdust, calcined coke, coke, and slag is effective in removing organics and sugar found in sugar wastes; for example, with sawdust, 96 per cent of organics and 50 per cent of sugar can be removed [1966, 9].
2. Total reuse of waste water is being included in the design of new plants [1964, 15; 1966, 21].
3. Aerobic oxidation ponds may give up to 90 per cent overall BOD reduction when they follow anaerobic digestion (of cane-sugar wastes) [1966, 2].
4. Activated-sludge treatment has given 95 per cent BOD removal in laboratory scale units [1963, 7], and 70 per cent in pilot-plant operations when mixed with sewage [1964, 16].
5. The following patented process [1965, 8] represents a novel approach. Highly alkaline wastes, separated from the solids, are sparged with CO_2, the precipitated $CaCO_3$ is separated, and the pH is further reduced with mineral acid. The BOD of the liquor after clarification is reduced by fermentation with yeast, the CO_2 evolved is used for $CaCO_3$ precipitation and excess yeast produced is used for forage.

References: Beet-Sugar Wastes

1. Allen, L. A., A. H. Cooper, A. Cairns, and M. C. Maxwell, "Microbiology of beet sugar production," *Sewage Ind. Wastes* **20**, 596 (1948).
2. Allen, L. A., A. H. Cooper, and M. C. Maxwell, "Microbiological problems in manufacture of beet sugar wastes," *Sewage Ind. Wastes* **21**, 942 (1949).
3. Ballou, F. H., "New Taber beet sugar factory," *Eng. J.* **33**, 791 (1950).
4. "Beet sugar waste, treatment by chlorination," *Chem. Eng. News* **28**, 1038 (1951).

5. Berlin, A. M., "Factory waste water disposal," *Sugar* **36**, 22 (1941).
6. Besselievre, E. B., *Industrial Waste Treatment*, McGraw-Hill Book Co., New York (1952).
7. Black, H. H., and G. N. McDermott, "Industrial waste guide," *Sewage Ind. Wastes* **24**, 181 (1952).
8. Brandon, T. W., "Waste waters from beet sugar factories," *Intern. Sugar J.* **49**, 98 (1947).
9. Brandon, T. W., "Reuse of process water," *Sewage Ind. Wastes* **20**, 360 (1948).
10. "British problem," *Sewage Ind. Wastes* **4**, 721 (1932).
11. Burdick, R. T., "Water, lifeblood of sugar beet," *Sugar* **42**, 28 (1947).
12. Carlton, "Holly's new plant in California," *Sugar* **42**, 29 (1948).
13. Daniels, E. M., "Screens for removing trash at beet receiving stations," *Sugar* **47**, 44 (1952).
14. Dennis, J. M., "Spring irrigation of food processing waste," *Sewage Ind. Wastes* **25**, 591 (1953).
15. Eldridge, E. F., and F. R. Theroux, "Chemical treatment," *Sewage Ind. Wastes* **7**, 769 (1935).
16. Eldridge, E. F., "Full-scale experimental plant results, Michigan," *Sewage Ind. Wastes* **9**, 531 (1937) and **10**, 913 (1938).
17. Eldridge, E. F., "Experimental plant, Monitor Sugar Co., Bay City, Mich.," *Sewage Ind. Wastes* **12**, 658 (1940) and **13**, 105 (1941).
18. Eldridge, E. F., *Industrial Waste Treatment Practice*, McGraw-Hill Book Co., New York (1942), p. 84.
19. Eldridge, E. F., and F. R. Theroux, "Michigan experiments, chemical and biological treatment," *Sewage Ind. Wastes* **5**, 881 (1933).
20. Eldridge, E. F., and F. R. Theroux, "Studies on the treatment of beet sugar factory wastes," *Mich Univ. Eng. Exp. Sta. Bull.* **51** (1933).
21. Eldridge, E. F., and F. R. Theroux, "Steffens waste, spray drying studies," *Sewage Ind. Wastes* **9**, 533 (1937).
22. Fleming, G., "Treatment of reused water in beet manufacture," *Sewage Ind. Wastes* **24**, 1383 (1952).
23. Garner, J. H., and J. M. Wishart, "Treatment and disposal methods, West Riding of Yorkshire," *Sewage Ind. Wastes* **4**, 384 (1932).
24. Garner, J. H., and J. M. Wishart, "Second supplementary report upon purification of waste waters from beet sugar factories," West Riding of Yorkshire Rivers Board, Report no. 169, 1931.
25. Griffin, A. E., "Treatment by chlorine of industrial wastes," *Eng. Contr. Record* **63**, 74 (1950).
26. Haupt, H., "By-product recovery," *Sewage Ind. Wastes* **8**, 350 (1936).
27. Hopkins, G., "Evaluation of broad field disposal of sugar beet wastes," *Sewage Ind. Wastes* **28**, 1466 (1956).
28. U.S. Public Health Service, *Industrial Waste Guide to Beet Sugar Industry,* Washington, D.C. (1950).
29. Jensen, L. T., "Recent developments in waste water treatment by the beet sugar industry," in Proceedings of 10th Industrial Waste Conference, Purdue University, May 1955.
30. Kalda, D. C., "Treatment of sugar beet wastes by lagooning," in Proceedings of 13th Industrial Waste Conference, Purdue University, May 1958.
31. Levine, M., and G. H. Nelson, "Stream-flow aeration purification studies," *Sewage Works J.* **1**, 40 (1928).
32. Levine, M., "Iowa factory, characteristics and disposal of wastes," *Sewage Ind. Wastes* **5**, 884 (1933).
33. Levine, M., "Purification of beet sugar wastes," *Am. J. Public Health* **23**, 585 (1951).
34. McDill, B. M., "Beet sugar industry," *Ind. Eng. Chem.* **39**, 657 (1947).
35. MacDonald, J. C., "Reuse of process waters in beet sugar factory," *Intern. Sugar J.* **46**, 208 (1944).
36. MacDonald, J. C., "Return of beet sugar factory waste waters," *Intern. Sugar J.* **47**, 100 (1945).
37. McGinnis, R. A., *Beet Sugar Technology*, Reinhold Publishing Corp., New York (1951).
38. O'Day, D., and E. Bartow, "Glutamic acid content of Steffens waste," *Sewage Ind. Wastes* **16**, 385 (1944).
39. American Public Health Association Committee Report, "Oxidation ponds," *Sewage Works J.* **20**, 1025 (1948).
40. Pailthorp, R. E., "Color problems with beet waste," *J. Water Pollution Control Federation* **32**, 1201 (1960).
41. Pearson, E., and C. N. Sawyer, "Recent developments in chlorination in the beet sugar industry," in Proceedings of 5th Industrial Waste Conference, Purdue University, November 1949, p. 110.
42. Pearson, E. A., and C. N. Sawyer, "Waste disposal," *Chem. Eng. Progr.* **46**, 337 (1950).
43. Pick, H., "Chlorination, Wischau, Germany," *Sewage Ind. Wastes* **2**, 465 (1930).
44. Porges, R., and G. Hopkins, "Broad field disposal of beet sugar waste," *Sewage Ind. Wastes* **27**, 1160 (1955).
45. Porter, L. B., "Beet juice purification by ion exchange," *Sugar* **42**, 22 (1947).

46. Department of Scientific and Industrial Research, *Purification of Waste Waters from Beet Sugar Factories,* Water Pollution Research Technical Paper no. 3, H.M.S.O., London (1933).
47. Richards, E. H., and D. W. Cutler, "British experiments, chemical and biological treatment," *Sewage Ind. Wastes* **5**, 877 (1933).
48. Riley, F. R., and W. E. Sanborn, "Ion exchange process has matured," *Sugar* **42**, 24 (1947).
49. Rogers, T. H., *Report of Technical Commission,* Upper Mississippi Board of Sanitary Engineering, February 1947.
50. Rogers, H. G., and L. Smith, "Beet sugar waste lagooning," in Proceedings of 8th Industrial Waste Conference, Purdue University, May 1953, p. 136.
51. Rudolfs, W., *Industrial Wastes,* Reinhold Publishing Corp., New York (1953), p. 473.
52. Sanborn, N. H., "Spray irrigation effluent disposal," *Sewage Ind. Wastes* **25**, 1034 (1953).
53. Sawyer, C. N., "Beet sugar treatment processes," *Sewage Ind. Wastes* **22**, 221 (1950).
54. Southgate, B. S., "Reuse of waste water," *Sewage Ind. Wastes* **20**, 169 (1948).
55. Southgate, B. S., *Treatment and Disposal of Industrial Waste Waters,* H.M.S.O., London (1948), p. 317.
56. Spengler, O., "German disposal practice," *Sewage Ind. Wastes* **7**, 131 (1935).
57. Stilz, W. P., "Vibrating screens in industrial waste water treatment," *Ind. Wastes* **1**, 68 (1955).
58. Stone, P., "Sugar cane process waste, origin and treatment," *Sewage Ind. Wastes* **23**, 1025 (1951).
59. Streeter, H. W., "Review of beet sugar wastes," *Sewage Ind. Wastes* **5**, 876 (1933).
60. National Lime Association, *The Use of Lime in Industrial Waste Treatment,* Trade Waste Bulletin no. 1, Washington, D.C. (1948), p. 9.
61. Wintzell, T., and T. Lauritzson, "Return of beet sugar factory waste water," *Sugar* **39**, 26 (1944) and **40**, 28 (1945).
62. Wood, A., J. H. Gorvin, and B. A. Forster, "British problem; treatment recommendations," *Sewage Ind. Wastes* **4**, 387 (1932).

Suggested Additional Reading

The following references have been published since 1960.

1. Banerjea, S., and M. P. Motwani, "Pollution of the Suvaon stream by the effluents of a sugar factory," *Indian J. Fisheries* **7**, 107 (1961).
2. Bhaskaran, T. R., and R. N. Chakrabarty, "Pilot plant for treatment of cane sugar waste," *J. Water Pollution Control Federation* **38**, 1160 (1966).
3. Bonomo, L., "Treatment of beet molasses distillery wastes," *Ing. Sanit.* **3**, 101 (1966); *Chem. Abstr.* **67**, 57073 (1967).
4. Carruthers A., et al., "The composition and treatment of sugar beet factory waste waters," Paper presented at 13th Technical Conference British Sugar Corp., 1960; *Water Pollution Abstr.* **34**, 343 (1961).
5. Chekurda, A. F., et al., "Improving the clarification of sugar industry wash waters," *Sakharn. Prom.* **35**, 28 (1961).
6. Congdon, J. V., and S. B. Zaiatz, "Development of a new autoanalyzer method for continuous measurement of sugar in refinery waste water in the presence of interfering agents," Annual Meeting Sugar Industry Technicians, Technical Paper no. 20, 1961, p. 30.
7. Dazai, M., Y. Yoshida, and H. Ono, "Treatment of industrial wastes by activated sludge. II: Treatment of beet sugar wastes," *Kogyo Gijutsuin, Hakko Kenkyusho Kenkyu Hokoku* **24**, 223 (1963); *Chem. Abstr.* **64**, 17242 (1966).
8. Ebara-Infilco Co., Ltd., "Procedure for the treatment of residual liquors from the production of beet sugar," French Patent no. 1398742, 14 May 1965; *Chem. Abstr.* **64**, 4786 (1966).
9. Gajban, C., "Absorbents for purification of waste water of sugar factories," *Ind. Aliment. Produse Vegetale* **17**, 372 (1966); *Chem. Abstr.* **68**, 2447f (1968).
10. Guzman, R. M., "Control of cane sugar wastes in Puerto Rico," *J. Water Pollution Control Federation* **34**, 1213 (1962).
11. Kramer, D., "Composition of waste waters from sugar factories and the necessity for and usual methods of treatment," *Wasserwirtsch.-Wassertech.* **8**, 499 (1958); *Water Pollution Abstr.* **34**, 278 (1961).
12. Leclere, E., and F. Edeline, "Study of the solution of organic material in the water circuit for carriage and washing of sugar beets in a sugar refinery," *Water Pollution Abstr.* **34**, 2 (1961).
13. Muehlpforte, H., "Recycling of flushing and washing water in sugar factories," *Wasserwirtsch.-Wassertech.* **11**, 262 (1961).
14. Muhlpforte, H., "Reuse of diffusion and press water in sugar factories," *Water Pollution Abstr.* **34**, 278 (1961).
15. Laughlin, J. E., "Waste treatment in new beet sugar plant," in Proceedings of 4th Texas Water Pollution

Control Association Industrial Water and Waste Conference, 1964, p. D2.
16. Shukla, J. P., and N. K. Varma, "Examination of activated sludge process for treatment of sugar factory effluents," *Proc. Ann. Conv. Sugar Technologists' Assoc. India* **32**, 62 (1964).
17. Shukla, J. P, et al., "Treatment of sugar factory effluent by high-rate trickling filter," *Proc. Ann. Conv. Sugar Technologists' Assoc. India* **28**, 231 (1960).
18. Skalski, K., "Waste-water economy of Austrian sugar factories," *Sugar Ind. Abstr.* **23**, 183 (1961).
19. Sladecek, V., Z. Cyrus, and A. Borovickova, "Hydrobiological investigations of a treatment of beet sugar factory's wastes in an experimental lagoon," *Sb. Vysoke Skoly Chem.-Technol. Praze Oddil Fak. Technol. Paliv. Vody* **2**, 7 (1958).
20. *Sugar Ind. Abstr.* **23**, 583 (1961).
21. Teranishi, I., "Sugar company gets new waste treatment facility," *Water Pollution Control* **104**, 30 (1966); *Eng. Index*, January 1967, p. 111.
22. Tuempling, W., and W. Mueller, "Influence of sugar factory sewage, pretreated in a settling basin, on the biological and chemical state of outfalls in the Unstret area," *Wasserwirtsch.-Wassertech.* **11**, 257 (1961).

MISCELLANEOUS FOOD-PROCESSING WASTES

Wastes from the manufacturing of such foods as coffee, rice, fish, pickles, beverages, baked goods, candy, and drinking water, are somewhat less prevalent than other food-processing wastes described in this chapter, but are nonetheless important and some study of them is advisable.

23.15 Coffee Wastes

The major part of our coffee comes from South America and most of it used to be processed there, but an increasing amount of raw coffee is being shipped to this country for processing. Since washed coffee of good quality receives the highest price in the United States, most of the coffee beans are washed before shipment; that is, the ripe berry is picked and milled in a process which requires the use of water. This process is differentiated from that of the "dry" coffee, in which the berry is picked from the tree and the hull removed by dry milling. The water requirement for washing is about 260 gallons per 100 pounds of finished coffee, so, especially if the washing is done in the same plant as the blending and roasting, a significant pollution problem may exist.

The coffee cherry (the de-hulled berry) is dumped into a receiving vat from which it is conveyed to the pulpers by water. During this conveyance, stones and other debris are separated by means of traps and floaters; unsound cherries are diverted to separate pulpers. The pulper removes the skin and a large proportion of the flesh from the coffee bean. The hulled bean is then transported by water to a fermentation vat, where it is allowed to remain and ferment in a moist state, the excess water being drained off and reused if necessary. The period of fermentation may be as short as twelve hours or as long as two days. Fermentation is necessary before all the flesh of the cherry can be removed from the parchment which immediately surrounds the silver skin of the coffee bean. The protopectin in the flesh is insoluble and cannot be washed or scrubbed off; no machine has been developed that can satisfactorily remove the unfermented flesh and undried parchment from the bean. Because the protopectin adheres to the tough parchment, the bean cannot be satisfactorily used unless fermentation of the bean itself occurs. The fermentation process makes it possible to dry the bean in a clean parchment which guarantees its purity.

The theory of fermentation in coffee processing is as follows:

1) Protopectin $\xrightarrow{\text{Protopectinase}}$ Pectin,

2) Pectin $\begin{cases} \xrightarrow{\text{Pectase}} \\ \text{Pectinase} \end{cases}$ $\begin{cases} \text{Pectic acid} + CH_3OH \\ \text{Galacturonic acid} \\ \quad + CH_3OH, \end{cases}$

3) Pectic acid + Calcium $\longrightarrow$ Calcium pectate (a soluble gel).

These products, some of which are soluble, are readily removed from the parchment of the coffee bean by washing. After fermentation, the beans are washed and conveyed to the drying patios, again by water. They are screened from the water and spread out to dry in the sun for several days; some plants use mechanical driers in addition to sun-drying. When the beans have dried sufficiently to ensure color and flavor, they are milled to remove the parchment, then graded, sacked, and shipped to the markets.

The main uses of water (and the origin of wastes) in coffee mills are (1) to transport beans to pulpers, (2) to transport pulp to a hopper or pile, (3) to transport beans to fermentation vats, (4) to wash fermented beans, (5) to transport fermented beans to drying patios, (6) miscellaneous uses, such as acting as a trap for stones and as a method of separating "floaters," for hydraulic classification of beans, and as boiler water.

The elements which make up the coffee cherry or bean are shown in Table 23.27. The four major wastes from processing the coffee bean are pulp, pulping

Table 23.27 Composition of coffee bean.

Component	Average %
Water	9–12
Ash	4
Nitrogen	12
Cellulose	24
Sugars	9
Dextrin	1–15
Fat	12
Caffetannic acid	8–9
Caffeine	0.7–1.3
Nitrogen-free extract	18
Essential oil	0.7
Water-soluble material	25.3

Table 23.28 Coffee wastes.

Source	BOD, ppm	Settleable solids, ppm	Total solids, ppm	Suspended solids, ppm
Pulp	47,000			
Fermentation wastes	1,250–2,220	660–700	4260	2060
Pulping waste	1,800–2,920	60–127	4960	848
Combined pulp and fermentation wastes	6,150–134,000	160	3220	

Table 23.29 Average results of examination of waste waters from processing of coffee (1946). (After Brandon [1].)

Waste water	Volume, gal/ton clean coffee	Proportion of total volume, %	BOD (3-day at 26.7°C), ppm	Proportion of total BOD, %
Pulping wastes				
Pulp water	4,490	34 ⎫	2400	45 ⎫
Main-tank effluent	2,220	17 ⎬ 57	3900	35 ⎬ 85
Repasser-tank effluent	840	6 ⎭	1450	5 ⎭
Tank-washing wastes				
First tank	280	2 ⎫	2800	4 ⎫
Second tank	270	2 ⎬ 5	1300	1 ⎬ 6
Repasser tank	165	1 ⎭	1900	1 ⎭
Channel washing				
Main	4,700	35 ⎫ 38	40	8 ⎫ 9
Repasser	440	3 ⎭		1 ⎭
Total	13,445*	100		100

*Usually 20,000 gal/ton of coffee.

Table 23.30 Coffee fermentation wastes.* (After Horton et al. [3].)

Characteristic	Minimum, ppm	Maximum, ppm	Mean, ppm
BOD	295	3600	1700
pH	4.1	5.5	4.5
Turbidity	250	4000	1750
Suspended solids	235	2385	900
Total solids	885	3140	2100

*Based on 30 different samples of effluents.

waste, fermentation wash water (tank water), and parchment.

The pulp is the waste that is potentially most troublesome, but it is generally recovered and used for fuel or fertilizer. When the fresh pulp is stored in open piles its sugar attracts flies, and when it begins to ferment a foul, repulsive odor emanates from it. The pulping water contains a relatively high amount of settleable solids and, since it contains sugar and other soluble materials, is highly pollutional. The fermentation (tank) wastes contain a great many colloidal gels of pectin and other products. This is a relatively weak waste, compared with the pulping waste water, and is relatively stable and inoffensive. The parchment from the dry milling of the dried bean has no significance as a waste product, since it is nearly pure cellulose and is generally utilized as a fuel for the steam boilers which provide power for the mills. Table 23.28 presents the average sanitary characteristics of wastes from three coffee plants using different amounts of water. In Table 23.29, the volumes of water and the BOD of coffee wastes is given [1]. Horton [3] reports the characteristics of fermentation wash-water wastes (Table 23.30) and depulping wastes (Table 23.31).

Horton [3] proved that biological filters with high rates of application and recirculation provided the most effective method of treating coffee wastes (Table 23.32) and proposed the effluent be used for irrigation. He also concluded that one hour of sedimentation reduced the BOD by only 16 to 29

Table 23.31 Coffee depulping wastes.* (After Horton et al. [3].)

Characteristic	Minimum, ppm	Maximum, ppm	Mean, ppm
BOD	3,280	15,000	9,400
pH	4.1	4.7	4.4
Turbidity	1,500	4,000	2,900
Suspended solids	625	1,055	790
Total solids	10,090	12,340	11,300

*Based on 12 samples on different days.

Table 23.32 Treatment of combined (fermentation and depulping) coffee wastes by recirculation through a biological filter.* (After Horton et al. [3].)

Rate of recirculation, mgd	Recirculations per hour	Settled waste, ppm	5-day BOD					
			After settling treatment for			Reduction after treatment for		
			2 hr, ppm	4 hr, ppm	6 hr, ppm	2 hr, %	4 hr, %	6 hr, %
20	5	2200	600	550	250	72.7	75.0	88.6
40	5	2450	920	650	420	62.4	73.5	82.9
40	8	2800	950	700	400	66.1	75.0	85.7
40	10	2951	900	700	450	69.5	76.3	84.7
60	5	2850	960	690	380	66.4	75.8	86.7

*Based on results of 5 tests.

Table 23.33 Comparative capital costs of alternative schemes for treatment of waste waters from processing of coffee. (After Heukelekian [1].)

Waste waters treated	Method of treatment	Reduction in polluting character, %	Reduction in total pollution, %	Capital cost for factory per ton of clean coffee per day, pounds sterling
Tank effluents and washings	Seepage pits		46	45
Pulp water and channel washings	Biological filtration	94	51	1000
Total			97	1045
Mixed wastes	Fermentation 12 days	85	85	580
	12 days plus biological filtration	94	14	480
Total			99	1060
Mixed wastes	Biological filtration to produce effluent with BOD of about 100 ppm	94	94	1850
	Biological filtration to produce effluent with BOD of about 40 ppm	98	98	2500

per cent, but the BOD of mixed coffee wastes can be reduced a maximum of 50 per cent by chemical coagulation. Brandon [1] presents results and costs for treatment of coffee wastes by various methods (Table 23.33) and states that in most coffee mills screening and primary sedimentation of the wastes is justified, since the dried pulp recovered is valuable as both a fuel and a fertilizer and has the following composition: protein, 1.3 per cent, fiber 19.7 per cent, nitrogen-free extract 50.1 per cent, ash 9.0 per cent.

References: Coffee Wastes

1. Brandon, T. W., "Treatment and disposal of waste waters from processing of coffee," *East African Agr. J.* **14**, 179 (1949).
2. Brandon, T. W., "Coffee waste waters, treatment and disposal," *Sewage Ind. Wastes* **22**, 142 (1950).
3. Horton, R. K., M. Pachelo, and M. F. Santana, "Study of the treatment of the wastes from the preparation of coffee," paper presented at Inter-American Regional Conference on Sanitary Engineering, Caracas, Venezuela, Sept. 26–Oct. 2, 1946.
4. "Interim report by Coffee Pollution Committee, Kenya Colony," *East African Agr. J.* **4**, 370 (1939).
5. Ward, P. C., "Industrial coffee wastes in El Salvador," *Sewage Works J.* **17**, 39 (1945).

23.16 Rice Wastes

In the preparation of edible rice, large volumes of waste are produced in the soaking, cooking, and washing processes. The volumes of waste produced average approximately 60,000 gallons per ton of raw rice handled. About 12 to 14 per cent of this volume comes from the soaking and an equal amount from the cooking process. The remaining 75 per cent results from washing and draining of the rice.

Heukelekian [1] gives an analysis based on the average of a number of composite samples (Table 23.34). Since most of the BOD is in the form of colloidal and soluble materials, settling effects only a 29 per cent reduction and is not recommended because of the thin and variable volumes of sludge. Sixty per cent BOD reduction has been obtained by using 2000 ppm of lime as a coagulant; digestion can yield a BOD reduction of over 90 per cent, at a loading of 0.02 to 0.10 pounds of BOD per cubic foot per day; and dispersed-growth aeration, with a waste-to-seed ratio of 14.5:1 and adequate nitrogen addition, has been found to produce more than 90 per cent BOD reduction.

Table 23.34 Characteristics of composite rice waste. (After Heukelekian [1].)

Characteristic	Amount or value
pH	4.2–7.0
Total solids, ppm	1460
Ash solids, %	20.5
Suspended solids, ppm	610
Ash in suspended solids, %	10.8
Total nitrogen, ppm	30
Phosphates, ppm	30
BOD, ppm	1065
Starch, ppm	1200
Reducing sugars, ppm	70

Reference: Rice Wastes

1. Heukelekian, H., "Treatment of rice water," *Ind. Eng. Chem.* **42**, 647 (1950).

23.17 Fish Wastes

The production of oil, meal, fish solubles, and other materials from fish constitutes a sizable industry in the United States. On the East and Gulf Coasts, these materials are primarily produced from menhaden, a nonedible fish, and on both the East and the West Coast they are by-products of the sardine-canning industry.

A typical flow sheet of a fish-processing plant is given in Fig. 23.8. (The Iw numbers refer to sources of waste waters.) (Iw1) Boat storage-compartment wastes occur when fish are pumped out of the fish-storage compartments of the boats. (Iw2) Spent fish-transfer water is kept in tanks of about 10,000-gallon capacity, which are emptied daily. Impurities in this waste water consist largely of blood, particulate, dissolved fish solids, fish scales, and oily scum. (Iw3) Raw-box-leakage waste occurs when liquids escape through cracks in the raw-fish storage box, because of the weight of the fish, and also when liquids escape at the point where the conveyors take out the fish. This waste has a high concentration of blood and dissolved solids. (Iw4) Dryer-kiln deodorizer spray water occurs when water is sprayed continuously during processing, to absorb fumes from the drier kilns. This spraying is a common practice in all plants and may be of importance in air-pollution control. (Iw5) "Stickwater" wastes have a BOD which varies from 33,800 to 112,000 and represents more than 90 per cent of the total BOD of the waste from a menhaden plant. Its quantity, however, is limited, and it is discharged early in the season when water temperatures are low. Essentially it is composed of the water-base reject liquid from the centrifuges. (Iw6) "Stickwater" storage-tank wastes consist of spillages due to overflows when filling the tanks and "bottoms" that must be disposed of when the tanks are cleaned. (Iw7) Evaporator wastes occur when the evaporators are boiled out with water, rinsed, boiled again with caustic soda, and rinsed again, normally after each run of fish. Cooling water and some condensate are also continuously discharged from the evaporators. (Iw8) Washing wastes occur from cleaning of presses, floors, tanks, centrifuges, and other equipment. Table 23.35 contains the average composition of these wastes [13].

Chemical coagulation, because of low BOD removals and excessive sludge formation, is not recommended for this waste. A satisfactory method of treatment has been to combine all individual wastes with the stickwater waste (Iw6) and evaporate them. Barging the wastes to sea has been suggested as an alternative to evaporation.

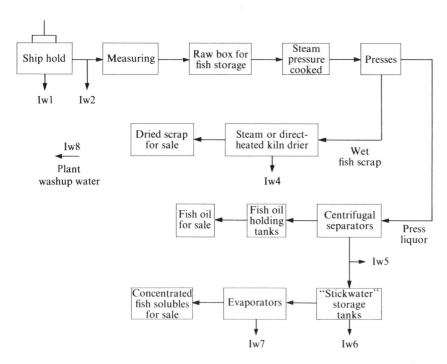

Fig. 23.8 Simplified flow sheet of menhaden fish-processing plant. The letters Iw refer to industrial waste. (After Paessler [13].)

Table 23.35 Composition of fish wastes. (After Paessler [13].)

Waste	BOD, ppm	Total solids, ppm	Total volatile solids, ppm	Grease, ppm
Iw1	42–265	15,576–20,606	2,489–3,394	
Iw2	3,050–67,205	18,421–64,857	5,912–46,907	1,314–17,234
Iw3	30,500–32,500	46,741–61,760	29,533–46,247	10,655
Iw4	120–300	14,171–18,949	1,906–7,957	45
Iw5	56,333–112,500	33,597–79,200	12,609–66,406	4,226–24,387
Iw6	47,063	52,998	45,483	18,157
Iw7	200–8,043	13,756–16,260	1,695–12,389	16–329

References: Fish Wastes

1. "British salmon fisheries, effect of wastes," *Sewage Ind. Wastes* **20**, 362 (1948).
2. Davis, H. C., "Fish cannery waste, effects on sewage treatment," *Sewage Ind. Wastes* **16**, 947 (1944).
3. "Fish products price cuts seen next year," *Reporter* **153**, 5 (1948).
4. Gallagher, F. S., "Fish meal fertilizer wastes," *Ind. Wastes* **4**, 87 (1959).
5. Goode, G. B., *A History of the Menhaden*, Orange Judd Co., New York (1880).
6. Hart, J. L., H. B. Marshall, and D. Beall, "The extent of the pollution caused by Pilchard reduction plant in British Columbia," *Biol. Board Can. Bull.* **39** (1933).
7. Hopkins, E. S., and J. Einarsson, "Water supply and waste disposal at a food processing plant," *Ind. Water Wastes* **6**, 5, 152 (1961).
8. Jensen, C. L., "Industrial wastes from seafood plants in the State of Alaska," in Proceedings of 20th

Industrial Waste Conference, Purdue University, May 1965.
9. Knowlton, W. T., "Fish cannery waste, effects on sewage treatment," *Sewage Ind. Wastes* **17**, 514 (1945).
10. Lassen, S., E. K. Bacon, and H. J. Dunn, "Fish reduction process," *Ind. Eng. Chem.* **43**, 2082 (1951).
11. Lewis, O., "U.S. fish research goals," *Oil, Paint, Drug Reptr.* **167**, 34 (1955).
12. Los Angeles Regional Water Pollution Control Board, *Los Angeles-Long Beach Harbor Pollution Survey*, no. 4 (1952).
13. Matusky, F. E., J. P. Lawler, and T. P. Quirk, "Preliminary process design and treatability studies of fish processing wastes," in Proceedings of 20th Industrial Waste Conference, Purdue University, May 1965.
14. "Menhaden's fame," *Chem. Week* **68**, 13 (1951).
15. "New hydro-vac lift speeds fish direct from boat to cannery," *Food Inds.* **21**, 1590 (1949).
16. Paessler, A. H., "Waste waters from menhaden fish oil and meal processing plants," in Proceedings of 11th Industrial Waste Conference, Purdue University, May 1956, p. 371.
17. "Recovering by-products from fish wastes," *Food Inds.* **22**, 96 (1950).
18. California Department of Public Health, *Report of a Study of Sardine Canning Wastes at San Carlos Cannery, Oxnard, Calif.* (1951).
19. California Department of Public Health, *Report on the Discharge of Waste from Sardine Canning and Reduction to Monterey Bay, Calif., for Regional Water Pollution Control Board*, Report no. 3 (1951).
20. Severance, H. D., "Fish canneries, odor control, Monterey, Calif.," *Sewage Works J.* **4**, 152 (1932).
21. Simmons, R. W., "By-product utilization brings hope to fish industry," *Chem. Eng. News* **31**, 1862 (1953).
22. "Sterile stickwater," *Chem. Week* **64**, 210 (1949).
23. "There's good hard cash in stickwater," *Food Inds.* **22**, 242 (1950).
24. Zimba, J., "Stickwater processing," *Chem. Eng.* **57**, 133 (1950).

23.18 Pickle Wastes

These wastes arise when cucumbers or other vegetables are converted—by a combination of aging, chemicals, and seasonings—to pickles. For a more detailed discussion of the processing and wastes involved, the reader is referred to the practical problem illustrated in Chapter 18. Since pickle factories produce either sweet or sour pickles, the wastes will therefore differ primarily in the degree of concentration of sugar. Four major wastes may be expected from a plant producing both types of pickle: (1) lime waste, from "sweetening" the wooden vats during the winter months; (2) brine waste, from fermenting the pickles for two to three months in a 10 to 20 per cent brine solution; (3) alum and turmeric waste, from the solution which causes swelling and coloring of the pickles; (4) syrup wastes, from the solution of vinegar and sugar used for seasoning and packing the pickles. Each of these contributes pollution of a different type. Combinations of two or more make the stream-pollution problem difficult to remedy.

The brine and the alum and turmeric wastes are normally discharged almost continuously in large plants. The intermittent discharge of the lime waste or the acid syrup waste determines the pH and organic-matter content of the final wastes. Screening, neutralization, and equalization are minimum treatments for pickle wastes. Good housekeeping practices are a necessity in the prevention of excessive pollution. Sudden shock loads on treatment plants or receiving streams should be avoided at all costs. The reader is urged to consult Chapter 18 for details of the characteristics and treatment of pickle wastes.

References: Pickle Wastes

1. Barnes, G. E., and L. W. Weinberger, "Internal housekeeping cuts waste treatment at pickle packing plants," *Wastes Eng.* **29**, 18 (1958).
2. Haseltine, T. R., "Biological filtration of kraut and pickle wastes," *Water Sewage Works*, **99**, 161 (1952).
3. Nemerow, N. L., *Pickle Waste Disposal*, private report to pickle company, 1952 (from files of the author).
4. Rice, J. K., "Water management reduces wastes and recovers water in plant effluents," *Chem. Eng.* **74**. no. 20 125 (1966).

23.19 Soft-Drink Bottling Wastes

Soft-drink bottling wastes result from the production of nonalcoholic beverages, both carbonated and noncarbonated. Wastes are produced from washing of bottles, production of syrup, treatment of water, and washing of floors. The wastes are usually highly alkaline, have a slightly higher BOD and suspended-solids content than domestic sewage, and are discharged to the sewer with or without screening.

The wastes from the bottle-washer are highly alkaline, since the washes consists of a series of alkaline detergent baths. Although, for reasons of economics as well as waste reduction, labels are used less than they were in the past, there are still large amounts of suspended solids resulting from straws, cigarette butts, paper, and other refuse left in the bottles. This foreign matter, plus leftover drinks in dirty bottles, is the major cause of the high BOD concentration. Wastes from the cleaning of floors, syrup-mixing and storage tanks, syrup filters, spillage, and so forth, are intermittent, and are not considered major sources of BOD and suspended solids. Wastes from water treatment will differ widely according to the quality required and the quality of the incoming water.

The characteristics of carbonated-beverage wastes, taken from Porges and Struzeski [6] and Besselievre [2] are presented in Table 23.36. Porges and Struzeski also observed [6] that, in 1954, 4643 bottling plants in the United States produced over one billion cases of soft drinks, valued at well over one billion dollars [3]. During that year, the per capita consumption in this country was more than 155 bottles. Analyses of several typical wastes from this giant industry are presented in Table 23.37.

Most soft-drink bottling plants are located near centers of population, so that discharge to the municipal sewer system appears to be the best means of waste disposal. Screening of wastes from the bottle-washer, to remove foreign matter left in bottles and labels if used, is sometimes practiced as a means of solids removal. To reduce the volume of waste, some plants reuse final rinse water from the bottle-washer for prerinsing the dirty bottles, or for other uses. Removal of waste drink and debris from the bottles and removal of labels before washing yields a pronounced reduction in BOD and suspended solids in the waste water. The small amounts of waste resulting from this operation can be disposed of in various other ways than letting them escape to the sewers. The remaining wastes, although they have a high pH and alkalinity, have little or no undesirable effect on most municipal sewage-treatment processes.

Table 23.36 Carbonated-beverage wastes.

Characteristic	Reference 6	Reference 2
pH	10.8	
Phenolphthalein alkalinity, mg/liter	150	
Total alkalinity, mg/liter	290	
5-day BOD, mg/liter	430	
Suspended solids, mg/liter	220	
Waste volume, gal/1000 cases	10,600	15,000
5-day BOD, lb/1000 cases		1,500
Suspended solids, lb/1000 cases		200

Table 23.37 Five-day BOD, total solids, acidity, and pH of carbonated beverages. (After Porges and Struzeski [6].)

Beverage	5-day BOD, ppm	Total solids, ppm	Acidity Mineral, ppm	Acidity Total, ppm	pH
Coca-Cola	67,400	114,900	244	1526	2.4
Pepsi-Cola	79,500	122,000	248	1466	2.5
Mission Orange	84,300	141,300	570	1579	3.0
Wagner Lift	64,600	110,800	316	2253	3.4
Tom Collins, Jr.	66,600	106,900	353	1246	3.2
Canada Dry quinine water	64,500	101,300	1181	3150	2.4
Average	71,200	116,200	490	1870	

References: Soft-Drink Bottling Wastes

1. Besselievre, E. B., "Industrial wastes, a community problem; the effects of certain types of wastes on city utilities," *Public Works Mag.* **83**, 74 (1952).
2. Besselievre, E. B., *Industrial Waste Treatment*, McGraw-Hill Book Co., New York, (1952), p. 107.
3. "Industrial statistics," in *1954 Census of Manufacturers*, Vol. 2, Bureau of Census, U.S. Dept. of Commerce, Washington, D.C.
4. Medbury, H. (ed.), *The Manufacture of Bottled Carbonated Beverages*, American Bottlers of Carbonated Beverages, Washington, D.C. (1945).
5. Morgan, R., *Beverage Manufacture*, Atwood and Co., London (1938).
6. Porges, R., and E. J. Struzeski, "Wastes from the soft drink bottling industry," *J. Water Pollution Control Federation*, **33**, 167 (1961); Proceedings of 15th Industrial Waste Conference, Purdue University, May 1960, p. 331.

23.20 Bakery Wastes

There are two types of bakery process. The first is a dry-baking operation, such as bread baking, where the only waste waters are the floor washings and some specialized machinery wastes. The mixing vats and baking pans and sheets are all cleaned dry, floors are swept, and breadcrumbs are recovered. The waste water is low in BOD and suspended solids, with the main contaminants being flour and some grease.

The second type of baking operation—the production of cakes, pies, doughnuts, cookies, etc.—is distinctly different in both operation and waste characteristics. Pans and trays have to be washed and greased after each baking, which results in a very strong waste, with BOD values from 3000 to 5000 ppm and suspended-solids content from 2000 to 3000 ppm. The major contaminants are grease, sugar, flour, fruit washings, and detergents.

Strong bakery wastes are biologically treatable, with activated sludge giving good results. Because of the high carbon character of the waste, bulking or a light filamentous sludge is often produced. Grease traps are often inefficient, because detergents emulsify the fats, and flotation is a recommended practice to remove suspended solids. Acid treatment of small-quantity, high-strength wash waters can reduce BOD values by as much as 80 per cent. To date, little has been done in the way of research on bakery wastes, but flotation and centrifugation have been used with apparently good results.

References: Bakery Wastes

1. Conner, W. R., and M. J. Perry, "Treats liquid wastes more efficiently," *Food Eng.* **40**, 92 (1968).
2. Mulligan, T., "Bakery sewage disposal," *Bakers Dig.* **41**, 81 (1967).
3. Grove, C. S., and D. Emerson, "Laboratory pilot-plant studies for treatment of bakery wastes," Report (part I) to the Ebinger Baking Company, Brooklyn, N.Y., Oct. 27, 1968.

23.21 Water-Treatment-Plant Wastes

In the past, the waste from water-treatment plants has not been considered an industrial waste. This may have had some basis in regard to municipal facilities, but certainly wastes from industry-owned water-treatment plants are industrial wastes. In addition, the water-treatment field is so large today that it should be considered an industry within municipal government.

Recent statements by both state and federal officials have indicated that even municipal plants will be required to treat their waste water, because federal and state legislation does not distinguish between sources of pollution. Cost of handling and treating wastes from water-treatment plants should be considered as a fundamental part of water-treatment costs. The sources of wastes in treatment plants are: (1) filter backwash water, (2) lime and lime soda sludge, (3) brine from cation exchange and sodium zeolite softeners, and (4) alum sludge. Common methods of treatment being used are:

1. Direct discharge to
    a) sanitary sewers (controlled (high flow), uncontrolled, or sludge discharge)
    b) deep well disposal
    c) lagoons

2. Reclamation and reuse
    a) recovery of alum by sulfuric acid

b) recalcination of lime
   c) recycle filter backwash after sedimentation to head of plant
3. Physical treatment methods
   a) centrifugation
   b) vacuum filtration
   c) sand drying beds
   d) flash drying
   e) thickening and sedimentation
   f) freezing

At present the most popular method is direct discharge to streams, with lagooning the most commonly used "treatment method."

References: Water-Treatment-Plant Wastes

1. Almquist, F. O. A., "Problems in disposal of sludge and waste water for Conn. waste filtration plants," *New Engl. Water Works Assoc. J.* **60**, 344 (1946).
2. Aultman, W. W., "Lime and lime soda sludge disposal," *J. Am. Water Works Assoc.* **39**, 1211 (1947).
3. Aultman, W. W., "Reclamation and reuse of lime in water softening plants," *J. Am. Water Works Assoc.* **31**, 640 (1939).
4. Aultman, W. W., "Waste disposal in water treatment plants," *J. Am. Water Works Assoc.* **58**, 1102 (1966).
5. Black, A. P., "Lime and lime-soda sludge disposal," *J. Am. Water Works Assoc.* **41**, 819 (1949).
6. Dean, J. B., "Disposal of wastes from filter plants and coagulation basins," *J. Am. Water Works Assoc.* **45**, 1226 (1953).
7. Dittoe, W. H., "Disposal of sludge at water purification and softening works at the Mahoning Valley Sanitary District," *J. Am. Water Works Assoc.* **25**, 1523 (1933).
8. Doe, P. W., "Sludge concentration by freezing," *Water Sewage Works* **112**, 401 (1965).
9. Gates, C. D., and R. F. McDermott, *Characteristics and Methodology for Measuring Water Filtration Plant Wastes*, New York State Department of Health, Research Report no. 14, May 1966.
10. Gordon, G. W., "Calcining sludge from a water softening plant," *J. Am. Water Works Assoc.* **36**, 1176 (1944).
11. Hall, H. R., "Disposal of wash water from purification plants," *J. Am. Water Works Assoc.* **39**, 1219 (1947).
12. Haney, P. D., "Brine disposal from cation exchange softeners," *J. Am. Water Works Assoc.* **41**, 829 (1949).
13. Haney, P. D., "Brine disposal from sodium-zeolite softeners," *J. Am. Water Works Assoc.* **39**, 1215 (1947).
14. Hoover, C. P., "Discussion of reclamation and reuse of lime in water softening," *J. Am. Water Works Assoc.* **31**, 640 (1939).
15. Howson, L. R., "Problem of pollution," *J. Am. Water Works Assoc.* **58**, 1104 (1966).
16. Kelley, E. M., "Discussion of reclamation and reuse of lime in water softening," *J. Am. Water Works Assoc.* **31**, 671 (1939).
17. Lyon, W. P., and D. A. Lazarchik, "Influence of state water quality standards," *J. Am. Water Works Assoc.* **58**, 1106 (1966).
18. Nelson, F. G., "Recalcination of water softening sludge," *J. Am. Water Works Assoc.* **36**, 1178 (1944).
19. O'Brien and Gere Consulting Engineers, *Waste Alum Sludge Characteristics and Treatment*, New York State Department of Health, Research Report no. 15, December 1966.
20. Pederson, H. V., "Calcining sludge from a softening plant," *J. Am. Water Works Assoc.* **36**, 1170 (1944).
21. Poston, H. W., "Federal pollution control," *J. Am. Water Works Assoc.* **58**, 1108 (1966).
22. Remus, G. S., "Detroit's waste water problem," *J. Am. Water Works Assoc.* **58**, 1112 (1966).
23. Roberts, J. M., and C. P. Ruddy, "Recovery and reuse of alum sludge at Tampa," *J. Am. Water Works Assoc.* **52**, 857 (1960).
24. Sheen, R. T., and H. B. Lammers, "Recovery of calcium carbonate or lime from water softening sludges," *J. Am. Water Works Assoc.* **36**, 1145 (1945).
25. Sowden, H. J., "Lime recovery from water softening sludges," *J. Am. Water Works Assoc.* **40**, 461 (1941).
26. Swab, B. H., "Pontiac recalcining plant," *J. Am. Water Works Assoc.* **40**, 461 (1948).

CHAPTER 24

THE MATERIALS INDUSTRIES

In this text, industries producing materials are differentiated from food-processing industries in that the products are nonedible; from the apparel industries in that the products are not articles of clothing; and from the chemical industries in that the products are not specific chemicals or associated products. These industries fall into four groups: (1) wood fiber (including pulp and paper wastes and photographic wastes); (2) metal (steel and other metal wastes, metal-plating and iron-foundry wastes); (3) liquid-processing (oil-refinery, rubber, glass-manufacturing, and naval-stores wastes); and (4) special materials (glue, wood preserving, candles, and plywood).

WOOD FIBER INDUSTRIES

The pulp- and paper-making industry is the fifth largest in our economy and ranks third in its rate of expansion [146]. The per capita demand for its products is nearly 400 pounds. More than 1000 mills in 40 states produce a wide variety of paper and paperboard products, which total more than 30 million tons annually. Water is used by this industry at a rate of 4000 mgd, a usage equivalent to that of a population of about 40 million. (These figures are as of 1956.)

The manufacture of paper, like the manufacture of textiles, can be divided into two phases: pulping the wood and making the final paper product. The raw materials generally used in the pulping phase are wood, cotton or linen rags, straw, hemp, esparto, flax, and jute or waste paper. These materials are reduced to fibers which are subsequently refined, sometimes bleached, and dried. At the paper mill, which is often integrated in the same plant with the pulping process, the pulps are combined and loaded with fillers; finishes are added and the products transformed into sheets. The fillers commonly used are clay, talc, and gypsum. The four main types of pulp used are groundwood, soda, kraft (sulfate), and sulfite.

Fiber industries, therefore, produce two main wastes, namely pulp-mill and paper-mill wastes. Pulp-mill wastes come from grinding, digester cooking, washing, bleaching, thickening, deinking, and defibering. These wastes contain sulfite liquor, fine pulp, bleaching chemicals, mercaptans, sodium sulfides, carbonates and hydroxides, sizing, casein, clay, ink, dyes, waxes, grease, oils, and fibers. Treatment of these wastes is primarily by recovery of chemicals, equalizing, sedimentation, controlled dilution, coagulation, lagooning, biological oxidation, evaporation, spray drying, burning, and some fermentation. Paper-mill wastes originate in water which passes through the screen wires, showers, and felts of the paper machines, beaters, regulating and mixing tanks, and screens. The paper-machine wastes (white waters) contain fine fibers, sizing, dye, and other loading material. Wastes are treated by recovery of the white water, settling and vacuum filtration, flotation, and chlorination. Figure 24.1 is a schematic flow diagram for pulp- and paper-mill processes.

24.1 Pulp- and Paper-Mill Wastes

Origin of pulp- and paper-mill wastes. The major portion of the pollution from papermaking originates in the pulping processes. Raw materials are reduced to a fibrous pulp by either mechanical or chemical means. The bark is mechanically or hydraulically removed from wood before it is reduced to chips for cooking. Mechanically prepared (groundwood) pulp is made by grinding the wood on large emery or sandstone wheels and then carrying it by water through screens. This type of pulp is low-grade, usually highly colored, and contains relatively short fibers; it is mainly used to manufacture nondurable paper products such as newspaper. The screened bark effluent contains fine particles of bark and wood and some dissolved solids. Additional sources of waste from

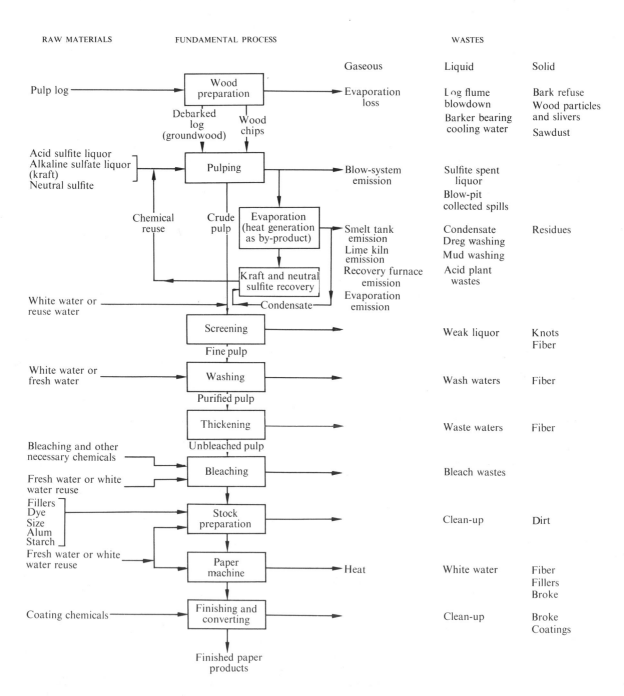

Fig. 24.1 Simplified diagram of fundamental pulp and paper processes. (Prepared for the F.W.P.C.A.)

wood preparation are the pressing of rejects prior to burning and floor drainings.

Chemically prepared pulps, as compared with mechanically prepared ones, are made by the soda, sulfate (kraft), or sulfite process; the semichemical process, which we will discuss later, is also used today. (Sulfate pulps, incidentally, do not bleach readily and therefore are generally used for brown, or other-colored, wrapping paper.) In all these methods the wood is prepared, as in the making of groundwood, by reduction to chips and screening to remove dust. The chemical processes differ from one another only in the chemical used to digest the chips.

Softwoods, such as poplar, are usually treated by the soda process. To a digester holding about four cords of chips, a mixture of soda ash (Na_2CO_3) and lime [$Ca(OH)_2$] is added and the total contents are boiled under steam pressure for about ten hours. This digestion decomposes or separates the binding, noncellulosic materials, such as lignins and resins, from the fiber, but it is a rather harsh treatment for the fibers and consequently weakens them. Coniferous woods are used in the preparation of both kraft and sulfite pulps. The sulfate process calls for a shorter digestion period of about five or six hours, with a mixture of sodium sulfide, hydroxide, sulfate, and carbonate. The lignin and noncellulosic materials are dissolved, leaving a stronger fiber for paper formation. Sulfite pulp is made by cooking with calcium bisulfite at over 300°F and 70 pounds of steam pressure.

After digestion, chemically prepared wood pulps are blown into a closed blow pit, where the black liquor is allowed to drain to the sewer or to recovery processes. The drained pulp is then washed. These wash waters may then be wasted, reused, or sent through recovery operations, while the washed pulp is passed through some type of refining machine to remove knots and other nondisintegrated matter. A cylindrical screen, called a decker, revolving across the path of the pulp partially dewaters it, after which it is passed to bleach tanks, where it is mixed in a warm, dilute solution of calcium hypochlorite or hydrogen peroxide. The dried, bleached pulp is then ready for sale or delivery to the paper mill.

The papermaking process involves first a selection of the appropriate mixture of pulps (wood, rag, flax, jute, straw, old newspaper, and so forth). The pulp mixture is disintegrated and mixed in a beater, to which are added various fillers and dyes, to improve the quality of the final paper product, and sizing, to fill the pores of the paper. The beater is essentially an oblong tank equipped with a rotating cylinder, to which are attached dull knives to break up the knotted or bunched fibers and cause a thorough mixing of the entire contents of the tank. Sometimes the pulps are washed in the "breaker beater" prior to the addition of the chemicals. The washing initially produces a rather strong waste, which is progressively diluted as the washing proceeds. After beating, the pulp is usually refined in a jordan, a machine that consists of a stationary hollow cone with projecting knives on its interior surface, fitted over a rapidly rotating adjustable cone having similar knives on its outside surface. This machine cuts the fibers to the final size desired. The pulp then passes to stuffing boxes, where it is stored, mixed, and adjusted to the proper uniform consistency for papermaking. Finally, the pulp is screened to remove lumps or slime spots, which would lower the quality of the final paper.

Next, the pulp is evenly distributed from a headbox over a traveling belt of fine wire screening, known as a fourdrinier wire, and carried to rolls. A small portion of the water contained in the pulp passes through the screen, while the longer fibers are laid down as a mat on the wire. A considerable portion of the fine fibers and some fillers also pass through the screen wire with the water. Because of its color, this waste water is called "white water." The paper mat passes through a series of rolls as follows: a screen roll to eliminate inequalities at the end of the wire, a suction roll to draw out more water, press and drying rolls to rid the paper of most of the remaining water, and finally finishing rolls (calenders), which produce the final shape of the paper. The final products are used for many purposes—printing paper, newspaper, wrapping paper, parchment paper, writing paper, blotting paper, tissue paper, and impervious food-wrapping paper.

The chief sources of waste at the pulp mills are the digester liquors and the chief sources at the paper mills are the beaters and paper machines. Fiber losses generally average 3 per cent or less. In so-called "closed systems," where white water is recirculated and reused, it is possible to reduce fiber losses to 0.1 per cent. However, even this low level of loss is significant, because of the large quantities of fibers processed per day. Table 24.1 gives some idea of the

Table 24.1 Average waste discharge per ton of paper product.*

Product	Waste, gal
Pulp mills	
Groundwood	5,000
Soda	85,000
Sulfate (kraft)	64,000
Sulfite	60,000
Miscellaneous paper	
No bleaching	39,000
With bleaching	47,000
Paperboard	14,000
Strawboard	26,000
Deinking used paper	83,000

*From *Industrial Waste Guide, Ohio River Pollution Control Survey*, Supplement D, U.S. Public Health Service (1943).

Characteristics of pulp- and paper-mill wastes. Since the four types of pulping produce somewhat different wastes, each should be considered separately. Gehm [273] presents the general characteristics of groundwood wastes (Table 24.2). The Ohio River study [145] shows typical analyses of all types of pulp and paper wastes (Table 24.3). Gehm [273] also gives volumes of wastes discharged per ton of product.

With softwood timber steadily becoming scarcer, the use of hardwoods for pulp has increased during recent years. Another pulping method, known as semichemical pulping, has been developed chiefly for hardwoods. It is carried out by digestion under relatively mild chemical conditions, which softens but does not fully pulp the wood, and subsequent actual reduction to pulp by mechanical means. The product is used mainly for container board and coarse wrapping paper. The name "semichemical" has been applied mainly to cooking methods employing neutral sodium sulfite. However, other cooks resulting in slightly acid or basic pH values and utilizing semisulfate and semisoda, are also sometimes referred to as semichemical pulping.

Table 24.2 Typical analysis of wood preparation wastes.

Characteristic	ppm
Total solids	1160
Suspended solids	600
Ash (suspended solids)	60
Dissolved solids	560
Ash (dissolved solids)	240
BOD, 5-day	250

Table 24.3 Typical analytical results for pulp- and paper-mill wastes.

Product	BOD, ppm	Suspended solids, ppm
Pulp		
Groundwood	645	
Soda	110	1720
Sulfate (kraft)	123	
Sulfite	443	
Miscellaneous paper		
No bleach	19	452
With bleach	24	156
Paperboard	121	660
Strawboard	965	1790
Deinking used paper	300	

Table 24.4 Characteristics of kraft-mill wastes.

Characteristic	Maximum	Minimum	Average
pH	9.5	7.6	8.2
Total alkalinity, ppm	300	100	175
Phenolphthalein alkalinity, ppm	50	0	0
Total solids, ppm	2000	800	1200
Volatile solids, %	75	60	65
Total suspended solids, ppm	300	75	150
Volatile solids, %	90	80	85
BOD, 5-day, ppm	350	100	175
Color, ppm	500	100	250

24.1 PULP- AND PAPER-MILL WASTES

the analyses of 24-hour composite samples of the combined effluent from modern unbleached kraft (sulfate) mills (Table 24.4).

Moggio [207] states that present-day kraft mills, efficiently operated, discharge an effluent containing no more than 100 pounds of the Na_2SO_4 equivalent of the cooking liquor per ton of pulp. He also claims water-use figures of 20,000 to 30,000 gallons for unbleached kraft and 40,000 to 60,000 gallons for bleached kraft (per ton of pulp). The effluent character varies somewhat, depending on bleaching practices: suspended solids range from 20 to 60 ppm and are primarily fiber (about 0.5 per cent of the total product); dissolved-solids concentrations range from

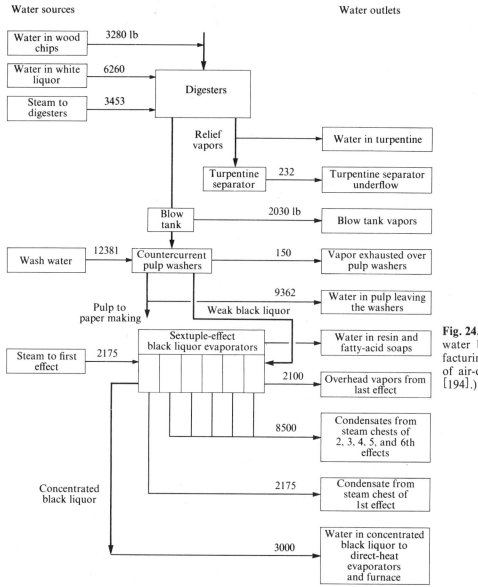

Fig. 24.2 Process flow diagram of water balance of kraft pulp manufacturing process (quantities in lb/ton of air-dry pulp). (After McDermott [194].)

1000 to 1500 ppm, of which about 60 per cent is ash; BOD values range from 100 to 200 ppm, or 20 to 40 pounds per ton of product. The effluent is coffee-colored and has a color value of about 500. Typical flow diagrams of kraft pulp-manufacturing processes and water balances are given in Figs. 24.2 and 24.3.

Gehm [273] states that, since the soda pulping process is very similar to the kraft one, the effluents are similar. The only essential difference is that either NaOH alone, or a lower concentration of Na_2S, is used in the soda process for cooking the wood. Soda cooks, then, will have a lower sulfur concentration.

Spent sulfite liquors average approximately 300 gallons per ton of pulp. Spent-sulfite-liquor waste has been described as a highly corrosive, dilute solution, containing about one-half of the solids in the pulp wood [28]. These solids may contain as much as 65 per cent lignosulfonic acid, 20 per cent reducing sugars, 8.4 per cent sugar-sulfur-dioxide derivatives, and 6.7 per cent calcium. The solids concentration of spent sulfite liquor drawn from the digesters may vary from 6 to 16 per cent and it may contain from 400 to 600 (or more) pounds of BOD per ton of pulp. The sulfur compounds possess an "immediate oxygen demand" which accounts for about 11 per cent of the 5-day BOD. The sugars (hexoses and pentoses) repre-

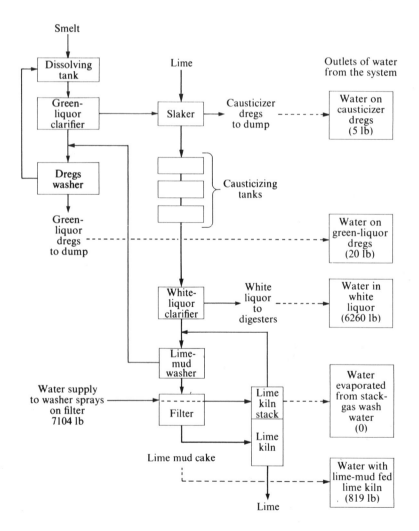

Fig. 24.3 Water balance in chemical reconstituting system (quantities in lb/ton of air-dry pulp). (After McDermott [194].)

Table 24.5 Relationship of major components of sulfite waste liquor [264].

Component	% Total solids, bone-dry basis
Lignin	51.6
Sugars	16.9
Sulfur	9.15
Calcium	4.5

sent about 65 per cent of the BOD. Although lignins account for over half the solids in this waste, they contribute little to the BOD. Haskins [cited in reference 264] describes sulfite waste liquor as a highly complex liquid, containing 10 to 12 per cent solids. The major components of the liquor are free and combined SO_2, volatile acids, alcohols, acetone, furfural, sugars, and lignin. A more detailed analysis of these wastes is presented in Tables 24.5 and 24.6.

An analysis of a typical semichemical-pulping mill waste was given by Crawford in 1947 [66]. This mill used the semisoda cook on pine and mixed hardwoods and the neutral cook on tannin-extracted chestnut chips (Table 24.7). Vilbrandt [266] reports a breakdown in the analysis of solids and BOD components of semichemical soda-pulping waste (Table 24.8). Rudolfs and Nemerow [281] present an analysis of typical white-water waste from chipboard mills (Table 24.9). The pulping of raw materials other than wood results in somewhat different waste analyses. Bloodgood [315] reports the analyses of strawboard-mill waste based on an eight-month daily test survey (Table 24.10).

Rudolfs and Nemerow [280] present the important sanitary characteristics of rag-, rope-, and jute-cooking liquors and of beater wash-water wastes (Table 24.11), and a further chemical analysis of the cooking liquors (Table 24.12). A Wisconsin survey of pulp and paper mills in 1946 [338] showed the wide variation in characteristics of all types of pulping and paper-machine wastes (Table 24.13). A recent survey [1968, 253] gives the solids, BOD, and pH waste loads and waste-water flows in various types of mill (Table 24.14).

Treatment of pulp- and paper-mill wastes. Pulp- and paper-mill wastes are treated in the following manner: (1) recovery, (2) sedimentation and flotation to remove suspended matter, (3) chemical precipitation to remove color, (4) activated sludge to remove oxygen-demanding matter, (5) lagooning, for purposes of storage, settling, equalization, and sometimes for biological degradation of organic matter.

Paper-mill waste treatment is still in its early stages,

Table 24.6 Composition of a typical sulfite pulp-plant waste.

Component	Digester liquor, ppm	Blow-pit liquor, ppm
Total solids	111,100	38,700
Volatile solids	101,000	34,000
Ash	10,100	4,700
Calcium	3,990	1,550
Total sulfate	31,200	8,620
BOD, 20-day	42,900	

Table 24.7 Major wastes in semichemical board production [66].

Characteristic	Globe-digester blow	Washer effluent	Machine effluent
Volume, gal/day	24,000	2,000,000	864,000
Total solids, ppm	102,300	4,593	658
Fixed solids, ppm	35,000	1,547	166
Volatile solids, ppm	67,300	3,046	492
Total solids, ton/day	10.8	38.3	2.4
Color, ppm	165,000	12,000	500
Per cent of total	13.8	84.7	1.5
BOD, 5-day, lb/day	1,940	16,440	230

Table 24.8 Solids and BOD analysis of semichemical soda-pulping waste [266].

Component	Total solids, %	BOD, %
Pentosans	5.7	18.9
$NaHCO_3$	11.5	0.0
$NaC_2H_3O_2$	26.1	56.7
Sodium formate	6.8	4.2
Lignin	14.9	—*
Hydrolyzable fraction of suspended solids other than pentosans	5.4	—*
Total	70.4	79.2

*Doubtful 5-day BOD value.

Table 24.9 Analysis of white-water waste from chipboard mill [281].

Characteristic	Value or concentration
pH	7.0
Alkalinity, ppm $CaCO_3$	118
Suspended solids, ppm	840
Fixed solids, %	11.9
Volatile solids, %	88.1
Total solids, ppm	2180
Ash, %	19.3
Volatile total solids, %	80.7
Dissolved solids, % of total	61.5
BOD, ppm*	100–400

*From Rudolfs and Axe [277].

Table 24.10 Analysis of total strawboard waste [315].

Characteristic	Maximum daily	Minimum daily	8-month average
Total solids, ppm	4900	2600	3691
Total volatile solids			
ppm	4100	1100	2502
%			67.7
Total suspended solids, ppm	2190	484	1369
Suspended volatile solids			
ppm	1590	320	909
%			66.5
BOD, ppm	1372	409	955

Table 24.11 Rag, rope, and jute-mill wastes [280].

Characteristic	Jute Cook	Jute Wash	Rope Cook	Rope Wash	Rag Cook	Rag Wash
pH	12	11.2	10.8	8.7	12	8.1
Alkalinity, ppm $CaCO_3$	2850	574	18,000	350	31,655	264
BOD, ppm	3381	385	12,125	1250	29,225	526
Total solids, ppm	7200	1300	45,000	1100	96,000	2200
Total volatile solids, %	56	56	56	64	64	54
Dissolved solids, %	82	65	95	59	93	73
Total nitrogen, ppm	126	6	157	21	1,270	44
Ratio of BOD to total nitrogen	2.7:1	65:1	88:1	60:1	23:1	12:1

Table 24.12 Breakdown of jute, rope, and rag cooking-liquor solids soluble in cold water [280].

Type of cook waste	Total solids in cold water, ppm	Volatile matter, %	Components of volatile matter in cold water, % of volatile matter					
			CO_2	Grease	Lignin	Volatile acids, as HAC	Complex polysaccharides	Protein, total N $\times$ 6.25
Jute	5,540	68.8	30.0	0	21.8	40.9	10.9	4.5
Rope	37,000	54.3	25.4	0	39.4	14.8	10.6	6.4
Rag	85,800	58.4	24.6	7.6	24.3	10.4	3.6	3.8

although a great deal of research and many pilot-plant investigations have been carried out and reported on. Actual treatment equipment is installed only after exhaustive study of all other possibilities, since the cost of treatment is considered high in relation to the cost of the product produced. Thus, economic limitations have forced the industry to place emphasis on recovery rather than treatment.

Recovery processes in the paper mill involve the use of "save-alls," in either closed or partly closed systems. These save-alls are installed not only as a waste-treatment measure, but also as a conservation measure to recover fibers and fillers. The main types are based on filtration, sedimentation, or flotation processes. Filtration devices are usually some variation of a revolving, cylindrical, perforated screen or filter that removes the suspended solids in the form of a mat, which is subsequently scraped off the drum and returned to the paper-making stock system. Conical or other sedimentation tanks are also often used to separate the suspended matter by difference in specific gravity. All the principles of sedimentation discussed in Chapter 10 apply to these treatment units.

Table 24.13 Results of 1946 survey of wastes at Wisconsin pulp and paper mills [338].

Type of mill	Waste, gal/ton of product	Solids, lb/ton of product			Fiber loss, % of production	5-day BOD*
		Fixed suspended	Volatile suspended	Total soluble		
Paper						
Book	13,071	28.8	22.5	44.6	0.9	40
Tissue	23,048	5.7	30.4	113.4	1.5	47
Wrapping	28,432	17.9	31.2	146.8	1.6	56
Bond	20,461	18.7	57.2	108.6	2.8	116
Glassine	48,727	5.3	32.2	111.4	1.6	59
Paperboard	5,600	0.9	10.9	26.3	0.5	11
Black wadding	57,400	13.2	86.3	232.9	4.3	158
Pulp						
Rag and deinked	49,796	157.6	188.2	911.6	6.1	403
Kraft	64,838	33.6	61.2	329.8	3.1	451
Sulfite	50,470	3.7	37.6	2413.8	2.0	2857
Groundwood	2,302	0.15	11.2	7.0	0.6	20

*Population equivalent, persons/ton of product.

Table 24.14 Waste loads and waste-water quantities in typical pulp and paper mills.

| Process | Waste load, in lb/ton of product ||||||||| Waste-water quantities, gal/ton ||
| | Suspended solids || Dissolved solids || Total solids || BOD || pH || | |
	Range*	Mean	Range	Mean	Range	Mean	Range	Mean	Range	Mean	Range	Mean
Wood preparation	1.9–40	9		4	4–50	13	2–10	3	6.5–8.0	7.0	1,000–10,000	3,400
Pulping												
Groundwood												
Sulfate (kraft)												
Blow tower	(3.7)	4		17	(21.0)	21	(1.3)	1	(12)	12.0	(1,000)	1,000
Dirty condensate	0–0.5	0.1		4	6–11	7	6.5–9.0	8	9.5–10	10.0	950–1,900	1,200
Evaporator ejector	0.06–0.2	0.1		2	1–3	2	1.6–4.5	3	9–10	9.5	290–640	300
Causticizing waste	2.2–5.7	5	(5)	96	46–240	101	8.0–10.5	9	9–11.0	10.0	600–9,600	2,500
Green dreg	(1.0)	1		21	(22)	22	(1.0)	1	(12)	12.0	(200)	200
Floor drain	0.5–10	6		1	11.0–11.5	11	0.3–1.7	1	11.6–12	12.0	340–580	400
Subtotal		17		141		164		23	9.5–12.0			5,600
Sulfite												
Blow tower	0.42–1.9	1		246	36–348	247	29–194	116	2.2–2.9	2.7	1,840–1,950	1,900
Condensate	0.05–0.2	0.1		47	18–87	47	48–71	66	2.3–3.1	2.6	750–1,700	1,100
Uncollected liquor	0.3–43	21		84	50–515	105	50–61	53	2.2–2.6	2.4	2,000–10,000	7,500
Acid plant wastes	(5)	5		5	(10)	10			(1.2)	1.2	(300)	300
Boiler blowdown	(2)	2			(22)	22	(0.05)	0.05		11.0	(100)	100
Subtotal		29		382		411		235	1.2–2.9			10,900
Semichemical												
Blow tower	(2)	2	(6)	6	(8)	8	(1)	1		4.0	(1,000)	1,000
Condensate	(0.1)	0.1	(2)	2	(2)	2	(3)	3		3.5	(2,000)	2,000
Recovery system	(9)	9	(111)	111	(150)	150	(8)	8			(2,000)	2,000
Uncollected liquor	(11)	11	(29)	29	(40)	40	(18)	18			(2,000)	2,000
Subtotal	(22)	22		148		200		30	2.5–4.0	2.5		7,000
Deinking (all sources)†							11–25				9,700–36,000	
Pulp screening												
Groundwood												
Sulfate (kraft)	5–8	4		58	60–63	62	10–18	14	9–10	10.0	900–9,600	3,600
Sulfite	1.7–14	8		19		27	22–10.7	8	5.4–5.7	5.6	1,700–14,300	6,000
Semichemicals												
Deinking												

Table 24.14 (continued)

	Range	Avg	Range	Avg	Range	Avg	Range	Avg	Range	Avg	Range	Avg
Pulp washing and thickening												
Groundwood (no washing)	9–14	11		44	51–107	75	22–46	33	5.0–6.25	6.0	4,800–10,000	7,500
Sulfate (kraft)	10–30	15		127	94–180	142	10–35	25	8.9–9.4	9.0	3,000–11,000	7,000
Sulfite	6.5–9.0	8		123	68–1037	131	7.4–34.0	18	2.4–3.9	2.9	1,800–15,000	7,500
Semichemical	0.9–6.0	3		90	42–141	93	10–42	24	7.0–7.9	7.4	2,400–7,800	5,400
Deinking												
Bleaching												
Groundwood												4,000
Sulfate (kraft)	14–124	60	92–280	180	216–294	240	8–88	30	2.9–6.8	2.9	12,000–32,000	19,000
Sulfite	4–44	15	126–409	205	131–415	220	17–44	25		3.8	9,000–30,000	15,000
Semichemical												
Deinking		6		119		125		12		2.2		5,500
Paper-making‡												
General	10–166	46	21–425	73	31–591	119	3–80	16	4.3–6.9	5	5,700–40,000	13,000
Related products	20–60	40					10–12	15			37,000	
Newsprint												
Uncoated groundwood												
Coated printing-paper												
Uncoated book paper	47–100	30		66		116	15–40	16			8,000–28,000	14,000
Fine paper	10–30	73		80		153	10–25	20			9,000–40,000	18,000
Coarse paper		20						15			2,000–29,000	10,000
Special industrial paper	200–400	300					140–170	155			20,000–100,000	
Sanitary and tissue paper	50–100	50		150		200	15–30	22			8,000–37,000	14,000
Total mill effluent (integrated pulp and paper mills)												
Bleached sulfate and paper	50–200	170	150–1130	640	200–1300	810	30–220	120			39,000–54,000	45,000
Unbleached sulfate and paper§		50		460		510						27,000
Bleached sulfite and paper	40–100	100	560–1600	1040	600–1700	1140	235–430	330			40,000–70,000	55,000

*Single pieces of data are entered under the "Range" column in parentheses. The mean values shown are not truly statistical averages; they are considered to be probable average values based on the available data.

†The deinking process includes pulping, screening, washing, and thickening.

‡Waste waters from paper-making include those from stock preparation, paper-machining, and finishing and converting operations.

§Data for integrated unbleached sulfate pulp and paper mills are generated by subtracting the data for bleaching from those for the integrated bleached sulfate pulp and paper mill.

In flotation recovery units, also discussed in Chapter 10, the suspended fibers and other solids are removed in the form of a mat floating on the surface of the tank. This is a very efficient method for certain fibers which have a natural tendency to float in a suspension, being buoyed up by minute bubbles of air dissolved in the fibrous waste. The air is usually forced into the waste water under a pressure of about 45 pounds per square inch and released in an open flotator tank under atmospheric pressure, or under a slight vacuum. Recovery efficiencies are often better than 95 per cent suspended solids recovered. Recovery of the clarifier white water is achieved by recirculating this water into the beaters, head boxes, and showers. One difficulty encountered with recirculation of clarified white water has been slime growths, both in the mixture and on equipment. This greatly reduces the paper machine rates and lowers the value of the paper produced. Chlorination, organic mercurials, and environmental controls (pH and temperature) are used to control these growths [281].

Sulfite pulp mills use various methods of recovery. Equipment has been developed in which the sulfite waste liquor can be burned to produce enough steam to run the evaporator. (This process does not produce a salable by-product, but merely eliminates the waste problem.) Complete evaporation of the sulfite waste liquor produces both a fuel which can be burned without an additional outside fuel supply and a salable by-product—used in making core binder, insecticides and fungicides, linoleum cement, road binder, roadbank stabilizer, ceramic hardener, boiler compounds, synthetic vanillin, and other useful by-products. However, the main problem associated with these by-products has been the fact that the market in this country cannot absorb more than 5 to 10 per cent of them. Since evaporation is costly, owing to the low initial concentration of solids in the sulfite waste liquor and the boiler-scaling difficulties encountered, operation of such a process is limited to mills close to users of waste liquors.

In addition to those by-products obtained by evaporating sulfite waste liquor, other valuable by-products are obtainable by other processes. The liquor may be fermented to produce ethyl alcohol; about 40 liters of alcohol can be produced per ton of dry solids. This process reduces the BOD of the liquor by utilizing the simple sugars alone. Acetone and butyl alcohol can also be produced from the waste, with an overall BOD reduction of about 82 per cent. However, in 1950 only one mill in the United States manufactured ethyl alcohol from sulfite waste liquor, and the situation remains similar in 1969. The drawback of this process is that it costs more to manufacture alcohol from the sulfite-liquor waste than from blackstrap molasses or ethylene.

Another product of fermenting the liquor is yeast for cattle feed. In 1948 a plant was built in Wisconsin to produce yeast fodder by this method and it achieved a 60 to 70 per cent BOD reduction. Unfortunately, the market was found to be very limited, because of competition with brewers' yeast. Laboratory experiments have been used to produce torula yeast for stock feed, with a resulting reduction of 40 per cent in BOD; 350 pounds of yeast are obtained from a ton of waste solids.

Recovery is also practiced in kraft mills. The black liquor (spent cooking liquor) is processed by evaporation and incineration, in order to recover chemicals and to utilize the heating value of the dissolved wood substances. During the recovery process, Na_2SO_4, with or without added sulfur, is added to replace the relatively small proportion of chemicals lost in the various steps of the process. Following these additions and the incineration, the smelt is dissolved in water to form "green liquor." The chemical compounds in the green liquor are converted to the desired cooking chemicals by the addition of lime, so as to form of "white liquor" and a lime mud consisting chiefly of $CaCO_3$. The white liquor is returned to the pulping operation as the cooking liquor and the lime mud is calcined to form calcium oxide, which is reused in converting other green liquor to white liquor. By-product recovery of turpentine, resin, and fatty acids also aids in reducing the strength of kraft waste-water effluents. Maximum recovery of these by-products may result in effluents in which chemical toxicants are no longer a significant factor, as far as stream pollution is concerned. The turpentine is recovered from the digester relief gases, which also contain small quantities of $(CH_4)_2S$, dimethyl sulfide, methyl mercaptan, and ketones. The black liquor also contains recoverable quantities of sodium salts, rosin, and fatty acids, which separate during the concentration and cooking of the black liquor. This material is called "crude sulfate soap"; after it has been skimmed from the black liquor, it is treated with acid to form tall oil. The resin and fatty acids are further refined and have a variety of applications in industry.

Sedimentation and flotation. These treatments are achieved through the use of save-alls, which have been described earlier. Although save-alls are used on paper-machine white waters primarily for purposes of clarification (for white-water reuse) and fiber recovery, they may still be considered as part of process equipment. Removal of the fiber naturally results in decreased loss of solids to the sewer, and therefore effluents of lower pollutional strength. Another type of equipment which has resulted in lower sewer losses of black liquor is the foam trap or foam breaker, which prevents foam from spilling over and carrying the entrained liquor into the sewer.

Sedimentation is the usual method of total and final treatment of paper-mill effluents, the save-alls being restricted to usage within the mill. In spite of the use of in-plant save-alls, there are losses to the drains which cannot be prevented. Modern concrete or steel sedimentation tanks, usually circular, have been installed at many paper mills, where they give good service. Diameters vary from about 20 to 120 feet; solids discharged have a concentration of 4 to 15 per cent, with an average of 6 to 8 per cent. The fillers present in the effluent make final clarification difficult; effluents usually contain at least 30 ppm suspended solids, unless upflow-type clarifiers are used.

Chemical precipitation to remove colloids and color. The use of chemicals to treat paper-mill waste has generally been avoided, since it increases the quantity of sludge which must be disposed of. However, some mills have used this method. An Indiana mill, using sulfite pulp, treats all wastes with alum. The sludge is dried on beds, and the effluent is recirculated to the process. A Michigan mill uses chemical precipitation and obtains a BOD reduction of 64 per cent—somewhat higher than usual.

The Howard process is a precipitation, with lime as a coagulant, in three stages to a final pH of 11. In the first stage, calcium sulfide settles out and is returned as a slurry to the cooking-liquor make-up. In the second stage, lignin is precipitated and converted to a cake on a rotary filter. In the third stage, settling removes any colloidal material that is left when the pH is raised to 11. About 40 per cent BOD removal is obtained, although claims for somewhat higher efficiencies have been made. The process is one of recovery and not actual treatment, even though chemical precipitation is used. The lignin is used both as a fuel and in the manufacture of plastics, the production of tannins, as an antiscale or antifoam agent in boilers, and in manufacturing synthetic vanilla. The Strellenert process is similar to the Howard process, except that: the precipitant is gypsum, and the liquors are heated to 160°C in pressure vessels; sulfur dioxide is evolved and reused in cooking liquors; lignin is precipitated and used for fuel, mainly in countries outside the United States.

The problem of color removal from kraft-mill effluents has become increasingly pressing. It is possible to remove color by treating the effluents with high dosages of hydrated lime, but this results in large volumes of hydrous sludges, which are extremely difficult to dewater. This has been a major hindrance to solving the color-removal problem. Because of the high lime dosages required, economic considerations have dictated the recovery of calcium from the sludges for reuse. A method of sludge treatment involving carbonation and heat is being investigated for calcium recovery.

A study of various methods of chemical treatment of rope-mill wastes [280] showed that, in general, most of the coagulants causing good clarification also produced excessive quantities of sludge and effected relatively little BOD reduction. A three-stage chemical treatment using $FeCl_3$, H_2SO_4, and alum in successive stages resulted in a total BOD reduction of 50 per cent, and an accumulated sludge volume of 27 per cent, with a turbidity reduction greater than 82 per cent, and a practically colorless effluent. All other single coagulants or combinations of chemical coagulants produced lower BOD reductions.

Activated-sludge treatment. Aerobic biological processes have been most successful on kraft-mill wastes. Whether this is due to the characteristics of this particular waste, or to the fact that the pollution problems of kraft mills are more pressing than those of other paper or pulp mills, or to more progressive viewpoints on the part of kraft-mill administrators, is still questionable; one modified version, dispersed-growth aeration, has been successful on a laboratory scale in treatment of specialty-paper-mill wastes. In recent years a promising accelerated treatment for kraft wastes has been developed, embodying the principle of activated sludge as used for domestic-sewage treatment, which requires the addition of nitrogen salt. Its greatest shortcomings are, however, the capital and operating costs involved when the process is applied continuously.

A Virginia kraft mill utilizes this process on its normal 16-mgd flow, containing an average BOD of 140 ppm. Aeration for three hours is provided, with a 25 per cent return of sludge. Based on average concentration, the BOD loading of the aerator is 56 pounds per 1000 cubic feet, and mixed-liquor concentrations vary between 0.2 and 0.3 per cent. Air requirements approximate one cubic foot per gallon of waste. Nitrogen and phosphorus nutrients are added as needed, to maintain one pound of available nitrogen and phosphorus for each 20 and 75 pounds of BOD respectively. During its initial operation, in June 1955, this plant accomplished an 85 per cent BOD reduction, while treating 60 per cent of the wastes of the kraft mill.

Experiments with aeration, in rag, rope, and jute mills, involved treating their cooking liquors and wash-water wastes with sludge and dispersed-growth seeds under optimum conditions, and showed these methods of treatment to be effective [280]. Aeration in the presence of sludge floc of mixtures of cooking liquors and the wash water from the first hour of washing appeared to be slightly more effective (but less dependable) than aeration in the presence of dispersed growth, as far as reduction of BOD was concerned. Important factors affecting the rate of oxidation of the wastes are pH, nutrients (N and P), seed adaptation, air supply, and temperature. At optimum conditions, BOD reduction of mixed wastes with a raw BOD concentration of about 2000 ppm, in the presence of dispersed-growth seed, was 78 to 96 per cent; with sludge floc, 90 to 98 per cent after 24 hours of aeration, depending primarily on speed and efficiency of adaptation of seed. Foaming of rag-mill wastes produced an operating difficulty during the aeration process.

Lagooning. The major method of pulp- and paper-mill final waste treatment is lagooning. Ponds, the most widely used form of lagoons, can be used for storage or as effluent-stabilization devices, with storage ranging from 10 days to 10 months. When ponds are used for stabilization purposes, retention periods of 10 to 30 days are common. Because of the tremendous size of ponds, their use depends on availability of land. Also, utilization of these ponds is usually seasonal and depends, as well, on the rate of flow of the receiving stream, with retention of wastes during low water flow and discharge during high flow.

A new kraft mill in Georgia handles its wastes in alternating settling basins, followed by two oxidation lagoons in series. The settling basins receive fiber-bearing waters after passage through flotation save-alls. Strong, nonfiber-bearing wastes are collected in an equalizing basin, from which they are pumped at controlled rates and blended with the effluent from the settling basins. The main object of this type of treatment is to remove the last traces of suspended matter and proportion the equalized waste to the receiving stream according to the flow. However, only limited BOD removals are obtained with lagoons, and odors and contamination of well waters must be taken into consideration.

High-pressure oxidation. The Zimmerman process (see Chapter 14) has been used successfully for treating sulfite-mill wastes in Norway [355], but economic factors must be studied carefully before such a high investment in capital equipment can be made in this country. The atomized suspension technique, utilizing a similar principle, has shown promise for treatment of concentrated, organic-waste liquors [104, 253]. Swedish mills appear to be more advanced in their use of this method of treatment than United States mills.

The current trends in waste treatment in the pulp and paper industry may be summarized in ten areas as follows:

1. In-plant changes have been successful in reducing both strength and quantity of waste. Extensive research is being undertaken on reuse and by-product recovery.
2. Disposal of spent liquor by deep well disposal or by conversion to salable products has been reported.
3. Sedimentation, flotation, and thickening: there has been considerable use of moving screens. In

all of these processes, more and more attention is being paid to fiber recovery.

4. Chemical coagulation: contact flocculation, alum or ferrous salts, and activated silica have proved effective.
5. Solids handling: vacuum filtration, centrifugation, thickening, straining, pressing, incineration, wet-air oxidation, and landfill have been used.
6. Treatment of receiving waters: the paper industry, perhaps more than any other, has been emphasizing consideration of such techniques as: impoundment, intermittent storage at low flow, diffusion of effluent, mechanical aerators within the stream, etc.
7. Biological treatment: activated sludge (and all modifications), trickling filtration, aerated lagoons, and anaerobic treatment are all being utilized, with the emphasis on activated sludge and aerated lagoons.
8. Irrigation disposal: excellent results have been reported on a number of crops.
9. Color removal: activated carbon adsorption.
10. Foam separation.

References: Pulp- and Paper-mill Wastes

1. "Activated sludge treatment of pulp mill wastes," *Sewage Ind. Wastes* **28**, 342 (1956).
2. "Alcohol fermentation of sulfite waste liquor," *Sewage Ind. Wastes* **18**, 168 (1946).
3. "Alcohol recovery from spent sulfite liquor," *Sewage Ind. Wastes* **25**, 705 (1953).
4. Allewelt, A. L., and W. R. Watt, "New methods for preparing cellulose thiourethanes," *Ind. Eng. Chem.* **39**, 69 (1957).
5. Amberg, H. R., "Factors affecting the lagooning of white water," in Proceedings of 6th Industrial Waste Conference, Purdue University, February 1951.
6. Amberg, H. R., "White water, factors affecting lagooning," *Sewage Ind. Wastes* **24**, 806 (1952).
7. Amberg, H. R., and J. F. Cormack, "Aerobic fermentation studies of spent sulfite liquor," *Sewage Ind. Wastes* **29**, 570 (1957).
8. Amberg, H. R., and R. Elder, "Intermittent discharge of spent sulfite liquor," *J. Sanit. Eng. Div. Am. Soc. Civil Engrs.* **82**, 929 (1956).
9. "Anaerobic digestion of paperboard mill white water," Technical Bulletin no. 36, National Council for Stream Improvement, New York (1950).
10. Applebaum, S. B., "Reclaimer treatment of white and waste water," *Tech. Assoc. Pulp Paper Ind.* **29** (1946).
11. Applebaum, S. B., and J. R. Rhinehart, "Waste water and fiber recovery with reclaimers of the precipitation type," *Tech. Assoc. Pulp Paper Ind.*, **28**, 125–126, February 1945.
12. Axe, E. J., "Paperboard mill white water treatment," in Proceedings of 3rd Industrial Waste Conference, Purdue University, May 1947.
13. Baecklund, G., "Methods of producing ethyl alcohol out of waste sulfite liquor in combination with evaporation," Canadian Patent no. 562250, 26 August 1958.
14. Barker, E. F., "How the Kalamazoo Valley deinking mills solved their waste disposal problems," *Paper Trade J.* **140**, 30 (1956).
15. Baum, M. A., and J. R. Salvessen, "Turbine aeration as a method of increasing the purification capacity of a stream," *Tech. Assoc. Pulp Paper Ind.* **41**, 800 (1958).
16. Baunick, H. F., and F. M. Mueller, "Spent sulfite liquor utilization," *Sewage Ind. Wastes* **24**, 1324 (1952).
17. Behn, V. C., "Mechanical dewatering of paper-mill sludges," *Water Sewage Works* **97**, 432 (1950).
18. Benson, H. K., and W. R. Benson, "Detection of sulfite waste liquor in sea water," *Sewage Ind. Wastes* **4**, 722 (1932).
19. Benson, H. K., and A. M. Partansky, "Anaerobic decomposition of sulfite waste liquor," *Sewage Ind. Wastes* **8**, 856 (1936).
20. Berger, H. F., "Solids removal practices in southern kraft and newsprint mills," *Paper Trade J.* **143**, 42 (1959).
21. Berger, H. F., "Summary of research on neutral sulphite semichemical wastes," *Paper Trade J.* **139**, 20 (1955).
22. Berman, R. I., and J. Osterman, "Dissolved air flotation for white water recovery," in Proceedings of 10th Industrial Waste Conference, Purdue University, May 1955.
23. Bialkowsky and Brown, "In-plant pollution control in practice, pulp and paper industry," *Chem. Eng. Progr.* **55**, 54 (1959).

24. Bialkowsky, H. W., and P. S. Billington, "Pilot studies, effect of waste discharge, prediction," *Sewage Ind. Wastes* **29**, 551 (1957).
25. Billings, R. M., "Stream improvement through spray disposal of sulphite liquor at the Kimberly-Clark Corp., Niagara, Wis.," Purdue University Engineering Extension Series, Bulletin no. 96, 1958, p. 71.
26. Bishop, F. W., and J. W. Wilson, "Integrated mill waste treatment and disposal, description," *Sewage Ind. Wastes* **26**, 1485 (1954).
27. Bjorkman, A., "Recovery in the cellulose industry," *Svensk Papperstid.* **61**, 760 (1958); *Inst. Paper Chem. Bull.* **29**, 877 (1959).
28. Black, H. H., "Spent sulfite liquor developments," *Ind. Eng. Chem.* **50**, 95A (1958).
29. Black, H. H., and V. A. Minch, "Industrial waste guide, wood naval stores," *Sewage Ind. Wastes* **25**, 462 (1953).
30. Blandin, H. M., et al., "Forum discussion, paper mill wastes," *Sewage Ind. Wastes* **19**, 1108 (1947).
31. Blandin, H. M., K. H. Holm, and B. F. Stahl, "Workshop on paper manufacturing wastes," in Proceedings of 2nd Industrial Waste Conference, Purdue University, January 1946, p. 155.
32. Bloodgood, D. E., "Development of a method for treating strawboard wastes," *Tech. Assoc. Pulp Paper Ind.* **33**, 317 (1950).
33. Bloodgood, D. E., "Disposal methods of strawboard mill wastes," *Sewage Ind. Wastes* **19**, 607 (1947).
34. Bloodgood, D. E., "Tenth Purdue Conference highlights industrial waste problems," *Ind. Wastes* **1**, 33 (1955).
35. Bloodgood, D. E., and G. Erganian, "Characteristics of strawboard mill wastes," *Sewage Ind. Wastes* **19**, 1021 (1947).
36. Bloodgood, D. E., et al., "Strawboard wastes, laboratory studies," *Sewage Ind. Wastes* **23**, 120 (1951).
37. Blosser, R. O., "BOD removal from deinking wastes," Purdue University Engineering Extension Series, Bulletin no. 96, 1958, p. 630.
38. Blosser, R. O., "Solids removal practices in the paper industry," in Proceedings of 7th Ontario Industrial Waste Conference, June 1960, p. 121.
39. "BOD reduction by heat hydrolysis," Sulfite Waste Research Report, Technical Bulletin no. 29, National Council for Stream Improvement, New York (1949).
40. Boyer, R. A., "Sodium-base pulping and recovery," *Tech. Assoc. Pulp Paper Ind.* **42**, 356 (1959).
41. Boyer, R. A., and S. R. Parsons, "Operation of the 1957 experimental sodium-base recovery plant at Consolidated Waste Power and Paper Company," *Tech. Assoc. Pulp Paper Ind.* **42**, 565 (1959).
42. Brookover, T. E., "A paper board mill's attack on stream pollution," *Tech. Assoc. Pulp Paper Ind.* **28**, 74–79 (1945).
43. Brookover, T. E., "The paper industry and stream pollution," *Paper Trade J.* **131**, 26 (1950).
44. Brown, H. B., "Water conservation," *Sewage Ind. Wastes* **29**, 1409 (1957).
45. Brown, H. B., "Effluent disposal problems and pollution abatement measures of the southern pulp and paper industry," *Southern Pulp Paper J.* **53**, 1–3, October 1955.
46. Brown, W. G., "Market potential for protein concentrate produced from fermentation of spent sulfite liquor," Technical Bulletin no. 110, National Council for Stream Improvement, New York (1958).
47. Brown, R. W., D. T. Jackson, and J. C. Tongren, "Semichemical recovery processes and pollution abatement," *Paper Trade J.* **143**, 28 (1959).
48. Buehler, H., Jr., "Waste treatment in a paper mill," in Proceedings of 12th Industrial Waste Conference, Purdue University, May 1957.
49. Burbank, N. C., and C. D. Eaton, "Pulp and paper mill waste treatment," *Tech. Assoc. Pulp Paper Ind.* **41** (Supplement 195A) 6 (1958).
50. Buswell, A. M., and F. W. Sollo, "Methane fermentation of a fiber board waste," *Sewage Works J.* **20**, 687 (1948).
51. Calise, V. J., and R. J. Keating, "Some economic aspects of white water treatment in pulp and paper mills," in Proceedings of 9th Industrial Waste Conference, Purdue University, May 1954.
52. Callaham, J. R., "Alcohol recovery from sulfite waste liquor," *Sewage Ind. Wastes* **16**, 388 (1944).
53. Carpenter, C., and C. C. Porter, "Waste water utilization by clarification," *Tech. Assoc. Pulp Paper Ind.*, **28**, 147–151 (1945).
54. Cawley, W. A., "Sphaerotilus in streams," *Sewage Ind. Wastes* **30**, 1174 (1958).
55. Cawley, W. A., and C. C. Wells, "Lagoon system for chemical cellulose wastes," *Ind. Wastes* **4**, 37 (1959).
56. Cederquist, K. N., "Some remarks on wet combustion of cellulose waste liquor," *Svensk Papperstid.* **61**, 114 (1958).

57. Chase, E. S., "Sulfite pulp wastes, control by sodium nitrate, Androscoggin river," *Sewage Ind. Wastes* **22**, 273 (1950).
58. "Chemical recovery from pulping liquors," Workbook Feature, *Ind. Eng. Chem.* **50**, 59A (1958).
59. Churchill, M., "Paper mill wastes, color reduction in streams," *Sewage Ind. Wastes* **23**, 661 (1951).
60. "Consolidated demonstrates its recovery of spent sulphite liquors," *Paper Trade J.* **143**, 50 (1959).
61. Coogan, F. J., "Waste control in a southern paper mill," *J. Water Pollution Control Federation* **32**, 853 (1960).
62. Cooke, W. B., "Some effects of spray disposal of spent sulfite liquor on soil mold populations," in Proceedings of 15th Industrial Waste Conference, Purdue University, May 1960, p. 35.
63. Copenhaver, J. E., Biggs, Boxley, and Wise, "Recovery of acetic and formic acids from black liquor," Canadian Patent no. 551736, January 1958.
64. Cormack, J. F., and H. R. Amberg, "The effect of biological treatment of sulphite waste liquor on the growth of sphaerotilus natans," in Proceedings of 14th Industrial Waste Conference, Purdue University, May 1959, p. 16.
65. Crawford, S. C., "Effects of kraft-mill wastes on oxygen balance in streams," *Sewage Ind. Wastes* **20**, 876 (1948).
66. Crawford, S. C., "Lagoon treatment of kraft-mill wastes," *Sewage Ind. Wastes* **19**, 621 (1947).
67. Crawford, S. C., "Spray irrigation of certain sulfate pulp mill wastes," *Sewage Ind. Wastes* **30**, 1266 (1958).
68. Crawford, S. C., "Stream pollution report to National Container Corp.," May 1947, Big Island, Virginia.
69. "Cyclator helps solve problem of stream improvement," *Paper Mill News* **80**, 16 (1957).
70. Darmstadt, W. J., "Neutral sulfite recovery process," *Tech. Assoc. Pulp Paper Ind.* **41**, 147 (1958).
71. Dedert and Brown, "Evaporation of neutral sodium sulfite concentrate spent liquor," *Paper Ind.* **39**, 914 (1958).
72. "Deep-water effluent dispersion," *Pulp Paper* **33**, 77 (1959).
73. "Deinking sludge, modified, utilization possibility," *Sewage Ind. Wastes* **25**, 627 (1953).
74. "Deinking wastes, biological treatment," *Sewage Ind. Wastes* **21**, 186 (1949).
75. "Development studies on the removal of color from caustic extract bleaching effluent by the surface reaction process," Technical Bulletin no. 107, National Council for Stream Improvement, New York (1958).
76. Dickerson, B. W., et al., "Treatment of industrial water from tall oil purification operations," in Proceedings of 13th Industrial Waste Conference, Purdue University, May 1958.
77. Diehl, W. F., "New process is key to expansion," *Pulp Paper* **32**, 54 (1958).
78. Dinsmore, R. F., "Long tube forced circulation evaporation of sulphite waste liquor," Maine Technology Experiment Station, Paper no. 74, April 1953; reprint from *Paper Mill News*, 14 February, 1953.
79. Doutt, F. V., "Pulp and paper mill wastes," in Proceedings of 5th Industrial Waste Conference, Purdue University, November 1949.
80. Doutt, F. V., "Pulp and paper wastes: problems, discussion," *Sewage Ind. Wastes* **23**, 699 (1951).
81. Doutt, F. V., "Unused water is no waste problem," in Proceedings of 4th Southern Municipal and Industrial Waste Conference, March 1955, p. 83.
82. Drummond, R. M., "Pulp waste reduction by mill and process improvement," *Sewage Ind. Wastes* **26**, 656 (1954).
83. Easton, P., and R. Baum, "White water recovery by a flotation method," *Tech. Assoc. Pulp Paper Ind.* **33**, 301 (1950).
84. Eden, G. E., et al., "Treatment of waste water from a paper mill by biological filtration," *Paper-Maker* **123**, 4 (1952).
85. Eden, G. E., E. E. Jones, and A. B. Wheatland, "Paper mill wastes, trickling filters, pilot study," *Sewage Ind. Wastes* **26**, 119 (1954).
86. "Edward Towgood and Sons install new effluent purification plant," *World Paper Trade Rev.* **148**, 1280 (1957).
87. "Effect of nutrients upon the rate of stabilization of spent sulfite liquor in receiving waters," Sanitary Engineering Division Research Committee, Stream Pollution Section, American Society of Civil Engineers, Research Report no. 3, October 1955. (From research data of H. R. Amberg.)
88. Ehemann, G. C., et al., "White water recovery, OCO system," *Sewage Ind. Wastes* **22**, 1573 (1950).
89. Eldridge, E. F., "Kalamazoo Valley, Mich., paper mill waste disposal," *Sewage Ind. Wastes* **16**, 1276 (1944).
90. Eldridge, E. F., "Monroe, Mich., fiberboard waste

treatment plant," *Sewage Ind. Wastes* **13**, 105 (1941).
91. Eldridge, E. F., and W. L. Mallman, "Digestion of strawboard mill waste with sewage sludge," *Sewage Ind. Wastes* **4**, 230 (1932).
92. Ergenien, G., and J. C. Hargleroad, "Strawboard waste disposal research," in Proceedings of 3rd Industrial Waste Conference, Purdue University, May 1947.
93. Evans, J. C. W., "A new sulphite pulping process and a new pulp," *Paper Trade J.* **143**, 42 (1959).
94. Evans, J. C. W., "Unique sulphite-soda pulping and recovery system used at Rayma," *Paper Trade J.* **143**, 50 (1959).
95. Evans, J. S., "Legal aspects of the effluent problem," *World Paper Trade Rev.* **150**, 1697 (1958); *Inst. Paper Chem. Bull.* **29**, 1367 (1959).
96. Felicetta, V. F., M. Lung, and J. L. McCarthy, "Spent sulphite liquor, sugar-lignin-sulphonate separations using ion exchange resins," *Tech. Assoc. Pulp Paper Ind.* **42**, 497 (1959).
97. Felicetta, V. F., and J. L. McCarthy, "The pulp mills research program at the University of Washington," *Tech. Assoc. Pulp Paper Ind.* **40**, 851 (1957).
98. "First large NSSC recovery system proves out," *Paper Trade J.* **143**, 22 (1959).
99. Fogler, H. H., et al., "Spray drying coefficients for sulfite waste liquor," Maine Technology Experiment Station, Paper no. 600, September 1949; reprint from *Tech. Assoc. Pulp Paper Ind.* **32**, 9 (1949).
100. Ganczarczyk, J., "Pulp waste treatment methods, review," *Sewage Ind. Wastes* **25**, 237 (1953).
101. Ganczarczyk, J., and J. Domanski, "Spent sulfite waste: coagulation, laboratory studies," *Sewage Ind. Wastes* **26**, 930 (1954).
102. Ganczarczyk, J., and J. Domanski, "White water: alkaline coagulation study," *Sewage Ind. Wastes* **26**, 1055 (1954).
103. Gaudy, A. J., "Wastes from pulp and paper processes," California State Water Pollution Control Board, Publication no. 17, 1957.
104. Gauvin, W. H., "Application of the atomized suspension technique," *Tech. Assoc. Pulp Paper Ind.* **40**, 866 (1957).
105. Gehm, H. W., "Activities of the pulp and paper industry in North Carolina on stream pollution control," in Proceedings of 1st Southern Municipal and Industrial Waste Conference, March 1952, p. 118.
106. Gehm, H. W., "Effects of fiberboard wastes on sewage treatment plants," *Sewage Ind. Wastes* **17**, 510 (1945).
107. Gehm, H. W., "Modern approaches to pulp and paper mill waste problems," *Sewage Ind. Wastes* **29**, 1370 (1957).
108. Gehm, H. W., "New and basic research approaches to liquid effluent treatment," *Paper Trade J.* **142**, 40 (1958).
109. Gehm, H. W., "New wrinkles for dewatering paper mill sludges," *Wastes Eng.* **30**, 256 (1959).
110. Gehm, H. W., "Physical factors involved in the clarification of paper effluents," *Tech. Assoc. Pulp Paper Ind.* **33**, 9 (1950).
111. Gehm, H. W., "Research developments in waste treatment," *Sewage Ind. Wastes* **19**, 827 (1947).
112. Gehm, H. W., "Research, National Council for Stream Improvement," *Sewage Ind. Wastes* **17**, 782 (1945).
113. Gehm, H. W., and V. Behn, "High-rate anaerobic digestion of wastes," *Sewage Works J.* **22**, 1034 (1950).
114. Gehm, H. W., and D. E. Bloodgood, "Problems in paper mill waste disposal," in Proceedings of 2nd Industrial Waste Conference, Purdue University, January 1946.
115. Gehm, H. W., and D. E. Bloodgood, "Waste disposal problems in paper mills," *Sewage Ind. Wastes* **19**, 1102 (1947).
116. Gehm, H. W., and N. J. Lardieri, "Waste treatment practices, paper industry," *Sewage Ind. Wastes* **28**, 287 (1956).
117. Gehm, H. W., and P. Morgan, "Paper and strawboard wastes, high-rate anaerobic waste treatment system," *Sewage Ind. Wastes* **21**, 851 (1949).
118. Gehm, H. W., "BOD test applications," *Sewage Ind. Wastes* **19**, 865 (1947).
119. Gellman, I., and R. Blosser, "Disposal of pulp and paper-mill waste by land application and irrigational use," in Proceedings of 14th Industrial Waste Conference, Purdue University, May 1959.
120. Giessen, J., "Process for hydrolyzing sulfite waste liquor," Canadian Patent no. 547591, 15 October 1957.
121. Gurnham, C. F., "Fourth Southern Waste Conference," *Ind. Wastes* **1**, 24 (1955).
122. Harmon, J. P., *Use of Lignin Sulfonate for Dust Control on Haulage Roads in Arid Regions*, Circular no. 7806, U.S. Bureau of Mines (1957).

123. Harris, E. L., et al., "Spent sulfite liquor for yeast growth," *Sewage Ind. Wastes* **24**, 684 (1952).
124. Haupt, H., "By-product recovery," *Sewage Ind. Wastes* **8**, 350 (1936).
125. Hawley, R. S., "Monroe, Mich., fiberboard recovery from wastes," *Sewage Ind. Wastes* **11**, 512 (1939).
126. Hearon, W. M., "The lignin dimethyl sulfide process," *Forest Prod. J.* **7**, 432 (1957).
127. Henley, E., *Interaction of Nuclear Radiation with Pulping Waste Products*, National Council for Stream Improvement, Technical Bulletin no. 109 (October 1958).
128. Herbet, A. J., and H. F. Berger, "A kraft bleach waste color reduction process integrated with the recovery system," in Proceedings of 15th Industrial Waste Conference, Purdue University, May 1960, p. 49.
129. Hickam, R., "Rubber reinforced by lignin complex," Canadian Patent no. 550293, 17 December 1957.
130. Hickerson, R. C., and E. K. McMahon, "Spray irrigation of wood distillation waste," *J. Water Pollution Control Federation* **32**, 55 (1960).
131. Hill, M. T., "Paper pulp digestion carbon-nitrogen ratio," *Sewage Ind. Wastes* **11**, 864 (1939).
132. Hodge, W. W., and P. F. Morgan, "Characteristics and methods for treatment of deinking wastes," *Sewage Works J.* **19**, 830 (1947).
133. Höhnl, W., "Fungi problems," *Sewage Ind. Wastes* **30**, 1078 (1958).
134. Holder, J. M., and W. A. Moggio, "Utilization of spent sulfite liquor," *J. Water Pollution Control Federation* **32**, 171 (1960).
135. Holderby, J. M., "Trickling filter experiments for sulphite waste liquor," *Sewage Ind. Wastes* **18**, 671 (1946).
136. Holderby, J. M., and W. A. Moggio, "The production of nutritional yeast from spent sulfite liquors," *Forest Prod. J.* **9**, 21A (1959).
137. Holderby, J. M., and W. A. Moggio, "Spent sulphite liquor treatment at Rhinelander Paper Company," in Proceedings of 14th Industrial Waste Conference, Purdue University, May 1959, p. 111.
138. Holderby, J. M., and A. J. Wiley, "Biological treatment of spent liquor from the sulfite pulping process," *Sewage Ind. Wastes* **22**, 61 (1950).
139. Howard, G. C., "Howard process for sulfite waste liquor disposal," *Sewage Ind. Wastes* **12**, 640 (1940).
140. "How National's 'nature ponds' work," *Pulp Paper*, **56**, 1–6, April 1956.
141. Hull, W. Q., et al., "Ammonia-base sulfite process, description," *Sewage Ind. Wastes* **26**, 1513 (1954).
142. Hull, W. Q., et al., "Magnesia base sulfite pulping developments," *Sewage Ind. Wastes* **24**, 1046 (1952).
143. Hutchinson, O. F., "Flotation processes: use and results in paper mill waste water clarification," *Tech. Assoc. Pulp Paper Ind.* **41**, (Supplement 158A), 7 (1958).
144. 78th U.S. Congress, First Session, House Document 266, *Industrial Guide to the Paper Industry*.
145. U.S. Public Health Service, *Industrial Waste Guide*, Ohio River Pollution Control Survey, Supplement D, Appendix X (1943), p. 1193.
146. "Industrial wastes forum," *Sewage Ind. Wastes* **28**, 654 (1956).
147. Inskeep, G. C., et al., "Spent sulfite liquor, yeast production," *Sewage Ind. Wastes* **24**, 687 (1952).
148. "Investigation of BOD reduction of waste sulfite liquor by heat hydrolysis," Technical Bulletin no. 53, National Council for Stream Improvement, New York (1952).
149. Jacobs, H. L., "Rayon wastes, recovery and treatment," *Sewage Ind. Wastes* **25**, 296 (1953).
150. Jahn, E. C., "Lignin recovery, pulp wastes," *Sewage Ind. Wastes* **13**, 835 (1941).
151. Jones, B. F., et al., "Avoidance reactions of salmonid fishes to pulp mill effluents," *Sewage Ind. Wastes* **28**, 1403 (1956).
152. Joseph, H. G., "Alcohol recovery from sulfite waste liquor," *Sewage Ind. Wastes* **19**, 60 (1947).
153. Jung, H., "Strawboard wastes, high-rate digestion," *Sewage Ind. Wastes* **25**, 627 (1953).
154. Keeth, G., "Howard process for sulfite waste liquor disposal," *Sewage Ind. Wastes* **12**, 643 (1940).
155. Kittrell, F., "Pulp and paper mill wastes, Tennessee Valley, effects," *Sewage Ind. Wastes* **22**, 1092 (1950).
156. Knack, M. F., "Board mill waste treatment; plant description," *Sewage Ind. Wastes* **23**, 120 and 1533 (1951).
157. Knack, M. F., "A waste treatment plant for board-mill wastes," in Proceedings of 4th Industrial Waste Conference, Purdue University, September 1948.
158. Kobe, K. A., "Utilization and treatment of sulfite waste liquor," *Sewage Ind. Wastes* **9**, 1019 (1937).
159. Kobe, K. A., and E. J. McCormack, "Pulping waste liquors, viscosity," *Sewage Ind. Wastes* **22**, 721 (1950).
160. Koch, H. C., "Paper mill waste disposal," *Sewage Ind. Wastes* **22**, 968 (1950).

161. Koch, H. C., and D. E. Bloodgood, "Experimental spray irrigation of paperboard mill wastes," *Sewage Ind. Wastes* **31**, 827 (1959).
162. Koch, H. C., and J. J. Lugar, "Addition of nitrate to paper mill wastes," in Proceedings of 13th Industrial Waste Conference, Purdue University Engineering Extension Series, Bulletin no. 96, 1958, p. 163.
163. Kominek, E. G., "Chemical treatment of white water," *Ind. Eng. Chem.* **42**, 616 (1950).
164. Kominek, E. G., "White water chemical treatment," *Sewage Ind. Wastes* **22**, 1386 (1950).
165. "Kraft mill waste, storage studies," *Sewage Ind. Wastes* **21**, 373 (1949).
166. Krancher, G. G., "Strawboard lagoon operation at Noblesville, Ind.," in Proceedings of 6th Industrial Waste Conference, Purdue University, February 1951.
167. Krancher, G. G., "Strawboard wastes, lagoons operation, Noblesville, Ind.," *Sewage Ind. Wastes* **24**, 925 (1952).
168. Kronbach, A. J., "Monroe, Mich., treatment with sewage," *Sewage Ind. Wastes* **11**, 518 (1939).
169. Kulkarni, G. R., and W. J. Nolan, "The mechanisms of alkaline pulping," University of Florida, vol. IX, no. 3, March 1955.
170. Kure, A. R., "Ethyl alcohol from sulphite waste liquor," in Proceedings of 3rd Ontario Industrial Waste Conference, June 1956, p. 1.
171. Lambe, T. W. and O. E. Anderson, "Impermeabilization of lagoon at International Paper Co., Chisholm, Me.," *Tech. Assoc. Pulp Paper Ind.* **38**, 39 (1955).
172. Lardieri, N. J., "The aerobic and benthol oxygen demand of paper mill waste deposits," *Tech. Assoc. Pulp Paper Ind.* **37**, 705 (1954).
173. Lardieri, N. J., "How to treat paper-mill effluent by controlled bio-oxidation methods," *Wastes Eng.* **28**, 456 (1957).
174. Lardieri, N. J., "Organic constituents of strawboard wastes," in Proceedings of 5th Industrial Waste Conference, Purdue University, November 1949.
175. Lardieri, N. J., "Strawboard waste organic constituents," *Sewage Ind. Wastes* **23**, 699 (1951).
176. Lasseter, F. P., and B. J. Queern, "Clariflocculation of mill wastes," *Paper Trade J.* **120**, 45 (1945).
177. Lawrance, W. A., and H. N. Fukui, "Calcium lignosulfonate, microbial oxidation," *Sewage Ind. Wastes* **28**, 1484 (1956).
178. Lawrance, W. A., "Androscoggin river, pollution by pulp and paper waste," *Sewage Ind. Wastes* **20**, 881 (1948).
179. Lawrance, W. A., and D. E. Bloodgood, "Paper mill wastes, oxygen demand, permanganate method," *Sewage Ind. Wastes* **22**, 792 (1950).
180. Lawrance, W. A., and W. Sakamoto, "The microbial oxidation of cellobiose, butyric and lactic acids in the presence of calcium lignosulfonate," *Tech. Assoc. Pulp Paper Ind.* **42**, 93 (1959).
181. Lawrance, W. A., "The addition of sodium nitrate to the Androscoggin river," *Sewage Works J.* **22**, 820 (1950).
182. Leonard, R. H., *et al.*, "Sulfite waste liquor, lactic acid from fermentation," *Sewage Ind. Wastes* **21**, 372 (1949).
183. Lebo, J. E., and J. W. Hassler, "Trickling filter studies," *Sewage Ind. Wastes* **29**, 170 (1957).
184. Leonard, A. G., and R. J. Keating, "Complete white water treatment in a tissue mill," Purdue University Engineering Extension Series, Bulletin no. 96, 1958, p. 286.
185. Lewis, H. F., "Fiber recovery, research, review," *Sewage Ind. Wastes* **23**, 1214 (1951).
186. Libby, C. E., "Introduction to the pulp and paper session," in Proceedings of the 3rd Southern Municipal and Industrial Waste Conference, March 1954, p. 86.
187. "Liquid waste, Central States Sewage and Industrial Wastes Association Meeting report," *Ind. Wastes* **2**, 152 (1957).
188. Logan, R. H., and H. Heukelekian, "Oxidation of waste sulfite liquor," *Sewage Works J.* **20**, 282 (1948).
189. Lucas, S. J., "Mechanical aspects of industrial waste treatment at Downington Paper Co.," *Tech Assoc. Pulp Paper Ind.* **42**, 117A (1959).
190. Lukemire, E. L., "Paper making with a closed system," *Inst. Paper Chem. Bull.* **29**, 482 (1959); Canadian Patent no. 562135.
191. Luner, Oubey, Boggs, and Wiley, "Upgrading spent sulfite liqour by sugar removal," *Forest Prod. J.* **8**, 82 (1958).
192. McCarthy, J. L., *et al.*, "Consideration of ammonium spent sulphite liquor processes using ion exchange resins," *Tech. Assoc. Pulp Paper Ind.* **42**, 379 (1959).
193. McCarthy, J. L., and L. Grimsbud, "Spent sulphite liquor, fanning friction factors and heat transfer

coefficients for concentrated spent sulphite liquors," *Tech. Assoc. Pulp Paper Ind.* **42**, 503 (1959).
194. McDermott, G. N., "Sources of wastes from kraft pulping and theoretical possibilities of reuse of condensates," in Proceedings of 3rd Southern Municipal and Industrial Waste Conference, March 1954, p. 105.
195. McIntosh, W. A., "Lignin disposal, asphalt emulsifier," *Sewage Ind. Wastes* **24**, 1552 (1952).
196. Maloney, T. E., "Utilization of sugars in spent sulfite liquor by a green alga, *Chlorococcum macrostigmatum*," *Sewage Ind. Wastes* **31**, 1395 (1959).
197. Marczek, E., and J. Zielinski, "Sulfite waste treatment, activated sludge trickling filter," *Sewage Ind. Wastes* **30**, 838 (1958).
198. Martin, G., and E. P. Miller, "BOD of sulfite waste liquor, effect of dilution water," *Sewage Ind. Wastes* **13**, 659 (1941).
199. Meighen, A. D., "Experimental spray irrigation of strawboard wastes," Purdue University Engineering Extension Series, Bulletin no. 96, 1958, p. 456.
200. Merrill, E. I., "Reduction of paper mill pollution through recirculation," in Proceedings of 5th Southern Municipal and Industrial Waste Conference, April 1956, p. 88.
201. Merryfield, F., "Utilization and treatment of sulfite waste liquor," *Sewage Ind. Wastes* **8**, 868 (1936).
202. Meyer, W. G., and J. G. Coma, "Dimethyl sulfide production from kraft pulp mill black liquor," *Chem. Eng. Progr.* **54**, 178 (1958).
203. Miller, H. E., and J. M. Kniskern, "Designs for control and treatment of wastes at West Virginia Pulp and Paper Co.," in Proceedings of 9th Industrial Waste Conference, Purdue University, May 1954, p. 531.
204. Miner, G. N., and R. J. Keating, "Recovery and reuse of boxboard mill effluent," 42nd Annual Meeting of Technical Association Pulp and Paper Industry, Technical Reprint, *T*-155, February 1957.
205. Miner, G. N., and R. J. Keating, "Recovery and reuse of boxboard mill effluent," *Can. Pulp Paper Ind.* **11**, 39 (1958).
206. Moggio, W. A., "Color removal from kraft mill wastes," in Proceedings of 9th Industrial Waste Conference, Purdue University, May 1954.
207. Moggio, W. A., "Control and disposal of kraft mill effluents," *Proc. Am. Soc. Civil Engrs.* **80**, separate no. 420 (1954).
208. Moggio, W. A., "Kraft mill effluent, control and disposal," *Sewage Ind. Wastes* **27**, 241 (1955).
209. Moggio, W. A., "Storage studies on kraft mill wastes," *Louisiana State Univ. Eng. Exp. Sta. Bull.* **13** (1948).
210. Moggio, W. A., Barnes, and Colmer, "Bacteriological studies of stored kraft paper mill wastes," *Louisiana State Univ. Eng. Exp. Sta. Bull.* **19** (1950).
211. Moggio, W. A., and H. W. Gehm, "Biological treatment of kraft mill wastes," *Sewage Ind. Wastes* **22**, 1326 (1950).
212. Moggio, W. A., and H. W. Gehm, "Kraft mill wastes, activated sludge pilot plant," *Sewage Ind. Wastes* **25**, 305 (1953).
213. Moggio, W. A., and H. W. Gehm, "Kraft mill wastes, biological treatment," *Sewage Ind. Wastes* **22**, 1326 (1950).
214. Moore, T. L., "An activated sludge plant for pulp and paper wastes," in Proceedings of 10th Industrial Waste Conference, Purdue University, May 1955.
215. Moore, T. L., "Applying the modified activated sludge process to the treatment of kraft wastes," in Proceedings of 4th Southern Municipal and Industrial Waste Conference, April 1955, p. 70.
216. Morgan, P., "Deinking wastes, pilot plant treatment," *Sewage Ind. Wastes* **21**, 512 (1949).
217. Morgan, P., "Deinking wastes, study of problem, Kalamazoo, Mich.," *Sewage Ind. Wastes* **22**, 140 (1950).
218. Morrison, H. A., "White water reuse and disposal," *Tech. Assoc. Pulp Paper Ind.* **29**, 157–159, 25 February 1946.
219. Murdock, H., "Pollutional wastes, status," *Sewage Ind. Wastes* **24**, 1443 (1952).
220. Murdock, H. R., "Pulp and paper industry," *Ind. Eng. Chem.* **44**, 507 (1952).
221. Murdock, H. R., "Pulp and paper industry symposium on liquid wastes," *Ind. Eng. Chem.* **44**, 507 (1952).
222. Murdock, H. R., "Waste treatment development, review," *Sewage Ind. Wastes* **26**, 71 (1954).
223. Murdock, H. R., "Wood pulp industries are reducing pollution loads on streams by modern methods and equipment," *Ind. Eng. Chem.* **45**, 102*A* (1953).
224. National Council for Stream Improvement Symposium on Waste Treatment, "Mill site selection," "Kraft mill process water," "Pollutional load factors," "Solids removal," "Biological treatment," "Sulfite liquor treatment," *Ind. Wastes* **1**, 177 (1956).
225. Neale, A. T., "Pulp and paper wastes in Washington," *Ind. Wastes* **3**, 29 (1958).
226. Nemerow, N. L., "Fiber losses, effect on streams and

227. Nethercut, P. E., "Pulp and paper," *Ind. Eng. Chem.* **51**, 47*A* (1959).
228. Nolan, W. J., "The alkaline pulping of bagasse for high strength papers and dissolving pulps," University of Florida, vol. IX, no. 3, March 1955.
229. Nolan, W. J., "Continuous kraft pulping, its sensitivity to the cooking variables," *Southern Pulp Paper J.* **19**, 6 (1956).
230. Nolan, W. J., "Increasing quality and production in high yield kraft and semichemical pulps," *Eng. Prog. Univ. Florida, Leaflet Ser.* **8** (1954); *Paper Ind.* **36**, 1954.
231. Nolan, W. J., "The sulphate process," *Eng. Prog. Univ. Florida, Tech. Paper Ser.* **102**, (1954).
232. Noordam-Geldewagen, M. A., *et al.*, "Sulfite influence on strawboard mill fermentation," *Sewage Ind. Wastes* **24**, 1437 (1952).
233. Opferkuch, R. E., Jr., "Combined treatment of sewage and pulp mill wastes," *Wastes Eng.* **22**, 658 (1951).
234. Opferkuch, R. E., Jr., and H. F. Berger, "Anaerobic digestion of spent semichemical pulping liquor," in Proceedings of 3rd Southern Municipal and Industrial Waste Conference, March 1954, p. 90.
235. Orlob, G. T., and E. F. Eldridge, "Pulp mill wastes, deep water disposal, Everett, Wash.," *Sewage Ind. Wastes* **26**, 520 (1954).
236. Othmer, D. F., "Chemicals recovery from pulping liquor," *Ind. Eng. Chem.* **50**, 60*A* (1958).
237. Palladino, A. J., "Deinking waste treatment plant, design," *Sewage Ind. Wastes* **23**, 699 (1951).
238. Palladino, A. J., "Deinking waste treatment plant design," *Sewage Ind. Wastes* **23**, 1419 (1951).
239. Palladino, A. J., "Design factors for primary treatment of pulp and paper mill waste," *Ind. Wastes* **1**, 267 (1956).
240. Palladino, A. J., "Design of deinking waste treatment plant," in Proceedings of 5th Industrial Waste Conference, Purdue University, November 1949.
241. Palladino, A. J., "Need for revised sewering techniques in paperboard mills," *Paper Trade J.* **141**, 26 (1957).
242. Palladino, A. J., "Operating results, demonstration plant for treatment of deinking wastes," in Proceedings of 7th Industrial Waste Conference, Purdue University, May 1952.
243. Palladino, A. J., "Treatment of pulp and paper mill effluents in the Kalamazoo river valley," in Proceedings of 10th Industrial Waste Conference, Purdue University, May 1955.
244. Palladino, A. J., and J. F. Baigas, "Deinking, two treatment plants, description," *Sewage Ind. Wastes* **28**, 924 (1956).
245. Palladino, A. J., and J. F. Baigas, "Deinking waste treatment plants at Plainwell, Mich., and Fitchburg, Mass," in Proceedings of 6th Industrial Waste Conference, Purdue University, February 1951.
246. "Paper mill waste treatment plant construction, 1954," *Sewage Ind. Wastes* **27**, 1368 (1954).
247. Pascoe, T. A., J. S. Buchanan, E. H. Kennedy, and G. Sivola, "The Sivola sulphite cooking and recovery process," *Tech. Assoc. Pulp Paper Ind.* **42**, 265 (1959).
248. Pearl, I. A., and H. K. Benson, "Catalytic oxidation by atmospheric oxygen," *Sewage Ind. Wastes* **14**, 1157 (1942).
249. Pearman, B. V., and O. B. Burns, "Activated sludge treatment of kraft and neutral sulfite mill wastes," *Sewage Ind. Wastes* **29**, 1145 (1957).
250. Pehrson and Rennerfelt, "Swedish pilot plants for biological treatment of waste water from the pulp and paper industry," *Svensk. Papperstid.* **61**, 879 (1958).
251. Perry, F. G., *et al.*, "Economics of furfural production from hardwood with reference to utilization of spent pickle liquors," in Proceedings of 14th Industrial Waste Conference, Purdue University, May 1959.
252. Peters, J. G., "Technical and biological aspects of waste water treatment at Downingtown Paper Co.," *Tech. Assoc. Pulp Paper Ind.* **42**, 175*A* (1959).
253. Pinder, K. L., and W. H. Gauvin, "Atomized suspension techniques," *Ind. Wastes* **4**, 26 (1959).
254. Porter, C. C., and F. W. Bishop, "Treatment of paper mill wastes in biochemical oxidation ponds." *Ind. Eng. Chem.* **42**, 102 (1950).
255. Pratt, S. K., "Accomplishments of mills on water quality control," *Tech. Assoc. Pulp Paper Ind.* **42**, 150*A* (1959).
256. Price, F. A., "The Lignosol saga: successful developments of sulfite liquor products," *Pulp Paper Mag. Can.* **59**, 88 (1958).
257. Priest, J. J., "Problems of paper mills when complex interests are involved," in Proceedings of 5th Southern Municipal and Industrial Waste Conference, April 1956, p. 179.
258. "Pulp and paper mill waste disposal by irrigation and land application," Technical Bulletin no. 124,

National Council for Stream Improvement, New York (1959).
259. "Pulp and paper waste," *Ind. Wastes* **3**, 63 (1958).
260. U.S. Bureau of Fisheries, "Pulp waste pollution of Puget Sound oyster beds," Report, *Sewage Ind. Wastes* **4**, 924 (1932).
261. Quirk, T. P., "Amenability of a mixture of sewage, cereal and board mill wastes to biological treatment," Purdue University Engineering Extension Series, Bulletin no. 96, 1958, p. 523.
262. Quirk, T. P., "Board mill wastes," in Proceedings of 7th Southern Municipal and Industrial Waste Conference, May 1958, p. 57.
263. Ratliff, F. C., "The use of a circulating lagoon in a paper mill effluent program," Purdue University Engineering Extension Series, Bulletin no. 94, 1957, p. 502.
264. "Rayonier's new sodium base process," *Paper Mill News* **82**, 12 (1959).
265. "Recover pulping chemicals this new way," *Chem. Eng.* **64**, 168 (1957).
266. National Council for Stream Improvement, *Report on Semichemical Wastes*, Report no. 24, New York (1949).
267. Robbins, M. H., "Activated sludge treatment of neutral sulfate semichemical waste," Technical Bulletin no. 103, National Council for Stream Improvement, New York (1958).
268. Robinson, L. E., "The control of paper machine stock losses," *Tech. Assoc. Pulp Paper Ind.* **42**, 158A (1959).
269. Robinson, L. E., and Harris, "Major savings for Megnefite," *Pulp Paper* **33**, 87 (1959).
270. Romano, A. H., "Fumaric acid fermentation of spent sulphite liquor," *Tech. Assoc. Pulp Paper Ind.* **41**, 687 (1958).
271. Ross, E. N., "Turbidity reduction in paper mill waste water," in Proceedings of 15th Industrial Waste Conference, Purdue University, May 1960, p. 240.
272. Rowley, F. B., R. C. Jordan, R. M. Olson, and R. F. Huettl, "Pulp, paper, and insulation mill waste analysis," *Minn. Univ. Eng. Exp. Sta. Bull.* **19**, (1942).
273. Rudolfs, W., *Industrial Waste Treatment*, Monograph Series no. 118, Reinhold Publishing Corp., New York (1953), p. 195.
274. Rudolfs, W., and H. R. Amberg, "White water treatment, effect of sulfides on digestion," *Sewage Ind. Wastes* **24**, 1278 (1952).
275. Rudolfs, W., and H. R. Amberg, "White water treatment, factors affecting anaerobic digestion," *Sewage Ind. Wastes* **24**, 1108 (1952).
276. Rudolfs, W., and H. R. Amberg, "White water treatment, factors affecting digestion efficiency," *Sewage Ind. Wastes* **24**, 1402 (1952).
277. Rudolfs, W., and E. J. Axe, "Clarification of paperboard mill white water," *Paper Trade J.* pages 1–6, April 1948.
278. Rudolfs, W., and E. J. Axe, "White water treatment and recovery," *Water Sewage Works* **95**, 219 (1948).
279. Rudolfs, W., and N. L. Nemerow, "Rag, rope, and jute wastes," *Sewage Ind. Wastes* **24**, 480 (1952).
280. Rudolfs, W., and N. L. Nemerow, "Rag, rope, and jute wastes from specialty paper mills," *Sewage Ind. Wastes* **24**, 661, 765, 882, and 1005 (1952).
281. Rudolfs, W., and N. L. Nemerow, "Some factors affecting slime formation and freeness in board-mill stock," *Tech. Assoc. Pulp Paper Ind.* **33**, 7 (1950).
282. Rue, J. D., "Waste disposal problem in pulp and paper industry," *Sewage Ind. Wastes,* **1**, 365 (1929).
283. Safford, T. H., "The location of wet process industries," in Proceedings of 10th Industrial Waste Conference, Purdue University, May 1955.
284. Salvesen, J. R., "From waste and nuisance to use and profit," *Ind. Wastes* **1**, 87 (1956).
285. Salvesen, J. R., "Utilizing lignosulfonates from spent sulfite liquor," *Forest Prod. J.* **9**, 6 (1959).
286. Sanborn, N., "Slime-producing bacteria in pulp and paper production," *Sewage Ind. Wastes* **17**, 405 (1945).
287. Savage, R. H., and H. C. Koch, "Waste disposal problems, evaluations," *Sewage Ind. Wastes* **22**, 1567 (1950).
288. Sawyer, C., "Activated sludge treatment of sulfite waste liquor with sewage," *Sewage Ind. Wastes* **13**, 625 (1941).
289. Shaffer, M. R., "Three commercial processes now available for NSSC recovery," *Paper Trade J.* **142**, 36 (1958).
290. Sheen, R. T., "White water treatment, an industrial waste problem," *Paper Trade J.* **126**, 106 (1948).
291. Shellow, W. S., and A. L. Pickens, "Soap collection in the sulfate process," *Southern Pulp Paper J.* **21**, 46 (1958).
292. Siekierzynska, H., and J. Zielinski, "Sulfate cellulose, process waste treatment," *Sewage Ind. Wastes* **26**, 931 (1954).
293. Simpson, R. W., "Tissue paper wastes treated by save-all closed system," *Sewage Works Eng.* **20**, 134 (1949).

294. Simpson, R. W., E. T. Duke, and K. Thompson, "Asbestos paper waste treatment," *Water Sewage Works* **99**, 286 (1952).
295. Skewes, T. J., and H. K. Benson, "Oxalic acid recovery from sulfite waste liquor," *Sewage Ind. Wastes* **11**, 921 (1939).
296. Skinner, H. J., "Waste problems in pulp and paper industry," *Sewage Ind. Wastes* **12**, 397 (1940).
297. Smith and Schirtzinger, "Prevention of fiber losses from paper machine system," *Tech. Assoc. Pulp Paper Ind.* **41**, 163A (1958).
298. Snell, F. D., "Characteristics of waste treatment," *Sewage Ind. Wastes* **9**, 327 (1937).
299. Southgate, B., "England, waste treatment and disposal practice," *Sewage Ind. Wastes* **18**, 340 (1946).
300. "Spent sulfite liquor, evaporation," 1951 Industrial Waste Forum, *Sewage Ind. Wastes* **24**, 870 (1952).
301. Sperry, W. A., "Paper and strawboard wastes, cooperative effort solves problem," *Sewage Ind. Wastes* **21**, 576 (1949).
302. Stahl, B. F., "Paper mill wastes, pilot plant experiments," *Sewage Ind. Wastes* **22**, 968 (1950).
303. Stanley, W. E., and R. D. Ellis, "Impregnated pulp pipe, waste treatment description," *Sewage Ind. Wastes* **26**, 991 (1954).
304. Steinschneider, "Germany, sources and characteristics of wastes," *Sewage Ind. Wastes* **7**, 586 (1935).
305. Stilz, W. P., "Vibrating screens in industrial waste water treatment," *Ind. Wastes* **1**, 68 (1955).
306. Straub, C. P., "Paper pulp digestion, effect of nitrogen," *Sewage Ind. Wastes* **16**, 30 (1944).
307. Straub, C. P., "Paper pulp digestion, loadings," *Sewage Ind. Wastes* **15**, 857 (1943).
308. "Stream improvement research," *Chem. Eng. News* **25**, 1579 (1947).
309. Suess, R. J., "Digester wastes, use as road binder," *Sewage Ind. Wastes* **27**, 1315 (1955).
310. "Sulfite wastes recovered, stream pollution reduced," (Editorial), *Wastes Eng.* **30**, 512 (1959).
311. Sullins, J. K., "Stream improvement program for a paper mill," *Sewage Ind. Wastes* **29**, 681 (1957).
312. "Survey of water usages in southern kraft industry," Technical Bulletin no. 97, National Council for Stream Improvement, New York (1957).
313. Sylvester, R. O., "Factors involved in the location of a proposed pulp mill on a tidal estuary," in Proceedings of 7th Industrial Waste Conference, Purdue University, May 1952.
314. Sylvester, R. O., "Pulp mill location study," *Sewage Ind. Wastes* **24**, 508 (1952).
315. National Council for Stream Improvement, *Treatment of Strawboard Wastes*, Report no. 15, New York (1947).
316. "Twelfth Purdue Industrial Waste Conference," *Ind. Wastes* **2**, 107 (1957).
317. Tyler, R. G., "BOD reduction by alcohol fermentation and fodder yeast recovery," *Sewage Ind. Wastes* **19**, 70 (1947).
318. Tyler, R. G., "Stream aeration for control of sulfite waste liquor," *Sewage Ind. Wastes* **14**, 834 (1942).
319. Tyler, R. G., and S. Gunter, "BOD of sulfite waste liquor," *Sewage Works J.* **20**, 709 (1948).
320. Tyler, R. G., and W. Maske, "Yeast production from sulfite waste liquor," *Sewage Ind. Wastes* **20**, 516 (1948).
321. Tyler, R. G., W. Maske, and R. Brewer, "High-rate filter treatment experiments with sulfite waste liquor," *Sewage Ind. Wastes* **18**, 1155 (1946).
322. Ulfsparre, S., "Evaporation and burning of sulfite waste liquor," *Svensk Papperstid.* **61**, 803 (1958).
323. Van Derveer, P. D., "How CZ gets profits out of sulfite and kraft mill liquors," *Paper Trade J.* **142**, 46 (1958).
324. Van Horn, W. M., "The effects of pulp and paper mill waste on aquatic life," in Proceedings of 5th Ontario Industrial Waste Conference, May 1958, p. 60.
325. Van Horn, W. M., "The relation of kraft pulping wastes to stream environment," in Proceedings of 6th Southern Municipal and Industrial Waste Conference, 1957, p. 187.
326. Van Horn, W. M., "Stream pollution abatement studies in the pulp and paper industry," *Trans. Wisconsin Acad. Sci.* **39**, 105 (1949).
327. Van Horn, W. M., and R. Balch, "Stream pollution aspect of slime control agents," *Tech. Assoc. Pulp Paper Ind.* **38**, 151 (1955).
328. Van Horn, W. M., *et al.*, "The effect of kraft pulp mill wastes on fish life," *Tech. Assoc. Pulp Paper Ind.* **35**, 5 (1950).
329. Vilbrandt, D. F., *et al.*, "Tall oil investigations," *Virginia Eng. Exp. Sta. Bull.* **89** (1953).
330. Vogler, J. F., and W. Rudolfs, "Factors involved in the drainage of white water sludge," in Proceedings of 5th Industrial Waste Conference, Purdue University, November 1949.
331. Vogler, J. F., and W. Rudolfs, "White water sludge

drainage factors," *Sewage Ind. Wastes* **23**, 699 (1951).
332. Waddell, J. C., "Application of the Ohio Boxboard Company water system in the recovery of paper-mill wastes," in Proceedings of 7th Industrial Waste Conference, Purdue University, May 1952.
333. Waldichuk, M., "Pollution survey, British Columbia, Canada," *Sewage Ind. Wastes* **28**, 199 (1956).
334. Waldemeyer, T., "Treatment of paper-mill wastes," *Papermaking* **76**, 6 (1957).
335. Wallerstein, J. S., *et al.*, "Protein recovery from sulfite waste liquor and fermentative wastes," *Sewage Ind. Wastes* **17**, 403 (1945).
336. Warrick, L. F., "Classification of wastes, characteristics, stream pollution aspects," *Sewage Ind. Wastes* **19**, 1098 (1947).
337. Warrick, L. F., "Howard process for sulfite waste liquor disposal," *Sewage Ind. Wastes* **12**, 641 (1940).
338. Warrick, L. F., "Pulp and paper industry wastes," *Ind. Eng. Chem.* **39**, 670 (1947).
339. Warrick, L. F., "Pulp waste treatment methods," *Sewage Ind. Wastes* **13**, 848 (1941).
340. Webber, H. A., "The production of oxalic acid from cellulose agricultural material," *Iowa State Univ. Eng. Exp. Sta. Bull.* **118** (1934).
341. Webster, W. T., "How natural purification facilities were developed for kraft mills," *Paper Trade J.* **139**, 25 (1955).
342. Webster, W. T., "Water and the paper industry," in Proceedings of 6th Southern Municipal and Industrial Waste Conference, April 1957, p. 227.
343. Wilcoxson, L. S., "Development of the magnesia base pulping process," in Proceedings of 6th Industrial Waste Conference, Purdue University, February 1951.
344. Wilcoxson, L. S., "Magnesia-base sulfite pulping developments," *Sewage Ind. Wastes* **24**, 803 (1952).
345. Wiley, A. J., *et al.*, "Alcohol fermentation of sulfite waste liquor," *Sewage Ind. Wastes* **13**, 840 (1941).
346. Wiley, A. J., *et al.*, "Sulfite waste liquor yeast growth," *Sewage Ind. Wastes* **23**, 1216 (1951).
347. Wiley, A. J., L. M. Whitmore, Jr., and L. A. Boggs, "Utilization of spent sulphite liquor carbohydrates," *Tech. Assoc. Pulp Paper Ind.* **42**, 14*A* (1959).
348. Wilson, J. N., "Kraft mill wastes; effect of wastes on stream bottom fauna," *Sewage Ind. Wastes* **25**, 1210 (1953).
349. Winget, R. L., "The present status of the pulp, paper, and paperboard industry's stream improvement program," *Tech. Assoc. Pulp Paper Ind.* series 29, page 175 (1946).
350. Winget, R. L., "Waste treatment water resources and expansion problems," *Ind. Wastes* **1**, 37 (1955).
351. Winget, R. L., and R. O. Blosser, "The pulp and paper industry pollution abatement program in the U.S.," in Proceedings of 4th Ontario Industrial Waste Conference, June 1957, p. 105.
352. Wisniewski, T. F., *et al.*, "Ponding and soil filtration for disposal of spent sulphite liquor in Wisconsin," in Proceedings of 10th Industrial Waste Conference, Purdue University, May 1955.
353. Zehnpfennig, R., and M. S. Nichols, "BOD determination, inoculating studies, pulp," *Sewage Ind. Wastes* **25**, 61 (1953).
354. Zieminski, S. A., F. J. Vermillion, and B. G. St. Ledger, "Final report on aeration development studies," Technical Bulletin no. 112, National Council for Stream Improvement, New York (1959).
355. Zimmerman, F. J., "New waste disposal process," *Chem. Eng.* **65**, 117 (1958).

Suggested Additional Reading

The following references have been published since 1962.
1. Abernathy, A. R., *et al.*, "Microbiological decomposition of sodium lignin sulfate," in Proceedings of 14th Southern Water Resources Pollution Control Conference, Chapel Hill, N.C., 1965.
2. Abernathy, A. R., *Measurement of Microbial Degradation of Sulfonated Lignin*, Purdue University, Lafayette, Ind. (1964), p. 602.
3. Adams, W. R., "Here's what the paper industry has done and is doing to solve the water pollution problem," *Am. Paper Ind.* **48**, 42 (1966).
4. Akaki, M., "Study of mixed yeast cultures in spent sulfite liquors," *Hakko Kogaku Zasshi* **48**, 365 (1965).
5. Amberg, H. R., "By-product recovery and methods of handling spent sulfite liquor," *J. Water Pollution Control Federation* **37**, 228 (1965).
6. Amberg, H. R., J. F. Cormack, and M. R. Rivers, "Slime growth control by intermittent discharge of spent sulfite liquor," *Tappi* **45**, 770 (1962).
7. Amberg, H. R., *et al.*, "Supplemental aeration of oxidation lagoons with surface aerators," *Tappi* **47**, 274 (1964).
8. Amberg, H. R., "Stream re-aeration using tonnage oxygen," Technical Bulletin no. 197, National

Council for Stream Improvement, New York (1966).
9. Andreev, K. P., "The basis for optimum process parameters in continuous cultivation of feed yeast," *Paper Chem.* **38**, 3172 (1965).
10. Andrews, G., *et al.*, "Effluent sludge dewatering as practiced by two pulp and paper mills of Mead Corporation," *Tappi* **50**, 99A (1967).
11. "Color removal and BOD reduction in kraft effluents by foam separation," Technical Bulletin no. 177, National Council for Stream Improvement, New York (1964).
12. "Mechanical pressing of primary dewatered paper mill sludges," Technical Bulletin no. 174, National Council for Stream Improvement, New York (1964).
13. "Glatfelter Company's effluent treatment," *Am. Paper Ind.* **67**, 72 (1966).
14. "Uses expanded for fluo solids system," *Chem. Eng. News* **43**, 43 (1965).
15. *Water in Industry*, National Association of Manufacturers, New York (1963).
16. "Waste water—unique disposal system at Howard Paper Mills," *Pulp Paper Mag. Can.* **67**, 94 (1966).
17. "Sludge thickening with a decanter type centrifuge," *Allgem. Papier Rundschau* **12**, 828 (1965).
18. "New fluidized bed system produces salt cake from spent NSSC liquor," *Paper Trade J.* **149**, 43 (1965).
19. "A survey of water use in the pulp and paper industry," *Pulp Paper Mag. Can.* **66**, 83 (1965).
20. Babcock & Wilcox Co., H. B. Markant, "Concentration of cellulose waste liquors," German Patent no. 1170235, 1964.
21. Bacher, A. A., "Advanced waste treatment processes for water renovation and reuse," in Proceedings of 4th Tappi Water Conference, Philadelphia, April 1967.
22. Bailey, A. C., "Waste water treatment plant," *Tappi* **47**, 165A (1964).
23. Baily, E. L., "Water supply and effluent treatment at Kamloops Pulp and Paper Co. Ltd.," *Pulp Paper Mag. Can.* **67**, 85 (1966).
24. Baker, M. C., "Test for dissolved oxygen correction to the Winkler method," *Tappi* **58**, 81A (1965).
25. Barnes, C. A., E. E. Collias, V. F. Fellicetta, *et al.*, "Standardized Pearl-Benson, or nitrose, method recommended for estimation of spent sulfite liquor or sulfite waste liquor concentration in waters," *Tappi* **46**, 347 (1963).
26. Bates, J. S., "Pulpmill water supply and pollution control in the Atlantic provinces," *Pulp Paper Mag. Can.* **67**, 103 (1966).
27. Beak, T. W., "Biological measurement of water pollution," *Chem. Eng. Progr.* **60**, 39 (1964).
28. Berger, H. F., H. W. Gehm, *et al.*, "Decolorizing kraft waste liquors," U.S. Patent no. 3120464, 4 February 1964.
29. Berger, H. F., "Evaluating water reclamation against rising cost of water and effluent treatment," *Tappi* **49**, 79A (1966).
30. "Big activated sludge plant for paper mill," *Eng. News Record* **167**, 34 (1961).
31. Biggs, W. A., J. T. Wise, *et al.*, "Commercial production of acetic and formic acids from NSSC black liquor," *Tappi* **44**, 385 (1961).
32. Billings, R. M., "Cooperation between industry and the state in the stream improvement program," *Tappi* **45**, 209A (1962).
33. Billings, R. M., and Q. A. Narum, "Criteria and operation of a liquid effluent treatment plant," *Tappi* **49**, 70A (1966).
34. Bloodgood, D. E., and A. S. Klaggar, "Decolorizing of semi-chemical bleaching wastes," in Proceedings of 16th Industrial Waste Conference, Purdue University Engineering Extension Series, Bulletin no. 109, 1967, p. 351.
35. Bloodgood, D. E., Vogel, *et al.*, "Spray irrigation of paper mill waste," Paper read at Oklahoma Industrial Waste Conference, November 1964.
36. Blosser, R. D., *et al.*, "Reuse of condensates from kraft recovery operations," Technical Bulletin no. 187, National Council for Stream Improvement, New York (1965).
37. Blosser, R. D., and A. L. Caron, "Recent progress in land disposal of mill efflents," *Tappi* **48**, 43A (1965).
38. Blosser, R. D., and E. L. Owens, "Irrigation and land disposal of pulp mill effluents," in Proceedings of Ontario Industrial Waste Conference, June 1964, p. 203; *Can. Munic. Utilities* **102**, 24 (1964).
39. Blosser, R. O., "Oxidation pond study for treatment of de-inking wastes," in Proceedings of 16th Industrial Waste Conference, Purdue University Engineering Extension Series, Bulletin no. 109, 1962, p. 87.
40. Blosser, R. O., "Practice in handling barker influent in mills in the United States," Technical Bulletin no. 194, National Council for Stream Improvement, New York (1966).
41. Boerger, H. E. A., "Prevention of water pollution by

burning sulfite spent liquor," *Papier* **18**, 267 (1964).
42. Borowski, W., "Effluents from the manufacture of fiberboards," *Wasserwirtschaft.-Wassertechn.* **16**, 234 (1965).
43. Bosko, K., "Self-purification of surface waters polluted with effluents from pulp manufacture," *Papir Celuloza* **20**, 101 (1965).
44. Bottenfield, W., and N. C. Burbank, "Putting industrial waste to work: Mead's new lime kiln recovers waste lime mud," *Ind. Water Wastes* **9**, 18 (1964).
45. Braicu, L., "Physiochemical purification of waste waters at the Braila Pulp and Paper Combine," *Celuloza Hirtie* **14**, 105 (1965).
46. Bratt, L. C., "Trends in production of silvichemicals in the U.S. and abroad," *Tappi* **48**, 46A (1963).
47. Brecht, W., et al., "Studies on the dewatering of waste water sludges from paper mills," *Papier* **18**, 741 (1964); *Paper Chem.* **36**, 340 (1965).
48. Brecht, W., and W. Merlau, "Possible ways to improve the performance of flotation savealls," *Papier* **19**, 704 (1965).
49. Brecht, W., "Veber die Entstoffung von Papiermaschinwassern, *Schweizer Archiv.* **29**, 8 (1963).
50. Britt, K. W., *Handbook of Pulp and Paper Technology*, Reinhold Publishing Corp., New York (1964).
51. Brown, R. W., and C. W. Spalding, "Deep well disposal of spent hardwood pulping liquors," *J. Water Pollution Control Federation* **38**, 1916 (1966).
52. Brown, R. W., and C. W. Spalding, "Deep well disposal at Hammermill," *Am. Paper Ind.* **48**, 64 (1966).
53. Burns, O. B., et al., *Biological Treatment of Pulp and Paper Mill Wastes*, Purdue University, Lafayette, Ind. (1963), p. 83.
54. Burns, O. B., et al., "Aeration improvements and adaptation of cooling tower to activated sludge plant," *Tappi* **48**, 96A (1965).
55. Burns, O. B., et al., "Pilot plant evaluation of plastic trickling filters in series with activated sludge," *Tappi* **48**, 42 (1965).
56. Butler, J., "Case history and evaluation of waste treatment problems at D. M. Bare Paper Co.," *Tappi* **47**, 82A (1964).
57. Carpenter, W. L., "Colorimetric measurements of resin acid soaps in pulp and paper mill wastes," *Tappi* **48**, 669 (1963).
58. Carpenter, W. L., et al., "Effects of polyelectrolytes on primary deinking and boardmill sludge and on effluent clarification of deinking effluent," in Proceedings of 19th Industrial Waste Conference, Purdue University Engineering Extension Series, Bulletin no. 117, 1964, p. 139.
59. Carpenter, W. L., "COD and BOD relationships of raw and biologically treated kraft mill effluents," Technical Bulletin no. 193, National Council for Stream Improvement, New York (1966).
60. Carpenter, W. L., "Foaming characteristics of pulping wastes during biological treatment," Technical Bulletin no. 195, National Council for Stream Improvement, New York (1966).
61. Cawley, W. A., et al., "Polyvinyl chloride for trickling filters," *Ind. Water Wastes* **7**, 111 (1962).
62. Cawley, W. A., et al., "Treatment of pulp and paper mill wastes," *Ind. Water Wastes* **8**, 12 (1963).
63. Class, C. P., "Vinyl lined lagoon solves Riegal's black liquor storage problem," *Paper Trade J.* **149**, 34 (1965).
64. Clement, J. L., "Magnesium oxide recovery system; design and performance," *Tappi* **127**, 34A (1966).
65. Cohn. M. M.,"Clean water: Challenge to pulp and paper," *Pulp Paper* **39**, 38 (1963).
66. Cohn, M. M., "Four 'R's of water and waste conservation," *Pulp Paper* **40**, 42 (1966).
67. Cohn, M. M., "Clean water: Challenge to pulp and paper. 6: Spent sulfite liquor," *Pulp Paper* **40**, 24 and 45 (1966).
68. Coogan, "Incineration of sludge from kraft pulp mill effluents," *Tappi* **46**, 44A (1965).
69. Cooke, W. B., "Fungi associated with spent sulfite liquor disposal in natural sand bed," *Tappi* **46**, 573 (1963).
70. Copeland, C. G., "Water reuse and black liquor oxidation by the container-Copeland process," in Proceedings of 19th Industrial Waste Conference, Purdue University Engineering Extension Series, Bulletin no. 117, 1964, p. 391.
71. Copeland, G. G., et al., "Elimination of sulfite mill wastes by fluid bed treatment," *Am. Paper Ind.* **49**, 41 (1967).
72. Copeland, G. G., and J. E. Hanway, Jr., "Treating waste NSSC liquors in fluidized-bed reactor," *Paper Trade J.* **45**, 40 (1963).
73. Coughlan, F. P., and A. E. Sparr, "Design and early operating experience of activated sludge plant for combined treatment of pulp, paper, and domestic wastes," in Proceedings of 16th Industrial Waste

Conference, Purdue University Engineering Extension Series, Bulletin no. 109, 1962, p. 375.
74. Davis, W. S., "A laboratory method for comparing the effects of starches in white water," *Tappi* **47**, 129*A* (1964).
75. Dietrich, K. R., "The protection of water with special regard to artificial aeration," *Inst. Paper Chem., Bull.* **33**, 317 (1962).
76. Dubey, G. A., et al., "Electrodialysis, a new unit operation for recovery of values from spent sulfite liquor," *Tappi* **48**, 95 (1965).
77. Dyck, A. W., "Whippany lets bugs do it," *Paper Ind.* **45**, 374 (1963).
78. Eckenfelder, W. W., Jr., "Design and performance of aerated lagoons for pulp and paper waste treatment," in Proceedings of 16th Industrial Waste Conference, Purdue University Engineering Extension Series, Bulletin no. 109, 1967, p. 115.
79. Eckenfelder, W. W., Jr., *Paper*, Purdue University, Lafayette, Ind. (1965), p. 115.
80. Eckenfelder, W. W., Jr., *Paper*, Purdue University, Lafayette, Ind. (1966), p. 105.
81. Eckenfelder, W. W., Jr., *Paper*, Purdue University, Lafayette, Ind. (1967), p. 83.
82. Edde, H., *High Rate Biological Treatment of Paper Mill Wastes*, Purdue University, Lafayette, Ind. (1964).
83. Ei, M. A., Dib, et al., "Characteristics of starch paperboard and gelatin wastes," *J. Water Pollution Control Federation* **38**, 46 (1966).
84. Enere, L., et al., "Study of waste waters from forest industries. 6: Bio-oxidation of forest industry effluents on a large laboratory scale," *Paper Chem.* **35**, 3724 (1965).
85. Enkvist, T., and T. Linders, "More organic chemicals from the spent liquors of the cellulose industry," *Finska Kemistsamfundets Medd.* **75**, 1 (1966).
86. Evilevich, M. A., and R. V. Tevilev, "Equipment for the biological purification of effluent at the Kotlas Combine," *Bumazhn. Prom.* **2**, 15 (1966).
87. Farin, W. G., "Exapex," *Am. Paper Ind.* **48**, 87 (1966).
88. Farin, W. G., "Short cuts to high efficiency collection of spent sulfite liquor," *Paper Trade J.* **150**, 38 (1966).
89. Farin, W. G., "Flambeau solves its sulfite waste liquor disposal," *Paper Trade J.* **150**, 44 (1966).
90. Fellicetta, V. F., et al., "Dissolved oxygen determination in waters containing spent sulfite liquor," *Tappi* **48**, 6362 (1965).
91. Fellicetta, V. F., and J. L. McCarthy, "Spent sulfite liquor. 10–11," *Tappi* **46**, 337 (1963).
92. Fenchel, V., "Filtration in fiber recovery from waste water," *Paper Chem. Abstr.* **36**, 4168 (1965).
93. Follett, R., and H. W. Gehm, "Manual of practice for sludge handling in the pulp and paper industry," Technical Bulletin no. 190, National Council for Stream Improvement, New York (1966).
94. Fraik, R. D., "Some elements of filler retention," *Tappi* **45**, 159*A* (1962).
95. Freeman, L., "Effluent treatment and abatement practices," Paper read at 4th Industrial Water and Waste Conference, University of Texas, January 1964.
96. Fuchs, R. E., "Decolorization of pulp mill bleaching effluents using activated carbon," Technical Bulletin no. 181, National Council for Stream Improvement, New York (1965).
97. Gafiteanu, M., et al., "Investigations on the treatment of waste waters from the manufacture of fiber building boards," *Water Pollution Abstr.* **38**, 1508 (1965).
98. Gaudy, A. F. Jr., "Shock loading activated sludge with spent sulfite pulp mill wastes," *J. Water Pollution Control Federation* **34**, 124 (1962).
99. Gavelin, G., "Leje and Thurne recovery and effluent treatment," *Paper Trade J.* **150**, 50 (1966).
100. Gehm, H. W., "Activated sludge process for pulp and paper mill effluents," *Ind. Water Wastes* **8**, 23 (1963).
101. Gehm, H. W., "Dissolved oxygen in surface waters," in Proceedings of 13th Southern Water Resources Pollution Control Conference, Durham, N.C. 1964.
102. Gehm, H. W., and I. Gellman, "Practical research and development in biological oxidation of pulp and paper mill effluents," *J. Water Pollution Control Federation* **37**, 1392 (1965).
103. Gellman, I., "Aerated stabilization basin," *Tappi* **48**, 106*A* (1965).
104. Gellman, I., and R. O. Blosser, "Disposal of pulp and papermill waste by land application," *Compost Sci.* **1**, 18 (1961).
105. Gettle, T. J., "Discharging molten black liquor from surface smelters during woodpulp manufacture," U.S. Patent no. 3122421, 25 February 1964.
106. Gloppen, R. C., et al., "Rating and application of surface aerators," *Tappi* **48**, 103*A* (1965).
107. Goeldner, R. W., et al., "A horizontal spray film

evaporator for sulfate liquor concentration," *Tappi* **47**, 185*A* (1964).
108. Golz, W. H., "New utilization products for sulfite liquor in the Federal Republic of Germany," *Tappi* **45**, 200*A* (1962).
109. Gravel, J. J., et al., "Coating and deinking—AST proving out for mill sludge disposal," *Pulp Paper* **39**, 28 (1965).
110. Gravel, J. J., and H. G. Barley, "Disposal of sewage and mill sludge by atomized suspension technique," *Pulp Paper Mag. Can.* **67**, *T*73 (1966).
111. Guerrieri, S. A., "Treating waste sulfite liquor," U.S. Patent no. 3133789, 19 May 1964.
112. Gunther, H. J., "In-plant recovery of paper machine fines and mill effluent water treatment by a Wabag contact flocculator," *Pulp Paper Mag. Can.* **67**, *T*91 (1966).
113. Han, S. T., "Engineering considerations for sulfite recovery," *Tappi* **48**, 66*A* (1965).
114. Hanousek, J., and V. Kamenick, "A laboratory study of the utilization of fibrous paper making sludges," *Papir Cellulosa* **21**, 33 (1966).
115. Haynes, D. C., "A survey of the pulp and paper industry," *Tappi* **49**, Supplement 51*A*, September 1966.
116. Hellstrom, B., "The SCA-Billerud recovery process," *Pulp Paper Mag. Can.* **66**, *T*289 (1965).
117. Hilgers, G., "The recovery of chemicals and energy in sulfate pulping mills," *Chem. Eng. Technol.* **36**, 23 (1964).
118. Howard, E. J., "Pollution and waste abatement in the European pulp and paper industry," *Pulp Paper Mag. Can.* **65**, 101 (1964).
119. Howard, T. E., et al., "Pollution and toxicity characteristics of kraft pulp mill effluents," *Tappi* **48**, 136 (1965).
120. Hurriet, B., "The effluents of the pulp and paper industry. II: Experimental studies," *Paper Chem. Abstr.* **36**, 3380 (1965).
121. Hurwitz, E., et al., "Degradation of cellulose by activated sludge treatment," in Proceedings of 16th Industrial Waste Conference, Purdue University Engineering Extension Series, Bulletin no. 109, 1962, p. 176.
122. Ignatenke, A. A., et al., "Deodorization of effluents from the manufacture of kraft pulp," *Paper Chem. Abstr.* **36**, 3380 (1965).
123. Ioffe, L., "The cation exchange method for the recovery of ammonia from spent sulfite liquor," *Zh. Prikl. Khim.* **39**, 1458 (1966).
124. Ivovacki, J., and J. Pidotek, "Plant investigations of biological and physico-chemical purification of effluents from the manufacture of kraft pulp," *Przeglad Papier.* **20**, 226 (1964) and **21**, 51 (1965).
125. Izumrudova, T. V., et al., "New drilling fluid additive," *Gidrolizn. i Lesokhim. Prom.* **19**, 25 (1966).
126. Jenkins, A., Jr., "Dewatering and disposal of paper mill wastes," *Paper Trade J.* **147**, 30 (1963).
127. Job, J. A., "The industrialist's viewpoint with respect to the treatment and disposal of industrial wastes," National Institute for Water Research, *C.S.I.R. (South Africa)*, 55, 1964, p. 240; *Water Pollution Abstr.* **38**, 1847 (1965).
128. Jones, P. H., "Experiments with bulking high carbohydrate wastes and sulfite liquor," *Pulp Paper Mag. Can.* **67**, *T*134 (1966).
129. Jorgenson, J. R., "Irrigation of slash pine with mill effluents," *Louisiana State Univ. Eng. Exp. Sta. Bull.* **80**, 90 (1965).
130. Julson, J. O., "Attitude—a management tool for resource protection," *Tappi* **48**, 120*A* (1965).
131. Kaila, E., "Effect of chemical pulp mills and settlement on the Lake Nasijarvi water system," *Paperi Puu* **46**, 1 (1964).
132. Klein, L., "Stream pollution and effluent treatment with special reference to textile and paper mill effluents," *Chem. Ind.* **21**, 866 (1964).
133. Kleinert, T. N., "Reuse of the impregnation liquor in continuous kraft pulping," *Tappi* **49**, 301 (1966).
134. Kleinert, T. N., "Lignin for spent sulfite liquor in plastic manufacturing," *Holzforsch. Holzverwert*, **17**, 69 (1965).
135. Klinger, L., "Dynamic activated sludge treatment," *Paper Ind.* **47**, 76 (1965).
136. Klinger, L. L., "Whippany completing final link in tri-mill waste treatment system," *Paper Trade J.* **146**, 36 (1962).
137. Klinger, L. L., "Whippany expands tri-mill treatment complex," *Water Works Wastes Eng.* **1**, 60 (1964).
138. Knauer, K. J., "White water and effluent systems for cylinder machine mills," *Paper Trade J.* **149**, 32 (1963).
139. Knowlton, D. C., "Protecting our water resources," *Tappi* **49**, 41*A* (1966).
140. Krenkel, P. A., et al., "The effect of impounding reservoirs on river waste assimilative capacity,"

J. Water Pollution Control Federation **37**, 1203 (1965).
141. Kretzschmar, G., "Producing yeast food from spent sulfite liquors—an example of the rational use of a waste product," *Paper Chem. Abstr.* **35**, 4620 (1965).
142. Krylova, T. B., *et al.*, "Effect of lignosulfonates on the biochemical processing of spent sulfite liquor," *Paper Chem. Abstr.* **36**, 4171 (1963).
143. Laumyanskas, G. A., *et al.*, "Pollution of the River Nyamunas with effluents from pulp and paper mills," *Tr. Akad. Nauk. Lit. SSR* **B3**, 121 (1965).
144. LeCompte, A. R., "Water reclamation by excess lime treatment of effluent," *Tappi* **49**, Supplement no. 121*A* (1966).
145. Lefrancois, L., and B. Reviz, "Yeast from SSL—a description of a recently constructed plant," *Ind. Aliment. Agr.* **81**, 1175 (1964).
146. Little, H. W., "Primary settling clarifier for treating mill effluent," *Paper Ind.* **47**, (1965).
147. Lindberg, A., "Determination of lignin and lignosulfonic acid," *Vattenhygien* **19**, 106 (1963).
148. Livar, A., and B. Foth, "Desulfonation and utilization of spent sulfite liquor as tanning material," *Papiripar* **8**, 5 (1964).
149. Lomova, M. A., "Microorganisms from activated sludge which purify effluent of pulp mills," *Nauchn. Issled. Inst. Tsellyuloz Bu. Prom.*, **51**, 63 (1965).
150. Lott, R. R., and H. F. Berger, "Neutralization of acid bleach effluents," Technical Bulletin no. 186, National Council for Stream Improvement, New York (1965).
151. Lure, Y. Y., L. A. Alferova, *et al.*, "Analysis of kraft pulping effluents," *Paper Chem. Abstr.* **35**, 6360 (1965).
152. Luthgens, M. W., "The new Sirola–Lurgi NSSC recovery plant," *Tappi* **45**, 837 (1962).
153. Maas, L., "Advanced effluent treatment plant," *Pulp Paper Intern.* **7**, 41 (1965).
154. Magne Grande, "Water pollution studies in the River Otra, Norway—effects of pulp and paper mill wastes on fish," *Air Water Pollution* **8**, 77 (1964).
155. Major, W. D., and O. B. Burns, "Protected improvements in Covington's waste treatment plant," *Am. Paper Ind.* **48**, 56 (1966).
156. Maloney, T. E., and E. L. Robinson, "Growth and respiration of green algae in spent sulfite liquor," *Tappi* **44**, 137 (1961).
157. Manocha, S. H., "Treatment of paper mill wastes," *Environ. Health* **6**, 37 (1964).
158. McCallion, J., and O. Mittlesteadt, "Centrifuge deftly removes solids from mill wastes," *Chem. Process.* **28**, 108 (1965).
159. McCormick, L. L., *et al.*, "Paper mill waste water for crop irrigation and its effect on the soil," *Louisiana Agr. Exp. Stat. Bull.* **604** (1965).
160. McKeown, J. J., "Control of Sphaerotilus natans," *Ind. Water Wastes* **8**, 19 (1963).
161. McKeown, J. J., "Comparative studies on the analysis for dissolved oxygen," Technical Bulletin no. 184, National Council for Stream Improvement, New York (1965).
162. McKeown, J. J., "Procedure for conducting mill effluent surveys," Technical Bulletin no. 183, National Council for Stream Improvement, New York (1965).
163. Minch, V. A., J. T. Eagan, *et al.*, "Design and operation of plastic filter media," *J. Water Pollution Control Federation* **34**, 459 (1962).
164. Minch, V. A., J. T. Eagan, *et al.*, "Plastic trickling filters—design and operation," *Paper Trade J.* **146**, 28 (1962).
165. Mohres, H. Z., "Utilization of secondary materials in paper board mills," *Tappi* **50**, supplement 109*A* (1967).
166. Masseli, J. W., *et al.*, *White Water from Paper and Paper Board Mills*, New England Interstate Water Pollution Control Commission, Boston, Mass., 1963.
167. Mataruev, K. U., "Purification of industrial effluents from the Baikal pulp mill," *Pererabotka Drevesiny Sb.* **31**, 3 (1964).
168. Meinhold, F. T., E. R. Hoppe, *et al.*, "Closed-loop water system pays out $100,000 a year," *Chem. Process.* **28**, 56 (1965).
169. Minch, V. A., "Some effects of impoundments upon the waste assimilation ability of the Coosa river," *Tappi* **48**, 49*A* (1965).
170. Morys, E., "Possibilities for the use of ion exchange resins in preliminary purification of effluents," *Papir Celulosa* **19**, 71 (1964).
171. Munteanu, A., "An experimental station for the treatment of waste waters from the manufacture of cellulose and viscose-artificial fibers," Institute of Hydrotechnological Research, Scientific Sessions, Section 4, 1964, p. 51; *Water Pollution Abstr.* **38**, 1507 (1965).

172. Murphy, N., and D. Gregory, *Removal of Color from Sulfate Pulp Wash Liquors*, Purdue University, Lafayette, Ind. (1965), p. 59.
173. "A process for removal of color from bleached kraft effluents through modification of the chemical recovery system," Technical Bulletin no. 157, National Council for Stream Improvement, New York (1962).
174. "Evaluation of fly ash as a filter aid for precoat vacuum filtration of papermill sludges," Technical Bulletin no. 158, National Council for Stream Improvement, New York (1962).
175. "Land disposal of evaporator condensates," Technical Bulletin no. 160, National Council for Stream Improvement, New York (1962).
176. Nelson, B. W., "Removal of kraft effluent from river water by suspended sediment," *Tappi* **46**, 277 (1963).
177. Nylander, G., "Waste water from forest products industries. V: Biochemical oxidation," *Svensk Papperstid.* **67**, 565 (1964).
178. O'Connor, R. E., "Increased water efficiency," *Paper Mill News* **85**, 21 and 26 (1962).
179. Okun, D. A., J. C. Lamb, *et al.*, "A waste control program for a river with highly variable flow," *J. Water Pollution Control Federation* **25**, 1025 (1963).
180. Orsler, R. J., and D. F. Packmar, "Determination of lignin in sulfite pulping liquors," *Paper Chem. Abstr.* **35**, 5421 (1963).
181. Palladino, A. J., "Reducing effluents for secondary treatment," *Tappi* **49**, 115A (1965).
182. Pearl, I. A., and D. L. Beyer, "The ether-insoluble water soluble components of several spent sulfite liquors," *Tappi* **47**, 779 (1964).
183. Petrowski, M., "The development and successful operation of a closed white water system at two Carolina paper board mills," Technical Bulletin no. 188, National Council for Stream Improvement, New York (1963).
184. Pirnie, M., Jr., and T. P. Quirk, "Design and cost considerations for treatment of deinking wastes," *Paper Trade J.* **146**, 30 (1962).
185. "Plastic grids help solve waste disposal problems," *Chem. Eng.* **6**, 70 (1961).
186. Plotz, L., "Waste water quality indices," *Papier* **18**, 676 (1964); *Paper Chem. Abstr.* **35**, 5582 (1965).
187. Pobis, J., "Biological final treatment of waste water from the sulfate pulping process," *Papir Celulosa* **19**, 9 (1964).
188. Pobis, J., "Research on activated sludge process as the second stage of sulfate pulp mill wastes treatment," *Vyzkum. Ustav Vodohospodarsky, Prace a Studie* **22**, 1963.
189. Polein, J., *et al.*, "Ultraviolet light absorption of lignosulfonic acids in spent sulfite liquors," *Paper Chem. Abstr.* **35**, 4623 (1965).
190. Potapenkse, A. B., "Biological purification of effluents at the Zhidachev Combine," *Brumazhn. Prom.* **4**, 15 (1965); *Paper Chem. Abstr.* **36**, 1928 (1965).
191. Quirk, T. P., R. C. Olson, and G. Richardson, "Biooxidation of concentrated board machine effluents," in Proceedings of 18th Industrial Waste Conference, Purdue University, 1963, p. 655; *J. Water Pollution Control Federation* **38**, 69 (1965).
192. Radej, Z., "Final neutralization of steam treated spent sulfite liquors before alcoholic fermentation," *Papir Celulosa* **20**, 168 (1965).
193. Rennerfelt, J., "Study of waste waters from forest industries. 8: Biological treatment of wall board mill effluents on a semi-industrial scale," *Paper Chem. Abstr.* **35**, 8117 (1965).
194. Riegal, P. S., "IWT at Downington Paper Co.," *Am. Paper Ind.* **48**, 82 (1966).
195. Roche, P., "Problems of disposal from an acid sulfite dissolving pulp mill," *Water Pollution Abstr.* **38**, 1847 (1965).
196. Ruus, L., "Study of waste waters of forest products industries—composition and BOD of sulfate pulp mill condensates," *Paper Chem. Abstr.* **35**, 4624 (1965).
197. Ruus, L., "Determination of resin acids in waste waters from forest products industries," *Paper Chem. Abstr.* **35**, 8118 (1965).
198. Ruzickova, D., "Problems with waste water purification in the Czech pulp and paper industry," *Paper Chem. Abstr.* **36**, 4176 (1965).
199. Sapotniskii, S. A., and R. M. Myasuikova, "Cultivation of yeast in spent sulfite liquor from hardwood pulping," *Paper Chem. Abstr.* **35**, 8120 (1965).
200. Sarnecki, K., and A. Szudzinska, "Effect of replacement of Ca base by Na base in the sulfite cooking liquor on the alcohol fermentation process," *Przeglad Papier.* **20**, 26 (1964).
201. Sarnecki, K., "Suitability of spent liquors from the manufacture of high yield sulfite pulps for processing into alcohol," *Przeglad Papier.* **21**, 110 (1965).

202. Sarnecki, K., et al., "Effect of the addition of birchwood in sulfite cooks on the composition of spent sulfite liquors and the course of alcoholic fermentation of the liquors," *Przeglad Papier.* **20**, 181 (1964).
203. Schonhuth, O., et al., "Fodder yeast from eucalyptus—derived spent sulfite liquor," *Papier* **20**, 171 (1966).
204. Schmidt, H., "Regeneration of sodium sulfite waste liquor by the gas fine distribution sulfitation process—principles and sani-technical tests," *Zellstoff Papier* **13**, 161 (1964).
205. Schmidt, I. J., and G. Weigt, "Biological purification of waste water in the pulp and paper industry. II: Laboratory studies of various aerating systems," *Zellstoff Papier* **13**, 360 (1964); *Paper Chem. Abstr.* **36**, 348 (1963).
206. Scott, R. H., "Disposal of high organic content wastes on land," *J. Water Pollution Control Federation* **34**, 932 (1962).
207. Seidl, J., "Separation of organic substances on ion exchange resins: sorption and desorption of lignosulfonic acids on anion exchange resins," *Chem. Prumysl.* **16**, 273 (1966).
208. Shell Internationale Research, "Slimicides," British Patent on. 952926, 19 March 1964.
209. Shepard, F. T., "Improving understanding of our industry: progress in waste treatment," *Paper Trade J.* **146**, 34 (1962).
210. Sherman, W. A., "Facts, not fiction concerning water cleanup," *Pulp Paper* **40**, 53 (1966).
211. Sitman, W. D., W. E. Hoover, et al., "Feed system adds nutrients for paper wastes treatment," *Wastes Eng.* **33**, 178 (1962).
212. Smith, L. L., and R. H. Kramer, "Some effects of paper fibers in fish eggs and small fish," in Proceedings of 19th Industrial Waste Conference, Purdue University Engineering Extension Series, Bulletin no. 117, 1964, p. 369.
213. Smith, L. L., R. H. Kramer, et al., "Effects of pulpwood fibers on fathead minnows and walleye fingerlings," *J. Water Pollution Control Federation* **37**, 130 (1965).
214. Sokolova, O. J., and V. F. Maksimov, "Treatment of effluents from the evaporation plant at the Sgeezha Combine," *Bumazhn Prom.* **1**, 15 (1965); *Paper Chem. Abstr.* **35**, 8122 (1965).
215. Soklova, O. J., and V. F. Maksimov, "Equilibrium concentration of sulfur containing gases in relation to deodorizing of foul smelling effluents from pulp mills." *Tr. Leningr. Tekhnol. Instit. Tzellyulozn. Bumazhn. Prom.* **17**, 122 (1963).
216. Solin, V., and J. Moravec, "Production of phenolic substance in the anaerobic decomposition of lignosulfonic acids," *Water Pollution Abstr.* **38**, 327 (1963).
217. Spalding, C. W., Halko, et al., "Deep well disposal of spent pulping liquors," *Tappi* **48**, 68A (1965).
218. Stanescu, N., and V. Sirbu, "Pilot plant for purification of waste waters of the pulp paper and synthetic fibre industries," *Celuloza Hirtie* **14**, 27 (1965).
219. Stefan, G., "Fermentation of spent bisulfite liquors for production of fodder yeast," *Celuloza Hirtie* **13**, 103 (1964); *Paper Chem. Abstr.* **35**, 5587 (1963).
220. Stovall, J. H., "Treatment of pulp and paper mill wastes," *Southwest Water Works J.* **46**, 19 (1964).
221. Sullins, J. K., "Developing patterns for efficiency water utilization: Pulp and paper industry," in Proceedings of 14th Southern Water Resources Pollution Control Conference, Chapel Hill, N.C., 1965, p. 73.
222. Swiezynska, H., "Selection of yeast strains for alcoholic fermentation of spent sulfite liquors," *Prace Inst. Lab. Badawczych Lesn.* **269**, 123 (1963).
223. Szudzinska, A., "Effect of phenol content in spent sulfite liquor on the alcoholic fermentation process," *Przeglad Papier.* **21**, 205 (1965).
224. Teletzke, G. E., "Zimmerman process," *Factory Plant* **52**, 62 (1964).
225. Toth, B., "Changes in spent sulfite liquors during yeast fermentation," *Acta Chim. Acad. Sci. Hung.* **47**, 431 (1966).
226. Twiss, R. H., "Water re-use and lagoon system solve waste treatment on small mill," *Am. Paper Ind.* **48**, 75 (1966).
227. Vamvakias, J., W. L. Carpenter, et al., "Temperature relationships in aerobic treatment and disposal of pulp and paper wastes," Technical Bulletin no. 191, Natural Council for Stream Improvement, New York (1966).
228. Van Luvan, A. L., "Kraft advanced design and techniques at Great Lakes," *Pulp Paper* **39**, 26 (1965).
229. Van Luven, A. L., "Primary treatment of effluent at Great Lakes Paper Co.," *Pulp Paper Mag. Can.* **67**, T99 (1966).
230. Vasseur, E. "Progress in sulfite pulp pollution abatement in Sweden," *J. Water Pollution Control Federation* **38**, 27 (1966).

231. Voci, J. J., et al., "Economics of spent liquor recovery," *Chem. Eng. Progr.* **61**, 110 (1965).
232. Waldichuk, M., "Dispersion of kraft mill effluent from a submarine diffuser in Stewart Channel, British Columbia," *J. Water Pollution Control Federation* **38**, 1484 (1966).
233. Waldichuk, M., "Effects of sulfite wastes in a partially enclosed marine system in British Columbia," *J. Water Pollution Control Federation* **38**, 1484 (1966).
234. Waldichuk, M., "Marine aspects of pulp mill pollution," *Can. Pulp Paper Ind.* **15**, 36 (1962).
235. Walter, J. H., and J. F. Byrd, "Disposal of paper wastes from a Sven-Pedersen flotation unit," in Proceedings of 16th Industrial Waste Conference, Purdue University Engineering Extension Series, Bulletin no. 109, 1962, p. 449.
236. Walter, L., "Clarifiers solve effluent problems," *Paper Ind.* **43**, 26 (1961).
237. Warner, H. L., and B. C. Miller, "Water pollution control by in-plant measures," *Tappi* **46**, 260 (1963).
238. Weinfurt, M., "Losses of suspended solids in the paper machine cycle and their reduction," *Papir Celulosa* **19**, 79 (1964).
239. Wells, C. C., "Better water monitoring is a result of Bowaters system," *Pulp Paper* **39**, 19 (1965).
240. Werner, A. E., "Suspended solids from mill effluents," *Can. Pulp Paper Ind.* **18**, 109 (1965).
241. Werner, A. E., "Sulfur compounds in kraft pulp mill effluents," *Can. Pulp Paper Ind.* **16**, 35 (1963).
242. Weston, R. F., and W. D. Rice, "Contact stabilization of activated sludge treatment for pulp and paper mill waste," *Tappi* **45**, 223 (1962).
243. Whalen, J. F., "How Combined Locks Paper handles deinking waste," *Paper Trade J.* **145**, 30 (1961).
244. White, M. T., "Surface aeration as secondary treatment," *Tappi* **48**, 128*A* (1965).
245. Whitney, R. L., S. T. Han, et al., "Bisulfate method of chemical recovery," *Tappi* **48**, 1 (1965).
246. Windons, G., et al., "Experiences with a completely biological method of waste water purification by use of the aero accelerator," *Papier* **19**, 696 (1965).
247. Woodard, F. E., and J. E. Etzel, "Coacervation and chemical coagulation of lignin from pulp mill black liquors," *J. Water Pollution Control Federation* **37**, 990 (1967).
248. Worley, J. L., T. F. Burgess, et al., "Identification of low flow augmentation requirements for water quality control by computer technique," *J. Water Pollution Control Federation* **37**, 659 (1965).
249. Young, B., "300 ft diffuser releases effluent along its length," *Can. Pulp Paper Ind.* **9**, 66 (1966).
250. Zieminski, S. A., and B. T. Coyle, "Biochemical oxygen demand reduction of sulfite waste liquor," *Tappi* **47**, 138 (1964).
251. Zieminski, S. A., and R. J. Martyn, "Reduction of the BOD of sulfite waste liquor by thermal and chemical methods of treatment," *Tappi* **45**, 878 (1962).
252. Zubranska, W., "Recovery of sodium base from spent sulfite liquors with the use of domestic cation exchange resins," *Przeglad Papier.* **21**, 350 (1965).
253. U.S. Department of the Interior, *The Pulp and Paper Industries*, Industrial Waste Profile, Vol. III, no. 3, Washington, D.C. (1967).

24.2 Photographic Wastes

Waste water from a large-scale film-developing and printing operation consists of spent solutions of developer and fixer, containing thiosulfates and compounds of silver. The solutions are usually alkaline and contain various organic reducing agents. Treatment usually consists of silver recovery carried out by the industry and subsequent treating of developer waste in combination with domestic sewage. Studies of two such plants* showed that the effect of developer waste on the treatment of sewage is insignificant if the ratio of this waste to sewage is comparatively low.

The Eastman Kodak Corporation† estimates the value of silver bullion recovered from photographic-processing waste at about $1.29 per troy ounce. It informs its processing plants that $800 worth of silver can be obtained by processing of the silver wastes from 100,000 rolls of black and white film and $2400 from a similar quantity of Kodacolor-X film. These monetary recoveries are based upon 100 per cent recovery efficiency and do not include any of the normal costs of waste treatment.

There are three methods generally available for silver recovery: metallic replacement, electrolysis,

---

*"Treatment Data for Photographic Wastes," *Public Works* **85**, 104 (1954).

†*Recovering Silver from Fixing Baths*, Pamphlet No. J-10 (3-66 minor revision), Eastman Kodak, Rochester, N.Y.

and chemical precipitation. Metallic replacement involves bringing spent hypo solutions into contact with a metal surface such as steel stampings, zinc, steel wool, and/or copper. After 66 to 99 per cent recovery of silver is completed, the metal is removed from the tanks, dried, and sold to the refiner. The electrolytic method consists of placing a cathode and an anode in the silver-bearing hypo solution. When an electric current is passed between these electrodes, silver plates out on the cathode. The permissible current density depends on whether sodium or ammonium hypo is used in the bath and on the acidity, silver, and sulfite concentration in the solution. Turbulence in the solution furnishes a continuous supply of silver-laden fixer solution to the cathode. To handle 50 gallons of solution containing 19 ounces of silver, 32 square feet of cathode area are required, along with compressed air agitation. Twenty hours of such operation will remove 98 per cent of the silver and can recover about 600 troy ounces of silver before the cathode has to be desilvered or replaced. Several compounds can be used in chemical precipitation. However, one common method utilizes a combination of sodium hydroxide and sodium sulfide to achieve the precipitation of silver sulfide. The sludge is dried and sold to the most convenient refiner. Typical quantities of precipitating chemical required are 1 ounce of sodium hydroxide (2 lb of $NaOH$ per gallon) and 1 ounce of sodium sulfide (2 lb of Na_2S per gallon) for each gallon of waste fixer.

METAL INDUSTRIES

Metal wastes include wastes from refining mills, plating mills, and parts washing and encompass a wide range of materials. For example, there are wastes not only from the manufacturing of steel, but from many other metals (copper and aluminum, to name two); wastes are produced from renewing surfaces on used metallic parts, such as airplane engines prior to their return to service; the coating of one metal with another, for protective purposes, for example, the plating of silverware or business machines, should be included as an intermediate process. The wastes from all three sources are similar, in that they possess various concentrations of metallic substances, acids, alkalis, and grease. They are characterized by their toxicity, relatively low organic matter, and greases.

24.3 Steel-mill Wastes

Origin of steel-mill wastes. Steel-mill wastes come mainly from the by-product coke, blast-furnace, rolling-mill, and pickling departments. Wastes contain cyanogen compounds, phenols, ore, coke, limestone, acids, alkalis, soluble and insoluble oils, and mill scale. Wastes are treated by recirculation, evaporation, benzol extraction, distillation, sedimentation, neutralization, skimming, flotation, and aeration.

The "by-product" coke process. Coal is heated in the absence of air to produce coke and other products. This may be done in an integrated steel mill or in a separate plant proximate to, but not at the same site as, the steel mill. The cooking process evolves a gas, the further processing of which leads to the major wastes from this process; tar and ammonia are its main constituents. The following products are obtained from the burning in retorts of a ton of coal: coke, 1300 to 1525 pounds; $(NH_4)_2SO_4$, 17 to 26 pounds; tar, 5 to 12 gallons; gas, 10,800 to 11,300 cubic feet; phenol, 0.1 to 2.0 pounds; light oil, 2 to 3 gallons; naphthalene, 0.5 to 1.2 pounds.

The major wastes from preparation of the coke product itself come from the quench tower, where the hot coke is deluged with water. The coke dust present in this quenching water is called "breeze" and is commonly recovered from the water. A schematic drawing of the coking process is presented in Fig. 24.4.

The blast furnace. The wet scrubbing of blast-furnace gas evolves waters laden with flue dust. The wet scrubbers are downflow water sprays which clean the dust from the upflowing gases, an operation which is usually an intermediate stage between dry (or cyclone) dust separation and final electrostatic precipitation of the remaining fine particles. Secondary gas washers or precipitators are periodically cleaned by flushing with water, thus adding to the flow of discolored water.

The pickling process. Before applying the final finish to steel products, the manufacturer must remove dirt, grease, and especially the iron-oxide scale which accumulates on the metal during fabrication. Normal-

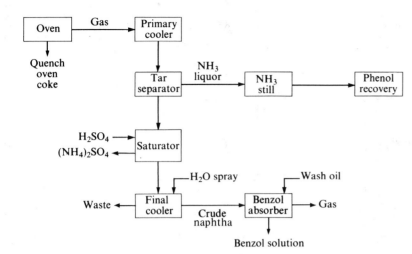

Fig. 24.4 Heating bituminous coal in the absence of air.

ly this is carried out by immersing the steel in dilute sulfuric acid (15 to 25 per cent by weight). This process, known as "pickling," produces a waste called "pickling liquor," composed mainly of unused acid and the iron salts of the acid (Fe^{+++} and Fe^{++}). The acid reacts with the iron salts, forming $FeSO_4$. As the acid is used, it becomes weaker and must be renewed. However, at a certain point, the concentration of $FeSO_4$ increases to such a degree that it inhibits the action of even a high concentration of sulfuric acid. At this point, the pickling liquor must be discharged and replaced by a fresh batch of sulfuric acid. It is this pickling-liquor waste which has received such wide publicity from an industrial-waste standpoint.

A recent report by the U.S. Department of the Interior [1967, 107] gives the following pertinent data on the steel industry. Gross water intake in 1964 was 3815 billion gallons; gross water use, including recirculation and reuse, was 5510 billion gallons. Overall water reuse was thus 42.1 per cent. Cooling water reuse is estimated to have been 52.2 per cent and process water reuse 18.1 per cent. Water uses by the industry are summarized in the following table, for the year 1964, in billions of gallons.

The general trend in the steel industry is toward the use of subprocesses which will produce products of lighter unit weight at increasingly high speeds with minimum manual operation. Production units tend to become larger in order to realize economies of scale.

Industry water uses in 1964, billions of gallons [107]		
Manufacturing process	Process water	Cooling water
Blast furnaces	276	586
Open-hearth furnaces	1	491
Basic oxygen furnaces	8	37
Electric furnaces	1	63
Hot-rolling mills and related	468	468
Cold mills and related	264	
Coke plants	6	632
Sanitary uses, boilers, etc.		254
Blowers, condensers, etc.		1955
	1024	4486

Characteristics of steel-mill wastes. Important wastes from the by-product coke phase of steel-mill operation come from the ammonia still, the final cooler, and from the pure still, where products such as benzene, toluene, and xylene are made from the crude naphthalene. Phenol and oxygen-demanding matter are the primary contaminants. A summary of the major constituents of these coke-plant wastes is given in Table 24.15.

The blast furnace wet-scrubber effluent contains flue-dust solids, from washing the gas, composed of iron oxide, alumina, silica, carbon, lime, and magnesia. The amount of each constituent, in comparison

Table 24.15 Analyses of by-product coke-plant wastes [1].

Characteristic	Source of wastes			
	Ammonia still	Final cooler*	Pure still	Combined
BOD, 5-day, 20°C	3974	218	647	53–125†
Total suspended solids, ppm	356		125	89‡
Volatile suspended solids, ppm	153		97	
Organic and NH_3-N, ppm	281	14	20	
NH_3-N, ppm	187		10	
Phenol, ppm	2057	105	72	6.4§
Cyanide, ppm	110			
pH	8.9		6.6	

*No recirculation.
†Depending on compositing technique.
‡Average of 11 daily 24-hour composites, including coke breeze.
§Single-catch sample.

with the total quantity of dust, varies with the type of ore used in the furnace, conditions of the furnace lining, the quality of coke used, the number of furnaces in blast, the amount of air being blown, and the regularity and thoroughness of dumping and flushing of dry dust catchers [33]. Fe_2O_3 comprises about 70 per cent, and silica about 12 per cent, of the flue-dust content. Some pertinent analyses of the physical characteristics of flue-dust wastes are given in Table 24.16 [33].

The amount of waste pickling liquor per ton of steel product depends on the size and type of plant. Steel production in the United States in 1948 was more than 11 million tons, with an estimated 600 million gallons of pickle-liquor waste [88], or about 55 gallons per ton of steel; whereas in Germany the figure ranges from 25 to 200 gallons per ton. One factor that increases the volume is that, since the steel products must be rinsed in water after they leave the pickling tank to remove all trace of acid, the rinse or wash water eventually becomes quite acidic and must also be discarded. The volume of rinse water is 4 to 20 times that of the actual pickling liquor, although naturally it is far more dilute. Wash waters contain from 0.02 to 0.5 per cent H_2SO_4 and 0.03 to 0.45 per cent $FeSO_4$, as compared with 0.5 to 2.0 per cent H_2SO_4 and 15 to 22 per cent $FeSO_4$ in the pickling liquors. Thus, H_2SO_4 and $FeSO_4$ in these ranges of concentration are the major contaminants in the wastes from pickling and washing of steel.

Treatment of steel-mill wastes. The primary method of treatment of by-product coke-plant wastes is to use recovery and removal units with high efficiencies, phenol being the main contaminant recovered. The BOD can be reduced by about one-third by the practice of recirculation and reuse of contaminated waters, and by-product recovery may be undertaken for profit in the case of such materials as ammonium sulfate, crude tar, naphthalene, coke dust, coal gas, benzene, toluene, and xylene. Quench water is usually settled to remove coke dust, and the supernatant liquor from

Table 24.16 Flue-dust content of wet-washer effluents [33]

Characteristic	Value or concentration
Suspended solids content	
Range, ppm*	500–4500
Per cent by weight passing 100-mesh sieve	86–99
Per cent by weight passing 200-mesh sieve	74–97
Temperature, °F	100–120
pH	6–8
Specific gravity	3–3.8

*1200 ppm is average at Fairless Steel plant.

the settling tanks is reused for quenching. Gravity separators are used to remove free oil from the wastes from benzol stills, since the emulsified oils are generally not treated and without separation the free portion of the oil would thus reach the sewers. Final cooler water is also recirculated, to reduce the amount of phenol being discharged to waste. Phenol is recovered primarily to prevent pollution of streams and to avoid the nuisance of taste in water supplies. Phenols may be removed by either conversion into nonodorous compounds or recovery as crude phenol or sodium phenolate, which have some commercial value. The conversion may be either biological (activated sludge or trickling filtration) or physical (ammonia-still wastes used to quench incandescent coke, a process which evaporates the NH_3. Although certain concentrations of phenol (0 to 25 ppm) may be handled by biological units, dilution with municipal sewage is a good idea, since this provides a buffering and diluting medium. The Koppers dephenolization process [1] lowers the phenol content by 80 to 90 per cent in ammonia-still wastes. The process, as shown in Fig. 24.5, is essentially a steam-stripping operation, followed by mixing in a solution of caustic soda and renewing pure phenol with flue gas.

In treating flue dust, sedimentation, followed by thickening the clarifier overflow with lime to encourage flocculation, has been found most effective for removing iron oxide and silica. Ninety to 95 per cent of the suspended matter settles readily and does so within a one-hour period, the resulting effluent having less than 50 ppm suspended solids. Primary and secondary (lime-coagulated) thickened sludges are also obtained, which can then be lagooned without creating nuisances. Henderson and Baffa [33] give details of a typical blast-furnace waste-treatment process (Fig. 24.6).

The treatment of pickling liquor is a problem of considerable magnitude. For most *small* steel plants, the recovery of by-products from waste pickling liquor is not economically feasible and they neutralize the liquor with lime. However, some companies do obtain by-products from this waste, namely: (1) copperas and $FeSO_4 \cdot H_2O$; (2) copperas and H_2SO_4; (3) $FeSO_4 \cdot H_2O$ and H_2SO_4; (4) $Fe_2(SO_4)_3$ and H_2SO_4; (5) Fe^{+++} and H_2SO_4; (6) iron powder; (7) Fe_3O_4 for polishing or pigments; (8) Fe_3O_4 and $Al_2(SO_4)_3$. These are described in more detail by Hoak [36].

The recently developed Blaw-Knox-Ruthner process for the recovery of sulfuric acid involves the concentration, by evaporation, of waste pickling liquor before it is discharged to a reactor, where anhydrous hydrogen chloride gas is bubbled through it, reacting with the ferrous sulfate to produce H_2SO_4 and $FeCl_2$. The ferrous chloride is separated from the sulfuric acid (which is returned to the pickling line) and is converted to iron oxide in a direct-fired roaster. This liberates HCl, which is recovered by scrubbing and stripping and is then recycled to the reactors. This process, shown in Fig. 24.7, has been successfully demonstrated in a pilot plant cooperatively run by many of the large steel producers [88].

Neutralization of pickle-liquor waste with lime is costly, because there is no saleable end-product and there is a voluminous, slow-settling sludge which is difficult to dispose of. Neutralization takes place in four stages: (1) formation of ferric hydrate with a pH below 4, (2) formation of acid sulfate, (3) formation of the ferrous hydrate with a pH between 6 and 8, and (4) formation of the normal sulfate. Calcium and dolomitic lime are the least expensive neutralizing agents, caustic soda and soda ash being too expensive for such a purpose. Even with the cheaper chemicals, Hoak [36] concludes that the overall cost of neutralization ranges from $5 to $10 per 1000 pounds of acid from the pickling operation. Much research has been done on reducing the cost of neutralization by increasing the basicity of the neutralizing chemicals and by various methods of decreasing sludge volumes. How-

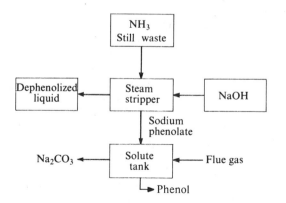

Fig. 24.5 The Koppers dephenolization process, using steam, caustic soda, and flue gas.

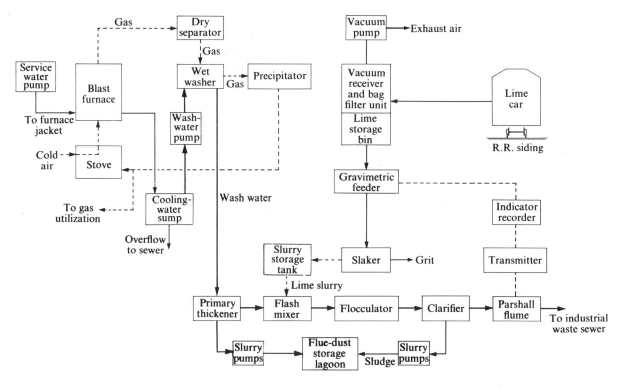

Fig. 24.6 Blast-furnace waste treatment process. (After Henderson and Baffa [33].)

ever, research should be directed toward finding new methods of treatment, rather than relying on neutralization only.

Three areas of change in the pickling-waste problem are: (1) improvements in the treatment of waste from pickling with H_2SO_4; (2) a new HCl pickling operation; and (3) a new dry descaling operation.

New treatment methods for H_2SO_4 pickling include *deep well disposal*, which costs $500,000 (on average) per installation and $1/1000 gallons to operate, with a removal efficiency of 85 per cent (based on the fact that the rinse water is not treated and a small percentage of the pickle liquor in the well may escape as a pollutant), and *ion exchange*, which has a removal efficiency of 80 per cent [1965, 87] and costs as yet undetermined.

Hydrochloric acid pickling differs from H_2SO_4 pickling in the basic chemistry of the pickling action. Hydrochloric acid readily dissolves all the various oxides of iron in the scale, yet reacts relatively slowly with the base metal. The dissolved solids in the HCl pickle liquor are far below saturation concentration and the steel is left clean and free of crystals or insoluble slime. Sulfuric acid on the other hand acts at a high reaction rate with the parent metal and "blows" off oxides on the strip. Because of this, more scale-breaking is required before pickling. The benefits of HCl pickling are: easier regeneration of acid; no over-pickling and more flexibility on the line; elimination of the secondary scale breaker; higher pickling speeds; and a 20 per cent reduction in waste-water volume. The one disadvantage is the increased cost of HCl over H_2SO_4; however, on the whole it is definitely more desirable. Hydrochloric acid wastes are treated by deep well disposal and by neutralization. The capital costs for neutralization are $1 million for a plant with a capacity of 100,000 gallons per day and the operating cost are $20 per 1000 gallons of waste neutralized, with a removal efficiency of 80 per cent (considering the calcium salt residual and the fact that

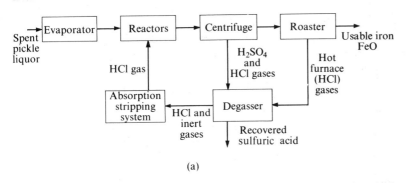

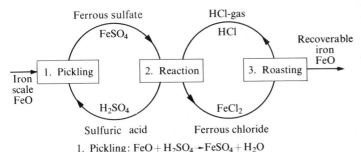

Fig. 24.7 Blaw-Knox-Ruthner process for recovery of acid from spent pickle liquor: (a) process flow diagram; (b) chemistry of process. (Courtesy Blaw-Knox Co.)

rinse waters are not neutralized). A third method of treatment is regeneration, which consists of the following processes: the pickle liquor is pumped to a spray roaster where water and free HCl in the pickle liquor are quickly driven off; the crystal descends inside the roaster while an increased temperature gradient roasts it, producing iron oxide and hydrogen chloride; the iron oxide is collected from the bottom of the roaster; some iron oxide is discharged in the gas and a cyclone is used to collect it; finally, the dry hydrogen chloride is recovered as HCl. For this method, the capital costs are $4 million for a plant with a capacity of 100,000 gallons per day; operation and maintenance costs are $8.80 per gallon of waste treated.

Abrasive descaling used on cold-rolled strip is done on two machines. The first uses steel spheres about 0.01 inch in diameter; the second uses even smaller angular grit. Both abrasives are cleaned continuously and recirculated [1966, 5].

References: Steel-Mill Wastes

1. American Iron and Steel Institute, *Annual Statistical Report* (1949).
2. Antwerpen, F. J., "Utilization of pickle liquor as a building material," *Sewage Ind. Wastes* **15**, 356 (1943).
3. Atwood, J. S., *et al.*, "Regeneration of waste pickle liquors to produce ferrous sulfate monohydrate," in Proceedings of 12th Industrial Waste Conference, Purdue University, May 1957.
4. Baffa, J. J., "Waste disposal at a steel plant: treatment of sewage," *Proc. Am. Soc. Civil. Engrs.* **80**, Separate no. 495 (1954).
5. Baffa, J. J., "Sanitary sewage treatment, steel mill," *Sewage Ind. Wastes* **27,** 990 (1955).
6. Barnhart, T. F., "Acid recovery from pickle liquor, Ruthner process," *Sewage Ind. Wastes* **30**, 296 (1958).
7. Bartholemew, F. J., "Recovery of waste iron sulfate—

sulfuric acid solutions," *Steel* **127**, 68 (1950) and *Chem. Eng.* **57**, 118 (1950).
8. Bartholomew, F. J., "Sulfuric acid recovery from pickle liquor," *Sewage Ind. Wastes* **24**, 1445 (1952).
9. Cameron, A. B., "Fostoria, Ohio, pickling liquor," *Sewage Ind. Wastes* **4**, 232 (1932).
10. Cassels, G. S., *et al.*, "Problem of water pollution in steel industry in Pennsylvania," *Iron Steel Eng.* **30**, 62 (1953).
11. "Chemical by-products from pickle liquor," *Sewage Ind. Wastes* **20**, 281 and 840 (1948).
12. Chivers, A. R. L., "Treatment of pickle liquors," *Sewage Ind. Wastes* **25**, 241 (1953).
13. Cohn, M. M., "A million tons of steel with sewage," *Wastes Eng.* **27**, 309 (1956).
14. Collier, J. R., "Elyria, Ohio, effect of steel mill wastes on sewage treatment," *Sewage Ind. Wastes* **18**, 528 (1946).
15. Cooper, J. E., "Waste treatment in modern steel mill," *Iron Steel Eng.* **28**, 82 (1951).
16. "Dunkirk, N. Y., pickling liquor," editorial report, *Sewage Ind. Wastes* **13**, 817 (1941).
17. Eden, G. E., and G. A. Truesdale, "Treatment of pickle liquors," *Sewage Ind. Wastes* **22**, 1099 (1950).
18. Fradkin, A. M., and E. B. Fooper, "Acid and iron recovery, ion exchange," *Sewage Ind. Wastes* **27**, 754 (1955).
19. Fuller, E., "Removal of oil and solids from rolling mill wastes," in Proceedings of 13th Industrial Waste Conference, Purdue University, May 1958.
20. Gehm, H. W., "By-products from pickle liquor," *Sewage Ind. Wastes* **14**, 1158 (1942).
21. Gehm, H. W., "Acid recovery from pickle liquor," *Sewage Ind. Wastes* **15**, 1248 (1943).
22. Gehm, H. W., "Neutralization with upflow expanded limestone bed," *Sewage Ind. Wastes* **16**, 104 (1944).
23. Gilbert, J. J., "Industrial waste treatment in Germany," in Proceedings of 2nd Industrial Waste Conference, Purdue University, January 1946.
24. Griffith, C. R., "Purification of wastes in a metal working plant by lagooning," Purdue University Engineering Extension Series, Bulletin no. 76, 1951, p. 421.
25. Griffith, C. R., "Lagoons for treating metal-working wastes," *Sewage Ind. Wastes* **27**, 180 (1955).
26. Groen, M. A., and J. E. Cooper, "Use of steel pickling liquor for sewage sludge conditioning," *Sewage Works J.* **21**, 1037 (1949).
27. Groen, M. A., "Pickle liquor use in sludge conditioning," *Sewage Ind. Wastes* **21**, 1037 (1949).
28. Guillot, E. F., "The use of vacuum filtration in pickle liquor disposal," in Proceedings of 6th Industrial Waste Conference, Purdue University, February 1951.
29. Guillot, E. F., "Vacuum filtration, pilot studies," *Sewage Ind. Wastes* **24**, 801 (1952).
30. Heise, L. W., and M. Johnson, "Disposal of waste pickle liquor by controlled oxidation and continuous lime treatment," in Proceedings of 9th Industrial Waste Conference, Purdue University, May 1954.
31. Heise, L. W., and M. Johnson, "Oxidation and continuous lime treatment of pickle liquor," *Sewage Ind. Wastes* **27**, 190 (1955).
32. Heise, L. W., and M. Johnson, "Practical development aspects of waste pickle liquor disposal," in Proceedings of 13th Industrial Waste Conference, Purdue University, May 1958.
33. Henderson, A. D., and J. J. Baffa, "Waste disposal at a steel plant: treatment of flue dust waste," *Proc. Am. Soc. Civil Engrs.* **80**, separate no. 494, (1954).
34. Hicks, R., "Pickle liquor in sludge conditioning," *Sewage Ind. Wastes* **21**, 591 (1949).
35. Hoak, R. D., "Pickle liquor treatment and disposal," *Sewage Ind. Wastes* **17**, 940 (1945).
36. Hoak, R. D., "New developments in the disposal and utilization of waste pickling liquors," in Proceedings of 2nd Industrial Waste Conference, Purdue University, January 1946.
37. Hoak, R. D., "Industrial wastes: waste pickle liquor," *Ind. Eng. Chem.* **19**, 614 (1947).
38. Hoak, R. D., "Lime treatment of pickle liquor," *Sewage Ind. Wastes* **19**, 945 (1947).
39. Hoak, R. D., "Pickle liquor waste problem and research," *Sewage Ind. Wastes* **19**, 1094 (1947).
40. Hoak, R. D., "Acid iron wastes, neutralization," *Sewage Ind. Wastes* **22**, 212 (1950).
41. Hoak, R. D., "Neutralization nomograph," *Water Sewage Works* **98**, 360 (1951).
42. Hoak, R. D., "Disposal of spent sulfate pickling solutions," Ohio River Valley Sanitation Commission, October 1952.
43. Hoak, R. D., "Disposal and utilization practices," *Sewage Ind. Wastes* **24**, 1444 (1952).
44. Hoak, R. D., "Liquid industrial wastes: steel industry," *Ind. Eng. Chem.* **44**, 513 (1952).
45. Hoak, R. D., *et al.*, "Neutralization studies of basicity

of limestone and lime," *Sewage Ind. Wastes* **16**, 855 (1944).
46. Hoak, R. D., et al., "Treatment of spent pickling liquors with limestone and lime," *Ind. Eng. Chem.* **37**, 553 (1945).
47. Hoak, R. D., et al., "Pickle liquor neutralization: economic and technological factors," *Ind. Eng. Chem.* **40**, 2062 (1948).
48. Hoak, R. D., and C. J. Sindlinger, "New technique for waste pickle liquor neutralization," *Ind. Eng. Chem.* **41**, 65 (1949).
49. Hodge, W. W., "Waste problems of the iron and steel industries," *Ind. Eng. Chem.* **31**, 1364 (1939).
50. Hodge, W. W., "Pickle liquor disposal," *Sewage Ind. Wastes* **14**, 736 (1942).
51. Horner, C., et al., "The electrolytic treatment of waste sulfate pickle liquor using permselective membranes," in Proceedings of 9th Industrial Waste Conference, Purdue University, May 1954, p. 423.
52. Horner, C., et al., "Electrolytic treatment of waste sulfate pickle liquor using anion exchange membranes," *Ind. Eng. Chem.* **47**, 1121 (1955).
53. Howell, G. A., "Water conservation practice," *Sewage Ind. Wastes* **24**, 1168 (1952).
54. Howell, G. A., "Disposal of steel production wastes at Fairless works," *Sewage Ind. Wastes* **26**, 286 (1954).
55. Howell, G. A., "Disposal of pickle liquor," *Sewage Ind. Wastes* **29**, 1278 (1957).
56. Imhoff, W. G., "Pickle liquor treatment plant," *Sewage Ind. Wastes* **17**, 656 (1945).
57. Imhoff, W. G., "Waste pickle liquor problem," *Wire Wire Prod.* **24**, 1040 and 1127 (1949).
58. U.S. Public Health Service, *Industrial Wastes, Ohio River Survey*, Supplement D (1943), p. 1092.
59. Johannes, C. A., "Iron ore mining and processing waste disposal," in Proceedings of 11th Industrial Waste Conference. Purdue University, May 1956.
60. Jones, E. M., "Acid wastes treatment," *Sewage Ind. Wastes* **22**, 224 (1950).
61. Jones, H. G., "Grit deposition from integrated steelworks," *Iron Coal Trades Rev.* **170**, 839 (1955).
62. Kinney, J. E., "Fusing the phenol frenzy," in Proceedings of 15th Industrial Waste Conference, Purdue University, May 1960, p. 28.
63. Kraiker, H., "Chemical recovery from pickle liquor," *Sewage Ind. Wastes,* **21**, 1105 (1949).
64. Lab, R. F., "Acid iron wastes neutralization plant," *Sewage Ind. Wastes* **22**, 222 (1950).
65. Lanphear, R. S., "Iron wastes, effect on Worcester, Mass, filters," *Sewage Ind. Wastes* **22**, 261 (1950).
66. Levine, R., *A Study of the Sludge Characteristics of Lime-Neutralized Pickling Liquor*, National Lime Association, Washington, D.C. (1948).
67. Levine, R., and W. Rudolfs, "Sludge characteristics of lime-neutralized pickling liquor," in Proceedings of 7th Industrial Waste Conference, Purdue University, May 1952.
68. Lewis, C. J., "Lime treatment of pickle liquor," *Sewage Ind. Wastes* **21**, 597 (1949).
69. McDermott, G., "Blast furnace industrial waste guide," *Sewage Ind. Wastes* **26**, 976 (1954).
70. MacDougall, H., "Sheet and tin mill wastes, treatment," *Sewage Ind. Wastes* **26**, 538 (1954).
71. MacDougall, H., "Waste disposal at a steel plant: treatment of sheet and tin mill wastes," *Proc. Am. Soc. Civil Engrs.* **88**, separate no. 493 (1954).
72. McGarvey, F. X., et al., "Liquid industrial wastes: brass and copper industry cation exchanges for metals concentration from pickle rinse waters," *Ind. Eng. Chem.* **44**, 534 (1952).
73. McNicholas, J., "Neutralization of acid wastes," *Sewage Ind. Wastes* **11**, 559 (1939).
74. Madarass, M. F., "Oil-removal facilities for steel mill wastes," in Proceedings of 10th Industrial Waste Conference, Purdue University, May 1955.
75. Morgan, L. S., "Treatment of steel mill wastes at Pittsburgh Mill," *Sewage Ind. Wastes* **14**, 404 (1942).
76. Nebolsine, R., "Waste disposal at a steel plant: general problems," *Proc. Am. Soc. Civil Engrs.* **80**, separate no. 492 (1954).
77. Nebolsine, R., "General problems, waste disposal, steel plant," *Sewage Ind. Wastes* **27**, 990 (1955).
78. Nusbaum, I., "Effect of industrial wastes on municipal sewage works at Detroit," *Sewage Ind. Wastes* **22**, 1538 (1950).
79. Petit, G., "Treating steel mill phenols and acid wastes," *Wastes Eng.* **24**, 560 (1953).
80. Queern, B. J., "Disposal of pickling liquors," *Wire Wire Prod.* **23**, 655 (1948).
81. "Recovery of steel mill wastes," Staff Report, *Sewage Ind. Wastes* **21**, 374 (1949).
82. Reed, T. F., "Disposal of waste pickling liquor," in Proceedings of 5th Industrial Waste Conference, Purdue University, November 1949.

83. Reed, T. F., "Disposal of pickle liquor," *Sewage Ind. Wastes* **23**, 698 (1951).
84. Reed, T. F., et al., "Treatment and disposal of pickle liquor," *Sewage Ind. Wastes* **24**, 66 (1952).
85. Reents, A. C., and F. H. Kahler, "Iron removal from pickling baths, ion exchange," *Sewage Ind. Wastes* **27**, 632 (1955).
86. Riegel, H. I., "Waste disposal at Fontana steel plant," *Sewage Ind. Wastes* **24**, 1121 (1952).
87. Rudolfs, W., *Industrial Waste Treatment*, Reinhold Publishing Corp., New York (1953), p. 272.
88. "Sanitary engineering division research report no. 22," *J. Sanit. Eng. Div. Am. Soc. Civil Engrs.* **SA3**, Paper 2031 (1959).
89. Shaw, J. A., "Coke plant wastes, oxygen consumed, test," *Sewage Ind. Wastes* **24**, 1202 (1952).
90. Siebert, C. L., "Treatment of steel industry wastes," *Sewage Ind. Wastes* **19**, 137 (1947).
91. Simpson, R. W., and W. Garlow, "Making mill effluent pay off," *Steel* **133**, 90 (1953).
92. Simpson, R. W., and J. L. Samsel, "Industry treats its sewage with pickling and rolling mill wastes," *Wastes Eng.* **27**, 583 (1956).
93. Smith, F., "Neutralization of pickle liquor," *Sewage Ind. Wastes* **15**, 157 (1943).
94. Smith, E. C., "What the steel industry is doing about stream pollution," *Ind. Wastes* **1**, 157 (1956).
95. Southgate, B. S., "Reuse of steel mill waste water," *Sewage Ind. Wastes* **20**, 169 (1948).
96. Strassburger, J. H., "Evaluation of the Blaw-Knox-Ruthner pilot plant program," Paper presented at General Meeting of American Iron and Steel Institute, New York, 21 May 1958.
97. Swindin, N., "Pickle liquor treatment, England," *Sewage Ind. Wastes* **16**, 1279 (1944).
98. Tänzler, K. H., "Germany, utilization of pickle liquor," *Sewage Ind. Wastes* **17**, 1052 (1945).
99. Townsend, J. W., "Handling of metal-bearing wastes at Erie, Pa.," in Proceedings of 6th Industrial Waste Conference, Purdue University, February 1951.
100. Van Voorhis, M. G., "Pipe mill waste water treatment," *Water Sewage Works* **101**, 242 (1954).
101. Woefle, A. H., "Dunkirk, N.Y., effect of steel mill wastes on sewage treatment," *Sewage Ind. Wastes* **14**, 402 (1942).
102. "Workshop on metal plating and steel mill wastes," in Proceedings of 2nd Industrial Waste Conference, Purdue University, January 1946.
103. "Zinc wastes and pickle liquor, recovery of by-products," *Sewage Ind. Wastes* **21**, 511 (1949).

Suggested Additional Reading

The following references have been published since 1962.

1. Anders, W., "Ion exchange for pickling waste waters," *Water Pollution Abstr.* **38**, 1705 (1965).
2. "New push for HCl in pickling," *Chem. Weekly* **99**, 101 (1966).
3. "Submerged combustion recovery plant for C. and W. Walker," *Chem. Abstr.* **60**, 650 (1964); *Water Pollution Abstr.* **38**, 708 (1965).
4. "Abrasives blast acid from pickling line," *Iron Age* **198**, 66 (1966).
5. "Trends in steel pickling and waste acid treatment," *33 Mag. Metal Prod.* **4**, 65 (1966).
6. Antoine, L., "Industrial effluents and water pollution," *Met. Constr. Mecan.* **98**, 471 (1966); *J. Iron Steel Inst.* **204**, 1281 (1966).
7. Asendorf, E., "Recovery of sulfuric acid from pickling solutions," *J. Iron Steel Inst.* **201**, 894 (1963).
8. Bahme, R. B., "Fertilizer from waste pickling liquor," *Chem. Abstr.* **64**, 14917 (1966).
9. Baker, W. M., "Waste disposal well completion and maintenance," *Ind. Water Wastes* **8**, 43 (1963).
10. Barraclough, J. T., "Waste injection into deep limestone in Northwestern Florida," *Ground Water* **4**, 22 (1966).
11. Bauerlein, C. R., et al., "Simple system treats metal finishing wastes," *Water Works Wastes Eng.* **1**, 56 (1964).
12. Blazenke, E. J., "Deep well disposal of liquid wastes," *33 Mag. Metal Prod.* **4**, 71 (1966).
13. Borgolte, T., "Pickling with HCl," *Galvanotechnik* **57**, 531 (1966); *Finishing Abstr.* **8**, 234 (1966).
14. Box, P., et al., "Improvements in and pertaining to the treatment of waste acid pickling liquor," *Water Pollution Abstr.* **39**, 1375 (1966).
15. Brwha, G., et al., "Regeneration of sulfuric acid pickling solution," *J. Iron Steel Inst.* **203**, 1168 (1965).
16. Buckley, "Liquor regeneration slashes cost of steel pickling," *Chem. Eng.* **74**, 56 (1967).
17. Cairns, D. F., "Stabilization lagoons successfully treat steel mill wastes," *J. Water Pollution Control Federation* **38**, 1645 (1966).

18. Crain, R. W., "A realistic approach to waste treatment," in Proceedings of 9th Ontario Industrial Waste Conference, 1962, p. 57.
19. Clough, G. F. G., "Biological oxidation of phenolic waste liquor," *Chem. Process Eng.* **42**, 11 (1961).
20. Czizi, G., "Electrolytic regeneration of pickling liquor," *Ind. Water Wastes* **6**, 191 (1961).
21. Dunk, G., H. Dembeck, A. Montens, P. Dickens, H. Ternes, E. Winkel, H. Winzer, and M. Hauke, "Behandlung von bekereiabwaessern und aufbereiten von beizloesungen," *Stahl Eisen* **83**, 833 (1963).
22. Eisenhauer, H. R., "Oxidation of phenolic wastes," *J. Water Pollution Control Federation* **36**, 1116 (1964).
23. Francis, C. B., "Recovery of by-products of waste pickle liquor," *Chem. Abstr.* **57**, 8378 (1962).
24. Funk, B. W., and R. C. Allen, "Process for reclamation of cold-strip rolling oils for reuse," *Iron Steel Eng.* **41**, 87 (1964).
25. Goerz, R., and G. Hoffman, "Development and use of rotary disk extractors in phenol-removal plants," *Chem. Tech.* **16**, 80 (1964).
27. Harsha, P., "Steel mill treats waste acids," *Water Sewage Works* **100**, 286 (1963).
28. Hoak, R. D., "Water resources and the steel industry," *Iron Steel Eng.* **41**, 87 (1964).
29. Jeffery, J., "Biological treatment of phenolic effluents," *Gas J.* **306**, 121 (1961).
30. Kestner Evaporator & Engineering Co., "Treating pickle liquors," *Water Pollution Abstr.* **35**, 745 (1962); British Patent no. 877066.
31. Kiheiji, T., "New method for treatment of waste pickling acid," *Natl. Metal Lab. Technicians J.* **6**, 45 (1964).
32. Mantell, C. L., and L. G. Grenni, "Iron from pickle liquor," *J. Water Pollution Control Federation* **34**, 95 (1962).
33. Monteus, A., "Processes for treating spent pickling solutions," *Batelle Tech. Rev.* **12**, 6707 (1963).
34. "Pickling process solves pollution problem," *Ind. Water Wastes,* **12**, 1963.
35. Roblin, J. M., and E. G. Bobalek, "Process reclaims waste pickle liquor," *Chem. Eng. News* **40**, 82 (1962).
36. Ruthner Electochemisch Metallurgische Industrieanlagen of Vienna, Austria, "Technical developments in waste treatment—vertical pickling tower," *Water Waste Treat. J.* **9**, 185 (1962).
37. Crouse, D. J., "Sulfuric acid recovery," *Finishing Abstr.* **7**, 280 (1965).
38. Dasher, J., et al., "Mixed acid pickling wastes," *Metal Finishing* **61**, 60 (1963).
39. Dean, B. T., "The design and operation of a deep well disposal system," *J. Water Pollution Control Federation* **32**, 2245 (1965).
40. Dembeck, H., "Ion exchange—a modern rinsing water treatment," *Chem. Abstr.* **64**, 17251 (1966).
41. De Poy, "Wire mill liquid waste disposal," *Wire Wire Prod.* **41**, 1614 (1966).
42. Dickens, P., "Disposal and utilization of spent sulphate acid pickling solutions and iron sulphate," *J. Iron Steel Inst.* **201**, 1069 (1963).
43. Drogon, J., et al., "Continuous electrolytic destruction of cyanide wastes," *Electroplating Metal Finishing* **18**, 310 (1965).
44. Fekete, L., "A new method for the graphical determination of the quantitative conditions in the continuous regeneration of sulfuric acid pickling baths," *J. Iron Steel Inst.* **204**, 413 (1966).
45. Fielder, H., "Treatment of pickling shop wastes by the SAG process," *J. Iron Steel Inst.* **203**, 1272 (1965).
46. Foulke, D. G., "Waste treatment and the metal finishing industry," *Plating* **53**, 1217 (1966).
47. Foulke, D. G., "Biography of metal finishing wastes for 1964," *Plating* **52**, 1114, 1965.
48. Ganz, S. N., et al., "Production of a nitrogen-iron fertilizer from waste pickling solutions," *Metal Finishing Abstr.* **202**, 949 (1964).
49. Ganz, S. N., et al., "Ferrous nitrate fertilizers from metal plating liquors," *J. Appl. Chem.* **15**, 337 (1965).
50. Ganz, S. N., et al., "Intensification of the process of neutralization of pickling solutions and acid rinsing liquors," *J. Iron Steel Inst.* **202**, 949 (1964).
51. Grieves, R. B., et al., "Flotation of the dichromate ion," *Nature* **205**, 1066 (1965).
52. Gross, J. A., et al., "Recovery of pickle liquor," *Finishing Abstr.* **7**, 280 (1965).
53. Hake, A., "The total regeneration of HCl pickling baths," *J. Iron Steel Inst.* **203**, 1272 (1965).
54. Hake, A., "Regeneration of HCl from iron solutions," *Chem. Abstr.* **64**, 9350 (1966).
55. Hann, V. A., "New developments eliminate waste pickle liquor problem," *Iron Steel Eng.* **42**, 167 (1965).

56. Harsha, P., "Steel mill treats waste acid," *Ind. Water Wastes* **8**, 9 (1963).
57. Hartman, C. H., "Deep well waste disposal at Midwest Steel," *Iron Steel Eng.* **43**, 118 (1966).
58. Hoak, R. D., "Pollution control in steel industry," *Chem. Eng. Progr.* **62**, 48 (1966).
59. Howell, G. A., "Reuse of water in the steel industry," *Public Works Mag.* **94**, 114 (1963).
60. Kobrin, C. L., "Getting in a pickle over HCl," *Iron Age* **198**, 68 (1966).
61. Kramer, A. E., and H. Nierstrasz, "Design of a treatment plant for metal finishing wastes," *Plating* **54**, 66 (1967).
62. Krikau, F. G., "Effective solids removal for pollution control," *Iron Steel Eng.* **43**, 14280 (1966).
63. Krofchak, D., "Steel pickling and pollution," *Wire Wire Prod.* **41**, 1603 (1966).
64. Krug, J., "Ion exchange," *Metal Finishing Abstr.* **6**, 33 (1964).
65. Kuntze, A., "Decontamination and neutralization of effluents produced in the metals industry," *J. Iron Steel Inst.* **204**, 1281 (1966).
66. Labergere, J. L. A., "Treating used metal pickling baths," *Chem. Abstr.* **64**, 18249 (1966).
67. Lancy, E., and R. Pinner, "Waste treatment and metal recovery in copper and copper alloy pickling plant," *Metallurgia* **73**, 119 (1966).
68. Lancy, L. E., "Neutralizing liquid wastes in metal finishing," *Metal Progr.* **91**, 82 (1967).
69. Leidner, R. N., "Bethlehem Steel Burns Harbor waste water treatment plant," *Water Sewage Works* **113**, 468 (1966).
70. Malz, F., "Application of treatment with ion-exchangers for treatment of waste waters," *Chem. Abstr.* **64**, 3197 (1966).
71. Mattock, G., "Automatic control in effluent treatment," *Trans. Soc. Instr. Tech.* **16**, 173 (1964).
72. McCurdy, F. H., et al., "Recent advancement in pickling technique with HCl compound," *Iron Steel Eng.* **42**, 144 (1965).
73. Merwin, R. F., "Efficient ways to recover tramp iron," *Mill Factory* **77**, 66 (1965).
74. Meuthen, B., et al., "Control and adjustments of acid pickling baths and their wastes," *J. Iron Steel Inst.* **203**, 547 (1965).
75. Miller, J. H., "Closed cycle systems as a method of water pollution control," *Iron Steel Eng.* **44**, 103 (1967).
76. Mittenberger, R. S., "The use of HCl in conventional facilities," in Yearbook of American Iron and Steel Institute, New York (1965).
77. Montens, A., "Processes for treating spent pickling solutions," *J. Iron Steel Inst.* **201**, 1069 (1963).
78. Molyneux, F., "Waste acid recovery," *Chem. Progr. Eng.* **45**, 485 (1964).
79. Nagendran, R., et al., "Electrochemical method of treatment of cyanide," *Plating* **54**, 179 (1967).
80. Nebolsine, R., "Steel plant waste water treatment and reuse," *Iron Steel Eng.* **44**, 122 (1967).
81. Noack, W., "Water problems connected with the sheet metal industry and ways for their practical solution," *J. Iron Steel Inst.* **202**, 4389 (1964).
82. Patterson, J. A., et al., "Crystallization apparatus including a swirling film evaporator on a cyclone separator," *Chem. Abstr.* **64**, 10504 (1966).
83. Perkins, E., et al., "HCl pickling and acid regeneration," *Iron Steel Eng.* **42**, 157 (1965).
84. Petlicka, J., et al., "The crystallization of iron. III. Sulfate monohydrate from spent pickling baths at higher temperatures and pressures," *J. Iron Steel Inst.* **203**, 1168 (1965).
85. Poliskin, J., "Recovery and reuse of waste acids," *Ind. Water Eng.* **20** (1965).
86. Poliskin, J., "Pickling acid regeneration," *Metal Finishing* **63**, 72 (1965).
87. Poole, D. E., "Republic's continuous reclamation of HCl pickling at Gadsden, Alabama," *Iron Steel Eng.* **42**, 4160 (1965).
88. Rathmell, R. K., "Pickle liquor treatment," *Finishing Abstr.* **8**, 289 (1966).
89. Robbins, J. L., et al., "Recent developments in hydraulic descaling," *Iron Steel Eng.* **42**, 167 (1965).
90. Rueb, F., "Waste disposal and treatment in metal finishing industry," *Metalloberflaeche (1965); Chem. Abstr.* **63**, 17676 (1965).
91. Schick, J. H., "The problems involved in the neutralization of acid waste water dehydration of sludge," *J. Iron Steel Inst.* **202**, 7627 (1964).
92. Skrylev, L. A., "Removal of zinc from waste waters," *Chem. Abstr.* **65**, 1950 (1966).
93. Spalding, C. U., et al., "Deep well disposal of spent pulping liquors," *Tappi* **48**, 68 (1965).
94. Spanier, G., "New process for purification and processing of pickling waste waters," *Chem. Abstr.* **59**, 4896 (1963).
95. Stratesleffen, E. O., "Completely automatic treat-

ment of plating and pickling waste waters," *Water Pollution Abstr.* **38**, 1876 (1965).
96. Symons, G. E., *The Present and Future of Industrial Waste Water Treatment*, Purdue University, Lafayette, Ind. (1962), p. 717.
97. Talbot, J. B., et al., "The deep well method of industrial waste disposal," *Chem. Eng. Progr.* **60**, 49 (1964).
98. Tomaki, K., "New method of treatment for waste pickling acid," *Water Pollution Abstr.* **38**, 303 (1965).
99. Tucker, F. E., "Recovery and disposal of HCl pickling waste," *Water Sewage Works* **113**, 272 (1966).
100. Ternes, H., et al., "Experience with a circulating system for HCl containing wash waters from the pickling plant of a sheet galvanizing shop," *J. Iron Steel Inst.* **201**, 1069 (1963).
101. Vought, J. H., "Approach to the prevention of water pollution by cyanide-bearing materials," *Plating* **54**, 63 (1967).
102. Vought, J. H., "Preventing stream pollution by cyanide-bearing materials," *Plating* **52**, 420 (1965).
103. Warner, D. L., "Deep well injection of liquid wastes—a review," U.S. Public Health Service, Publication no. 999-*WP*-21, U.S. Government Printing Office, Washington, D.C. (1965).
104. Warner, D. L., "Deep well waste injection—reaction with aquifier water," *J. Sanit. Eng. Div. Am. Soc. Civil Engrs.* **92**, 45 (1966).
105. Williams, A. L., Jr., et al., "Injection well drilling and injection experiments," in Proceedings of 8th Texas Water Pollution Control Association Industrial Water and Waste Conference, 1965.
106. Zabban, W., "Treatment and recovery of effluents from metal treating and metal fabrication operations," in Proceedings of 23rd Annual Water Conference Engineers' Society, Western, Pa., 1962, p. 143.
107. U.S. Department of the Interior, *Blast Furnaces and Steel Mills*, Industrial Waste Profile, Vol. III, no. 1 Washington, D.C. (1967).

The following references on coke plants have been published since 1962.
1. Albrecht, K. H., and F. Lindauer, "Removal of waste liquors containing phenols," East German Patent no. 52, 5 December 1966; *Chem Abstr.* **66**, 794289 (1967).
2. Ashmore, A. G., J. R. Catchpole, and R. L. Cooper, *Water Res.* **1**, 605 (1967); *Chem. Abstr.* **68**, 42950 (1968).
3. Bell, W. P., and G. Jones, "Biological degradation of phenolic wastes from a refinery: influence of various contaminants," *Erdoel Kohle* **18**, 462 (1965); *Chem. Abstr.* **63**, 8021g (1966).
4. Biczysko, J., "Biological purification of phenol-containing effluents," *Prace Inst. Hutniczych* **17**, 241 (1965); *Chem. Abstr.* **63**, 16022d (1966).
5. Biczysko, J., S. Mielus, and J. Sobota, "Removal of phenols from sewage by sorption on dust," *Przemysl Chem.* **45**, 454 (1966); *Chem. Abstr.* **65**, 16665d (1967).
6. Biczysko, J., "Biological purification of phenol waste waters on tower beds within the thermophic range," *Gaz, Woda Tech. Sanit.* **34**, 90 (1965); *Chem. Abstr.* **63**, 4008e (1966).
7. Biczysko, J., "Phenolic sewage treatment in circulation ditches," *Gaz, Woda Tech. Sanit.* **401**, 307 (1966); *Chem. Abstr.* **66**, 79366t (1967).
8. Borkowski, B., "Catalytic oxidation of phenols in gas works effluents," in Proceedings of 9th International Gas Conference, The Hague, 1964, *IGU-B*10–64; *Chem. Abstr.* **65**, 6908e (1967).
9. Brebion, G., R. Carbridenc, and M. Rogeon, "Laboratory study of mixed purification, of urban and industrial effluents containing phenol," *Centre Belge Etudes Doc. Eaux Tribune* **19**, 406 (1966); *Chem. Abstr.* **66**, 49090f (1967).
10. Chen, N. G., and L. I. Gerasyutina, "Effect of aeration on the protective properties of phenol-containing waters of the coke-byproduct industry," *Koks i Khim.*, 1965, p. 54; *Chem. Abstr.* **63**, 5366b (1966).
11. Chambers, C. W., and P. W. Kabler, "Biodegradability of phenols as related to chemical structure," *Develop. Ind. Microbiol.* **5**, 85 (1963); *Chem. Abstr.* **64**, 3191e (1966).
12. Chetverikov, D. I., "Purification of phenol-containing waste water," *Gidrolizn. i Lesokhim. Prom.* **17**, 17 (1964); *Chem. Abstr.* **64**, 13904e (1966).
13. Choulat, C., "New ways for treating the ammonia liquor produced in coke oven plants," *Intern. Z. Gas-Waerme* **15**, 116 (1966); *Chem. Abstr.* **65**, 5208 (1967).
14. Cooke, R., and P. W. Graham, "The biological purification of the effluent from a large plant gasifying

bituminous coals," *Intern. J. Air Water Pollution* **9**, 97 (1965); *Public Health Eng. Abstr.* **45**, Sept. 1965.
15. Dzhobzdze, S. A., and V. A. Perenyshlin, "Use of diisopropyl ether as solvent for the removal of phenols from waste water," *Gaz. Prom.* **7**, 14 (1962); *Chem. Abstr.* **63**, 9655*f* (1966).
16. Eisenhauer, H. R., "Oxidation of phenolic wastes. I: Oxidation with hydrogen peroxide and a ferrous salt reagent. II: Oxidation with chlorine," *J. Water Pollution Control Federation* **36**, 116 (1964).
17. Evans, C. G. T., and S. Kite, "Further experiments on the treatment of spent liquors by homogenous continuous culture (metabolism by bacteria of phenols in phenolic waste waters produced during making of coke)," in *Continuous Cultivation of Microorganisms*, Academic Press, New York (1964); *Biol. Abstr.* **46**, 104455 (1965).
18. Fairall, J. M., "Biological treatment of certain organic industrial wastes," in Proceedings of 10th Ontario Industrial Waste Conference, 1963, p. 113.
19. Fedorenko, Z. P., "The effect of biochemical purification of the waste water of a by-product coke plant on the 3–4–benzpyrene content," *Gigiena i Sanit.* **3**, 17 (1964); *Biol. Abstr.* **48**, 2892 (1967).
20. Fedorenko, Z. P., "Carcinogenic properties of sewage from by-product coke industry," *Vopr. Gigiery Noselen Mestorozhd. Kiev Sb.* **5**, 101 (1964); *Chem. Abstr.* **64**, 15561*f* (1966).
21. Fukuoka, S., H. Eto, and H. Ono, "Microbial purification of some specific industrial wastes. 1: Treatment of industrial phenolic wastes with activated sludge," *Kogyo Gitjutsuin, Hakko Kenkyusho Kenkyu Hokoku* **28**, 107 (1965); *Chem. Abstr.* **64**, 17242*d* (1966).
22. Glabisz, U., "Chlorine dioxide action on phenol wastes," *Przemysl Chem.* **45**, 211 (1966); *Chem. Abstr.* **65**, 10319*b* (1967).
23. Glabisz, U., "Reaction of chlorine dioxide with the components of phenolic waste waters. I: Reaction of chlorine dioxide with phenol," *Chem. Stosowana* **A10**, 211 (1966); *Chem. Abstr.* **66**, 31844*v* (1967).
24. Glabisz, V., "Reaction of chlorine dioxide with the components of phenolic waste waters. II: Reaction of chlorine dioxide with dehydroxyphenols," *Chem. Stosowana* **A10**, 221 (1966); *Chem. Abstr.* **66**, 31845*W* (1967).
25. Grieves, R., and R. C. Aronica, "Foam separation of phenol with a cationic surfactant," *Air Water Pollution* **10**, 31 (1966); *Chem. Abstr.* **64**, 19180*h* (1966).
26. Grossman, A., and J. Pasynkiewica, "Oxidation of phenols in water and wastes," *Koks, Smola, Gaz* **9**, 55 (1964); *Chem. Abstr.* **63**, 1159*c* (1966).
27. Guillaume, F., "Treatment of industrial wastes by means of the oxidation ditch," in Proceedings of 12th Ontario Industrial Waste Conference, 1965, p. 58; *Chem. Abstr.* **66**, 79392*y* (1967).
28. Hall, D. A., and G. R. Nellist, "Phenolic effluents treatment," *Chem. Trade J. Chem. Eng.* **156**, 786 (1965); *Chem. Abstr.* **66**, 58686*g* (1967).
29. Hsu, C., W. Yang, and C. Weng, "Phenolic industrial wastes treatment by a trickling filter," *K'uo Li Taiwan Ta Hsueh Kung Cheng Hsueh Kan* **10**, 162 (1966); *Chem. Abstr.* **67**, 193793 (1967).
30. Irlenbusch, J., "Studies about the use of ammoniacol waste waters in winter stubble-cultures," *Wiss. Z. Karl-Marx-Univ. Leipzig, Math. Naturw. Reihe* **12**, 725 (1963); *Chem. Abstr.* **65**, 15063*b* (1967).
31. Jackson, M. I., and P. W. Graham, "Biological oxidation of waste water from gas works," *Gas Rome* **31**, B61 (1963); *Chem. Abstr.* **63**, 12861*d* (1966).
32. Jankovic, S., and S. Dakovic, "Improvement of absorption capacities of some kinds of Yugoslavian semi-cokes by a chemical treatment and their use for dephenolation of industrial sewage," *Glasnik Hem. Drustrva, Beograd* **28**, 573 (1963); *Chem. Abstr.* **64**, 4784*b* (1966).
33. Jones, G. I., and J. M. Miller, "Biological purification of coking plant waste waters," *Berbautechnik* **14**, 544 (1964); *Chem. Abstr.* **64**, 13903*b* (1966).
34. Karkowski, J., "Problems in phenolic wastes," *Chemik* **19**, 130 (1966); *Chem. Abstr.* **65**, 8554*e* (1967).
35. Knybel, F., "Balance parameters determining the substance outlet of monofunctional phenols, cyanides, ammonia, and chemical oxygen demand in corrosive waste water of coke plants," *Vodni Hospodarstvi* **17**, 23 (1967); *Chem. Abstr.* **67**, 47004*c* (1967).
36. Kostovetskii, Y. I., and E. M. Yoroyskaya, "Hygenic evaluation of total clarification of sewage from coke-chemical industry by a microbe method," *Vopr. Gigieny Noselen Mestorozhd. Kiev Sb.* **5**, 97 (1964); *Chem. Abstr.* **64**, 13903*c* (1966).
37. Kotulski, B., I. Kohut, and K. Susul, *Gaz, Woda Tech. Sanit.* **41**, 304 (1967); *Chem. Abstr.* **68**, 24445 (1968).

38. Kresta, V., and M. Koubik, "Determination of efficiency of the phenolic waste water treatment plants," *Vodni Hospodarstvi* **14**, 103 (1964); *Biol. Abstr.* **47**, 13683 (1966).
39. Kupryakhina, K. Z., P. R. Zimtsev, and A. T. Ivashckenko, "Decontamination of waste waters with ion-exchange resins," *Koks i Khim.*, 1965, p. 46; *Chem. Abstr.* **63**, 9655h (1966).
40. Levin, E. D., N. A. Chuprova, Z. P. Belikova, and T. F. Kandalintseva, "Contents of phenol and cresols in pyroligneous waters," *Chem. Abstr.* **64**, 5208e (1966).
41. Maly, F., "Treatment and utilization of phenolic waste water adsorption and thermodesorption," *Vodni Hospodarstvi* **17**, 379 (1967); *Chem. Abstr.* **68**, 6250h (1968).
42. Malz, W., "Determination of phenols and detergents in waters and in waste waters," *Foederation Europaeischer Gerwaesserschuts Informationasbl.* **11**, 19 (1964); *Chem. Abstr.* **64**, 1808h (1966).
43. Medrea, F., "Purification of phenolic water: general processes of dephenolation," *Rev. Chim.* **17**, 299 (1966).
44. Metsik, R., and A. Tomberg, "Corrosion of wastewater dephenolation equipment," *Khim. i Tekhnol. Goryuch. Slantsev i Produktov ikh Pererabotki* **13**, 229 (1964); *Chem. Abstr.* **64**, 10915e (1966).
45. Misek, T., and B. Rozkos, "Dephenolization of waste water in a rotating-disk contactor (RDC)," *Chem. Prumysl* **15**, 450 (1965); *Chem. Abstr.* **64**, 1807f (1966).
46. Mueller, W., and W. V. Tuempling, "Degradation of phenols in the oxidation ditch," *Wasserwirtsch.-Wassertech.* **15**, 269 (1965); *Chem. Abstr.* **64**, 1805e (1966).
47. Nellist, G. R., "Biological treatment of coke-works liquor," *J. Inst. Sewage Purif.*, 1965, p. 461; *Chem. Abstr.* **65**, 11969f (1967).
48. Oberlaender, G., and W. Funke, "Properties of waste waters from gas works," *Energietechnik* **15**, 329 (1965); *Chem. Abstr.* **63**, 17677b (1966).
49. Pasynkiewicz, J., "Electrochemical treatment of phenolic sewage," *Gaz, Woda, Tech. Sanit.* **41**, 331 (1967); *Chem. Abstr.* **68**, 24446 (1968).
50. Pawlaczyk, M., "Effect of glucose and urea on the rate of phenol degradation (purification of phenol wastes) by *Pseudomonas fluorescens*," *Acta. Microbiol. Polon.* **14**, 207 (1965); *Biol. Abstr.* **47**, 28472 (1966).
51. Rashkevich, I. I., and V. I. Dol, "Detoxication of waste waters of coke-chemical plants by sulfonated coal," *Ukr. Khim. Zh.* **33**, 205 (1967); *Chem. Abstr.* **66**, 118622v (1967).
52. Richter, H., "The efficiency of intensive biological methods based on tests made on a laboratory and technical scale to purify waters of brown coal high-temperature coking subjected to a preliminary destruction of phenols," *Wasserwirt.-Wassertech.* **16**, 25 (1966); *Chem. Abstr.* **66**, 5580w (1967).
53. Richter, H., and H. J. Zuehlke, "Experimental investigations for the final purification of lignite-high temperature coking waste water," *Fortschr. Wasserchem. Grenzg.* **2**, 143 (1965); *Chem. Abstr.* **67**, 46997s (1967).
54. & 55. Rokyta, M., and V. Korinek, "Removing volatile substances, especially univalent phenols, ammonia, and hydrogen sulfide, from crude phenol waters," Czechoslovak Patent no. 114811, 15 May 1965; *Chem. Abstr.* **64**, 15566d (1966).
56. Schmidt, K. H., "Processing water containing high amounts of cyanogen compounds in a coke plant by a new method," *Air Water Pollution* **9**, 753 (1965); *Chem. Abstr.* **64**, 6296a (1966).
57. Schnitzspan, H. G., and I. Neff, "Possibilities for the purification of high polluted phenol-containing organic waste water," *Industrieabwaesser,* June 1966, p. 28; *Chem. Abstr.* **66**, 49091g (1967).
58. Schulmann, J., and P. Fiechs, "Post-purification of phenol-containing wastes," *Fortschr. Wasserchem. Grenzg.* **2**, 175 (1965); *Chem. Abstr.* **68**, 15874 (1968).
59. Sherwood, P. W., "Biological processing of aqueous gas processing wastes," *Gas- Wasserfach* **106**, 853 (1965); *Chem. Abstr.* **63**, 9655g (1966).
60. Shevchenko, M. A., E. M. Kaliniichuk, and R. S. Kasyanchuk, "Purification of water from phenols and petroleum products by oxidization," *Ukr. Khim. Zh.* **30**, 527 (1964); *J. Appl. Chem.* **15**, i–339 (1965); *Public Health Eng. Abstr.* 1965, p. 1569.
61. Solin, V., D. Volkmanmova, and J. Erlebach, "Purification of gas generator effluents on slag filters: removal of cresols on slag," *Sb. Vysoke Skoly Chem. Technol. Praze, Technol Vody* **5**, 43 (1962); *Chem. Abstr.* **63**, 2736d (1966).
62. Smith, R. M., "Some systems for the biological oxidation of phenol-bearing waste waters," in *Symposium on the Microbial Decomposition of Wastes and Detergents,* New York, 1963; *Biotechnol. Bioeng.*

514, 275 (1963); *Biol. Abstr.* **45**, 52221 (1964).
63. Stoneburner, G., "Method of removing phenolic compounds from waste water," U.S. Patent no. 3284337, 8 November 1966; *Chem. Abstr.* **66**, 22007x (1967).
64. Weber, R., "Disposal of phenosolvan waste liquor in the coke conditioning installation of the Espenhain low-temperature carbonizing plant," *Freiberger Forschungsh.* **298**A, 29 (1965); *Chem. Abstr.* **64**, 1807c (1966).
65. Wurm, H. J., and K. Graewe, "Treatment of nitrogenous waste water with special consideration being given to ammonia containing waste water of coal-refining industries," *Wasser, Luft Betrieb* **10**, 381 (1966); *Chem. Abstr.* **65**, 13398a (1967).
66. Wurm, H. J., and K. Graewe, "Sulfur removal from phenolate liquors in dephenolizing plants," *Gas-Wasserfach* **106**, 379 (1965); *Chem. Abstr.* **63**, 2734g (1966).
67. Yurovskaya, E. M., "Resistance of phenol-decomposing microbes to certain variation in the chemical composition of sewage from gas plants," *Vopr. Sanit. Bakter. i Virusol.* **S6**, 155 (1965); *Chem. Abstr.* **64**, 15562a (1966).
68. Zvengintseva, G. B., B. G. Ginzburg, E. Va. Korchilava, E. I. Davydova, A. B. Davankov, and L. B. Zubakova, "Extraction with pyrindine containing an ion exchanger of phenol from industrial caustic sulfates from a sulfonation process," *Ionoobmen. Tekhnol. Akad. Nauk. SSSR. Inst. Fiz. Khim.*, 1965, p. 223; *Chem. Abstr.* **63**, 5367b (1966).
69. Zenin, A. A., and G. S. Konovolov, "Some processes taking place in river water polluted with effluents from (coal) mines," *Gidrokhim. Materialy* **36**, 56 (1964); *Chem. Abstr.* **63**, 2736c (1966).
70. Zdybiewska, M., M. Trawinski, and Z. Paluch, "Full-scale tests on the biological purification of coking plant wastes," *Koks, Smola, Gaz.* **10**, 22 (1965); *Chem. Abstr.* **63**, 12864f (1966).

24.4 Other Metal-Plant Wastes

Processors of several other metals besides steel are significant waste contributors. Among these are brass, copper, gold, and aluminum plants, which are similar to steel mills in that impure metal is purified, worked, and fabricated into final usable products.

Brass and copper. The brass and copper industry produces plate, sheets, and strips by rolling operations, rods and wire by extrusion and drawing operations, and tubes by piercing or extrusion and drawing. Among the principal alloys used in these processes is what is called normal brass ($\frac{2}{3}$ copper and $\frac{1}{3}$ zinc, with small amounts of tin and lead). The molten metal comes from an electric furnace and is poured into molds of various sizes and shapes, to produce billets or bar castings for further operations. The bar castings are rolled into plates, sheets, and strips. After a certain amount of rolling, the metal becomes "hard" and must be annealed before further rolling, after which pickling is required to remove the resulting oxide scale or stain. Similar procedures must be carried out with rods, in order to form wire and tubes from billets. All processes require annealing followed by pickling.

Annealing is done in oil-fired furnaces; the alternate heating and cooling causes a rather heavy oxide scale to form on the surface of the metal. This scale must be removed before the rods, wire, and tubes are drawn, to prevent damage to the dies and sheets and so that the scale is not embedded in the final product. This is done by pickling in a bath of 5 to 10 per cent H_2SO_4 by volume. Stains, particularly on the finished product, are removed in a "bright dip" solution of 5 to 10 per cent H_2SO_4 and up to 0.5 pound per gallon of sodium dichromate. The metal leaving the pickling bath and bright-dip tank is washed with fresh water, which eventually overflows to waste. When the concentration of dissolved metal is too great for economical operation, the bath is dumped as waste. The frequency depends on the metal composition, length of pickling time, and amount of metal pickled; bright dips may be dumped daily, while pickle baths are usually dumped monthly. These two wastes constitute the main waste problems of this industry. Wise *et al.* [16] give the composition of samples collected from

Table 24.17 Pickle-bath wastes [14].

Acid and metal	gm/liter
H_2SO_4	59.7–163.5
Cu	4.0–22.6
Zn	4.3–41.4
Cr	0–0.56
Fe	0.1–0.21

eight pickle tubs (Table 24.17) and bright-dip tanks (Table 24.18) during a brass-plant survey.

The most important methods of treatment appear to be precipitation of the metals as hydroxides in alkaline solutions or utilizing ion exchangers to recover valuable metals. Electrolysis is also sometimes used to recover or regenerate pure metals. The Bureau of Mines reports that high-grade alumina can be recovered economically from mineral-waste solutions such as copper-mine waste water. This method eliminates the cost of mining, crushing, and leaching the rock matter; thus, by-product recovery not only helps to solve a waste problem but also makes the new product more economical.

Gold. Figure 24.8 illustrates the processing of gold bullion from gold-rich ores and the associated wastes.

Aluminum. The production of various aluminum products by the Alcoa Company* illustrates typical processes of the aluminum industry and their associated wastes.

---

*The author wishes to thank the Alcoa Company for the information in this section.

Table 24.18 Bright-dip wastes [14].

Acid and metal	gm/liter
H_2SO_4	5.6–85.8
Cu	6.9–44.0
Zn	0.2–37.0
Cr^{+6}	4.3–19.1
Cr (total)	13.5–47.7
Fe	0.03–0.36

At the mines, power shovels scoop bauxite (the raw material of aluminum) out of the ground and dump it into railroad cars or trucks that haul it to a shipping point or directly to an alumina plant. Some bauxites are crushed, washed to remove some of the clay and sand waste, and dried in rotary kilns prior to shipment. Other bauxites are crushed and dried only, while bauxite from the Caribbean Islands is usually only dried. Bauxite delivered directly to an alumina plant is generally transported in its raw state.

In one manufacturing method, the Alcoa-Bayer process, the finely ground bauxite is fed into a steam-heated unit called a digester and a caustic solution

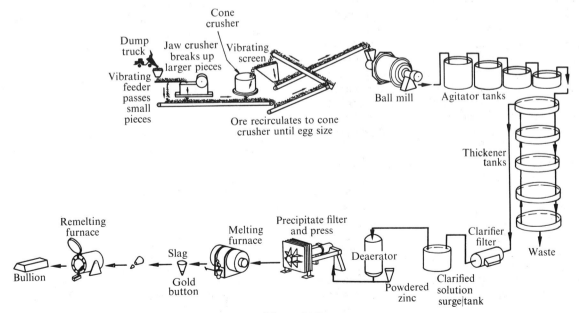

Figure 24.8

made from soda ash and ground lime is added. This mixture is heated under pressure, whereupon the alumina dissolves, but the impurities do not. The mixture then flows through pressure-reducing tanks into a filter press, where cloth filters hold back the solids but allow the alumina-containing solution to pass through. The remaining solids, known as "red mud," are discarded. The liquid solution next passes through a cooling tower and into tall, silo-like tanks, called precipitators. Small amounts of alumina in crystal form are stirred into the liquid to serve as "seed particles" that stimulate the precipitation of solid alumina as the solution cools. This is aluminum oxide, chemically combined with water, and in this form is known as hydrated alumina. The hydrated alumina crystals that have formed in the solution are settled out and removed and the weakened caustic soda solution is returned to the digesters, where it is strengthened, mixed with new bauxite, and used over again.

The ore used in this process should be of high quality, which means that it should contain a low percentage of silica. Silica is a problem in bauxite refining because it combines with alumina and soda to form an insoluble compound that is filtered out with the red mud; thus valuable quantities of alumina and soda are lost. To recover this unintentionally discarded alumina and soda, and to make possible the use of ores containing substantial percentages of silica, the Alcoa Combination process was invented. In this process, the filtered-out red mud is treated to recover the alumina and soda that have combined with the silica. Limestone and soda ash are added and the mixture is heated to form a clinker-like "sinter." This sinter is treated with water which dissolves out the alumina–soda compound produced in the sintering step. The solution of alumina and soda in water is then returned to the digesters at the start of the Alcoa-Bayer process.

In the final stages of alumina production, the hydrated alumina from the precipitators is filtered, washed, and dried in a long, slowly-revolving kiln at 1800°F to remove both the moisture associated with the hydrated alumina and the chemically combined water that is part of its crystalline structure. Hydrated alumina goes into the kiln with about 35 per cent combined water content, and comes out as a dry, white powder—commercially pure alumina.

Ingot, and scrap aluminum are converted into mill products—sheet, plate, bar, rod, wire, tube, extruded, and rolled shapes—that become the raw material for the finished goods manufactured by thousands of other companies.

References: Other Metal-Plant Wastes

1. Czensny, R., "Copper, toxicity to fish," *Sewage Ind. Wastes* **7**, 760 (1935).
2. Griffith, C. R., "Lagoons for treating metalworking wastes," *Sewage Ind. Wastes* **27**, 180 (1955).
3. Hill, H., "Effect on activated sludge process, nickel," *Sewage Ind. Wastes* **22**, 272 (1950).
4. Hupfer, M. E., "Brass mill waste treatment," *Sewage Ind. Wastes* **29**, 45 (1957).
5. Mitchell, R. D., *et al.*, "Brass and copper wastes, effect on sewage plant design, Waterbury, Conn.," *Sewage Ind. Wastes* **23**, 1001 (1951).
6. McDermott, G. N., A. W. Moore, M. A. Post, and M. B. Ettinger, "Effects of copper on aerobic biological sewage treatment," *J. Water Pollution Control Federation* **35**, 227 (1963).
7. McGarvey, F. X., R. E. Tenhoor, and R. P. Nevers, "Cation exchangers for metals concentration from pickle rinse waters," *Ind. Eng. Chem.* **44**, 534 (1952).
8. Pomelee, C. S., "Beryllium production wastes, toxicity studies," *Sewage Ind. Wastes* **25**, 1424 (1953).
9. Pirk, G. W., "Copper reclamation and water conservation," *Sewage Ind. Wastes* **29**, 805 (1957).
10. Rudgal, H. T., "Copper, effects on sludge digestion, Kenosha, Wis.," *Sewage Ind. Wastes* **18**, 1130 (1946).
11. Sanderson, W. W., and A. M. Hanson, "Colorimetric determination," *Sewage Ind. Wastes* **29**, 422 (1957).
12. Ullrich, H., "Treatment of waste-waters from aluminum surface treatment plants," *Chem. Abstr.* **58**, 7713 (1963).
13. Wise, W., "Treatment of metal processing wastes," *Sewage Ind. Wastes* **18**, 761 (1946).
14. Wise, W., "Character, treatment and disposal of wastes," *Sewage Ind. Wastes* **20**, 96 (1948).
15. Wise, W., *et al.*, "Composition of wastes, treatment," *Sewage Ind. Wastes* **20**, 772 (1948).
16. Wise, W., B. F. Dodge, and H. Bliss, "Brass and copper industry," *Ind. Eng. Chem.* **39**, 632 (1947).

24.5 Metal-plating Wastes

Origin of metal-plating wastes. After metals have been fabricated into the appropriate sizes and shapes to meet customers' specifications, they are finished to final product requirements. Finishing usually involves stripping, removal of undesirable oxides, cleaning, and plating. In plating, the metal to be plated acts as the cathode while the plating metal in solution serves as the anode. The total liquid wastes are not voluminous, but are extremely dangerous because of their toxic content. The most important toxic contaminants are acids and metals, such as chromium, zinc, copper, nickel, tin, and cyanides. Alkaline cleaners, grease, and oil are also found in the wastes.

There are two main sources of waste from plating operations, each one distinctive in its volume and chemical nature: (1) batch solutions; and (2) rinse waters, including both nonoverflowing reclaimable rinses and continuous overflow rinses. Various stripping and cleaning operations may precede the plating processes. The Ohio River Valley Water Sanitation Commission (ORSANCO) [156] lists the major types of wastes originating from the stripping, cleaning, and plating of metal parts as follows:

1. *Proprietary solutions.* Most metal-finishing plants use solutions prepared according to manufacturers' formulas. These are mainly cleaners or plating-process accelerators of various types. The exact chemical composition of each solution should be obtained from the manufacturer.
2. *Cyanide concentrates.* This includes cyanide plating solutions and cyanide dips with relatively high concentrations of cyanide. Since this chemical is one of the most toxic to both fish and other aquatic life, as well as man, even low concentrations in wastes are extremely dangerous and to be avoided at all costs.
3. *Cyanide rinse water* originates from the rinsing of cyanide-plated or -dipped metal parts.
4. *Concentrated acid and pickling wastes* originate primarily from stripping and cleaning of metal.
5. *Strong acid rinse waters* arise from rinsing after acid dips, pickling solutions, and strong acid process solutions.
6. *Chromates* originate from both plating and rinsing of metals that have been treated with chromate solutions to give them a durable protective finish. Since chromium, like cyanide, is toxic even in very low concentrations, chromium wastes are segregated and treated to remove all the chromium.
7. *Concentrated alkalis* are found in spent alkaline cleaning solutions—usually containing soaps, oils, and suspended solids—which are dumped periodically.
8. *Other wastes requiring treatment.* In most metal-finishing plants there are wastes which contain metal compounds, oils, soaps, and suspended solids, which can be treated by chemical precipitation and pH adjustment.
9. *Waste waters not requiring treatment.* These include cooling water and other waters unchanged in quality, which may be discharged without treatment.

Characteristics of metal-plating wastes. Most stripping baths are acidic in nature and consist of solutions of sulfuric, nitric, and hydrochloric acid, but alkaline baths containing sodium sulfide, cyanide, and hydroxide may also be used. Usually the chemicals in the stripping solution are present in concentrations of less than 10 per cent. Cleaning is carried out by organic solvents, pickling, or alkaline cleaning compounds. The organic-emulsion cleaners are petroleum or coal-tar solvents coupled with an emulsifier. Alkaline cleaners consist of sodium hydroxide, orthophosphate, complex phosphates, silicates, carbonates, some organic emulsifiers, and synthetic wetting agents.

Burford and Masselli [218] present a table showing the concentration of chemicals used in common plating baths (Table 24.19) and a flow chart for the most common plating baths (Table 24.20).

Cyanide salts are desirable, since they are good oxide solvents and in zinc plating they yield a brighter, less porous, galvanized plate. However, acid zinc sulfate is also being used in plating baths because it is

Table 24.19 Common plating baths. (After Burford and Masselli [218].)

Bath formulas	Metallic + cyanide concentrations, ppm	Rinse concentration, ppm	
		0.5 gph drag-out*	2.5 gph drag-out*
Nickel			
40 oz/gal nickel sulfate	82,000 Ni	171 Ni	855 Ni
8 oz/gal nickel chloride			
6 oz/gal boric acid			
Chromium			
53 oz/gal chromic acid	207,000 Cr	431 Cr	2155 Cr
0.53 oz/gal sulfuric acid			
Copper (acid)			
27 oz/gal copper sulfate	51,500 Cu	107 Cu	535 Cu
6.5 oz/gal sulfuric acid			
Copper (cyanide)			
3.0 oz/gal copper cyanide	12,400 Cu	2.8 Cu	14 Cu
4.5 oz/gal sodium cyanide	28,000 CN	58 CN	290 CN
2.0 oz/gal sodium carbonate			
Copper (pyrophosphate)			
4 oz/gal copper (as proprietary mix)	30,000 Cu	62 Cu	310 Cu
29 oz/gal sodium pyrophosphate			
0.4% ammonia (by volume)			
Cadmium			
3.5 oz/gal cadmium oxide	23,000 Cd	48 Cd	240 Cd
14.5 oz/gal sodium cyanide	57,700 CN	120 CN	600 CN
Zinc			
8 oz/gal zinc cyanide	33,800 Zn	70 Zn	350 Zn
5.6 oz/gal sodium cyanide	48,900 CN	102 CN	510 CN
10 oz/gal sodium hydroxide			
Brass			
4 oz/gal copper cyanide	21,000 Cu	44 Cu	220 Cu
1.25 oz/gal zinc cyanide	5,250 Zn	11 Zn	55 Zn
7.5 oz/gal sodium cyanide	47,500 CN	99 CN	495 CN
4 oz/gal sodium carbonate			
Tin (alkaline)			
16 oz/gal sodium stannate	53,000 Sn	110 Sn	550 Sn
1 oz/gal sodium hydroxide			
2 oz/gal sodium acetate			
Silver (cyanide)			
4 oz/gal silver cyanide	24,600 Ag	51 Ag	255 Ag
4 oz/gal sodium cyanide	21,800 CN	45 CN	225 CN
6 oz/gal sodium carbonate			

*Drag-out is the amount of solution carried out of the bath by the material being plated and the racks holding the material. Rinse rate is assumed to be 4 gpm.

Table 24.20 Flow chart for some common plating baths. (After Burford and Masselli [218].)

Copper plating	Nickel plating	Chrome plating	Zinc plating
Electrocleaner (cathodic)	Electrocleaner (cathodic)	Electrocleaner (cathodic)	Electrocleaner (cathodic)
Running rinse →	Electrocleaner (anodic)	Running rinse →	Running rinse →
Hydrochloric acid dip (5%)	Running rinse →	Sulfuric acid dip	5% sulfuric acid dip
Running rinse →	5% sulfuric acid dip	Running rinse + spray →	Running rinse →
Copper cyanide "strike"	Running rinse →	Chrome solution	Zinc cyanide solution
Running rinse →	Bright nickel solution	Recovery rinse	Running rinse →
Running rinse →	Running rinse →	Mist spray rinse	Spray rinse →
Copper pyrophosphate solution	Soap dip	Running rinse →	Brightener still dip (HNO_2)
Running rinse →	Hot running rinse →	Hot still dip	Running rinse →
Hot rinse (slow overflow) →	Drying oven	Running rinse →	Running rinse →
Drying oven		Hot rinse (slow overflow) →	Hot water dip (slow overflow) →
		Drying oven	Drying oven

*Flow sheets for common types of conveyorized electroplating. (Wastes overflowing to final effluent are indicated by an arrow.)

said to conduct the current with less resistance than zinc cyanide.

The character and strength of plating wastes vary considerably, depending on plating requirements and type of rinsing used. The total plant waste may be either acidic or alkaline, depending on the type and quantity of baths used. A preponderance of cyanide or alkaline cleaning baths is likely to result in a highly alkaline pH, while the opposite may be true for chromate baths. The amount of stripping done is also important, since this operation contributes a highly acid waste to the plant mixture. Table 24.21 presents typical plating-waste concentrations obtained from seven different plants. The total volume of wastes from metal-plating plants, usually expressed as gallons per finished number of metallic units, varies even more than the characteristics. In most metal-finishing plants, the volume of waste is less than 0.5 mgd. Since most plants use excessive chromates for plating, the concentration of chromium in chromium-plating bath waste will usually be several times the concentration of other metals in other baths (Table 24.19).

Treatment of metal-plating wastes. The methods used for disposal of waste from plating operations can be divided into two classes: (1) modifications in design and/or operation within the manufacturing process to minimize or eliminate the waste problem; (2) installation of a chemical (sometimes physical) treatment plant to destroy or remove toxic and objectionable materials in plating-room effluents.

Many recommendations for modifications in design and operation to reduce wastes have been suggested. The Ohio River Valley Water Sanitation Commission [197] has published a guide for these practices. Addi-

Table 24.21 Plating-waste concentrations. (From Burton and Masselli [218].)

Plant	pH	Cu, ppm	Fe, ppm	Ni, ppm	Zn, ppm	Chromium, ppm +6	Chromium, ppm Total	Cn, ppm
A	3.2	16	11	0	0	0	1.0	6
A	10.4	19	3	0	0	0	0.5	14
B	4.1	58	1.2	0	0	204	246	0.2
C	2.8	11		0.2		3	7	1.2
D	2.0	300	10	0	82	0	0	0.7
E	2.4	35	8			555	612	1.2
E	10.7	14	4	19		32	39	2.0
F	10.5	6	2	25	39			10
G	11.3	18	18	26		36		15
G	11.9	23	21	32		95		13

tional modifications include: (1) installing a gravity-fed, nonoverflowing emergency holding tank for toxic metals and their salts; (2) eliminating breakable containers for concentrated material; (3) designing special drip pans, spray rinses, and shaking mechanisms; (4) reducing spillage, drag-out leak to the floor, or other losses, by curbing the area and discharging these losses to a holding tank; (5) using high-pressure fog rinses rather than high-volume water washes; (6) reclaiming valuable metals from concentrated plating-bath wastes; (7) evaporating reclaimed wastes to desired volume and returning to plating bath at rate equal to loss from bath; and (8) recirculating wet-washer wastes from fume scrubbers.

Treatments of plating wastes by chemical and physical means are designed primarily to accomplish three objectives: removal of cyanides, removal of chromium, and removal of all other metals, oil, and greases.

The treatment of cyanides, although mostly accomplished by alkaline chlorination, is being carried out by no less than ten methods [156]: (1) chlorination (gas), (2) hypochlorites, (3) ClO_2, (4) O_3 (ozonation), (5) conversion to less toxic cyanide complexes, (6) electrolytic oxidation, (7) acidification, (8) lime-sulfur method, (9) ion exchange, and (10) heating to dryness.

Chromium-bearing plating wastes are normally segregated from cyanide wastes, since they must be reduced and acidified (to convert the hexavalent chromium to the trivalent stage) before precipitation can occur. Although it is possible to precipitate the chromium directly in the hexavalent form with barium chloride, this method is not widely used. The removal of other metals such as Cu, Zn, Ni, Fe, and greases is usually accomplished by neutralization followed by chemical precipitation. To give the reader a better working knowledge of the processes, the three most widely used treatment methods are discussed in the following paragraphs.

Alkaline chlorination. The treatment of cyanide-bearing wastes by alkaline chlorination involves the addition of a chlorine gas to a metal-plating waste of high pH. Sufficient alkalinity, usually $Ca(OH)_2$ or NaOH, is added prior to chlorination to bring the waste to a pH of about 11, thus ensuring the complete oxidation of the cyanide. Violent agitation must accompany the chlorination, to prevent the cyanide salt of sodium or calcium from precipitating out prior to oxidation. The presence of other metals may also interfere with cyanide oxidation, because of the formation of metal cyanide complexes. Extended chlorination may be necessary under these conditions. The probable reaction with excess chlorine in the presence of caustic soda has been expressed as

$$2NaCN + 5Cl_2 + 12NaOH \rightarrow N_2 + 2Na_2CO_3 + 10NaCl + 6H_2O.$$

About 6 pounds each of caustic soda and chlorine are normally required to oxidize one pound of CN to N_2. Sometimes a full 24-hour chlorination period may be required to effect complete oxidation. Schematic

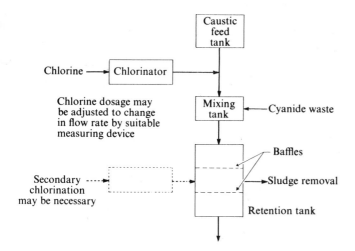

Fig. 24.9 Continuous chlorination of cyanide wastes. (After ORSANCO [156].)

drawings of both continuous and batch treatment of cyanide wastes are presented in Figs. 24.9 and 24.10.

Reduction and precipitation. Chromium-plating-waste treatment by reduction and precipitation involves reducing the hexavalent chromium (Cr^{+6} as chromic acid or chromates) in the waste to the trivalent stage (Cr^{+++}) with reducing agents such as $FeSO_4$, SO_2, or $NaHSO_3$. Sufficient free mineral acid should also be present to combine with the reduced chromium and to maintain a residual pH of 3.0 or lower, which will ensure complete reaction. When the reduction is complete, an alkali (usually lime slurry) is added, to neutralize the acid and precipitate the trivalent chromium. The following chemical reactions, using ferrous sulfate, illustrate this method of treatment:

$$H_2Cr_2O_7 + 6FeSO_4 + 6H_2SO_4$$
$$\rightarrow Cr_2(SO_4)_3 + 3Fe_2(SO_4)_3 + 7H_2O,$$
$$Cr_2(SO_4)_3 + 3Ca(OH)_2 \rightarrow 2Cr(OH)_3 + 3CaSO_4,$$
$$Fe_2(SO_4)_3 + 3Ca(OH)_2 \rightarrow 2Fe(OH)_3 + 3CaSO_4.$$

Figure 24.11 diagrams these reactions. One part per million of chromium usually requires about 16 ppm of copperas, 6 ppm of sulfuric acid, and 9.5 ppm of lime and produces about 2 ppm of chromic hydroxide and 0.4 ppm ferric hydroxide sludges, as well as almost 2 ppm of calcium sufate (some of which is also precipitated).

Neutralization. Treatment of other metal, oil, and grease-bearing wastes by neutralization and precipitation usually involves recombining the wastes with previously oxidized cyanide and reduced chromium wastes for subsequent and final treatment. If the combined waste is acid, an alkali (usually 5 to 10 per cent lime slurry) is added to neutralize and precipitate the metals. The floc produced is large and quite heavy, and hence the velocity of flow is decreased after adequate flocculation has occurred. The waste is then allowed to settle. Sludge is removed and usually lagooned, since this is the most economical treatment for the slow-drying, relatively innocuous, metal

24.10 Batch chlorination of cyanide wastes. (After ORSANCO [156].)

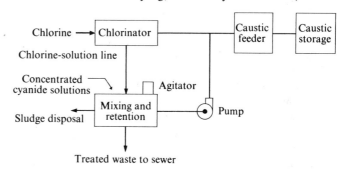

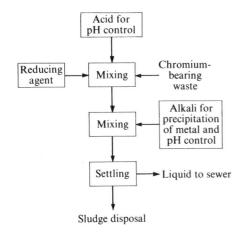

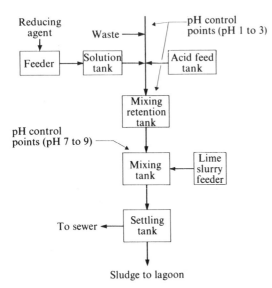

Fig. 24.11 Reduction and precipitation of chromium. (After ORSANCO [156].)

sludges. The processes involved in treating the final wastes are shown in Fig. 24.12.

Recovery practices are mainly those involving ion exchange and evaporation. The use of ion exchangers is only an application of water-softening methods, and its best application is in the treatment of rinse water following plating operations, so that little or no foreign contamination other than the recoverable metal is present. Rinse water is passed through beds of

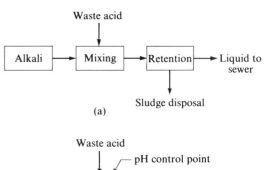

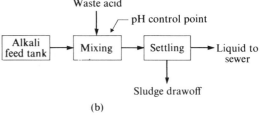

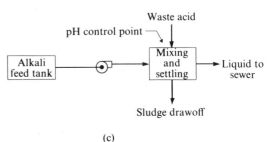

Fig. 24.12 (a) Acid neutralization; (b) continuous acid neutralization; (c) batch acid neutralization. (After ORSANCO [156].)

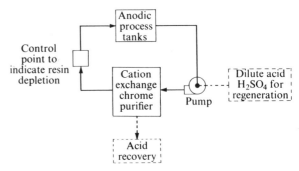

Fig. 24.13 Chrome purifier and recovery system. (After ORSANCO [156].

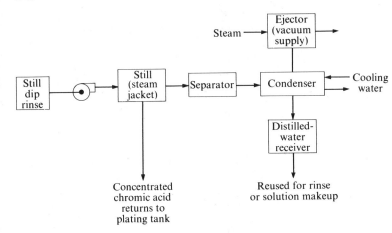

Fig. 24.14 Chrome acid recovery by vacuum evaporation. (After ORSANCO [156].)

cationic and anionic resins selected for the particular application, and the deionized water is recycled through the rinse tank. The ion beds must be regenerated periodically. The regenerating solution containing the concentrated metal salts may require further treatment prior to its reuse in plating operations. Figure 24.13 illustrates the use of an ion exchanger for chromium recovery.

Chrome, nickel, and copper acid-type plating solutions may be reclaimed from the rinse tank by evaporation in glass-lined equipment; the concentrated solution is then returned to the plating systems. The water condensed from the steam is reused in the rinse tank following the plating tank, to eliminate build-up of natural water salts. This process has proved effective for recovering valuable metal salts. The high initial cost for equipment is more than recovered, not only in the value of waste treatment, but also by recovery of metals, especially when volumes of metallic wastes are large. ORSANCO [156] shows a schematic arrangement for evaporating chromium wastes (Fig. 24.14).

References: Metal-Plating Wastes

1. Wallace & Tiernan Co., "Alkaline chlorination for cyanide waste treatment," 25 Main St., Belleville 9, N.Y. (reprinted 1952).
2. Allen, L. A., *et al.*, "Cyanogen chloride formation during chlorination of certain wastes, toxicity to fish," *Sewage Ind. Wastes* **21**, 189 (1949).
3. American Electroplating Society Research Report no. 10, Progress Report, Sterling Laboratory, Yale University, 6 December 1951.
4. Anderson, E. F., "Eliminating acid disposal in metal cleaning operation," *Ind. Wastes* **2**, 52 (1957).
5. Barnes, G. E., "Treatment works for plating wastes containing toxic metals and cyanides," *Water Sewage Works* **94**, 267 (1947).
6. Barnes, G. E., and M. M. Braidech, "Treating pickling liquors for removal of toxic metals," *Eng. News Record* **129**, 86 (1942).
7. Barnes, G. E., and L. W. Weinberger, "Complex metal finishing wastes licked by effective chemical treatment," *Wastes Eng.* **28**, 124 (1957).
8. Barnes, G. E., and L. W. Weinberger, "Waste disposal problems in the metals finishing industry," in Proceedings of 5th Southern Municipal and Industrial Waste Conference, April 1956, p. 201.
9. Baxter, R. R., "Cyanide wastes destroy fish, Anderson, Ind.," *Sewage Ind. Wastes* **18**, 585 (1946).
10. Balden, A. R., "Treatment of industrial process waste at Chrysler Corp.," *Sewage Ind. Wastes* **31**, 934 (1959).
11. Beohner, H. L., and A. B. Mindler, "Ion exchange in waste treatment," *Ind. Eng. Chem.* **41**, 448 (1949).
12. Besselievre, E. B., "A realistic approach to the treatment of plating wastes," in Proceedings of 5th Ontario Industrial Waste Conference, May 1958, p. 90.
13. Billings, N., "Ground water pollution in Michigan," *Sewage Ind. Wastes* **22**, 1596 (1950).
14. Bleiweis, J., "Plating waste disposal," *Iron Age* **163**, 78 (1948).
15. Bloodgood, D. E., "Tenth Purdue Conference highlights industrial waste problems," *Ind. Wastes* **1**, 33 (1955).

16. Bloodgood, D. E., and F. J. Looson, "Removal of toxic substances from metal plating wastes by ion exchange," in Proceedings of 3rd Industrial Waste Conference, Purdue University Engineering Extension Series, Bulletin no. 64, 1947, p. 196.
17. Bloodgood, D. E., and A. Strickland, "Characteristics of chromium wastes," *Sewage Ind. Wastes* **23**, 697 (1951).
18. Bloodgood, D. E., and A. Strickland, "Peculiar characteristics of chromium," in Proceedings of 5th Industrial Waste Conference, Purdue University, November 1949.
19. Brink, R. J., "Systems for the destruction of cyanide wastes at Buick Motor Division," in Proceedings of 8th Industrial Waste Conference, Purdue University, May 1953.
20. Brink, R. J., "Chromic acid recovery by ion exchange," *Sewage Ind. Wastes* **26**, 197 (1954).
21. Burdick, G. E., and M. Lipschuetz, "Toxicity of ferro and ferricyanide solutions to fish, and determination of the cause of mortality," *Trans. Am. Fisheries Soc.* **78**, 192 (1950).
22. Carlson, P. R., "Cyanide waste disposal survey," *Sewage Ind. Wastes* **24**, 1541 (1952).
23. Carmichael, D. C., "Cyanide wastes at Dupont," *Consulting Engr.*, December 1952.
24. Carmichael, D. C., "A continuous method for treatment of cyanide wastes," in Proceedings of 7th Industrial Waste Conference, Purdue University, May 1952.
25. Carran, J. W., "The flue gas method of treating chrome plating waste," in Proceedings of 14th Industrial Waste Conference, Purdue University, May 1959.
26. Chamberlin, N. S., and R. V. Day, "Technology of chrome reduction with sulphur dioxide," in Proceedings of the 11th Industrial Waste Conference, Purdue University, May 1956.
27. Chamberlin, N. S., and H. B. Snyder, Jr., "Technology of treating plating wastes," in Proceedings of 10th Industrial Waste Conference, Purdue University, May 1955.
28. Chamberlin, N. S., and H. B. Snyder, Jr., "Treatment of cyanide and chromium wastes," in Proceedings of Conference on Industrial Health, Houston, Texas, 27–29 September 1952.
29. Chester, A. E., "The importance of mineral-free water in the plating industry," in Proceedings of 3rd Industrial Waste Conference, Purdue University, May 1947.
30. Christie, A. A., *et al.*, "The colorimetric determination of cadmium, chromium, copper, iron, lead, manganese, nickel and zinc in sewage and industrial wastes," Water Pollution Research Laboratory, Reprint no. 309; *Analyst* **82**, 974 (1957).
31. "Chromium salvage," *Chem. Inds.* **68**, 21 (1951).
32. Cooper, J. E., and W. S. Wise, "Plating and metallurgical wastes, problems and plants," *Sewage Ind. Wastes* **22**, 796 (1950).
33. Corcoran, L. M., "Anodizing wastes, treatment, ion exchange," *Sewage Ind. Wastes* **27**, 1259 (1955).
34. Corcoran, A. N., "Treatment of cyanide wastes," *Sewage Ind. Wastes* **22**, 228 (1950).
35. Costa, R. L., "Regeneration of chromic acid solutions by cation exchange," *Ind. Eng. Chem.* **42**, 308 (1950).
36. Cotton, D. A., and A. W. Neel, "Cyanide waste treatment methods," *Sewage Works J.* **19**, 1108 (1947).
37. Cox, I. D., "A combined water softening and plating waste treatment plant operation," in Proceedings of 7th Industrial Waste Conference, Purdue University, May 1952.
38. Cox, I. D., "Practical results of alkali chlorine treatment of cyanide wastes," *Sewage Ind. Wastes* **24**, 1312 (1952).
39. "Curtis Wright plant treats cyanide wastes," *Chlorination Topics*, no. 4, Wallace & Tiernan Co., Newark, N.J. (1940), p. 45.
40. David, M. M., and H. Bliss, "The application of ion exchange to the treatment of dilute brass mill wastes," *Trans. Indiana Inst. Chem. Engrs.* **5**, serial no. 214 (1952).
41. Davids, H. W., and M. Lieber, "Underground water contamination by chromium wastes," *Water Sewage Works* **98**, 528 (1951).
42. Delos, J. S., "Hexavalent chromium: its presence in treated cyanide rinse waters," in Proceedings of 12th Industrial Waste Conference, Purdue University, May 1957.
43. DeWitt, C. C., *et al.*, "Synthetic iron oxide pigments," *Mich. State Univ. Eng. Exp. Sta. Bull.* **110** (1952).
44. Dickerson, B. W., and R. M. Brooks, "Neutralization of acid wastes," *Ind. Eng. Chem.* **42**, 599 (1950).
45. Steel Industry Action Committee of the Ohio River Valley Water Sanitation Commission, *Disposal of Spent Sulfate Pickling Solutions* (1952).
46. Dobson, J. G., "Cyanide waste treatment by chlori-

nation," *Sewage Ind. Wastes* **19**, 1007 (1947).
47. Dobson, J. G., "Disposal of cyanide wastes," *Metal Finishing* **45** (2), 78 and (3), 68 (1947).
48. Dobson, J. G., "Treatment of cyanide wastes by chlorination," in Proceedings of 3rd Industrial Waste Conference, Purdue University, May 1947.
49. Dobson, J. G., "Chlorination of cyanide wastes," *Sewage Ind. Wastes* **22**, 141 (1950).
50. Dodge, B. F., and D. C. Reams, "Disposal of plating room wastes," American Electroplating Society Research Report no. 14, *Plating* **36**, 8 (1949).
51. Dodge, B. F., and D. C. Reams, "Disposal of plating room wastes; a critical review of the literature pertaining to the disposal of waste cyanide solutions," *Plating* **36**, 463 (1949).
52. Dodge, B. F., and W. Zabban, "Disposal of plating room wastes; cyanide wastes: treatment with hypochlorites and removal of cyanates," *Plating* **38**, 561 (1951) and **39**, 385 (1952).
53. Dodge, B. F., and W. Zabban, "How small electroplater can treat cyanide plating waste solutions with hypochlorites," *Plating* **42**, 71 (1955).
54. D'Orazio, A. J., "Pretreatment of cyanide and chromium wastes plus treatment of oily wastes," in Proceedings of 14th Industrial Waste Conference, Purdue University, May 1959.
55. D'Orazio, A. J., "Treatment of chromate waste using liquid sulfur dioxide," in Proceedings of 15th Industrial Waste Conference, Purdue University, May 1960.
56. D'Orazio, A. J., "Vacuum evaporation and deionization for the recovering of plating material," in Proceedings of 9th Industrial Waste Conference, Purdue University, May 1954.
57. Dvorin, R., "Variations in the design of plating waste treatment systems," *Plating* **45**, 827 (1958).
58. Dvorin, R., "Water and waste treatment, a review of methods for the metal finishing industry," in *Metal Finishing Guidebook Directory*, 28th ed. (1960).
59. Eden, G. E., et al., "Chlorination of cyanide wastes," *Sewage Ind. Wastes* **23**, 1213 (1951).
60. Eden, G. E., B. L. Hampson, and A. B. Wheatland, "Destruction of cyanide in waste waters by chlorination," *J. Soc. Chem. Ind.* **69**, 244 (1950).
61. Edwards, G. P., and F. E. Nussberger, "Chromium wastes, effects on activated sludge, Tallmans Islands, New York," *Sewage Ind. Wastes* **19**, 598 (1947).
62. Ehlen, W. J., "Character of plating wastes," *Sewage Ind. Wastes* **19**, 1103 (1947).
63. Ehlen, W. J., "The composition of metal plating wastes," in Proceedings of 2nd Industrial Waste Conference, Purdue University, January 1946.
64. Eldridge, E. F., "Cyanide removal methods," *Sewage Ind. Wastes* **5**, 897 (1933).
65. Eldridge, E. F., *Industrial Waste Treatment Practice*, McGraw-Hill Book Co., New York (1942), p. 289.
66. Eldridge, E. F., "The removal of cyanide from plating room wastes," *Mich. State Univ. Eng. Exp. Sta. Bull.* **52** (July 1933).
67. Erganian, G. K., "The effect of chrome plating wastes on the Warsaw, Indiana, sewage treatment plant," in Proceedings of 14th Industrial Waste Conference, Purdue University, May 1959.
68. Fadgen, T. J., "A study of the possible use of ion exchange for the recovery of metal from electroplating wastes," *Purdue Univ. Eng. Bull. Ext. Ser.* **79**, 24 (1952).
69. Fadgen, T. J., "Metal recovery by ion exchange," *Sewage Ind. Wastes* **24**, 1101 (1952).
70. Fadgen, T. J., "Electroplating wastes, ion exchange units, operation," *Sewage Ind. Wastes* **27**, 206 (1955).
71. Fair, C. M., "Economies in metal finishing wastes management," *J. Water Pollution Control Federation* **32**, 632 (1960).
72. Fales, A. L., "Analysis of plating waste disposal problems," *Sewage Ind. Wastes* **20**, 857 (1948).
73. Fisco, R., "A review of the process of hexavalent chromium reduction utilizing waste flue gas," in Proceedings of 15th Industrial Waste Conference, Purdue University, May 1960.
74. Fisher & Porter Co., "Instrumentation for Control of Cyanide and Chrome Waste Treatment Processes Application," Bulletins no. 10, 90, and 242, Warminster, Pennsylvania, March 1955.
75. Floe, C. F., and A. E. Drucker, "Electro-hydrometallurgical process for copper flotation concentrate," *Wash. State Coll. Monthly Bull.* **14**, 5 (1931).
76. "For effective neutralization of industrial wastes (micromax automatic pH control)," Bulletin no. *ND-96-708*, Leeds & Northrup Co., Philadelphia (1949).
77. Foulke, D. G., and R. F. Ledfora, "Plating wastes, review of research," *Metal Finishing* **53**, 67 (1955).
78. Fradkin, A. M., and E. B. Tooper, "Treatment of spent sulfuric acid pickling liquors," *Ind. Eng. Chem.* **47**, 87 (1955).
79. Friel, F. S., and G. T. Weist, "Cyanide removal from

metal finishing wastes," *Water Sewage Works* **92**, 97 (1945).
80. Gard, C. M., C. A. Snavely, and D. J. Lemon, "Design and operation of a metal waste treatment plant," *Sewage Ind. Wastes* **23**, 1429 (1951).
81. Gallman, H. A., and W. W. Hodge, "Vacuum carbonate process for recovery of hydrogen sulphide and cyanide compounds," in Proceedings of 10th Industrial Waste Conference, Purdue University, May 1955.
82. Garrett, R. L., et al., "How Trans-World Airlines treats plating shop wastes," *Plating* **45**, 847 (1958).
83. Gray, A. G., "Finishing clinic," *Prod. Finishing* **12**, 74 (1948).
84. Gray, A. G., "Practical methods for treatment of plating room wastes," *Prod. Finishing* **14**, 68 (1950).
85. Gray, A. G., "Use of limestone beds to neutralize waste from acid dipping and pickling operations," *Prod. Finishing* **12**, 78 (1947).
86. Greenberg, L., "Plating and waste treatment," *Ind. Eng. Chem.* **52**, 83A (1960).
87. Grindley, J., "Treatment and disposal of waste waters containing chromate," *J. Soc. Chem. Ind.* **64**, 339 (1945).
88. Grindley, J., "Chromium waste treatment and disposal," *Sewage Ind. Wastes* **18**, 1244 (1946).
89. Gruner, C. T., "Incineration as a means for the disposal of solid cyanide," in Proceedings of 11th Industrial Waste Conference, Purdue University, May 1956.
90. Gurnham, C. F., "Current trends in plating waste abatement," in Proceedings of 4th Ontario Industrial Waste Conference, June 1957, p. 8.
91. Gurnham, C. F., "Cyanide destruction on trickling filters," in Proceedings of 10th Industrial Waste Conference, Purdue University, May 1955.
92. Gurnham, C. F., "Disposal of metal finishing wastes," *Metal Finishing Guidebook*, 19th ed. (1950), p. 26.
93. Gurnham, C. F., "Plating room waste disposal, preliminary planning," *Sewage Ind. Wastes* **22**, 721 (1950).
94. Gurnham, C. F., "Stream pollution and the plating industry," *Prod. Finishing* **14**, 26 (1950).
95. Gurnham, C. F., "Waste disposal studies by the American Electroplaters Society," in Proceedings of 4th Industrial Waste Conference, Purdue University, September 1948.
96. Gurnham, C. F., and C. T. Guner, "Solid cyanide waste incineration," *Ind. Wastes* **3**, 137 (1958).
97. Hanson, W. H., and W. Zabban, "Design and operation problems of a continuous automatic plating waste treatment plant at the Data Processing Division, IBM, Rochester, Minn.," in Proceedings of 14th Industrial Waste Conference, Purdue University, May 1959.
98. Haseltine, T. R., "Determination of cyanides," *Water Sewage Works* **94**, 187 (1947).
99. Hart, W. B., "Specialized biological treatment opens new possibilities in treatment of industrial wastes," *Ind. Eng. Chem.* **48**, 93A (1956).
100. Hathaway, C. W., "The vacuum evaporation of chromic acid rinse water," in Proceedings of 7th Industrial Waste Conference, Purdue University, May 1952.
101. Hathaway, C. W., R. E. Harvie, and D. J. Flynn, "Treatment of gray iron foundry waste water," *Ind. Waste* **1**, 166 (1956).
102. Hauri, C. F., "Cyanide disposal after plating cycle," in Proceedings of 6th Industrial Waste Conference, Purdue University, February 1951.
103. Hauri, C. F., "Reduction of plating waste losses by reclaim tanks," *Sewage Ind. Wastes* **25**, 586 (1953).
104. Heidon, R. F., and H. W. Keller, "Methods for disposal and treatment of plating room solution," in Proceedings of 13th Industrial Waste Conference, Purdue University, May 1958.
105. Herda, N., "Cyanide and acid plating waste treatment, Willow Run, Mich.," *Sewage Ind. Wastes* **18**, 499 (1946).
106. Hesler, J. C., "Practical methods for treatment of metal finishing wastes," *Plating* **42**, 1019 (1955).
107. Hoak, R. D., "Steel industries," *Ind. Eng. Chem.* **44**, 513 (1952).
108. Hoak, R. D., "Waste pickle liquor," *Ind. Eng. Chem.* **39**, 614 (1947).
109. Hoak, R. D., C. J. Lewis, C. Sindlinger, and B. Klein, "Lime treatment of waste pickle liquor," *Ind. Eng. Chem.* **39**, 131 (1947).
110. Hoak, R. D., C. J. Lewis, and W. W. Hodge, "Treatment of spent pickling liquors with limestone and lime," *Ind. Eng. Chem.* **37**, 553 (1945).
111. Hoover, C. R., and J. W. Masselli, "Chromium waste disposal," *Sewage Ind. Wastes* **13**, 835 (1941).
112. Hoover, C. R., and J. W. Masselli, "Disposal of waste liquors from chromium plating," *Ind. Eng. Chem.* **33**, 131 (1941).
113. Hoppl, T. C., and W. L. Casper, "Plating waste treat-

ment and water reclamation for the Maytag Company," in Proceedings of 9th Industrial Waste Conference, Purdue University, May 1954.
114. Hosman, L. J., and R. J. Keating, "Chromium and cyanide use of ion exchange," *Sewage Ind. Wastes* **29**, 34 (1957).
115. Hosman, L. J., and R. J. Keating, "Design of industrial waste treatment facilities at IBM, Endicott, N.Y.," Technical Reprint no. *T*-150, Graver Water Conditioning Co., New York; Proceedings of 11th Industrial Waste Conference, Purdue University, May 1956.
116. Horsman, L. J., and R. J. Keating, "Treatment of plating wastes from computer manufacturing," *Sewage Ind. Wastes* **29**, 34 (1957).
117. Howard, F. S., "Airplane washing waste treatment," Infilco Technical Data Bulletin no. *TE*-23-91-3, Chicago, Ill., August 1954.
118. Howell, G. A., "Water conservation in steel mills," *Sewage Ind. Wastes* **24**, 1368 (1952).
119. Huke, B. T., R. P. Selm, and G. E. Summers, "Control of metal finishing wastes using ORP," *J. Water Pollution Control Federation* **20**, 975 (1960).
120. Ingols, R. S., and E. S. Kirkpatrick, "Toxicity study of chromium wastes," *Sewage Ind. Wastes* **25**, 747 (1953).
121. Ingols, R. S., "The toxicity of chromium," in Proceedings of 8th Industrial Waste Conference, Purdue University, May 1953.
122. Jenkins, S. H., and C. W. Hewitt, "Chromium wastes, effects on activated sludge," *Sewage Ind. Wastes* **14**, 1358 (1942).
123. Jenkins, S. H., and C. W. Hewitt, "Chromium wastes, effects on trickling filter," *Sewage Ind. Wastes* **12**, 646 (1940).
124. Kallin, J. F., "Industrial waste control in the Ford Motor Company," *Sewage Ind. Wastes* **31**, 1059 (1959).
125. Keating, R. J., et al., "Application of ion exchange to plating plant problems," in Proceedings of 9th Industrial Waste Conference, Purdue University, May 1954.
126. Keating, R. J., et al., "Plating waste solutions: recovery and disposal," in Proceedings of 10th Industrial Waste Conference, Purdue University, May 1955.
127. Kelch, J. L., and K. A. Graham, "Electrometric system for continuous control of reduction of hexavalent chromium in plant wastes," *Plating* **36**, 1028 (1949).
128. Keller, F. R., C. C. Cupps, and R. E. Shaw, "Recovery of chromic acid from plating operations," *Plating* **39**, 152 (1952).
129. Kempson, N. W., "Alkaline chlorination of metal finishing waste waters," *Wastes Eng.* **22**, 646 (1951).
130. Kessler, R. L., and R. W. Oyler, "Disposal methods for cyanide wastes," *Sewage Ind. Wastes* **22**, 1381 (1950).
131. Kessler, R. L., and R. W. Oyler, "Methods of disposing of cyanide wastes," in Proceedings of 4th Industrial Waste Conference, Purdue University, September 1948.
132. Kittrell, F. W., "Metal plating waste in municipal sewerage," in Proceedings of 5th Southern Municipal and Industrial Waste Conference, April 1956, p. 216.
133. Klassen, C. W., et al., "Toxicity of chromium waste to fish," *Sewage Ind. Wastes* **22**, 1381 (1950).
134. Klassen, C. W., W. A. Hasfurther, and M. F. Young, "The toxicity of hexavalent chromium to sunfish and bluegills," in Proceedings of 4th Industrial Waste Conference, Purdue University, September 1948.
135. Kline, H. S., "Methods for treating metal finishing wastes," in Proceedings of 8th Industrial Waste Conference, Purdue University, May 1953.
136. Kominek, E. G., "Treatment of plating wastes," *Metal Finishing* **47**, 56 (1949).
137. Langford, J. M., "The rocket propulsion industry," *Ind. Wastes* **2**, 42 (1957).
138. Ledford, R. F., "Solids-liquid separation in the treatment of metal finishing wastes," *Plating* **42**, 1030 (1955).
139. Ledford, R. F., and J. C. Hesler, "Chromic acid and copper recovery by ion exchange," *Sewage Ind. Wastes* **27**, 754 (1955).
140. Lewis, C. J., "Dry lime treatment of waste pickle liquor," *Iron Age* **163**, 48 (1949).
141. National Lime Association, *Lime Handling, Application and Storage*, Washington, D.C. (1949).
142. "Liquid waste; Central States Sewage and Industrial Waste Association Meeting report," *Ind. Wastes* **2**, 15 (1957).
143. Lockett, W. T., and J. Griffiths, "Cyanides in trade effluents and their effect on the bacterial purification of sewage," *J. Proc. Inst. Sewage Purif.* Part 2 (1947), p. 121.
144. Lockett, W. T., and J. Griffiths, "Cyanide wastes,

effects on sewage treatment," *Sewage Ind. Wastes* **20**, 357 (1948).
145. Losson, F. J., Jr., and D. E. Bloodgood, "Cyanide waste treatment by ion exchange methods," *Purdue Univ. Eng. Bull. Ext. Ser.* **68**, 314 (1949).
146. Ludzack, F. J., "Biochemical oxidation of some commercially important organic cyanides," in Proceedings of 13th Industrial Waste Conference, Purdue University, May 1958.
147. Ludzack, F. J., et al., "Effect of cyanide on biochemical oxidation in sewage and polluted water," *Sewage Ind. Wastes* **23**, 1928 (1951).
148. Ludzack, F. J., et al., "Experimental treatment of organic cyanides by conventional sewage disposal processes," in Proceedings of 14th Industrial Waste Conference, Purdue University, May 1959.
149. Ludzack, F. J., W. A. Moore, and C. C. Ruchhoft, "Analysis of cyanide in water and waste samples," Standard Methods Committee of Federation of Sewage and Industrial Wastes Association, March 1953.
150. McCormick, C. D., "How can the waste from a plating room be decreased?" in Proceedings of 3rd Industrial Waste Conference, Purdue University, May 1947.
151. McElhaney, H. W., "Metal-finishing wastes treatment at the Meadville, Pa., plant of Talon, Inc.," *Sewage Ind. Wastes* **25**, 475 (1953).
152. McGarvey, F. X., "The application of ion exchange resins to metallurgical waste problems," in Proceedings of 7th Industrial Waste Conference, Purdue University, May 1952.
153. McGarvey, F. X., R. E. Tenhoor, and R. P. Nevers, "Brass and copper industry—cation exchange for metals concentration from pickle rinse waters," *Ind. Eng. Chem.* **44**, 534 (1952).
154. McNicholas, J., "Neutralization of acid wastes," *Sewage Ind. Wastes* **11**, 559 (1939).
155. Marks, H. C., and J. S. Chamberlin, "Determination of residual chlorine in metal finishing wastes," *Anal. Chem.* **24**, 1885 (1953).
156. Metal-Finishing Industry Action Committee, *Methods for Treating Metal Finishing Wastes*, Ohio River Valley Water Sanitation Commission, January 1953.
157. Miller, P. E., "Accidental discharges of cyanide wastes," *Sewage Ind. Wastes* **22**, 1381 (1950).
158. Miller, P. E., "Accidents with cyanide plating solutions," in Proceedings of 4th Industrial Waste Conference, Purdue University, September 1948.
159. Milne, D., "Chemistry of waste cyanide treatment," *Sewage Ind. Wastes* **23**, 174 (1951).
160. Milne, D., "Control and treatment of metal finishing wastes," *Metal Finishing Guidebook Directory* **21**, 104 (1953).
161. Milne, D., "Disposal of cyanides by complexation," *Sewage Ind. Wastes* **22**, 1192 (1950).
162. Milne, D., P. W. Uhl, C. F. Hauri, and E. J. Roy, "Experiences with chlorination of cyanides in the General Motors Corp.," *Sewage Ind. Wastes* **23**, 64 (1951).
163. Milne, D., "Organization for liquid waste control in General Motors Corp.," *Sewage Ind. Wastes* **31**, 447 (1959).
164. Mindler, A. B., and C. Buettman, "Rinse water reuse by ion exchange," *Plating* **42**, 1012 (1955).
165. Mitchell, R. D., et al., "Effect of metal wastes on sewage treatment plant design at Waterbury, Conn.," *Sewage Ind. Wastes* **23**, 1001 (1951).
166. Mohler, J. B., "Control of metal plating rinse waters," *Ind. Wastes* **1**, 77 (1955).
167. Monk, H. E., "Chromium wastes, chemical and bacteriological properties," *Sewage Ind. Wastes* **11**, 1099 (1939).
168. Monk, H. E., and J. H. Spencer, "Chromium waste treatment," *Sewage Ind. Wastes* **11**, 920 (1939).
169. Moore, W. A., et al., "The effect of chromium on the activated sludge process of sewage treatment," in Proceedings of 15th Industrial Waste Conference, Purdue University, May 1960.
170. Neben, F. W., and W. F. Swanton, "Recovery of chromic acid from plating rinse waters," *Plating* **38**, 457 (1951).
171. Neel, A. W., D. A. Cotton, and T. J. Fadgen, "Workshop on metal plating and steel mill wastes," in Proceedings of 2nd Industrial Waste Conference, Purdue University, January 1946, p. 161.
172. Neil, J. H., "Toxicity of cyanides to fish," in Proceedings of 3rd Ontario Industrial Waste Conference, June 1956, p. 125.
173. Nesbitt, J. B., et al., "The aerobic metabolism of potassium cyanide," in Proceedings of 14th Industrial Waste Conference, Purdue University, May 1959.
174. Nyquist, O. W., and H. R. Carroll, "Design treatment of metal processing waste waters," *Sewage Ind. Wastes* **31**, 941 (1959).

175. Oeming, L. F., "Stream pollution problems in plating industry," *Sewage Ind. Wastes* **18**, 678 (1946).
176. O'Kane, G. J., "Planning and operating an industrial waste disposal plant for a new plating facility," *General Motors Eng. J.* **5**, 8 (1958).
177. Oyler, R. W., "Disposal of waste cyanides by electrolytic oxidation," *Plating* **36**, 341 (1949).
178. "Ozone counters waste cyanides lethal punch," *Chem. Eng.* **65**, 63 (1958).
179. Pagano, J. F., et al., "Effect of chromium wastes on methane formation of acetic acid," *Sewage Ind. Wastes* **22**, 336 (1950).
180. Patton, W. G., "Dual disposal system fully neutralizes plating wastes," *Iron Age* **175**, 102 (1955).
181. Paulson, C. F., "Chromate recovery by ion exchange," in Proceedings of 7th Industrial Waste Conference, Purdue University, May 1952.
182. Paulson, C. F., "Plating waste treatment," in Proceedings of 8th Industrial Waste Conference, Purdue University, May 1953.
183. Paulson, C. F., "Profits from metal wastes," *Water Sewage Works* **99**, 199 (1952).
184. Paulson, C. F., "Wastes recovery by ion exchange," *Wastes Eng.* **23**, 208 (1952).
185. Pennsylvania State Health Department, *Pennsylvania Clean Streams*, Harrisburg, Pa., June 1960, p. 3.
186. Pennsylvania State Health Department, "Pennsylvania clean streams," Harrisburg, Pa., December 1948, p. 3.
187. Pennsylvania Sewage and Industrial Waste Association, "Tannery waste plus pickle liquor," *Ind. Wastes* **3**, 18 (1958).
188. Pettet, A. E. J., *Treatment of Plating Wastes and Cyanides*, Institute of Sewage Purification, Midland Branch (England), April 1957.
189. Pettet, A. E. J., "Plating shop effluents, disposal," *Sewage Ind. Wastes* **23**, 1062 (1951).
190. Pettet, A. E. J., "The treatment of electroplating wastes," *Prod. Finishing* **19** (7), 56 and (8), 57 (1955).
191. Pettet, A. E. J., and E. V. Mills, "Biological treatment of cyanides with and without sewage," *J. Appl. Chem.* **4**, 434 (1954).
192. Pettet, A. E. J., and H. N. Thomas, "The effect of cyanides on treatment of sewage in percolating filters," *J. Proc. Inst. Sewage Purif.* Part 2 (1948), p. 61.
193. Pettet, A. E. J., and G. C. Ware, "Disposal of cyanide wastes," *Chem. Ind.* **33**, 1232 (1955).
194. Pettit, G. A., "Mill scale waste treatment," *Ind. Wastes* **3**, 133 (1958).
195. Pinner, W. L., "Metal finishing industry action council of Ohio River Valley Water Sanitation Commission," in Proceedings of 7th Industrial Waste Conference, Purdue University, May 1952.
196. Pinner, W. L., "Practical methods for in-plant reduction of metal plants," *Sewage Ind. Wastes* **24**, 1432 (1952).
197. Ohio River Valley Water Sanitation Commission, *Plating Room Controls for Pollution Abatement*, (1951).
198. "Plating waste treatment plant, a highly efficient unit," *Sewage Works Eng.* **20**, 131 (1949).
199. Poole, B. A., et al., "Control of accidental discharge of cyanide solutions," *Sewage Ind. Wastes* **26**, 1382 (1954).
200. Pratt, M. A., "Extended use of oil emulsions to minimize disposal problems," *Sewage Ind. Wastes* **22**, 331 (1950).
201. Quinlan, E. J., "Plating waste treatment and chrome recovery," *Ind. Wastes* **1**, 19 (1955).
202. Quinlan, E. J., R. F. Keating, and A. L. Wilcox, "Plating waste treatment," *Ind. Wastes* **1**, 19 (1955).
203. Reed, A., "Treatment of plating wastes at Electric Auto-Lite Plant, Lockland, Ohio," *Sewage Ind. Wastes* **22**, 1338 (1950).
204. Reents, A. C., and D. M. Stromquist, "Recovery of chromate and nickel ions from rinse water by ion exchange," in Proceedings of 7th Industrial Waste Conference, Purdue University, May 1952.
205. Reidl, A. L., "Limestone used to neutralize acid wastes," *Chem. Eng.* **54**, 100 (1947).
206. Reidl, A. L., "Neutralization with upflow limestone bed," *Sewage Ind. Wastes* **19**, 1093 (1947).
207. Report of the Water Pollution Research Board, Department of Scientific and Industrial Research, London, 1939–45, p. 49; available from British Information Services, 30 Rockefeller Plaza, New York 21, N.Y.
208. Rhame, G. A., "Treatment of a strong industrial waste containing carbohydrates and chromium," in Proceedings of 9th Industrial Waste Conference, Purdue University, May 1954.
209. Rhame, G. A., "Treatment of a strong carbohydrate plus chromium waste," *Water Sewage Works* **102**, 405 (1955).
210. Ridenour, G. M., "Effects of cyanide wastes on sludge

digestion," *Sewage Ind. Wastes* **20**, 1059 (1948).
211. Ridenour, G. M., *et al.*, "Effects of cyanide wastes on sludge digestion," *Sewage Ind. Wastes* **17**, 966 (1945).
212. Ridenour, G. M., and J. Greenbank, "Preliminary report on effect of cyanide case hardening, copper and zinc plating wastes on activated sludge sewage treatment," *Sewage Works J.* **16**, 774 (1944).
213. Robbins, B. H., "Practical methods of disposing of cyanide wastes," in Proceedings of 12th Industrial Waste Conference, Purdue University, May 1957.
214. Rothstein, S., "Five years of ion exchange; service experience in plating department chemical waste treatment," *Plating* **45**, 835 (1958).
215. Roy, E. J., "Use of hydrochlorine solution as a treatment agent for cyanide," in Proceedings of 7th Industrial Waste Conference, Purdue University, May 1952.
216. Rudgal, H. T., "Brass waste detrimental to digesters at Kenosha, Wis.," *Sewage Works Eng. Munic. Sanit.* **18**, 626 (1947).
217. Rudgal, H. T., "Effects of copper bearing wastes on sludge digestion," *Sewage Works J.* **18**, 1130 (1946).
218. Rudolfs, W., *Industrial Waste Treatment*, Reinhold Publishing Corp., New York (1953), p. 289.
219. Sanders, F. A., "Combating oil and metal plating waste problems at Kelly Air Force Base," in Proceedings of 7th Industrial Waste Conference, Purdue University, May 1952.
220. Southwick, D. E., and J. F. Ryan, "Automatic controls improve cyanide waste treatment," *Wastes Eng.* **30**, 490 (1959).
221. Sellers, W. W., "Caterpillar tractor plant solves waste problems," *Ind. Wastes* **1**, 153 (1956).
222. Serota, L., "Science for electroplaters; cyanide disposal methods," *Metal Finishing* **55**, 75 (1957).
223. Sheets, W. D., "Toxicity of metal-finishing wastes," *Sewage Ind. Wastes* **29**, 1380 (1957).
224. Simpson, R. W., and K. Thompson, "Chlorine treatment of cyanide wastes," Builders Iron Foundry Industries, Providence, R. I, reference no. 100-G5.
225. Small, H. M., and W. C. Graulich, "Plating waste disposal by precipitation and vacuum filtration," *Ind. Wastes* **2**, 75 (1957).
226. Snyder, H. B., "Problems of metal finishing wastes in municipal treatment," in Proceedings of 4th Southern Municipal and Industrial Waste Conference, March–April 1955, p. 88.
227. *Solution of Some Industrial Water and Waste Problems*, Technical Paper no. 64, Milton Roy Company, 1300 East Mermaid Lane, Philadelphia.
228. Sperry, L. B., and M. R. Caldwell, "Destruction of cyanide copper by hot electrolysis," *Plating* **36**, 343 (1949).
229. "Steel mill reclaims process wastes in chemical treatment plant," *Wastes Eng.* **28**, 36 (1957).
230. Stromquist, D. M., and A. C. Reents, "Chromic acid solutions, removal of cations," *Sewage Ind. Wastes* **24**, 807 (1952).
231. Stromquist, D. M., and A. C. Reents, "Removal of cations from chromic acid solutions," in Proceedings of 6th Industrial Waste Conference, Purdue University, February 1951.
232. Sussman, S., F. C. Nachold, and W. Wood, "Metal recovery by anion exchange," *Ind. Eng. Chem.* **37**, 618 (1945).
233. Sweylar, C., "Plating solutions," *Ind. Wastes* **4**, 40 (1959).
234. Tadsen, V. S., "Lime neutralization of metal bearing acid wastes," in Proceedings of 7th Ontario Industrial Waste Conference, June 1960, p. 63.
235. Tarman, J., and M. Priester, "Treatment and disposal of cyanide bearing wastes," *Water Sewage Works* **97**, 385 (1950).
236. Tarzwell, C. M., "The use of bio-assays in the safe disposal of electroplating wastes," in Proceedings of 45th Annual Convention of Electroplaters' Society, 1958; *Tech. Proc. Am. Electroplaters' Soc.*, **44**, 1958.
237. *The Use of Lime in Industrial Trade Waste Treatment*, Trade Waste Bulletin no. 1, National Lime Association, Washington, D.C. (1948).
238. *Treatment of Cyanide and Chromium Wastes*, Wallace & Tiernan Co., *RA*-2120-*C*.
239. *Treatment of Trade Wastes with Dolomitic Lime*, Lime Finishing Association, Bulletin no. 2.
240. Tupholme, C. H. S., "Cyanide wastes destroy fish," *Sewage Ind. Wastes* **5**, 893 (1933).
241. "Twelfth Purdue Industrial Waste Conference," *Ind. Wastes* **2**, 107 (1957) and **3**, 133 (1958).
242. Tyler, R. G., *et al.*, "Treatment of chromium wastes by ion exchange," *Sewage Ind. Wastes* **23**, 1032 (1951).
243. Tyler, R. G., *et al.*, "Treatment of chromium wastes by ion exchange," in Proceedings of 6th Industrial Waste Conference, Purdue University, February 1951.
244. Tyler, R. G., W. Maake, M. J. Westin, and W.

Matthews, "Ozonation of cyanide wastes," *Sewage Ind. Wastes* **23**, 1150 (1951).
245. Unwin, H. D., "Metallurgical plant wastes and their treatment," *Water Sewage Works* **96**, 399 (1949).
246. Vogel, A., and D. E. Bloodgood, "Cyanide poisoning," in Proceedings of 6th Industrial Waste Conference, Purdue University, February 1951.
247. Vrÿburg, R., "Effect of chromium wastes on sewage plant processes," *Sewage Ind. Wastes* **25**, 240 (1953).
248. Waite, C. F., "Acid, cyanide, chromium wastes, treatment," *Sewage Ind. Wastes* **23**, 697 (1951).
249. Waite, C. F., "Treatment of acid, cyanide and chromium wastes," in Proceedings of 5th Industrial Waste Conference, Purdue University, November 1949.
250. Walker, C. A., and W. Zabban, "Disposal of plating room wastes; treatment of cyanide waste solutions by ion exchange," *Plating* **40**, 165 (1953).
251. Walker, C. A., and W. Zabban, "Disposal of plating room waste solutions with ozone," *Plating* **40**, 777 (1953).
252. Walker, D. J., "Treatment for disposal of spent and contaminated soluble oil mixtures," in Proceedings of 5th Industrial Waste Conference, Purdue University Engineering Extension Series, Bulletin no. 72, November 1949, p. 63.
253. Walker, C. A., and P. W. Eichenlaub, "Disposal of electroplating wastes by Oneida, Ltd.," *Sewage Ind. Wastes* **26**, 843 (1954); Walker, C. A., B. F. Dodge, and J. Madden, **26**, 1002 (1954); Eichenlaub, P. W., and J. Cox, **26**, 1130 (1954).
254. Ware, G. C., "Effect of temperature on the biological destruction of cyanide," *Water Waste Treat. J.* **5**, (1958).
255. Washburn, G. N., "Toxicity of plating wastes to fish," *Sewage Ind. Wastes* **20**, 1074 (1948).
256. "Waste disposal," Editorial, *Chem. Eng.* **56**, 96 (1949).
257. "Waste treatment at Douglas Aircraft," *Plating* **42**, 58 (1955).
258. Watson, K. S., and C. M. Fair, "Electric appliance manufacturing wastes," *Sewage Ind. Wastes* **28**, 49 (1956).
259. Weisberg, L., and E. J. Quinlan, "Recovery of plating wastes," *Plating* **42**, 1012 (1955).
260. Wells, W. N., "Chromium wastes, effects on activated sludge, Grand Prairie, Texas," *Sewage Ind. Wastes* **15**, 798 (1943).
261. "What the steel industry is doing about stream pollution," *Ind. Wastes* **1**, 157 (1956).
262. Williams, R., "Disposal plant description for cyanide wastes," *Sewage Ind. Wastes* **22**, 590 (1950).
263. Wischmeyer, W. J., and J. T. Chapman, "Nickel, effects on sludge digestion," *Sewage Ind. Wastes* **19**, 790 (1947).
264. Wise, W. S., et al., "Industrial wastes, brass and copper industry," *Ind. Eng. Chem.* **39**, 632 (1947).
265. Wise, W. S., "Character and disposal of brass and copper plating wastes," *Sewage Ind. Wastes* **20**, 96 (1948).
266. Young, M. K., "Anionic and cationic exchange for recovery and purification of chrome from plating process waste water," in Proceedings of 11th Industrial Waste Conference, Purdue University, May 1956.

Suggested Additional Reading

The following references have been published since 1962.
1. Albright, P. N., "Method of cyanide waste disposal," *Water Pollution Abstr.* **39**, 1186 (1966).
2. Lawrence, A. W., and P. L. McCarty, "The role of sulfide in preventing heavy metal toxicity in anaerobic treatment," *J. Water Pollution Control Federation* **37**, 392 (1965).
3. "Chemical control of effluents at new plating plant," *Water Pollution Abstr.* **39**, 1187 (1966).
4. Ayers, K. C., et al., "Toxicity of copper to activated sludge," in Proceedings of 20th Industrial Waste Conference, Purdue University, 1965, p. 516.
5. Barth, E., et al., "Summary report on the effect of heavy metals on the biological treatment processes," *J. Water Pollution Control Federation* **37**, 86 (1965).
6. Bloodgood, D. E., "Twenty years of industrial waste treatment," in Proceedings of 20th Industrial Waste Conference, Purdue University, 1965, p. 183.
7. Bon, J. J., "Waste water arising from the treatment of metallic surfaces," *Water Pollution Abstr.* **39**, 779 (1966).
8. Bucksteeg, W., "Decontamination of cyanide wastes by methods of catalytic oxidation and adsorption," in Proceedings of 21st Industrial Waste Conference, Purdue University, 1966, p. 688.
9. Csizi, G., "Possibilities in the use of electrolytic regeneration of waste waters," *Chem. Ingr. Tech.* **36**, 686 (1964).
10. Culotta, J. M., "Treatment of cyanide and chromic acid plating wastes," *Water Pollution Abstr.* **39**, 447 (1966).

11. Cupps, C. C., "Treatment of wastes for auto. bumper finishing," *Ind. Water Wastes* **6**, 11 (1961).
12. Dart, M. C., J. D. Gentiles, and D. C. Trenton, "Electrolytic oxidation of strong cyanide wastes," *J. Appl. Chem.* **13**, 55 (1963).
13. Dart, M. C., "Electrolytic treatment of cyanides," *Chem. Abstr.* **65**, 480 (1966).
14. Demidov, V. I., and V. S. Volodin, "Sorption method of removing cyanide compounds from industrial waste waters," *Tsvetn. Metal.* **37**, 5 (1964).
15. Finkenwirth, W., "Die Aufbereitang saereholtiger galvanischer abwaesser voit ionenaustauschern," *Elektrie* **15**, 92 (1961).
16. Foulke, D. G., "Bibliography of metal finishing wastes for 1962," *Plating* **50**, 1114 (1963).
17. Fromm, P., and R. Stokes, "Assimilation and metabolism of chromium by trout," *J. Water Pollution Control Federation* **34**, 1151 (1962).
18. Geyer, L. C., "Treatment of bearing manufacturing waste," *Ind. Water Waste* **6**, 33 (1961).
19. Grieves, R. B., et al., "Ion flotation for industrial wastes: Separation of hexavalent chromium," in Proceedings of 20th Industrial Waste Conference, Purdue University, 1965.
20. Haider, G., "Heavy metal toxicity to fish," *Chem. Abstr.* **64**, 18084 (1966).
21. Hartung, K. H., "Electroplating waste water," *Galvanotechnik* **55**, 123 (1964).
22. Herpers, H., et al., "Influence of metal salts and other compounds on sewage gas production," *Chem. Abstr.* **65**, 13395 (1966).
23. Hesler, J. C., "Recovery and treatment of metal finishing wastes by ion exchange," *Ind. Water Wastes* **6**, 75, 93, and 138 (1961).
24. Hethwer, E. W., "Plating waste treatment," in Proceedings of 9th Ontario Industrial Waste Conference, 1962, p. 87; *Chem. Abstr.* **64**, 1808 (1966).
25. Howe, R. H. L., "Disposal of toxic chemical wastes having a high concentration of cyanide ion," Eli Lilly & Co.; U.S. Patent no. 3145166, 18 August 1964.
26. Howe, R., "Biological treatment for cyanide and plating waste," *Intern. J. Air Water Pollution* **9**, 463 (1965).
27. Husmann, W., "Purification and removal of mining waste waters from anthracite mines," *Chem. Abstr.* **64**, 435 (1966).
28. Ibl, N., and A. M. Frei, "Electrolytic reduction of chrome in waste water," *Galvanotech. Oberflaechenschutz* **5**, 117 (1964).
29. "Incinerating cyanide slurries," *Wastes Eng.* **32**, 342 (1961).
30. Kantawala, D., and H. D. Tomlinson, "Comparative study of recovery of zinc and nickel by ion exchange media and chemical precipitation," *Water Sewage Works* **111**, 192 (1964).
31. Kongiel-Chablo, I., "Behavior of complex cyanides in natural water with high rate of contamination," *Chem. Abstr.* **65**, 6914 (1966).
32. Krug, J., "Ion exchanger—its suitability and economy for the purification of galvanic wastes," *Galvanotechnik* **55**, 32 (1964).
33. Krusenstjern, A., and L. Axmacher, "Neutralization of plating wastes with calcium and sodium hydroxide," *Metalloberflaeche* **18**, 65 (1964).
34. Lakin, J., "Effluent treatment for small plater," *Electroplating Metal Finishing* **14**, 89 (1961).
35. Lancy, L. E., and W. Zabban, "Analytical methods and instrumentation for determining cyanogen compounds," *Am. Soc. Testing Mater. Spec. Tech. Publ.* **337**, 32 (1963).
36. Lancy, L. E., "Conditioning cyanide compounds," *Water Pollution Abstr.* **39**, 138 (1966).
37. Lancy, L. E., and W. Zabban, "Die Bexiehungen zwischen analyse und behandlung von cyanid haltigem abwasser," *Metalloberflaeche* **17**, 65 (1963).
38. Lieber, H. W., "Water supply of electrodeposition plants," *Metalloberflaeche* **18**, 611 (1964).
39. Ludzack, F. J., and R. B. Schaffer, "Activated sludge treatment of cyanide, cyanatic, and thiocyanate," *J. Water Pollution Control Federation* **34**, 320 (1962).
40. McDermott, G. N., et al., "Nickel in relation to activated sludge and anaerobic digestion process," *J. Water Pollution Control Federation* **37**, 163 (1965).
41. Michalson, A. W., "What's new, practical and important in ion exchange," *Chem. Eng.* **70**, 163 (1963).
42. Murphy, R. S., and J. B. Nesbitt, "Biological treatment of cyanide wastes," *Penn. State Univ. Eng. Res. Bull.* **B-88** (1964).
43. Nehring, D., "Experiments on the toxicological effect of thallium ions on fish and fish-food organisms," *Water Pollution Abstr.* **39**, 813 (1966).
44. "New way to incinerate cyanide wastes," *Air Eng.* **4**, 36 (1962).
45. Nohse, W., et al., "Removal of toxic effluents from the rinsing solution of surface treatment of metal

articles," *Water Pollution Abstr.* **39**, 446 (1966).
46. Ochme, F., et al., "Comparison of the important conventional processes for chemical removal of cyanides," *Chem. Abstr.* **65**, 356 (1966).
47. Page, L. J., "Waste disposal at the Dana Corporation Plant, Toledo, Ohio," in Proceedings of 21st Industrial Waste Conference, Purdue University, 1966, p. 33.
48. Pickering, Q. H., "The acute toxicity of some heavy metals to different species of warm water fishes," in Proceedings of 19th Industrial Waste Conference, Purdue University, 1964.
49. Poston, W. H., "The battle to save Lake Michigan," *Civil Eng.* **37**, December 1967.
50. & 51. Rincke, G., "Pickling and plating waste treatment in particular plants in West Germany," in Proceedings of 21st Industrial Waste Conference, Purdue University, 1966, p. 72.
52. Rothkegel, J., "The treatment of effluents from non-electrolytic wet-chemical metal finishing processes," *Metalloberflaeche* **18**, 97 (1964).
53. Sanders, F. A., "From wings to wastes," *Wastes Eng.* **32**, 180 (1961).
54. Schlegel, H., "Die hydroxidfaellung der schwermetalle in galvanischen abwaessern," *Metalloberflaeche* **17**, 129 (1963).
55. Sobota, J., "Elimination of cyanide compounds from the circulation water as a secondary problem connected with wet dedusting of blast furnaces, gas," *Chem. Abstr.* **65**, 8427 (1966).
56. Salotto, B. V., "Organic load and the toxicity of copper to the activated-sludge process," in Proceedings of 21st Industrial Waste Conference, Purdue University, 1966, p. 1025.
57. Salotto, B. V., et al., "Organic load and the toxicity of copper to the activated sludge process," in Proceedings of 19th Industrial Waste Conference, Purdue University, 1964, p. 1025.
58. Sondak, N. E., and B. F. Dodge, "Oxidation of cyanide bearing plating wastes by ozone," *Plating* **48**, 173 (1961).
59. Spanier, G., "The placement of ion exchange processes in galvanizing operations," *Galvanotechnik* **55**, 409 (1964).
60. Tallmadge, J. A., "Ion exchange treatment of mixed electroplating waste," American Chemists' Society, Division Waste Water Chemistry, Preprints no. 4, 2, 61 (1964).
61. Tallmadge, J., and R. Mattson, "Natural convection and modified diffusion in metal plating," *J. Water Pollution Control Federation* **34**, 723 (1962).
62. Tallmadge, J. A., Jr., and B. A. Buffham, "Rinsing effectiveness in metal finishing," *J. Water Pollution Control Federation* **33**, 817 (1961).
63. Sondak, N. E., and B. F. Dodge, "The oxidation of cyanide-bearing plating wastes, by ozone," *Plating* **48**, 173 (1961); *Water Pollution Abstr.* **34**, 1713 (1961).
64. Tixerant, M., "Recirculation, an answer to effluent problems," *Prod. Finishing* **17**, 83 (1964).
65. Unwin, H. D., "Industrial waste treatment programs and facilities," *Heating Ventilation* **60**, 61 (1963).
66. Vought, J. H., "Monitoring and treatment of cyanide-bearing plating wastes," *J. Water Pollution Control Federation* **39**, 19 (1967).
67. Woodley, R. A. Jr., "Cyanide control experience: Indiana Water Pollution Control Board," in Proceedings of 19th Industrial Waste Conference, Purdue University, 1964.
68. "Literature Review," *J. Water Pollution Control Federation* **38**, 886 (1965).

24.6 Motor Industry Wastes

Origins of motor industry wastes. Stamping plants, body, and final assembly operations contribute 23 per cent of the water intake and use 70 per cent of the manpower of the motor vehicle equipment industry [1967, 1]. In the stamping operation, which produces major body parts, the metal (normally strip or sheet steel) is cut to size and then stamped into the desired shape by large hydraulic presses. Portions of the stamped parts are normally welded together in the stamping operation. From here the parts are sent to the body manufacturing facility. In conventional industry terminology "body" refers to the passenger enclosure from the fire wall back and does not include the front-end parts such as the front fenders and hood. In the body-assembly plant, the body is first constructed from stamped metal parts in the body shop and is then treated and painted in the paint shop. The exterior and interior trim, produced in parts plants, are added in the trim shop.

From the body-assembly operations, the completed body goes to the assembly plant. First, the chassis, wheels, and power train (engine, transmission, etc.) are assembled from parts produced elsewhere; this assembled chassis is joined to the already assembled body; finally, the front-end parts (fenders, hood, etc.) are added. These last items are stamped in a conventional stamping plant and are normally painted in the final assembly plant. The stamping plants are separate from the body and final assembly operations.

Characteristics of motor industry wastes

Stamping plants. These operations produce no significant liquid processing wastes, since only small amounts of water are used directly in processing. However, large amounts of oils (both lubricating and hydraulic) are used, and in many cases some of these find their way into the sewer system. Because these oils originate from a variety of points in the plant and because their introduction into the sewer system is of a miscellaneous nature, the concentration in the plant effluent can vary widely. Amounts of extractable material from 50 mg/liter to several thousand mg/liter have been encountered. The flow of contaminated process water will be, as noted above, quite small, varying from about 2000 to 10,000 gallons per day.

Large amounts of cooling water are used in welding systems. Recirculation is widely practiced, so in most cases the discharge will be restricted to cooling system blowdown. This can, however, represent the major part of the plant water discharge. For typical stamping plants the cooling-system blowdown can vary from 25 to 150 gallons per minute. Powerhouse water (boiler blowdown, boiler-water pretreatment system blowdown, etc.) also represents a source of contaminants. Processes and significant wastes from body stamping and assembly are shown in Fig. 24.15.

Assembly plants. The waste waters discharged from final assembly plants, body-assembly plants, or combined operations are of the same general type, that is, organic waste waters containing suspended solids.

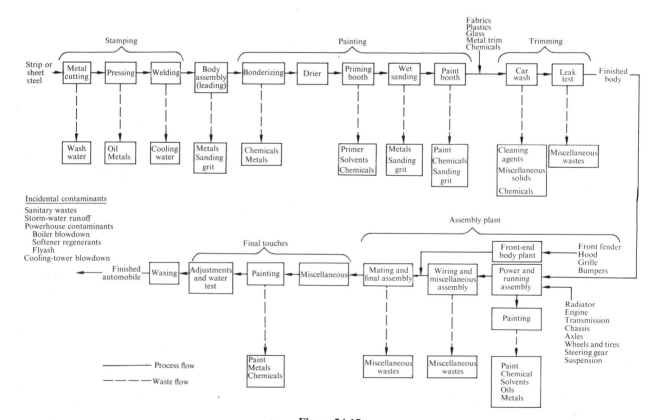

Figure 24.15

These solids originate primarily from the painting and paint-sanding operations. In addition, heavy metals such as zinc and chromium that originate in metal-treating (bonderizing) operations may be present, as well as powerhouse and cooling water, although the amount of this latter waste relative to the overall plant discharge will be considerably less than with stamping plants.

Because the waste waters are primarily organic in nature and contain suspended solids they are similar to sanitary waste water, but the organic content (as measured by BOD and COD tests) and suspended solids will normally be higher than in typical sanitary wastes.

The processes which produce liquid wastes are essentially uniform throughout the industry and do not vary between small and large plants. There are no new processes anticipated that will materially add to the pollution load, but one new process, electrostatic painting, may in the future actually effect a reduction in pollution loading. In a typical painting operation, a water curtain is used in the paint booth to entrap overspray, which otherwise could present an environmental pollution problem. This water is discharged (after a significant amount of reuse) and is a major source of contaminants, primarily organic materials and suspended solids. With electrostatic painting there is less overspray, in fact the water curtain can sometimes be eliminated entirely. It appears unlikely that this technique will be widely used in the foreseeable future to replace existing equipment, but it will probably be tried where new facilities are constructed. Thus, it will probably not make a significant contribution within the next decade, but may become increasingly important thereafter.

The average amount of waste produced per 100 cars for body assembly and final assembly is summarized as follows:

	lb/100 cars
Chemical oxygen demand (COD)	1007.77
Biochemical oxygen demand (BOD)	322.33
Hexavalent chromium (CrO_4)	4.50
Trivalent chromium (CrO_4)	2.08
Zinc (Zn)	1.12
Suspended solids	360.30
	gal/100 cars
Flow	201,958

Treatment of motor industry wastes

Stamping plants. Stamping-plant treatment systems are of several types. If the cooling water and powerhouse water do not require treatment and if the concentrated waste can be collected separately, the treatment system will usually consist of a batch system for removal of oil and suspended solids. Alternatively, incineration has been used for the concentrated, oil-containing waste water.

If, on the other hand, the cooling water and powerhouse water require treatment or if the concentrated waste water cannot be separated from the general plant collection system, an end-of-line facility for removal of suspended solids and oil is dictated. The overall efficiency for the removal of suspended solids and oil is in the range of 85 to 95 per cent.

Assembly plants. Typical waste-treatment facilities for assembly plants incorporate chemical clarification followed by conventional biological treatment, such as the activated-sludge process. Provision for reducing hexavalent chromium is incorporated in the chemical treatment, and trivalent chromium and other heavy metals are removed in the clarification step. The removal of heavy metals including chromium is essentially complete. Normal efficiencies for removal of organic material (as measured by BOD and COD tests) are in the range of 80 to 95 per cent and for suspended solids in the range of 85 to 95 per cent. Several modifications of this general approach can be used and these are detailed in a recent report.*

Since the waste waters from assembly operations are basically organic in nature, they can be sent to municipal sewage-treatment facilities. However, it is usually necessary first to adjust the pH and remove the heavy metals, including chromium. Also, since the concentrations of COD, BOD, and suspended solids are usually higher than in conventional sanitary waste, pretreatment to remove excess suspended solids (including some organic material) is normally provided. After such pretreatment, it is the general practice of the industry to discharge to municipal systems where possible rather than to provide secondary biological treatment on the plant site.

---

*_The Cost of Clean Water_, Motor Vehicles and Parts, Industrial Waste Profile no. 2, Vol. III, U.S. Department of the Interior, Washington, D.C. (1967).

24.7 Iron-Foundry Wastes

The waste from most small gray-iron foundries is dry, being composed of solids from molding and core sands and fly ash. Foundries produce castings from molten metal, which are then machined to final specifications. The major waste resulting from the formation of rough castings from molten metal is used sand. Often the sand is conveyed to a disposal site by flushing with water, in a procedure similar to that used in conveying flue dust from steel mills and coal dust from coal-mining operations. Disposing of the used sand is a difficult problem, since it requires considerable land area and working time. Foundry sand, normally new Ottawa sand, possesses specific sieve analyses and is quite expensive, so its recovery ought to be considered. The waste material from used molds consists of about 85 to 90 per cent sand, the remainder being clay, sea coal, and so forth. The suspended solids concentration varies between 2500 and 5000 ppm.

Most waste-treatment measures include some method of sand reclamation. One processor [5] recommends a filtering unit to remove excess water from the sand. The sand then enters a drier, which contains a firing unit at the end where the sand enters. The processor states that the performance of the reclamation unit has been most satisfactory, not only from an operating standpoint, but also as an investment, showing immediate returns. Typical screenings of the sand as purchased and the sieve analyses of the sand after reclamation are given in Table 24.22. The analyses of the reclaimed sand after screening show that it is comparable to the new sand.

Another waste-treatment system, set up according to the flow diagram in Fig. 24.16, consists of primary sedimentation prior to reuse of the water. The excess over that required for reuse is further clarified by chemical treatment and sedimentation. This effluent is subjected to a flotation treatment before final disposal, and the sludge is dewatered by filtration. With this method, it was found that, although the sand waste contained 3760 ppm suspended solids, this figure was reduced to 363 ppm after primary settling, 30 ppm after the clarifier, and 10 ppm after flotation. The 13 per cent solids sludge, filtered at a rate of 13 pounds per square foot per hour, produced a cake of $\frac{1}{4}$-inch thickness and a filtrate with 900 ppm suspended solids.

Bader [1] describes a new foundry with a modern industrial-waste treatment system utilizing gravity separation of oils, a holding basin, air flotation, and sludge disposal (Fig. 24.17). The air-flotation equip-

Table 24.22 Comparative sieve analyses of new and reclaimed sand [5].

Sieve mesh	Ottawa sand, new	Reclaimed sand
20		0.08
30	0.04	0.30
40	2.04	5.84
50	32.30	41.08
70	44.90	36.44
100	16.80	13.26
140	2.84	1.68
200	0.20	0.68
270	0.30	0.02
Pan	0.04	0.02

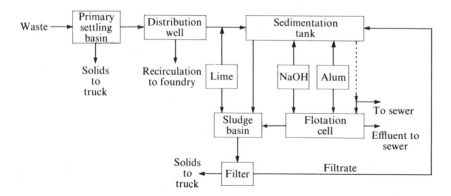

Fig. 24.16 Flow diagram of recovery of water and treatment of residual gray-iron foundry waste.

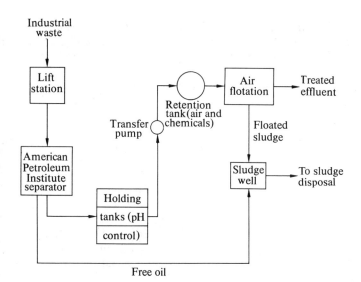

Fig. 24.17 Schematic plan of industrial waste system and description of various stages of waste treatment. This system handles all the liquid wastes except sanitary, dust collection, and cooling.

ment provides the secondary treatment necessary because the separator (designed by the American Petroleum Institute) does not completely remove emulsified oil, small particles of free oil, and suspended oil. The system pressurizes the entire waste flow and consists of a transfer pump, retention tank, flotation unit, and related air-supply and chemical-feed accessories. Compressed air, alum, and a coagulant are added to the retention tank to improve removal efficiencies. After treatment, floated sludge is discharged to a sludge well and treated liquid is drawn from the bottom of the flotation compartment and either discharged to the spent-cooling-water sewer or returned to the holding tank for additional treatment.

References: Iron-Foundry Wastes

1. Bader, A. J., "Complete waste treatment system designed for new foundry," *Plant Eng.* 118 (1968).
2. Chapple, H., "Wet foundry sand reclamation," *Ind. Wastes* **1**, 121 (1911).
3. Hartman, C. D., "Waste water control at a midwest finishing mill near Lake Michigan," in Proceedings of the 11th Ontario Industrial Waste Conference, June 1964, p. 85.
4. Hartman, C. D., et al., "Waste water control at Midwest Steel's new finishing mill," *Iron Steel Eng.* **40**, 182 (1963).
5. "Industrial waste treatment in foundry industry," *Mod. Castings* **44**, 471 (1963).
6. Hathaway, C. W., et al., "Treatment of machine shop and foundry wastes," *Sewage Ind. Wastes* **26**, 1363 (1954).
7. Hathaway, C. W., et al., "Treatment of gray iron foundry waste water," *Ind. Wastes* **1**, 166 (1956).

LIQUID-MATERIALS INDUSTRIES

The major liquid-materials industries are here categorized as oil, rubber, and glass manufacturing. We shall discuss the waste problems of each one separately in the pages which follow.

24.8 Oil-Field and Refinery Wastes

Origin of oil-field and refinery wastes. Oil wastes can be classified into those originating from (1) oil production and (2) oil refining. Wastes result from pumping, desalting, distilling, fractionation, alkylation, and polymerization processes; they are of large volume and contain suspended and dissolved solids, oil, wax, sulfides, chlorides, mercaptans, phenolic compounds, cresylates, and sometimes large amounts of dissolved iron. Treatment is by scrubbing with flue gas, evaporation, flotation, mixing, aeration, biological oxidation, coagulation, centrifugation, and incineration.

Generally, petroleum contains about 85 per cent carbon and 12 per cent hydrogen. The remaining 3 per cent is composed of small amounts of oxygen, nitrogen, and sulfur. Some products and by-products of oil refining are gasoline, kerosene, lubricants, gas oil and fuel oil, wax, asphalt, petroleum coke, and miscellaneous materials such as petrolatum and insecticides.

Crude oil is refined by fractional distillation to separate the various hydrocarbons, by application of heat and pressure (with or without catalysts) to alter the molecular structure of some of the distillation products, and by chemical and mechanical treatment of various fractions or products to remove impurities. The crude oil is first passed through a pipe still into a fractionating tower where the lighter products—gasoline, kerosene, and gas oil—are taken off and condensed. The gasoline and kerosene then pass through tanks in which sulfuric acid, caustic soda, plumbite, and water washes are applied to remove impurities. The gas oil, after certain proprietary treatment, is stored for sale as light fuel oil. Distillation of the remainder is continued in the fractionating tower, to take off lubricating or wax distillates. After all products have been taken off as distillates, the residue in the vacuum chamber is used for the manufacture of asphalt.

Wastes from oil fields are drilling muds, salt water, free and emulsified oil, tank-bottom sludge, and natural gas. Of these, salt water (brine) presents the most difficulty. Many oil-bearing strata have brine-bearing formations directly over or under them. Pumping rates are controlled and some of the areas of the well are sealed off to prevent these briny waters from seeping into the oil. However, this can never be completely accomplished, and brine is often pumped out of the well with the oil. The brine and oil must then be separated by gravity and the brine disposed of.

Wastes from oil refineries include free and emulsified oil from leaks, spills, tank draw-off, and other sources; waste caustic, caustic sludges, and alkaline waters; acid sludges and acid waters; emulsions incident to chemical treatment; condensate waters from distillate separators and tank draw-off; tank-bottom sludges; coke from equipment tubes, towers, and other locations; acid gases; waste catalyst and filtering clays; special chemicals from by-product chemical manufacture; and cooling waters. Oils from leaks and spills can amount to as much as 3 per cent of the total crude oil treated. The treatment of oils with alkaline reagents to remove acidic components and the sweetening processing of oils to convert or remove mercaptans produce a series of alkaline wastes which give off obnoxious odors and present difficult, and costly, waste-disposal problems; as do the acid sludges resulting from sulfuric-acid treatment of oils. A typical flow sheet of the petroleum-refining process, showing the above wastes, is given in Fig. 24.18.

Characteristics of oil-refinery wastes. Like most industries, oil refineries use enormous quantities of water. Virtually every refinery operation, from primary distillation through final treatment, requires large volumes of process and cooling waters. The demand is estimated by the National Association of Manufacturers to be 770 gallons per barrel of crude oil. In 1952 Giles [48] estimated that the oil refineries used 5400 mgd of water (based on a national crude-oil figure of 7 million barrels per day). This volume of water used was second only to that of the steel industry and represented about 20 per cent of the total industrial consumption in the United States and slightly less than 50 per cent of municipal needs.

Giles also gave data on waste characteristics of a typical oil refinery (Table 24.23). Analyses of oil-field brine wastes in four major oil-producing states are given in Table 24.24. The Ohio River Survey [69] gives a more complete analysis of the mineral content of five Kansas oil-field brines (Table 24.25).

A recent study by the U.S. Department of the Interior [1967, 65] describes the petroleum refinery "as a complex combination of interdependent processes and operations, many of which are complex in themselves." The reader is referred to Fig. 24.18 for a flow diagram of a typical refinery and to this report for complete and detailed descriptions.

Although the sheer volume of wastes from oil refineries looms as so large a problem, the American Petroleum Institute reports that 80 to 90 per cent of the total water used by the average refinery is for cooling purposes only and is not contaminated except by leaks in the lines. The combined refinery wastes, however, may contain crude oil, and various fractions thereof, and dissolved or suspended mineral and organic compounds discharged in liquors and sludges from the various stages of processing. The oil may appear in waste waters as free oil, emulsified oil, and as a coating or suspended matter, though ordinarily not in proportions greater than 100 ppm. However,

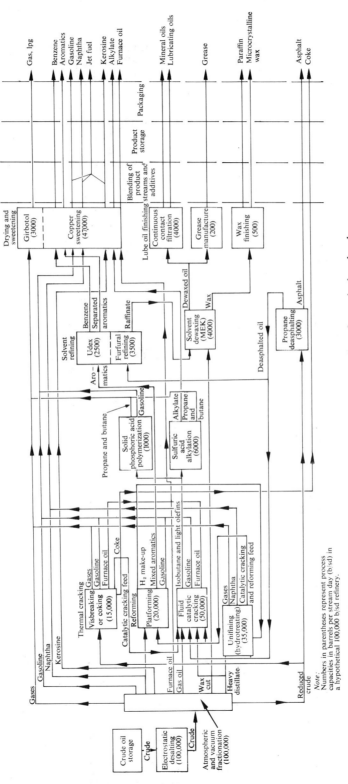

Fig 24.18 Subprocess series representative of a typical technology.

Table 24.23 Data for typical oil refinery. (After Giles [48].)

Production, water, and waste character	Value
Crude run, bbl/day	20,500
Water usage, mgd	16
Effluent water quality	
Oil content, bbl/day	16
Phenolics, lb/day	380
BOD, population equivalent	31,500

Table 24.24 Oil-field brines.

State	Total solids, ppm	Ca and Mg, ppm	Na and K, ppm	Cl, ppm
Illinois	113,000	50,000	38,000	69,000
Kentucky	251,000	220,000	57,000	137,000
Oklahoma	236,000	130,000	77,000	145,000
Texas	69,000	1,800	24,000	40,000

Table 24.25 Mineral analyses of oil-field brine samples [69].

Radical	Concentration range, ppm
Calcium	1507–12,888
Magnesium	346–4290
Sodium	8260–63,275
Bromide	32–633
Carbonate	0
Bicarbonate	43–644
Sulfate	0–1578
Chloride	12,750–127,220
Total solids	25,210–248,600

floating oil is visible even when present in very small concentrations, because of its ability to spread in very thin (0.000003 inch), nondestructible layers. The American Petroleum Institute has classified waste constituents according to the refinery units from which they are released (Table 24.26). Weston [112] gives the sanitary characteristics of 20 typical refinery wastes (Table 24.27).

Treatment of oil-field and refinery wastes. Several methods of disposing of brine wastes have been used, including solar evaporation of impounded brine, controlled diversion of brine into surface waters, recovery of mineral salts, and injection of the brine into subterranean formations; in spite of all these, there is sometimes no alternative to the payment of damages to downstream riparian owners through claim-adjustment associations. The only procedure finding wide use and acceptance for other than coastal fields is disposal by injection. A drawback is that brines must be pretreated prior to subterranean injection to prevent equipment from being corroded and to prevent plugging of the sand; it often becomes apparent that the pretreatment required may be so extensive and costly that it is better to discharge the brines elsewhere after some other pretreatment. Chemical coagulation of brine wastes with alum is also practical prior to discharging them into surface waters.

The main remedial measures for reducing refinery wastes are: (1) reduction of oil leakage by preventive maintenance of pipelines and equipment; (2) preventing formation of oil emulsions or, where these exist, isolation and separate treatment; (3) removal of floating oil in separators located as close to the original source of waste as possible; (4) isolation and separate treatment of objectionable wastes, e.g. with American Petroleum Institute (API) separators, which reduce the BOD to 5 to 10 ppm, provided emulsions are not present.

Acidification of caustic wastes (with H_2SO_4) removes some of the objectionable compounds. Acid sludges may be used as a source of fuel or to produce by-products such as oils, tars, asphalts, resins, fatty acids, and chemicals. Some refineries recover sulfuric acid from the acid sludges for their own use. Giles [48] gives the reduction of sanitary constituents of refinery wastes after chemical and biological treatment (Table 24.28).

Weston [140] points out that the method of handling petroleum-industry product losses and by-products is an economic problem in which, in some cases, the choice will lie between utilization of the wastes for by-products and proper industrial-waste treatment prior to disposal.

Beychok in an excellent new text [95] on refinery wastes presents a considerable amount of basic

Table 24.26 Oil refinery waste: classification of substances found in wastes from the oil refinery industry as prepared by the American Petroleum Institute.*

Refinery unit	Native solutes and those present in end-products	Solutes resulting from chemical reactions	Naturally occurring emulsoids and suspensoids, and those persisting in end-products	Suspensoids and emulsoids resulting from chemical and physical actions
1. Oil storage	Organic sulfur compounds† Acids: H_2S, CO_2, organic acids Inorganic salts: NaCl, $MgCl_2$, Fe and Al compounds, $CaCl_2$, $(NH_4)_2S$, etc.		Suspended matter in tank bottoms Insoluble salts, SiO_2, $Al_2(SiO_3)_3$, S, finely divided substances Asphaltic compounds (in some cases)	
2. Distillation A. Straight distillation	Organic nitrogen compounds‡	Inorganic salts, sulfites, acid sulfites, Na_2CO_3, $(NH_4)_2S$, Na_2S, sulfates, acid sulfates	Insoluble organic and inorganic salts, S compounds, sulfonic and naphthenic acids, and insoluble mercaptides	Oil-water emulsions from steam in towers, etc.
	Organic sulfur compounds†	Acids and alkalis: H_2S, NaOH, NH_4OH, $Ca(OH)_2$		Soaps
	Phenol and like compounds Naphthenic acids	$(NH_4)_2SO_4$, $(NH_4)_2S$, NH_4Cl		Waxy emulsions Oxides of metals
B. Cracking and distillation	Same as 2.A	Same as 2.A with the addition of phenols and phenolic compounds	Suspended coke Insoluble salts, FeS, and SiO_2	Same as 2.A
3. Treating A. Sweetening, sulfuric acid, neutralization	Organic sulfur compounds† Organic nitrogen compounds‡ Naphthenic acids	Organic sulfur compounds Phenolic and sulfonate compounds Weak H_2SO_4 and other acid solutions	Suspended matter: Pbs, S, S compounds Acid and alkaline sludges Polymers and resins	Waxy emulsions Oil-water emulsions Inorganic salts: PbS, $CaSO_4$, $CaHPO_4$

Table 24.26 (*continued*) Oil refinery waste: classification of substances found in wastes from the oil refinery industry as prepared by the American Petroleum Institute.*

Refinery unit	Native solutes and those present in end-products	Solutes resulting from chemical reactions	Naturally occurring emulsoids and suspensoids, and those persisting in end-products	Suspensoids and emulsoids resulting from chemical and physical actions
	Phenylates	Weak alkaline solutions		Soaps
		Soaps		
		Inorganic salts: $CaCl_2$, Na_2CO_3, Na_2SO_4, NaCl		Oxides: PhO, Fe_2O_2
		Oxides dissolved in alkaline solutions as PbO, CuO, etc.		
B. Clay		Organic sulfur compounds	Suspended clay, earth	Suspended clay, earth
	Same as 3.A	Phenolic and sulfonate compounds	Polymers and resins	SiO_2, H_2SiO_3, $Al(OH)_3$
		Weak H_2SO_4 and other acid solutions		
		Inorganic salts: $CaCl_2$, Na_2CO_3, Na_2SO_4, NaCl		
4. Recovery A. Gas purification and recovery	Organic sulfur compounds†	Inorganic salts: sulfates, acid sulfates, sulfites, acid sulfites, FeS, $(NH_4)_2S$, Na_2CO_3, Na_2S	Insoluble S compounds and mercaptides	Suspended Fe and S compounds
	Organic nitrogen compounds‡	Mercaptides		

24.8 OIL-FIELD AND REFINERY WASTES 441

B. Acid recovery	Sulfonates	Inorganic salts, and H_2SO_4, SO_2, SO_3	Acid sludges
	Mineral acids	Organic esters	
	Organic nitrogen compounds‡		
5. Miscellaneous			
A. Cooling- and boiler-water treating		Inorganic salts: $BaCl_2$, $NaCl$, $NaHCO_3$, Na_2SiO_3, $CaCl_2$, $MgCl_2$, Na_2CO_3, Na_2SO_4, Na_2HPO_4, $CaHPO_4$, etc.	Insoluble and colloidal compounds: $CaCO_3$, $BaCO_3$, $Ca(OH)_2$, $Mg(OH)_2$, $Ba(OH)_2$, $Ca_3(PO_4)_2$
B. Fire protection		Inorganic salts: $NaHCO_3$, Na_2SO_4, $Al_2(SO_4)_3$ Organic compounds	

* Asphalt and lubricating-oil units have not been included in the separate headings. Straight-run distillation involves the heavier oils, also sulfuric-acid treating embraces that of lubricating oils as well as light oils.
† Under this caption are included mercaptans, dialkyl sulfides, sulfonates, sulfonic acids, some alkyl and aryl sulfides, etc. Only a few of these compounds will be found in any one type of oil.
‡ Under this caption are included amines, some amides, quinolines, and pyridines. Only a few of these compounds will be found in any one type of oil.
§ Some alkyl sulfides, thiophenes, etc. Not all of these compounds are found in any one type of oil.
** Quinolines, some amides, pyridines, and some aryl amines. Not all of these compounds are found in any one type of oil.

Table 24.27 Characteristics of typical refinery wastes. (After Weston [112].)

Characteristic	Type of waste					
	Water layer	Water layer	Milk-water emulsion	Emulsion	Emulsion	Emulsion
Source of waste	Slop oil treatment	Slop oil treatment	Water wash	Bar condenser	Jet vacuum pump	Desalting
Type of unit	Spent caustic and heat, plant scale	Vacuum pre-coat filter, pilot scale	Treating	Combination unit	Lube oil vacuum still	Desalting unit
Quantity of waste			1 bbl/bbl product			
Acidity, ppm						
Alkalinity, ppm	5660–14,440	77–153	15,313	59.5	520	739
Ammonia, ppm				4.1	225	2.0
BOD, ppm	22,000–56,000	500–1360	7,900		425	404
COD, ppm			86,775	72.3		865–3031
Odor threshold				1.0		1.8
Oil, ppm	4900–10,300	37–130	31,600	236	94.3	32–713
pH	10–10.2	6.9–7.7	9.79	7.22	7.03	9.26
Phenol, ppm				2.3		4.1
Sulfide, ppm				1.3		
Suspended solids, ppm	60–940	30–139				
TLM, 24-hr*						

Table 24.27 (*Continued*)

Characteristic	Type of waste					
	Condensate	Condensate	Condensate	Acid	Acid	Spent caustic
Source of waste	Low sulfur gas separator	Viscous breaker gas separator	Light oil recovery	Unit sewer	Unit sewer	Alkylate wash
Type of unit	Combination unit	Combination unit		Alkylation unit	H_2SO_4 sludge conversion unit	Chemical manufacturing
Quantity of waste						0.21 lb NaOH used per lb product
Acidity, ppm		2963	69–13,175	1,105–12,325	1140–10,050	
Alkalinity, ppm	1518	130–999	3–8350			46,250
Ammonia, ppm	500	3040	55–9500	1.2	2–13	
BOD, ppm	408	7239	214–16,255	31	10–272	256
COD, ppm	1204			1251	910	3,230
Odor threshold	2.5	4.5	3.5	3.7	3–5	1.5
Oil, ppm	3	2.5	6–230		124	10
pH	8.5	7.85	5.0–9.2	131.5	1.71	12.8
Phenol, ppm	0.06	156	0–213	0.6–1.9		50
Sulfide, ppm	600	1500	T–5000			2
Suspended solids, ppm			14			253
TLM, 24-hr*				0.4	3	

Table 24.27 (continued) Characteristics of typical refinery wastes. (After Weston [112].)

Characteristic	Spent caustic	Spent caustic	Carbolate	Alkaline	Special chemicals	Special chemicals	Refinery separator inlet
Source of waste					Unit sewer	Unit sewer	
Type of unit	Catalytic polymerization	Naphtha wash Treating	Naphtha wash Fluid catalytic cracking unit	Shut-down Caustic methanol sweetening	Alkylation sulfonation	Detergent drying	
Quantity of waste	0.8 lb NaOH used per bbl product						9.6 mgd
Acidity, ppm							0–188
Alkalinity, ppm	209,330	80,020	247,900		18–245	150	0–92
Ammonia, ppm					1.8	0.2–12.6	
BOD, ppm	8,440	51,154	363,600		28–151	8–1,180	96–501
COD, ppm	50,350	144,120	901,200	371–299,000	259–5382	2,585–51,350	
Odor threshold					1.2		3.25–5.32
Oil, ppm	T–12	13.4	13 +		13–1310	40–7,750	3900–6500
pH	12.9			8–58	4.6	7.37–9.31	2.4–6.2
Phenol, ppm	22.2	23,312	309,300	9.5–12.5	0–7.4		0.5–7.9
Sulfide, ppm	3,060		0–3,380				
Suspended solids, ppm	54–279		678				131–678
TLM, 24-hr		0.04					33

*Approximate 24-hour median tolerance limit for bluegill sunfish, expressed as percentage concentration of waste.
†Waste represents combined discharges from a complete refinery, with the exception of naphtha-treating wastes.

Table 24.28 Reduction in refinery effluent contaminants by secondary treatment [48].

Characteristic	Reduction, %	
	Chemical flocculation	Biological treatment
Odor	95	90
Turbidity	93	85
Oil	90	75
Suspended solids	65	45
BOD	50	85
Phenolics	0	90

theories and practical formulations, tables, and graphs, as well as calculations to enable the engineer to predict composition and quantities of contaminants in various wastes. He discusses in detail three major methods of treatment: (1) in-plant pretreatment, (2) API and similar separators, and (3) secondary treatment following the use of separators. He also includes a much-needed chapter on cost data. The reader is urged to refer to this text for a comprehensive treatment of petrochemical processing and wastes.

The U.S. Department of the Interior [1967, 65] reports that the waste-treatment methods applicable to petroleum refineries can be divided into five types: physical, chemical, biological, tertiary, and special in-plant methods. (1) Physical methods include gravity separators, air flotation (without chemicals), and evaporation. Gravity separators (API and earthen basins), which are used in practically all refineries, are designed primarily for removal of floatable oil and settleable solids. They remove 50 to 99 per cent of the separable oil and 10 to 85 per cent of the suspended solids, as well as BOD, COD, and phenol, at times to a substantial degree depending on the influent waste-water characteristics. Air flotation without chemical addition obtains comparable results. Pollutant removals by evaporation ponds are very high, but application of this method is severely limited by location, climate, and land-availability considerations. (2) Chemical methods (coagulation-sedimentation and chemically assisted air flotation) are more effective in oil and solids removal, particularly in respect to emulsified oil. (3) Biological methods include activated sludge, trickling filters, aerated lagoons, and oxidation ponds. In general these treatment processes require waste-water pretreatment to remove oil and remove or control other conditions (such as pH and toxic substances). The activated-sludge process is the most effective for removal of organic materials (the main purpose of biological treatment); expected removal efficiencies are 70 to 95 per cent for BOD, 30 to 70 per cent for COD, and 65 to 99 per cent for phenols and cyanides. (4) Tertiary treatment to date has been limited to activated carbon and ozonation, which are effective in removing taste and odor and refractory organic substances from biologically treated waste waters. (5) The most important in-plant treatment methods are sour water stripping, neutralization and oxidation of spent caustics, ballast-water treatment, slop-oil recovery, and temperature control. These measures substantially reduce the waste loadings in the influent to general refinery treatment facilities and are necessary to ensure reasonable performance of these facilities.

References: Oil-Refinery Wastes

1. Aeschliman, P. D., R. P. Selm, and J. Ward, *Study of a Kansas Oil Refinery Waste Problem*, McPherson Refinery, National Cooperative Refinery Association (1957).
2. American Petroleum Institute, *Chemical Wastes*, Vol. III of *Manual on Disposal of Refinery Wastes*, 2nd ed. (1951).
3. American Petroleum Institute, *Waste Water Containing Oil*, Vol. I of *Manual on Disposal of Refinery Wastes*, 5th ed. (1953).
4. Anderson, C. O., "Oil refinery waste treatment as practiced by Rock Island Refinery Corp., at Indianapolis, Ind.," *Ind. Wastes* **2**, 1959 (1957).
5. Austin, R. J., "Sulfur dioxide and coke recovery from acid refinery sludge," *Sewage Ind. Wastes* **19**, 1108 (1947).
6. Austin, R. J., et al., "Treatment of oil-containing waste waters, trickling filter studies," *Sewage Ind. Wastes* **26**, 1057 (1954).
7. Austin, R. J., W. F. Meehan, and J. D. Stockham, "Operations of experimental trickling filters on oil-containing waste waters," in Proceedings of 8th Industrial Waste Conference, Purdue University, May 1953.

8. Austin, R. J., and E. H. Vause, "Chemical flocculation of refinery waste water," Purdue University Engineering Extension Series, Bulletin no. 76, 1951, p. 272.
9. Baker, R. A., and R. F. Weston, "Biological treatment, refinery wastes," *Sewage Ind. Wastes* **28**, 58 (1956).
10. Barr, B. A., "Reclamation of oil in a large industry," in Proceedings of 4th Industrial Waste Conference, Purdue University, September 1948.
11. Bertram, F. W., *et al.*, "Chemical oxygen demand of petrochemical wastes, modification of the standard catalytic reflux procedure," *Anal. Chem.* **30**, 1482 (1958).
12. Biehl, J. A., "Reduction of phenol in wastes from catalytic petroleum processes," in Proceedings of 7th Industrial Waste Conference, Purdue University, May 1952.
13. Bland, W. F., "Laboratory waste control, Atlantic Refining Company," *Sewage Ind. Wastes* **19**, 132 (1947).
14. Bloodgood, D. E., "Petroleum wastes problem," *Sewage Ind. Wastes* **19**, 607 (1947).
15. Bloodgood, D. E., and W. F. Kelleher, "Fundamental studies on the removal of emulsified oil by chemical flocculation," in Proceedings of 7th Industrial Waste Conference, Purdue University, May 1952.
16. Borroughs, L. C., "Segregation and treatment," *Sewage Ind. Wastes* **30**, 57 (1958).
17. Bowers, F. J., *et al.*, "Oil-water separation, electrical method," *Sewage Ind. Wastes* **3**, 286 (1931).
18. Brady, S. D., "Effluent improvement program at Humble's Baytown refinery," in Proceedings of 9th Industrial Waste Conference, Purdue University, May 1954.
19. Brandel, R. J., "Cooling water recirculation," *Sewage Ind. Wastes* **29**, 1409 (1957).
20. Brannan, T. F., J. E. Etzel, and D. E. Bloodgood, "Uses of lime in treating oil wastes," in Proceedings of 9th Industrial Waste Conference, Purdue University, May 1954.
21. Brink, R. J., "The new Buick waste disposal and oil reclamation system," in Proceedings of 12th Industrial Waste Conference, Purdue University, May 1957.
22. Brown, G. W., and J. E. Sublett, "Union Oil Company builds new waste water facilities," *Ind. Wastes* **2**, 6 (1957).
23. Brown, K. N., and S. Kurtis, "Difficult waste disposal problem solved," *Sewage Works J.* **22**, 1507 (1950).
24. Buck, W. B., "Progress made by the oil industry of Oklahoma in the disposal of brine," in Proceedings of 13th Industrial Waste Conference, Purdue University, May 1958.
25. Calise, V. J., "Removing oil from water by flocculation and filtration," *Power* **98**, 112 (1954).
26. Carbone, W., L. T. Hartman, and J. J. Lawton, "Continuous chemical purification of coke oven light oil, including thiophene removal," in Proceedings of 6th Ontario Industrial Waste Conference, June 1959, p. 265.
27. Coulter, J. B., *et al.*, "Emulsified oil waste—an Air Force problem," in Proceedings of 11th Industrial Waste Conference, Purdue University, Engineering *Bulletin* no. 41, 1957, p. 99; Engineering Extension Series no. 91.
28. Crosby, E. S., W. Rudolfs, and H. Heukelekian, "Biological growths in petroleum refinery waste waters," *Ind. Eng. Chem.* **46**, 296 (1954).
29. Daugherty, F. M., "Oil well drilling chemicals, effects on marine animals," *Sewage Ind. Wastes* **23**, 1282 (1951).
30. Dickerson, B. W., W. T. Laffey, and R. O. McNeil, "Treatment of waste from tall oil refining," *Sewage Ind. Wastes* **31**, 141 (1959).
31. Elkin, H. F., "Activated sludge, applications," *Sewage Ind. Wastes* **28**, 1122 (1956).
32. Elkin, H. F., "Biological oxidation of oil refinery wastes in cooling tower systems," in Proceedings of 4th Ontario Industrial Waste Conference, June 1957, p. 62.
33. Elkin, H. F., "Condensates, quenches and wash waters as petrochemical waste sources," *Sewage Ind. Wastes* **31**, 836 (1959).
34. Elkin, H. F., "Successful initial operation of water reuse at refinery," *Ind. Wastes* **1**, 75 (1955).
35. Elkin, H. F., "Waste treatment at the Sun Oil Company refinery, Sarnia, Ontario," in Proceedings of 3rd Ontario Industrial Waste Conference, June 1956, p. 80.
36. Elkin, H. F., and W. E. Soden, "Gravity separation of oil," *Sewage Ind. Wastes* **26**, 854 (1954).
37. Ettinger, M. B., K. J. Leshka, and R. C. Kroner, "Persistence of pyridine bases in polluted water," *Ind. Eng. Chem.* **46**, 791 (1954).
38. Etzel, J. E., T. F. Brannan, and D. E. Bloodgood, "New economies possible in treating oil emulsion

wastes by chemical coagulation," *Ind. Wastes* **1**, 237 (1956).
39. Fellows, F. G., "Removing oils, tars, alkalis from refinery waste," *Wastes Eng.* **22**, 468 (1951).
40. Fiske, C. E., "Air flotation for refinery wastes," *Sewage Ind. Wastes* **27**, 1317 (1955).
41. Fowler, I. A., "The waste-disposal program of the Sinclair Refining Company, East Chicago, Ind.," in Proceedings of 5th Industrial Waste Conference, Purdue University, November 1949.
42. Frame, J., "Approach to treatment of refinery wastes," in Proceedings of 5th Ontario Industrial Waste Conference, May 1958, p. 59.
43. Frame, J. D., "From refinery wastes to pure water at the Cities Service Trafalgar refinery," in Proceedings of 6th Ontario Industrial Waste Conference, June 1959, p. 171.
44. Frame, J. D., "Refining refinery wastes," *Wastes Eng.* **30**, 38 (1959).
45. Garrett, J. T., "Tars, spent catalysts, and complexes as petrochemical waste sources," *Sewage Ind. Wastes* **31**, 841 (1959).
46. Giles, R. N., "Oil-water separation," *Sewage Ind. Wastes* **24**, 801 (1952).
47. Giles, R. N., "Oil-water separation," in Proceedings of 6th Industrial Waste Conference, Purdue University, February 1951.
48. Giles, R. N., "A rational approach to industrial waste disposal problems," *Sewage Ind. Wastes* **24**, 1495 (1952).
49. Gilliam, A. S., and F. C. Anderegg, "Biological disposal of refinery wastes," in Proceedings of 14th Industrial Waste Conference, Purdue University, May 1959.
50. Gothard, N. J., and J. A. Fowler, "Liquid industrial wastes: petroleum refineries," *Ind. Eng. Chem.* **44**, 503 (1952).
51. Gothard, N. J., and J. A. Fowler, "Petroleum refineries," *Ind. Eng. Chem.* **44**, 503 (1952).
52. Gleason, R. J., "Waste-oil treatment and disposal practices in the 13th naval district," in Proceedings of Conference on Industrial Wastes, Engineering Experiment Station, University of Washington, April 1949, p. 48.
53. Granville, M. F., "Brine disposal, East Texas," *Sewage Ind. Wastes* **17**, 1295 (1945) and **18**, 589 (1946).
54. Harris, T. R., "Disposal of refinery waste sulphuric acid," *Ind. Eng. Chem.* **50**, 81A (1958).
55. Hart, W. B., "A case history of a refinery waste," *Ind. Eng. Chem.* **49**, 107A (1957).
56. Hart, W. B., "Flocculation as a treatment for petroleum refinery waste water," *Ind. Eng. Chem.* **49**, 77A (1957).
57. Hart, W. B., "Flocculation of oil refinery wastes," *Ind. Eng. Chem.* **49**, 63A (1957).
58. Hart, W. B., "Pollution abatement in the petroleum industry," *Ind. Wastes* **1**, 110 (1956).
59. Hart, W. B., "Refinery waste disposal," *Sewage Ind. Wastes* **18**, 1245 (1946), **19**, 133, 541, 708, 956, and 1111 (1947), **20**, 363 and 596 (1948).
60. Hart, W. B., "Treatment of oil industry wastes," *Sewage Ind. Wastes* **17**, 307 (1945).
61. Henderson, G. R., "Waste disposal at Sun Oil Co., Ltd., Sarnia, Ontario, refinery," in Proceedings of 2nd Ontario Industrial Waste Conference, June 1955, p. 1.
62. Hill, J. B., "Refinery waste problems," *Sewage Ind. Wastes* **12**, 401 (1940).
63. Hill, J. B., "Waste problems in the petroleum industry," *Ind. Eng. Chem.* **31**, 1361 (1939).
64. Hodgkinson, C. F., "Oil refinery waste treatment in Kansas," *Sewage Ind. Wastes* **31**, 1304 (1959).
65. Howland, W. E., "Oil separation from refinery wastes," *Sewage Ind. Wastes* **23**, 121 (1951).
66. Howland, W. E., "Oil separation in the disposal of refinery wastes," in Proceedings of 4th Industrial Waste Conference, Purdue University, September 1948.
67. Humphreys, W., "Production and refinery waste disposal, California," *Sewage Ind. Wastes* **16**, 936 (1944).
68. Huntress, C. O., "Treatment of petrochemical wastes," in Proceedings of 8th Industrial Waste Conference, Purdue University, May 1953.
69. 78th U.S. Congress, First Session, House Document 266, "Industrial waste guide to the oil industry," *Ohio River Pollution Control Survey*, Supplement D, Appendix IX (1943), p. 1175.
70. Johnston, C. B., "Underground water pollution by oilfield brines," *Public Works Mag.* **85**, 93 (1954).
71. Jones, O. S., "Brine disposal, subsurface," *Sewage Ind. Wastes* **20**, 184 (1948).
72. Juterbock, E. E., and B. H. Weil, "Petroleum," *Ind. Eng. Chem.* **51**, 40A (1959).
73. Kaplovsky, A. J., "Evaluation of treatment methods at Tidewater's Delaware refinery, waste control and

treatment," *Sewage Ind. Wastes* **31**, 432 (1959).
74. Kleber, J. P., "Treatment of flooding waters and disposal brines with Calgon," *Water Pollution Abstr.* **18**, 64 (1955).
75. Koshen, A., "Esso refinery reconstructs industrial waste sewer," *Water Sewage Works* **100**, 126 (1953).
76. Lee, J. A., "Brines, well disposal," *Sewage Ind. Wastes* **23**, 326 (1951).
77. Lewis, A. W., "Facilities and treatment methods at Tidewater's Delaware refinery, waste control and treatment," *Sewage Ind. Wastes* **31**, 424 (1959).
78. "Liquid waste," Central States Sewage and Industrial Waste Association Meeting Report, *Ind. Wastes* **2**, 152 (1957).
79. Lucas, W. R., "Direct oxygen regeneration of spent petroleum caustics," Workbook Feature, *Ind. Eng. Chem.* **51**, 84A (1959).
80. Ludzack, R. J., et al., "Observation and measurement of refinery wastes," *Sewage Ind. Wastes* **30**, 662 (1958).
81. Ludzack, R. J., W. M. Ingram, and M. B. Ettinger, "Characteristics of a stream composed of oil refinery and activated sludge effluents," *Sewage Ind. Wastes* **29**, 1177 (1957).
82. Ludzack, R. J., and D. Kinkead, "Oil wastes, persistence in water," *Sewage Ind. Wastes* **28**, 827 (1956).
83. McKinney, R. E., et al., "Aromatic compounds, activated sludge treatment," *Sewage Ind. Wastes* **28**, 547 (1956).
84. McRae, A. D., "Disposal of alkaline wastes in the petrochemical industry," *Sewage Ind. Wastes* **31**, 712 (1959).
85. McRae, A. D., "Modern waste disposal and recovery in a petroleum refinery," in Proceedings of 9th Industrial Waste Conference, Purdue University, May 1954, p. 440.
86. McRae, A. D., F. H. Griffiths, and R. G. Lane, "Detection and monitoring of phenolic waste water," in Proceedings of 7th Ontario Industrial Waste Conference, June 1960, p. 56.
87. Madarasz, M. F., "Pilot plant recovery of an ion coagulant from oil waste treatment sludge," in Proceedings of 14th Industrial Waste Conference, Purdue University, May 1959.
88. Madarasz, M. F., "Treatment of oil wastes from machining plants," *Lubrication Eng.* **14**, 145 (1958).
89. Marsh, G. A., "Brines, dissolved oxygen determination, portable meter," *Sewage Ind. Wastes* **24**, 1046 (1952).
90. Mau, G. E., "Trickling filter treatment, pilot plant studies," *Sewage Ind. Wastes* **26**, 1236 (1954).
91. Milne, D., "Industrial waste control at General Motors Corporation," in Proceedings of 5th Ontario Industrial Waste Conference, May 1958, p. 111.
92. Morgan, M. J., "Collection and segregation of wastes," *Sewage Ind. Wastes* **19**, 1107 (1947).
93. Morris, W. S., "Brine disposal, East Texas," *Sewage Ind. Wastes* **23**, 697 (1951).
94. Morris, W. S., "Disposal of oil field salt waters," in Proceedings of 5th Industrial Waste Conference, Purdue University, November 1949.
95. Musante, A. F. S., "Determination of oil in refinery wastes," *Sewage Ind. Wastes* **24**, 1046 (1952).
96. Neumann, E. D., et al., "Modern waste-disposal facilities at Shell's Ancortes refinery," *Oil, Gas J.* **56**, 124 (1958).
97. Niegowski, S. J., "Phenol waste, ozone destruction," *Sewage Ind. Wastes* **28**, 1266 (1956).
98. "Oil pollution and refinery wastes," Committee Report, *Sewage Ind. Wastes* **7**, 104 (1935).
99. Pomeroy, R., "Oil and grease, floatability and separation design," *Sewage Ind. Wastes* **25**, 1304 (1953).
100. Pomeroy, R., "Oil field wastes, disposal, California coastal counties," *Sewage Ind. Wastes* **26**, 59 (1954).
101. Pratt, N. A., "Oil emulsions in disposal problems," *Sewage Works J.* **22**, 331 (1950).
102. Randolph, J. W., "Refinery waste disposal," *Sewage Ind. Wastes* **19**, 1107 (1947).
103. Rather, J. B., "Determination of oil in refinery effluents," *Sewage Ind. Wastes* **30**, 1325 (1958).
104. "Refinery removes oil wastes from 45,000-barrel-daily operations," *Wastes Eng.* **28**, 182 (1957).
105. Reid, G. W., R. Daigh, and R. L. Wortman, "Phenolic wastes from aircraft maintenance," *J. Water Pollution Control Federation* **32**, 383 (1960).
106. Rody, J. J., "Brine treatment and disposal," *Sewage Ind. Wastes* **13**, 848 (1941).
107. Rohlich, G. A., "Air flocculation, application to refinery wastes," *Sewage Ind. Wastes* **26**, 1056 (1954).
108. Rohlich, G. A., "Pilot plant studies of air flotation of oil refinery waste water," in Proceedings of 8th Industrial Waste Conference, Purdue University, May 1953, p. 368.
109. Rook, C. G., "Refinery waste effluent utilization," *Sewage Ind. Wastes* **18**, 582 (1946).
110. Ross, L., "Sohio separator, unusual design," *Sewage*

Ind. Wastes **20**, 956 (1948).
111. Ross, W. K., and A. A. Sheppard, "Biological oxidation of petroleum phenolic waste waters," in Proceedings of Conference on Biological Waste Treatment, Manhattan College, New York, April 1955.
112. Rudolfs, W., *Industrial Waste Treatment*, Reinhold Publishing Corp., New York (1953), p. 427.
113. Ruggles, W. L., "Basic petrochemical process as waste source," *Sewage Ind. Wastes* **32**, 274 (1960).
114. Ryan, B. J., "Spent caustic disposal and petrochemical industries," in Proceedings of 6th Ontario Industrial Waste Conference, June 1959, p. 179.
115. Sabina, L. R., and H. Pivnick, "Oil, emulsion, oxidation," *Sewage Ind. Wastes* **29**, 841 (1957).
116. Salem, E. D., "Development, construction and operation of an oily waste treatment plant," in Proceedings of 7th Industrial Waste Conference, Purdue University, May 1952.
117. Salvatorelli, J. J., "Aircraft engine waste treated in continuous 'flow-through' plant," *Wastes Eng.* **30**, 310 (1959).
118. Sanders, F. A., "Combating oil and metal plating waste problems at Kelly Air Force Base," in Proceedings of 7th Industrial Waste Conference, Purdue University, May 1952.
119. Sawyer, C. N., *et al.*, "Brines, iodine recovery," *Sewage Ind. Wastes* **22**, 583 (1950).
120. Schindler, H., "Chemical treating plant for refinery waste water from white oils and petroleum sulfonates," in Proceedings of 6th Industrial Waste Conference, Purdue University, February 1951.
121. Schindler, H., "Refinery wastes chemical treating plant," *Sewage Ind. Wastes* **24**, 811 (1952).
122. Schindler, H., "Sulfonates, disposal," *Sewage Ind. Wastes* **22**, 1255 (1950).
123. Stein, M., "Oil well brine problem of the cannery drainage system," in Proceedings of 12th Industrial Waste Conference, Purdue University, May 1957.
124. Sheets, W. D., *et al.*, "Phenolic refinery wastes, treatment, pilot filter studies," *Sewage Ind. Wastes* **26**, 862 (1954).
125. Sherwood, P. W., "Modern American methods of acid sludge disposal," *Petroleum* **18**, 224 (1955).
126. Sinard, R. G., *et al.*, "Infrared spectrophotometric determination of oil and phenols in water," *Anal. Chem.* **23**, 1384 (1951).
127. Stormont, D. H., "Air flotation used to separate oil at Richfield's new waste water plant," *Oil, Gas J.* **54**, 82 (1956).
128. Strong, E. R., and R. Hatfield, "Petrochemical wastes, treatment high-rate activated sludge, pilot studies," *Sewage Ind. Wastes* **26**, 1057 (1954).
129. Sukes, G. L., "Soluble oil and industrial cleaning waste treatment," in Proceedings of 9th Industrial Waste Conference, Purdue University, May 1954, p. 477.
130. "Sulfuric acid recovery from refinery sludge," *Sewage Ind. Wastes* **13**, 1010 (1941).
131. Swarthout, Z., "Is that refinery a stinker or a boom?" *Miami News*, 12 July 1955.
132. "Symposium of paper on oil refinery wastes," in Proceedings of Oklahoma Industrial Wastes Conference, October 1955; *Oklahoma Eng. Exp. Stat. Publ.* **97** (1955).
133. "Symposium of nine technical papers on waste disposal in the petroleum industry," *Ind. Eng. Chem.* **46**, 283 (1954).
134. Sylvester, R. O., *et al.*, "A study of the pre-operational environment in the vicinity of a new oil refinery," in Proceedings of 11th Industrial Waste Conference, Purdue University, May 1956.
135. Turnbull, H., *et al.*, "Toxicity of refinery wastes to fresh water fish," *Sewage Ind. Wastes* **26**, 1057 (1954).
136. Umbach, R. D., "How one refinery is handling its waste treatment problem," in Proceedings of 14th Industrial Waste Conference, Purdue University, May 1959.
137. Vrablek, E. R., "An evaluation of circular gravity-type separators and dissolved-air flotation for treating oil refinery waste water," in Proceedings of 12th Industrial Waste Conference, Purdue University, May 1957.
138. Walker, D. J., "Treatment for disposal of spent and contaminated soluble oil mixtures," in Proceedings of 5th Industrial Waste Conference, Purdue University, November 1949.
139. Weston, R. F., "Separation of oil refinery waste waters," *Ind. Eng. Chem.* **42**, 607 (1950).
140. Weston, R. F., "Waste disposal and utilization problems of the petroleum industry," in Proceedings of 1st Industrial Waste Conference, Purdue University, November 1944, p. 98.
141. Weston, R. F., "Waste disposal and utilization," *Sewage Ind. Wastes* **18**, 763 (1946).
142. Weston, R. F., and W. B. Hart, "Water pollution

abatement problems of the petroleum industry," *Water Works Sewerage* **88**, 208 (1941).
143. Weston, R. F., and R. G. Merman, "Chemical flocculation of refinery waste," reprinted from *Petrol. Engr.*, by Builders' Iron Foundry Industries, May 1954.
144. Weston, R. F., R. G. Merman, and J. G. DeMann, "Coagulation of refinery wastes by twin pilot-plant units," in Proceedings of 4th Industrial Waste Conference, Purdue University, September 1948, p. 290.
145. Weston, R. F., et al., "Coagulation of refinery wastes by coagulation by twin pilot plant unit," *Sewage Ind. Wastes* **23**, 121 (1951).
146. Williams, J. T., "Methods adopted by Imperial Oil, Ltd., to prevent phenol contamination of the St. Clair River in the event of an accident," in Proceedings of 3rd Ontario Industrial Waste Conference, June 1956, p. 111.
147. Williamson, A. E., "Land disposal of refinery wastes," in Proceedings of 13th Industrial Waste Conference, Purdue University, May 1958.
148. Wright, E. R., "Secondary petrochemical process as waste source," *Sewage Ind. Wastes* **31**, 574 (1959).
149. Zeien, J. T., "Reduction and control of wastes in a new refinery," in Proceedings of 10th Industrial Waste Conference, Purdue University, May 1955.

Suggested Additional Reading

The following references have been published since 1962.
1. "Air monitoring device helps keep water clean," *Chem. Eng.* **73**, 96 (1966).
2. Aizenshtain, P. G., et al., "Chemical and electrical flotation methods for sewage purification," *Neftepererabotka i Neftekhim., Nauchn. Tekhn. Sb.* **3**, 18 (1964).
3. "Air bubbles polish refinery waste water," *Oil Gas J.* **64**, 194 (1966); *J. Water Pollution Control Federation* **39**, 887 (1966).
4. "New plant makes H_2SO_4 from wastes," *Oil, Gas J.* **60**, 128 (1962).
5. "Plastic media effective in waste treatment," *Petrol. Management* **35**, 164 (1963).
6. "Three methods of combating oil pollution," *Petroleum* **25**, 148 (1962).
7. Ball, H. K., "New approach to leaded gasoline sludge disposal," *Hydrocarbon Process. Petrol. Refiner* **42**, 147 (1963).
8. Ball, H. K., "Methods of disposing of sludge from leaded gasoline storage tanks," *Proc. Am. Petrol. Inst.* **43**, 302 (1963).
9. Becksmann, E., "Das Verhalten der Erdoelderivate im Grundwasser," *Gas- Wasserfach* **104**, 689 (1963).
10. Benger, M., "The disposal of liquid effluents from oil refineries," *Fluid Handling* **15**, 270 (1964).
11. Benger, M., "The disposal of liquid and solid effluents from oil refineries," in Proceedings of 21st Industrial Waste Conference, Purdue University, May 1966, p. 759.
12. Beychok, M. R., *Aqueous Wastes from Petroleum and Petrochemical Plants*, John Wiley & Sons, New York (1967).
13. "Biofilter cleans waste water," *Petrol. Management* **35**, 197 (1963).
14. Bozeman, H. C., "Longest polypropylene line handles salt-water disposal," *Oil, Gas J.* **61**, 163 (1963).
15. Bozeman, H. C., "Shell's filter system borrows from nature," *Oil, Gas J.* **62**, 92 (1964).
16. Brown, K. M., W. K. T. Gleim, and P. Urben, "Sulfur containing refinery waste water," U.S. Patents nos. 3029021 and 3029022.
17. Bryan, E. H., "Two-stage biological waste treatment," *Ind. Water Wastes* **8**, 31 (1963).
18. Brunsmann, J. J., J. Cornelissen, and H. Eilers, "Improved oil separation in gravity separators," *J. Water Pollution Control Federation* **34**, 44 (1962).
19. Busch, A. W., "Process kinetics as design criteria for bio-oxidation of petrochemical wastes," *J. Eng. Ind.* **85**, 163 (1963).
20. Campbell, C., and G. R. Scoullar, "How Shell treats Oakville effluent," *Hydrocarbon Process. Petrol. Refiner* **43**, 137 (1964).
21. Campbell, G. C., and G. R. Scoullar, "Effluent water treating facilities at Oakville refinery," *Proc. Am. Petrol. Inst.* **44**, 83 (1964).
22. Copeland, B. J., and T. C. Dorris, "Effectiveness of oil refinery effluent holding ponds," *Kansas State Univ. Eng. Architectural Bull.* **51**, 8 (1966).
23. Copeland, B. J., and T. C. Dorris, "Photosynthetic productivity in oil refinery effluent holding ponds," *J. Water Pollution Control Federation* **34**, 1104 (1962).
24. Cornelissen, J., and W. Steck, "Deoiling of refinery effluents," *J. Appl. Chem.* **12**, 621 (1962).
25. Cunningham, R. J., "Effluent problems in an edible oil refinery and margarine factory," *Chem. Ind.* 21 August 1965, p. 1481.
26. Davis, R. W., J. A. Riehe, and R. M. Smith, "Pollution control and waste treatment at inland refinery,"

Purdue University Engineering Extension Series, Bulletin no. 117, 1965, p. 126.
27. Douglas, N. H., and W. H. Irwin, "Relative resistance of fish species to petroleum refinery wastes," *Water Sewage Works* **100**, R-246 (1963).
28. Dorris, T. C., D. Patterson, and B. J. Copeland, "Oil refinery effluent treatment in ponds," *J. Water Pollution Control Federation* **35**, 932 (1962).
29. Dungs, G., "Electrical equipment for effluent and sludge treatment plant in B.P. Ruhr refinery," *Am. Gas Elec. Progr.* **3**, 260 (1963).
30. Easthagen, J. H., and V. Skrylov, "Development of refinery washwater control at Pascagoula, Miss.," *J. Water Pollution Control Federation* **37**, 467 (1965).
31. Emde, W. von der, "Die Neue Abwasserreinigungsanlage der Raffinerie Hamburg-Harburg der Deutschen Shell AG," *Gas- Wasserfach* **104**, 94 (1963).
32. Engelbrecht, R. S., *et al.*, "Diffused air stripping of volatile waste components of petrochemical wastes," *J. Water Pollution Control Federation* **33**, 127 (1961).
33. Engelbrecht, R. S., and B. B. Ewing, "Treatment of petrochemical wastes by activated sludge process," *Ind. Water Wastes* **8** (3), 36 (1963) and (4), 34 (1963).
34. Enright, R. J., "Oil field pollution and what's being done about it," *Oil, Gas J.* **61**, 76 (1963).
35. Fiske, C. E., and J. E. Garner, "Stabilization pond treatment of American oil refinery effluent," *Water Sewage Works* **103**, R-277-280 (1966).
36. Forbes, M. C., and P. A. Witt, "Estimate cost of waste disposal," *Hydrocarbon Process. Petrol. Refiner* **44**, 8 (1965).
37. Gaudy, A. F., Jr., *et al.*, "Stripping kinetics of volatile components of petrochemical waste," *J. Water Pollution Control Federation* **33**, 382 (1961).
38. Gloyna, E. F., and J. F. Malina, "Petrochemical waste effects on water," *Ind. Water Wastes* **8**, 14 (1963).
39. Gloyna, E. F., and J. F. Malina, "Petrochemical wastes effect on water," *Water Sewage Works* **100**, R-262 (1963).
40. Gossom, W. J., "New waste treatment unit follows 12-year study," *Oil, Gas J.* **60**, 82 (1962).
41. Halladay, W. B., and R. H. Crosby, "Current techniques of treating recovered oils and emulsions," *Proc. Am. Petrol. Inst.* **44**, 68 (1964).
42. Harris, A. J., "Water pollution control activities of the central Ontario lakeshore refineries," *J. Water Pollution Control Federation* **35**, 1154 (1963).
43. Hatch, L. F., "Can Syndet shift beat foam problem," *Hydrocarbon Process. Petrol. Refiner* **41**, 146 (1962).
44. "How Shell will clean up its waste water," *Oil, Gas J.* **59**, 72 (1961).
45. "Humble uses ponds to improve waste-water quality," *Oil, Gas J.* **62**, 96 (1964).
46. Karavaev, I. I., and F. Reznik, "Purification of waste water from petroleum products by flotation," *Chem. Abstr.* **57**, 2001 (1962).
47. Karapaev, B. I., "Vliyanie nchk na ochistku stochnyku vod ot nefti filtratsiei," *Neft. Khoz.* **6**, 38 (1963).
48. Karelin, Y. A., *et al.*, "Investigation of the industrial waste waters of an oil refinery, and their biochemical purification," *Khim i Tekhnol. Topliv i Masel* **9**, 29 (1964).
49. Katz, W. J., A. Geinopolis, and J. L. Mancini, "Concepts of sedimentation applied to design," *Water Sewage Works* **109**, 162 (1962).
50. Kingsbury, A. W., "Development of an oily water separator," *J. Water Pollution Control Federation* **38**, 236 (1966).
51. Kissling, L. F., "Process for removing phenol," U.S. Patent no. 2911363.
52. Lamkin, J. C., and L. V. Sorg, "Aerated lagoon aids disposal of waste water," *Oil, Gas J.* **62**, 84 (1964).
53. Lamkin, J. C., and L. V. Sorg, "American oil cleans up wastes in aerated lagoons," *Hydrocarbon Process. Petrol. Refiner* **43**, 133 (1964).
54. Lamkin, J. C., and L. V. Sorg, "Treatment of refining wastes in aerated lagoons," *Proc. Am. Petrol. Inst.* **44**, 113 (1964); *Oil Gas J.* **62**, 84 (1964).
55. Martin, J. D., and L. D. Levanas, "Air oxidation of sulfide in process water," *Proc. Am. Petrol. Inst. Refining* **42**, 392 (1962).
56. Martin, J. D., and L. D. Levanas, "New column removes sulfide with air," *Hydrocarbon Process. Petrol. Refiner* **41**, 149 (1962).
57. McKinney, R. E., "Biological treatment of petroleum refinery wastes," American Petroleum Institute Publication (1963).
58. McKinney, R. E., "Mathematics of complete-mixing activated sludge," *J. Sanit. Eng. Div. Am. Soc. Civil Engrs.* **88**, 87 (1962).
59. McPhee, W. T., and A. R. Smith, "From refining waste to potable water," *Civil Eng.* **31**, 64 (1961).
60. Medem, M., *et al.*, "Biochemical process for oxidizing petroleum fractions," *Chem. Technol. Fuels Oils* **4**, 287 (1965).

61. Merman, R. C., P. J. Ferrall, and G. T. Foradori, "Sludge disposal at Philadelphia refinery," *J. Water Pollution Control Federation* **33**, 1153 (1961).
62. "Mobile oil decontamination equipment," *Engrs.* **222**, 395 (1966).
63. Munson, E. D., "New concepts in industrial sewage collection," *J. Water Pollution Control Federation* **36**, 1146 (1964).
64. "Navy saves the day," *Water Sewage Works* **113**, 72 (1966).
65. "Oil pollution of water supplies," *J. Am. Water Works Assoc.* **58**, 3 (1966).
66. "Plastic media effective in waste treatment," *Petrol. Management* **35**, 164 (1963).
67. Prather, B. V., and A. F. Gaudy, Jr., "Combined chemical, physical, and biological processes in refinery waste water purification," *Proc. Am. Petrol. Inst. (Refining)* **44**, 105 (1964).
68. Prather, B. V., "Development of a modern petroleum refinery waste treatment program," *J. Water Pollution Control Federation* **36**, 96 (1964).
69. Prather, B. V., and J. W. Gaudy, "Purifying refinery waste water," *Oil, Gas J.* **62**, 96 (1964).
70. Prather, B. V., "Will air flotation remove chemical oxygen demand of refinery waste water?" *Petrol. Refiner* **40**, 177 (1961).
71. Pyplant, H. S., "Here's how petrochemical companies dispose of waste," *Oil, Gas J.* **61**, 118 (1963).
72. Quigley, R. E., and E. L. Hoffman, "Flotation of oily wastes—a refineries approach to waste water treatment," *Ind. Water Eng.* **4**, 22 (1967).
73. Rabb, A., "Sludge disposal—growing problem," *Hydrocarbon Process. Petrol. Refiner* **44**, 149 (1965).
74. Reed, R. D., "Waste gas disposal—a problem?" *Oil, Gas J.* **60**, 92 (1962).
75. Sadow, R. D., "Waste treatment at large petrochemical plant," *J. Water Pollution Control Federation* **38**, 548 (1965).
76. Shannon, E. S., "Handling and treating petrochemical plant wastes: a case history," *Water Sewage Works* **111**, 240 (1964).
77. Shaver, R. A., "Waste water treatment at Regent's Port Credit refinery," in Proceedings of 10th Ontario Industrial Waste Conference, June 1963, p. 21.
78. Sherwood, P. W., "Activated sludge treatment of industrial waste waters," *Water Water Eng.* **65**, 66 (1961).
79. Sherwood, P. W., "Disposal of refinery acid sludge," *Petroleum* **24**, 22 and 65 (1961).
80. Slyker, J. V., "Recent developments in the clarification of oil field waters," *Producers' Monthly* **28**, 8 (1964).
81. Simonsen, R. N., "Oil removal by air flotation at Sohio refineries," *Proc. Am. Petrol. Inst. (Refining)* **42**, 399 (1964); *Hydrocarbon Process. Petrol. Refiner*, (1962).
82. Simonsen, R. N., "Remove oil by air flotation," *Hydrocarbon Process. Petrol. Refiner* **41**, 145 (1962).
83. Smith, R. T., "Cooling towers used for waste treatment," *Oil, Gas J.* **62**, 115 (1964).
84. Sontheimer, H., "Waste-water treatment in petroleum refineries," *Oil, Gas J.* **61**, 159 (1963).
85. "Spot shutdown seen in Shakleford Co. pollution case," *Oil, Gas J.* **64**, 68 (1966).
86. Talbot, J. S., and P. Beradon, "Deep well method of industrial waste disposal," *Chem. Eng. Progr.* **60**, 49 (1964).
87. Tassoney, J. P., E. B. Stuart, and R. L. Albright, "Batch process removes oil, fat, grease," *Water Works Wastes Eng.* **1**, 38 (1964).
88. Taylor, J. C., "The recovery and containing of oil spills," *Petrol. Times* **66**, 292 (1962).
89. "Texas aims broad attack to clean up pollution," *Oil, Gas J.* **61**, 121 (1963).
90. Thomas, A. M., "Thermal decomposition of sodium carbonate solutions," *J. Chem. Eng. Data* **8**, 51 (1963).
91. U.S. Department of the Interior, *Petroleum Refineries*. Industrial Waste Profile no. 5, Washington, D.C. (1967).
92. Van Wyk, J. W., "Design for portable waste treatment plant study of flocculation," *Proc. Am. Petrol. Inst. (Refining)* **42**, 407 (1962).
93. Vaughn, S. H., "Oil retention and removal facilities at Ford Motor Company's Rouge manufacturing complex," in Proceedings of 21st Industrial Waste Conference, Purdue University, 1966, p. 639.
94. Voege, F. A., and D. R. Stanley, "Industrial waste stabilization ponds in Canada," *J. Water Pollution Control Federation* **35**, 1019 (1963).
95. Wallace, A. T., *et al.*, "The effect of inlet conditions on oil—water separations at Sohio's Toledo refinery," in Proceedings of 12th Industrial Waste Conference, Purdue University Engineering Extension Series, Bulletin no. 118, May 1965, p. 618.
96. Beychok, M. R., *Aqueous Wastes from Petroleum and Petrochemical Plants*, London: John Wiley.

24.9 Fuel-Oil Wastes

Origin of wastes. Fuel-oil wastes are also a problem at the distribution stage. Two major types normally occur in metropolitan areas.

The first fuel-oil waste is found at refueling stations for diesel trains. A typical diesel engine may contain several fuel tanks on each side. These tanks are filled by 3-inch hose-pipes under pressure. Although most hose-pipes are fitted with automatic shutoff valves that close when each fuel tank is full, this procedure is usually too slow and inefficient for the operators. In some cases the hoses are not removed from the tanks until the fuel oil has reached the very top and overflows. The oil spills onto the surrounding ground and over a period of time saturates the soil. Subsequent rains may carry this oil into a nearby receiving stream.

The second oil waste originates from automobile and truck crankcase oil. The dirty crankcase oil drained out at garages or terminals was formerly collected and resold to refineries for reprocessing. Since the Federal Government recently eliminated its financial support to the refineries for fuel-oil reprocessing, most refineries are not too enthusiastic about collecting, transporting, and reprocessing this oil. In fact, garages, instead of receiving some payment for the old oil, are now often forced to pay the refineries to take the oil away. In some instances this cost may force collectors of crankcase oil to let these wastes find their way to drainage ditches and creeks.

Waste characteristics. Although the wastes are obviously various dilutions of oil in water, no specific analyses have been officially published.

Waste treatment. The only effective treatment is retention of the oily wastes for shipment to refineries. When the oils have been contaminated with water, skimming often precedes collection for transportation away from the site.

24.10 Rubber Wastes

Origin of rubber wastes. Rubber is not one substance, but many. The principal types of rubber are: (1) natural rubber—all rubberlike materials produced by coagulation of the rubber-plant sap (latex); (2) synthetic rubber, made by copolymerization of butadiene and styrene (GR-S) or isoprene and butadiene with small amounts of isobutylene (GR-1) for non-oil-resisting rubbers and neoprene-type oil-resisting rubbers (polymers of chloroprene); (3) scrap rubber, a mixture of discarded rubber items and residues from manufacturing processes; (4) rubberlike plastics, including a group of nonrigid ones, both thermoplastic and thermosetting. Rubber latex is usually distributed to manufacturers of finished rubber products as a milky colloidal emulsion of sugars, resins, and protein constituents. The composition of rubber varies widely with the source, but the useful constituent is rubber hydrocarbon, which makes up 90 per cent of plantation rubber. It is now universally accepted that the unit structural group in the rubber molecule is $(C_5H_8)_n$,

$$\begin{array}{ccc} H & & H \quad H \\ C=C & - & C=C, \\ H \quad | & & H \\ \quad HCH & & \\ \quad H & & \end{array}$$

with the configuration which is 2-methyl-butadiene-1, 3, commonly known as isoprene. Naturally, the most important characteristic of rubber is its high modulus of elasticity, the combination of strength, flexibility, and resilience.

Wastes from the production of rubber have a high BOD, taste, and odor; the problems they present vary considerably, depending on the plant site, the raw material used, and the number of intermediary products. Rubber-manufacturing wastes may be divided into four general classes: (1) steel-products wastes, (2) rubber-commodities wastes, (3) reclaimed-rubber wastes, and (4) synthetic-rubber wastes.

To explain these classifications: the same company that produces the rubber product may also manufacture steel rims for wheels, metal parts for mechanical rubber goods, stainless-steel-lined beverage containers, and so forth. As a result, these rubber wastes would include zinc- and brass-plating wastes, and other metal-manufacturing wastes.

The making of rubber commodities involves washing, compounding, calendering, and curing processes, followed by the actual manufacturing of all kinds of rubber products. The wastes include a large volume of washing waters with the impurities exuded by the crude rubber.

Reclamation of used rubber is accomplished by shredding the old rubber and discharging it on a travel-

Table 24.29 Characteristics of reclaimed and synthetic rubber wastes. (After Sechrist and Chamberlin [38].)

Characteristic	Reclaimed rubber	Synthetic rubber
Total solids, ppm	16,800–63,400	1900–9600
Suspended solids, ppm	1,000–24,000	60–2700
Oxygen consumed, ppm	3,600–13,900	75–4500
BOD, ppm	3,500–12,500	25–1600
Chlorides, ppm	130–2,000	90–3300
Hydroxide alkalinity, ppm	0–2,700	
pH	10.9–12.2	3.2–7.9

ling belt, where it passes under a magnet which removes any bits of metal. The ground rubber and fabric, freed of metal, are subjected to a caustic treatment under high temperatures for several hours, which destroys the fabric and frees the rubber. The recovered rubber is then washed, dried, milled, strained, and refined for reuse.

Synthetic rubber evolves from a process whereby butadiene is mixed with some other monomer such as styrene or acrylonitrile plus a catalyst, in a soap solution, to produce synthetic latex. Coagulation of the latex either in an acid-brine solution or with alum follows, after which the latex is washed, dried, and baled. Waste from the synthetic-rubber plant consists of whatever coagulated rubber escapes, plus the acid and saline liquid, and occasional batches of materials that will not polymerize properly.

Characteristics of rubber wastes. Schatze [38] presents some of the sanitary characteristics of reclaimed and synthetic rubber wastes (Table 24.29). Black [6] and Rostenbach [34] give other analyses of typical rubber wastes (Tables 24.30 to 24.32). Study of these tables reveals that rubber wastes are highly objectionable because of their high BOD and odiferous nature. The odors are detectable even at extremely low concentrations and make water unpalatable for several hundred miles downstream from a rubber plant.

Treatment of rubber wastes. The most common means of treatment practiced today are aeration, chlorination, sulfonation, and biological methods; coagulation, ozonation, and treatment with activated carbon are also used.

Black [6] presents the results of aeration achieved by spraying rubber wastes at a pressure of 10 psi (Table 24.33). Increase in aeration pressure above 10 psi

Table 24.30 Analysis of typical wastes from the synthetic-rubber industry. (After Black [6].)

Plant area	Approximate daily discharge, mgd	pH	Total solids, ppm	Suspended solids, ppm	5-day BOD, ppm	Odor concentration
1. Butadiene waste	1.90	2.8	300	27.6	2550	16,100
2. Styrene waste	4.62	6.2	150	4.5	180	690
Copolymer plant						
3. Process waste	2.34	4.3	5580	12.3	69	62
4. Recovery and reactor	0.39	8.0	570	23.6	492	8,760
5. Main sewer*	2.63	7.0	6530	46.0	168	930
6. Main outfall, institute†	119.40	5.4	270	15.5	81	460

*Represents the mixture of 3 and 4.
†Represents the mixture of 1, 2, and 5, plus large quantities of condenser water.

24.10   RUBBER WASTES   455

Table 24.31 Wastes from combined operations of three plants: butadiene, GR-1, and RD-S [34].

Characteristic	Value or concentration
pH	7.6
Oil, ppm	9
Suspended solids, ppm	79
BOD, ppm	78
Dissolved oxygen, ppm	2.0
Flow, gpm	2000

Table 24.32 Wastes from combined operations of three plants: butadiene, styrene, and GR-S [34].

Characteristic	Value or concentration
pH	8.6
Oil, ppm	18
Settleable solids, ml/liter	1.4
Odor threshold	16
Flow, gpm	
Styrene plant	190
GR-S plant	735
Butadiene plant	1120
Total	2045

improved the efficiency. Spray aeration was less effective than diffused-air aeration, and both methods were less effective with styrene wastes than with butadiene wastes. Chlorination has been used to reduce the phenolic constituents of rubber wastes [6, 39], and sulfonation of styrene waste can yield an almost odorless waste [6] if sufficient time is given for the reaction (Table 24.34). In comparison with aeration, chlorination, and sulfonation, which are valuable mostly for their capability as to odor removal, biological treatment of rubber wastes is a method which affords the greatest reduction of BOD (Tables 24.35 and 24.36). However, addition of nitrogen and phosphorus, or mixture with domestic sewage in a ratio of one to three, is required for efficient biological treatment of rubber wastes.

In certain cases rubber wastes are treated by

Table 24.33 Effect of spray aeration on butadiene and styrene wastes and their dilutions [6].

Material	Concentration, %	Temperature of treatment, °C	Odor concentration		
			Initial	After bubble aeration	After spray aeration
Butadiene	100	24	1200	100	250
	4*	25.5	64		16
	4*	50	64		16
	4	50	16		8
	4	50	8	1	8
	2	24	32		8
	1	50	2	None	2
	1	24	4		4
	1	50	8		4
	1*	50	16		8
	1*	24	16		8
	0.5	24	8		4
Styrene	100	24	600	500	600
	4	24	64	4	16
	1	24	16		4
	0.5	50	8		4

*Duplicates except for temperature of treatment.

Table 24.34 Odor concentrations obtained after treating butadiene and styrene wastes with Na_2SO_3 and Na_2S [6].

Waste and original odor level	Days treated	Treated with Na_2SO_3			Treated with Na_2S		
		100 ppm	250 ppm	500 ppm	100 ppm	250 ppm	500 ppm
Butadiene (4100)	1	4100	4100	4100	4100	4100	4100
	3	4100	4100	4100	8200	8200	8200
	4	4100	4100	4100	4100	8200	8200
	5	4100	4100	4100	4100	4100	4100
	6	4100	4100	4100	4100	4100	4100
	7	4100	4100	4100	4100	4100	4100
	10	1000	2000	2000	4100	4100	4100
	13	500	1000	1000	2000	2000	2000
	17	250	250	250	500	500	500
Styrene (1000)	1	128	64	64	65	6	128
	2	64	32	32	32	32	64
	3	32	32	32	32	32	32
	5	16	16	16	8	16	32
	6	16	16	16	8	8	16
	7	8	16	16	8	8	8
	8	4	16	8	4	4	8
	9	4	8	8	4	4	4

Table 24.35 Treatment of composite rubber waste by the use of an experimental trickling filter. (After Black [6].)

Feed applied to filter	Total time, hr	Total amount applied		pH		5-day BOD, ppm		BOD removed, %
		liters	mgd	Influent	Effluent	Influent	Effluent	
50% composite rubber waste and 50% sewage	0	0	0	6.3		340		
	16.5	14.0	1.49	6.9	7.8	340.5	25.3	92.6
	20.0	16.0	1.01	6.9	7.7	404	18.2	95.5
	23.0	18.0	1.16	6.9	7.8	398	39.6	90.1
	39.5	42.0	2.56	6.7	7.9	359	35.2	90.2
	42.5	43.0	0.59	6.9	7.8	252	28.8	88.6
	45.5	44.0	0.59	6.9	7.7	436	46.2	89.4
	68.0	58.0	1.10	6.9	7.7	195	32.5	83.3
	72.0	60.0	0.88		7.7	230	19.3	91.6
Fresh sewage only	24		1.0	7.4	7.5	292	205.5	29.6
	48		1.0	7.7	7.5	189	26.9	85.8
	72		1.0	7.7	7.6	263*	34.5	86.9
100% neutralization of composite rubber waste	2			7.3	7.9	328	72	78.0
	5			7.2	7.9	392	35	91.1
	7	4.0	1.01	7.3	8.4	385	71.2	78.9
	24	15.0	0.81	7.4	8.3	445	35	92.1
	31	18.0	0.99	7.4	8.4	484	43.8	91.0

*4-day BOD.

Table 24.36 Effects of adding butadiene and styrene wastes to activated sludge. (After Black [6].)

Type of waste	Concentration of waste, %	Feed*	pH	Initial Suspended solids, ppm	Initial BOD, ppm	4-hr effluent BOD, ppm	4-hr effluent Reduction, %	24-hr effluent BOD, ppm	24-hr effluent Reduction, %
Neutralized	3.0	C	7.7	2364	320.0	27.2	92	10.7	96
		B	7.7	2196	371.1	40.3	91	13.6	96
		S	7.7	2244	310.0	47.4	85	12.2	96
Neutralized	6.0	C	7.5	2176	167.2	50.8	70	12.9	92
		B	7.7	2128	405.3	53.1	81	12.4	95
		S	7.4	2176	161.3	69.1	57	9.5	94
Neutralized	12.0	C	7.1	1828	185.3	59.2	68	13.1	93
		B	7.7	2128	405.3	131.4	67	20.0	95
		S	7.1	1940	168.4	95.6	43	8.8	95
Neutralized	24.0	C	7.1	1712	167.4	53.5	68	17.8	89
		B	7.7	2080	674.7	412.5	39	53.8	92
		S	7.3	1668	187.1	146.8	22	15.6	92
Sample and fresh-filtered sewages	0.0	C	7.7	1720	301.7	108.1	64	6.7	98
		B	8.5	2384	307.6	72.8	76	9.5	97
		S	7.7	1672	301.2	91.9	70	63	98
Sample and fresh-filtered sewages	0.0	C	7.5	1696	151.7	69.5	54	11.8	92
		B	8.4	2416	152.6	78.4	49	6.7	96
		S	7.5	1748	151.5	93.8	38	7.5	95
Unneutralized	3.0	C	7.3	1644	189.3	45.0	76	15.5	92
		B	6.6	2320	259.2	26.7	90	5.7	98
		S	7.1	1716	195.1	44.6	77	11.2	94
Unneutralized	6.0	C	6.9	1492	71.2	27.9	61	25.3	65
		B	4.7	2244	220.6	29.7	86	11.2	95
		S	6.7	1576	93.8	35.0	37	16.2	83

*C, control; B, butadiene; S, styrene.

trickling filtration. Culver [1963, 2] stated that combined treatment of rubber wastes with municipal sewage could be achieved by the activated-sludge process and a BOD removal of 85 per cent was obtained. Morzycki [1966, 8] reported that sewage containing synthetic latexes was purified by coagulation with CaC_2, $MgSO_4$, $Al_2(SO_4)_3$, $FeCl_3$, and $Fe_2(SO_4)_3$; optimum results were obtained in experiments involving $Al_2(SO_4)_3$ and iron salts and BOD was reduced by 84 to 87 per cent. Ruchhoft [37] mentioned that the coagulation method is of little value because less removal of taste and odor is obtained; ozone treatment is also not valuable, but activated carbon completely removes taste and odor, provided 100 ppm of carbon is applied.

References: Rubber Wastes

1. Adams, C. D., "Control of taste and odor from industrial wastes," *J. Am. Water Works Assoc.* **38**, 702 (1946).
2. American Society Testing Materials Committee, *American Society Testing Materials Standards on Rubber Products*, Philadelphia (1944), p. 424.
3. Barron, H., *Modern Synthetic Rubber*, 2nd ed., Hutchinson Scientific Publications, London (1944).
4. Barron, H., *Modern Synthetic Rubber*, 3rd ed., Chapman & Hall, London (1954), p. 636.
5. Barron, H., *Modern Rubber Chemistry*, Hutchinson Scientific Publications, London (1948).
6. Black, O. R., "Study of wastes from synthetic

rubber industry," *Sewage Works J.* **18**, 1169 (1946).
7. Bridgewater, E. R., "Neoprene as a construction material for the chemical industry," *Trans. Am. Inst. Chem. Eng.* **35**, 435 (1939).
8. Bushee, R. J., "Economics in sewage treatment construction," *Am. City* **69**, 83 (1954).
9. Catton, N. L., *The Neoprenes*, E. I. DuPont de Nemours & Co., Wilmington, Del. (1953), p. 246.
10. Cox, J. T., "Rubber in 1954," *Chem. Eng. News* **33**, 107 (1955).
11. Davis, C. C., and J. T. Blake, *Chemistry and Technology of Rubber*, Reinhold Publishing Corp., New York (1937).
12. Dougan, L. D., and J. C. Bell, "Synthetic rubber wastes disposal," *Munic. Utilities* **88**, 62 (1950).
13. Dougan, L. D., and J. C. Bell, "Synthetic rubber waste disposal," *Eng. Contract Record* **63**, 67 (1950).
14. Dougan, L. D., and J. C. Bell, "Waste disposal at synthetic rubber plant," *Sewage Ind. Wastes* **23**, 181 (1951).
15. Drakeley, T. J. (ed.), *Annual Report on the Progress of Rubber Technology*, Vol. 18, W. Heffer & Sons, Ltd., Cambridge (1954).
16. Fieser, L. F., *Organic Chemistry*, 3rd ed., D.C. Heath Co. and Reinhold Publishing Corp., New York (1956), p. 322.
17. Firestone, H. S., Jr., *The Romance and Drama of the Rubber Industry* and *Rubber, Its History and Development*, Firestone Tire and Rubber Co., Akron, Ohio (1922).
18. "First commercial phenol destruction unit," *Eng. News Record* **148**, 30 (1952).
19. Fontana, M. G., "Review of synthetic rubber and plastics as materials of construction in the chemical industry," *Trans. Am. Inst. Chem. Engrs.* **42**, 359 (1946).
20. *Handbook of Molded and Extruded Rubber*, 1st ed., The Goodyear Rubber Co., Inc., Ohio (1949).
21. Hauser, E. A., "Latex," *The Chemical Catalog*, New York (1930).
22. Hebbard, J. M., "Rubber industry," *Ind. Eng. Chem.* **39**, 589 (1947).
23. Hebbard, J. M., "Synthetic rubber waste treatment," *Sewage Ind. Wastes* **19**, 951 (1947).
24. Jones, S. A., "Historical review of synthetic rubbers," *Chem. Ind.*, March 1954, p. 343.
25. Keshen, A. A., "Solid rubber wastes," *Ind. Wastes* **4**, 14 (1959).
26. Marchionna, F., *Butalastic Polymers*, Reinhold Publishing Corp., New York (1946), p. 11.
27. Martin, A. A., and R. E. Rostenbach, "Industrial waste treatment and disposal," *Ind. Eng. Chem.* **45**, 9 (1953).
28. Meakes, R. C. W., and W. C. Wake (eds.), *Rubber Technology*, Butterworth's Scientific Publications, London (1951).
29. Memmler, K. (ed.), *The Science of Rubber*, Reinhold Publishing Corp., New York (1934).
30. Mills, R. E., "Progress report on the bio-oxidation of phenolic and 2,4-D waste waters," in Proceedings of 4th Ontario Industrial Waste Conference, June 1957, p. 30.
31. Nicolai, A. L., W. W. Eckenfelder, and D. G. Gardner, "Effluent treatment study for a rubber research lab," *Ind. Wastes* **1**, 136 (1956).
32. Noble, B. J., "Latex in industry," in *Rubber Age*, 2nd ed., New York (1953).
33. *Organic Chemistry*, Vol. 4, John Wiley & Sons, New York (1953), p. 717.
34. Rostenbach, R. E., "Status report on synthetic rubber wastes," *Sewage Ind. Wastes* **24**, 1138 (1952).
35. "Rubber," *Information Please Almanac*, Macmillan Co., New York (1956), p. 708.
36. *Rubber Technology*, Academic Press, New York (1951).
37. Ruchhoft, C. C., et al., "Synthetic rubber waste disposal," *Sewage Ind. Wastes* **20**, 180 (1948).
38. Schatze, T. C., "Effect of rubber wastes on sewage treatment processes," *Sewage Works J.* **17**, 497 (1945).
39. Sechrist, W. D., and N. S. Chamberlin, "Chlorination of phenol-bearing rubber wastes," in Proceedings of 6th Industrial Waste Conference, Purdue University, November 1951, p. 396.
40. Stern, H. J., *Practical Latex Work*, 2nd ed., Blackfriars Press, Ltd., London (1947).
41. Stevens, H. P., and W. H. Stearns, *Rubber Latex*, 4th ed., Rubber Growers' Association, London (1936), p. 224.
42. American Society Testing Materials Committee, *Symposium on the Application of Synthetic Rubbers*, Philadelphia (1944), p. 134.
43. Talalay, A., *Synthetic Rubber from Alcohol*, Interscience Publishers, New York (1945), p. 298.
44. *The Vanderbilt Latex Handbook*, Vanderbilt Co., New York (1954).
45. Wertheim, E., *Textbook of Organic Chemistry*, 3rd ed., Blackiston Co., New York (1951).

46. Whalley, M. E., *Abstracts on Synthetic Rubber*, 2 vols., Ottawa (1943).
47. Wilson, C. M., *Trees and Test Tubes*, Henry Holt & Co., New York (1943), p. 352.
48. "World rubber output," *Chem. Ind.*, April 1954, p. 466.

Suggested Additional Reading

The following references have been published since 1962.

1. "Advances in natural rubber technology," *Chem. Ind.*, 12th September 1964, p. 1573.
2. Culver. R. H., "Pilot-plant testing for domestic and industrial waste treatment," *J. Water Pollution Control Federation* **35**, 1455 (1963).
3. Drugov, Yu. S., "Use of spectrometry for analysis of sewage from synthetic rubber production," *Chem. Abstr.* **62**, 7496 (1965).
4. Iranov, V. A., and M. V. Pogorelova, "Erythrocytometry for a hygenico–toxicological evaluation of sewage from synthetic rubber production," *Chem. Abstr.* **60**, 5197 (1964).
5. Iranov, V. A., "Hygenic evaluation of the BOD of waste waters from synthetic rubber production," *Chem. Abstr.* **61**, 1600 (1964).
6. Iranov, V. A., "Toxicohygenic characterization of waste water from new synthetic rubber production," *Chem. Abstr.* **61**, 432 (1964).
7. Kotulski, B., "Effluents from synthetic rubber production," *Chem. Abst.* **61**, 4056 (1964).
8. Morzycki, J., et al., "Purification of sewage contaminated with latex," *Chem. Abstr.* **64**, 3192 (1966).
9. Orestova, N. N., "Local purification and use of waste waters from the polymerization shop in the Voronezh Synthetic Rubber Plant," *Chem. Abstr.* **62**, 11524 (1965).
10. Strukov, F. I., "Use of coagulation and extraction for purifying synthetic rubber production waste waters," *Chem. Abstr.* **62**, 11522 (1965).

24.11 Glass-Industry Wastes

In the manufacture of optical and other specialty glass, polishing produces a waste containing detergents, finely divided iron from the polishing process, and a considerable quantity of glass particles, many of which are microscopic in size. These materials form a quasi-emulsion with the following characteristics: (1) brick-red to scarlet color, (2) low BOD, (3) alkaline reaction, (4) nonsettleable solids, (5) resistance to acid cracking, and (6) resistance to alum coagulation.

McCarthy [2] showed that coagulation with calcium chloride produced a clear supernatant, when 250 ppm were used. This method reduced the total solids from 1080 to 3 ppm, the BOD from 40 to 28 ppm, and the color from 900 to 35 ppm. No other coagulant produced comparable results.

References: Glass-Industry Wastes

1. Durham, R. W., "Disposal of fission products in glass," Paper no. 57-*NESC*-54, read at 2nd Nuclear Engineering and Science Conference, March 1957, at Philadelphia, American Society of Mechanical Engineers, New York.
2. McCarthy, J. A., "Coagulating rouge with calcium chloride," *Public Works Mag.* **85**, 170 (1954).

24.12 Naval-Stores Wastes

Naval stores are manufactured by refining oleoresinous materials from pine wood. Among the products are wood rosin, wood turpentine, pine oil, dipentene, and other monocyclic hydrocarbons. Gum naval stores are obtained by simple steam distillation of resinous material taken from living pine trees. Wood naval stores, obtained from cut pine wood and stumps, can be subdivided into products obtained by (1) destructive distillation, (2) sulfate distillation or (3) steam distillation. Because of the nature of the raw material, gum refining is carried out on a seasonal basis. Wood (southern pine), on the other hand, is refined all year round.

Stumps that have been rotting on the ground for many years are conveyed to a washer where dirt, sand, and loose or rotted wood are removed. The clean stumps are fed directly to grinding equipment, which delivers wood chips about one inch long and $\frac{3}{16}$ inch thick for extraction. The extraction process is carried out in large pressure vessels at about 280°F and at a pressure of 80 psi. A solvent, usually a narrow-range petroleum fraction (200 to 400°F naphtha), is pumped into the bottom of the first extractor. The solvent flows out of the top of the first extractor and into the bottom of the second, and so on, until about 15,000 gallons of solvent have been pumped into 10 tons of pine chips. The spent chips are used for fuel, and the crude extract solution of rosin, solvent, and terpene oils, known as "drop solution," is pumped to tanks where water sprays wash out color bodies. A heavy, dark resin (nigre) separates from the wash water and is drained off. The remaining solution is evaporated to recover the solvent, and a series of successive equilibrium distillations remove the terpene oils, leaving a marketable residue known as wood rosin. The terpene oils are further fractionated by batch distillation into turpentine, pine oil, dipentene, and other terpene derivatives. Black [1] presents a simplified flow diagram of these processes (Fig. 24.19).

For each ton of pine stumps, 100 to 600 gallons of extraction solvent and 2000 to 5000 gallons of water

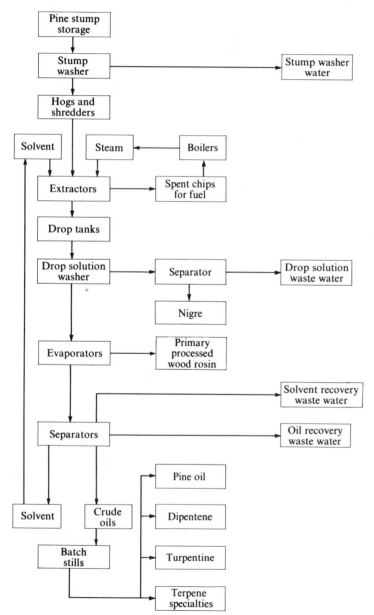

Fig. 24.19 Flow diagram for wood naval-stores plant, using naphtha as the solvent for the primary extraction process. (After Black [1].)

Table 24.37 Representative values of character and volume of primary process naval-stores manufacturing wastes. (After Black [1].)

Primary process waste source	Discharge, gal*		Oxygen consumed (dichromate), ppm		Total solids, ppm		pH range		Phenols, ppm		BOD, 5-day, 20°C				Suspended solids				Population equivalent‡	
											ppm		lb*		ppm		lb*			
	N†	B†	N	B	N	B	N	B	N	B	N	B	N	B	N	B	N	B	N	B
Stump washer	350–3500		5700–16,400		11,000–20,000		4.5–6.7		0		50–1500		0.05–9.63		1220–16,900		2.34–105		0.3–58	
Drop solution wash	180		5200		1900		4.5–6.1		0		3130		4.34		150		0.22		29	
Solvent recovery	730	150	1100	2,800	285	150	6.1–7.2	3.3–3.5	0		550	2,100	3.5	2.5		16			20	16
Oil recovery	250	1170	4000	1,360	310	30	4.3–6.9	4.1–5.2	0		2170	60	5.7	0.58		65	0.63		34	4
Furfuraldehyde recovery		12		14,500		150		3.4–3.5				10,700		1.07		8				6

*Per ton of stumps.
†N, naphtha extraction; B, benzene extraction.
‡Per ton of stumps, based on 0.167 lb of 5-day BOD per capita per day.

Table 24.38 Representative values of character and volume of secondary process naval-stores manufacturing wastes. (After Black [1].)

Secondary process waste source	Unit	Discharge, gal*	Oxygen consumed (dichromate), ppm	Total solids, ppm	pH range	Phenols, ppm	BOD		Suspended solids		Population equivalent*†
							ppm	lb*	ppm	lb*	
Pale rosin refining											
Solvent recovery	1 ton stumps	490			3.6+	0	820	3.35	12	0.05	20
Alcohol recovery		485			5.8–6.2	0	385	1.55	24	0.10	9.30
Polymerization											
Solvent recovery	1000 lb rosin	2100	430	1,530	1.7–3.2		75	1.33	28	0.5	8
Acid recovery		750	30		2.5–3.5		17	0.11			0.6
Disproportionation											
Condenser water	1000 lb rosin	3900	590	150	5.9–6.5	{2.0}	260	8.45	41	1.33	50
Alkaline wash water		140	79,000	14,600	8.5–11.9		38,100	44.4	3,350	3.90	265
Hydrogenation											
Waste rosin oils	1000 lb rosin	1	960,000	410,000	12.2+		382,000	3.2	26,500	0.22	19
Isomerization											
(wash water)	1000 gal pine oil	500	17,000	128,000	12.2+	1970	14,800	49.3	75	0.25	295
Synthesization											
(condenser water)	1000 lb rosin	6000	830	415	9.5–10.5	0.2	270	13.5	130	6.5	80

*Per unit.
†Based on 0.167 lb 5-day BOD per capita per day.

are used. Actually the solvent is recovered and reused (see Fig. 24.19) so that only about one gallon of solvent is wasted per ton of chips. Secondary refining processes are carried out by some naval-stores plants, in which the rosins and oils obtained from primary processing are utilized as raw materials. Black [1] presents the characteristics of primary-processed wastes of wood naval stores (Table 24.37) and also gives representative values of character and volume of secondary-processed wastes (Table 24.38).

A few remedial measures to handle naval-stores wastes have been proposed, although little, if any, waste treatment is currently practiced. The suggested methods include elimination (process change or stopping of waste discharge), by-product recovery (such as terpene hydrate), equalization of flow, recirculation and reuse, and waste treatment where necessary; pilot-plant results have shown that trickling filtration produces an 83 per cent BOD reduction.

References: Naval-Stores Wastes

1. Black, H. H., and V. A. Minch, "Industrial waste guide, wood naval stores," *Sewage Ind. Wastes* **25**, 462 (1953).
2. "1951 Industrial wastes forum," *Sewage Ind. Wastes* **24**, 869 (1952).
3. Palmer, R. C., "Producing naval stores from waste pine wood," *Chem. Met. Eng.* **37**, 140, 289, and 422 (1930).
4. Shantz, J. L., and T. Marvin, "Waste utilization," *Ind. Eng. Chem.* **31**, 585 (1939).

SPECIAL-MATERIALS INDUSTRIES

The major special-materials industries are glue manufacturing, wood preserving, candle manufacturing, and plywood making.

24.13 Animal-Glue Manufacturing Wastes

Sources of wastes. The manufacturing process may be divided into three areas; the mill house, where stock is made amenable to glue extraction by washing and milling; the cook house, where the stock is cooked and the light liquor is extracted; and the drying rooms, where the light liquor is condensed and dried to the final product (Fig. 24.20).

The bulk of water-borne wastes come from the mill house and include lime from soaking, acids, wash water, dirt, hair, and miscellaneous solids brought in with the stock. Three types of stock (raw material) are used—green fleshings, hide stock, and chrome-split stock—and processing is somewhat different for each. Fleshings, the substance which joins animal skins to the carcass, is the most perishable raw material. They are washed for approximately 12 hours, acidulated, then soaked, before being brought to the cook house. Hides are first desalted and cleaned by milling for 6 hours, then limed and soaked for 80 to 90 days in large vats. After this soaking period they are washed free of lime, acidulated, and sent to the cook house. Chrome splits are first chopped up mechanically and limed for approximately 5 hours. The lime is washed out, sulfuric acid is added, and the washing begins. After the acid has been washed out, the stock soaks in dilute acid for approximately 8 hours before neutralization is provided. After a final hour of soaking the stock is sent to the cook house.

Cook-house operations consist of loading the prepared stock into large kettles and cooking it with steam at temperatures ranging from 160° to 200°F. After about 6 hours the light liquor is drained off. The process is repeated three more times, each one yielding a lower grade of glue. The material left after the final cook cycle is called tankage. This is further treated to yield products such as grease, hair, and fertilizer.

The wastes generated have been shown to be amenable to biological treatment; methods such as foam separation, flotation, and sedimentation seem valid primary processes. Table 24.39 gives the unit waste loads recently found by the author in an animal-glue manufacturing plant.

Table 24.39 Summary of unit waste loads of an animal-glue manufacturing plant.*

Stock	Waste, gal	BOD, lb	Suspended solids, lb	COD, lb
Fleshings	50.5	2.5	4.25	4.8
Hide	54.7	0.58	1.92	1.42
Chrome stock	51	0.28	0.40	0.65

*All quantities are those produced per pound of glue.

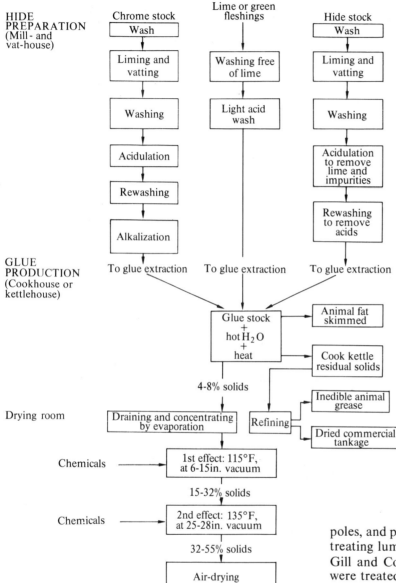

Fig. 24.20 Glue manufacturing process flow chart.

24.14 Wood-Preservation Wastes

There are about 400 wood-preserving plants in the United States. Almost all use pressure processes to preserve wood with either creosote or pentachlorophenol or both. Both preservatives are increasing in total use, the creosote chiefly to preserve crossties, poles, and piling and pentachlorophenol primarily for treating lumber, poles, and cross-arms. According to Gill and Corey [2], 262,000,000 cubic feet of wood were treated with preservatives and fire retardants in 1965 in the United States.

Origin of wastes. Air-dried timber (or sometimes properly conditioned green wood) is first steamed at about 20 psi for up to 12 hours while the condensate is continually removed. The second step involves a relatively short period of evacuation before the pressure is raised with air to 30 to 90 psi. During this latter period the retort is filled with either creosote or pentachlorophenol oil. Air is vented while the preservative is added until all the air has been displaced by the chemical. A maximum pressure of about 200 psi is used for a period of 2 to 8 hours. As the pressure is then released, air trapped in the wood bubbles out, bringing any excess preservative with it. The preservative is usually too valuable to waste and is returned to the storage tank. A final vacuum is applied to remove any excess oil from the surface of the wood.

The condensate removed during the first steaming process appears to be the major source of waste, since it contains dissolved preservative left on the walls of the retort in previous cycles. An attempt is usually made to recover preservatives by pumping the condensate to sumps for differential gravity separation. Creosote, being denser than water, is drawn off the bottom of the sump and returned to storage tanks or directly to the retort. Further recovery is obtained by using additional settling tanks in series. Pentachlorophenol oil, being lighter than water, is removed from the top of the sump and returned to storage tanks.

The water and material extracted from the wood during steaming (the other major source of waste) contain a considerable portion of contaminating matter and are usually treated similarly to the steam condensates. After all the differential separation has taken place, the residual water-bearing wastes are discharged to the appropriate watercourse.

Characteristics of the waste. Middlebrooks [3] gives the following characteristics of the total waste:

Constituent	Raw waste concentration (mg/liter)
COD	11,500–19,600
BOD	2800–5000
Total solids	6340
Total volatile solids	5730
Phenols	17–85
pH	5.2
Suspended solids	1420
Volatile suspended solids	1100
COD from raw waste filtrate	7080
Nitrogen (total) as N	89
Nitrogen (NH_3) as N	32
Nitrogen (organic) as N	57
Phosphates (total)	< 5.0

Although Middlebrooks does not give unit volumes of waste, he notes a volume of about 3000 gallons per day. He recognizes that "although the waste water from the wood preservation process is highly contaminated, the volume is relatively small."

Treatment of wastes. Both Middlebrooks [3] and Gaudy *et al.* [1] found that this waste was amenable to biological treatment. The former also found that flocculation with lime achieved a significant reduction in COD and BOD. Chlorination is relatively ineffective in reducing the COD. However, chemical flocculation followed by chlorination reduced the COD of the waste by 85 per cent and reduced the phenol concentration to less than 0.5 mg/liter. Gaudy found considerable success with oxidation-pond treatment.

References: Wood-Preservation Wastes

1. Gaudy, A. G., Jr., R. Scudder, M. M. Neely, and J. J. Perot, "Studies on the treatment of wood preserving wastes," Presented at Symposium on Selected Papers in Water and Wastewater Technology, 55th Meeting of the American Institute of Chemical Engineers, Houston, Texas, Feb. 7–11, 1965.
2. Gill, T., and E. A. Corey, *Wood Preservation Statistics—1965*, Forest Service, U.S. Department of Agriculture, Washington, D.C. (1966).
3. Middlebrooks, E. J., "Wastes from the preservation of wood," *J. Sanit. Eng. Div. Am. Soc. Civil Engrs.* **SA1**, No. 5785, 41 (1968).

24.15 Candle-Manufacturing Wastes

The candle industry has moved with the times.* From the old, white, hand-made product has evolved the modern, colored, mass-produced, long-lasting, and dripless candle. Quality candles always incorporate stearic acid, although the most important single ingredient in candle manufacturing is still paraffin wax. The two types of candle are named after these ingredients. The paraffin candle is the less expensive one; it melts easily and burns down rapidly. Because

---

*"Lighting the way," *Wallace and Tiernan Topics* **20**, 1 (1966).

they also drip, paraffin candles are used for the multicolor drip effects popular today. The stearic-acid candle burns longer, stays cleaner, and does not drip, because of its 20 to 30 per cent stearic-acid content (the remainder being mainly paraffin wax).

Whatever their composition, candles are made in only two ways, either dipped or molded. In the dipping method, racks of suspended wicks are repeatedly dipped into a vat of molten wax. With each dip a new layer is built onto the candle until it is complete. In the molding method, molten wax is poured into molds around already-positioned wicks. The wax ingredients are carefully controlled to ensure that the candles shrink after cooling and do not stick to the molds. High-quality candles are finally dipped into a bath of Cenwax G (hydrogenated castor oil) to give them a hard shell. When the candle is lit, this shell collects the molten wax and prevents dripping.

It is reported* that little waste is evolved from this industry, because any spilled wax is simply recovered, melted, and reused. However, when a candle manufacturer makes his own stearic acid (in an "integrated" plant) there is a possibility of some waste being produced. Stearic acid is made from the high-pressure steam-splitting of tallow, with glycerine and oleic acid separated and sold as by-products. The only published treatment for wastes from an "integrated" candle-manufacturing plant is anaerobic digestion.

24.16 Plywood-Plant Glue Wastes

Plywood plants are divided into two parts: the *green end*, designed to store, debark, veneer, and prepare the wood laminates for sheet formation; and the *glue end*, where the laminates are glued, pressed, trimmed, sanded, and packaged for shipment. The complete plant processes are shown schematically in Fig. 24.21.

Major wastes from these plants originate from the glue kettles and the glue spreaders. Bodien† reports that a typical plant, producing 100 million square feet of plywood per year (one-half interior and one-half exterior grade), makes about 11 batches of interior glue and 9 batches of exterior glue a day (equivalent to 400,000 and 350,000 pounds, respectively, per month). Washing the glue-mixing equipment produces a small volume of highly concentrated waste. A typical plant also has six periods of use of glue-spreaders per day; the spreaders are usually washed down once a shift when interior glue is used and at least once a day for exterior

---

*"Lighting the way," *Wallace and Tiernan Topics* **20**, 1 (1966).

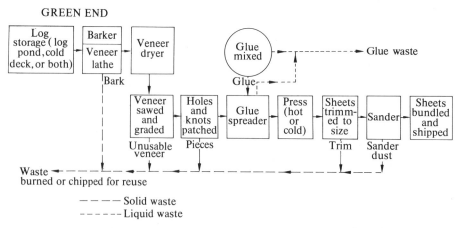

Fig. 24.21 Plywood plant flow diagram. (After Bodien [1968].)

Table 24.40 Typical plywood-plant waste characteristics. (After Bodien.†)

Plant no.	1966 production, ft² (⅜-inch basis)	Average discharge, gpm	pH	COD, mg/l	Total PO₄, mg/l	Total Kjeldahl nitrogen, mg/l	Phenol, mg/l	Suspended solids, ppm	Total solids, ppm
1	100,000,000	18.2							
2	135,000,000	30.2	11.6	1814	15	110	1667	148	1627
3	100,000,000	21.6	9.4	1917	9	64	1790	356	1814
4	70,000,000	54.0	10.8	1621	12	3	222	330	1120

†Bodien, D. G., *Progress Report, Plywood Plant Glue Waste Disposal*, Report no. PR-2. United States Department of Interior, Technical Projects Branch, Northwest Region, Pacific Northwest Water Laboratory, Corvallis, Oregon, February 1968.

glue. The difference in frequency is due to the relatively short pot life of interior glues (6 to 8 hours).

Exterior glues contain a furfural extraction of corncobs and oat hulls, as well as wheat flour, phenolic formaldehyde resin, and caustic soda and ash. Interior glues contain dried blood, soya flour, lime, caustic soda, and sodium silicate. In addition, formaldehyde and pentachlorophenol may be added to produce a glue toxic to insect pests. Typical waste characteristics, derived from Bodien's study, are shown in Table 24.40.

Treatment. Most plywood plants practice settling, either in ponds or septic tanks. This treatment removes some of the glue solids and wood chips, but achieves little reduction in phenols, phosphates, or total Kjeldahl nitrogen. Some plants discharge their glue wastes to municipal treatment systems, where high pH, glue solids, and wood chips have been known to cause difficulty. Because of the high organic content of the relatively low-volume glue wastes, properly controlled incineration offers a potential solution to glue-waste disposal.

CHAPTER 25

CHEMICAL INDUSTRIES

Once again we must draw a very thin, and often questionable, distinction between the chemical industries described in this chapter and the materials industries of the previous chapter. Undoubtedly, there are some industries which could be classified in either chapter. In general, however, the chemical industry encompasses smaller plants which produce basic chemicals and raw materials to be used by other manufacturers, whereas the materials industry consists of larger plants producing materials for direct public use.

Chemical wastes are produced by plants that manufacture acids, bases, detergents, cornstarch, powder and explosives, insecticides and fungicides, fertilizers, silicones, plastics, resins, synthetics, and other substances which are often used as raw materials for further manufacturing processes. Chemical processes vary greatly, according to the nature of the substance being manufactured; they include chemical reactions at high temperatures and pressures with or without catalyst and flux, separation of a liquid from a gas, solid, or other liquids, and so forth. Some of the manufacturing methods are sedimentation, flotation, evaporation and distillation, washing, filtering, electrolysis, burning, centrifugal separating, absorption, crystallization, and screening.

Chemical wastes include acids, bases, toxic materials, and matter high in BOD, color, and inflammability (phosphorus) and low in suspended solids. Often chemical wastes require neutralization, as in the manufacture of silicones, smokeless powder, TNT, insecticides, and herbicides which are acid in character. Many can be treated by some biological-oxidation method such as trickling filters, activated sludge, or lagooning. It has been found that if cornstarch waste is mixed with an equal volume of domestic sewage, it can be treated by either activated sludge or trickling filters.

Coagulation is necessary for some wastes, such as those containing phosphorus. Wastes with a complex molecular structure are often separated by some physical process. Detergents, which present the problem of foaming in aeration tanks, have been found amenable to biological degradation, although some of the so-called "hard" syndets are resistant to this. Lake studies have shown that the advent of detergents in domestic sewage has increased the residual soluble inorganic phosphorus by as much as 100 per cent. This increase was directly related to frequent nuisance blooms of blue-green algae.

25.1 Acid Wastes

Acid wastes may be discharged by any of the plants mentioned in the apparel, materials, and chemical industries. Since most states have laws relating to stream standards which require that the pH of a receiving stream be maintained between 6.0 and 9.0, acid wastes usually cannot be allowed to flow untreated into our watercourses. In this chapter we are especially concerned with acids arising from chemical plants producing primary raw materials such as dyes, explosives, pharmaceuticals, and silicone resins. The most important are the dilute wastes of hydrochloric, sulfuric, and sometimes nitric acid. Because of the varied uses of acids, the origin of any one of these acid wastes may bear little or no resemblance to the origin of another.

Regardless of the degree of acidity, the main method of treatment of acid wastes is neutralization (see Chapter 8). Gehm [8] describes a method of

25.1　　　　　　　　　　　　　　　　　　　　　　　　　　　ACID WASTES　　469

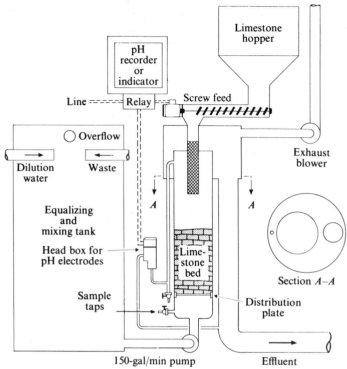

Fig. 25.1 Proposed design of an upflow neutralizing bed, capable of handling 100,000 gal/day of nitrocellulose waste containing from 10,000 to 15,000 ppm mineral acidity. (After Gehm [8]).

neutralizing acid wastes by means of an upflow limestone bed, which handled wastes with up to 10,000 ppm of mineral acidity in a bed capable of receiving 0.1 mgd of waste (Fig. 25.1). Shugart [23] describes a process utilizing lime for automatically neutralizing acid citrus wastes, employing Beckman pH electrodes for control (Fig. 25.2).

As mentioned in Chapter 24, oil refineries use sulfuric acid for desulfurization, improvement of color, refining of lubricating oils, as a catalyst in alkylation, and for other miscellaneous purposes. Two processes are commonly used by oil refineries for disposal of sulfuric-acid wastes: spray-burning and indirect combustion. Spray-burning involves spraying waste acid into a hot combustion chamber (1700 to 2000°F) with small amounts of excess air added to oxidize hydrocarbons. Sulfur is converted to SO_2 and hydrocarbons to CO_2 and H_2O, the hot gases are cooled and dried, and the SO_2 is absorbed to make new sulfuric acid. The principal reaction of the second method, indirect

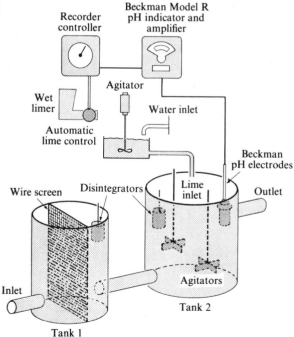

Fig. 25.2 Diagram showing equipment layout for treatment of orange and lemon wastes [22].

combustion, is the reduction of the sulfuric acid in the sludge by the hydrocarbons which are present. Granular by-product coke is recirculated through a mixer, acid sludge is added to this circulating stream, and heat is applied in the decomposing chamber. Flow diagrams of these processes are presented in Fig. 25.3.

Dickerson and Brooks [2] describe a process using lime neutralization for a mixed nitric and sulfuric acid waste from a nitrocellulose manufacturing plant. Although the acids varied widely in volume, concentration, and ratio, neutralization was accomplished effectively in a multiple-unit reaction chamber provided with two-point pH-controlled addition of dolomitic lime slurry.

Tully [27] also describes an upflow limestone bed acid-neutralizing unit for treating a mixture of hydrochloric and sulfuric acids in varying concentrations—wastes from the manufacture of certain resins. The wastes were diluted until they reached a concentration of less than 1 per cent and then passed upward through a 3-foot expanded limestone bed at an average rate of 20 to 30 gallons per minute per square foot of bed area. The effluent pH averaged 4.6, and the 1958 operating costs were about $0.49 per ton of 1 per cent acid neutralized. The installation is shown in Figs. 25.4 and 25.5.

References: Acid Wastes

1. Cooper, J. E., "How to dispose of acid wastes," *Chem. Ind.* **66**, 684 (1950).
2. Dickerson, B. W., and R. H. Brooks, "Neutralization of acid wastes," *Ind. Eng. Chem.* **42**, 599 (1950).
3. "Disposal at sea, National Lead Company," *Sewage Ind. Wastes* **18**, 1250 (1946) and **19**, 1111 (1947).
4. Faust, S. D., "Sludge characteristics resulting from

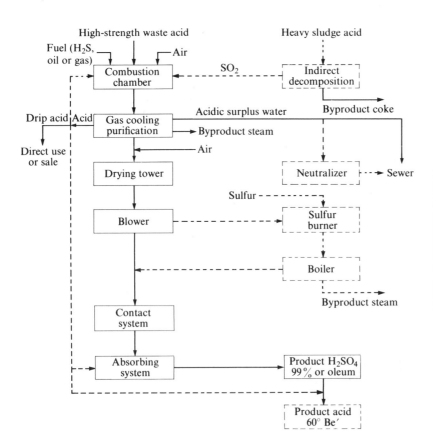

Fig. 25.3 Flow diagram showing (solid lines) basic elements in the spray-burning type of acid recovery plant, (dotted lines) supplementary features which may be used, depending on particular requirements, and the necessary features present in a combination of the spray-burning and indirect-combustion processes. (Courtesy T. R. Harris, Monsanto Chemical Co.)

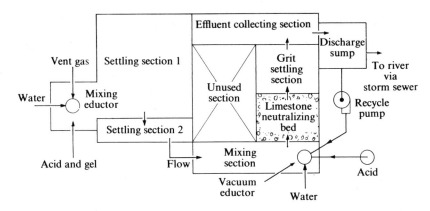

Fig. 25.4 Schematic plan and flow diagram of existing acid neutralization system [26].

lime neutralization of dilute sulfuric acid wastes," in Proceedings of 13th Industrial Waste Conference, Purdue University, May 1958.
5. Faust, S. D., H. E. Orford, and W. A. Parsons, "Control of sludge volume following live neutralization of acid wastes," *Sewage Ind. Wastes* **28**, 872 (1956).
6. Faust, S. D., and H. E. Orford, "Crystal seeding by returned sludge," *Ind. Wastes* **2**, 36 (1957).
7. Gehm, H. W., "Neutralization of acid waste water with an up-flow expanded limestone bed," *Sewage Works J.* **16**, 104 (1944).
8. Gehm, H. W., "Up-flow neutralization of acid wastes," *Chem. Met. Eng.* **51**, 124 (1944).
9. Gross, C. D., and C. Lee, "Collection and treatment of acid runoff from coal gob-pile-storage areas," in Proceedings of 6th Industrial Waste Conference, Purdue University, February 1951, p. 10.
10. Jacobs, H. L., "Acid wastes, neutralization," *Sewage Ind. Wastes* **23**, 900 (1951).
11. Jones, E. M., "Acid wastes, treatment studies," *Sewage Ind. Wastes* **22**, 224 (1950).
12. Keating, R. J., and R. Dvorin, "Dialysis for acid recovery," in Proceedings of 15th Industrial Waste Conference, Purdue University, May 1960, p. 567.
13. Ledford, R. F., and J. C. Hesler, "Treatment of chromic acid wastes," *Ind. Eng. Chem.* **47**, 83 (1955).
14. Levine, R. Y., and W. Rudolfs, "Sludge characteristics of lime neutralized pickling liquor," in Pro-

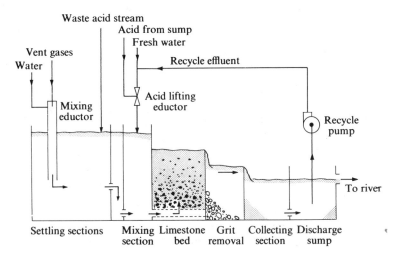

Fig. 25.5 Schematic cross section of existing acid neutralization system [26].

ceedings of 7th Industrial Waste Conference, Purdue University, May 1952.
15. Lewis, C. J., and L. J. Yost, "Acid wastes, lime treatment," *Sewage Ind. Wastes* **22**, 893 (1950).
16. The Finishing Lime Association of Ohio, *Lime, The Low-Cost Way to Prevent Stream Pollution*, (1949).
17. Minnick, L. J., and C. Presgrave, "Physico-chemical characteristics of liming materials as related to neutralization of mineral acids," in Proceedings of 9th Industrial Waste Conference, Purdue University, May 1954, p. 284.
18. Parsons, W. A., "After-precipitation resulting from lime neutralization," in Proceedings of 6th Industrial Waste Conference, Purdue University, February 1951.
19. Powell, S. T., "Acid wastes and temperature of surface and ground waters are important considerations to many industries," *Ind. Eng. Chem.* **46**, 97A (1954).
20. Rudolfs, W., "Neutralization with lime," *Sewage Ind. Wastes* **15**, 590 (1943).
21. Rudolfs, W., "Pretreatment of acid wastes," *Sewage Ind. Wastes* **15**, 48 (1943).
22. "Shortcut to neutral water (ion exchange)," Editorial, *Chem. Week* **90**, 69 (1962).
23. Shugart, P. L., "Automatic pH control replaces manual operation for acid waste treatment," *Public Works Mag.* **85**, 67 (1954).
24. Stone, T., "Studies of acid precipitation at Salford," *J. Inst. Sewage Purif.* (1952), p. 361.
25. Tatlock, M. W., "Treatment of acid wastes for Emery Industries, Inc., Cincinnati, Ohio," in Proceedings of 14th Industrial Waste Conference, Purdue University, May 1959.
26. Temple, K. L., and A. R. Colmer, "The formation of acid lime drainage," in Proceedings of 6th Industrial Waste Conference, Purdue University, February 1951.
27. Tully, T. J., "Waste acid neutralization," *Sewage Ind. Wastes* **30**, 1385 (1958).
28. National Lime Association, "The use of lime *vs.* caustic soda and soda ash as acid neutralizing agents," Trade Waste Bulletin no. 2, 1 October 1948.
29. Waite, C. F., "Treatment of acid cyanide, acid chromium wastes," in Proceedings of 5th Industrial Waste Conference, Purdue University, November 1949.
30. Wing, W. E., "A pilot plant for lime neutralization studies," in Proceedings of 5th Industrial Waste Conference, Purdue University, November 1949.

Suggested Additional Reading

The following references have been published since 1962.
1. "Vat evaporator cuts acid recovery cost," *Chem. Eng.* **67**, 66 (1960).
2. Bray, L. A., and E. C. Martin, "Use of sugar to neutralize nitric acid waste liquors," U.S. Atomic Energy Commission, Invention Report, *HW*-75562 (1962).
3. Baffa, J., "Design of neutralization basins for acid wastes," *Ind. Water Wastes* **5**, 107 (1960).
4. Crits, G. J., and C. M. C. McKoewn, "Neutralize acid and alkaline wastes," *Power* **106**, 182 (1962).
5. Eifert, G., and E. Mueller, "Automatic neutralization of waste waters," *Regelungstech. Praxis* **2**, 61 (1960); *Water Pollution Abstr.* **34**, 152 (1961).
6. Greuter, E., "Practical experience in the region of automatic neutralization of waste waters," *Lit. Bver. Wasser Abwasser Luft Boden* **8**, 119 (1959).
7. Huggett, R. E., and W. H. Toller, "Automatic continuous acid neutralization," in Proceedings of 16th Industrial Waste Conference, Purdue University Engineering Extension Series, no. 109, May 1961, p. 438.
8. Krofchak, D., "Submerged combustion evaporation of acid wastes," *Ind. Water Wastes* **7**, 63 (1962).
9. Krofchak, D., "Treatment of acid wastes and recovery of valuable by-products using submerged combustion," in Proceedings of 8th Ontario Industrial Waste Conference, 1961.
11. Stein, V., U. Esche, and R. Miehels, "Treatment of sulfuric acid waste," Belgian Patent no. 633219, November 1963.
12. Willy, C. F., and Buesching, "Reclaiming free sulfuric acid from wastes," *Buesching & Co., Ingenieurbauer*, 3 January 1963.
13. Wiest, G. J., "Lime as a neutralizing agent," *Ind. Water Wastes* **7**, 14 (1962).
14. Wheatland, A. B., and B. J. Borne, "Neutralization of acidic waste waters in beds of calcined magnesite," *Water Waste Treat. J.* **8**, 650 (1962).
16. Vieweq, R., "Purification of waste HC—an example of steam distillation of trace components," *Chem. Tech.* **14**, 728 (1962).

25.2 Cornstarch-Industry Wastes

Although this industry might be included under food processing in Chapter 23, it is described here because cornstarch products are so widely used in the chemical and materials field. This industry processes corn to produce starch, oil, and feed. A bushel of corn weighs on average about 56 pounds [19] and by the wet-milling process this yields about 32 pounds of pearl starch (used in the manufacture of textiles), 1.6 pounds of oil, and 13 to 14 pounds of feed. The composition of the corn kernel is approximately as follows: carbohydrates, 80 per cent; protein, 10 per cent; oil (fat), 4.5 per cent; fiber, 3.5 per cent; and minerals, 2 per cent. In the early 1930s industry began a waste-water reuse program which reduced plant losses to less than 0.5

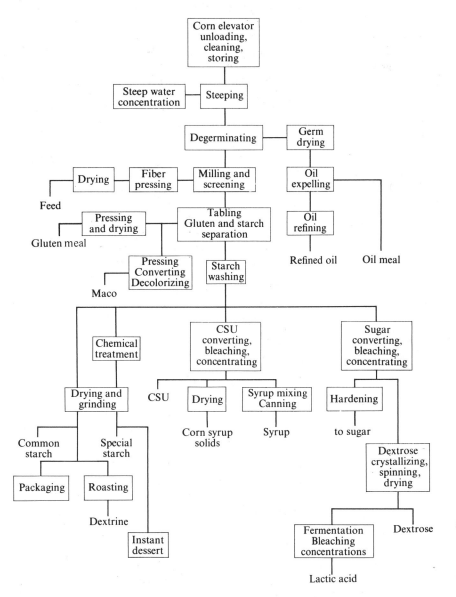

Fig. 25.6 Flowsheet of wet-milling process for corn. (After Van Patten and McIntosh [24].)

per cent of the dry corn raw material. This has become known as the "bottled-up" system.

Clean corn is steeped in a dilute solution of sulfuric acid in order to loosen the hull, soften the gluten, and dissolve minerals or organic matter in the kernel. Next the corn is ground to free, but not crush, the germ; the ground corn is mixed with water and placed in settling tanks. When the germs float to the top, they are skimmed off, and oil is pressed or extracted from them. The kernel residues are ground finely, to separate the soluble starch and gluten from the fiber and hull, which are known as "grits and bran" and are used as feed additives. The starch is separated from the gluten by settling, centrifuging, and countercurrent washing. The gluten is added to feed, and the starch is filtered and washed on vacuum filters and dried. Then it is either marketed as starch, or modified starch, or hydrolyzed into corn syrup or corn sugar.

Both the feed-producing and starch-manufacturing processes evolve a process-water waste containing about 3 per cent of the corn in soluble form. Van Patten and McIntosh [24] present a flow sheet for wet milling of corn (Fig. 25.6), and Hatfield [19] presents a flow sheet of both the wet-milling process and "bottling-up" reuse of process waters in cornstarch manufacture (Fig. 25.7).

Even before waste-treatment practices were utilized in the industry, the "bottling-up" process was common, since it was introduced to abate stream pollution, and it can now be considered as an actual part of the cornstarch-plant process. The wastes from this industry consist of residues and leaks from the reuse processes. The bottling-up process has been described [19] as: (1) recirculation of process water, (2) evaporation of a portion of this recirculated water as steep water, and (3) addition of all dried organic res-

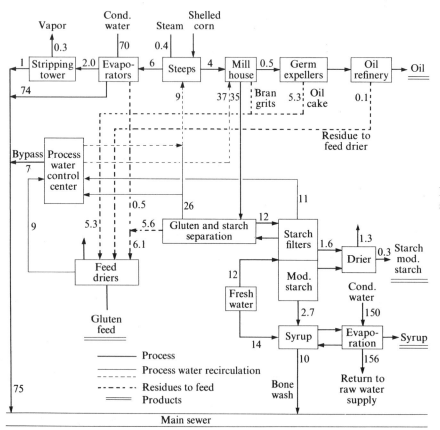

Fig. 25.7 Flow sheet showing cornstarch production. (After Hatfield [19].)

idues to the gluten to make an improved cattle feed.

Major wastes from the cornstarch plants are: (1) volatile organics entrained in the evaporator condensate, (2) syrup from final wastes, and (3) wastes from bottling-up processes—arising primarily because of an imbalance between the amount of fresh water added, the amount of recirculated water, and the amount of steep water taken to the evaporators. Hatfield [19] lists the population equivalents of these wastes (Table 25.1). For each bushel of corn processed, 40 gallons of water are used directly in the process and 100 to 200 gallons per bushel for other purposes, although much of this is reused [21].

Table 25.1 Population equivalents of cornstarch wastes [19].

Waste source	Population equivalent*
Steep-water evaporators	30,000
Light bone wash	8,000
Cleanup, etc.	12,000
Total	50,000

*Per 50,000-bushel grind.

Van Patten and McIntosh [24] describe waste reduction at the American Maize Products Company, which manufactures starches, sugars, syrups, and fermentation products from corn. In 1951, the company spent about $850,000 to replace equipment and install modifications in production procedures that would reduce its waste BOD load by 92 per cent. Since the residual cornstarch-plant wastes generally contain organic solubles from whole-kernel corn and corn syrup, they are particularly amenable to biological treatment, especially when mixed with domestic sewage. The wastes are normally quite hot, and all factors directly attributable to this characteristic must be considered; for example, oxygen solubility is lowered, but at the same time digestion is enhanced. Settling is hindered by the heat of the wastes, and sewer lines tend to clog as the starch wastes cool. Equalization, to effect both cooling and homogeneity of the wastes, should thus be practiced prior to biological treatment, either at the industrial-plant site or in the municipal treatment plant.

Hatfield *et al.* [9] used trickling filters with a 5:1 recirculation ratio to effect a 90 per cent BOD reduction, with an influent BOD concentration of 1400 ppm or less and a loading of about 4 pounds of BOD per cubic yard per day. Control of pH and addition of nitrogen and phosphorus were necessary.

References: Cornstarch-Industry Wastes

1. Adinoff, J., "Disposal of organic chemical wastes to underground formations," *Ind. Wastes* **1**, 40 (1955).
2. Douglass, I. B., "By-products and waste in potato processing," in Proceedings of 15th Industrial Waste Conference, Purdue University, May 1960, p. 99.
3. Fergason, A. A., "Bacterial utilization of potato starch wastes," in Proceedings of 15th Industrial Waste Conference, Purdue University, May 1960, p. 258.
4. Greeley, S. A., and W. D. Hatfield, "The sewage disposal works of Decatur, Ill.," *Trans. Am. Soc. Civil Engrs.* **94**, 544 (1930).
5. Greenfield, R. E., *et al.*, "Cornstarch waste treatment with sewage, Decatur, Ill.," *Sewage Ind. Wastes* **19**, 951 (1947).
6. Greenfield, R. E., G. N. Cornell, and W. D. Hatfield, "Cornstarch wastes," in Proceedings of 3rd Industrial Waste Conference, Purdue University, May 1947, p. 360.
7. Greenfield, R. E., G. N. Cornell, and W. D. Hatfield, "Industrial wastes: cornstarch processes," *Ind. Eng. Chem.* **39**, 583 (1947).
8. Gurnham, C. F. (ed.), *Principles of Industrial Waste Treatment*, John Wiley & Sons, New York (1955), p. 375.
9. Hatfield, R., E. R. Strong, F. Heinsohn, H. Powell, and T. G. Stone, "Treatment of wastes from a corn industry by pilot-plant trickling filters," *Sewage Ind. Wastes* **28**, 1240 (1956).
10. Hatfield, W. D., "Operation of the pre-aeration plant at Decatur, Ill.," *Sewage Works J.* **3**, 621 (1931).
11. Hatfield, W. D., "Special test for corn starch wastes," *Sewage Ind. Wastes* **22**, 1381 (1951).
12. Hatfield, W. D., "Cornstarch processes," in *Industrial Wastes, Their Disposal and Treatment*, ed. W. Rudolfs, Reinhold Publishing Corp., New York (1953), p. 132.
13. Haupt, H., "By-product recovery from starch wastes," *Sewage Ind. Wastes* **8**, 350 (1936).
14. Hussman, W., "Starch waste treatment experiments," *Sewage Ind. Wastes* **6**, 342 (1934).

15. Mohlman, F. W., "Treatment of packing-house, tannery, and corn-products wastes," *Ind. Eng. Chem.* **18**, 1076 (1926).
16. Mohlman, F. W., "Utilization and disposal of industrial wastes," in Proceedings of 1st Industrial Waste Conference, Purdue University, November 1944, p. 43.
17. Mohlman, F. W., and A. J. Beck, "Disposal of industrial wastes," *Ind. Eng. Chem.* **21**, 205 (1929).
18. Pulfrey, A. L., R. W. Kerr, and H. R. Reintjes, "Wet milling of corn," *Ind. Eng. Chem.* **32**, 1483 (1940).
19. Rudolfs, W., *Industrial Waste Treatment*, Reinhold Publishing Corp., New York (1953), p. 132.
20. Sjostrom, O. A., "Treatment of waste water from a starch and glucose factory," *Ind. Eng. Chem.* **3**, 100 (1911).
21. Corn Industry Research Foundation, *The Story of Corn and Its Products*, New York (1952).
22. Van Patten, E. M., and G. H. McIntosh, "Waste savings at American Maize Products Company," in Proceedings of 6th Industrial Waste Conference, Purdue University, February 1951, p. 344.
23. Van Patten, E. M., and G. H. McIntosh, "Corn products manufacture, waste load reduction," *Sewage Ind. Wastes* **24**, 1443 (1952).
24. Van Patten, E. M., and McIntosh, G. H., "Liquid industrial wastes, corn products manufacture," *Ind. Eng. Chem.* **44**, 483 (1952).
25. Wagner, T. B., "Disposal of starch factory wastes," *Ind. Eng. Chem.* **3**, 99 (1911).

Suggested Additional Reading

The following references have been published since 1961.
1. Buzzell, J. C., A. L. Caron, S. J. Rychman, and O. J. Sproul, "Biological treatment of protein water from the manufacture of potato starch. I," *Water Sewage Works* **V-3**, 327 (1964).
2. Hemens, J., P. G. J. Meiring, and G. J. Stander, "Full-scale anaerobic digestion of effluents from the production of maize-starch," *Water Waste Treat. J.* **9**, 16 (1962).
3. Ling, J. T., "Pilot investigation of starch-gluten waste treatment," in Proceedings of 16th Industrial Waste Conference, Purdue University Engineering Extension Series, no. 109, May 1961, p. 217.
4. Vennes, J. W., and E. G. Olmstead, "Stabilization of potato wastes," *Official Bulletin* (Official Organ of the North Dakota Water and Sewage Works Conference) **29**, 10 (1961).

25.3 Phosphate-Industry Wastes

Two billion years ago, when the earth was first formed, the molten rocks cooled and solidified into rocks containing small amounts of apatite, a tricalcium fluorophosphate mineral. Exposed to the elements, these rocks slowly weathered, were washed into streams, and eventually dissolved in the ocean. Species of sea life withdrew these minute forms of phosphorus, now combined with calcium, limestone, quartz, sand, and so forth, to built their shells and bodies. These various forms of sea life eventually died, settled to the bottom of the ocean, and formed thick layers of deposits containing phosphorus. Such a deposit is now being commercially mined mainly in Florida, which 10 million years ago was at the bottom of the ocean [20]. About 70 per cent of the world's supply of phosphate rock comes from an area in central Florida approximately 50 miles in diameter, centered in the small city of Bartow.

The phosphate rock in this region is found in the form of small pebbles embedded in a matrix of phosphatic sands and clays. These beds, sedimentary in origin, are overlaid with nonphosphate sands and lime rock of more recent origin. Although mining is seldom carried out below 60 feet, the phosphate rock is found at varying depths down to several hundred feet. The overburden is first stripped off with large drag lines and the exposed matrix is excavated in strips; this excavated matrix is dropped into a previously prepared pit and mixed with water from hydraulic guns that wash the rock into a pump sump. From there, the mixture of phosphate matrix and water is carried in pressure pipelines to a washer plant and the larger phosphate-rock particles are removed by screens, shaker tables, and hydraulic sizing cones. All the particles retained on a 200-mesh screen are recovered, by hydraulic sizing in large clarifiers and by a flotation process in which the phosphate particles are selectively coated (with a material such as tall or rosin oil) after conditioning with NaOH for pH control.

The nonphosphatic sands are removed by another flotation process in which, after dewatering and treatment with sulfuric acid, the silica material is selectively coated with an amine and floated off, thus concentrating the phosphate content in the tailings sufficiently for commercial purposes. The finished

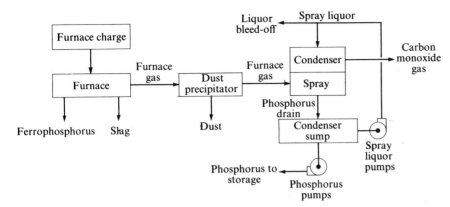

Fig. 25.8 Electric furnace for conversion of phosphate rock into phosphorus [10].

product is dried in kilns and either sold directly as fertilizer or further processed to superphosphate or triple phosphate. It may also be burned in electric furnaces to produce elemental phosphorus or phosphoric acid.

The electric-furnace conversion process (Fig. 25.8) is described by Horton et al. [10]. Phosphate rock is blended with coke, the reducing agent, and silica, which acts as a flux, to form the furnace charge. In the furnace, the charge is smelted by electrical energy to liberate phosphorus as a gas, leaving a dense slag—a mixture of calcium silicate and ferrophosphorus—which is tapped periodically. The furnace gases (with temperatures of 500 to 800°F) consist of phosphorus, carbon monoxide, and various other gaseous impurities in small amounts. The gases are cleaned, usually by electrostatic precipitators, and condensed by water to a heavy liquid; after which the water and heavy liquid phosphorus are separated by sedimentation.

Wastes from the phosphate industry arise from (1) mining the rock and (2) processing the rock to elemental phosphorus and other pure chemicals. In mining, the main wastes originate from the washer plant, where phosphate rock is separated from the water solution, and the flotators, where phosphate particles are separated from the impurities retained on the screens. In processing the phosphate, the major source of water-borne waste is the condenser water bleed-off from the reduction furnace (see Fig. 25.8).

The wastes from phosphate-rock mining are high in volume. Flotation plants [5] commonly use and discharge approximately 30,000 gpm of waste, containing fine clays and colloidal slimes, as well as some tall oil (a resinous by-product from the manufacture of chemical wood pulp) or rosin oil from the flotators. The condenser water bleed-off from phosphorus refining varies in volume from 10 to 100 gpm and has as its most important ingredient the elemental phosphorus in colloidal form, which may ignite if allowed to dry out in ditches. Another significant component of the waste is fluorine, which is also present in the furnace gases. Horton [10] gives the general characteristics of phosphorus wastes (Table 25.2).

To treat these wastes, land areas are provided for tailings, storage, and settling of slime. These methods, and the use of modern flotators, comprise the bulk of phosphate-rock waste-treatment measures. However, it is also common to use mechanical clarifiers for removing sand tailings and to store waste water in lagoons prior to reuse [36].

Table 25.2 Phosphorus wastes [10].

Characteristic	Amount or value
pH	1.5–2.0
Temperature	120–150°F
Elemental phosphorus	400–2500 ppm
Total suspended solids	1000–5000 ppm
Fluorine	500–2000 ppm
Silica	300–700 ppm
P_2O_5	600–900 ppm
Reducing substances as (I_2)	40–50 ppm
Ionic charge	Predominantly positive (+)

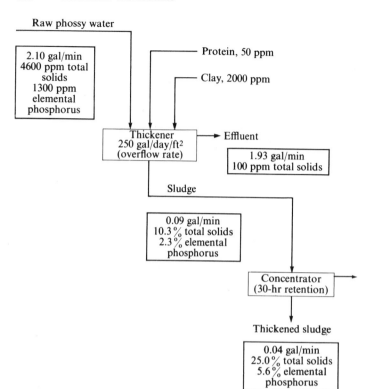

Fig. 25.9 Summary of pilot plant for the treatment of wastes from phosphorus refining. (After Horton et al. [10].)

Among the treatment or disposal methods for phosphorus-refining plants that have been investigated or attempted are not only the above-mentioned lagooning, oxidizing, and settling (with or without prior chemical coagulation), but also filtering and centrifuging. Horton et al. [10] state that coagulation and settling appear to offer the best solution. They present their pilot-plant results in schematic form (Fig. 25.9) and state that concentrations of 40-fold (25 per cent solids) were obtained by a simple coagulation and sedimentation process.

References: Phosphate-Industry Wastes

1. Bixler, G. H., J. Work, and R. M. Lattig, "Elemental phosphorous," *Ind. Eng. Chem.* **48**, 2 (1956).
2. Bowles, O., "Florida phosphate holds attention of American Institute of Mining Engineers," *Rock Products* **53**, 164 (1950).
3. Breton, E., Jr., and N. H. Waggaman, "Calcium fluoride," U.S. Patent no. 2410043, 29 October 1946.
4. Bridgers, G. L., J. W. Moore, and H. M. McLeod, Jr., "Phosphatic animal-feed supplement," *Ind. Eng. Chem.* **41**, 1391 (1949).
5. Fuller, R. B., "The position of the pebble phosphate industry in stream sanitation," in Proceedings of 1st Annual Public Health Engineering Conference, University of Florida, November 1948; Bulletin no. 26, April 1949.
6. Fuller, R. B., "The position of the pebble phosphate industry of Florida in stream sanitation as of November, 1949," in Proceedings of 5th Industrial Waste Conference, Purdue University, November 1949.
7. Fuller, R. B., "Phosphate industry position in Florida," *Sewage Ind. Wastes* **23**, 700 (1951).
8. Hall, J. P., and N. A. Hodges, "Recovery of phosphate fines," U.S. Patent no. 2113727, 12 April 1938.
9. Hignett, T. P., and M. R. Siegel, "Recovery of fluorine from stack gases," *Ind. Eng. Chem.* **41**, 2493 (1949).
10. Horton, J. P., J. D. Molley, and H. C. Bays, "Processing of phosphorus furnace wastes," *Sewage Ind. Wastes* **28**, 70 (1956).
11. Lea, W. L., and G. A. Rohlich, "Removal of phosphate

from treated sewage," *Sewage Ind. Wastes* **26**, 261 (1954).
12. Lenhart, W. B., "Latest recovery methods highlight Florida phosphate plant," *Rock Products* **54**, 74 (1951).
13. Lenhart, W. B., "Developments in producing of phosphates," *Rock Products* **55**, 80 (1952).
14. McCarthy, J. A., and W. E. Cassidy, "Viability of bacteria in the presence of phosphates," *J. New England Water Works Assoc.* **57**, 287 (1943).
15. Maynard, P., "Lightweight aggregate from phosphate slimes," *Eng. Mining J.* **152**, 92 (1951).
16. Murdock, H. R., "Industrial waste," *Ind. Eng. Chem.* **43**, 89A (1951).
17. Murdock, H. R., "Industrial wastes," *Ind. Eng. Chem.* **44**, 115A (1952).
18. Orago, A., "Three new steps in treating Florida phosphate rock," *Eng. Mining J.* **151**, 78 (1950).
19. Owen, R., "Removal of phosphorus from sewage effluent with lime," *Sewage Ind. Wastes* **25**, 548 (1953).
20. "Phosphate, the servant of mankind," *Oil Power* **26**, 3 (1951).
21. State Water Pollution Control Board, "Phosphates," in *Water Quality Criteria*, Publication no. 3, Sacramento, California (1952), p. 322.
22. Rudolfs, W., "Phosphates in sewage and sludge treatment," *Sewage Ind. Wastes* **19**, 43 (1947).
23. Sawyer, C. N., "Some new aspects of phosphates in relation to lake fertilization," *Sewage Ind. Wastes* **24**, 768 (1952).
24. Seyfried, W. R., "Recovery of values from phosphate rock," U.S. Patent no. 2152364, 28 March 1939.
25. Specht, R. C., "Effect of waste disposal of the pebble phosphate rock industry in Florida on conditions of receiving streams," *Mining Eng.* **187**, 779 (1950).
26. Specht, R. C., "Phosphate waste studies," *Florida Univ. Eng. Exp. Sta. Bull.* **32**, February 1950.
27. Specht, R. C., "Disposal of wastes from the phosphate industry," *J. Water Pollution Control Federation* **32**, 969 (1960).
28. Specht, R. C., and W. E. Herron, Jr., "Lightweight aggregate from phosphate slimes," *Rock Products* **53**, 96 (1950).
29. Swainson, S. J., "Washing and concentrating Florida pebble phosphate," *Mining Met.* **25**, 454 (1944).
30. "Symposium on phosphates and phosphorus," *Ind. Eng. Chem.* **44**, 1519 (1952).
31. Thompson, D., "Ultrasonic coagulation of phosphate tailing," *Bull. V. Polytech. Inst. Eng. Exp. Sta. Ser.* **75**, 5 (1950).
32. "Twelfth Purdue Industrial Waste Conference," *Ind. Wastes* **3**, 48 (1957).
33. Waggaman, W. H., and R. E. Bell, "Factors affecting development, Western Phosphate," *Ind. Eng. Chem.* **42**, 269 (1950).
34. Waggaman, W. H., and E. R. Ruhlman, "Conservation problems of the phosphate industry," *Ind. Eng. Chem.* **48**, 360 (1956).
35. Wakefield, J. W., "Semi-tropical industrial waste problems," in Proceedings of 7th Industrial Waste Conference, Purdue University, May 1952, p. 503.
36. Wakefield, J. W., "As Florida grows, so does its industrial waste problem," *Wastes Eng.* **24**, 495 (1953).
37. Williams, D. E., F. L. MacLeod, E. Murrell, and H. Patrick, "Animal feeding test," *Ind. Eng. Chem.* **41**, 1396 (1948).
38. Wright, D. M., "Dewatering materials such as Florida phosphate rock," U.S. Patent no. 2158169, 16 May 1939.

Suggested Additional Reading

The following references have been published since 1962.
1. Chodak, M. E., "Chemical industry solids disposal problems," *J. Proc. Inst. Sewage Purif.*, Part 5 (1962), p. 431.
2. Jones, W. E., and R. L. Olmsted, "Waste disposal at a phosphoric acid and ammonium phosphate fertilizer plant," Purdue University Engineering Extension Series, Bulletin no. 112, 1962, p. 198.
3. Knopsack-Griesheim, A. G., "Treating waste waters obtained in the production of phosphorus," British Patent no. 888085, 24 January 1962.
4. Patton, V. D., "Phosphate mining and water resources," *Ind. Water Wastes* **8**, 24 (1963).

25.4 Soap- and Detergent-Industry Wastes

The soap and detergent industry produces relatively small volumes of liquid wastes directly, but causes great public concern when its products are discharged after use in homes and factories.

In soap manufacturing, the waste waters discharge into trap tanks on skimming basins, where floatable fatty acids are recovered. The recovered fatty acids may not only pay all costs of operation, but also amortize the treatment-plant investment. Gibbs [9] successfully treated soap-plant wastes by flotation with fine air bubbles for a retention period of 40 minutes. The floated sludge was skimmed into a receiving tank, from which it was periodically pumped back to the soap factory for reprocessing or recovery.

Detergents are a class of surface-active compounds used as cleansing agents. They can be grouped as anionic, cationic, and nonionic agents. The synthetic detergents are effective in wide pH ranges and do not form insoluble precipitates in hard water, as many soaps do. Use of synthetic organic detergents in place of, and in addition to, soap has increased at a rapid rate, the production in the United States growing from about 28 million pounds in 1941 to more than 2 billion pounds in 1955 [32]. These synthetic compounds are used not only in households, but also increasingly in the textile, cosmetic, pharmaceutical, metal, paint, leather, paper, and rubber industries, because of their properties of dispersing, wetting, and emulsifying [20].

Many difficulties are reportedly being caused by these residual, water-soluble detergents, such as: (1) interference with oxygen transfer in activated-sludge treatment and in receiving streams, (2) excessive frothing, (3) toxicity to fresh-water game fish, and (4) difficulty of removal at water-treatment plants. The main concern of the public appears to be the presence of these compounds in water supplies, which implies not only that we may be forced to drink a small amount of detergent in our water, but also that other pollution, of a more dangerous type, could conceivably find its way into the water supply. Bogan [2] found that, although all detergents (Table 25.3) decompose somewhat by biological attack, the rate of decomposition is related to chemical structure. Branching in the alkyl group of the alkyl-aryl detergent types causes a definite retarding of oxidation of the alkyl-benzene-sulfonate detergents. Susceptibility of a nonionic substance to biochemical oxidation, other things being equal, decreases as the size of the polyoxyethylene hydrophilic group increases.

Research is currently being carried out on the replacement of polyphosphates in detergents with organic acids to reduce the algae problem. Studies have shown that both alkyl-benzene-sulfonate and linear alkyl-sulfate can be removed by percolation through sand–silt-type soils.

References: Soap- and Detergent-Industry Wastes

1. Bell, C. E., "Assimilation of hydrocarbons by microorganisms," *Advan. Enzymol.* **10**, 443 (1950).
2. Bogan, R. H., "The biochemical oxidation of synthetic detergents," in Proceedings of 10th Industrial Waste Conference, Purdue University, 1955, p. 231.
3. Bogan, R. H., and C. N. Sawyer, "Biochemical degradation of synthetic detergents," *Sewage Ind. Wastes* **26**, 1069 (1954).
4. Borden, L., and P. C. F. Isaac, "Effects of synthetic detergents on the biological stabilization of sewage," *Surveyor* **115**, 915 (1956).
5. Committee Report, Association of Soap and Glycerine Producers, "Determinations of orthophosphate, hydrolyzable phosphate and total phosphate," *J. Am. Water Works Assoc.* **50**, 1563 (1958).
6. Degens, P. N., Jr., H. Van der Zee, and J. D. Kommer, "Effects of synthetic detergents on the settling of suspended solids," *Sewage Ind. Wastes* **26**, 1081 (1954).
7. Downing, A. L., and L. J. Scragg, "The effect of synthetic detergents on the rate of aeration in diffused-air activated sludge plants," *Water Waste Treat. J.* **7**, 102 (1958).
8. Fairing, J. D., and F. R. Short, "Spectrophotometric determination of alkyl benzene sulfonate detergents in surface water and sewage," *Anal. Chem.* **28**, 1827 (1956).
9. Gibbs, F. S., "The removal of fatty acids and soaps from soap-manufacturing waste waters," in Proceedings of 5th Industrial Waste Conference, Purdue University, 1949, p. 400.
10. Gibbs, F. S., "Soap manufacturing wastes, removal of fatty acids and soaps," *Sewage Ind. Wastes* **23**, 700 (1951).
11. House, R., "Analytical development work for detergent ABS determination in waste waters," *Sewage Ind. Wastes* **29**, 1225 (1957).
12. Lammana, C., and M. F. Mallett, *Basic Bacteriology*, Williams & Wilkins Co., Baltimore (1953).
13. McGauhey, P. H., and S. A. Klein, "Removal of ABS by sewage treatment," *Sewage Ind. Wastes* **31**, 877 (1959).
14. McKinney, R. E., "Syndets and waste disposal," *Sewage Ind. Wastes* **29**, 654 (1957).
15. McKinney, R. E., and E. J. Donovan, "Bacterial

Table 25.3 Biochemical oxidation of the principal detergent types. (After Bogan [2].)

Synthetic detergent, class and type	5-day 20°C BOD				BOD, Warburg method (acclimated seed)	
	Sewage seed		Acclimated seed			
	ppm $\times 10^3$	%*	ppm $\times 10^3$	%*	ppm $\times 10^3$	%*
Anionic						
Akyl benzene sulfonate						
n-dodecyl	237	10.0	1046	44.2	300	12.7
Keryl	0	0	535	21.6	240	10.1
Tetrapropene	0	0	81	3.4	40	1.7
Alkyl sulfate						
n-dodecyl	1307	57.2	1362	59.7	930	40.7
Dupanol C	1250	4.8	1330	57.4	920	45.2
Sulfonated ester						
Igepon AP-78	1315	59.8	1460	66.5	990	45.0
Sulfonated amide						
Igepon T-77	1443	52.0	1558	56.0	1560	56.0
Nonionic						
Alkyl phenoxy polyoxyethylene						
Neutronyx 600	0	0	116	5.4	254	11.8
Igepal CA 630	132	6.1	124	5.7	440	20.4
Rohm and Hass						
OPE-5	310	13.1	390	16.4	170	7.2
Polyethoxy amide						
Ethomid HT/15	996	47.5	880	42.0	412	19.5
Ethomid HT/60	29	1.5	310	15.9	16	8.2
Polyethoxy ester						
Ethofat C/15	1000	46.5	880	42.0	640	29.1
Ethofat C/60	220	11.8	240	12.8	160	8.5
Polyglycol ethers						
Pluronic F68	124	6.0	20	1.1	67	3.6

*Based on theoretical amount of oxygen required for complete conversion to CO_2 and H_2O.

degradation of ABS," *Sewage Ind. Wastes* **31**, 690 (1959).
16. McKinney, R. E., and J. M. Symons, "Bacterial degradation of ABS," *Sewage Ind. Wastes* **31**, 549 (1959).
17. Maloney, G. W., and W. D. Sheets, "Detergent builders and BOD," *Sewage Ind. Wastes* **29**, 263 (1957).
18. Manganelli, R., H. Heukelekian, and C. N. Henderson, "Persistence and effect of alkyl aryl sulfonate in sludge digestion," in Proceedings of 15th Industrial Waste Conference, Purdue University, May 1960, p. 199.
19. Munro, L. A., and M. Yatabe, "Frothing of synthetic sewages," *Sewage Ind. Wastes* **29**, 883 (1957).
20. Niven, W. W., *Fundamentals of Detergency*, Reinhold Publishing Corp., New York (1950).
21. Oldham, L. W., "Investigations into the effects of a non-ionic synthetic detergent on biological percolating filters," *J. Inst. Sewage Purif.*, Part 2 (1958), p. 136.

22. Porter, J. R., *Bacterial Chemistry and Physiology*, John Wiley & Sons, New York (1946).
23. Raybould, R. D., and L. H. Thompson, "Tide and santomesse in the sewage works," *Surveyor* **115**, 41 (1956).
24. Roberts, F. W., "The removal of anionic syndets by biological purification processes—observations at Luton and Letchworth," *Water Waste Treat. J.* **6**, 302 (1957).
25. Roberts, F. W., and G. R. Lawson, "Some determinations of the synthetic detergent content of sewage sludge," *Water Waste Treat. J.* **7**, 14 (1958).
26. Sawyer, C. N., "Effects of synthetic detergents on sewage treatment processes," *Sewage Ind. Wastes* **30**, 757 (1958).
27. Sawyer, C. N., R. H. Bogan, and J. R. Simpson, "Biochemical behavior of synthetic detergents," *Ind. Eng. Chem.* **48**, 236 (1956).
28. Sawyer, C. N., and D. W. Ryckman, "Anionic synthetic detergents and water supply problems," *J. Am. Water Works Assoc.* **49**, 480 (1957).
29. Sheets, W. D., and G. W. Maloney, "Synthetic detergents and the BOD test," *Sewage Ind. Wastes* **28**, 10 (1956).
30. Stanier, R. Y., "Problems of bacterial oxidative metabolism," *Bacterial. Rev.* **14**, 179 (1950).
31. Soap & Detergent Association, *Synthetic Detergents in Perspective*, compiled by the Association, 295 Madison Ave., New York, N.Y. (1962).
32. Truesdale, G. A., "Foaming of liquids containing synthetic detergents," *Water Waste Treat. J.* **7**, 108 (1958).
33. U.S. Tariff Commission Reports, "Synthetic organic chemicals, U.S. production and sales," 2nd Series, Government Printing Office, Washington, D.C.
34. Weaver, P. J., "Determination of trace amounts of alkyl benzene sulfonates in water," *Anal. Chem.* **28**, 1922 (1956).

Suggested Additional Reading

The following references have been published since 1962.

1. Abrams, I. M., and S. M. Lewon, "Removal of ABS from water by chloride cycle—anion exchange," *J. Am. Water Works Assoc.* **54**, 43 (1962).
2. Anderson, D. A., "Growth responses of certain bacteria to ABS and other surfactants," in Proceedings of 19th Industrial Waste Conference, Purdue University, 1964, p. 592.
3. "Biodegradability of detergents: a story about surfactants," *Chem. Eng. News* **41**, 102 (1963); *Textile Technicians' Dig.* **20**, 2272 (1963).
4. "Armour solves problems of waste water treatment," *Soap Chem. Specialties* **43**, 47 (1967).
5. Barth, E. F., and M. B. Ettinger, "Anionic detergents in waste water received by municipal treatment plants," *J. Water Pollution Control Federation* **39**, 815 (1967).
6. Basu, A. K., "Treatment of effluents from the manufacture of soap and hydrogenated vegetable oil," *J. Water Pollution Control Federation* **39**, 1653 (1967).
7. Bock, K. J., "Biological investigations of detergents," *Textil-Rundschau* **17**, 87 (1962); *Chem. Abstr.* **56**, 15303 (1962).
8. Buescher, C. A., and D. W. Rychman, "Reduction of foaming of ABS by ozonation," in Proceedings of 16th Industrial Waste Conference, Purdue University, May 1961, p. 251.
9. Burgess, S. G., and L. B. Wood, "Removal and disposal of synthetic detergents in sewage effluents," *J. Proc. Inst. Sewage Purif.* (1962), p. 158.
10. Chadwell, J. H., "Soap plant waste treatment," in Proceedings of 13th Ontario Industrial Waste Conference, June 1966, p. 231.
11. Eden, G. E., et al., "The destruction of ABS in sewage treatment processes," *Water Sewage Works* **108**, 275 (1961).
12. Eldridge, E. F., "Irrigation as a source of water pollution," *J. Water Pollution Control Federation* **35**, 614 (1963).
13. Eisenhauer, H. R., "Chemical removal of ABS from wastewater effluents," *J. Water Pollution Control Federation* **37**, 1567 (1965).
14. Eldib, I. A., "Testing biodegradability of detergents," *Textile Technicians' Dig.* **20**, 4372 (1963).
15. El'kun, D. I., and D. B. Maeve, "Waste waters from production of synthetic detergents as a source for acetic acid and its homologs," *Inst. Lesokhim. Prom.* **14**, 85 (1961).
16. Evans, F. L., and D. W. Rychman, "Ozonated treatment of wastes containing ABS," in Proceedings of 18th Industrial Waste Conference, Purdue University, 1963, p. 141.
17. Feng, T. H., "Exploration of sludge adsorption of syndets," *Water Sewage Works* **109**, 183 (1962).
18. Gates, W. E., and J. A. Borchardt, "Nitrogen and phosphorus extraction from domestic waste water treatment plant effluent by controlled algal culture,"

J. Water Pollution Control Federation **36**, 443 (1964).
19. Hlavsa, E., "Treating waste water containing synthetic detergents," Czechoslovak Patent no. 105679, 1962.
20. "Ion exchange systems may remove detergents wastes," *Eng.* **32**, 369 (1961).
21. Jackson, D. F., "The effects of algae on water quality," in Proceedings of 1st Annual Water Quality Research Symposium, Albany, N.Y., New York State Health Department, *FGB*-20, 1964.
22. Isaac P. C. G., and S. H. Jenkins, "A laboratory investigation of the breakdown of some of the newer synthetic detergents in sewage treatment," *J. Proc. Inst. Sewage Purif.*, part 3, p. 314 (1960).
23. Jenkins, S. H., N. Harkness, A. Lennon, and K. James, "The biological oxidation of synthetic detergent in recirculating filters," *Water Res.* **1**, 31 (1967).
24. Johnson, W. K., and G. J. Schroepfer, "Nitrogen removal by nitrification and denitrification," *J. Water Pollution Control Federation* **36**, 1015 (1964).
25. Kaufmann, H. P., and F. Malz, "The adsorptive precipitation and the skimming of detergents from aqueous solutions," *Fette, Seifen, Anstrichmittel* **62**, 1024 (1960); *Chem. Abstr.* **55**, 9739 (1961).
26. Koefer, C. E., "The syndet problem after five years of progress," *Public Works Mag.* **95**, 82 (1964).
27. Klein, S. A., and P. H. McGauhey, "Detergent removal by surface stripping," *J. Water Pollution Control Federation* **35**, 100 (1963).
28. Klein, S. A., "Effect of ABS on digester performance," *Water Sewage Works* **109**, 373 (1962).
29. Kling, W., "Surface action agents in effluents," *J. Textile Inst. Proc. Abstr.* **54**, *A*140 (1963).
30. Knesta, V., "Elimination of synthetic detergents from industrial waste waiters," *Chem. Prumysl* **13**, 281 (1963).
31. "Knocking out ABS by biodegradation wastes," *Eng.* **34**, 41 (1963).
32. Kucharski, J., "Production of synthetic detergents and problems in industrial waste disposal," *Chemik* **15**, 94 (1962).
33. Kumke, G. W., and C. E. Renn, "LAS removal across an institutional trickling filter," *Am. Oil Chem. Soc. J.* **43**, 92 (1966).
34. Lashen, E. S., and K. A. Booman, "Biodegradability and treatability of alkylphenol ethoxylates," Proceedings 22nd Industrial Waste Conference, Purdue University, Eng. Ext. Series 129, p. 211, 1967.
35. Lockwood, J. C., "Detergent industry and clean waters," *Soap Chem. Specialties* **41**, 67 (1965).
36. Ludzack, F. J., and M. B. Ettinger, "Controlling operation to minimize activated sludge effluent nitrogen," *J. Water Pollution Control Federation* **34**, 920 (1962).
37. Manay, K. H., W. E. Gates, J. D. Eye, and P. K. Deb, "The adsorption kinetics of ABS on fly ash," in Proceedings of 19th Industrial Waste Conference, Purdue University, 1964, p. 146.
38. Mackenthum, K. M., "A review of algae lake weeds and nutrients," *J. Water Pollution Control Federation* **34**, 1077 (1962).
39. Mackenthum, K. M., "Public Health Service research on algae and aquatic nuisances," in Proceedings of 1st Annual Water Quality Research Symposium, Albany, N.Y., New York State Health Department, 20 February 1964.
40. Maloney, T. E., "Detergent phosphorus effect on algae," *J. Water Pollution Control Federation* **38**, 38 (1966).
41. McGauhey, P. H., and S. A. Klein, "Degradable pollutants—a study of new detergents," unpublished data.
42. McGauhey, P. H., and S. A. Klein, "Travel of synthetic detergents with percolating water," in Proceedings of 19th Industrial Waste Conference, Purdue University, 1964, p. 1.
43. Metzgen, A., "Washing and scanning agents in waste water," *Spinner Weber Textilveredl.* **78**, 541 (1960).
44. Nemerow, N. L., and M. C. Rand, "Algal nutrient removal from domestic waste waters," in Proceedings of 1st Annual Water Quality Research Symposium, Albany, N.Y., New York State Health Department, 20 February 1964.
45. O'Neill, R. D., "Exotic chemicals," in Proceedings of 1st Annual Water Quality Research Symposium, Albany, N.Y., New York State Health Department, 20 February 1964.
46. Perlman, J. L., "Detergent biodegradability," *Textile Technicians' Dig.* **20**, 5583 (1963).
47. "Phosphates may lose detergent markets," *Chem. Eng. News* **45**, 18 (1967).
48. Pitter, P., and J. Chudoba, "Removing soaps by coagulation," *Vodni Hospodarstvi* **12**, 164 (1962).
49. Renn, E. E., W. A. Kline, and G. Orgel, "Destruction of linear alkylate sulfonates in biological waste treatment by field test," *J. Water Pollution Control Federation* **36**, 864 (1964).
50. Reymonds, T. D., "Pollutional effects of agricultural

insecticides and synthetic detergents," *Water Sewage Works* **109**, 352 (1962).
51. Samples, W. R., "Removal of ABS from waste-water effluent," *J. Water Pollution Control Federation* **34**, 1070 (1962).
52. Sawyer, C. N., "Causes, effects and control of aquatic growth," *J. Water Pollution Control Federation* **34**, 279 (1962).
53. Schoen, H. M., et al., "Foam separation," *Ind. Water Wastes* **6**, 71 (1961).
54. Sengupta, A. K., and W. W. Pipes, "Foam fractionation–the effect of salts and low molecular weight organics on ABS removal," in Proceedings of 19th Industrial Waste Conference, Purdue University, 1964, p. 81.
55. Soap & Detergent Association Technical Advisory Council, *Synthetic Detergents in Perspective*, New York (1962).
56. Spade, J. F., "Treatment methods for laundry wastes," *Water Sewage Works* **109**, 110 (1962).
57. Sweeney, W. A., "Note on straight chain ABS removal by adsorption during activated sludge treatment."
58. Swisher, R. D., "Biodegradation of ABS in relation to chemical structure," *J. Water Pollution Control Federation* **35**, 877 (1963).
59. Swisher, R. D., "Chemical mechanism of straight chain ABS biodegradation," *Textile Technicians' Dig.* **20**, 4374 (1963).
60. Swisher, R. D., "LAS major development in detergents," *Chem. Eng. Progr.*, 41 (1964).
61. Swisher, R. D., "Transient intermediate in the biodegradation of straight chain ABS," *J. Water Pollution Control Federation* **35**, 1557 (1963).
62. Symons, J. M., and L. A. Del Valle-Rivera, "Metabolism of organic sulfonates by activated sludge," in Proceedings of 16th Industrial Waste Conference, Purdue University, May 1961.
63. "Symposium-field evaluation of LAS and ABS treatability," in Proceedings of 20th Industrial Waste Conference, Purdue University, 1965, p. 724.
64. Tatsumi, C., and M. Nakiagawa, "Production of fatty microorganisms. IV: Production of fat yeast from soap waste liquor," *Hakko Kyokaishi* **19**, 396 (1961).
65. Waymon, C. H., and J. B. Robertson, "Adsorption of ABS on soil minerals," in Proceedings of 18th Industrial Waste Conference, Purdue University, 1963, p. 253.
66. Waymon, C. H., J. B. Robertson, and C. W. Hall, "Biodegradation of surfactants under aerobic and anaerobic conditions," in Proceedings of 18th Industrial Waste Conference, Purdue University, May 1963, p. 578.

25.5 Explosives-Industry Wastes

The explosives industry is concerned with three major processes servicing the public and private interests of our society: (1) manufacture of TNT, (2) manufacture of smokeless powders, and (3) manufacture of small-arms ammunition. Although this industry is particularly active during wartime, the peacetime uses of guns, ammunition, and explosives are also significant, because of hunting for sport, mining operations, construction projects which require blasting, and the manufacture of such items as fireworks and cap pistols.

In the manufacture of TNT (trinitrotoluene), toluene is mixed with a solution of nitric and sulfuric acids, under proper temperature conditions, and nitrate groups are gradually added, one by one, until the product is primarily trinitrotoluene, which is then washed free of residual acid, crystallized, and purified with sodium sulfite. The impure beta and gamma trinitrotoluenes are washed out of the alpha trinitrotoluenes as soluble sulfonates, after which the purified product is finally remelted, flaked, and packaged.

In the manufacture of smokeless powder, purified cotton, known as cotton linters, is treated with a mixture of nitric and sulfuric acids, producing cellulose nitrate (guncotton). Various methods of purifying this product are used, including boiling, macerating, and washing; it is then mixed with ether–alcohol and a stabilizer to render it colloidal. The powder is granulated by pressing it through steel dies, the solvent is recovered, and the powder dried and blended for shipping.

In the manufacture of small-arms ammunition, the brass case is made first, then the projectile, after which the percussion cap is filled, and finally all the parts, including the smokeless powder, are assembled. The major wastes originate from making the cartridge cases and projectile jackets; this process consists of extruding and annealing metals, pickling them in acids, washing them with detergents, and lubricating the dies. Some wastes also come from the lead shop,

Table 25.4 Quantities of TNT wastes [12].

Characteristic	Waste per 100,000 lb of explosive produced (TNT and DNT)			
	Plant A	Plant B	Plant C	Average
Flow, million gallons	1.17	1.08		1.12
Free mineral acid as H_2SO_4, lb	2,070	1,210	3140	2140
Sulfates, lb	5,560	5,450	2840	4620
NH_3 nitrogen, lb	49.7		27.2	38.5
NO_2 nitrogen, lb	140	179	60	116
NO_3 nitrogen, lb	1,062		302	684
Oxygen consumed, lb	8,360	4,990	1055	4800
Total solids				
Volatile, lb	9,460	6,180	6440	7360
Ash, lb	12,240	10,220	4980	9150
Suspended solids				
Volatile, lb	200	118	170	163
Ash, lb	1,380	130	6	505

where lead ingots are drawn into wire and shaped into projectiles.

The characteristics of TNT wastes are that they are generally clear, highly colored, strongly acid, have a high percentage of volatile solids, a "chemical" odor and an acid taste, and are quite resistant to alteration after they reach the receiving waters. The acid wash waters from the washing after nitration and the sulfite-purification wash water (red water) are the two major wastes. The nitration wash water is acid and yellow, while the red water is alkaline and so intensely red that it appears almost black. Smith and Walker [12] presents typical results of two TNT-plant wastes (Tables 25.4 and 25.5).

The four principal smokeless-powder wastes are (1) acid residue remaining after nitrating the cotton and purifying the product, (2) guncotton lost in water in which cellulose nitrate is boiled, (3) ether–alcohol lost from the solvent-recovery system, and (4) aniline from the manufacture of diphenylamine. One estimate [12] of the quantities of waste from manufacturing 100,000 pounds of powder gives 89,500 pounds

Table 25.5 Average of analytical results of TNT wastes [12].*

Plant	pH	Color	Odor	Acidity		Oxygen consumed	SO_4	Nitrogen			Total solids		Suspended	
				Methyl red	Phenolphthalein			NH_2	NO_2	NO_3	Volatile	Ash	Volatile	Ash
A	2.4	7,100	70	291	485	795	672	5.3	15	107	1004	1273	22	144
B	2.7	6,300	16	134	178	551	604		20		686	1123	14	15
Average A and B	2.6	6,700	43	212	332	673	638	5.3	18	107	850	1198	18	80
C (no cooling water)	1.2	34,000	11	3230	3460	1057	2923	2.8	62	310	5490	4685	17	0

*All results are given in parts per million, except for the pH, color, and odor concentration.

Table 25.6 Average analytical results for wastes from three powder plants [12].*

Plant	pH	Color	Odor	Acidity		5-day BOD	Oxygen consumed	SO$_4$	Nitrogen		Total solids		Suspended solids		Soap hardness
				Methyl red	Phenolphthalein				NO$_2$	NO$_3$	Volatile	Ash	Volatile	Ash	
A	<1.6			1860	1990	49.1	76.2	1280	2.7	530			29	24	
B	<1.6	53	21	1540	1610	47.6	99.8	1100	1.9	470	687	354	54	136	241
C (combined flow)		52		2820	2970	62.9	91.4	1512	2.4	650	2645	229	29	13	368
C (pyrocotton)	1.1	36	4	4250	4400	52.3	106.0	2265	2.6	970	3930	346	37	7	526
C (finish)	8.2	104	29			83.9	62.0	1	2.3	9	69	122	13	26	50

*All results are given in parts per million, except for the pH, color, and odor concentration.

of mixed acids, 2500 pounds of alcohol, and 125 pounds of cotton, all in a total of 8.3 million gallons of waste water. Smith and Walker [12] also present the average analytical results from three powder plants (Table 25.6) which produce a large volume of strongly acid liquid waste, containing high concentrations of sulfates and nitrates.

The predominant characteristics of small-arms ammunition wastes are that they are turbid, greenish-gray in color, and may have an oily or soapy odor, contain copper and zinc from acid pickling baths, and grease from the cutting oils and soaps. The average waste characteristics for three typical plants [12] are given in Tables 25.7 and 25.8, which show that a flow of 36,000 gallons contains 100 pounds of grease, 12 pounds of copper, 95 pounds of volatile suspended solids, and a 5-day BOD population equivalent of 300 people for each 100,000 rounds of mixed ammunition.

Smith and Walker [12] describe a plant treating ammunition wastes by means of grease flotation and chemical precipitation (Table 25.9), which removed 84 per cent of the BOD and 94 per cent of the suspended solids. The pH was also raised from 3.5 to 6.9 and the odor reduced about 90 per cent.

Dickerson [1] concluded that coagulation and sedimentation are not effective methods of treatment for powder-manufacturing wastes and that chlorination, though effective, is too costly. He did state, however, that these wastes are amenable to decom-

Table 25.7 Average analytical results, waste flows from three small-arms ammunition plants [12].*

Plant	Acidity (methyl red)	Alkalinity (methyl orange)	Grease	Copper	5-day BOD	Oxygen consumed	Total solids			Suspended solids			Sulfate
							Total	Volatile	Ash	Total	Volatile	Ash	
A		213	184	31	141	67	1577	445	1132	276	205	71	244
B	6		419	22	138	103	1092	505	587	408	372	36	397
C	60		502	86	295	163	1753	675	1078	408	374	34	647
Average			368	46	191	111	1474	542	932	364	317	47	309

*All results are given as parts per million.

Table 25.8 Waste products per 100,000 rounds of ammunition [12].*

Product and plant	Million gallon flow	Alkalinity	Mineral acid	SO$_4$	Grease	Copper	BOD	Oxygen consumed	Total solids		Suspended solids	
									Volatile	Ash	Volatile	Ash
0.50-caliber ammunition												
A	0.0623	131		265	284	29.7	159	64	31.4	504	252	116
B	0.0975	20		212	475	16.3	144	137	567	474	442	39.7
Average	0.0799	76		239	380	23.0	152	67	299	489	347	78
0.30-caliber ammunition												
A	0.0252	58		37	46	7.9	42	22	114	141	69	19
B	0.0242		9.40	73	40	5.4	17	14	58	120	34	3
Average	0.0242			55	43	6.7	30	18	86	131	52	11
Combined output												
A	0.0417	74		84	64	10.5	49	23	155	392	71	24.7
B	0.0408		3.0	104	144	7.5	47	40	174	198	129	12.2
C	0.0258		12.9	140	109	18.6	64	35	146	232	81	7.4
Average	0.0361			76	106	12.2	53	33	158	274	94	14.8

*All results are given in pounds.

Table 25.9 Average analytical results, treatment plants for ammunition wastes [12].*

Sampling	pH	Acidity		Alkalinity	Grease	Copper	Total solids			Suspended solids			5-day BOD	Odor	
		Methyl red	Phenol-phthalein				Total	Volatile	Ash	Total	Volatile	Ash		Concentration	Type
Raw waste	3.5	60	239		709	86	1763	684	1079	413	379	34	298	174	Oily
After grease removal	3.6	55	211		54	82								55	Oily
Influent (final settling)	6.9			333		64				867	358	509		48	Soapy
Final effluent	6.9			43	8	9	1510	302	1208	25	13	12	47	14	Soapy

*All results are given in parts per million, except for pH and odor.

position by both aeration and biological oxidation.

Ruchhoft et al. [7] concluded that black garden soil was best for filtering TNT wastes and that such soils can remove an amount of TNT up to about 0.1 per cent of their weight. Dosing and mixing activated carbon into the waste were more effective in removing TNT than filtration through carbon.

More recently, Edwards and Ingram [2] stated that the color of TNT wastes is very resistant to biological and chemical attack, since reducing substances such as SO_2, $FeSO_4$, $Na_2S_4O_7$, and hydrosulfites are not effective and adsorption, on materials such as clays and activated carbon, may remove much color, but is impractical because of cost. It is true that all the color can be removed by ion exchange, but short cycles and difficulty in regeneration make this method uneconomical. Oxidizing agents seem to be most promising, but only chlorine can remove color at a reasonable cost. For example, fresh TNT stellite waste can be chlorinated directly in a column (Fig. 25.10) and more than 90 per cent of the color can be removed if one uses about 9000 ppm of chlorine—a considerable dosage.

References: Explosives-Industry Wastes

1. Dickerson, B. W., "Treatment of powder plant wastes," in Proceedings of 6th Industrial Waste Conference, Purdue University, February 1951, p. 30.
2. Edwards, G., and W. T. Ingram, "The removal of color from TNT wastes," *J. Sanit. Eng. Div. Am. Soc. Civil Engrs.* **81**, separate no. 645 (1955).

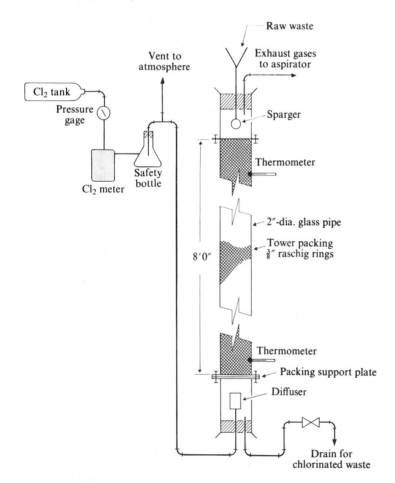

Fig. 25.10 Chlorination tower for treating TNT wastes. (After Edwards and Ingram [2].)

3. Edwards, G. P., and W. T. Ingram, "TNT wastes, description of treatment methods," *Sewage Ind. Wastes* **26**, 1484 (1954).
4. Mohlman, F. W., "Industrial wastes in wartime," *Sewage Works J.* **15**, 1164 (1943).
5. Morris, R. L., and J. D. Dougherty, "Infrared identification of degraded TNT wastes in streams and shallow wells," in Proceedings of 15th Industrial Waste Conference, Purdue University, May 1960, p. 281.
6. Ruchhoft, C. C., et al., "TNT wastes, color reactions and disposal procedure," *Sewage Ind. Wastes* **18**, 339 (1946).
7. Ruchhoft, C. C., M. LeBosquet, Jr., and W. G. Meckler, "TNT wastes from shell-loading plants," *Ind. Eng. Chem.* **37**, 937 (1945).
8. Ruchhoft, C. C., and W. G. Meckler, "Colored TNT derivative and alpha TNT in colored aqueous alpha-TNT solutions," *Ind. Eng. Chem.* (analytical edition) **17**, 430 (1945).
9. Ruchhoft, C. C., and W. G. Meckler, "TNT waste treatment studies," *Sewage Ind. Wastes* **18**, 779 (1946).
10. Ruchhoft, C. C., and F. J. Noms, "Estimation of ammonium picrate in wastes from bomb and shell loading plants," *Ind. Eng. Chem.* (analytical edition) **18**, 480 (1946).
11. Schott, S., C. C. Ruchhoft, and S. Megregian, "TNT waste," *Ind. Eng. Chem.* **35**, 1122 (1943).
12. Smith, R. S., and W. W. Walker, "Surveys of liquid wastes from munitions manufacturing," U.S. Public Health Service, Reprint no. 2508, Public Health Reports nos. 58, 194, 1365, and 1393.
13. Wilkinson, R., "Treatment and disposal of waste waters containing picric acid and dinitrophenol," *Ind. Chem.*, Jan.–Feb. 1951.
14. Wilkinson, R. W., "Shell-filling plant waste treatment and disposal," *Sewage Ind. Wastes* **20**, 590 (1948).

Suggested Additional Reading

The following references have been published since 1958.
1. Madera, Solin, Vucka, "The biochemical reduction of TNT—the course and intermediary substances of reduction of 2,4,6-TNT," *Sb. Vysoke Skoly Chem. Technol. Praze, Oddil Fak. Technol. Paliv. Vody* **3**, 129 (1959).
2. Schuster, G., "Possible biochemical decomposition of hard-to-handle constituents of individual waste water," *Math.-Naturw. Reihe* **11**, 179 (1962).
3. Sercu, C., "Chemical plant waste incinerator," *Natl. Fire Protect. Quart.* **55**, 90 (1961).
4. Solin, V., and M. Kustka, "The treatment of waste waters containing TNT by sprinkling on ashes," *Sb. Vysoke Skoly Chem. Technol. Praze, Oddil Fak. Technol. Paliv. Vody* **2**, 247 (1958).
5. U.S. Department of Commerce, "Air and stream pollution control: preliminary survey of thermal methods for TNT red water disposal," Pamphlet no. 556, Washington, D.C., Office of Technical Service (1961), p. 810.

25.6 Formaldehyde Wastes

This one chemical, formaldehyde, warrants individual consideration in this chapter for several reasons. First, it is used in numerous industries such as plastics, leather, and antibiotics. Second, by virtue of its special chemical nature, it has long been considered an effective antiseptic agent, and treatment of a biologically inhibiting agent presents difficult or, at least, unique problems. The toxicity problem is illustrated in the work of Gellman and Heukelekian [5], in which they proved that a formaldehyde concentration somewhere between 130 and 175 ppm was lethal to bacteria in sewage (Table 25.10) and that even in small concentrations a lag period occurs, owing to the action of sublethal doses of formaldehyde on the biological flora in sewage. Another valuable conclusion evolved from their research, which they summarized as follows: "By a process of adaptation and selection of bacteria, the oxidizable concentration of formal-

Table 25.10 Effects of formaldehyde concentrations in sewage on oxidation of formaldehyde. (After Gellman and Heukelekian [4].)

Days of oxidation	Formaldehyde concentration, ppm			
	45	90	130	175
1	0	0	0	0
2	15	0	0	0
3	21	72	43	0
$3\frac{1}{4}$	23	81	92	0
4	25	89	112	0
7	28	90	135	0

Table 25.11 Trickling filtration of formaldehyde waste. (After Dickerson [2].)

Concentration of formaldehyde, ppm	Removal of formaldehyde, lb/yd³	Removal efficiency, %
110	1.12	23
184	1.25	16
266	1.75	15
300	3.10	23
360	3.45	28

dehyde was increased from 135 ppm to 1750 ppm." This means that even a toxic chemical can serve as food for bacteria, *provided* that the bacterial species is carefully acclimated to the food.

Dickerson [2] reported at about the same time on a synthetic-resin plant producing a total BOD load of 2000 pounds per day, with individual BOD values ranging from 300 to 10,000 ppm, and formaldehyde concentrations up to 5000 ppm (90 per cent of the total BOD load) in a volume of about half a million gallons of water. He found that a high-rate trickling filter can be used for treatment of formaldehyde, organic oils, and organic acids and can provide satisfactory reduction, as shown in Table 25.11. He advised holding a uniform pH, but said that adjustment is not necessary as long as the pH stays somewhere between 4.5 and 8.5. His report also showed that toxicity is only relative and that long periods of pilot-plant operation are necessary to ascertain the correct degree of oxidation required [2].

In later studies, Dickerson *et al.* [4] report on a complete treatment plant for formaldehyde wastes involving aeration and activated sludge along with trickling filtration, for treating about 4700 pounds of BOD, with a removal efficiency of approximately 90 per cent (see Fig. 25.11). Operating costs (1952–1954) were estimated at about 2 cents per pound of BOD treated.

References: Formaldehyde Wastes

1. Dickerson, B. W., "A high-rate trickling filter pilot plant for certain chemical wastes," *Sewage Works J.* **21**, 685 (1949).
2. Dickerson, B. W., "High-rate trickling filter operation on formaldehyde wastes," *Sewage Ind. Wastes* **22**, 536 (1950).
3. Dickerson, B. W., C. J. Campbell, and M. Stankard, "Further operating experiences in biological purification of formaldehyde wastes," in Proceedings of 9th Industrial Waste Conference, Purdue University, May 1954, p. 331.
4. Gellman, I., and H. Heukelekian, "Biological

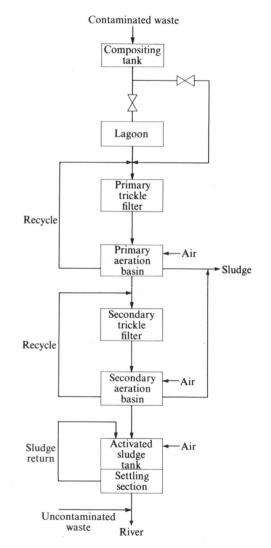

Fig. 25.11 Flow diagram of complete treatment plant for formaldehyde wastes [3].

oxidation of formaldehyde," *Sewage Ind. Wastes* **22**, 1321 (1950).
5. Ragan, J. L., and R. H. Maurea, "Industrial waste disposal by solar evaporation," *Ind. Water Wastes* **8**, 37 (1963).
6. Sakharnov, A. V., et al., "Purification of waste waters from formaldehyde and phenol in phenol-formaldehyde resin production," U.S.S.R. Patent no. 141814, Oct. 16, 1961; application December 1960.
7. "Treatment of trade water containing formaldehyde," *Effluent Water Treat. J.* **3**, 88 (1963).
8. Waldemeyer, T., "Treatment of formaldehyde wastes by activated sludge," *Surveyor* **111**, 445 (1952).

25.7 Pesticide Wastes

The most troublesome waste from the production of chemicals used to make insecticides, herbicides, and pesticides is that arising from production of 2,4-dichlorophenoxyacetic acid (2,4-D). The chemical actually reaching the waste stream is dichlorophenol (DCP).

Mills [7] originally treated this waste by passing the waste water through a filter bed of activated carbon, at a cost of about $6 per pound of dichlorophenol removed, but the method was soon abandoned because of the cost. The alkaline chlorination process then came into favor, which gives 95 to 98 per cent destruction of dichlorophenol, although the effluent still contains about 25 ppm of dichlorophenol and the cost is somewhat more than $1 per pound of dichlorophenol destroyed. In his search for an economical method of removing the remaining DCP, Mills concluded that, although it is a phenolic compound, it cannot be biologically oxidized under the same conditions as phenol.

Another major insecticide, DDT, is manufactured by the reaction of monochlorobenzene and chloral alcohol in the presence of H_2SO_4 containing 20 per cent free SO_3. The crude DDT is purified either by washing with large volumes of water or by neutralization with Na_2CO_3 after draining off the spent acid. The wastes contain a great deal of acid; for example, the neutralization process results in 500 gallons of waste acid per ton of DDT manufactured. The acid waste contains 55 per cent H_2SO_4, 20 per cent ethyl hydrogen sulfate, and 20 per cent chlorobenzene sulfuric acid. In addition, this method produces about 800 gallons of wash water per ton of DDT, containing 2 to 6 per cent acid, and 90 gallons of wash water from centrifuges. Other wastes arise from the manufacture of chloral alcohol [4]. The only waste-treatment method used has been dumping the wastes into sufficient dilution water, such as the ocean. Treatment by municipalities in biological plants is possible, but is greatly hampered by the toxicity of most of the constituents of the wastes.

Increased interest is being shown in the effect of new organic insecticides on fish and wildlife. Rachel Carson's book aimed at the layman, *Silent Spring*, which first appeared in the *New Yorker* magazine in June 1962 and was published in book form by Houghton Mifflin, is an example of current public interest in this subject. Aldrin, toxaphene, rotenone, and dieldrin are a few of the more toxic sprays recently reported to be responsible for large-scale destruction of aquatic life in many streams and farm ponds [3, 5, 6, 10, 15]. These insecticides have been found to be lethal even in very small concentrations. For example, Burdick [1] found that 0.05 to 0.40 ppm of rotenone was lethal to brown trout. The toxicity of these chemicals varies according to the type of fish and increases with an increase in the temperature of the water.

At the present time, little treatment of these wastes (other than those described by Mills) is being practiced. However, there is increasing concern over keeping these chemicals out of the watercourses.

Stutz [1966, 10] reports successful biological oxidation of parathion wastes after nine years of experimentation. He characterized these wastes as very strong, highly mineralized, and acid in nature. Liquid wastes are prechlorinated, lagooned, neutralized with limestone, adjusted for pH control with soda ash, treated by activated sludge for 5 to 9 days at a suspended-solids concentration of 18,000 ppm in mixed liquor, clarified, and discharged to the city sewer. Influent and effluent analyses are given in Table 25.12.

References: Pesticide Wastes

1. Burdick, G. E., H. J. Dean, and E. J. Harris, "Toxicity of emulsifiable rotenone to various species of fish," *New York Fish Game J.*, January 1955, p. 36.
2. Chanin, G., and R. P. Dempster, "A complex chemical

Table 25.12 Sanitary characteristics of raw and treated parathion wastes. (After Stutz [10].)

Characteristic	Raw wastes*	Plant effluent*
COD	3,000	100
Total solids	27,000	18,000
Volatile solids	25%	0.1%
pH	2.0	7.0
Acidity	3,000	
Sodium	6,000	
Chlorides	7,000	
Phosphates	250	
Nitrogen	20	
Sulfates	3,000	
Parathion		< 0.1
p-Nitrophenol		< 1.0

*All results are given in mg/liter unless otherwise indicated.

waste," *Ind. Wastes* **3**, 155 (1958).
3. Doudoroff, P., M. Katz, and C. M. Tarzwell, "Toxicity of some organic insecticides to fish," *Sewage Ind. Wastes* **25**, 840 (1953).
4. Grindley, J., "Effluent disposal in DDT manufacture," *Ind. Chem.* **26**, November 1950.
5. Hoffman C. H., and A. T. Drooz, "Effects of a C-47 airplane application of DDT on fish food organisms in two Pennsylvania watersheds," *Am. Midland Naturalist* **50**, 175 (1953).
6. Ingram, W. M., and C. M. Tarzwell, "Selected bibliography of publications relating to undesirable effects upon aquatic life by algicides, insecticides, and weedicides," Publication no. 400, U.S. Public Health Service (1954).
7. Mills, R. E., "Development of design criteria for biological treatment of an industrial effluent containing 2,4-D waste water," in Proceedings of 14th Industrial Waste Conference, Purdue University, May 1959, p. 340.
8. "Report on New Jersey Sewage and Industrial Wastes Association meeting, industrial waste problems," *Ind. Wastes* **3**, 72 (1958).
9. Tarzwell, C. M., "Disposal of toxic wastes," *Ind. Wastes* **3**, 48 (1958).
10. Tarzwell, C. M., and C. Henderson, "Toxicity of dieldrin to fish," *Trans. Am. Fisheries Soc.* **86**, (1956).
11. Warrick, L. F., "Blitz on insects creates water problems," in Proceedings of 6th Industrial Waste Conference, Purdue University, February 1951, p. 455.
12. Warrick, L. F., "Fish kills by leaching of insecticides," *Sewage Ind. Wastes* **24**, 924 (1952).
13. Weiss, C. M., "Response of fish to sub-lethal exposures of organic phosphorus insecticides," *Sewage Ind. Wastes* **31**, 580 (1959).
14. Wilson, I. S., "The Monsanto plant for the treatment of chemical wastes," *J. Inst. Sewage Purif.*, Midland Branch, 18 March 1954.
15. Young, L. A., and H. P. Nicholson, "Stream pollution resulting from the use of organic insecticides," *Progressive Fish Culturist* **13**, 193 (1951).

Suggested Additional Reading

The following references have been published since 1961.
1. Buescher, C. A., and J. H. Dougherty, "Chemical oxidation of selected organic pesticides," *J. Water Pollution Control Federation* **36**, 1005 (1966).
2. Coley, G., and C. N. Stutz, "Treatment of parathion wastes and other organics," *J. Water Pollution Control Federation* **38**, 1345 (1966).
3. Lutin, P. A., and J. J. Cibulka, "Oxidation of selected carcinogenic compounds by activated sludge," Purdue University Engineering Extension Series, Bulletin no. 118, 1965, p. 131.
4. Morris, J. C., and W. J. Weber, Jr., "Adsorption of biochemically resistant material from solution," U.S. Public Health Service, Environmental Health Series Water Supply and Pollution Control, 999-WP-33, March 1966, p. 108.
5. Oshina, I. A., and N. K. Tyurina, "Clarification of 2,4,D production by adsorption," *Chem. Abstr.* **63**, 12860 (1965).
6. Pitter, P., and F. Tucek, "Influence of waste water from chlorophenol production on biological purification," *Chem. Abstr.* **65**, 8556 (1966).
7. Pitter, P., and J. Chudoba, "Purification of waste effluents from the industrial production of the fungicides Kaptan and Faltan," *Chem. Abstr.* **63**, 12864 (1965).
8. Randall, C. W., and R. A. Lauderdale, "Biodegradation of malathion," *J. Sanit. Eng. Div. Am. Soc. Civil Engrs.* **93**, 145 (1967).
9. Riklis, S. G., and A. R. Perkins, "The hydrolysis of chlorobenzenesulfonic acid, a by-product of DDT

manufacture," *Khim. Prom.*, 1961, p. 461.
10. Stutz, C. N., "Treating parathion wastes," *Chem. Eng. Progr.* **62**, 82 (1966).
11. Teasley, J. I., "Identification of cholinesterase inhibiting compound from industrial effluent," *Environ. Sci. Technol.* **1**, 411 (1967).
12. Wilroy, R. D., "Industrial wastes from scouring rug wools and the removal of dieldrin," Purdue University Engineering Extension Series, Bulletin no. 115, 1963, p. 413.
13. Winar, R. M., "The disposal of waste water underground," *Ind. Water Eng.* **4**, 21 (1967).
14. Woodland, R. G., M. C. Mall and R. R. Russell, "Process for disposal of chlorinated organic residues," *J. Air Pollution Control Assoc.* **15**, 56 (1965); *Chem. Abstr.* **62**, 12890 (1965).

25.8 Plastic and Resin Wastes

Plastics and resins are chainlike structures known chemically as polymers. All polymers are synthesized by one or more of the following processes: bulk, solution, emulsion, and suspension. A typical production reaction requires the addition of a free radical initiator and modifiers to the monomer, the building block of the polymer. This polymerization process creates relatively little water-borne waste, compared with other chemical manufacturing processes. In most cases, the preliminary step—the synthesis of the monomer—creates considerably more waste than the production of the polymer from the monomer.

The U.S. Department of the Interior Profile [1] separated these chemical industries into nine subdivisions, which represent over 85 per cent of all plastic production: (1) cellulosics, (2) vinyls, (3) styrenes, (4) polyolefins, (5) acrylics, (6) polyesters and alkyds, (7) urea and melamine resins, (8) phenolics, and (9) miscellaneous resins. This profile reports that the current total production of 14.25×10^9 pounds per year generates 113×10^6 pounds of waterborne waste, expressed as 5-day BOD. Since that is roughly equivalent to the waste discharged by 1,800,000 persons, we cannot overlook this industry as a major contributor to the organic waste loads on our streams and lakes. It is further reported that of the BOD generated, 55 per cent is removed by treatment-plant systems, while the remaining 45 per cent (51×10^6 pounds per year) is discharged to watercourses. The assumption is that 1 pound (dry weight) of industrial waste generates 0.75 pounds of BOD. From Table 25.13 it can be seen that the polyolefins, vinyls, and styrenes make up about two-thirds of all production, while all divisions except the cellulosics have made substantial production gains during the 5-year period from 1962 to 1967.

Each subdivision of this industry manufactures its product in a different manner from any other. It would be very difficult for the author to present, or the reader absorb, all the details of each subdivision's processes and wastes. However, brief descriptions

Table 25.13 Production of plastics and resins.

Division	Production, 10^9 lb/year		Increase, %	Percentage of total production in 1967
	1962	1967		
Cellulosics	0.48	0.50	4	3.5
Vinyls	1.55	2.80	80	19.5
Styrenes	1.25	2.25	80	16.0
Polyolefins	2.00	4.35	82	30.5
Acrylics	0.18	0.29	61	2.0
Polyesters and alkyds	0.68	1.10	62	7.5
Urea and melamines	0.49	0.66	35	4.5
Phenolics	0.69	1.05	52	7.5
Miscellaneous resins	0.65	1.25	93	9.0
Total	7.97	14.25	78	100.0

are given to help the reader in understanding the industry.

1. Cellulosics. Cellulosics, which are plastic materials produced from cellulose, range from regenerated cellulose, or cellophane, to the nitrocellulose in guncotton. Purified wood pulp is the main raw material. In the United States, purified cellulose (the major product) is made by the xanthate process; in Europe the cuprammonium process is widely used. In the xanthate process cellulose is treated in a solution of NaOH and CS_2. The resultant cellulose xanthate solution is coagulated, and cellulose is regenerated in the form of a continuous film by acidification (Fig. 25.12).

The wastes contain much biodegradable cellulosic materials, sulfates, and heavy metals. The level of waste generated has been reported to range from 0.015 to 0.10 pounds per pound of product manufactured. The bulk of the remainder of the waste is reported to be H_2SO_4, $NaSO_4$, and heavy metals.

Neutralization of acids with caustic is a common practice. In some operations, such as biological treatment, CS_2 may be toxic. Scrap cellophane from production processes is also a waste that requires disposal; usually it is buried or incinerated.

2. Vinyl resins. The mono- and copolymers of vinyl chloride are among the oldest and most versatile thermoplastic resins. Production is essentially in batch operations, and suspension polymerization is the most widely used process in terms of the variety and quantity of products; it accounts for 85 to 90 per cent of the total vinyl resins produced. In this method, the monomer is dispersed as small droplets in a stabilized suspending medium consisting of water and 0.01 to 0.50 per cent by weight of the suspending agents, such as polyvinyl alcohol, gelatin, and cellulose ethers. The suspension is then heated in a reactor in the presence of catalysts such as benzoyl, lauroyl, and *tert.*-butyl peroxides in order to initiate polymerization. When polymerization is complete, the polymer suspension is taken to a blowdown tank or stripper, where residual unreacted monomer is recovered. The stripped polymer is transferred to a blend tank and mixed with sufficient other batches to form a lot. Finally, the polymer slurry mix is pumped to centrifuges, where it is washed and dewatered, and dried in rotary driers. A simplified flow diagram of the suspension process for polyvinyl chloride is shown in Fig. 25.13.

The effluent from the centrifugation step contains most of the contaminants from these plants—suspending agents, surface-active agents, catalysts, small amounts of unreacted monomer, and significant amounts of very fine particles of the polymer product. It has been reported for an average plant, producing 100 million pounds of vinyl resins a year, that the waste contains 1 million pounds of BOD, 150,000 pounds of suspended solids and 100 to 200 × 10^6 gallons of waste.

Primary-sedimentation and activated-sludge treatments have been reported to produce less than 1 per cent reduction in BOD or COD and 98 per cent re-

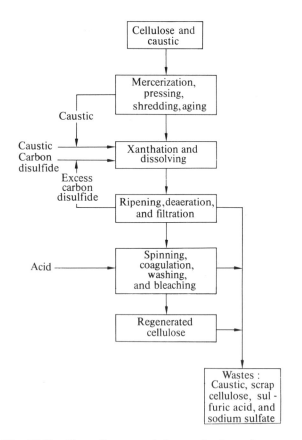

Fig. 25.12 Flow diagram of the production of regenerated cellulose.

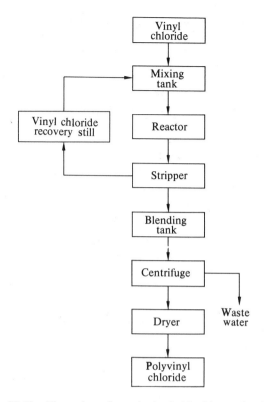

Fig. 25.13 Flow chart for polyvinyl chloride production.

moval of suspended solids and 89 per cent removal of BOD and COD and 98 per cent removal of suspended solids, respectively.

3. Polystyrene resins and copolymers. Polystyrene's combination of physical properties and ease of processing (by injection molding and extrusion) makes it a unique thermoplastic. The crystal-clear product has excellent thermal and dimensional stability, high flexural and tensile strength, and good electrical properties.

The fundamental manufacturing process for polystyrene resins and copolymers is a batch process that uses a combination of both bulk (mass polymerization) and suspension methods. The styrene monomer, or mixtures of monomers, is purified by distillation or caustic washing to remove inhibitors. The purified raw materials, together with an initiator, are charged into stainless-steel or aluminum polymerization vessels, which are jacketed for heating and cooling and contain agitators. Polymerization of the monomer is carried out at about 90°C to approximately 30 per cent conversion, at which stage the reaction mass is syrupy. During this prepolymerization step, water is used only as a heat-exchange medium. Since it does not come into contact with the product and is therefore not contaminated, it can be recirculated. The prepolymer, or partially polymerized mass, is then transferred to suspension-polymerization reactors containing water and proprietary suspending and dispersing agents. The reactors are usually jacketed, and the contents stirred in stainless-steel vessels. The syrupy mass is broken up into droplets by means of the stirrer and held in suspension in the aqueous phase. Temperature is a critical variable in the further polymerization of the product. After completion of polymerization, the polymer suspension is sent to a blowdown tank where any unreacted monomer is stripped. The stripped batch is centrifuged, and the polymer product is filtered, washed, and dewatered. A flow chart for this process is shown in Fig. 25.14.

Reaction water (suspension medium) and wash water are the two significant sources of waste water in the production of polystyrene. Some cooling water is lost through evaporation; however, the amount lost is insignificant, compared with the primary sources of water waste. Approximately 1.5 gallons of water, other than cooling water, is used for each pound of polymer product. The pollutional character of the effluent is slight, because of the small quantities of additives (catalyst and suspending agents), used in suspension polymerization; and the low reaction-medium temperatures required (120 to 180°F). The catalysts are generally of the peroxide type; the suspending agents may be methyl or ethyl cellulose, polyacrylic acids, polyvinyl alcohol, and numerous other naturally occurring materials such as gelatin, starches, gums, casein, zein, and alginates. Inorganic materials such as calcium carbonate, calcium phosphate, talc, clays, and silicates may also be present in effluent reaction water.

No plants employing typical technology have waste-treatment facilities. It is estimated that over 90 per cent of the waste is discharged to municipal sewers.

4. Polyolefins (polyethylenes). The polyethylenes produced today run the gamut of molecular weight from

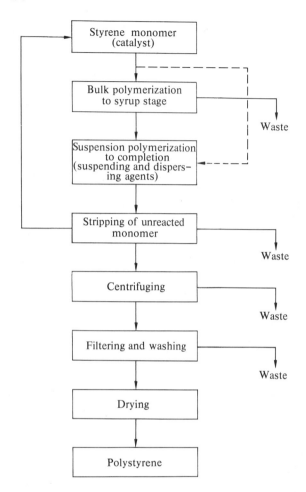

Fig. 25.14 Flow chart for polystyrene production.

pressure and passed through a reactor (of a tubular or an autoclave design) in the presence of free radical initiators. The resulting polymer-monomer mixture is separated by pressure reduction into a monomer-rich and a polymer-rich stream. The monomer-rich stream can be cleaned and recycled as feed to the reactor or used in another process. The polymer-rich stream is usually further concentrated by a second separation step and then extruded into ribbons or strands for pelletizing. Another process is the low-pressure method (producing high-density polyethylene), of which the most common type is the Phillips process, which uses a supported catalyst of

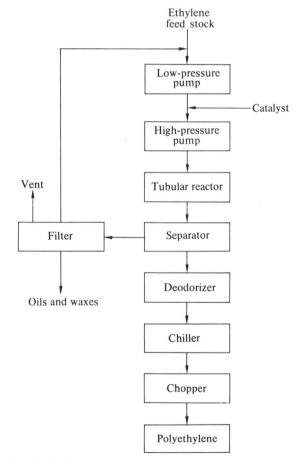

Fig. 25.15 Tubular-reactor process for low-density polyethylene production.

waxes of a few thousand to polyethylenes of several million molecules. In addition to this range in molecular weights, an equally wide range in stiffness is available. In decreasing order of utilization, polyethylene is used for: film and sheet, injection molding, blow-molded bottles, cable insulation, coatings, pipe, and all other uses. Since this represents about 30 per cent of the plastic and resin industry, its production is significant.

One fundamental process for manufacturing polyethylene is the high-pressure method. The final product is often called low-density polyethylene. A high-purity ethylene stream is elevated to a suitable

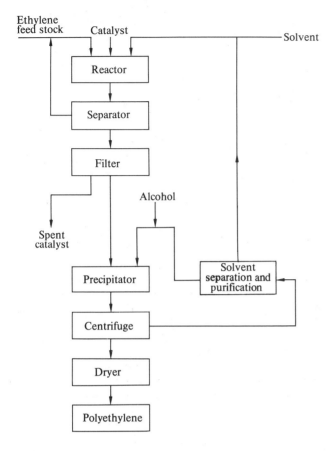

Fig. 25.16 Phillips process for high-density polyethylene production.

chrome and alumina. Figures 25.15 and 25.16 depict the high-pressure and low-pressure methods of producing polyethylene.

Low-density polyethylene processes create no significant water wastes. Water contacts the product only at the chiller-chopper step. Analysis of a typical highly recirculated chill water revealed a very low total organic carbon of 0.4 ppm. The high-density polyethylene processes also produce no significant water wastes. Typical process waste waters contain a BOD of less than 10 ppm. Potential hazards that might generate water-borne wastes are improper operation, spills, and washdown of equipment and facilities.

5. Acrylics. Acrylic resins are made by three processes: bulk, solution, and emulsion polymerization. Bulk polymerization is used for cast sheets and molding and extruding powders. Solution polymerization is used to produce coatings for industrial sales, including automobile paints and fabric coatings. Emulsion polymerization is used mainly to produce coatings for trade sales, such as home paints. About 40 per cent of acrylic resins are made by this last process, which is a batch operation. The monomer is combined with the catalyst, water, and surfactant in a large vat. Polymerization and emulsification are carried out simultaneously. Some of the water is removed and additional surfactant added. Lumps are removed by either filtration or centrifugation, and the emulsion is placed in storage. The final product contains about 50 per cent acrylic resin. Figure 25.17 presents a flow diagram of the emulsion process.

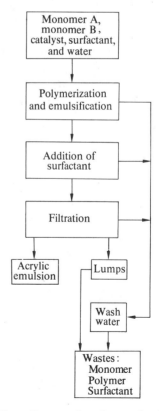

Fig. 25.17 Flow diagram for the production of acrylic emulsion.

The bulk and solution polymerization methods create very little waste. Virtually all wastes from the manufacture of acrylic resins are from emulsion polymerization. This process creates a concentrated water waste from washing the vat between batches and from lumps that are filtered or centrifuged out of the emulsion mix. The waste contains acrylic monomer, acrylic polymer, emulsifying agents, and catalyst; it is white, highly turbid, and has a high suspended-solids content. It is reported that for every pound of product there are 0.125 gallons of water and 0.0015 pounds of BOD in the waste.

Acrylic plants generally feed their waste into municipal systems. This has created a number of problems from the start and a variety of solutions. These range from a self-contained, closed-end, waste-treatment plant, in which the water is continuously recycled, to a wash-water treatment plant that also treats the waste for a municipality in a joint operation. One particular plant, where some waste-removal-efficiency data were available, reports an 85 per cent BOD removal by the activated-sludge treatment.

6. Polyester and alkyd resins. Alkyds and polyesters are characterized by great variation in their formulations, not only according to the class of resin produced, such as oil-modified polyesters (alkyds), unsaturated polyesters (laminates amd molding compounds), and linear polyesters (films and fibers), but also within each group. The fundamental manufacturing process for alkyd and polyester resins is batch-type condensation polymerization of a dibasic acid and a polyfunctional alcohol. Polymerization in the presence of oils or fatty acids results in a complicated polymer known as an alkyd resin. The polymerization process is illustrated in Fig. 25.18.

The significant wastes associated with the production of alkyd and polyester resins are (1) unreacted volatile fractions of raw materials, which either appear in the withdrawn water of esterification and in the water used in scrubbers or are vented to the atmosphere, and (2) residue in kettles cleaned out with either caustic solutions or solvents.

Flotation and land disposal are the only two methods of waste treatment known to be utilized, other than discharge to a municipal sewer system. It is maintained that there is no adequate treatment process for waste water from polyester production.

7. Urea and melamine resins. Urea and melamine resins, which can be used interchangeably, compete with phenolics. Their superior tensile strength and modulus of rupture command a higher price than phenolics, and their electrical properties are outstanding. The fundamental manufacturing process for urea and melamine resins is batch-condensation polymerization of the urea or melamine with formalin (a 40 per cent solution of formaldehyde in water). The raw materials, urea or melamine and formalin, together with catalysts, miscellaneous additives, and modifiers of a proprietary nature, are charged into a jacketed reaction vessel and heated to initiate the reaction. Once initiated, the reaction is exothermic, and the heating steam is shut off. Cooling water is introduced into the jacket to control the reaction

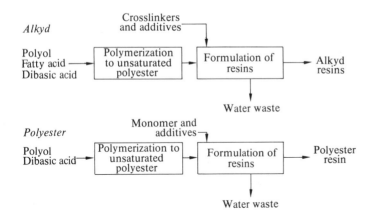

Fig. 25.18 Flow chart for polyester production.

temperature. The mixture is refluxed until the proper degree of polymerization takes place. The resin is soluble in water, so no separation occurs as in the case of phenolic–formaldehyde condensation resins. The urea and melamine resins are vacuum-dehydrated until the solids content is 50 to 60 per cent and then are either sold as a solution or spray-dried and sold as a solid product. A flow diagram for the manufacturing processes is shown in Fig. 25.19.

The significant water wastes from the production of urea and melamine resins are: water introduced with the raw materials, water formed as a product of the condensation reaction, caustic solutions used for cleaning the reaction kettles, and blowdown from cooling towers. Quantities of waste can be computed

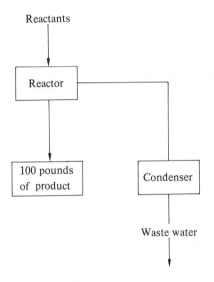

Fig. 25.19 Flow chart for the production of urea, melamine, and phenolic resins. To produce 100 lb of urea resin, 75 lb of urea and 188 lb of formalin are put into the reactor; the waste water contains 133 lb H_2O from formalin, 43 lb H_2O from the reaction, 3.5 lb urea, and 3.5 lb formaldehyde. For 100 lb of melamine resin, the quantities are 52 lb melamine and 182 lb formalin; the waste water contains 109 lb H_2O from formalin, 21 lb H_2O from the reaction, 2 lb melamine, and 2 lb formaldehyde. For 100 lb of phenolic resin, the quantities are 92 lb of phenol, 73 lb of formalin, and 0.3 lb of catalyst; the waste water contains 44 lb H_2O from formalin, 17 lb H_2O from the reaction, 3 lb phenol, 1 lb formaldehyde, and 0.3 lb nonvolatile matter, with a pH of 6.5.

from Fig. 25.19. The waste water discharged from resin plants has a temperature of about 85 to 90°F.

Treatment is carried out by lagooning and thermal incineration, as well as discharge to municipal plants. Lagoons are designed with no overflows, so that evaporation and seepage are the only means of volume reduction. Malodors have been observed with this method. Incineration can also lead to air pollution problems, depending upon the fuel used and the operating efficiency.

8. Phenolic resins. The phenol-derived resins are the oldest family of resins. A wide variety of products and uses make up the distribution pattern. The major uses are for inexpensive casting and plywood bonding. The fundamental manufacturing process for phenolic resins is batch-condensation polymerization of phenolics with formalin. The phenolics and formalin, together with catalysts and miscellaneous additives and modifiers, are charged into a jacketed reaction kettle and heated to initiate the reaction. Once initiated, the reduction is exothermic; heating is terminated, and cooling water is introduced into the jacket to control the reaction temperature. The mixture is refluxed until the contents separate into two, a heavy viscous resin layer and an aqueous one. At this point, a vacuum is applied and the temperature is raised to remove the water. The molten resin is drained into a pan where it solidifies on cooling. The manufacturing process is illustrated in Fig. 25.19.

The significant water wastes from the production of phenolic resins are: water introduced with the raw materials, water formed as a product of the condensation reaction, caustic solutions used for cleaning the reaction kettles, and blowdown from cooling towers. The process water resulting from resin production is about 7.3 gallons per 100 pounds of resin. About 600 gallons of cooling water are required to control the reaction: 42 gallons are discharged to the sewers as cooling-tower blowdown and 558 gallons are recirculated. The waste waters discharged from resin plants have a temperature of about 85 to 90°F.

Phenolic-resin wastes are treated by lagooning, phenol extraction, and thermal incineration, and also are discharged to municipal sewage-treatment plants. Normally a single-stage phenol-extraction plant will remove about 96 per cent of the phenols and 100 per cent of the formaldehyde. One report of the character of waste from such plants is as follows:

Phenol	1,600 ppm
BOD	11,500 ppm
Chlorine demand	68 ppm
Total solids	500 ppm
Volatile solids	250 ppm
Total suspended solids	40 ppm
Volatile suspended solids	20 ppm
pH	6.4

Municipal sewage authorities indicate a high degree of accommodation for phenolic waste. Studies have shown that both phenolics and formaldehyde are biodegradable in conventional biological sewage-treatment processes, providing the concentration is maintained below toxic levels. The difficulties that do arise are due to fluctuation in pH. Therefore, combination of phenolics and municipal wastes is technically feasible, and pretreatment is generally not required.

9. Miscellaneous resins. The miscellaneous resins are chemically unrelated but have one common feature—low-volume production. They include polyurethanes, epoxy, acetal, polycarbonates, silicone, nylon 6, and coumarone-indene. Several of the resins generate no water wastes. Some specialty resins are manufactured by just a few companies. Accurate reports of treatment practices are not available.

References: Plastic and Resin Wastes

1. *The Cost of Clean Water*, Vol. III, Industrial Waste Profile no. 10: Plastic Materials and Resins, U.S. Department of the Interior, Washington, D.C. (1967).
2. Mayo Smith, W., *Manufacture of Plastics*, Vol. 1, Reinhold, New York (1964).
3. Raff, R. A., and J. B. Allison, *Polyethylene*, Interscience, New York (1956).

CHAPTER 26

ENERGY INDUSTRIES

Many industrial activities that are seldom considered as contributing to stream pollution may nonetheless, alone or in combination with others, create pollution problems of considerable magnitude. Among these is the generation of electrical power by steam, which is carried out by central utility plants located throughout the country. We shall describe in this chapter the wastes that arise in the operations of industries involved, either directly or indirectly, in the production of power for public and/or private users. All these industries have one trait in common—the use of large volumes of water, mainly for cooling purposes.

Another difficult waste problem arises in the mining of coal, a material frequently used for energy production. Although the problems of acid mine drainage and coal-preparation wastes (the main sources of pollution in mining operations) could have been considered in Chapter 24 on "Materials Industries," they will be treated here, because coal is used almost exclusively for the production of power, whereas oil, as shown previously, has a multitude of other uses.

Steam is also produced by the heat generated by nuclear fission, so that nuclear wastes rightfully belong in this chapter. However, since the operation of nuclear power plants gives rise to a unique and extremely significant disposal problem, the discussion of radioactive power-plant wastes will be postponed to Chapter 27, which presents a unified treatment of the most important disposal problems arising in connection with radioactive wastes regardless of their origin.

26.1 Steam Power Plants

The operation of steam power plants involves the generation of heat from coal, oil, or other fuel to produce steam from exceptionally pure water. The steam is used to drive turbines, which in turn are coupled to generators. After driving the turbine, the steam is condensed and reused as boiler-water feed. Some steam is lost, and hence make-up water is required to balance the water cycle. Fresh water is generally used to cool the steam, but this water is heated by its contact with the steam and is therefore discharged to the receiving steam at an elevated temperature. The nature and extent of the problems connected with discharge of cooling waters vary, depending on the location, the availability, and the type of waterway into which the liquids may be discharged. When the location of generating stations permits disposal of wastes into salt water, the problems are minimized, but not completely eliminated, since the discharge of certain types of contaminated waste waters into coastal waters is prohibited by federal or state statutes.

Thermal pollution. Davidson and Bradshaw [1967, 21] report that in coal-fired steam plants 6000 Btu's of heat must be dissipated by means of cooling water in heat-exchangers for every kilowatt-hour of electricity generated. The cooling-water discharges are often from 11° to 17°F higher than the temperature of the water in the stream. Cooling towers are now being required to prevent thermal pollution.

Cadwallader [1965, 16] reports that a single-pass condenser will limit the temperature use to 11°F while a two-pass condenser will give a temperature use of 16°F. Condenser design is dependent on the quantity of cooling water available and the permissible temperature of water returning to the watercourse. He predicts the cooling-water requirements

Table 26.1 Total water circulated through condensers.* (After Cadwallader [1965, 16].)

Year	Rivers	Lakes	Brackish and sea water	Cooling towers	Artificial reservoirs or ponds	Total
1959	12,428	3,254	7,820	2,755	556	26,813
1970	23,842	8,168	16,455	6,635	2185	57,285
1980	42,754	15,260	28,015	15,487	4893	106,409

*Data presented as billions of gallons circulated annually.

in the United States for the period 1959 to 1980 (Table 26.1). The combined increase in demand for river water, lake water, brackish, and sea water for cooling purposes will be approximately 300 per cent, whereas, for the same period, it is predicted that water circulated through cooling towers will increase almost 500 per cent and the use of artificial cooling reservoirs and ponds will increase about 800 per cent.

Skanks [1966, 48] reports that, for the steel industry, mechanical-draft cooling towers provide more efficient and less expensive cooling water than spray ponds and atmospheric towers. Berg [1963, 15] makes a cost evaluation of cooling towers and recommends that, if the costs of once-through and conservation methods differ by only about 30 per cent, a thorough study of cooling-water needs is justified.

Wastes other than cooling waters. Powell [24] presents ten classes of waste materials other than heated cooling water that must be disposed of by power stations:

1) hot, concentrated water salines from boiler and evaporator blowdown;
2) acid and alkaline chemical solutions used in cleaning power-plant equipment;
3) water from blowdown of cooling towers, containing minerals in high concentrations;
4) waste water from washing stack gases;
5) acid and caustic solutions resulting from the regeneration of ion-exchange softeners and demineralization water-treatment plants which are used for conditioning feed water for boilers and evaporators;
6) hot alkaline water from blowing down chemical softening plants;
7) acid water drainage from coal storage;
8) drainage from cinder and ash dumps;
9) sanitary sewage;
10) oil, greases, and miscellaneous solid and liquid wastes.

All boilers must be cleaned both before and during operation. Acid and alkaline solutions as well as special detergents are used as cleaning agents, and are periodically discharged as waste. Powell [24] gives the typical dosage of chemicals used for boiling out a new unit with a capacity of 15,000 gallons of water at the operating level (Table 26.2). Since many boilers are cleaned two or more times each year to remove scale from tubes and boiler-drum surfaces, central stations with many banks of boilers may have a frequently recurring acid-waste problem. Neutralization procedures, such as described in Chapter 25 for acid-waste treatment, are sometimes required for these steam-plant wastes. In normal operation, many boilers are blown down daily, or at least weekly, to eliminate precipitated sludge from the mud drum. This process constitutes a major share of the wastes from power plants.

Table 26.2 Boiler compounds for cleaning new boilers. (After Powell [24].)

Chemical	Dosage
Trisodium phosphate (Na_3PO_4)	600 lb
Sodium carbonate (Na_2CO_3)	250 lb
Sodium hydroxide (NaOH)	250 lb
Sodium sulfite (Na_2SO_3)	38 lb
Sodium nitrate ($NaNO_3$)	160 lb
Detergent	10 gal

Forbes [1967, 24] describes pollutants from cooling tower blowdowns and makes suggestions for the control of microbiological as well as chemical agents that contribute to pollution.

Cadwallader [1965, 16] points out that steam-electric generating technology has rapidly advanced since World War II, steam pressures have jumped from 1200 to 3500 psi, and boiler-water quality demands have changed from tolerating contamination in parts per million to parts per billion. Demineralizing ion-exchange resins require acid and alkali regeneration every few days. He suggests treating by neutralization and then controlled release at low rates. Very high-pressure boilers must be cleaned on the average of once a year. The chemical cleaning solutions most commonly used are hydrochloric acid, acetic acid, potassium bromate, ammonia, corrosion inhibitors, detergents, and phosphates. He suggests discharge to an equalization and neutralization basin and slow, controlled-rate discharge into the condenser cooling water as it leaves the station.

Eckenfelder [1966, 23] describes industrial powerhouse water as having the following concentrations (in milligrams per liter): COD, 66; BOD, 10; suspended solids, 50; and condenser water as: COD, 59; BOD, 21; suspended solids, 24.

Rice [1966, 44] proposes using boilers and cooling towers as waste evaporators. He shows that it would cost $121.26 per million lb ($1.01/1000 gallons) to treat boiler-feed water for complete reuse in low-pressure boilers, whereas complete treatment of discharge water would cost $207.33 per million lb ($1.73/1000 gallons). Other cost comparisons are given by Rice.

Hoppe [1964] describes the chemical treatment of cooling-tower water, listing the chemicals used and the waste problem they create, if any. Controlled bleed-off of cooling-tower water will help the waste problem. He suggests regulating bleed-off with a conductivity meter, with the number of cycles of cooling water varying from six to two depending on the hardness of the make-up water.

Some cooling processes require that the water serving as cooling agent be cooled itself prior to use. This precooling is not usually a prerequisite for the condensation of steam from boilers, but cooling waters for air conditioning or other uses within a plant may have to be precooled. Eventually the cooling water will be discharged or it will be reused to a point where it must be treated before further use. According to Riley [15], cooling mechanisms are usually made of wood and designed to produce intimate contact between a down-flowing stream of water and an up-flowing stream of air. The wooden structures contain fans, which move a current of air up through the cooling tower, and pumps, which lift the water to the top of the tower to supply the down-flowing stream of water. (The circulation of these waters through an industrial plant poses two major problems: the loss of water and the excessive quantities of chemicals required for treating these waters.) Cooling towers are primarily used in the absence of an adequate water supply which would make it possible to accomplish the cooling in one single pass (a procedure used by most power plants). The quality of this water must be controlled, for protection against slime and algae growths, corrosion, infiltration of foreign contaminants, and wood deterioration.

Fly ash. Another water-pollution problem is presented by fly ash, a solid waste which results from the use of pulverized coal. Fly ash weighs about 30 lb/ft^3 and constitutes about 10 per cent of the coal burned. These wastes are sometimes collected and deposited in, or on the banks of, races or receiving streams, to be carried away at high water. Jacobs [18] estimates that in 1948 public utilities in the United States discharged a total of about 3,000,000 tons of fly ash. For each carload of coal used per day, the storage area for fly ash amounts to about 2.1 acre-ft per year.

A common method of fly-ash removal from steam power-plant flue gases is by use of a wet scrubber, an electrostatic precipitator, or cyclone separators. In all cases, the fly ash may be transported by pumping in a water slurry. The common method of separation is settling in a decanting basin. Eckenfelder [1966, 23] describes an industrial power plant with waste water containing 6750 gm/liter of fly-ash and suggests lagooning as the method of treatment. An August 1969 article in *Power* reported that a dyke of a fly-ash settling pond of the 700 megawatt Appalachian Power Company station at Carbo, Virginia, broke, releasing a large quantity of alkaline liquor to the Clinch River, with the result that all aquatic life was killed for 125 miles downstream.

A new method of SO₂ removal from coal-fired steam power stations utilizes powdered limestone or dolomite that is injected into the combustion chamber. It reacts with the SO₂ in the hot gas stream to form calcium sulfate ($CaSO_4$) or magnesium sulfate ($MgSO_4$). For a coal with 3 per cent sulfur, the ratio of coal to limestone would be 10:1. A 500-megawatt plant burning 5000 tons of coal per day would produce 465 tons of fly ash per day. Removal of the $CaSO_4$ or $MgSO_4$ from the solution is difficult and researchers suggest reuse of the supernatant liquor from the decanting basin in the wet scrubber. A wet scrubber using a sodium-carbonate solution removes 98 per cent of the particulate matter and 91 per cent of the SO₂. The cost of a 25-megawatt pilot plant is $10 per kilowatt and the operating cost is $1.17 per ton of coal. The main problem with the system is disposal of large quantities of sludge.

Waste treatment. Bennett [18] describes the procedure of disposing of fly ash as follows. Cyclone separators separate the fly ash from gases, water is added to the collected fly ash, and the slurry is pumped to settling or decanting basins. The ash settles, and the water is decanted to the river. Powell [24] disposed of blowdown wastes by constructing a 14-mile sewer line from the plant to the county sewage system. Since the waste was highly acid and alkaline, caused corrosion of equipment, and interfered with the efficiency of biological-treatment units, all waste waters were discharged first into a holding tank and then released to the sewer at a uniform and low rate of flow. Little evidence is currently available of the treatment of other power-plant wastes.

Bender [1967, 13] reports that 20 million tons of fly ash were produced in the United States in 1966, of which 1.25 million tons (about 6 per cent) were put to a useful application. The 45 million tons of fly ash available in 1980 will contain 5 million tons of pure aluminum metal, 8 million tons of iron pellets with 65 per cent iron, 21 million tons of silica and sand, and lesser quantities of oxides of titanium, potassium, sodium, and phosphorus, also uranium and germanium.

Fly-ash may be used as a pozzolana ingredient in Portland cement, for making lightweight aggregate, as a mineral filler in asphalt pavement, as a cement replacement in concrete, as a pozzolana in soil stabilization, as a fill for land development, as a grout in oil wells, as a soil conditioner in agriculture, for the making of bricks, as a filter aid in the vacuum filtration of industrial sludges, as a "choking" material for large, heavy slag roadway base courses, in asphalt roofing and siding materials and as a coagulant in sewage treatment.

Waste characteristics. Gibsen [15] defines "boiler blowdown" as the water containing impurities in considerable concentration which is removed from the boiler circulatory system in order to maintain concentration control. In Gibsen's study, the water removed had a pH of around 11 and total-solids content of about 6000 ppm; however, these characteristics vary with the type of boiler, the operating pressure, and the type of boiler feed used. The blowdown water usually contains some other agents, such as antifoam materials or some type of high-phosphate organic agent. Powell [24] presents a formula and diagrams to show that the concentrations reached may be sufficiently high to constitute a disposal problem of considerable magnitude (Figs. 26.1 and 26.2). He also presents a detailed analysis of raw and concentrated tower-blowdown water (Table 26.3). Fly

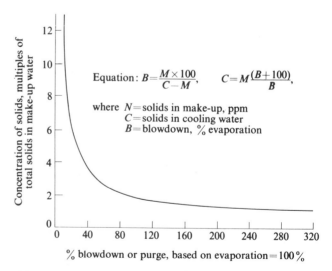

Fig. 26.1 Concentration of solids in water circulated over cooling tower as function of total losses due to windage, leakage, and blowdown. (After Powell [24].)

Equation: $B = \dfrac{M \times 100}{C - M}$, $C = M\dfrac{(B + 100)}{B}$,

where N = solids in make-up, ppm
C = solids in cooling water
B = blowdown, % evaporation

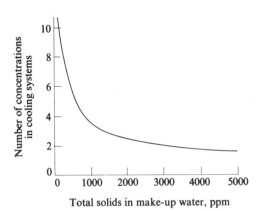

Fig. 26.2 Estimated allowable concentration of make-up water in cooling system for various amounts of total solids to make-up water. Chart based on use of acid when necessary and maintaining a Langelier index of 0.6 ± 0.1. (After Powell [24].)

Table 26.3 Analyses of raw make-up water and concentrated circulating water over cooling towers after corrective treatment with sulfuric acid.

Constituents	As	Make-up water*	Circulating water*
Calcium	Ca	84.0	345.0
Magnesium	Mg	30.0	123.0
Sodium	Na	107.0	438.0
Chloride	Cl	85.0	348.0
Sulfate	SO_4	417.5	1710.0
Nitrate	NO_3	0.2	0.8
Bicarbonate	HCO_3	16.0	65.0
Silica	SiO_2	8.3	35.0
Fluoride		0.5	2.0
Boron		0.1	0.4
Total		748.80	3067.2
Hardness, as calcium carbonate			
Calcium		210.0	863.0
Magnesium		124.0	505.0
Total hardness		334.0	1368.0
Methyl orange alkalinity		13.0	53.0
No. concentrations			4.1

*All results are given as parts per million.

ash contains carbon (1 to 60 per cent by weight), iron, aluminum, calcium, magnesium, silica, sulfur, titanium, and phosphorus. The particle-size distribution depends on the manner in which the fuel is burned. Normally, with ordinary stokers, 20 to 40 per cent of the ash has a diameter of less than 10 microns, and 80 to 90 per cent of the ash has a diameter of less than 200 microns.

Effects of power-plant wastes on receiving waters. The present author [1968, 40A] has summarized some of the major effects on receiving streams of wastes from steam-generating plants, powered by either nuclear or fossil fuel. These effects are as follows:

1. Increase in temperature, which
    a) lowers the amount of oxygen which can be dissolved in water;
    b) increases bacterial as well as aquatic vertebrate activity, which rapidly diminishes already lowered oxygen resources;
    c) increases the growth rate of microscopic plant life and fish;
    d) may cause fish to hatch earlier in spring, ahead of the availability of organisms on which they feed;
    e) increases sensitivity of aquatic life to toxic elements;
    f) decreases value of water for drinking-water use;
    g) may kill small aquatic crustaceans with the sudden temperature rise as they pass through condensers.

2. Addition of salts to water from boiler compounds (mainly phosphates, carbonates, sulfates, and certain organic compounds), which can
    a) stimulate algae growth;
    b) decrease evaporation rates slightly;
    c) increase hardness of water;
    d) make water more corrosive to boats and home water equipment.

3. Addition of disinfectants, such as chlorine and copper sulfate, to decrease slime formation in cooling waters, which
    a) adds color and taste to receiving water;
    b) reduces bacteria population level.

4. Possible addition of radioactive matter to receiving water, which

   a) may concentrate and thus increase in fish and human food;
   b) may introduce trace amounts in water supply which may be harmful in the long term to people drinking water.

5. Reduction of the deeper, cold-water (hypolimnion) layer of water in a lake which serves as a cooling-water supply, which may

   a) cause mixing of upper layer (epilimnion) and lower layer (hypolimnion), with an overall decrease or increase in oxygen depending on whether the initial concentration of oxygen in the deep hypolimnion layer is withdrawn, heated, and subjected to some aeration before being spread out on the surface;
   b) withdraw stored-up algal nutrients (salts such as phosphates and nitrates) and put them into the epilimnion, where they become readily available for algae and other plant growth;
   c) withdraw accumulated dead, organic matter from the bottom and remove a certain amount of it by entrainment on the pump suction inlet screens (some passing the screens may be oxidized by the chlorine compounds added by plant to control slime growth).

In general, there are five methods of solving thermal pollution problems posed by power plants:

1) management of the resource, by dilution, dispersion, increasing turbulence to increase aeration, and cooling, as by using cooling-water storage ponds;
2) improving efficiency of thermal electric plants, as by the use of closed-circuit (evaporative) cooling;
3) utilization of waste heat, as in process heating, desalting water, heating building, etc.
4) disposal of waste heat to atmosphere, in spray cooling towers or in diversion channels;
5) using new methods of electric power generation, such as non-steam-driven turbines or air-cooled condensers.

References: Steam Power Plants

1. Bachmair, A., "Dust from flue gas in public thermal power station," *Atmospheric Pollution Bull.* **24**, December 1956.
2. Basse, B., "Gases cleaned by use of scrubbers," *Blast Furnace* **44**, 1307 (1956).
3. Branin, M. L., "Control of airborne waste," *Coal Util.* **12**, 18 (1958).
4. Brown, D. J., "Keep plant air clean electrically," *Iron Age* **180**, 100 (1957).
5. Craig, O. L., "Abatement of dust from power plant stacks," in Proceedings of U.S. Technical Conference on Air Pollution, 1952, p. 443.
6. Dallavalle, J. M., "Methods of control of atmospheric pollution," *Air Conditioning, Heating, Ventilating* **53**, 75 (1956).
7. Danis, A. L., "Effect of local weather on air pollution," *J. Sanit. Eng. Div. Am. Soc. Civil Engrs.* **83**, 1463 (1957).
8. Drinker, P., and T. Hatch, *Industrial Dust*, 2nd ed., McGraw-Hill Book Co., New York (1954).
9. Duncan, D. M., "Making choice of dust collectors," *Design Eng.* **4**, 34 (1958).
10. Farquhar, J. C., "Combustion efficiency, flue gas analyzer and high pressure," *Fuel Econ. Rev.* **34**, 73 (1956).
11. Gallaer, C. A., "How to measure dust in stacks," *Power Eng.* **101**, 88 (1957).
12. Geisheker, B. J., "Fly ash emission limits," *Public Works Mag.* **88**, 152 (1957).
13. Given, M. D., "Flue gas solves lime waste treatment problem," *Power Eng.* **60**, 93 (1956).
14. Glaysher, E. A., "Institute for smoke prevention in small boiler plants," *Air Pollution Bull.* **24**, 49 (1956).
15. Hedgepeth, L. L., "1957 industrial wastes forum—solving the cooling tower blowdown pollution problem," *Sewage Ind. Wastes* **30**, 539 (1958).
16. Heinrich, R. F., "Problems of electrical cleaning of flue gas," *Air Pollution Bull.* **24**, 52 (1957).
17. Hilst, G. N., "Dispersion of stack gases in stable atmosphere," *J. Air Pollution Control Assoc.* **7**, 205 (1957).
18. Jacobs, H. L., "Fly ash disposal," *Sewage Ind. Wastes* **22**, 1207 (1950).
19. Kamps, R., "Gas filters," *Atmospheric Pollution Bull.* **24**, 50 (1956).
20. Kane, J. M., "Public nuisance problems from invisible stack effluent," *Air Pollution Control Assoc. News* **5**, 4 (1957).

21. Kennedy, H. W., "Fifty years of air pollution law," *J. Air Pollution Control Assoc.* **7**, 125 (1957).
22. Kingsley, K., "Air pollution review, 1956–1957," *Ind. Eng. Chem.* **50**, 1175 (1958).
23. Lerner, L. E., "Conference on smokeless charging of coke ovens," *Atmospheric Pollution Bull.* **24**, 51 (1956).
24. Powell, S. T., "Power generating stations can develop stream pollution problems," *Ind. Eng. Chem.* **46**, 112A (1954).
25. Richter, E., "Smoke nuisance from power plant chimneys," *Fuel Abstr.* **22**, 167 (1957).
26. Russell, H. H., "Fly ash aid to operating efficiency," *District Heating* **42**, 53 (1956).
27. Stamfield, J. P., "Dust collecting equipment," *Cost Eng.* **2**, 106 (1957).
28. Staniar, W. J., *Plant Engineering Handbook*, McGraw-Hill Book Co., New York (1951), p. 1573.
29. Sutton, G., "Dispersal of airborne effluents from stacks," *Air Pollution Control Assoc. Abstr.* **2**, 3 (1956).
30. Tanner, E., "Disposal and utilization of ash from large boiler plants," *J. Appl. Chem.* **6**, page 1 (1956).
31. Taylor, R. A., "A modern development in fly ash handling," *Fuel Econ. Rev.* **35**, 83 (1957).
32. Watson, J. W., "New incineration designed to reduce fly ash," *Public Work Mag.* **89**, 97 (1958).
33. Williams, T. H., "Purification of coke oven effluents," *Chem. Abstr.* **50**, 13346 (1956).
34. Woodall, J. D., "Atmospheric pollution and demand for smokeless fuels," *Heating Air Treatment Eng.* **20**, 298 (1957).
35. Wordley, W. A., "Economics of boiler flue cleaning," *Cheap Steam* **40**, 68 (1956).
36. Wurtz, C. B., and T. Dolan, "A biological method used in the evaluation of effects of thermal discharge in the Schuylkill river," in Proceedings of 15th Industrial Waste Conference, Purdue University, May 1960, p. 461.
37. Yano, N., "Radioactive dust in the air," *Chem. Abstr.* **51**, 100 (1957).

Suggested Additional Reading

The following references have been published since 1962.
1. "Nuclear industry plans to control its thermal pollution," *Environ. Sci. Technol.* **2**, 165 (1968).
2. "Limestone for SO_2 capture," *Environ. Sci. Technol.* **1**, 9 (1967).
3. "Fly ash—SO_2 scrubber," *Environ. Sci. Technol.* **1**, 13 (1967).
4. "Boiler cleaning—survey of off load methods," *Eng. Boiler House Rev.* **81**, 286 (1966); *Eng. Index*, 1967, p. 21.
5. "From the frying pan into the fire," *Power* **111**, 5 (1967).
6. "Do additives help control pollution?" *Power* **111**, 96 (1967).
7. "Fly ash used in concrete at Lansing," *Power* **111**, 4 (1967).
8. "Coal ash slag may make fine mineral wool," *Power* **111**, 4 (1967).
9. "Clinch river, Va., polluted by dyke break," *Power* **111**, 126 (1967).
10. "KLP to inject dolomite into boilers," *Power* **122**, 108 (1968).
11. "Reviewed interest in fly ash brick," *Power* **111**, 4 (1967).
12. Arnold, G. E., "Thermal pollution of surface supplies," *J. Am. Water Works Assoc.* **54**, 1332 (1962).
13. Bender, R. J., "Fly ash utilization makes slow progress," *Power* **111**, 116 (1967).
14. Berg, B., R. W. Lane, and T. E. Larson, "Water use and related costs with cooling towers," *J. Am. Water Works Assoc.* **56**, 311 (1964); *Eng. Index*, 1964, p. 2202.
15. Berg, B., R. W. Lane, and I. E. Larson, "Water use and related costs with cooling towers," Illinois State Water Survey, Circular no. 86, 1963; *Eng. Index*, 1965, p. 2619.
16. Cadwallader, W. W., "Industrial wastewater control," in *Power*, ed. C. F. Gurnham, Academic Press, New York (1965).
17. Coutant, C. C., "The effect of a heated water effluent upon the macroinvertebrate riffle fauna of the Delaware river," *Proc. Penn. Acad. Sci.* **36**, 58, 1963.
18. Dalton, T. F., "Planning a cooling tower treatment program for waste and pollution control," in Proceedings of 11th Ontario Industrial Waste Conference, 1964, p. 49; *Chem. Abstr.* **66**, 79383w (1967).
19. Dalton, T. F., "Planning a cooling water treatment program for waste and pollution control," in Proceedings of 11th Ontario Industrial Waste Conference, June 1964.
20. Daniels, R. C., "Heat recovery cuts cooling water costs," *Power* **111**, 60 (1967); *Eng. Index*, August 1967, p. 230.
21. Davidson, B., and R. W. Bradshaw, "Thermal pollution of water systems," *Environ. Sci. Technol.* **1**, 619 (1967).

22. Dryer, W., and N. G. Benson, "Observations on the influence of the new Johnsonville steam plant on fish and plankton populations," in Proceedings of 10th Annual Conference Southeastern Association of Fish and Game Commissioners, 1957, p. 85.
23. Eckenfelder, W. W., Jr., *Industrial Water Pollution Control*, McGraw-Hill Book Co., New York (1966) p. 25.
24. Forbes, M. C., "Cooling towers not unmixed blessing in pollution control," *Oil, Gas J.* **65**, 88 (1967); *Eng. Index*, December 1967, p. 292.
25. Frankenberg, T. T., "Praktische Erfahrungen zur SO$_2$ Beseitigung aus Abgasen in Versuchs—und Betriebsanlagen," *Staub*, **25**, 1965, p. 429.
26. Hertel, W., "Rauchgas und Aschenabfuehrung sowie Haupttragkonstruktion eines Blockkraftwerks," *Brennstoff-Waerme-Kraft* **19**, 245 (1967); *Eng. Index*, November 1967, p. 240.
27. Hoak, R. D., "The thermal pollution problem," *J. Water Pollution Control Federation* **33**, 1267 (1961).
28. Hoppe, T. C., "Industrial cooling water treatment for minimum pollution from blowdown," *Proc. Am. Power Conf.* **28**, 719 (1966); *Chem. Abstr.* **67**, 25252 (1967).
29. Katell, S., and K. D. Plants, "Here's what SO$_2$ removal costs," *Hydrocarbon Process. Petrol Refiner* **46**, 161 (1967); *Eng. Index*, December 1967, p. 7.
30. Kovats, A., "Economics of condenser circulating water supply in power stations," in Proceedings of American Society of Mechanical Engineers, Paper no. 67-*WA/PWR*-2, 12 November 1967; *Eng. Index*, January 1968, p. 273.
31. Krumholz, L. A., and W. L. Minckley, "Changes in the upper Ohio river following temporary pollution abatement," *Trans. Am. Fisheries Soc.* **93**, 1 (1964).
32. Laberge, R. H., "A critical problem in stream pollution comes from thermal discharges," *Water Sewage Works* **106**, 536 (1959).
33. Landers, W. S., "Trends in steam station design affecting air pollution," in Proceedings of American Society of Mechanical Engineers, Paper no. 66-*PUR*-1, 18 September 1966; *Eng. Abstr.*, April 1967, p. 226.
34. Magnus, M. N., "History of fly ash collection of South Charleston Plant Union Carbide Corporation—chemicals division," *J. Air Pollution Control Assoc.* **15**, 149 (1965); *Eng. Index*, 1966, p. 1024.
35. Magnus, M. N., "History of fly ash collection of South Charleston Plant Union Carbide Corporation—chemicals division," *J. Air Pollution Control Assoc.* **15**, 149 (1965); *Eng. Index*, 1966, p. 1024.
36. Mapstone, G. E., "Control cooling tower blowdown," *Hydrocarbon Process. Petrol. Refiner* **46**, 155 (1967); *Eng. Index*, September 1967, p. 251.
37. Markowski, S., "The cooling water of power stations: a new factor in the environment of marine and freshwater invertebrates," *J. Animal Ecol.* **28**, 243 (1959).
38. Markowski, S., "Observations on the response of some benthonic organisms to power station cooling water," *J. Animal Ecol.* **29**, 349 (1960).
39. Moehle, F. W., "Fly ash aids in sludge disposal," *Environ. Sci. Technol.* **1**, 374 (1967); *Eng. Index*, November 1967, p. 226.
40. Mumby, K., "Moving five million cubic yards of earth in five months," *Mech. Handling* **50**, 596 (1963); *Eng. Index*, 1964, p. 96.
41. Nemerow, N. L., "Some major effects of nuclear powered steam generating plant wastes on receiving lakes and streams," Paper presented at New York State Society of Professional Engineers' Meeting, Ithaca, N.Y., 15 November 1968.
42. Paige, P. M., "Costlier cooling towers require new approach to water-systems design, *Chem. Eng.* **74**, 93 (1967); *Eng. Index*, January 1968, p. 113.
43. Palaty, J., "Purification of cyanide waste waters on black coal slag," *Sb. Vysoke Skoly Chem. Technol. Praze, Oddil. Fak. Technol. Paliv Vody*, **5**, 7 (1961); *J. Water Pollution Control Federation* **37**, 748 (1965); *Water Pollution Abstr.* **37**, 1499 (1964).
44. Phelps, A. H., "What doesn't go up must come down," *Chem. Eng. Progr.* **62**, 37 (1966); *Eng. Abstr.*, May 1967, p. 105.
45. Rice, J. K., "Water management reduces waste and recovers water in plant effluents," *Chem. Eng.*, 26 September 1966, p. 125.
46. Ritchings, F. A., "Closed cooling water cycle saves water," *Elec. West* **131**, 44 (1964); *Eng. Index*, 1965, p. 2383.
47. Schaffer, S. G., and R. W. Noble, "Techniques for utilization of fly ash," *Proc. Am. Chemists' Soc. Div. Fuel Chem.* **10**, 73 (1966); *Eng. Index*, 1966, p. 1024.
48. Schmidt, K. G., "Nasswaschgeraete aus der Sicht des Betriebsmannes," *Stabu* **24**, 485 (1964); *Eng. Index*, 1965, p. 571.
49. Shanks, R. I., "Water conservation in heavy industry," *Steel Times* **193**, 682 (1966); *Eng. Index*, June 1967, p. 247.
50. Siebert, O. W., and W. C. Engman, "Case history on

economics of chemical treatment of recirculating water cooling tower," *Protection* **3**, 20 (1964); *Eng. Index*, 1965, p. 2645.
51. Snyder, M. J., "Properties and uses of fly ash," *Batelle Tech. Rev.* **13**, 14 (1964); *J. Water Pollution Control Federation* **37**, 748 (1965).
52. Snyder, J. A., A. J. Roese, R. I. Hunter, and P. Gluck, "Progress report on fly ash utilization research program," *Edison Elec. Inst. Bull.* **34**, 38 (1966); *Eng. Index*, 1966, p. 1024.
53. Squires, A. M., "An introduction to the top neat cycle: a means for eliminating sulfur from power station effluent while improving heat rate," *Proc. Am. Power Conf.* **28**, 1966, p. 505.
54. Sprague, J. B., "Resistance of four freshwater crustaceans to lethal high temperature and low oxygen," *J. Fisheries Res. Board Can.* **20**, 387 (1963).
55. Stevens, H. A., "Water conservation saves $250,000," *Power* **122**, 76 (1968).
56. Stairmand, C. J., "Removal of grit, dust and fume from exhaust gases from chemical engineering processes," *Chem. Eng.* **194**, CE310–326 (1965); *Eng. Index*, 1966, p. 1340.
57. Trembley, F. J., "Effects of cooling water from steam-condenser power plants on stream biota," in *Biological Problems in Water Pollution*, Transactions of 1962 Seminar, Robert A. Taft Sanitary Engineering Center, Cincinnati, Ohio; U.S. Public Health Service Publication no. 999-*WP*-25 (1965), p. 334.
58. Snyder, M. J., and H. W. Nelson, "Critical review of technical information on utilization of fly ash," Battelle Memorial Institute Research, Report no. 902, 1962; *Eng. Index*, 1964, p. 675.
59. Van Vliet, R., "Effect of heated condenser discharge upon aquatic life," in Proceedings of American Society of Mechanical Engineers, Paper no. 57–PWR-4, p. 1.
60. Warinner, J. E., and M. L. Brehmer, "Effects of thermal effluents on marine organisms," *Air Water Pollution* **10**, 277 (1966); *Eng. Index*, February 1967, p. 265.
61. Welmer, J., "By-product gypsum," German Patent no. 1224190, 1 September 1966; *Chem. Abstr.* **65**, 19824 (1966); *J. Water Pollution Control Federation* **39**, 878 (1967).
62. Winkelman, F. W., "Ash-handling problems and practices in large thermal power stations," in Proceedings of American Society of Mechanical Engineers, Paper no. 65-*WPR*-13, September 1965; *Eng. Index*, 1965, p. 117.

26.2 The Coal Industry

There are two major wastes associated with the production of coal: coal-preparation wastes (coal washeries) and coal-mine drainage wastes (acid mine drainage). Since these wastes and their subsequent effects and treatment are so different, they are usually considered as separate categories; however, in the present context, they will be considered as two different wastes associated with one industry.

Origin of Wastes

Coal-preparation wastes. After coal is mined and brought to the surface, it is processed by "breakers," or preparation plants, where the impurities are removed. Coal solids from a diameter of $3\frac{1}{4}$ inches down to $\frac{3}{64}$ inch are now used by the public; they are described, in order of decreasing size, as egg, stove, chestnut, pea, buckwheat, rice, barley, and No. 4 and No. 5 buckwheat. Both Rickert [179] and Parton [166] describe in considerable detail the series of crushing, screening, classifying, and washing processes involved in the cleaning of coal and the separating of the various sizes. During the cleaning operation, the fine coal becomes suspended in the large quantity of water required for the cleaning. One plant [179] discharges 9000 gpm, approximately 4 per cent of its volume being solids.

Acid mine-drainage wastes. Acid mine-drainage wastes result from the passage of water through mines where iron disulfides, usually pyrites, are exposed to the oxidizing action of air, water, and bacteria. Coal and adjacent rock strata buried in the earth contain sulfur in the form of various compounds. In the process of mining, the sulfuritic materials are uncovered and exposed to air and moisture, with the result that the sulfide oxidizes to ferrous sulfate (copperas) and sulfuric acid, according to the equation

$$2FeS_2 + 7O_2 + 2H_2O \rightarrow 2FeSO_4 + 2H_2SO_4. \quad (1)$$

As the mine is drained, the oxides form additional sulfuric acid in the water. Water finds its way into a strip mine primarily through surface runoff during periods of rainfall. (Ash [11] found that up to 97 per cent of the rainfall on an extensively mined area was pumped out as mine drainage.)

Since the ferrous sulfate in the presence of sulfuric acid is quite resistant to oxidation, the leaching action of the mine water carries off most of the iron and sulfur in this form. After a time, however, there is further oxidation by atmospheric oxygen and from the products of the reaction in Eq. (1), we have

$$2FeSO_4 + O + 2H_2SO_4 \rightarrow Fe_2(SO_4)_3 + H_2O + H_2SO_4. \quad (2)$$

When reactions (1) and (2) are considered to be occurring simultaneously, we have

$$4FeS_2 + 15O_2 + 2H_2O \rightarrow 2Fe_2(SO_4)_3 + 2H_2SO_4. \quad (3)$$

When the sulfuric acid concentration is diluted by receiving streams, the ferric sulfate hydrolyzes as follows:

$$Fe_2(SO_4)_3 + 6H_2O \rightarrow 2Fe(OH)_3 + 3H_2SO_4. \quad (4)$$

It is apparent that all the sulfur which, at the start, was present in the ground as an insoluble sulfate is now dissolved in the receiving stream as sulfuric acid.

There is also much evidence that bacterial activity plays an important role in acid formation [48, 220, 221]. The sulfur-oxidizing bacterium, *thiobacillus thiooxidans*, an autotrophic bacterium, uses inorganic sulfur, thiosulfate, or tetrathionate for food and manufactures sulfate, which is eventually converted to sulfuric acid. It thus appears that the formation of acid mine water is both a chemical and a biological reaction and may occur in almost any mine where oxidizable sulfur comes into contact with air and water.

Character of Wastes

Coal-preparation wastes. Richert [179] presents typical size and ash analyses of the solids which are discharged from a coal-cleaning plant (Table 26.4). These data indicate that 7.4 per cent of the solids are retained on a 28-mesh sieve (openings of 0.023 inch). The refuse content increases progressively with the increase in the size of the particles and the proportion of clay present. The solids consist of coal, shale, clay, sandstone, bone, and bony coal. In addition to clay, relatively small amounts of other minerals are also present, the more common ones being calcite, gypsum, kaolin, and pyrite. The clays and shales break down into fine particles when subjected to the action of water or when immersed in it, and thus are

Table 26.4 Hydroclassifier operating data. (After Richert [179].)

Discharge from cleaning plant	Feed	Underflow	Overflow
Waste water, gpm	9000	560	8400
Solids, %	4.1	35.0	1.0
Solids, tons/hr	91.0	60.0	31.0
Recovery, % waste water	100.0	67.0	33.0
Ash, %	32.3	25.4	42.9
Mesh	Screen analysis, washout		
+10	0.2	0.2	0
+14	0.2	0.3	0
+20	2.4	2.0	0
+28	4.6	5.0	0
+35	7.6	10.0	0.2
+48	9.4	14.5	0.2
+65	9.6	15.0	0.2
+100	10.8	10.5	0.2
+150	8.6	14.5	0.4
+200	7.0	9.0	1.8
+270	2.6	4.0	2.0
−207	37.2	6.0	95.0

responsible for the large quantities of semicolloidal particles present in suspension in the washing water. Hall [1965, 74] reports that coal-washing water has as its principal pollutant suspended solids and may contain calcium and magnesium sulfates and iron. It is sometimes acid but is usually kept alkaline to minimize corrosion of processing equipment. He describes the black solid discharge from coal-preparation plants as finely divided clay, black shale, and other minerals.

Acid mine-drainage wastes. The concentration of acid in mine waters varies widely, from less than 100 ppm to nearly 50,000 ppm, with typical values of about 100 to 6000 ppm of H_2SO_4, 10 to 1500 ppm of $FeSO_4$, 0 to 350 ppm of $Al_2(SO_4)_3$, and 0 to 250 ppm of $MnSO_4$. Mine drainage presents numerous anomalies, and the acidity of the drainage cannot be predicted with certainty. For example, a study of two mines located within a few miles of each other showed that the drainage from the mine whose coal contained 3 per cent of sulfur was 200,000 gal/day of water, with an alkalinity of 170 ppm, whereas the other mine,

Table 26.5 Analysis of two different mine drainage waste waters [4].

Mine drainage analyses	Mine A	Mine B
Drainage flow, gpm	2500	900
pH	3.7	6.2
Free acid, ppm	124	4
Total acid, ppm	466	13
SiO_2, ppm	14	9.6
Al, ppm	17	1.9
Fe, ppm	22*	0.2
Mn, ppm	10	4.6
Ca, ppm	95	34
Mg, ppm	55	12
SO_4, ppm	746	172
Cl, ppm	9	2.4

*Apparently, a considerable amount of iron originally in solution has been precipitated out in suspended form.

whose coal had a sulfur content of 2.6 per cent, discharged 130,000 gal/day of water with an acidity of 30,000 ppm [117].

The Bureau of Mines [4] also provides data illustrating the variability of mine-drainage waters (Table 26.5). Mine A has a highly acid drainage, containing relatively large quantities of iron, other metals, and sulfates, while the drainage of mine B exhibits primarily the characteristics of typical ground water. Hinkle [108, 109] provides representative analyses of mine waters from a West Virginia area that is badly polluted with acid from the softcoal country (Table 26.6).

Acidic stream water to be used for industrial or domestic purposes will require treatment that will vary depending on the eventual use. Acidity can be overcome by the addition of alkaline materials, but the resultant water is usually "hard" or contains an excess of alkaline materials. Although this water may be softened or treated for the excess of alkaline compounds, the problem is not always eliminated, because an amount of "foaming" and priming occurs as a result of the high chemical content [122]. Table 26.7 shows the characteristics of a Pennsylvania surface-water stream that receives acid mine drainage.

Table 26.6 Analysis of some West Virginia mine water. (Hinkle [108])

Analysis*	Source of water			
	Pittsburgh seam	Sewickley seam	Upper Freeport	Roof drips, Pittsburgh
pH	2.2	7.7	4.3	8.3
Acidity				
Phenolphthalein	3880	12	40	8
Methyl orange	1420	−440 (alk)	10	−800 (alk)
Methyl red	2420	−408 (alk)	30	−
Total solids	9362	2466	733	3105
Organic and volatile solids	4616	270	69	149
Fixed mineral matter	5746	2196	664	2956
Insoluble matter	−	25	47	14
SiO_2	78	19	42	7
Fe_2O_3	1520	3	8	35
Al_2O_3	231	7	12	4
Mn_2O_3	+	5	0	147
CaO	589	121	156	199
MgO	268	53	60	24
Na_2O	754	829	24	1130
Sulfates as SO_3	4760	987	369	1315
Chloride	28	24	7	6

*All results are given as ppm, except for the pH.

Table 26.7 Chemical analyses* of a Pennsylvania stream† containing acid mine drainage [122].

Mean discharge, ft³/sec	Temperature, °F	Color	pH	Specific conductance at 25°C, μmhos	Silica	Aluminum	Iron	Manganese	Calcium	Magnesium	Combined sodium and potassium	Bicarbonate	Sulfate	Chloride	Fluoride	Nitrate	Dissolved solids	Hardness as CaCO₃ Total	Hardness as CaCO₃ Noncarbonate	Total acidity as H₂SO₄
40		2	3.70	1840	23	77	0.83	15	139	94	16	0	1170	12	0.3	1.0	1750	1200	1200	580
73.2	48	1	3.60	1430								0	798					660	660	322
59.3	58	2	3.30	1430	17	37	0.57	7.2	117	74	13	0	829	7.0	0.1	0.2	1230	844	844	308
54.2	52	1	3.70	1690								0	980		0.1			900	900	384
25.1	66	1	3.30	1670								0	996		0.4			716	716	536
29.2	67	1	3.25	1340	13	37	1.2	7.2	1000	67	3.0	0	744	7.0	0.2	0.4	1120	779	779	312
41.7	67	4	3.20	1870	15							0	1140	8.0	0.1			1010	1010	
32.4	62	2	2.60	1930	21							0	1300	12	0.0	0.5		587	587	690
43.6	62	1	3.50	1470	14	34	0.85	5.7	144	86	6.6	0	899	6.0	0.0	0.1	1430	931	931	248
30.4	60	7	3.30	1590	14							0	1090	4.0	0.0	3.5		650	650	496
22.3	68	0	3.15	1940	25	64	1.6	8.2	167	107	1.2	0	1190	21	2.0	1.1	1690	1270	1270	436

*All results are given in parts per million unless otherwise specified.
†Panther Creek at Tamaqua, Pa.

From this table the reader may note the following: low pH; high specific conductance; high concentration of iron, aluminum, manganese, and magnesium; high sulfate concentration; high dissolved solids; high degree of hardness, most of which is of noncarbonate origin, which makes its removal by chemicals more expensive; and high acidity. All these observations serve to emphasize the great polluting potential of acid mine waters and the subsequent difficulty for any users of stream water receiving such waste waters.

Treatment of Coal-industry Wastes

Coal-preparation wastes. Parton [166] describes three methods of treatment or developments that have helped to reduce the amount of solids reaching the streams:

1) installation of settling and impounding facilities for collecting the fine-size anthracite (silt) in the waste water discharged from the wet washing of coal;
2) increased demand for fine coal, formerly discarded as refuse, which reduces the quantity of solid reaching the waste;
3) adoption of more efficient methods of cleaning the ultrafine sizes, e.g. froth flotation was developed as a method of solving this problem.

Richert and Bishop [179] show a sketch of the treatment plant at the Tamaqua Colliery, which has made use of all three developments (Fig. 26.3). The froth-flotation process is used to clean the coal sludge; i.e., the mineral is floated to the surface by an artificially induced froth. Since it is important that the bubbles be strong enough to support the weight of the floated coal, pine oil—an organic polar compound capable of forming a stable froth on the surface—is used as the frother. Operating results for the primary and secondary (scavenger) flotation units are given in Tables 26.8 and 26.9 respectively.

Sometimes rivers must be dredged to reclaim the "culm," or deposited river coal. Van Ness [232] describes an operation for recovering riverborne anthracite from the Susquehanna River. When the reclaimed coal is cleaned, it becomes steam-size coal, containing 16 per cent ash and 18 per cent moisture and supplying 10,000 Btu's of heat per pound.

Coal-mining wastes. Since 1962, twelve methods of

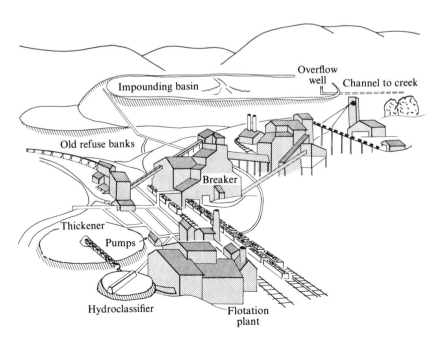

Fig. 26.3 Waste-water treatment at Tamaqua Colliery [179].

Table 26.8 Operating results, primary flotation circuit [179].

	Feed	Coal	Refuse
Gal/min	1000	280	720
Solids, %	10.0	40.6	9.6
Solids, tons/hr	52.0	33.5	18.5
Recovery, %	100.0	64.5	35.5
Ash, %	30.0	13.0	60.8

Screen and ash analyses

Mesh size	Wash-out	Ash, %	Wash-out	Ash, %	Wash-out	Ash, %
+28	3.7	11.3	5.7	8.8	1.5	14.2
+48	20.7	18.2	21.4	9.2	10.9	32.9
+100	34.0	27.9	34.0	12.8	26.5	60.5
+200	23.2	33.4	24.7	13.9	24.8	65.8
−200	18.4	46.4	14.2	19.5	36.3	70.0
Total	100.0		100.0		100.0	

coal-mining wastes treatment have received most consideration.

1. Thickener. The first step in clarification of coal-washery waste waters is to concentrate the material with hydraulic cyclones or thickeners. The clarified water is returned to the coal-washing circuit, while the high-solids underflow from the thickener may be passed through a drum-type filter or pumped to settling ponds. Clarified water from the pond may be returned to the coal-preparation circuit or discharged to a stream. Many clarification systems are overloaded and the immediate and economic solution is to minimize the amount of water necessary for proper coal preparation. In some cases the fine coals recovered from washery water have economic value but normally such water is disposed of in a permanent impoundment or a specially constructed pit.

2. Settling ponds. Deane [1966] states that waste water from a strip-mine coal-cleaning plant should be cycled through two settling ponds in series and the effluent from the second pond should be reused. Woodley [1967] advocates using large slurry ponds for wash water with reuse. Abandoned strip-mining excavations are excellent slurry ponds. Hummer [1965] describes minus 58-mesh refuse disposal in storage ponds and dam design. Cooley [1944] reports that Virginia limits discharge from coal-preparation plants to 2 mg/liter of settleable solids and no discoloration. Several hours of sedimentation in a lagoon usually effects removal.

3. Coagulation. Hrebacka [1965] reports that the new mechanical mining techniques produce a great amount of fines (small unusable coal dust) and an increased amount of sludge. He discusses sedimentation with polyelectrolytes and sludge disposal. Fomenko et al. [1965] show that clarifier efficiency may be markedly increased by determining the optimum water-slurry system. Rozgaj [1965] found that organic and inorganic coagulants were unsatisfactory when used alone, but combinations of the two gave good settling in 30 minutes. The inorganic coagulants were H_2SO_4, $Ca(OH)_2$, $Al_2(SO_4)_3$, $CaCl_2$, and $NaOH$. The organic coagulants (polyacrylamides) were Separan 2610, Separan NP1020, and Aerofloc 550. Bochkaren and Baryshnikov [1963] and Olfert [1965] also found polyacrylamide to be an effective coagulant. Green [1966] used a mixture of clay and polyelectrolyte blended in ratios from 10:1 to 2:1 respectively in concentrations of 0.25 to 15 mg/liter to achieve settling of finely divided solids. Johnson

Table 26.9 Operating results, scavenger flotation circuit [179].

	Feed	Coal	Refuse
Gal/min	360.0	25.0	335.0
Solids, %	15.4	41.5	11.7
Solids, tons/hr	15.2	3.1	12.1
Recovery, %	29.2	6.0	23.2
Ash, %	61.4	24.1	70.7

Screen and ash analyses

Mesh size	Wash-out	Ash, %	Wash-out	Ash, %	Wash-out	Ash, %
+28	3.5	16.9	2.5	14.9	3.5	16.4
+48	14.7	46.4	17.8	15.6	14.2	48.4
+100	31.2	66.1	31.9	24.6	32.2	73.7
+200	28.1	66.6	29.9	26.3	26.7	78.1
−200	22.5	72.1	17.9	30.6	23.4	79.4
Total	100.0		100.0		100.0	

[1964] found that free silica and hydrated lime (1 to 5 lb/ton of suspended solids) achieved good coagulation. Kumanomedo and Ogeshima [1966] used alum and either cationic, anionic, or nonionic polymers to coagulate coal-washery wastes. Day [1965] found that water-soluble polymeric chemicals aid solids settling of coal tailings and clay and effectively produce a reusable water.

4. *Froth flotation.* Hamilton et al. [1967] summarize improved treatment of coal fines by froth flotation and discuss flotation reagents, bubble contact, cell design, and kinetics studies. Adamson [1965] describes fines treatment by flocculation or froth flotation, depending on the character of the coal and catering to the fraction under 0.5 mm in a 400 to 600 ton/hr coal-cleaning plant. Mistrek [1966] reports using a product of olefin hydroformylation in amounts of 1 to 2.5 kg/ton as a flotation agent. Negulescu et al. [1964] report that a flotation agent rendered coal-washing water too high in oily matter and phenols for discharge and instead it was effectively reused. Roe [1964] advocates recycling flotation agents to decrease dosages required and recommends pressure rather than vacuum filtration of the froth-flotation tailings. Bucklen and Smith [1964] attain 97 per cent coal recovery in an air-lift froth-flotation cyclone. Ferney [1964] reports on reclamation of fine coal from settling ponds using cyclone froth flotation or centrifuging. Florin [1964] describes drying and sintering of froth-flotation tailings in a shaft furnace. Sorokin and Tsiperovich [1964] found that methylcyclohexane was an outstanding reagent in coal flotation. Flotation-concentrated coal fines are in one case sent to a dryer. The dryer exhaust goes to a cyclone unit and an impingement wet scrubber to remove ultra-fines.

5. *Thickeners and hydrocyclones.* Smidt [1967] reports on clarifying and thickening of coal slurry in a new Dorr-type thickener. Visman [1967] describes a multicyclone which produces semi-solid lumps from coal slurries, which may be further compacted into pellets with vibratory screens.

6. *Centrifuges.* Smidt [1965] reports on the use of a shell, solid-bowl centrifuge, and flocculation agents to dewater flotation refuse mechanically and so to reduce water consumption by 70 per cent. Iwaski et al. [1967] describe a new vacuum and stream centrifuge used to dewater fine suspensions. Llewellyn [1964] found that a continuous solid-bowl centrifuge is practical and economical in dewatering froth-flotation concentrates. Bruk [1965] defined working and construction parameters for coal-washing centrifuges. Eveson and Pickin [1966] use polyelectrolyte coagulant aids for settling bentonite and shale and centrifuged the sludge to produce a cake with 36 per cent moisture.

7. *Filtration.* Berger [1965] studied continuous rotary vacuum filtration combined with froth flotation for fines separation. Heertjes [1964] studied parameters affecting filtration including weight of particles, their concentration, resistance to filtration, and the use of filter aids. Washburn [1963] describes a closed circuit plant which handles 5000 to 7000 tons per shift. Flotation and flocculants are not used. Underflow from the thickener at 50 per cent solids is dewatered on two 1000-ft^2 dish filters. Hill et al. [1962] discuss use of the filter press to dewater froth-flotation tailings. They suggest using a maximum chamber size of 60 in.^{2} and a ram-type pump for filling and bringing the press up to pressure.

8. *Burial.* May and Berg [1966] report that segregation and burial of bone and rider coal wastes from strip mining, which contain very acid materials, limit bad effects.

9. *"Gob" piles.* Davison and Jefferies report that the stability of pit waste heaps is influenced by vegetative cover which depends on nutrition factors. The recovery of quality coal from old anthracite tailings banks eliminates regional eyesores, acid drainage, fills abandoned strip mines, and provides jobs for coal workers [Coal Age **70**, 1965]. To recover coal from spoil banks, they must be 19 per cent coal by weight [*Tech. Digests* **9**, 1967].

10. *Atomized water.* Brady et al. [1967] experimented with removing fines of minus 60-mesh anthracite from a slurry by atomizing water and removing mist in an air current. This process recovers 67 to 79 per cent.

11. *Dry cleaning.* The use of an electropneumatic concentrating separator to clean all coal particles greater than $\frac{5}{8}$ inch without water would eliminate stream pollution and cleaning of sludge ponds [*Coal*

Age **71**, 1967]. Thoma and Geppert [1967] describe a dry-cleaning process for use at the start of the coal-preparation process.

12. Trough separation. Tanaka *et al.* [1966] found that a trough separation unit was effective in separating shale and sandstone in coal refuse.

In a discussion of the problems associated with the treatment of acid mine drainage, Hert [106] recommended the following five control measures:

1) drainage control and diversion of water, to prevent water from entering the mining area, and rapid removal of any water present;
2) proper disposal of sulfur-bearing materials, to ensure that none of the "gob" (sulfuric refuse) comes in contact with water;
3) elimination of slug effects of pumping, i.e., to equalize loading to treatment plants by distributed pumping;
4) sealing terminal activities, actually a process of sealing up abandoned mines to prevent water from entering the sulfur-bearing soil;
5) treatment of mine drainage and in certain circumstances chemical treatment of controlled quantities of drainage from workings, to protect water quality.

Hert also points out that experience has indicated how mine waste-water drainage can be prevented to a great degree by proper advance planning; for example, the control of surface drainage, by protective measures such as barrier dams and so forth, should be worked out prior to opening a mine. After closing down operations, proper measures should be taken to prevent drainage, such as covering stretches of gob roads with packed clay material.

Neutralization is always a possibility, but has thus far been found too costly and presents many difficulties from an operating standpoint. The problem of collecting water for treatment is especially difficult in strip mines, where volumes and flow channels are constantly changing. Theoretically, one ton of either soda ash or limestone, or $\frac{3}{4}$ ton of hydrated lime, is required to neutralize one ton of H_2SO_4 in acid mine water. Braley [27] presents some results of a pilot-plant neutralization treatment. He concluded that, although neutralization is theoretically and physically possible, it is impractical.

After reviewing the research and plant studies, the author concludes that the formation of acid mine water can be prevented by observing the following general rules: keep water out, keep drainage moving, segregate sulfuritic materials, and neutralize acid pools. These four objectives can be realized by the following means. The major method of keeping surface water out is the construction of drainage ditches along the bank above the high wall of the stream. Underground water can be collected in a sump and pumped outside the stripping area. Proper grading keeps water moving so that it will not have a long contact time with the sulfuritic material. The segregation of sulfuritic material, in order to prevent it from contributing to acid mine drainage, can be achieved by keeping floors clean, burying acid-forming refuse, and not leaving coal exposed. A proper job of backfilling or grading the land so that it will shed water will permanently stop or reduce both the formation of acid and the total amount of drainage discharged.

Since 1962, fifteen major methods of preventing and treating acid mine drainage have been proposed.

1. Reducing water inflow. Hall [1965, 176] feels that the most effective method is to minimize contact (both in time and quantity) between water and acid-producing materials, by control of water flow. Flowing streams and surface run-off may be diverted around and away from mining operations by means of high-wall diversion ditches, drains, and conduits to carry flowing waters through or around the mining operation, and by rechanneling or diverting streams away from the mine. Water that does enter should be removed from the mine as quickly as possible. In underground mining, contact between flowing water and acid-producing materials can be minimized by sealing off the surface of the earth above the mine to close cracks, fissures, sinkholes, and other openings when they can be detected, by picking up water as close as possible to the point of entry in the mine, and by conducting it through and out of the mine either in closed conduits or in ditches or sewers that prevent further contact of the water with acid-producing materials.

Conrad [1966] stresses improved mining practices and mine layout to prevent and reduce water flow in mines. Deane [1966] says that water encountered in strip-mine operations should be intercepted by a high wall and drained to diversion ditches. Dye may be used to locate underground seepage [*Coal Age* **69**, 1965].

2. *Flooding abandoned mines.* Hall [1965, 176] states that flooding abandoned mines is effective in excluding oxygen when the mine is below drainage level and the water does not jeopardize active mines. Deane [1966] suggests that the final strip-mine cut be covered with water. Water eliminates air and prevents oxidation and acid formation. Several flooded pits in southern Indiana went from pH 2.8 to 5.8 and these waters were used for municipal supplies. Woodley [1967] also proposed flooding Indiana strip mines.

3. *Lagooning of acid mine water.* Hall [1965, 178] suggests that surface flows from strip mines in areas where annual rainfall and evaporation are equal may be impounded with no discharge to streams. Prolonged lagooning will cause a reduction in the suspended ferric hydroxide and the ferrous sulfate will be oxidized and precipitate out. These lagoons eventually fill with deposits and must be cleaned out or a new lagoon constructed.

Campbell *et al.* [1964] report variable recovery from acid mine drainage in shallow strip-mine lakes. Some in Missouri took 45 years. Alkaline farm run-off and location of "gob" piles were important.

4. *Proportioning of acid drainage to streams.* Hall [1965, 178] points out that equalization basins and the proportioning of discharge with stream flow helps to minimize downstream effects.

5. *Acid mine water used for coal washing.* Dillon [1967] describes a coal-cleaning plant where acid mine water is used to wash coal. The raw coal has 30 per cent $CaCO_3$ and $MgCO_3$. The plant treats 600 tons/hr of raw coal using an average of 225 gpm of mine water (about 23 gals/ton). Mine water at pH 3 with 4340 ppm acidity as $CaCO_3$ and 551 ppm Fe (approximately 50 per cent ferrous) is discharged as effluent from the process with a pH of 6.7 to 7.1 and 0 to 1 ppm Fe. Lovell and Reese [1966] studied the use of acid mine water for coal washing and found it amenable in the froth-flotation, flocculation, and agglomeration processes of coal preparation.

6. *Neutralization.* Deul and Mihok [1967] experimented with neutralization using limestone and lime. They found that waters with low to moderate Fe could be treated by mixing with limestone, which produced a pH of 7 to 8 and < 7 ppm Fe. Waters high in ferrous Fe required a reaction time of less than 15 minutes and second-stage treatment with lime was necessary for short reaction times. They found that the sludges compacted well.

Gerard [1966] in a pilot-plant study shows that lime-neutralization, aeration, sedimentation, and dewatering produces an effluent containing less than 5 ppm Fe and with a pH of 7 to 7.5 but the process is expensive and has high operating costs. Matasov [1967] found that aeration helped the lime-neutralization of H_2SO_4 by promoting the formation of heavier particles of $Fe(OH)_3$ instead of $Fe^-(OH)_2$. Conrad [1967] proposes using an automatic lime feeder with continuous mixing controlled by a pH analyzer. Lunney [1964] stresses strict pH control in lime-neutralization to comply with effluent standards and to minimize the neutralizing agents, undesirable sludges, and plant facilities required. Denby [1965] describes lime-neutralization and sludge removal. Girard and Kaplan [1967] report the costs of lime-neutralization as ranging from $0.07 to $1.09 per 1000 gal or 5.2 cents to $3.25 per ton of coal produced. Hall [1965, 178] reports that neutralization with lime or related alkalis is expensive and waters containing iron must be intimately mixed to prevent deactivation of the alkali particles with an iron-hydroxide precipitate covering.

The Pennsylvania State Health Department has awarded a 55 million dollar contract to neutralize acid mine water from abandoned coal mines near Wilkes Barre [*Power* **112**, 1968].

There is available an automatic limer, giving flows from $\frac{1}{2}$ to 100 gpm, which requires no attention other than periodic filling of the lime hopper. Water enters the unit through a 4-inch pipe and is discharged into an overshot water wheel that drives a stirrer in a lime-feed tank and operates a vibrating device to prevent blockage of the lime-feed chamber.

7. *Sealing of abandoned mines.* Hall [1965] feels that sealing a mine to prevent the entry of air is difficult or impossible because of the permeability of the overlying strata caused by cracking during settlement. Porges *et al.* [1966] also discuss mine sealing. Moebs [1966] reports that the U.S. Bureau of Mines is sealing

an abandoned mine with a highly acid drainage, on a trial basis.

8. Covering strip mines and "gob" piles with earth. Woodley [1967] suggests that coal-haulage roads made of acid-producing materials should be covered. He notes that pollution from "gob" piles is extensive. In one 18-acre site with 30,000 gpd seepage, the following ranges in water quality were measured over a two-year period: pH, 2.8 to 3.6; methyl orange acidity, 284 to 7200 mg/liter; phenolphthalein acidity, 2450 to 12,600 mg/liter; sulfates, 7900 to 30,000 mg/liter; Fe, 760 to 4240 mg/liter; and Mg, 40 to 50 mg/liter. Covering the site with two feet of soil has not decreased either the acidity or the seepage. Hall [1965] states that covering acid-forming materials with earth in strip mines is effective if the materials and earth are compacted to a sufficient density. Deane [1966] suggests that "gob" piles should be covered with soil and planted or, better yet, the spoil should be returned to the active or a nearby inactive mine. He suggests controlling silting by leaving the strip-mine area ungraded to get higher percolation rates and better vegetative growth.

9. Deep-well disposal. Linden and Stefanko [1966] describe subsurface disposal and point out that this method requires a knowledge of the geology of the area and of the chemical and physical properties of the water. The problem of chemical precipitation in the disposal formation must be considered. Dutcher [1966] reports on deep-well disposal in sandstone.

10. Aeration. Rhodes [1967] has patented a process of extended aeration with metallic Fe, which neutralizes the H_2SO_4 and forms insoluble Fe sulfates which are removed by sedimentation. He suggests using metallic Fe from old automobile bodies.

11. Ion exchange. Pollio and Kunin [1967] found that an anion-exchange resin, Amberlite, IRA-68, followed by aeration and clarification was efficient in treating 3000 ppm acidity.

12. Activated sludge. Klappach [1965] found that activated sludge provides adequate treatment.

13. Radiation. Steinberg *et al.* [1968] propose removal of Fe (II) by neutralization with limestone followed by gamma-radiation from Co^{60}. At pH 5.7, with aeration, Fe (II) went from 409 to less than 1 ppm in a short time, since gamma-radiation acts as a catalyst and crystalline precipitate is formed. The process also removes organic pollutants since gamma-radiation oxidizes the material to CO_2 and H_2O. The cost range is \$0.05 to \$0.25 per 1000 gal using Co^{60} or Cs^{137} as the radiation source.

14. Flash distillation. Westinghouse has a 5-mgd demineralization plant using flash distillation to make "ultra-pure" water from acid mine water [*Environ. Sci. Technol.* **1**, 600, 1967].

15. Modern coal mining. Hall [1965, 180] points out that acid mine drainage from modern coal mining is decreasing, because strip mines are becoming depleted and underground mines are in deeper seams where less water is encountered. The present underground full-seam mining techniques are less likely to cause acid drainage. The mine roof is deliberately collapsed after the coal is removed and subsidence occurs quickly with less void space, causing less breakage of acid-forming materials and generally compacting the earth above the mine. Today most acid-forming materials are brought to the surface with the coal, are separated in the coal-cleaning plant, and properly disposed of in coal-refuse piles.

References: Coal-Industry Wastes

1. "Acid mine drainage, control methods," *Sewage Ind. Wastes* **26**, 1190 (1954).
2. U.S. Public Health Service, "Acid mine drainage studies," *Ohio River Pollution Survey, Final Report to the Ohio River Committee*, Supplement C (1942).
3. "Acid mine drainage, summary report," *Sewage Ind. Wastes* **26**, 655 (1954).
4. U.S. Bureau of Mines, "Acid mine water in the anthracite region of Pennsylvania," Technical Paper no. 710.
5. Adler, O., "Biologische Untersuchungen von Natürlichen Eisenwassern," *Deutsche Med. Wochschr.* **26**, 431 (1901).
6. Alexander, L. J., "Control of iron and sulfur organisms by superchlorination and dechlorination," *J. Am. Water Works Assoc.* **32**, 1137 (1940).
7. Allen, E. T., J. L. Crenshaw, and J. Johnson, "The mineral sulfides of iron," *Am. J. Sci.* **33**, 169 (1912).
8. Allen, E. T., and J. L. Crenshaw, "The Stokes method for the determination of pyrite and marcasite," *Am. J. Sci.* **38**, 371 (1914).

9. Allen, E. T., and J. Johnson, "The exact determination of sulfur in pyrite and marcasite," *Ind. Eng. Chem.* **2**, 196 (1910).
10. Arbeiter, E., *Mineralogisch-Chemische Untersuchungen an Markasit*, Pyrit und Magnethies Dissertation, University of Breslau (1913).
11. Ash, S. H., "Acid mine drainage problems," U.S. Bureau of Mines, Bulletin no. 508.
12. Ash, S. H., "Disposal of sulfur waters," *Mining Met.* **22**, 167 (1941).
13. Ash, S. H., et al., "Acid mine drainage, Pennsylvania anthracite region, problems," *Sewage Ind. Wastes* **25**, 630 (1953).
14. Ash, S. H., R. E. Doherty, and P. S. Miller, "Cost of an acid mine drainage tunnel," U.S. Bureau of Mines, Bulletin no. 513, 1952.
15. Baas, L., and G. M. Becking, "Studies on the sulphur bacteria," *Ann. Botany* **39**, 613 (1925).
16. Baas, L., G. M. Becking, and G. S. Parks, "Energy relations in the metabolism of autotrophic bacteria," *Physiol. Rev.* **7**, 85 (1927).
17. Bach, H., "The disposal of coal mine liquid wastes," in Proceedings of Third International Conference on Bituminous Coal, Carnegie Institute of Technology, Pittsburgh, **2**, 1931, p. 924.
18. Bain, G. W., "Pyrite oxidation," *Econ. Geol.* **30**, 166 (1935).
19. Bandy, M. C., "Mineralogy of three sulphate deposits of northern Chile," *Am. Mineralogist* **23**, 669 (1938).
20. Beal, G. B., Common Fallacies about Acid Mine Water, Pennsylvania Board of Health, Harrisburg, Pa.
21. Beckwith, T. D., and P. F. Bovard, "Bacteria destroy pipe line in California," *Chem. Met. Eng.* **40**, 530 (1933).
22. Beckwith, T. D., and P. F. Bovard, "Bacterial disintegration of sulphur-containing sealing compounds in pipe joints," *Univ. Calif., L.A. Publ. Biol. Sci.* **1**, 121 (1937).
23. Bilharz, O. W., "Experiences with acid mine water drainage in tristate field," American Institute of Mining and Metallurgical Engineers, Technical Publication no. 2267, Class A, Mining Technology, 1947, p. 1.
24. Black, H. H., et al., "Industrial waste guide—by-product coke," in Proceedings of 11th Industrial Waste Conference, Purdue University, May 1956.
25. Braley, S. A., "Recent research as to the effect of coal mine drainage on the clean stream program," Paper presented before Second Annual Meeting of the Pennsylvania Section of American Water Works Association, 1950.
26. Braley, S. A., "Acid drainage from coal mines," American Institute of Mining and Metallurgical Engineers, Technical Publication no. 3098-F; *Mining Eng.* **3**, 703 (1951).
27. Braley, S. A., *A Pilot Plant Study of the Neutralization of Acid Drainage from Bituminous Coal Mines*, Pennsylvania Health Department (1951).
28. Braley, S. A., "Experimental strip mines show no stream pollution," *Mining Congress J.*, September 1952, p. 50.
29. Bunker, H. F., "A review of the physiology and biochemistry of the sulfur bacteria," Chemical Research Special Report no. 3, Department of Scientific and Industrial Research, London (1936).
30. Burke, S. P., and W. R. Downs, "Oxidation of pyritis sulfur in coal mines," *Trans. Am. Inst. Mining Engrs. Coal Div.* **130**, 425 (1938).
31. Cady, G. H., "The Illinois pyrite inventory of 1918," *Illinois State Geol. Surv. Bull.* **38**, 427 (1922).
32. Carpenter, L. V., "Pollution in the Monongahela river basin and its effect on water supplies," *West Va. Univ. Eng. Exp. Sta. Tech. Bull.* **2**, 27 (1928).
33. Carpenter, L. V., "Wanted: more research on acid mine waters," *Coal Age* **35**, 406 (1930).
34. Carpenter, L. V., and A. H. Davidson, "Developments in the treatment of acid mine drainage," *Proc. West Va. Acad. Sci.* **4**, 93 (1930).
35. Carpenter, L. V., and L. K. Herndon, "Report on pollution survey of the Cheat river basin," Special Report of West Virginia Water Commission (1929).
36. Carpenter, L. V., and L. K. Herndon, "Acid mine drainage from bituminous coal mines," *West Va. Univ. Eng. Exp. Sta. Res. Bull.* **10**, 38 (1933).
37. Carpenter, L. C., and E. T. Roetman, "The sterilizing effect of acid mine drainage," in Proceedings of Pennsylvania Water Association, 1932.
38. Chubb, R. S., and P. D. Merkel, "Effects of acid wastes on natural purification of the Schuylkill river," *Sewage Works J.* **18**, 692 (1946).
39. "The clarification of coal washery slimes," *Chem. Abstr.* **32**, 3580 (1938).
40. West Virginia Geological Survey, *Coal in Monongahela County* (1932).
41. "Coal processing waste problem, report of West

Riding, Yorkshire, River Board," *Sewage Ind. Wastes* **21**, 193 (1949).
42. "Coal washery waste study, West Virginia," *Sewage Ind. Wastes* **22**, 1040 (1950).
43. U.S. Public Health Service, "Coal washery wastes," *Ohio River Survey*, Supplement D (1943), p. 1073.
44. U.S. Bureau of Mines, "Coal washing investigation, methods and tests," Bulletin no. 300.
45. Crohurst, H. R., "A study of the pollution and natural purification of the Ohio river," U.S. Public Health Service, Bulletin no. 204, 1933.
46. Cole, V. W., "Lime treatment of lake reduces acid mine waste pollution," *Ind. Wastes* **2**, 100 (1957).
47. Collins, C. P., "Pollution of water supplies by coal mine drainage," *Eng. News Record* **91**, 638 (1923).
48. Colmer, A. R., and M. E. Hinkle, "The role of microorganisms in acid mine drainage: a preliminary report," *Science* **106**, 253 (1947).
49. Colmer, A. R., and M. E. Hinkle, "Acid mine drainage, microorganisms," *Sewage Ind. Wastes* **20**, 177 (1948).
50. Colmer, A. R., K. L. Temple, and M. E. Hinkle, "An iron oxidizing bacterium from the acid drainage of some bituminous coal mines," *J. Bacteriol.* **59**, 317 (1950).
51. Pennsylvania Board of Health, "Control of acid drainage from coal mines," Harrisburg, Pa.
52. Cooper, J. E., "How to dispose of acid waste," *Chem. Ind.* **66**, 5 (1950).
53. Crichton, A. B., "Neutralization of acid mine drainage," *Mining Congress J.* **12**, 418 (1926).
54. Crichton, A. B., "Disposal of drainage from coal mines," *Trans. Am. Soc. Civil Engrs.* **92**, 1332 (1928).
55. Davidson, A. H., *Studies on Acid Drainage from Bituminous Coal Mines*, Engineering Thesis, West Virginia University (1931).
56. Davis, D. E., "Stream pollution," *Eng. News Record* **117**, 586 (1936).
57. Dexter, G. M., "Municipal water needs versus strip coal mining," American Institute of Mining and Metallurgical Engineers, Technical Publication no. 2570-F; *Mining Eng.* **1**, 37 (1949).
58. Dinegar, R. H., R. H. Smellie, and V. K. LaMer, "Kinetics of the acid decomposition of sodium thiosulfate in dilute solutions," *J. Am. Chem. Soc.* **73**, 2050 (1951).
59. Downs, W. S., *The Natural Oxidation of Pyritic Sulfur in Coal*, M.S. Thesis, Chemistry, West Virginia University (1936).
60. Downs, W. S., "A survey of culverts in West Virginia," *West Va. Univ. Eng. Exp. Sta. Res. Bull.* **13**, (1954).
61. Drake, C. F., "Effect of acid mine drainage on river water supply," *J. Am. Water Works Assoc.* **23**, 1474 (1931).
62. Drake, C. F., "Water-purification problems in mining and manufacturing districts," *J. Am. Water Works Assoc.* **23**, 1261 (1931).
63. Drewes, K., "Mikrobiologische Untersuchungen Eines Stark Sauren Moorbodens," *Z. Bakteriol. Parasitenk. II*, **76**, 114 (1928).
64. Eavenson, H. N., *Coal Mines Drainage*, Isaac Walton League of America, Chicago (1927).
65. Eavenson, H. N., "Surplus water in the bituminous coal fields," *Coal Age* **40**, 195 (1935).
66. Edwards, A. B., and G. Baker, "Some occurrences of supergene iron sulfides in relation to their environments of decomposition," *J. Sediment. Petrology* **21**, 34 (1051).
67. "Efficiency in cleaning," *Coal Age* **44**, 44 (1939).
68. Farrell, M. A., *Living Bacteria in Ancient Rocks and Meteorites*, American Museum of Natural History, American Museum Navitiates, New York (1933).
69. Farrell, M. A., and H. G. Turner, "Bacteria in anthracite coal," *Fuel* **11**, 229 (1932).
70. Felegy, E. W., et al., "Acid mine drainage, Pennsylvania," *Sewage Ind. Wastes* **20**, 1146 (1948).
71. Felegy, E. W., L. H. Johnson, and J. Westfield, "Acid mine water in the anthracite region of Pennsylvania," U.S. Bureau of Mines, Technical Paper no. 710, 1948.
72. Fish, E. L., L. A. Turnbull, A. L. Toenges, and I. Hartman, "A study of summer air conditioning with water sprays to prevent roof falls at the Beech Bottom coal mine, West Virginia," U.S. Bureau of Mines, Reports of Investigations no. 3775, 1944.
73. Foster, W. D., and F. L. Feicht, "Mineralogy of concretions from Pittsburgh coal seam with special reference to analcite," *Am. Mineralogist* **31**, 357 (1946).
74. Frederick, L. R., *Studies of Some Factors Affecting Bacterial Oxidation of Sulfur with Particular Consideration of Inhibitors*, M.S. Thesis, Rutgers University (1947).
75. Frederick, L. R., and R. L. Starkey, "Bacterial

oxidation of sulfur in pipe sealing mixtures," *J. Am. Water Works Assoc.* **40**, 729 (1948).
76. Frost, W. H., E. J. Theriault, H. W. Streeter, and J. K. Hoskins, "Study of the pollution and natural purification of the Ohio river. I, II, III," U.S. Public Health Service Bulletin no. 171, 1932, p. 198.
77. Fujishige, H., and K. Haga, "Thiosulfate and trithionate dehydrase in chemoautotrophic sulfur bacteria," Imperial University, Tokyo, *Acta Phytochim.* **14**, 141 (1944).
78. Galle, E., "Über Selbstentzündung der Steinkohle," *Z. Bakteriol. Parasitenk. II,* **28**, 461 (1910).
79. Gidley, L. P., *Treatment of Acid Mine Drainage*, Undergraduate Chemistry Thesis, West Virginia University (1928).
80. Gifford, R. D., "Colliery waste treatment," *Sewage Ind. Wastes* **6**, 1185 (1934).
81. Gillenwater, L. E., "Coal washery wastes, West Virginia," *Sewage Ind. Wastes* **23**, 869 (1051).
82. Gleen, H., "Biological oxidation of iron in soil," *Nature* **166**, 871 (1950).
83. Gleen, H., "Some aspects of the metabolism of a new group of iron-oxidizing microorganisms in soil," *J. Gen. Microbiol.* **5**, 3 (1951).
84. Gleen, H., and J. H. Quastel, "Sulphur metabolism in soil," *Appl. Microbiol.* **1**, 70 (1953).
85. Gottschalk, V. A., and H. A. Buehler, "Oxidation of sulfides," *Econ. Geol.* **7**, 15 (1912).
86. Grawe, O. R., "Pyrite deposits of Missouri," *Missouri Geol. Surv. Water Resources* **30**, 1945.
87. Graf, G., "New wash water clarification process in coal mining," *Sewage Works J.* **4**, 925 (1932).
88. Gross, C. D., and C. Lee, "Collection and treatment of acid runoff from coal gob-pile-storage areas," in Proceedings of 6th Industrial Waste Conference, Purdue University, February 1951.
89. Gross, C. D., and C. Lee, "Gob-pile-storage areas, acid runoff, collection and treatment," *Sewage Ind. Wastes* **24**, 801 (1952).
90. Gutzeit, G., E. J. Lyons, and D. C. MacLean, "Treatment of phenolic wastes," *Ind. Wastes* **4**, 57 (1959).
91. Hall, G. L., "The control of stream pollution in Maryland from acid coal mine drainage," in Proceedings of 12th Annual Conference of the Maryland—Delaware Water and Sewage Association, **83**, 1938, p. 95.
92. Halvorson, H. O., "Studies on the transformation of iron in nature; the effect of CO_2 on the equilibrium in iron solution," *Soil Sci.* **32**, 141 (1931).
93. Harder, E. C., "Iron depositing bacteria and their geologic relations," U.S. Geological Survey, Professional Paper no. 113, 1919.
94. Harris, S., *Controlling the Acidity of Mine Water by Sealing*, Indiana Coal Mining Institute (1935).
95. Hart, E. J., "Radiation chemistry of ferrous sulfate solutions," *J. Am. Chem. Soc.* **73**, 1891 (1951).
96. Hatch, B. F., "Sealing abandoned coal mines," *Water Works Sewerage* **81**, 99 (1934).
97. Hebley, H. F., "Neutralizing acid mine water," *Mining Congress J.* **36**, 62 (1950).
98. Hebley, H. F., "Stream pollution by coal mine waste," *Mining Eng. Trans.* **5**, 410, April 1953.
99. Hebley, H. F., "Coal washery wastes, treatment," *Sewage Ind. Wastes* **21**, 592 (1949).
100. Henson, E. B., *Bottom Fauna in Relation to the Hydrogen Ion Concentration of Lower Deckers Creek Basin*, M.S. Thesis, Biology Department, West Virginia University (1950).
101. Herndon, L. K., "Survey of mine drainage in the West Fork basin," in Proceedings of 6th Annual West Virginia Conference on Water Purification; *West Va. Univ. Eng. Exp. Sta. Tech. Bull.* **4**, 115 (1931).
102. Herndon, L. K., *Acid Mine Drainage from West Virginia Mines*, Chemical Engineering Thesis, West Virginia University (1934).
103. Herndon, L. K., "Sanitary survey of Ohio river in West Virginia during 1932," *Proc. West Va. Acad. Sci.* **7**, 63 (1934).
104. Herndon, L. K., and W. W. Hodge, "Coal seams of West Virginia and their drainage," *Proc. West Va. Acad. Sci.* **9**, 39 (1936).
105. Herndon, L. K., and W. W. Hodge, "West Virginia coal seams and their drainage," *West Va. Univ. Eng. Exp. Sta. Res. Bull.* **14**, 1936.
106. Hert, O. H., "Practical control measures to reduce acid mine drainage," in Proceedings of 13th Industrial Waste Conference, Purdue University, May 1958, p. 189.
107. Hert, O. H., "Mine drainage control in Indiana," *J. Water Pollution Control Federation* **32**, 505 (1960).
108. Hinkle, M. E., and W. A. Koehler, "Investigations of coal mine drainage," West Virginia University, Engineering Experiment Station, Bureau of Coal

Research Fellowship, Annual Report, 1944, and Semi-Annual Report, 1946.
109. Hinkle, M. E., and W. A. Koehler, "The action of certain microorganisms in acid mine water," American Institute of Mining and Metallurgical Engineers, Technical Publication no. 2381, 1948.
110. Hoak, R. D., "A rational examination of stream pollution abatement," *Science* **101**, 523 (1945).
111. Hodge, W. W., "Pollution of streams by coal mine drainage," *Ind. Eng. Chem.* **29**, 48 (1937).
112. Hodge, W. W., "The effect of coal mine drainage on West Virginia rivers and water supplies," *West Va. Univ. Eng. Exp. Sta. Bull.* **18**, 1938.
113. Hodge, W. W., "Mine drainage pollution," *Sewage Ind. Wastes* **10**, 168 (1938).
114. Hodge, W. W., and M. E. Hinkle, "Role of microorganisms," Paper presented at the Spring meeting of the American Chemical Society, Division Water, Sewage, and Sanitation Chemists, April 1946.
115. Hodge, W. W., and R. Newton, "Studies on Morgantown water supplies, especially their variations in mineral content," *West Va. Univ. Eng. Exp. Sta. Tech. Bull.* **7**, 52 (1934).
116. Hodge, W. W., and E. J. Niehaus, "Ohio river water in the Wheeling district and its treatment for use in boilers," in Proceedings of 10th West Virginia Conference on Water Purification; *West Va. Eng. Exp. Sta. Tech. Bull.* **8**, 41 (1936).
117. Hoffert, J. R., "Acid mine drainage," *Ind. Eng. Chem.* **39**, 642 (1947).
118. Hoffert, J. R., "Acid mine drainage," *Sewage Ind. Wastes* **19**, 1095 (1947).
119. Jensen, H. J., "Vorkommen von Thiobacillus Thiooxidans in Dänischem Boden," *Z. Bakteriol. Parasitenk. II*, **72**, 242 (1927).
120. Joffe, J. S., "Preliminary studies on the isolation of sulfur-oxidizing bacteria from sulfur float soil compost," *Soil Sci.* **13**, 161 (1922).
121. Johnson, L. H., "Treatment of acid mine water for breaker use in the anthracite region of Pennsylvania," U.S. Bureau of Mines, Information Circular no. 7382, 1946.
122. Jones, D. C., "Acid mine water, its control reduces stream pollution," *Mechanization* Part I, **15**, 10 (Oct. 1951) and Part II, **15**, 11 (Nov. 1951).
123. Joseph, J. M., "Microbiological study of acid mine waters: preliminary report," *Ohio J. Sci.* **53**, 123 (1953).
124. Kappen, H., and E. Quensell, "Über die Umwandlungen von Schwefel und Schwefelverbindungen im Ackerboden, ein Beitrag zur Kenntnis des Schwefelkreislaufs," *Landwirtsch. Versuchssta.* **86**, 1 (1915).
125. Kaplan, B. B., "The recovery of marketable by-products from acid mine water," *Proc. West Va. Acad. Sci.* **4**, 90 (1930).
126. Kaplan, B. B., and D. B. Reger, "Process of purifying water," U.S. Patent no. 1878525, 1932.
127. Kilpatrick, M., and M. L. Kilpatrick, "The stability of sodium thiosulfate solutions," *J. Am. Chem. Soc.* **45**, 2132 (1923).
128. Lackey, J. B., "The flora and fauna of surface waters polluted by acid mine drainage," *Public Health Rept.* **53**, 1499 (1938).
129. Leathen, W. W., and S. A. Braley, "Bacterial activity on sulfuritic constituents associated with coal," *Proc. Soc. Am. Bacteriologists*, 1950, p. 21.
130. Leathen, W. W., and S. A. Braley, "The effect of iron oxidizing bacteria on certain sulfuritic constituents of bituminous coal," *Proc. Soc. Am. Bacteriologists*, 1951, p. 21.
131. Leathen, W. W., S. A. Braley, and L. D. McIntyre, "The role of bacteria in formation of acid from certain sulfuritic constituents associated with bituminous coal. I: Thiobacillus thiooxidans," *Appl. Microbiol.* **1**, 61 (1953).
132. Leathen, W. W., S. A. Braley, and L. D. McIntyre, "The role of bacteria in the formation of acid from certain sulfuritic constituents associated with bituminous coal. II: Ferrous iron oxidizing bacteria," *Appl. Microbiol.* **1**, 65 (1953).
133. Leathen, W. W., L. D. McIntyre, and S. A. Braley, "A medium for the study of the bacterial oxidation of ferrous iron," *Science* **14**, 280 (1951).
134. Leathen, W. W., and K. M. Madison, "The oxidation of ferrous iron by bacteria found in acid mine water," *Proc. Soc. Am. Bacteriologists*, 1949, p. 64.
135. Leitch, R. D., "Observations on acid mine drainage in Western Pennsylvania," U.S. Bureau of Mines, Reports of Investigations, no. 2725, January 1926.
136. Leitch, R. D., "Acid mine drainage in Western Pennsylvania," U.S. Bureau of Mines, Reports of Investigations, no. 2889, September 1928.
137. Leitch. R. D., "Observations in acid mine drainage in western Pennsylvania," *Mining Congress J.* **14**, 835 (1928).
138. Leitch, R. D., "Abandoned mines," U.S. Bureau of

Mines, Reports of Investigations, no. 3098, April 1931.
139. Leitch, R. D., "The acidity of Bennet branch of Sinnemahoning creek, Pennsylvania, during low water," U.S. Bureau of Mines, Reports of Investigations, no. 3097, July 1931.
140. Leitch, R. D., "The acidity of Black Lick, Two Lick and Yellow creeks, Pennsylvania, during low water in 1930," U.S. Bureau of Mines, Reports of Investigations, no. 3102, July 1931.
141. Leitch, R. D., "Acidity of drainage from high pyritic coal areas in Pennsylvania," U.S. Bureau of Mines, Reports of Investigations, no. 3146, January 1932.
142. Leitch, R. D., "The acidity of several Pennsylvania streams during low water," U.S. Bureau of Mines, Reports of Investigations, no. 3119, September 1931.
143. Leitch, R. D., and W. P. Yant, "A comparison of the acidity of waters from some active and abandoned coal mines," U.S. Bureau of Mines, Reports of Investigations, no. 2895, October 1928.
144. Leitch, R. D., W. P. Yant, and R. R. Sayers, "Effect of sealing on acidity of mine drainage," U.S. Bureau of Mines, Reports of Investigations, no. 2994, April 1930.
145. Leitch, R. D., and W. P. Yant, "Sealing old workings prevents acid formation and saves pipes and streams," *Coal Age* **35**, 78 (1930).
146. Lewis, W. M., and C. Peters, "Coal mine slag drainage," *Ind. Wastes* **1**, 145 (1956).
147. Lockett, W. T., "Oxidation of thiosulfates on bacterial filters," *J. Soc. Chem. Ind.* **32**, 573 (1913).
148. Lockett, W. T., "Oxidation of thiosulfate by certain bacteria in pure culture," *Proc. Roy. Soc. (London), Ser. B*, **87**, 441 (1914).
149. Lutz, W. F., and J. T. Griffith, "Effects of rainfall in mining area," *Trans. Am. Inst. Mining Met. Engrs.* **168**, 145 (1946).
150. Lyon, E. W., "Report on sealing abandoned coal mines to the State Planning Board of West Virginia," State Health Department, Division of Sanitary Engineering, 1936.
151. McGauhey, P. H., "A study of the stream pollution problem in the Roanoke, Va., metropolitan district," *Bull. Va. Polytech. Inst.* **35**, 10 (1942).
152. MacIntire, W. H., W. M. Shaw, and J. B. Young, "The oxidation of pyrite and sulfur as influenced by lime and magnesia—a twelve year lysimeter study," *Soil Sci.* **30**, 443 (1930).
153. MacLean, H. C., "The oxidation of sulfur by microorganisms in its relation to the availability of phosphate," *Soil Sci.* **5**, 251 (1918).
154. Morgan, L. S., "Acidity and hardness at Monongahela river plants," *Eng. News Record* **106**, 850 (1931).
155. Morgan, L. S., "Acid mine waste, treatment and disposal," *Sewage Ind. Wastes* **14**, 404 (1942).
156. Morgan, L. S., "Treatment of acid mine water," Paper presented at Annual Meeting of Pennsylvania Sewage and Industrial Waste Association, August 1952.
156. A. Moulton, E. Q., "The acid mine-drainage problem in Ohio," *Ohio State Univ. Eng. Exp. Sta. Bull.* **166**, (1957).
157. Mumford, E. M, "A new iron bacterium," *Chem. Soc.* **103**, 645 (1913).
158. Murdock, H. R., "Preventing the formation of acid mine drainage in Pennsylvania strip coal mines," *Ind. Eng. Chem.* **45**, 101A (1953).
159. Newhouse, W. H., "Some forms of iron sulfide occurring in coal and other sedimentary rocks," *J. Geol.* **35**, 73 (1927).
160. Nusbaum, I., "Effects of industrial waste on municipal sewage works at Detroit," *Sewage Ind. Wastes* **22**, 1583 (1950).
161. Ohio River Committee, *Ohio River Pollution Control*, U.S. Government Printing Office, Washington, D.C (1944).
162. Olsen, E., and W. Szybalski, "Aerobic microbiological corrosion of water pipes," *Acta Chem.* **3**, 1094 (1949).
163. Parker, C. D., "The corrosion of concrete; the isolation of a species of bacterium associated with the corrosion of concrete exposed to atmosphere containing hydrogen sulfide," *Australian J. Exptl. Biol. Med. Sci.* **23**, 81 (1945).
164. Parker, C. D., "Species of sulphur bacteria associated with the corrosion of concrete," *Nature* **159**, 439 (1947).
165. Parr, S. W., and A. R. Powell, "A study of the forms in which sulphur occurs in coal," *Univ. Illinois Eng. Exp. Sta. Bull.* **111** (1919).
166. Parton, W. J., "Coal washery plants," *Ind. Eng. Chem.* **39**, 646 (1947).
167. Permutit Company, *Graphical Studies of the Hardness and Other Properties of the Rivers of the Pittsburgh District, Pittsburgh, Pa.*, New York (1929).
168. Plummer, C. W., *Constitution of Marcasite and Pyrite*,

Chemistry Thesis, University of Pennsylvania (1910–11).
169. Pochon, J., and O. Coppier, "Role of sulfate-reducing bacteria in the biologic alteration of stones of monuments," *Comptes Rendus* **231**, 1584 (1950).
170. Pochon, J., and T. Yao-Tseng, "Role of sulphur bacteria in disintegration of building stone," *Comptes Rendus* **226**, 2188 (1948).
171. Powell, A. R., "Methods of analyzing coal and coke," U.S. Bureau of Mines, Technical Paper no. 254, 1921.
172. Pringsheim, E. G., "The filamentous bacteria Sphaerotilus, Leptothrix, Caldothrix, and their relation to iron and manganese," *Trans. Roy. Soc. (London), Ser. B*, 233 (1949).
173. Pringsheim, E. G., "Iron bacteria," *Biol. Rev.* **24**, 200 (1949).
174. Prins, K., "Coal preparation plants," *Mining Eng.* **4**, 572 (1952).
175. "Public water supplies of West Virginia," West Virginia State Department of Health, Bulletin no. 18, 1931, p. 42.
176. Purby, W. C., "A study of the pollution and natural purification of the Ohio river," U.S. Public Health Service, Bulletin no. 131, 1923, and Bulletin no. 198, 1930.
177. "Purifying coal washing slurry liquor," *Chem. Abstr.* **30**, 8580 (1936); British Patent no. 448593, 11 June 1936.
178. Quispel, A., G. W. Harmsen, and D. Otzen, "Contribution to the chemical and bacteriological oxidation of pyrite in soil," *Plant Soil* **4**, 43 (1952).
179. Rickert, E. E., and W. T. Bishop, "Wash water treatment and fine coal recovery," *Ind. Eng. Chem.* **42**, 626 (1950).
180. Rickert, E. E., and W. T. Bishop, "Colliery waste treatment," *Sewage Ind. Wastes* **22**, 1379 (1950).
181. Roberts, P. T., "Acids in the Monongahela river," *Eng. News Record* **34**, 26 (1911).
182. Roetman, E. T., *The Sterilization of Sewage by Acid Mine Waters*, M.S. Thesis, Chemical Engineering, West Virginia University (1932).
183. Roetman, E. T., "A further study of acid mine drainage with relation to stream pollution," West Virginia University Library, Bulletins nos. 628 and 541, R 62.
184. Ross, G. A., "Preparation of anthracite," *Coal Mining J.* **31**, 32 (1945).
185. Rudolfs, W., "Oxidation of iron pyrites by sulfur oxidizing organisms and their use for making mineral phosphates available," *Soil Sci.* **14**, 135 (1922).
186. Sanitary Water Board, *Control of Acid Drainage from Coal Mines*, Pennsylvania Department of Health (1952).
187. Sayers, R. R., et al., "A general review of the U.S. Bureau of Mines stream pollution investigations," U.S. Bureau of Mines, Reports of Investigations, no. 3098, April 1931.
188. Selvig, W. A., and W. C. Ratliff, "The nature of acid water from coal mines and the determination of acidity," *Ind. Eng. Chem.* **14**, 125 (1922).
189. Seymour, R. B., W. Pascoe, W. J. Eney, A. C. Loewer, R. H. Steiner, and R. D. Stout, "Performance studies on sulfur jointing compounds," *J. Am. Water Works Assoc.* **43**, 1001 (1951).
190. Seymour, R. B., W. Pascoe, and R. D. Stout, "Corrosion studies of iron in the presence of sulfur," *Corrosion* **7**, 263 (1951).
191. Sheen, R. T., and W. H. Kohler, "Direct titration of sulfates," *Ind. Eng. Chem.* (analytical edition) **8**, 127 (1936).
192. Siebert, C. L., Jr., "Colliery silt, discharge prevention," *Sewage Ind. Wastes* **24**, 809 (1952).
193. Sijderius, R., *Heterotrophe Bacterien, Die Thiosulfaat Oxydeeren*, Doctoral Dissertation, University of Amsterdam (1946).
194. Sinnatt, F. S., A. Grounds, and F. Bayley, "The inorganic constituents of coal with special reference to Lancashire seams," *J. Soc. Chem. Ind.* **40**, 1 (1921).
195. Sinnatt, F. S., and N. Simpkin, "The inorganic constituents of coal with special references to Lancashire seams. II: The iron in coal," *J. Soc. Chem. Ind.* **41**, 164 (1922).
196. Smyth, C. H., *On the Genesis of the Pyrite Deposits of St. Lawrence County, Albany, N.Y.*, New York State Education Department (1912).
197. Sneedon, R., "Casing failures traced to bacterial action," *Petrol. Engr.* **23**, 7 (1951).
198. Snyder, R. H., "Effect of coal strip mining upon water supplies," *J. Am. Water Works Assoc.* **39**, 751 (1947).
199. Snyder, R. H., "Strip mine drainage effect on water supplies," *Sewage Works J.* **20**, 178 (1948).
200. Spencer, K. A., "Pyrite recovery from coal mine refuse," in *American Mining Congress Yearbook* (1940), p. 97.
201. American Public Health Association, *Standard Methods of Water Analysis*, 9th ed. (1946), p. 246.

202. Starkey, R. L., "Concerning the carbon and nitrogen nutrition of Thiobacillus thiooxidans, an autotrophic bacterium oxidizing sulfur under acid conditions," *J. Bacteriol.* **10**, 165 (1925).
203. Starkey, R. L., "Cultivation of organisms concerned in the oxidation of thiosulfate," *J. Bacteriol.* **28**, 365 (1934).
204. Starkey, R. L., "Products of the oxidation of thiosulfate by bacteria in mineral media," *J. Gen. Physiol.* **18**, 325 (1934).
205. Starkey, R. L., "Isolation of some bacteria which oxidize thiosulfate," *Soil Sci.* **39**, 197 (1935).
206. Starkey, R. L., "Precipitation of ferric hydrate by iron bacteria," *Science* **102**, 532 (1945).
207. Starkey, R. L., "Transformation of iron by bacteria in water," *J. Am. Water Works Assoc.* **37**, 963 (1945).
208. Starkey, R. L., and H. O. Halvorson, "Studies on the transformation of iron in nature. II: Concerning the importance of microorganisms in the solution and precipitation of iron," *Soil Sci.* **24**, 381 (1927).
209. Starkey, R. L., and K. M. Wight, "Anaerobic corrosion of iron in soil, with particular consideration of the soil redox potential as an indicator of corrosiveness," *Am. Gas Assoc. Monthly* **28**, 108 (1946).
210. Starkey, R. L., and K. M. Wight, "Anaerobic corrosion of iron in soil," Bulletin of Technical Section American Gas Association, 1945; abstracted in *J. Am. Water Works Assoc.* **38**, 1210 (1946).
211. Stevenson, W. L., "Stream pollution due to acid mine waters," *Gas World (Coking)* **21**, (1921).
212. Stevenson, W. L., "Coal mine drainage disposal," in Proceedings of 3rd International Conference on Bituminous Coal, Carnegie Institute of Technology, **2**, 1931, p. 912.
213. Stewart, A. H., "Stream pollution control in Pennsylvania," *Sewage Works J.* **17**, 586 (1945).
214. Stokes, H. N., "On pyrite and marcasite," Bulletin no. 186, U.S. Geological Survey, 1901.
215. Stokes, H. N., "Experiments on the action of various solutions on pyrite and marcasite," *Econ. Geol.* **2**, 14 (1907).
216. "Stream pollution by mine wastes," *Sewage Ind. Wastes* **4**, 915 (1932).
217. "Strip mining, acid control," *Sewage Ind. Wastes* **24**, 1553 (1952).
218. Temple, K. L., "A modified design of the Lees soil percolation apparatus," *Soil Sci.* **71**, 209 (1951).
219. Temple, K. L., "Acid mine drainage," *Ind. Wastes* **1**, 114 (1956).
220. Temple, K. L., and A. R. Colmer, "The autotrophic oxidation of iron by a new bacterium: Thiobacillus ferrooxidans," *J. Bacteriol.* **62**, 605 (1951).
221. Temple, K. L., and A. R. Colmer, "The formation of acid mine drainage," *Mining Eng.* **3**, 1090 (1951).
222. Temple, K. L., and A. R. Colmer, "Formation of acid mine drainage," *Sewage Ind. Wastes* **24**, 810 (1952).
223. Temple, K. L., and W. A. Koehler, "Drainage from bituminous coal mines," *West Va. Univ. Bull.* **25**, 1954.
224. Tisdale, E. S., and E. W. Lyon, "Acid mine drainage control on upper Ohio river tributaries," *J. Am. Water Works Assoc.* **27**, 1186 (1935).
225. Thiessen, R., "Occurrence and origin of finely disseminated sulphur compounds in coal," *Trans. Am. Inst. Mining Met. Engrs.* **63**, 913 (1920).
226. Thiessen, G., "Forms of sulfur in coal," in *Chemistry of Coal Utilization*, John Wiley & Sons, New York (1945), p. 1.
227. Tracy, L. D., "Mine water neutralization at the Calumet mine," *Mining Met. Eng.* **161**, 29 (1920).
228. Trax, E. C., "A quarter century of progress in the purification of acid waters," *West Va. Univ. Eng. Exp. Sta. Tech. Bull.* **6**, 1933, p. 5.
229. Tucker, W. M., "Pyrite deposits in Ohio coal," *Econ. Geol.* **14**, 198 (1919).
230. Tyner, E. H., R. M. Smith, and S. L. Galpin, "Reclamation of strip-mined areas in West Virginia," *J. Am. Soc. Agron.* **40**, 313 (1948).
231. Umbreit, W. W., H. R. Vogel, and K. G. Vogler, "The significance of fat in sulfur oxidation by Thiobacillus thiooxidans," *J. Bacteriol.* **43**, 141 (1942).
232. Van Ness, B., Jr., "Recovery of river coal," *Ind. Wastes* **1**, 232 (1956).
233. Vishniac, W., *On the Metabolism of the Chemolithoautotrophic Bacterium Thiobacillus Thioparus Beijerinck*, Ph.D. Thesis, Stanford University (1949).
234. Vishniac, W., "The metabolism of Thiobacillus thioparus; the oxidation of thiosulfate," *J. Bacteriol.* **64**, 363 (1952).
235. Vogler, K. G., G. A. LePage, and W. W. Umbreit, "Studies on the metabolism of autotrophic bacteria; the respiration of Thiobacillus thiooxidans on sulfur," *J. Gen. Physiol.* **26**, 89 (1942).
236. Vogler, K. G., and W. W. Umbreit, "The necessity for

direct contact in sulphur oxidation by Thiobacillus thiooxidans," *Soil Sci.* **51**, 331 (1941).
237. Waksman, S. A., "Microorganisms concerned in the oxidation of sulfur in the soil; media used for the isolation of sulfur bacteria from the soil," *Soil Sci.* **13**, 329 (1922).
238. Waksman, S. A., "Microorganisms concerned in the oxidation of sulfur in the soil; a solid medium for the isolation and cultivation of Thiobacillus thiooxidans," *J. Bacteriol.* **7**, 605 (1922).
239. Waksman, S. A., and J. S. Joffe, "Microorganisms concerned in the oxidation of sulfur in the soil; Thiobacillus thiooxidans, a new sulfur-oxidizing organism isolated from the soil," *J. Bacteriol.* **7**, 239 (1922).
240. Waksman, S. A., and R. L. Starkey, "Carbon assimilation and respiration of autotrophic bacteria," *Proc. Soc. Exp. Biol. Med.* **20**, 9 (1922).
241. Waksman, S. A., and R. L. Starkey, "On the growth and respiration of sulfur-oxidizing bacteria," *J. Gen. Physiol.* **5**, 285 (1923).
242. Weigmann, D. H., "Utilization of coal wastes," *Sewage Ind. Wastes* **8**, 351 (1936).
243. "West Virginia coal seams and their drainage," *West Va. Univ. Eng. Exp. Sta. Res. Bull.* **14**, 1936.
244. White, L. C., "Geographic distribution of sulphur in West Virginia coal beds," *Trans. Am. Inst. Mining Met. Engrs.* **43**, 1919.
245. Whyte, R. O., and J. W. B. Sisam, "The establishment of vegetation on industrial waste land," Aberystwyth and Oxford (England), Commonwealth Agricultural Bureaux, Joint Publication no. 14, 1949.
246. Wilson, A. W. G., *Pyrite in Canada, Its Occurrence, Exploiting, Dressing, and Uses*, Government Printing Office, Ottawa (1912).
247. Winchell, A. N., "The oxidation of pyrite," *Econ. Geol.* **2**, 290 (1907).
248. Winogradsky, S., *Beitrag zur Morphologie und Physiologie der Bakterien. Heft I. Zur Morphologie und Physiologie der Schwefelbakterien*, A. Felix, Leipzig (1888).
249. Winogradsky, S., "Eisenbakterien als Anorgoxydanten," *Zentr. Bakteriol. Parasitenk.* **57**, 1 (1922).
250. Yancey, H. F., "Some chemical data on coal pyrite," *Chem. Met. Eng.* **22**, 105 (1920).
251. Yancey, H. F., and T. Fraser, "The distribution of the forms of sulphur in the coal bed," *Univ. Illinois Eng. Exp. Sta. Bull* **125**, 11 (1921).
252. Young, C. M., "Pollution of river water in Pittsburgh district," *J. Am. Water Works Assoc.* **8**, 201 (1921).

Suggested Additional Reading

Numerous important references have been published since 1962.
1. Adamson, G. F. S., "Some modern aspects of coal cleaning and their influence on avoidance of river pollution," *Effluent Water Treat. J.* **5**, 143 (1965); *Eng. Index*, 1965, p. 1081.
2. Adema, D., and R. Tietema, "Activated sludge biological treatment of coke wastes with greatly varying pollution," *Gas-Wasserfach* **103**, 617 (1962).
3. "Acid mine-drainage control—principles and practices guide," *Coal Age* **69**, 81 (1964); *Eng. Index*, 1964, p. 287.
4. "Acid mine-drainage—control case histories," *Coal Age* **69**, 72 (1964); *Eng. Index*, 1964, p. 287.
5. "Solving problem of acid mine drainage," *Coal Age* **70**, 72 (1965); *Eng. Index*, 1965, p. 358.
6. "New reclaim-type plant produces quality coal, provides backfilling," *Coal Age* **70**, 118 (1965); *Eng. Index*, 1965, p. 378.
7. "Acid mine drainage programs set," *Environ. Sci. Technol.* **1**, 600 (1967).
8. "Streamlined refuse disposal," *Coal Age* **70**, 106 (1965); *Eng. Index*, 1965, p. 378.
9. "Accentrifloc system of water clarification," *Colliery Guardian* **209**, 799 (1944); *Eng. Index*, 1965, p. 2645.
10. "Mine drainage can be costly," *Power* **112**, 102 (1968).
11. "Stabilization of surfaces of mine tailings dumps," *S. African Mining Eng. J.* **75**, 1021 (1964); *Eng. Index*, 1965, p. 1431.
12. "Coal saved, pollution controlled by Guyan 1 fine-coal plant," *Coal Age* **69**, 87 (1964); *Eng. Index*, 1964, p. 311.
13. "Coal preparation of Seaham," *Colliery Eng.* **41**, 185 (1964); *Eng. Index*, 1964, p. 306.
14. "Automatic limer prevents pollution," *Coal Age* **71**, 120 (1966); *Eng. Index*, 1966, p. 3207.
15. "P. & M. reclamation—forests, lakes, recreation centers," *Coal Age* **71**, 100 (1966); *Eng. Index*, 1967, p. 31.
16. "Stripping and reclamation at Robert Bailery Coal

Co.," *Coal Age* **71**, 66 (1966); *Eng. Index*, 1967, p. 31.
17. "Coal waste processing plant profitably recovers coal and building materials," *Tech. Dig.* **9**, 45 (1964); *J. Water Pollution Control Federation* **38**, 880 (1966).
18. "Chinook modernizes—digs deeper, up-grades preparation," *Coal Age* **71**, 82 (1967); *Eng. Index*, 1967, p. 31.
19. "Electropneumatic coal cleaning," *Coal Age* **71**, 76 (1966); *Eng. Index*, 1967, p. 33.
20. Barnes., I., W. T. Stuart, and D. W. Fisher, "Field investigation of mine waters in northern anthracite field, Pa." *Eng. Index*, 1964, p. 288.
21. Bellano, W., "An operator looks at acid mine drainage," *Mining Congr. J.* **50**, 66 (1964); *Eng. Index*, 1964, p. 288; *J. Water Pollution Control Federation* **37**, 748 (1965).
22. Berger, O., "Filtration in coal processing," *Chem. Process Eng.* **46**, 617 (1965); *J. Water Pollution Control Federation* **28**, 880 (1966).
23. Bochkarev, G. R., and F. A. Barysnikov, "Coagulating coal slurries of different material composition with polyacrylamide," *Koks i Khim.*, no. 4, 1963, p. 10; *Eng. Index*, 1964, p. 306.
24. Braley, S. A., "The oxidation of pyritic industrial wastes," **5**, 89 (1960).
25. Brady, G. A., H. H. Griffiths, and J. W. Echerd, "Dewatering anthracite slurry," U.S. Bureau of Mines, Report Investigations, no. 7012, 1967; *Eng. Index*, 1967, p. 35.
26. Bramer, H. C., and R. D. Hoak, "Measuring sedimentation—flocculation efficiencies," *Ind. Eng. Chem., Process Design Develop.* **5**, 316 (1966); *J. Water Pollution Control Federation* **39**, 878 (1967).
27. Barthauer, G. L., "Mine drainage treatment—fact and fiction," *Cool Age* **71**, 79 (1966); *Eng. Index*, 1966, p. 441.
28. Brown, D. J., "A photographic study of froth flotation," *Fuel Soc. J., Univ. Sheffield* **16**, 22 (1965); *Chem. Abstr.* **64**, 4665 (1966); *J. Water Pollution Control Federation* **39**, 877 (1967).
29. Bruk, O. L., "Counterflow process of washing precipitates in sedimentation washing centrifuges," *Khim. Prom.* **41**, 778 (1965); *Chem. Abstr.* **64**, 6123 (1966); *J. Water Pollution Control Federation* **39**, 879 (1967).
30. Bucklen, O. B., and J. W. Smith, "A simplified device for the froth flotation of fine coal," *Trans. Met. Soc. Am. Inst. Mining Met. Petrol. Engrs.* **229**, 373 (1964); *Battelle Tech. Rev. Abstr.* **14**, 2168 (1965); *J. Water Pollution Control Federation* **38**, 880 (1966).
31. Buscavage, J. J., "Research and demonstration projects in abatement of acid mine drainage," Purdue University Engineering Extension Series, Bulletin no. 118, 1965, p. 664.
32. Campbell, R. S., O. T. Lind, W. T. Geiling, and G. L. Harp, "Recovery from acid pollution in shallow strip-mine lakes in Missouri," Purdue University Engineering Extension Series, Bulletin no. 117, 1964, p. 17.
33. Cann, E. D., "Recovery of flocculating agents from waste waters," U.S. Patent no. 3268443, 23 August 1966; *Chem. Abstr.* **65**, 15067 (1966); *J. Water Pollution Control Federation* **39**, 878 (1967).
34. Cavallaro, J. A., and A. W. Deurbrouck, "Reclaiming magnetite in dense medium circuits by froth flotation," U.S. Bureau of Mines, Report Investigation no. 6821, 1966; *Eng. Index*, April 1967, p. 35.
35. Chmielowski, J., and A. Skowronek, "Autotrophic activated sludge for the biochemical oxidation of some sulfur compounds in mine water," *Zeszyty Nauk. Politech. Slask., Inzh. Sanit.* **8**, 123 (1965); *Chem. Abstr.* **67**, 36217y (1967).
36. Clark, C. S., "Oxidation of coal mine pyrite," *J. Sanit. Eng. Div. Am. Soc. Civil Engrs.* **92**, 1271 (1966); *Public Health Eng. Abstr.* **46**, 2012 (1966); *J. Water Pollution Control Federation* **39**, 877 (1967); *Eng. Index*, 1966, p. 428.
37. Coal Industry Advisory Committee, "Acid mine-drainage: control case histories," *Coal Age* **69**, 72 (1964); *J. Water Pollution Control Federation* **37**, 748 (1965).
38. Cooley, C. E., "Cooperative approach to solving difficult industrial waste problems (coal mining)," in Proceedings of 11th Ontario Industrial Waste Conference, June 1964, p. 256.
39. Conrad, J. W., "Solving the problem of acid mine water pollution," *Analyzer* **8**, page 16, 1967.
40. Conrad, J. W., "Control of acid mine wastes," *Proc. Illinois Mining Inst.* **74**, 1966, p. 104; *Chem. Abstr.* **67**, 25259n (1957).
41. Corbett, D. M., *Summary of Facts—Runoff from Cast-over Burdens Formed by Surface Mining of Coal, Pike County, Indiana*, Water Resources Research Center, Indiana University (1964).
42. Corbett, R. G., and D. J. Growitz, "Composition of water discharged from bituminous coal mines in

northern West Virginia," *Econ. Geol.* **62**, 848 (1967).
43. Cummins, D. G., W. T. Plass, and C. E. Gentry, "Properties and plantability of East Kentucky spoil banks," *Coal Age* **71**, 82 (1966); *Eng. Index*, 1967, p. 31.
44. Davison, A., and B. J. Jefferies, "Some experiments on the nutrition of plants growing on coal mine waste heaps," *Nature* **210**, 649 (1966); *Biol. Abstr.* **47**, 95725 (1966).
45. Day, R. F., "Use of organic polymers in the treatment of industrial wastes," in Proceedings of 12th Ontario Industrial Waste Conference, June 1965, p. 89.
46. Deane, J. A., "How strip mining improves Mid-West water supplies," *Coal Age* **71**, 66 (1966); *Eng. Index*, 1966, p. 446.
47. Deane, J. A., "Control of water at strip mining operations through sound mining and reclamation practices," Purdue University Engineering Extension Series, Bulletin no. 121. 1966, p. 1.
48. Denby, H. G., "Milk of lime plant using bulk lime," *Mining Mag.* **113**, 284 (1965); *J. Water Pollution Control Federation* **38**, 880 (1966).
49. Denne, A., and G. Derenk, "Removal of acidity, iron and manganese from drinking and industrial waters," German Patent no. 1202737, 7 October 1965.
50. Deul, M., and E. A. Mihok, "Mine water research—neutralization," U.S. Bureau of Mines, Report Investigation no. 6987, 1967; *Chem. Abstr.* **66**, 1186262 (1967).
51. Dillon, K. E., "Waste disposal made profitable," *Chem. Eng.* **74**, 146 (1967); *Chem. Abstr.* **67**, 14698*d* (1967).
52. Dru, B., "Some relevant factors in flocculation of shale waters from washing coal," *Rev. Ind. Minerale* **47**, 374 (1965); *Chem. Abstr.* **64**, 9420 (1966); *J. Water Pollution Control Federation* **39**, 878 (1967).
53. Dunning, R. "Coal handling and preparation plant at Newcastle steelworks—fundamental design considerations," B.H.P. Bulletin no. 10, July 1966, p. 16; *Eng. Index*, November 1967, p. 395.
54. Dutcher, R. R., E. B. Jones, H. L. Lovell, R. Parizek, and R. Stefanko, "Mine drainage. I: Abatement, disposal, treatment," *Mineral Ind., Penn. State Univ.* **36**, 1 (1966); *Chem. Abstr.* **66**, 108078*h* (1967).
55. Eichholtz, K. "Planung, Bau und Betrieb von zwei Lagern zum Vergleichmaessigen der Rohwaschkohle vor der Aufbereitung," *Glueckauf* **103**, 206 (1967); *Eng. Index*, December 1967, p. 40.
56. Eveson, G. F., and G. A. Pickin, "Recovery of finely divided solids from suspensions," British Patent no. 1019035, 2 February 1966; *Chem. Abstr.* **64**, 13974 (1966); *J. Water Pollution Control Federation* **39**, 878 (1967).
57. Fairbanks, H. V., and R. E. Cline, "Acoustic drying of coal," *IEEE Trans. Sonics Ultrasonics*, **SU-14**, 175 (1967).
58. Ferney, F. X., "Reclaiming fine coal from settling ponds," *Mining Congress J.* **50**, 48 (1964); *Eng. Index*, 1964, p. 311.
59. Florin, G., "Ein neues Verfahren zum Trocknen feinstkoerniger Flotations-abgaenge," *Bergbau Arch.* **25**, 65 (1964); *Eng. Index*, 1966, p. 457.
60. Flowers, A. E., "Effective reclamation sets stage for industrial development," *Coal Age* **71**, 66 (1966).
61. Fomenko, T. G., and V. S. Butovetskii, "Choosing and justifying a rational water-slurry flowsheet for coal cleaning plants," *Koks i Khim.* **3**, 40 (1966); *J. Water Pollution Control Federation* **39**, 878 (1967).
62. Fomenko, T. G., E. M. Pogartseva, A. M. Kotkin, and V. S. Butovetskii, "New schemes for clarifying contaminated waters," *Koks i Khim.* **7**, 18 (1965); *Pollution Control Federation* **38**, 880 (1966).
63. Freke, A. M., and D. Tate, "The formation of magnetic iron sulphide by bacterial reduction of iron solutions," *J. Biochem. Microbiol. Technol. Eng.* **3**, 29 (1961).
64. Gartrell, F. E., and J. C. Barber, "Pollution control interrelationships," *Chem. Eng. Progr.* **62**, 44 (1966); *Eng. Index*, April 1967, p. 104.
65. Guard, L., "Design and economics of an acid mine drainage treatment plant operation Yellow Bay," *Am. Chemists' Soc., Div. Fuel Chem. (Preprints)* **10**, 107 (1966); *Chem. Abstr.* **66**, 49077*g* (1967).
66. Girard, L., and R. Kaplan, "Operation Yellow Bay—treatment of acid mine drainage," *Coal Age* **72**, 72 (1967); *Eng. Index*, August 1967, p. 388.
67. Gluskoter, H. J., "Composition of ground water associated with coal in Illinois and Indiana," *Econ. Geol.* **60**, 614 (1965); *Eng. Index*, 1965, p. 2617.
68. Gisler, H. J., and C. T. Beardsley, "Closed cycle and washing," *Ind. Water Wastes* **6**, 163 (1961).
69. Green, J., "Coagulating composition," U.S. Patent no. 3130167, 21 April 1964; *Water Pollution Abstr.* **39**, 1360 (1966); *J. Water Pollution Control Federation* **39**, 878 (1967).
70. Hall, E. P., "Coal mining," in *Industrial Waste Water*

Control, Academic Press, New York (1965), p. 169.
71. Hall, E. P., "Long-range look at acid mine drainage," *Mining Eng.* **18**, 61 (1966); *Eng. Index*, 1966, p. 441.
72. Hamilton, W., J. R. Norman, and J. S. Ratcliffe, *Australian Chem. Eng.* **7**, 3 (1966); *Eng. Index*, May 1967, p. 35.
73. Hanna, B. P., Jr., "Relation of water to strip-mine operation," *Ohio J. Sci.* **64**, 120 (1964); *Eng. Index*, 1966, p. 441.
74. Hassett, N. J., "Graphical analysis of filtration processes," *But. Chem. Eng.* **9**, 753 (1964); *J. Water Pollution Control Federation* **38**, 880 (1966).
75. Heertjes, P. M., "Filtration," *Trans. Inst. Chem. Engrs.* **42**, 266 (1964). *J. Water Pollution Control Federation* **38**, 880 (1966).
76. Hill, N. W., P. L. Hughes, H. Ryder, and A. E. Whittle, "Disposal of froth flotation tailings," in Proceedings of 4th International Coal Preparation Congress, Harrogate, England, 1962, p. 413; *Eng. Index*, 1964, p. 311.
77. Hrebacka, J., "Treatment of waste water from collieries," *Vodni Hospodarstvi* **15**, 208 (1965); *Biol. Abstr.* **47**, 98369 (1966).
78. Hummer, E. D., "Minus 48-mesh refuse disposal at U.S. Steel's Gary center coal preparation plant," *Mining Eng.* **17**, 63 (1965); *Eng. Index*, 1965, p. 378; *J. Water Pollution Control Federation* **38**, 880 (1966).
79. Iwasaki, J., E. Oi, and H. Hirano, "Moisture reduction in rewatering coal slime by new vacuum and steam centrifuges," *Sentan* **17**, 140 (1967); *Eng. Index*, November 1967, p. 35.
80. Johnson, G. E., L. M. Kunka, and J. H. Field, "Use of coal and fly ash as adsorbents for removing organic contaminants from secondary municipal effluents," *Ind. Eng. Chem. Process Design Develop.* **4**, 323 (1965); *J. Water Pollution Control Federation* **38**, 880 (1966).
81. Johnson, T. A., "Method of flocculating suspended solids," U.S. Patent no. 2890608 (to U.S. Steel Corp.); *Water Pollution Abstr.* **37**, 1911 (1964); *J. Water Pollution Control Federation* **37**, 748 (1965).
82. Kelsey, G. D., "Continuous slurry dewatering system in coal preparation," *Effluent Water Treat. J.* **2**, 273 (1963).
83. Klappach, G., "Significance of the chemical reaction technique in the microbiological purification of coal processing waste water," *Fortschr. Wasserchem. Ihre Grenzg.* **2**, 134 (1965).
84. Kloessel, E., "Effluent problems at the Rammelsberg mines," *Z. Erzbergbau Metallhuettenw.* **19**, 243 (1966); *Chem. Abstr.* **64**, 3551b (1966).
85. Kumanomido, K., and K. Ogishima, "Application of flocculants to treatment of coal preparation plant waste water," *Sentan* **16**, 284 (1966).
86. Lehmann, H., Pietzner, and H. Werner, "Wasserchemismus und Mineralneubildung in der Grube Luederich," *Eng. Index*, 1967, p. 136.
87. Leszczynska, H., M. Kaczorek, R. Sobiesiak, B. Witkowska, B. Hoffman, and A. Pfeffer, "Development of a method for the removal of hydrogen sulfide from mine waters," *Freiberger Forschungsh.* **350A**, 183 (1965); *Chem. Abstr.* **64**, 4783 (1966); *J. Water Pollution Control Federation* **39**, 877 (1967).
88. Linden, K. V., and R. Stefanko, "Subsurface disposal of acid mine water," *Am. Chemists' Soc., Div. Fuel Chem. (Preprints)* **10**, 101 (1966); *Chem. Abstr.* **66**, 40581j (1967).
89. Llewellyn, R. L., "Dewatering flotation slurries in centrifuges at Stotesbury," *Coal Age* **69**, 88 (1964); *Eng. Index*, 1964, p. 311.
90. Lovell, H. L., and R. D. Reese, "Interactions between coal and water which change water quality," *Am. Chemists' Soc., Div. Fuel Chem. (Preprints)* **10**, 117 (1966); *Chem. Abstr.* **66**.49080c (1967).
91. Lunney, R. H., "Precise pH control in the treatment of effluents," *Effluent Water Treat. J.* **4**, 468 (1964); *Public Health Eng. Abstr.* **45**, 577 (1965); *J. Water Pollution Control Federation* **38**, 880 (1966).
92. Maneval, D. R., and H. B. Charmbury, "A mobile demonstration plant to combat acid mine drainage," *Water Sewage Works* **112**, 268 (1965); *Water Pollution Abstr.* **39**, 1974 (1966); *J. Water Pollution Control Federation* **39**, 878 (1967).
93. Matasov, V. L., "Purification of sulfuric acid-iron containing washing discharges," *Byul. Tekhn.-Inform., Gos Nauchn.- Issled. Inst. Nauchn. Tekhn. Inform.* **20**, 3 (1967); *Chem. Abstr.* **68**, 62501b (1968).
94. May, R. F., "Surface-mine reclamation—continuing research challenge," *Coal Age* **69**, 98 (1964); *Eng. Index*, 1964, p. 294.
95. May, R. F., "Strip-mine reclamation research—where are we?" *Mining Congr. J.* **51**, 52 (1965); *Eng. Index*, 1965, p. 365.
96. May, R. F., and W. A. Berg, "Overburden and bank acidity—Eastern Kentucky strip mines," *Coal Age* **71**, 74 (1966); *Eng. Index*, 1966, p. 446.

97. McIlhenny, W. F., "Recovery of additional water from industrial waste water," *Chem. Eng. Progr.* **63**, 76 (1967).
98. Meerman, P. G., "Moeglichkeiten zur Abwasserbehandlung in der Steinkohlenauf-bereitung, dargestellt am Beispiel der niederlaendischen Staats-gruben," *Glueckauf* **100**, 562 (1964); *Eng. Index*, 1964, p. 2207.
99. Mehsen, J. J., "Microbiological study of acid mine waters, preliminary report," *Ohio J. Sci.* **53**, 123 (1953).
100. Michel, G., and K. H. Rueller, "Hydrochemische Untersuchungen des Grubenwassers der Zechen der Huettenwerk Oberhausen AG," *Bergbau Arch.* **25**, 21 (1964); *Eng. Index*, 1966, p. 441.
101. Mistrick, E. J., "Flotation of coal and minerals," Czechoslovak Patent no. 117661, 15 March 1966; *E. European Sci. Abstr.* **2**, 3285 (1966); *J. Water Pollution Control Federation* **38**, 878 (1967).
102. Moebs, N. N., "Air sealing as means of abating acid mine drainage pollution," *Am. Chemists' Soc., Div. Fuel Chem.* **19**, and Preprints, **10**, 1966, p. 93; *Eng. Index*, 1966, p. 1779; *Chem. Abstr.* **66**, 40582 (1967).
103. Monroe, H. L., "Design of mine water settler at Pea Ridge," *Engineering* **17**, 81 (1965); *Eng. Index*, 1966, p. 1779.
104. Negulescu, M., V. Ghederim, and L. Mihailescu, "Investigations on the treatment of waste waters from the preparation of coal in the Jiu valley," *Stud. Prot. Epur. Apel. (Bucharest)* **5**, 65 (1964); *J. Water Pollution Control Federation* **38**, 880 (1966).
105. Olfert, A. I., "The performance of the iron and steel industry's coal cleaning plants during 1964," *Koks i Khim.* **4**, 10 (1965); *J. Water Pollution Control Federation* **38**, 880 (1966).
106. Pollio, F., and R. Kunin, "Treatment of acid mine waters," *Water Wastes Eng.* **4**, 1 August 1967.
107. Pollio, F., and R. Kunin, "Ion exchange processes for the reclamation of acid mine drainage waters," *Environ. Sci. Technol.* **1**, 183 (1967).
108. Pollio, F., and R. Kunin, "Ion exchange processes for the reclamation of acid mine drainage waters," *Environ. Sci. Technol.* **1**. 235 (1967).
109. Porges, R., L. A. Vandenberg, and D. G. Ballinger, & "Reassessing an old problem—acid mine drainage,"
110. *J. Sanit. Eng. Div. Am. Soc. Civil Engrs.* **92**, 69 (1966); *Chem. Abstr.* **64**, 3551a (1966); *Public Health Eng. Abstr.* **46**, 1071 (1966); *J. Water Pollution Control Federation* **39**, 878 (1967).
111. Radek, O., "Technological concepts of coal preparation plants in Ostrava—Karvina coal district in Czechoslovakia," *Czech. Heavy Ind.* **11**, 23 (1966); *Eng. Index*, August 1967, p. 32.
112. Reed, W., "The use of water in coal preparation," *Effluent Water Treat. J.* **2**, 263 (1962).
113. Reese, R. D., and H. L. Lovell, "Some interactions between coal and water which change water quality," *Am. Chemists' Soc., Div. Fuel Chem.*, **10**, 117 (1966); *Eng. Index*, 1966, p. 1780.
114. Rhodes, J. C., "Purifying mine drainage water," U.S. Patent no. 3347787, 17 October 1967; *Chem. Abstr.* **68**, 43016 (1968).
115. Roe, L. A., "Needed: more attention on recovery and recycling of flotation reagents," *Eng. Mining J.* **165**, 86 (1964); *Battelle Tech. Rev. Abstr.* **14**, 1304 (1965); *J. Water Pollution Control Federation* **38**, 880 (1966).
116. Rogers, T. O., and H. A. Wilson, "pH as a selecting mechanism of the microbial flora in waste water—polluted acid mine drainage," *J. Water Pollution Control Federation* **38**, 990 (1966).
117. Rozgaj, S., "Coagulation of minute particles in coal separation waste water," *Khim. Prom.* **14**, 134 (1965); *Chem. Abstr.* **64**, 4787b (1966); *J. Water Pollution Control Federation* **38**, 880 (1966).
118. Schaffer, R. B., "Polyelectrolytes in industrial waste treatment," *Ind. Water Wastes* **8**, 34 (1963); *J. Water Pollution Control Federation* **37**, 748 (1967).
119. Smidt, O., "Mechanical dewatering of flotation dirt in solid-bowl shell centrifuges," *Glueckauf* **101**, 41 (1965); *Battelle Tech. Rev. Abstr.* **14**, 2857 (1965); *J. Water Pollution Control Federation* **39**, 878 (1966); *Eng. Index*, 1965, p. 378.
120. Smidt, O., "Klaeren und Eindecken von Feinst Korn—Kohlenschlaemmen in einen neuceriten Runderindecker," *Glueckauf* **103**, 784 (1967); *Eng. Index*, February 1968,
121. Soboleva, I. M., and S. V. Peltikhin, "Installation for purifying mine water," *Ochistka i Ispol'z. Stochn. Vod i Prom. Vybrosov*, 1964, p. 111; *Chem. Abstr.* **63**, 12864 (1965); *J. Water Pollution Control Federation* **38**, 880 (1966).
122. Sorokin, A. F., and M. V. Tsiperovich, "The flotation properties of some carbocyclic compounds," *Koks i Khim.* **2**, 5 (1964); *J. Water Pollution Control Federation* **37**, 738 (1968).
123. Steinberg, M., J. Pruzansky, L. R. Jefferson, and B.

Manowitz, "Removal of iron from mine drainage waste with the aid of high energy radiation," U.S. Atomic Energy Commission, *BNL*-11576, 1967; *Chem. Abstr.* **68**, 33012 (1968).

124. Steinberg, M., and B. Manowitz, "Purification of mine drainage water by a radiolytically induced chain process," U.S. Atomic Energy Commission, *BNL*-11220, 1967; *Chem. Abstr.* **68**, 33016 (1968).

125. Steinman, R. E., "Controlled mine water drainage," *Ind. Water Wastes* **5**, 201 (1960).

126. Struthers, P. H., "Chemical weathering of strip-mine spoils," *Ohio J. Sci.* **64**, 125 (1964); *Eng. Index*, 1966, p. 446.

127. Sullivan, G. D., "Current research trends in mined-land conservation and utilization," *Mining Eng.* **19**, 63 (1967); *Eng. Index*, August 1967, p. 144.

128. Tanaka, T., H. Hirano, S. Vinori, S. Arai, and K. Nakamura, "Study on utilization of coal refuse—2," *Eng. Index*, February 1967, p. 35.

129. Tanabe, S., and R. Kawanami, "Fundamental studies of production of lightweight aggregate—4," *Sentan* **16**, 198 (1966); *Eng. Index*, February 1967, p. 35.

130. Thoma, K., and H. Geppert, "Ein neuer Trockensortierer zum Ausscheiden von Bergen vor den Setzmaschinen," *Glueckauf* **103**, 18 (1967); *Eng. Index*, December 1967, p. 40.

131. Thomas, C. M., "The application of pressure filters in coal preparation," *Effluent Water Treatment J.* **2**, 266 (1962).

132. Visman, J., "Stripping solids from effluents with slugging cyclone," *Can. Mining Met. Bull.* **60**, 553 (1967); *Eng. Index*, October 1967, p. 40.

133. Walter, G., "Ecological studies on the effect of waste water from lignite surface mining and containing Fe-II on the outfall organisms," *Wiss. Z. Karl-Marx-Univ., Leipzig, Math. Naturw. Reihe* **15**, 247 (1966).

134. Washburn, H. L., "Closed circuit systems for stream pollution," *Proc. West Va. Coal Mining Inst.*, 1963, p. 54; *Eng. Index*, 1965, p. 2646.

135. Wilson, H. A., J. L. Hipke, and T. O. Rodgers, "Sewage decomposition in acid mine drainage water," Purdue University Engineering Extension Series, Bulletin no. 117, 1964, p. 272.

136. Woodley, R. A., and S. L. Moore, "Pollution control in mining and processing of Indiana coal," *J. Water Pollution Control Federation* **39**, 41 (1967); *Eng. Index*, August 1967, p. 112.

137. Yamanouchi, T., "Utilization of refuse in coal mine for subbase of pavements," *Sentan* **16**, 186 (1966); *Eng. Index*, 1967, p. 35.

CHAPTER 27

RADIOACTIVE WASTES

Radioactive waste materials are the unwanted by-products formed during nuclear fission, which is induced by bombarding the nuclei of heavy elements (those with a mass number greater than 230) with neutrons. The resulting fission products, isotopes of approximately 30 elements, have mass numbers in the range of 72 to 162, are for the most part solids, and emit beta particles, together with electromagnetic radiation (gamma rays). The emission in turn produces further changes in the identity of the isotopes. This so-called "decay" proceeds according to the laws of monomolecular theory and is measured in terms of half-lives (the time required for half of the radioactive atoms to disentegrate), which vary from seconds to milleniums.

Unlike the alpha and beta emissions, the gamma rays ionize only a little and hence are exceedingly penetrating. Since these radiations are capable of altering the atoms of matter through which they pass and have a cumulative effect in nature, they may cause irreparable damage to living tissue.

Fission is the basic process used in the operation of nuclear reactors (piles), in the production of plutonium for military purposes, or in the conversion of nuclear energy into electricity. Radioactive waste matters are produced regardless of whether fission is controlled, as in the reactor of a power station, or occurs explosively, as in the atom bomb. The chemical separation of fissioned products and their conversion to nuclear fuel elements are the most important sources of radioactive wastes, in terms of level of activity and frequency of occurrence. To eliminate any radiation hazard to man, the disposal of these wastes must follow one of these two general principles: (1) concentrate and contain (applicable to highly active wastes) or (2) dilute and disperse (suitable for large volumes of wastes exhibiting a relatively low level of radioactivity).

To help the reader comprehend the unique problems that have to be solved by industries handling radioactive materials, we shall digress momentarily to present some basic quantitative information on radiation.

The unit frequently encountered in studies of radioactivity is the curie,* which is defined as the quantity of radioactive matter giving 3.70×10^{10} disintegrations per second. For mixed radioisotopes, the maximum permissible concentration (MPC) of radioactive matter in water is set by the Atomic Energy Commission at 10^{-7} microcurie/milliliter, or one tenbillionth of a curie per liter. Since it is expected that by 1980 the activity of fission products produced each year will amount to approximately 100 billion curies, the reader will readily comprehend the magnitude of the problem posed by the treatment of radioactive wastes.

Industries using radioactive matter must accept the responsibility for protecting our biological environment. Waste, whether solid, liquid, or gaseous, which may eventually come in contact with human beings must be rendered harmless to man's biological system. Unfortunately, this is not easily accomplished. Whereas industrial wastes lend themselves to stabilization by chemical, physical, or biological processes, radioactive wastes do not respond sufficiently to any of these methods of treatment. Only time can render radioactive wastes inactive, and in

---

*Until 1948, the curie was officially defined as the quantity of radium emanation (radon) in equilibrium with one gram of radium.

some cases hundreds of years must pass before the wastes are safe for discharge to the environment. Thus, at present, storage appears to be the only way of successfully solving the disposal problem. But if storage, particularly in liquid form, should continue to be the only means of effective disposal, the accumulated solutions of wastes could amount to 200 million gallons by 1980 and more than 2000 million gallons by the year 2000. Although many improved storage systems have been developed through research, only a few systems have proved both economical and practical.

As mentioned before, radioactive wastes may be liquid, gaseous, or solid; furthermore, they may exhibit high or low levels of activity. In the following sections, we shall discuss the origin and disposal of the three types of waste, as well as the different methods of treatment that may be indicated, depending on the amounts of radioactive materials present.

27.1 Origin of Wastes

The most frequent sources of radioactive wastes are the following.

Processing of uranium ores (in mining plants in the southwestern United States) produces considerable volumes of alpha emitters, mainly radium. Storage of these wastes will usually remove suspended radioactive materials.

Laundering of contaminated clothes usually produces large volumes of water containing decreasing concentrations of radioactive materials as the laundry progresses through a series of cycles. The initial wash water generally requires treatment.

Research-laboratory wastes contain all the materials prevalent in chemical, metallurgical, and biological research operations. Some of the wastes are in the form of stable isotopes of the radioactive chemicals. An acceptable method of disposal mixes the particular radioisotope with a quantity of stable isotopes of the same element in a certain ratio. This reduces the danger of contamination, since any living organism will take up the various isotopes of an element in the proportion in which they are present.

Hospitals contribute radioactive wastes from diagnostic and therapeutic uses. Iodine-131 and phosphorus-32 are the radioisotopes which predominate in hospital wastes. Fortunately these possess short half-lives, and simple detention tanks can render them inactive.

Fuel-element processing produces high-level wastes; however, the waste products of secondary operations may exhibit low levels of radioactivity and it is usually preferable to separate the two types. Fuel reprocessing, which also produces high-level wastes, is normally done only at AEC installations. Industry, therefore, will not be involved in this waste-disposal problem. However, the future policy of the government or the AEC may change, and private industry may be allowed to reprocess its own fuel.

Power-plant cooling waters may acquire some radioactivity. The contamination may be due to leaks in the piping caused by corrosion or to neutron bombardment of salts present in the cooling water. The water is stored for some time to permit decay and is then monitored before discharge to a river at a controlled rate.

27.2 Power-Plant Wastes

We shall now describe in some detail the wastes encountered and the treatment needed in nuclear power plants and in nuclear fuel processing.

Falk [63] describes a boiling-water reactor which will produce steam at both 1000 psig and 500 psig (Figs. 27.1 and 27.2). The radioactive wastes encountered in reactor operations are to a large extent due to leakage, blowdown, maintenance, refueling, and other mechanisms. Falk states that corrosion products formed in the system (circulating reactor water is used as a source of heat) are the primary source of radioactive isotopes in the reactor water. Those formed outside the reactor vessel, particularly those in the circulating and feed water systems, may be carried into the vessel and become radioactive, along with corrosion products formed within the vessel. Hence it is mandatory that the water used for cooling purposes, as well as that which is used as a source of steam, be ultra-pure. Any salts or other impurities in the water may capture neutrons and become radioactive. Another potential source of radioisotopes in the reactor water is the fission products formed within the fuel elements. It may be possible, but fortunately

534 RADIOACTIVE WASTES

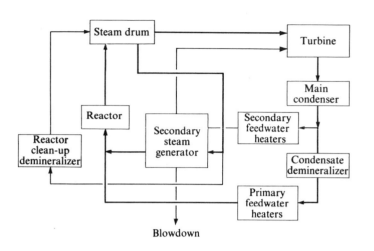

Fig. 27.1 Schematic flow diagram of power plant cycle. (After Falk [63].)

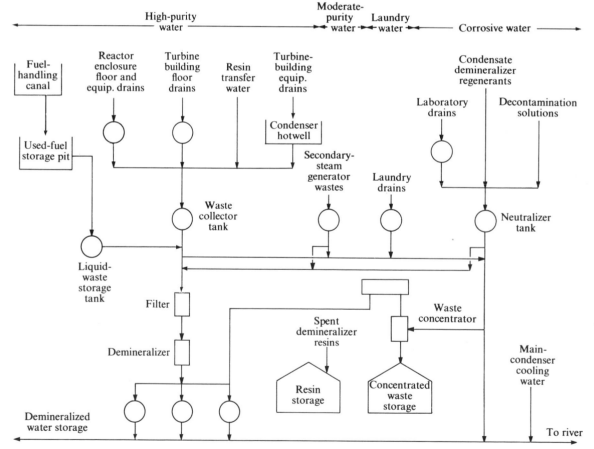

Fig. 27.2 Schematic flow diagram of radioactive waste disposal system. (After Falk [63].)

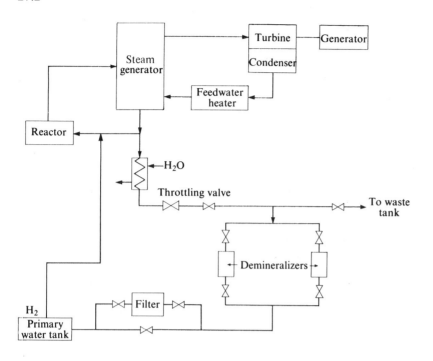

Fig. 27.3 Schematic diagram showing by-pass purification using mixed-bed demineralizers required to maintain high-purity water in a primary system. (After Medin [141].)

not probable, that holes in the cladding of some of the fuel elements could occur, with subsequent release of some fission products to the reactor water. The quantity of radioactive isotopes in the reactor water thus depends on corrosion rates, frequency of failure of fuel-element cladding, and rate of removal by condensate and reactor clean-up demineralizers.

The possible presence of radioactive isotopes in the water necessitates waste-treatment precautions. Medin [141] describes a pressurized-water package power reactor that the army has recently installed at Fort Belvoir, Virginia. In the primary system, great care is taken to maintain the water at a high level of purity to minimize build-up of excessive radioactivity caused by either impurities or corrosion products. No primary water is wasted, but a portion is removed, purified, and recirculated. The purification is accomplished with demineralizers (mixed ion exchangers), a micrometallic filter, and a hold-up tank (Fig. 27.3). Because of the danger of stress corrosion, it is essential that the boiler water contain neither oxygen nor chlorides in any perceptible concentration. Raw water is deaerated and evaporated to reduce the content of chloride and oxygen, which should not exceed 0.3 and 0.03 ppm, respectively.

Table 27.1 Characteristics of solvent extraction processes. (After Blomeke [17].)

Process	Application	Solvent	Salting agent	Approximate volume of untreated high-activity waste*
Purex	Pu, natural	TBP in hydrocarbon	HNO_3	990 gal/metric ton U
Redox	Enriched uranium	Hexone	$Al(NO_3)_3$	1000 gal/metric ton U
Hexone-25	U-Al alloys	TBP in hydrocarbon	$Al(NO_3)_3$	700 liters/kg enriched U
TBP-25	Th-U^{233}	TBP in hydrocarbon	$HNO_3 + Al(NO_3)_3$	670 liters/kg enriched U
Thorex			$Al(NO_3)_3$	1360 gal/metric ton Th

*Volumes refer to the wastes as they come from the extraction column.

27.3 Fuel-processing Wastes

Irradiated reactor fuels are chemically processed to reclaim the unburned nuclear fuel and recover the transmutation products, such as Pu^{239} or U^{233}, from mixtures of the fission products and inert components of the fuel [17]. Several processes in current use are described by Blomeke et al. [17] and are listed in Table 27.1. All involve solvent extraction of the type illustrated in Fig. 27.4.

Generally speaking, all fuel-processing methods are based on similar principles. Solid fuels, which have been stored for about three to four months, are dissolved in HNO_3 and fed to the extraction column (Fig. 27.4), where uranium and plutonium are extracted with the particular solvent chosen. To remove traces of fission products, the extracted uranium and plutonium are scrubbed by an aqueous salt solution added at the top of the extraction column. The waste solutions, which contain more than 99.9 per cent of the total fission products and inerts in the feed and scrub, leave at the bottom of the column. The actual concentration of both inert and radioactive elements found in fuel-processing wastes depends on the operating characteristics of the particular chemical plant and the particular extraction process used. Blomeke et al. [17] give the characteristics of the high-activity wastes produced by the five common extraction processes (Table 27.2). In addition to the concentration of inert chemicals, all the wastes con-

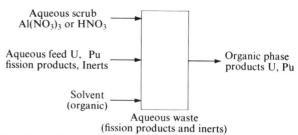

Fig. 27.4 Solvent extraction separation column. (After Blomeke et al. [17].)

Table 27.2 Characteristics of current high-activity wastes. (After Blomeke [17].)

Characteristic*	Purex	Redox	Hexone 25	TBP-25	Thorex
H, M	0.93	− 0.3	− 0.2	1.33	− 0.05
Al, M		1.08	1.6	1.63	0.62
Na, M		0.23			
NH_4, M			1.4		
Hg, M			0.01	0.01	0.01
NO_3, M	0.93	3.05	6.0	6.2	1.8
F, M					0.039
Cr_2O_7, M		0.06			
NH_2SO_3, M				0.04	
Fe, Ni, Cr, gm/liter	< 1	< 1	< 1	< 1	< 1
SiO_2			< 1	< 1	
PO_4, SO_4					< 1
Volume, untreated	990 gal/ton U	1000 gal/ton U	700 liters/kg U	670 liters/kg U	1360 gal/ton Th
Specific gravity	1.03	1.16	1.25	1.25	1.10
Boiling point, °C	101	108	105	105	101
Freezing point, °C	− 3	− 18	− 24	− 24	− 15
Specific heat	0.97	0.78	0.7	0.7	0.85
Volume after evaporation	60 gal/ton U	490 gal/ton U	510 liters/kg U	500 liters/kg U	380 gal/ton Th
Volume after neutralization	80 gal/ton U	830 gal/ton U	860 liters/kg U	840 liters/kg U	640 gal/ton Th

*Chemical composition is exclusive of fission products and heavy elements.

tain fission products and lesser amounts of uranium, plutonium, and other heavy elements. The presence of these heavy elements is attributable mainly to process losses, which usually approximate 0.1 per cent. The wastes obtained by the Purex process [97] are mainly solutions of fission products in nitric acid and their physical properties are essentially those of 1 M HNO_3. Trace constituents of iron, nickel, and chromium are present, from corrosion of stainless-steel equipment. These wastes lend themselves quite readily to concentration by evaporation. The fission product ruthenium begins to oxidize as the waste is concentrated. All other extraction-process wastes contain various amounts of aluminum. The Redox waste [120] contains some sodium and some dichromate. Hexone-25 and TBP-25 wastes [206] contain mercury, and the Thorex waste [24] has both mercury and fluoride. The trace constituents in these wastes include those from products of corrosion, as in the Purex wastes, and also small amounts of SiO_2 and other impurities originally present in the fuel.

Blomeke [17] has also tabulated the characteristics of three high-activity power-reactor fuel-processing wastes (Tables 27.3 and 27.4). These wastes, in addition to being radioactive, are corrosive and possess peculiar chemical properties. None of them can be neutralized without precipitation of dissolved salts. For every megawatt-day of irradiation, about 1.1 gm of fission products (total stable and radioactive) is formed. Consequently, in most of the untreated wastes, the total content of fission products, consisting of isotopes of some 37 elements having atomic

Table 27.3 Characteristics of high-activity power reactor fuel-processing wastes. (After Blomeke [17].)

Characteristic*	Zirconium, TBP-25	Stainless steel, TBP-25	Darex, TBP-SS-28
H, M	1.0	3.4	2.0
Al, M	0.75		
Zr, M	0.55		
Fe, M		0.07	0.18
Cr, M		0.02	0.05
Ni, M		0.007	0.018
NO_3, M	2.3	2.7	2.3
F, M	3.2		
SO_4, M		0.5	
Cr_2O_7, M	0.01		
Fe, Ni, Cr, gm/liter	< 1		
Cl, gm/liter			< 1
Mn, P, Si, gm/liter		< 1	< 1
Volume, untreated	330 liters/kg U	330 liters/kg U	1760 gal/ton U
Specific gravity	1.2	1.1	1.2
Boiling point, °C	101	106	107
Freezing point, °C	Metastable < 25°C	−6	−22
Specific heat		0.84	0.75
Viscosity, cp	2	1.2	1.3
Volume after evaporation			300 gal/ton U (80 gm stainless steel/liter)

*Chemical composition is exclusive of fission products and heavy elements.

Table 27.4 Power reactor fuel-recovery processes. (After Blomeke [17].)

Process	Fuel	Process description
Aqueous HF – TBP-25	High Zr, enriched U	HF dissolution; Zr and F complexing with Al; oxidation of Pu with Cr_2O_7; extraction with TBP
Sulfuric acid – TBP-25	Stainless steel, enriched U	H_2SO_4 dissolution of stainless steel; HNO_3 dissolution of U; extraction with TBP
Darex – Purex	Stainless steel, natural U	Dilute aqua regia dissolution; distillation of HCl; extraction with TBP

weights near that of zinc, amounts to 1 gm/liter. Mixtures of fission products are characterized by the great amount of heat generated.

Culler [42] also lists the physical, chemical, and radiochemical natures of high-level fuel-processing wastes (Table 27.5), and Silverman [15] tabulates some values of inert and radioactive aerosols arising from ore processing (Table 27.6).

Table 27.5 Physical, chemical, and radiochemical characteristics of high-level fuel-processing wastes. (After Culler [42].)

Characteristic	Natural U-Pu		Enriched U-235, diluent-salted
	Salted	HNO_3-salted	
Physical			
Boiling point, °C	120	112	103
Specific gravity	1.18	1.24	1.23
Concentration possible	3	3	2
U^{235} burned (unneutralized), gal/gm	2.5	0.5	5.0
U^{235} burned (neutralized with NaOH), gal/gm	10.0	1.5–2.0	20.0
Chemical			
Acidity, N	−0.3 (basic)	8.0	−0.2
Total salts, M			
Unneutralized	1.2	0.2	2.0
Neutralized	6.0	8.2	Acid storage
Solids stability	Unstable basic	Stable	Unstable basic
Radiochemical			
Curies/gal of acid	80	400	2000
Curies/gal of base	20	200	Acid storage
Lead shielding for 1 cm³ acid (in.)*	3.5	4	4
Lead shielding for 500 gal/acid (in.)*	11.5	12	12
Watts/gal of acid	0.3	1.2	5.4

*Inches of lead shielding required for protection.

Table 27.6 Particle size and distribution of some inert and radioactive aerosols. (After Blatz [15].)

Source[†]	Aerosol	Geometric* mean diameter, Mg	Mass median* diameter, $M'g$	Standard geometric diameter, σ_g
a	Atmospheric dust from 14 U.S. cities, average	0.54 μ	0.97 μ	1.56
b	Atmospheric dust as measured by electron microscope (Knolls Atomic Power Laboratory)	0.028	1.12	3.05
c	Beryllium fluoride fume, BeF_2, from furnace-pouring operation, 10 ft from furnace	0.36	2.3	2.2
d	Iron oxide fume from open-hearth furnace			
	Before waste-heat boiler	0.047	0.65	2.55
	After waste-heat boiler	0.057	0.82	2.60
e	Fission-product source pilot plant			
	Radioactive material measured by light microscope	0.47	2.25	2.17
	Radioactive material measured by electron microscope	0.014	0.14	2.39
	Nonradioactive material measured by light microscope	0.42	11.30	2.86
f	Uranium oxide fume produced by burning scrap	0.12	8.11	3.29
g	Separations process stack effluent during dissolving[‡]	0.2	3.1	2.6
h	Sodium oxide from burning Na metal[§]	0.04	0.17	2.0
i	Air-borne dust from incinerator ash disposal	0.43	101	3.86
j	Fume from ferrosilicon electric furnace during tapping	0.43	2.77	2.2

*Conversion to mass or geometric size based on mathematical conversion $\log Mg = \log M'g - 6.91 \log 2\sigma_g$.
[†]The following references were used:
a. J. E. Ives, et. al., *Atmospheric Pollution of American Cities for the Years 1931 to 1933*, Public Health Bulletin no. 224, U.S. Government Printing Office, Washington, D.C. (1936).
b. J. J. Fitzgerald and C. G. Detwiler, "Collection Efficiency of Air Cleaning and Air Sampling Media," *Am. Ind. Hyg. Assoc. Quart.* **16**, 123 (1955).
c. A. J. Vorwald (ed.), *Pneumoconiosis*, p. 378, Hoeber, New York (1950).
d. C. E. Billings, W. D. Small, and L. Silverman, "Pilot-plant Studies of Continuous Slag-wool Filter for Open-hearth Fume," *J. Air Pollution Control Assoc.* **5**, 159 (1955).
e. J. J. Fitzgerald and C. G. Detwiler, "Size Distribution of Particles Produced by Fission Product Source Pilot Plant," *KAPL* 1232, General Electric Co., Schenectady, N.Y., 1954.
f. E. W. Conners, Jr., and D. P. O'Neil, "Efficiency Studies of a High Efficiency, High Temperature Filter against Freshly Generated Uranium Oxide Fume," *ANL* 5453, Argonne National Laboratory, Lemont, Ill., June 1954.
g. J. J. Fitzgerald, "Evaluation of *KAPL* Separations Process Stack Effluent," *KAPL* 1015, General Electric Co., Schenectady, N.Y., 1952.
h. R. C. Lumatainen and W. J. Mechan, "Removal of Halogens, Carbon Dioxide and Aerosols from Air in a Spray Tower," *ANL* 5429, Argonne National Laboratory, Lemont, Ill., February 1955.
i. W. H. Megonnell, J. H. Ludwig, and L. Silverman, "Dust Exposures during Ash Removal from Incinerators," *Arch. Ind. Health*, **15**, 215 (1957).
j. L. Silverman and R. A. Davidson, "Electric Furnace Ferro-silicon Fume Collection," *J. Metals* (*Trans. AIME*) **203**, 1327 (1955).
[‡]Obtained by light microscopy; electron microscopy showed a mean size of 0.05.
[§]Light-field microscopy gave a mean size of 0.2μ (magnification $\times$ 900).

27.4 Treatment of Radioactive Wastes

Any discussion of the treatment of radioactive wastes must be preceded by some information about the extent to which the waste must be treated. The waste engineer is, at this point, well acquainted with the tolerance limits of streams and municipal treatment plants with respect to "usual" types of industrial waste. Because radiation is a unique phenomenon with which the engineer is not too familiar, our attention will be directed to tolerance values pertaining to radioactive wastes. The main threat to human beings arises by contamination (in excess of the critical tolerance value) of the water used for human consumption or recreation, the fish harvested as food, or the edible plants which, irrigated with the contaminated water, have absorbed some of the radioactive substances [56]. The International Commission on Radiation Protection [204] has established limits of maximum permissible concentration (MPC) for the United Nations, and the National Committee for Radiation Protection and Measurement has established similar values for the United States. The latter group has set the MPC for the general population at five roentgens over a 30-year period. According to Eliassen [56], one roentgen (r) is the quantity of radiation which causes one gram of tissue to absorb 97 ergs of energy.

Because of their sensitivity to radiation, the reproductive organs are most seriously affected by exposure to radioactivity. Beadle [10] estimates that the dosage mentioned above will produce 33 mutations (births of abnormal children) per 10,000 births. The MPC's in air and in domestic water supply for most radioelements are listed in *Handbook 52*, from the National Bureau of Standards [138]. *Handbook 59* [139], a supplement to *Handbook 52*, contains data concerning the allowable internal exposure for radiation workers and advises a safety factor of ten for the general population outside the controlled area. The AEC provides the following regulations [139] concerning the disposal of radioactive wastes by discharge into sanitary sewage systems and burial in soil.

20.303 *Disposal by release into sanitary sewerage systems*. No licensee shall discharge licensed material into a sanitary sewerage system unless:
(a) It is readily soluble or dispersible in water; and
(b) The quantity of any licensed or other radioactive material released into the system by the licensee in any one day does not exceed the larger of subparagraphs (1) or (2) of this paragraph:
 (1) The quantity which, if diluted by the average daily quantity of sewage released into the sewer by the licensee, will result in an average concentration equal to the limits specified in Appendix B [Table 27.8] Table I, Column 2 of this part; or
 (2) Ten times the quantity of such material specified in Appendix C [Table 27.9] of this part; and
(c) The quantity of any licensed or other radioactive material released in any one month, if diluted by the average monthly quantity of water released by the licensee, will not result in an average concentration exceeding the limits specified in Appendix B [Table 27.8] Table I, Column 2 of this part; and
(d) The gross quantity of licensed and other radioactive material released into the sewerage system by the licensee does not exceed one curie per year. Excreta from individuals undergoing medical diagnosis or therapy with radioactive material shall be exempt from any limitations contained in this section.

20.304 *Disposal by burial in soil*. No licensee shall dispose of licensed material by burial in soil unless:
(a) The total quantity of licensed and other radioactive materials buried at any one location and time does not exceed, at the time of burial, 1000 times the amount specified in Appendix C [Table 27.9] of this part; and
(b) Burial is at a minimum depth of four feet; and
(c) Successive burials are separated by distances of at least six feet and not more than 12 burials are made in any year.

The *Federal Register* specifies items of records, exceptions, and enforcement, and reprints the Appendixes A, B, and C, containing maximum permissible limits referred to in AEC regulations 20.303 and 20.304 (Tables 27.7 through 27.9). Falk [63] also presents a table of MPC values in water for several typical radioisotopes and adds for comparison the equivalent concentrations expressed in more conventional units (Table 27.10). Lundgren [128] concludes that the doses of radiation allowed in sewers by the AEC will have relatively little effect on bacteria and other microorganisms, or on the functioning of their enzyme systems in waste-disposal plants. Considering the structural organization of disposal plants, the amount of radiation present, and the dosage of radiation that will affect microorganisms, we see that no immediate problem will arise due to hazards of radiation.

Table 27.7 Permissible weekly dose of radiation. (From *Federal Register*, Jan. 29, 1957.)

Conditions of exposure		Dose in critical organs (mrem)			
Parts of body	Radiation	Skin, at basal layer of epidermis	Blood-forming organs	Gonads	Lens of eye
Whole body	Any radiation with half-value layer greater than 1 mm of soft tissue	600*	300*	300*	300*
Whole body	Any radiation with half-value layer less than 1 mm of soft tissue	1500	300	300	300
Hands and forearms, or feet and ankles, or head and neck	Any radiation	1500†			

*For exposures of the whole body to X- or gamma rays up to 3 mev, this condition may be assumed to be met if the "air dose" does not exceed 300 mr, provided the dose to the gonads does not exceed 300 mrem. "Air dose" means that the dose is measured by an appropriate instrument in air in the region of highest dosage rate to be occupied by an individual, without the presence of the human body or other absorbing and scattering material.

†Exposure of these limited portions of the body under these conditions does not alter the total weekly dose of 300 mrem permitted to the blood-forming organs in the main portion of the body, to the gonads, or to the lens of the eye.

Treatment of liquid radioactive wastes. Two main procedures have been used for the disposal of liquid wastes: concentration and storage, with subsequent burial, or dilution and dispersal, with subsequent discharge to sewers or streams. A typical treatment procedure is illustrated in Fig. 27.5, which shows in schematic fashion the handling of all liquid wastes. It is interesting to note that the ultimate disposition of all liquid wastes is accomplished by either burying in the earth or dumping in the ocean. Although the ocean was once considered adequate to handle all the radioactive materials we could manufacture, it has recently been proved woefully inadequate as a safe receptacle for the radioactivity we will be discharging by the year 2000.

For example, the AEC, despite recent findings that ocean disposal areas for radioactive wastes have so far been safe, warns that extreme care must be taken in the future, in particular with respect to possible damage to the containers. For that reason, the AEC refused at one time to grant that portion of a license to the Industrial Waste Corporation of Houston, Texas, which would have permitted the organization to use sea disposal for radioactive wastes from hospitals, research facilities, and industry.

A combination of dilution and dispersal is the usual method of disposing of liquid wastes of low activity. Dilution may have to be preceded by temporary storage to allow sufficient time for the reduction of radioactivity by natural decay. Temporary storage is usually restricted to short-lived isotopes (Na^{22}, P^{32}, I^{131}). Liquid wastes of very low activity are held for some time to permit decay and are then discharged directly to the watercourse. Generally, wastes having activity concentrations greater than 20 $\mu c/cm^3$ are considered too radioactive for temporary storage. There are practical limits to the amount of radioactive waste that can be treated by dilution and discharged. Lieberman [125] recommends the following limits of radioactivity discharged per million gallons of sewage effluent: (1) 1 millicurie of Sr^{90} or Pu^{210}, (2) 100 millicuries of any radioactive material having a half-life of less than 30 days, (3) 10 millicuries of any other radioactive material.

It is common practice to treat solid wastes, as well as high-level radioactive liquid wastes, by concentration and burial. A small volume of waste is less difficult to monitor, package, transport, and dispose of than a larger volume with the same activity. However, the cost of concentration must not exceed that saved by the concentration process. From Fig. 27.5, the three most widely used methods of concentration are coprecipitation, evaporation, and ion exchange.

Table 27.8 Permissible concentrations in air and water above natural background. (From *Federal Register*, Jan. 29, 1957.)

Material	Table I (Restricted areas)		Table II (Unrestricted areas)	
	Column 1, air*	Column 2, water†	Column 1, air*	Column 2, water†
A^{41}	1.6×10^{-6}	1.4×10^{-3}	5×10^{-8}	5×10^{-5}
Ag^{105}	3.6×10^{-5}	5	1.2×10^{-6}	1.6×10^{-1}
Ag^{111}	1×10^{-4}	13	3×10^{-6}	4×10^{-1}
Am^{241}	8×10^{-11}	4×10^{-4}	2×10^{-12}	1.3×10^{-5}
As^{76}	7×10^{-6}	6×10^{-1}	2×10^{-7}	2×10^{-2}
At^{211}	9×10^{-10}	6×10^{-6}	3×10^{-11}	2×10^{-7}
Au^{198}	3.4×10^{-7}	9×10^{-3}	1.1×10^{-8}	3×10^{-4}
Au^{199}	8×10^{-7}	2×10^{-2}	2.5×10^{-8}	7×10^{-4}
$Ba^{140} + La^{140}$	2×10^{-7}	6×10^{-3}	6×10^{-9}	2×10^{-4}
Be^{7}	1.3×10^{-5}	3	4×10^{-7}	1×10^{-1}
C^{14}	1.4×10^{-6}	1×10^{-2}	5×10^{-8}	3.6×10^{-4}
Ca^{45}	9×10^{-8}	1.5×10^{-3}	3×10^{-9}	5×10^{-5}
$Cd^{109} + Ag^{109}$	2×10^{-7}	2×10^{-1}	7×10^{-9}	7×10^{-3}
$Ce^{144} + Pr^{144}$	2×10^{-8}	1×10^{-1}	7×10^{-10}	3.6×10^{-3}
Cl^{36}	1×10^{-6}	7×10^{-3}	4×10^{-8}	2.4×10^{-4}
Cm^{242}	5×10^{-10}	2.7×10^{-3}	1.8×10^{-11}	1×10^{-4}
Co^{60}	3.4×10^{-6}	5×10^{-2}	1.2×10^{-7}	1.8×10^{-3}
Cr^{51}	2.4×10^{-5}	1.4	8×10^{-7}	5×10^{-2}
$Cs^{137} + Ba^{137}$	6×10^{-7}	4.5×10^{-3}	2×10^{-8}	1.5×10^{-4}
Cu^{64}	2×10^{-5}	2.5×10^{-1}	6×10^{-7}	8×10^{-3}
Eu^{154}	2×10^{-8}	1×10^{-1}	6×10^{-10}	3×10^{-8}
F^{18}	3.5×10^{-4}	2.6	1.2×10^{-5}	9×10^{-2}
Fe^{55}	1.8×10^{-6}	1.3×10^{-2}	6×10^{-8}	4×10^{-4}
Fe^{59}	5×10^{-8}	3.3×10^{-4}	1.5×10^{-9}	1.1×10^{-5}
Ga^{72}	1×10^{-5}	26	3.4×10^{-7}	9×10^{-1}
Ge^{71}	1×10^{-4}	27	3.6×10^{-6}	9×10^{-1}
H^{3} (HTO or T_2O)	7×10^{-5}	5×10^{-1}	2.5×10^{-6}	1.6×10^{-2}
Ho^{166}	1×10^{-5}	70	3×10^{-7}	2.3
I^{131}	9×10^{-9}	9×10^{-5}	3×10^{-10}	3×10^{-6}
Ir^{190}	2.2×10^{-6}	4×10^{-2}	7×10^{-8}	1.3×10^{-3}
Ir^{190}	1.5×10^{-7}	2.7×10^{-3}	5×10^{-9}	9×10^{-5}
K^{42}	6×10^{-6}	4×10^{-2}	2×10^{-7}	1.4×10^{-3}
La^{140}	4×10^{-6}	3.4	1.4×10^{-7}	1.1×10^{-1}
Lu^{177}	1.5×10^{-5}	70	5×10^{-7}	2.4
Mn^{56}	8×10^{-6}	5×10^{-1}	3×10^{-7}	1.5×10^{-2}
Mo^{99}	5×10^{-3}	40	1.8×10^{-4}	1.4
Na^{24}	5×10^{-6}	2.4×10^{-2}	1.6×10^{-7}	8×10^{-4}
Nb^{95}	1.3×10^{-6}	1.2×10^{-2}	4×10^{-8}	4×10^{-4}
Ni^{59}	5×10^{-5}	7×10^{-1}	1.6×10^{-6}	2.5×10^{-2}
P^{32}	4×10^{-7}	6×10^{-4}	1.4×10^{-8}	2×10^{-5}
Pb^{203}	2×10^{-5}	4×10^{-1}	6×10^{-7}	1.4×10^{-2}
$Pb^{103} + Rh^{103}$	2×10^{-6}	3×10^{-2}	7×10^{-8}	1×10^{-3}
Pm^{147}	6×10^{-7}	3	2×10^{-8}	1×10^{-2}
Po^{210} (soluble)	6×10^{-10}	9×10^{-5}	2×10^{-11}	3×10^{-6}
Po^{210} (insoluble)	2×10^{-10}		7×10^{-12}	
Pr^{143}	2.3×10^{-6}	1	7×10^{-8}	3.6×10^{-2}

27.4 TREATMENT OF RADIOACTIVE WASTES

Table 27.8 (*continued*)

Material	Table I (Restricted areas)		Table II (Unrestricted areas)	
	Column 1, air*	Column 2, water†	Column 1, air*	Column 2, water†
Pu^{239} (soluble)	6×10^{-12}	4.5×10^{-6}	2×10^{-13}	1.5×10^{-7}
Pu^{239} (insoluble)	6×10^{-12}		2×10^{-13}	
$Ra^{226} + \frac{1}{2}$ dr	2.4×10^{-11}	1.2×10^{-7}	8×10^{-13}	4×10^{-9}
Rb^{86}	1.1×10^{-6}	9×10^{-3}	4×10^{-8}	3×10^{-4}
Re^{183}	2.4×10^{-5}	2.4×10^{-1}	8×10^{-7}	8×10^{-3}
Rh^{105}	3×10^{-6}	5×10^{-2}	1×10^{-7}	1.6×10^{-3}
$Rh^{222} + $ dr	1×10^{-7}	6×10^{-6}	3.3×10^{-9}	2×10^{-7}
$Ru^{106} + Rh^{106}$	8×10^{-8}	4×10^{-1}	2.6×10^{-9}	1.3×10^{-2}
S^{35}	3×10^{-6}	1.5×10^{-2}	1×10^{-7}	5×10^{-4}
Sc^{46}	2×10^{-7}	1	7×10^{-9}	3.6×10^{-2}
Sm^{151}	4×10^{-8}	6×10^{-1}	1.3×10^{-9}	2×10^{-2}
Sn^{113}	1.7×10^{-6}	5×10^{-1}	6×10^{-8}	1.6×10^{-2}
Sr^{89}	6×10^{-8}	2×10^{-4}	2×10^{-9}	7×10^{-6}
$Sr^{90} + Y^{90}$	6×10^{-10}	2.4×10^{-6}	2×10^{-11}	8×10^{-8}
Tc^{90}	8×10^{-6}	8×10^{-2}	3×10^{-7}	3×10^{-3}
Te^{127}	3×10^{-7}	8×10^{-2}	1×10^{-8}	3×10^{-3}
Te^{129}	1.2×10^{-7}	3.3×10^{-2}	4×10^{-9}	1.1×10^{-3}
Th^{234}	2×10^{-6}	10	6×10^{-8}	3×10^{-1}
Th-natural (soluble)	5×10^{-11}	1.5×10^{-6}	1.7×10^{-12}	5×10^{-8}
Th-natural (insoluble)	5×10^{-11}		1.7×10^{-12}	
Tm^{170}	1.5×10^{-7}	8×10^{-1}	5×10^{-9}	2.5×10^{-3}
U-natural (soluble)‡	5×10^{-11}	2×10^{-4}	1.7×10^{-12}	7×10^{-6}
U-natural (insoluble)‡	5×10^{-11}		1.7×10^{-12}	
U^{233} (soluble)	4×10^{-10}	4.5×10^{-4}	1×10^{-11}	1.5×10^{-5}
U^{233} (insoluble)	5×10^{-11}		1.6×10^{-12}	
V^{48}	3×10^{-6}	1.5	1×10^{-7}	5×10^{-2}
Xe^{133}	1.3×10^{-5}	1.3×10^{-2}	4×10^{-7}	4×10^{-4}
Xe^{135}	5×10^{-6}	4×10^{-3}	1.7×10^{-7}	1.4×10^{-4}
Y^{91}	1.2×10^{-7}	6×10^{-1}	4×10^{-9}	2×10^{-2}
Zn^{65}	6×10^{-6}	2×10^{-1}	2×10^{-7}	6×10^{-3}
Unidentified beta or gamma emitters or any undetermined mixtures of beta or gamma emitters			1×10^{-9}	1×10^{-7}
Unidentified alpha emitters or any undetermined mixtures of alpha emitters			5×10^{-12}	1×10^{-7}

*Air concentrations are given in microcuries per milliliter of air.
†Water concentrations are given in microcuries per milliliter of water. These figures also apply to foodstuffs in microcuries per gram (wet weight).
‡For enriched uranium the same radioactivities per unit volume as for natural uranium are applicable. It should be noted that the contribution of U^{234} to the gross activity of enriched uranium is 20 to 40 times that of U^{235}.

Table 27.9 Permissible quantities of radioactive material. (From *Federal Register*, Jan. 29, 1957.)

Material	Micro-curies	Material	Micro-curies	Material	Micro-curies
Ag^{105}	1	K^{42}	10	Sn^{113}	10
Ag^{111}	10	La^{140}	10	Sr^{89}	1
As^{76}, As^{77}	10	Mn^{52}	1	$Sr^{90} + Y^{90}$	0.1
Au^{198}	10	Mn^{56}	50	Ta^{182}	10
Au^{199}	10	Mo^{99}	10	Tc^{96}	1
$Ba^{140} + La^{140}$	1	Na^{22}	10	Tc^{99}	1
Be^7	50	Na^{24}	10	Te^{127}	10
C^{14}	50	Nb^{95}	10	Te^{129}	1
Ca^{45}	10	Ni^{59}	1	Th (natural)	50
$Cd^{109} + Ag^{109}$	10	Ni^{63}	1	Tl^{204}	50
$Ce^{144} + Pr^{144}$	1	P^{32}	10	Tritium, see H^3	250
Cl^{36}	1	$Pd^{103} + Rh^{103}$	50	U (natural)	50
Co^{60}	1	Pd^{109}	10	U^{233}	1
Cr^{51}	50	Pm^{147}	10	U^{234}-U^{235}	50
$Cs^{137} + Ba^{137}$	1	Po^{210}	0.1	V^{48}	1
Cu^{64}	50	Pr^{143}	10	W^{185}	10
Eu^{154}	1	Pu^{239}	1	Y^{90}	1
F^{18}	50	Ra^{226}	0.1	Y^{91}	1
Fe^{55}	50	Rb^{86}	10	Zn^{65}	10
Fe^{59}	1	Re^{186}	10	Unidentified radioactive materials or any of the above in unknown mixtures	0.1
Ga^{72}	10	Rh^{105}	10		
Ge^{71}	50	$Ru^{106} + Rh^{106}$	1		
H^3(HTO or H_2^3O)	250	S^{35}	50		
I^{131}	10	Sb^{124}	1		
In^{114}	1	Sc^{46}	1		
Ir^{192}	10	Sm^{153}	10		

Table 27.10 (After Falk [63].)

Radioisotope	Half-life	Maximum permissible concentration in H_2O, $\mu c/cm^3$	Equivalent concentration, gm/cm^3
Mn^{56}	2.6 hr	3×10^{-3}	1.4×10^{-16}
Cu^{64}	12.8 hr	5×10^{-3}	1.3×10^{-15}
Fe^{59}	45 days	10^{-4}	2.0×10^{-14}
Zr^{95}	65 days	6×10^{-4}	2.8×10^{-14}
Cs^{137}	33 yr	2×10^{-3}	2.6×10^{-11}
Sr^{90}	19.9 yr	8×10^{-7}	4.0×10^{-15}
Unknown mixture		10^{-7}	10^{-15} to 10^{-16}

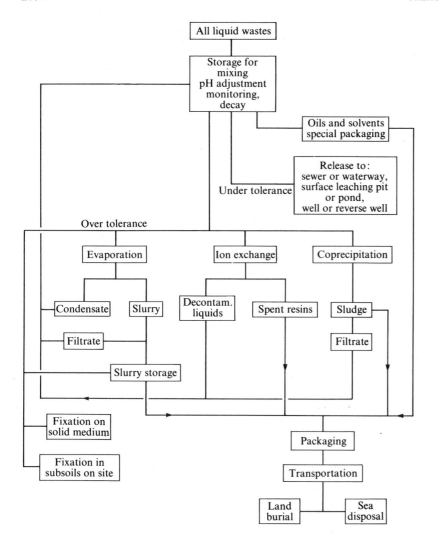

Fig. 27.5 Stages in processing liquid radioactive waste.

Coprecipitation. This is the precipitation of an otherwise soluble (because of its low concentration) material along with an insoluble precipitate. The insoluble precipitate acts as a "scavenger" or "carrier" to coprecipitate radioactive ions from solution. Usually the pH is adjusted to an alkaline condition prior to precipitation to form metal hydroxide flocs, which aid in the scavenging process. After the supernatant settles, the sludge is removed, packaged for shipment, and disposed of by burial. Further concentration, however, may be effected by drying the sludge prior to burial. Coprecipitation is known to yield decontamination factors of 200 to 1000, which is less than that realized by evaporation; however, the process is much less expensive.

Evaporation. This is at present the most widely used concentration process for water solutions containing a mixture of wastes of low-level activity. The wastes are boiled, the resultant vapor is condensed, and the condensate is subsequently released to the sewer, provided there is no significant radioactivity. The high-activity concentrate is usually a sludge that solidifies upon cooling and is then transferred to polyethylene-lined 55-gallon drums for ultimate burial. Foaming and priming with resultant carry-over occur and present operational difficulties. Although evaporation is expensive, it usually is very effective and provides decontamination factors of 100,000 to 1,000,000.

Ion exchange. In this process the radioactive ions in the waste are exchanged for stable ions of the exchanger. Cation exchange is the principal ion-exchange process used to concentrate radioactive waste. Both synthetic resins and Montmorillonite clay serve as exchangers. This method is customarily reserved for small volumes of solutions containing low concentrations of solids and exhibiting low levels of radioactivity. With these wastes, the exchangers may be used for long periods before regeneration becomes necessary.

Eliassen [1964, 147] reports that high-level aluminum-bearing fission-product wastes can be incorporated into ceramic glazes (glasses) by the addition of silica and fluxes to calcined wastes and firing the mixture at 1450°C. A suggested flowsheet is presented in Fig. 27.6.

Treatment of solid radioactive wastes. Solid wastes may be disposed of by burial, by incineration if combustible, or by remelting if metallic. Typical stages in processing solid wastes are presented in Fig. 27.7.

Burial. The AEC operates several burial areas in this country. In selecting these sites, the AEC has been guided by one imperative consideration, namely, to reduce environmental hazards below maximum tolerance levels. Constant surveillance should be possible, and the land should be forbidden to the public for generations to come. Burial grounds, as established by the AEC, should: (1) be not less than 10 acres in area; (2) be accessible; (3) have greater than 15, and preferably greater than 20, feet of unconsolidated sedimentary overburden on the bedrock; (4) the overburden should be sufficiently coherent that the vertical, or nearly vertical, walls of a burial trench will stand up for short periods of time; (5) areas should not be located directly upstream from a ground-water course, from existing or potential plant sites, or nearby populated areas.

Radioactive wastes are usually buried in narrow,

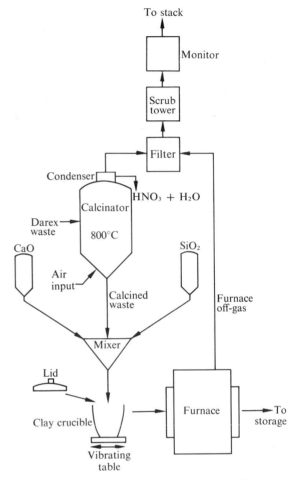

Fig. 27.6 Suggested flow sheet for fixation of high-level radioactive waste in glass. (After Eliassen [145].)

deep trenches excavated by means of a back hoe and backfilled with several feet of earth cover compacted by bulldozers. Burial in containers is, as one would expect, expensive, and one must also consider the cost of excavating the trenches. At present, a major portion of the disposal problem is connected with the ultimate disposal of low-volume, high-level liquid wastes resulting from the chemical processing of spent reactor-fuel elements. Since the effective life of the fission-product mixtures making up these wastes is on the order of several hundred years, and since the future portends great volumes of wastes, the present

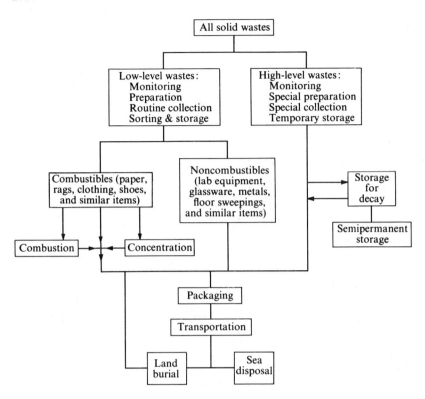

Fig. 27.7 Stages in processing solid radioactive waste.

practice of storing the wastes in tanks cannot be considered a permanent solution to the problem. Rich [168] predicts that for at least 100 years mined-out salt mines will provide enough storage space for anticipated waste production. The states of New York, Michigan, Ohio, and Kansas are ideally suited to this type of disposal. Rich and Rodger [169] conclude that high-level wastes can be concentrated and converted into solids at a cost of no more than 10 cents per gallon. Containers can be fabricated (probably) out of $\frac{1}{8}$-inch aluminum bronze at a cost of $50 per cubic foot of capacity. The cost of storage in mined-out salt deposits is estimated at $1 per square foot of floor space. Rich concludes that if one could succeed in developing an economical method of concentrating and fixing the radioactive waste in a nonleachable solid, a system of isolation would thus be available for the ultimate, safe, and economical disposal of high-level wastes.

According to recent reports, storage of radioactive wastes in glass appears to be an effective, safe, and inexpensive method of atomic-waste disposal. The British Atomic Energy Authority at Harwell has developed a method by which radioactive waste materials may be converted into insoluble glasslike particles. A mixture of silica and borax is added to a nitric-acid solution of the concentrated liquid wastes. The solution is then treated until the liquid evaporates, and the remainder becomes red hot, sinters, and melts. On cooling, the meltage solidifies to a glass of predetermined composition, containing from 20 to 30 per cent of waste oxides.

Incineration. The burning of combustible wastes has not been adopted on any large scale, but many experiments have been carried out. The by-products of active-waste incineration (stack gases and ashes) are also radioactive. The problem of radioactive gases will probably become more acute as atomic-energy installations increase. This was brought home dramatically when an overheated nuclear reactor at Windscale (Cumberland, England) spewed radioactive iodine[131] over an area of 200 square miles on October 10, 1957. As a consequence, all milk produced in the area was declared unsafe, barred from shipment, and dumped.

Filters, electrostatic precipitators, scrubbers, settling chambers, inertial collectors, or wet collectors can be used in the stacks to reduce or eliminate air contamination. Persons handling radioactive ashes must be careful to avoid inhalation or direct exposure to concentrated radioactive material. Volume reduction by incineration may exceed a ratio of 50:1, but the problem of contamination control, as well as high initial and operating costs, discourage the use of this method of solid-waste treatment.

Remelting. Contaminated iron and steel remelted in an induction furnace produces a slag which carries off a major portion of uranium originating from the processing of natural uranium ores. The content of radioactive uranium in the melted steel is reduced by a factor of 40 to 1.

Treatment of gaseous radioactive wastes. Radioactive atmospheric contaminants can be divided into two main categories, gases and particulate matter. These wastes result mainly from mining and ore processing, and also from incineration of contaminated combustibles. The principal method of controlling such radioactive materials at source has been the application of local exhaust ventilation. The principles of local exhaust-hood design apply equally well to the handling of radioactive contaminants. Features of ventilation for radioactive laboratories (Silverman [15]) which may differ from those governing ventilation in general are:

1. Sufficient air supply must be provided to equal that exhausted by the hoods.
2. No recirculation of air shall be permitted in ventilating radioactive laboratories.
3. It is essential to maintain the laboratories in which active materials are handled at a slight negative pressure with respect to adjacent nonactive chemical or physical laboratories, corridors, or offices.
4. Supply air should always be taken from the outside or from uncontaminated areas.
5. Adequate by-passes, crossovers, and auxiliary equipment should be installed to provide exhaust-air flow in the event of shutdown or power failures.
6. For highly contaminated working areas, a parallel system for stand-by or emergency operation is essential in case of power or exhauster failure.
7. Converging velocities from exhaust ducts should range from 2000 ft/min for gases to 5000 ft/min for particulates of heavy metals.
8. Air- and gas-cleaning equipment should meet design decontamination factors and be readily accessible or removable without exposing maintenance personnel to undue radiation danger. The equipment should also be provided with efficient monitoring devices, as well as continuous resistance- or pressure-loss indicators or recorders.

From an economic standpoint, the disposal of radioactive gases should, wherever permissible, utilize the atmospheric dispersion and dilution obtaining for tall stacks. Other procedures which are sometimes used are: (1) containment by ventilating the gases to a detention chamber, where they are stored for an adequate period to permit decay before release to the atmosphere; (2) scrubbing and adsorption at normal or extremely subnormal temperatures; (3) adsorption on special materials; (4) reaction with solid materials; and (5) combustion and special reactions.

Gases created by incineration or combustion of solid wastes are not a significant problem since these processes do not create large amounts of radioactive gases. Where gases have been involved, and dilution by the atmosphere is not adequate, ethanolamine scrubbing towers for C^{14} or S^{35} have been used. For the removal of particulates created by nuclear-energy operations, inertial collectors, high-efficiency filters, or electrostatic precipitators are used.

27.5 Cost of Radioactive-Waste Treatment

Eliassen *et al.* [15] present a breakdown of costs for radioactive waste treatment in various locations:

Large-diameter cyclone	$0.10–0.25 per cfm
Inertial scrubbers (power-driven)	$0.15–0.25 per cfm
High-efficiency cellulose asbestos filters	$0.04–0.06 per cfm
Single-stage electrostatic precipitator	$0.50–2.00 per cfm

Cost for off-site disposal of solid wastes from Rocky Flats, Colo, facility, April 1954–Sept. 1955	$2.95 per hundredweight $1.02 per ft^3 $0.08 per ton-mile* (750 mi)
Estimated cost of burial of solid radioactive waste, National Reactor Testing Station, 1955	$9.11 per cubic yard
Estimated cost (capital) for burial of concentrated liquid wastes in trenches	$0.37–2.00 per gallon
Total evaporation costs (Argonne National Laboratory)	$0.12 per gallon
Total evaporation costs (Knolls Atomic Power Laboratory)	$0.138 per gallon
Total annual operating cost of Los Alamos waste-treatment plant	$5.86 per 1000 gallons
Initial capital cost of Los Alamos waste-treatment plant, plus laboratory facilities	$350,000

References: Radioactive Wastes

1. Amphlett, C. B., *Treatment and Disposal of Radioactive Wastes*, Pergamon Press, New York (1961), p. 289.
2. Atkins, C. H., "Treatment of liquid radioactive wastes," *J. Sanit. Eng. Div. Am. Soc. Civil Engrs.* **85**, 61 (1959).
3. "Atomic Energy Commission, waste disposal symposium," *Nucleonics* **4**, 9 (1949).
4. "Atomic Energy Program, research," *Sewage Ind. Wastes* **23**, 1059 (1951).
5. "Atomic wastes disposal adds to biological hazards of radiation," *Wastes Eng.* **27**, 374 (1956).
6. Ayres, J. A., "Treatment of radioactive waste by ion exchange," *Ind. Eng. Chem.* **43**, 1526 (1951).
7. Ayres, J. A., "Ion exchange for waste treatment," *Sewage Ind. Wastes* **23**, 1217 (1951).
8. Ayres, J. A., "The fuel situation," *Sci. Am.* **195**, 43 (1956).
9. Barton, G. B., et al., "Chemical processing wastes—recovering fission products," *Ind. Eng. Chem.* **50**, 213 (1958).
10. Beadle, G. W., "Ionizing radiation and the citizen," *Sci. Am.* **201**, 219 (1959).
11. Beers, N. R., "Discussion," *Sewage Ind. Wastes* **23**, 703 (1951).
12. Bidwell, R. W., "A biological investigation of the fate of strontium in a diffused filter bed," U.K. Atomic Energy Commission, *IGR*-658, 1957.
13. Biladeau, A. L., "Trickling filter sewage plant built to remove radioactive wastes," *Civil Eng.* **23**, 546 (1953).
14. Biladeau, A. L., "Trickling filter for waste treatment," *Sewage Ind. Wastes* **25**, 1482 (1953).
15. Blatz, H. (ed.), *Radiation Hygiene Handbook*, McGraw-Hill Book Co., New York (1959).
16. Blew, J. B., "Disposal of radioactive wastes by non-Atomic-Energy-Commission users," *Nucleonics* **5**, 22 (1949).
17. Blomeke, J. O., E. D. Arnold, and A. K. Gresky, "Characteristics of reactor fuel process wastes," Nuclear Engineering and Science Conference, Preprint no. 44, Session XVI, March 1958.
18. Blomeke, J. O., E. D. Arnold, and A. K. Gresky, "Characteristics of reactor fuel process wastes," *J. Sanit. Eng. Div. Am. Soc. Civil Engrs.* **85**, 1 (1959).
19. Blomeke, J. O., and M. F. Todd, "U^{235} fission product production as a function of thermal neutron flux irradiation time, and decay time," U.S. Atomic Energy Commission, *ORNL*-2127, 19 August 1957.
20. Boegly, W. J., R. L. Bradshaw, F. M. Empson, and F. L. Parker, "Disposal for radioactive waste in natural salt-field experiments," in Proceedings of 15th Industrial Waste Conference, Purdue University, May 1960, p. 577.
21. Brockett, T. W., and O. R. Placak, "Removal of radioisotopes from waste solution by soils—soil studies with Conosauga shale," in Proceedings of 8th Industrial Waste Conference, Purdue University, May 1953.
22. Browder, F. N., "Liquid waste disposal at Oak Ridge," *Ind. Eng. Chem.* **43**, 1502 (1951).
23. Browder, F. N., "Oak Ridge National Laboratory, liquid waste disposal," *Sewage Ind. Wastes* **23**, 1463 (1951).
24. Bruce, F. R., "The Thorex process," U.S. Atomic Energy Commission, *TID*-7534, May 1957, p. 180.
25. Burns, R. H., "Disposal of industrial atomic waste products," *Soc. Chem. Ind.* **33**, 143 (1956).
26. Burris, L., Jr., and I. G. Dillon, "Estimation of fission product spectra in discharged fuel from fast reactors,"

---

*One ton of solid waste carried one mile for disposal.

U.S. Atomic Energy Commission, *ANL*-5742, July 1957.
27. Biershcenk, W. H., "Hydrological aspects of radioactive waste disposal," *J. Sanit. Eng. Div. Am. Soc. Civil Engrs.* **84**, 1835 (1958).
28. Butrico, F., "Massachusetts General Hospital, disposal study," *Sewage Ind. Wastes* **23**, 123 (1951).
29. Carter, M. W., "Iodine, radioactive, removal by trickling filter," *Sewage Ind. Wastes* **25**, 560 (1953).
30. Chadwick, R. C., "Disposition of airborne radioactive vapor," *Nucleonics* **11**, 22 (1953).
31. Chamberlain, A., "Behavior of radioactive iodine and strontium in food," in Proceedings of Conference on Peaceful Uses of Atomic Energy, 1956, p. 360.
32. Christenson, C. W., "Removal of plutonium from laboratory wastes," *Ind. Eng. Chem.* **43**, 1509 (1951).
33. Christenson, C. W., "Plutonium, removal from laboratory wastes," *Sewage Ind. Wastes* **23**, 1463 (1951).
34. Christenson, C. W., et al., "Radioactive wastes disposal," *Sewage Ind. Wastes* **23**, 861 (1951).
35. Christenson, C. W., et al., "Laboratory studies on removal of plutonium from laundry wastes," *Ind. Eng. Chem.* **43**, 1516 (1951).
36. Coats, H. B., "Recovery of spent fuel is important," *Ind. Wastes* **2**, 49 (1957).
37. Coleman, R. D., and L. Silverman,"Control of radioactive airborne wastes," in Proceedings of 8th Industrial Waste Conference, Purdue University, May 1953.
38. U.S. Atomic Energy Commission, "Concentration of Purex waste," Progress Report on Waste Processing Development Project, *BNL*-266, December 1953.
39. Cowan, F. P., "Sensitivity of the evaporation method of liquid waste monitoring," *Nucleonics* **7**, 39 (1950).
40. Cowser, K. E., and R. J. Morton, "Radioactive contaminant removal from waste water, evaluation of performance," *J. Sanit. Eng. Div., Am. Soc. Civil Engrs.* **85**, 55 (1959).
41. Culbreath, M. C., "Radioactive contaminant removal from waste water, engineering design features," *J. Sanit. Eng. Div. Am. Soc. Civil Engrs.* **85**, 41 (1959).
42. Culler, F. L., "Notes on fission product wastes from proposed power reactors," U.S. Atomic Energy Commission, Report no. CF 55-4-25, Oak Ridge National Laboratory, 1955.
43. Davey, H. G., "Some problems in the maintenance of a nuclear reactor," *J. Brit. Nucl. Energy Conf.* **1**, 174 (1956).
44. Davis, J. J., "Accumulation of radioactive substances in aquatic forms," in Proceedings of Conference on Peaceful Uses of Atomic Energy, 1956, p. 280.
45. "Disposal," *Sewage Ind. Wastes* **21**, 1106 (1949).
46. "Disposal: Atomic Energy Commission Interim Recommendations on the Disposal of Radioactive Wastes by non-A.E.C. Users," *Sewage Ind. Wastes* **22**, 574 (1950).
47. "Disposal problems," *Sewage Ind. Wastes* **21**, 603 (1949).
48. Dunn, C. G., "Treatment of water and sewage by ionizing radiation," *Sewage Ind. Wastes* **25**, 1277 (1953).
49. Dunster, H. J., "Preliminary estimate of safe daily discharge of radioactive effluent," in Proceedings of International Conference on Peaceful Uses of Atomic Energy, Geneva, 1956, p. 419.
50. Dunster, H. J., "The disposal of radioactive liquid wastes into coastal waters," in Proceedings of 2nd U.N. International Conference on Peacetime Uses of Atomic Energy, Vol. 18, 1958, p. 390.
51. Eaton, S. E., and R. J. Bowen, "Decontaminable surfaces for millicuric level laboratories," *Nucleonics* **8**, 27 (1951).
52. Edelmann, A., "Some possible applications of nuclear energy to problems of disposal of industrial waste," in Proceedings of 13th Industrial Waste Conference, Purdue University, May 1958.
53. Eden, G. E., G. H. I. Elkins, and G. A. Travesdale, "Removal of radioactive substances from water by biological treatment processes," *Atomics* **5**, May 1954; Water Poll. Res. Lab. Reprint no. 249, Stevenage, England.
54. Eisenbud, M., "The Atomic Energy Commission fallout monitoring network," American Institute of Chemical Engineers, Preprint no. 192, 12 Dec. 1955.
55. Eisenbud, M., and Harley, "Radioactive fallout in the U.S.," *Science* **121**, 677 (1955).
56. Eliassen, R., "Atomic wastes disposal: an international problem," Paper presented at Conference on Biological Waste Treatment, Manhattan College, April 1960.
57. Eliassen, R., W. J. Kaufman, et al., "Studies on radioisotopes, removal by water treatment processes," *J. Am. Water Works Assoc.* **43**, 615 (1951).
58. Eliassen, R., and R. A. Lauderdale, "Removal of radioactive fallout from water by municipal water

treatment plants," U.S. Atomic Energy Commission, *TID*-7517, October 1956, p. 19.
59. Eliassen, R., and R. A. Lauderdale, "Liquid and solid waste disposal," in *Handbook of Radiation Hygiene*, McGraw-Hill Book Co., New York (1958).
60. "Estimated future power requirements for the United States," U.S. Federal Power Commission, 1955.
61. Etzel, J. E., "Atomic wastes problem," *Ind. Wastes* **1**, 148 (1956).
62. Evans, R. D., "Determination of the thorium content of air and its bearings on lung cancer hazards in industry," *J. Ind. Hyg.* **22**, 89 (1957).
63. Falk, C. F., "Radioactive liquid waste disposal from the Dresden nuclear power station," Nuclear Engineering and Science Conference, Reprint no. 102, Session XVI, March 1958.
64. U.S. Atomic Energy Commission, *Federal Register*, May 1957, p. 3389.
65. Finch, J., "Report from abroad, atomic energy wastes," *Water Sewage Works* **102**, 86 (1955).
66. Foster, R. F., and J. J. Davis, "Accumulation of radioactive substances in aquatic forms," in Proceedings of International Conference on Peaceful Uses of Atomic Energy, Geneva, Vol. 13, Paper no. 280, August 1955.
67. Gifford, F., "Meteorological parameters in waste disposal," U.S. Atomic Energy Commission, *TID*-7517, October 1956, p. 57.
68. Gloyna, E. F., "Radioactive contaminated laundry waste treatment," *Sewage Ind. Wastes* **26**, 777 and 869 (1954).
69. Goodall, R. L., "The protection of workers against ionizing radiation," *Occupational Safety Health* **5**, 61 (1955).
70. Goodman, E. I., "Selection and handling of radioactive tracer for studying sewer distribution," *Ind. Eng. Chem.* **50**, 210 (1958).
71. Gorman, A. E., "Nuclear fission operations and the sanitary engineer," *Sewage Works J.* **21**, 63 (1949).
72. Gorman, A. E., "Sanitary engineering and waste disposal problems in the atomic energy industry," in Proceedings of 6th Industrial Waste Conference, Purdue University, February 1951.
73. Gorman, A. E., "Disposal problems review," *Sewage Ind. Wastes* **24**, 808 (1952).
74. Gorman, A. E., "Disposal of wastes from atomic energy industry," *Wastes Eng.* **24**, 528 (1953).
75. Gorman, A. E., "Disposal of atomic energy wastes," *Ind. Eng. Chem.* **45**, 2672 (1953).
76. Gorman, A. E., "Waste disposal as related to site selection," *Proc. Am. Inst. Chem. Eng.*, 12 December 1955, preprint 3, Cleveland, Ohio.
77. Gorman, A. E., "Sanitary engineering objectives in the atomic energy industry," U.S. Atomic Energy Commission, *TID*-7517, October 1956, p. 4.
78. Gorman, A. E., and A. Nolman, "Nuclear fission operations and the sanitary engineer," *Sewage Works J.* **21**, 63 (1949).
79. Grune, W. N., "Radiological laboratory," *Mod. Sanit.*, May 1950.
80. Grune, W. N., "Studies of the effect of radioactive phosphorus on biochemical oxidation of sewage," *Sewage Ind. Wastes* **23**, 141 (1951).
81. Grune, W. N., "Sanitary engineering laboratory design to permit use of radioactive isotopes," *Nucleonics* **9**, 59 (1951).
82. Grune, W. N., "Radioactive effects on the BOD of sewage," *Sewage Ind. Wastes* **25**, 882 (1953).
83. Hanson, W. C., "Radioactivity in terrestrial animals near an atomic energy site," in Proceedings of International Conference on Peaceful Uses of Atomic Energy, Geneva, 1956, p. 281.
84. Harmeson, R. H., "Effect of radioactive substance on sludge digestion," *Univ. Illinois, Eng. Exp. Sta. Bull.* **441**, January 1957.
85. Harward, E. D., "Accident preparedness in reactor waste treatment," *J. Sanit. Eng. Div. Am. Soc. Civil Engrs.* **86**, 1 (1960).
86. Hatch, L. P., "Processes for high-level radioactive waste disposal," in *United States National Bureau of Standards Handbook*, Washington, D.C. (1954), p. 58.
87. Hatch, L. P., J. J. Martin, and W. S. Ginell, "Ultimate disposal of radioactive wastes," U.S. Atomic Energy Commission, *BNL*-1781, February 1954.
88. Hatch, L. P., W. H. Ragen, B. Manowitz, and F. Hitman, "Processes for high-level active waste disposal," in Proceedings of International Conference on Peaceful Uses of Atomic Energy, Geneva, Paper no. 553, August 1955.
89. Hayner, J. H., "Atomic energy industry," *Ind. Eng. Chem.* **44**, 472 (1952).
90. Hayner, J. H., "Waste treatment status," *Sewage Ind. Wastes* **24**, 1442 (1952).
91. Hembree, H. G., "Atomic Energy Commission Safety research program reactor design factors," *Ind. Eng. Chem.* **51**, 82A (1959).

92. Hermann, E. R., and E. R. Gloyna, "Summary of investigations on the removal of radioisotopes from waste water by oxidation ponds," U.S. Atomic Energy Commission, *TID*-7517, October 1956, p. 27.
93. Heroy, W. B., "Disposal of radioactive wastes in salt cavities," Report prepared for National Research Council, 20 July 1956.
94. Herrington, A. C., "Economic evaluation of permanent disposal of radioactive wastes," *Nucleonics* **11**, 34 (1953).
95. "Hospital wastes, studies," *Sewage Ind. Wastes* **23**, 1122 (1951).
96. Ingersoll, L. R., O. J. Zobel, and A. C. Ingersoll, *Heat Conduction*, University of Wisconsin Press, Madison (1954), p. 245.
97. Irish, E. R., and W. H. Reas, "The Purex process—a solvent extraction reprocessing method for irradiated uranium," Symposium on the Reprocessing of Irradiated Fuel, Brussels, U.S. Atomic Energy Commission, *TID*-7534, May 1957, p. 83.
98. Irwin, A. E., *Salt, an Industrial Potential for Kansas*, University of Kansas Research Foundation, Lawrence (1951).
99. Jackson, N. P., "Atomic energy and the American economy," *J. Am. Water Works Assoc.* **47**, 1139 (1955).
100. Jensen, J. H., "National Committee on Radiation Protection," *Ind. Eng. Chem.* **43**, 1499 (1951).
101. Jenson, J. H., "Radiation protection, National Committee activities," *Sewage Ind. Wastes* **23**, 1463 (1951).
102. Jonke, A. A., "A fluidized-bed technique for treatment of aqueous nuclear wastes by calcination to oxides," U.S. Atomic Energy Commission, *TID*-7517, October 1956, p. 374.
103. Joseph, A. B., "The status of land disposal of atomic reactor wastes," American Institute of Chemical Engineers, Reprint no 185, 12 December 1955.
104. Kahn, B., and S. A. Reynolds, *Determination of Radionuclides in Low Concentrations of Water*; reprinted from *J. Am. Water Works Assoc.* **50**, 613 (1958).
105. Kennedy, W. J., "Disposal of radioactive wastes," *Ind. Wastes* **4**, 44 (1959).
106. Kenny, A. W., "The behavior of radioisotopes in sewage treatment," *Proc. Inst. Civil Engrs. (London)* **7**, 326 (1957).
107. Kenny, A. W., "Radioactive discharge to sewers and rivers," Paper presented at Annual Conference of Institute of Sewage Purification, June 1957.
108. Kittrell, F. W., "Radioactive waste disposal to public sewers," *Sewage Ind. Wastes* **24**, 985 (1952).
109. Koltzbach, R. J., "Problems in radioactive wastes disposal," *Ind. Wastes* **5**, 21 (1960).
110. Krieger, H. L., A. S. Goldin, and C. S. Straub, "Laboratory studies of removal and segregation of fission products from reactor wastes," *J. Water Pollution Control Federation* **32**, 495 (1960).
111. Krumholz, L. A., "A summary of findings of the ecology of White Oak Creek, Tenn.," U.S. Atomic Energy Commission, *ORO*-132, 1954.
112. Kunin, R. P., "Ion exchange in the atomic energy program," *Ind. Eng. Chem.* **48**, 30A (1956).
113. Lacy, W. J., "Water decontamination with clay slurry," *Sewage Ind. Wastes* **26**, 1305 (1954).
114. Lackey, J. B., "Turbidity effects in natural waters in relation to organisms and the uptake of radioisotopes," in Proceedings of 6th Ontario Industrial Waste Conference, June 1959, p.37.
115. Langford, J. M., "Here's a new waste problem—the rocket propulsion industry," *Ind. Wastes* **2**, 42 (1957).
116. La Pointe, J. R., W. J. Hahn, and E. D. Harward, Jr., "Waste treatment at the Shippingport reactor," *J. Sanit. Eng. Div. Am. Soc. Civil Engrs.* **86**, 129 (1960).
117. Lauderdale, R. A., "Decontamination of small volume of radioactive water," *Nucleonics* **8**, 21 (1951).
118. Lauderdale, R. A., "Treatment of radioactive water by phosphate precipitation," *Ind. Eng. Chem.* **43**, 1538 (1951).
119. Lauderdale, R. A., and A. H. Emmons, "A method of decontaminating small volumes of radioactive wastes," *Sewage Ind. Wastes* **24**, 1322 (1952).
120. Lawroski, S., and M. Levenson, "Redox process—a solvent extraction reprocessing method for irradiated uranium," U.S. Atomic Energy Commission, *TID*-7534, May 1957, p. 45.
121. Lewis, C. J., "Treatment of uranium mill wastes," in Proceedings of 14th Industrial Waste Conference, Purdue University, May 1959.
122. Lewis, J. G., "Effects produced from gamma ray," *Ind. Eng. Chem.* **49**, 1197 (1957).
123. Love, S. K., "Natural radioactivity of water," *Sewage Ind. Wastes* **23**, 1217 (1951).
124. Lieberman, J. A., "Disposal of radioactive wastes—A growing problem," *Civil Eng.* **25**, 44 (1955).
125. Lieberman, J. A., "Engineering aspects of the disposal of radioactive wastes from peacetime ap-

plications of nuclear technology," *Am. J. Public Health* **47** (1957).
126. Lieberman, J. A., and A. E. Gorman, "Treatment and disposal of atomic energy industry wastes," *Proc. Am. Soc. Civil Engrs.* **80**, Separate no. 422 (1954).
127. Lackey, J. B., "Microscopy in radioactive waste treatment by dilution," Paper presented at a National Nuclear Institute Conference, Atlanta, 12 April 1957.
128. Lundgren, D., "Effect of radiation on cells and their bacterial enzymes," Paper presented at Syracuse University Civil Engineering Seminar, 1960.
129. McBay, T. N., R. L. Hamner, and M. P. Haydon, "Fixation of radioactive wastes in clay flux mixes," U.S. Atomic Energy Commission, *TID*-7517, October 1956, p. 335.
130. McCullough, G. E., "Concentration of radioactive liquid waste by evaporation," *Ind. Eng. Chem.* **43**, 1505 (1951).
131. McCullough, G. E., "Concentration by evaporation," *Sewage Ind. Wastes* **23**, 1463 (1951).
132. McKay, H. A., "Safety criteria in radioactive water monitoring," *Nucleonics* **5**, 12 (1949).
133. McKay, H. A. C., and G. N. Walton, "Water quality criteria, monitoring," *Sewage Ind. Wastes* **22**, 144 (1950).
134. MacIntosh, A. D., "The radiochemical laboratory, an architectural approach," *Nucleonics* **5**, 48 (1949).
135. McVay, T. J., "High-level radioactive waste disposal problem," *Proc. Am. Inst. Chem. Engrs.*, 12 December 1955.
136. Marley, W. G., "Problems of waste disposal in the wide-scale use of radioactive isotopes," UNESCO Conference, 1957.
137. Mawson, C. A., "Report of waste disposal systems at Chalk River," U.S. Atomic Energy Commission, *CRB*-658, 1956.
138. "Maximum permissible amounts of radioisotopes in the human body, and maximum permissible concentrations in air and water," in *United States National Bureau of Standards Handbook*, no. 52, U.S. Government Printing Office, Washington, D.C. (1953).
139. "Maximum permissible radiation exposures to man," Insert to accompany *United States National Bureau of Standards Handbook*, no. 59, U.S. Government Printing Office, Washington, D.C. (1957).
140. Medin, A. L., "Nuclear reactor wastes from small power reactors," *Ind. Wastes* **1**, 278 (1956).
141. Medin, A. L., "Army package power reactor water treatment and waste disposal," *Ind. Eng. Chem.* **50**, 989 (1958).
142. "Meteorology and atomic energy," U.S. Atomic Energy Commission, *AECU*, 3066, 1955.
143. Miller, H. S., et al., "Survey of radioactive waste disposal practices," *Nucleonics* **12**, 68 (1954).
144. "Monitoring, methods and instruments," *Sewage Ind. Wastes* **24**, 1446 (1952).
145. Morgan, G. W., "Surveying and monitoring of radiation from radioisotopes," *Sewage Ind. Wastes* **22**, 144 (1950).
146. Mortland, H. S., "General view of hazards caused by ingestion of luminous paint," *J. Am. Med. Assoc.* **92**, 466 (1929).
147. Newell, J. F., and C. W. Christenson, "Radioactive waste disposal," *Sewage Ind. Wastes* **23**, 861 (1951).
148. Newell, J. F., et al., "Plutonium, removal from laundry wastes," *Sewage Ind. Wastes* **23**, 1464 (1951).
149. Odum, E. P., "Consideration of the total environment in power reactor waste disposal," in Proceedings of International Conference on Peaceful Uses of Atomic Energy, Geneva, 1955.
150. Ophel, I. L., "The disposal of radioactive wastes," in Proceedings of 2nd Ontario Industrial Waste Conference, June 1955, p. 83.
151. Ophel, I. L., "Recent developments in the handling of radioactive wastes," in Proceedings of 7th Ontario Industrial Waste Conference, June 1960, p. 17.
152. Palange, R. C., "Radioactivity as a factor in stream pollution," *Proc. Am. Inst. Chem. Engrs.*, 12 December 1955, p. 190.
153. Parker, H. M., "Health protection in chemical processing plants," in Proceedings of International Conference on Peaceful Uses of Atomic Energy, Geneva, 1956, p. 354.
154. Parker, H. M., "Radioactive waste management operations at the Hanford plant," Hearings of Joint Committee on Atomic Energy, 1959, Vol. 1, p. 394.
155. Parkhurst, J. D., "Handling radioactive wastes in sewers," *Sewage Ind. Wastes* **22**, 1073 (1950).
156. Patrick, W. A., "Use of artificial clays for removal and fixation of radioactive nuclides," U.S. Atomic Energy Commission, *TID*-7517, October 1956, p. 368.
157. Pearce, P. W., and J. F. Honstead, "Radioactive waste disposal at Hanford," in Proceedings of

13th Industrial Waste Conference, Purdue University, May 1958.
158. Phalen, W. C., "Salt resources of the United States," U.S. Geological Survey, Bulletin no. 669, 1919.
159. Powell, C. C., and H. L. Andrews, "Sea disposal," *Sewage Ind. Wastes* **25**, 313 (1953).
160. "Progress report, Johns Hopkins University, Sanitary Engineering Department," *Sewage Ind. Wastes* **23**, 122 (1951).
161. International Atomic Energy Agency, *Provisional Report on Radioactive Wastes Disposal to the Sea*, Vienna (1959).
162. "Radioisotopes, bibliography," *Sewage Ind. Wastes* **21**, 673 (1949).
163. "Radioisotopes, permissible concentration, handbook," *Sewage Ind. Wastes* **25**, 1367 (1953).
164. Reading, L. M., *et al.*, "Laundry waste, radioactive, activated sludge treatment," *Sewage Ind. Wastes* **25**, 1414 (1953).
165. Renn, C. E., "Ultimate disposal of radioactive reactor wastes in the oceans," U.S. Atomic Energy Commission, *TID*-7517, October 1956, p. 53.
166. Rhame, G. A., "Nuclear fission operators and the waste treatment operator," *Sewage Works J.* **22**, 1404 (1950).
167. Rice, T. R., "Accumulation and exchange of strontium by marine planktonic algae," *Limnol. Oceanog.* **1**, 1 (1956).
168. Rich, C. G., "The disposal of high-level radioactive wastes in salt formations," in Proceedings of 13th Industrial Waste Conference, Purdue University, May 1958, p. 581.
169. Rodger, W. A., "Radioactive wastes—treatment, use, disposal," *Chem. Eng. Progr.* **50**, 263 (1954).
170. Rodger, W. A., and P. Fineman, "A complete waste disposal system for a radio chemical laboratory," *Nucleonics* **9**, 51 (1951).
171. Rodger, W. A., and P. Fineman, "Ultimate disposal of radioactive wastes," Paper presented at the Symposium on Chemical Processing, Brussels, May 1957.
172. Rodger, W. A., P. Fineman, and H. G. Swope, "Disposal of radioactive waste at Argonne National Laboratory," in Proceedings of 8th Industrial Waste Conference, Purdue University, May 1953.
173. Ruchhoft, C. C., "Biological treatment methods, possibilities of disposal of radioactive wastes," *Sewage Ind. Wastes* **21**, 877 (1949).
174. Ruchhoft, C. C., "Disposition of radioactive sources," University of Michigan lecture, 5 February 1951.
175. Ruchhoft, C. C., "Activated sludge from foods for treatment of radioactive wastes," *Ind. Eng. Chem.* **43**, 1520 (1951).
176. Ruchhoft, C. C., "Estimate on the concentration of radioactive iodine in sewage and sludge from hospital wastes," *Nucleonics* **6**, 29 (1951).
177. Ruchhoft, C. C., "Wastes containing radioactive isotopes," *Ind. Eng. Chem.* **44**, 545 (1952).
178. Ruchhoft, C. C., "Activated sludge from foods for treatment of radioactive wastes," *Sewage Ind. Wastes* **25**, 48 (1953).
179. Ruchhoft, C. C., *et al.*, "Los Alamos treatment plant, description," *Sewage Ind. Wastes* **24**, 1445 (1952).
180. Ruchhoft, C. C., *et al.*, "Waste treatment by activated sludge," *Sewage Ind. Wastes* **25**, 1465 (1953).
181. Ruchhoft, C. C., A. E. Gorman, and C. W. Christenson, "Wastes containing radioactive isotopes," *Ind. Eng. Chem.* **44**, 545 (1952).
182. Ruchhoft, C. C., F. I. Norris, and L. R. Setter, "Activated sludge from foods for treatment of radioactive waste," *Ind. Eng. Chem.* **43**, 1520 (1951).
183. Ruchhoft, C. C., and E. R. Setter, "Application of biological methods in treatment of radioactive wastes," *Sewage Ind. Wastes* **25**, 48 (1953).
184. Ryan, J. L., and E. J. Wheelwright, "Recovery and purification of plutonium by anion exchange," *Ind. Eng. Chem.* **51**, 60 (1959).
185. Sackheim, G. J., *Chemical Calculations*, 6th ed., (1956), p. 171.
186. Saddington, K., and W. L. Templeton, *Disposal of Radioactive Waste*, George Newnes, London (1958).
187. Schauer, P. J., "Organic radioactive material, removal from gas," *Sewage Ind. Wastes* **23**, 1216 (1951).
188. Schauer, P. J., "Removal of submicron aerosol particles from moving gas stream," *Ind. Eng. Chem.* **43**, 1532 (1951).
189. Schechter, R. S., and E. F. Gloyna, "Thermal considerations on the storage of radioactive wastes," *Sewage Ind. Wastes* **31**, 1165 (1959).
190. Schmidt, W. C., "Treatment of gaseous effluents of irradiated fuels," U.S. Atomic Energy Commission, *TID*-7534, 1957.
191. Scott, K. G., "Radioactive waste, how will it affect man's economy?" *Nucleonics* **6**, 18 (1950).

192. Scott, K. G., "Disposal, economic effect," *Sewage Ind. Wastes* **22**, 972 (1950).
193. Seligman, H., "Discharge of radioactive waste products in the Irish Sea," in Proceedings of International Conference on Peaceful Uses of Atomic Energy, Part I, 1956, p. 418.
194. Setter, L. K., "Radioactive waste disposal problems as related to water supply," Paper presented at Ohio section of American Water Works Association, 3 November 1949.
195. Setter, L. K., and A. S. Goldin, "Radioactive fallout in surface water," *Ind. Eng. Chem.* **48**, 251 (1956).
196. Shannon, R. L., "Bibliography," *Sewage Ind. Wastes* **23**, 1059 (1951).
197. Shannon, R. L., "Radioactive waste disposal," U.S. Atomic Energy Commission, *TID*-375, August 1950.
198. Simpson, *et al.*, *Geneva Papers UN-A Conference*, 8/P/815 (June 1955).
199. Skrinde, R. T., and C. N. Sawyer, "Application of the Warburg respirometer to industrial waste analysis, special studies on radioactive wastes," in Proceedings of 7th Industrial Waste Conference, Purdue University, May 1952.
200. Smith, A. L., "Safe disposal of radioactive wastes," *Ind. Wastes* **4**, 23 (1959).
201. Sodd, V. J., "Determination of radioactivity in saline waters," *Anal. Chem.* **32**, 25 (1960).
202. Somerville, A., "Waste control at the General Motors Research Isotope Laboratory," in Proceedings of 15th Industrial Waste Conference, Purdue University, May 1960, p. 341.
203. "A staff report, automotive, chemical, and atomic energy," *Ind. Wastes* **4**, 11 (1959).
204. Federal Register of Atomic Energy Commission, *Standards for Protection against Radiation*, Part 20, Title 10, Code Federal Regulations, February 1957.
205. Stephenson, R., *Introduction to Nuclear Engineering*, McGraw-Hill Book Co., New York (1954).
206. Stevenson, C. E., "Solvent extraction processes for enriched uranium," U.S. Atomic Energy Commission, *TID*-7534, May 1957, p. 152.
207. Straub, C. P., "Atomic energy program (Part III)," *Ind. Wastes* **3**, 60 (1958).
208. Straub, C. P., "Atomic energy program," *Ind. Wastes* **3**, 91 (1958).
209. Straub, C. P., *et al.*, "Studies of removal of radioactive contaminants from water," *J. Am. Water Works Assoc.* **43**, 773 (1941).
210. Straub, C. P., "Removal of radioactive wastes from water," *Nucleonics* **10**, 40 (1952).
211. Straub, C. P., "Waste removal from water," *Sewage Ind. Wastes* **25**, 629 (1953).
212. Straub, C. P., "Observations on the removal of radioactive materials from waste solutions," *Sewage Ind. Wastes* **23**, 118 (1955).
213. Straub, C. P., and H. L. Krieger, "Removal of radioisotopes from waste solutions; soil suspension studies," in Proceedings of 8th Industrial Waste Conference, Purdue University, May 1953.
214. Straub, C. P., R. J. Morton, and O. R. Placak, "Oak Ridge reports results on water decontamination study," *Eng. News Record* **147**, 38 (1951).
215. Sutton, O. G., *Micrometeorology*, McGraw-Hill Book Co., New York (1953).
216. Swope, H. G., "Radioactive wastes handling," *Sewage Ind. Wastes* **31**, 1191 (1959).
217. Swope, H. G., and E. Anderson, "Cation exchange removal of radioactivity from waste," *Ind. Eng. Chem.* **47**, 78 (1955).
218. Teletzke, G. H., "The 1957 Nuclear Congress on Radioactive Wastes," *Ind. Wastes* **2**, 62 (1957).
219. Teletzke, G. H., "Radioactive waste," *Ind. Wastes* **3**, 99 (1958).
220. Terrill, J. G., "Fission products from nuclear reactors," *Proc. Am. Soc. Civil Engrs.*, **81**, 643, March 1955.
221. Terrill, J. G., Jr., and M. D. Hollis, "Sanitary engineering and reactor waste disposal," *J. Sanit. Eng. Div. Am. Soc. Civil Engrs.* **83**, 1407 (1947).
222. Thomas, H. A., Jr., "Disposal of low-level liquid radioactive wastes in inland waterways," U.S. Atomic Energy Commission, *TID*-7517, October 1956, p. 457.
223. Thomas, H. A., Jr., "Public health implications of radioactive fallout in water supplies," *Am. J. Public Health* **46**, 1266 (1956).
224. Thurston, W. R., "Summary of Princeton conference on disposal of high-level radioactive waste products in geologic structures," U.S. Atomic Energy Commission, *TID*-7517, October 1956, p. 47.
225. Thurston, W. R., "Summary of Princeton conference on disposal of high-level radioactive waste products in geologic structures," U.S. Atomic Energy Commission, *TID*-7517, October 1956, p. 374.
226. Tompkins, P. C., O. M. Bizzell, and C. P. Watson, "Working surface for radiochemical laboratories," *Ind. Eng. Chem.* **42**, 1475 (1950).

227. Tsivoglou, E. D., et al., "Uranium ore, refinery wastes," *Sewage Ind. Wastes* **30**, 1012 (1958).
228. Tsivoglou, E. D., and W. W. Towne, "Sources and control of radioactive water pollutants," in Proceedings of Pennsylvania Sewage and Industrial Wastes Association, University Park, Pa., August 1956, p. 143.
229. Tsivoglou, E. D., D. M. Harward, and W. M. Ingram, "Stream surveys for radioactive waste control," American Society of Mechanical Engineers, March 1957; A.S.M.E., New York, no. 57–NESC–21.
230. Tsivoglou, E. D., M. Stein, and W. W. Towne, "Control of radioactive pollution of the Animas river," *J. Water Pollution Control Federation* **32**, 262 (1960).
231. Van Kleeck, L. W., "Radiological problems, sewage works aspects," *Sewage Ind. Wastes* **23**, 1311 (1951).
232. Van Rozen, W., and O. Bowles, *Atlas of the World's Resources. Vol. II: The Mineral Resources of the World*, Prentice-Hall, Englewood Cliffs, N.J. (1952).
233. Walters, W. R., D. W. Weiser, and L. J. Marek, "Concentrations of radioactive aqueous wastes," *Ind. Eng. Chem.* **47**, 61 (1955).
234. Warde, J. M., and T. N. McVay, "High-level radioactive waste disposal problems," American Institute of Chemical Engineers, Reprint no. 189, 12 December 1955.
235. Watson, C. D., "Asphalt lining of radiochemical waste storage basins," *Ind. Eng. Chem.* **50**, 87A (1958).
236. Western, F. O., "Problems of radioactive waste," *Nucleonics* **3**, 43 (1948).
237. Western, F. O. "Disposal problems," *Sewage Ind. Wastes* **21**, 1107 (1949).
238. Wilcox, W. G., "Salt production," *Sci. Monthly* **61**, 157 (1950).
239. Wilson, W. L., "'Hot stuff,' big hurdle for atomic power," *Business Week*, July 1952, p. 72.
240. Wilson, W. L., "Radioactive waste disposal in the ocean," in *United States National Bureau of Standards Handbook*, Washington, D.C. (1954), p. 58.
241. Wilson, W. L., "Recommendations of the International Committee on Radiological Protection," *Brit. J. Radiol.*, Supplement 6, 1954.
242. Wilson, W. L., "Atoms and man, radiation hazard," *Ind. Eng. Chem.* **48**, 17A (1956).
243. Wilson, W. L., "Design and construction of handling and treatment systems for radioactive wastes," *Civil Eng.* **4**, 20 (1958).
244. Wilson, W. L., "Atomic power rides out stormy year," *Business Conditions* **35**, 10 (1958).
245. Wilson, W. L., "Regulation of radiation exposure by legislative means," U.S. Department of Commerce, House Bill 61, 1958.
246. Wolman, A., A. E. Gorman, and J. A. Lieberman, "Disposal of radioactive wastes in the U.S. Atomic Energy Program," U.S. Atomic Energy Commission, *WASH*-408, May 1956.
247. Wolman, A., and A. E. Gorman, "Radioactive waste disposal," *Chem. Eng. Progr.* **51**, 471 (1955).

Suggested Additional Reading

Since 1962 many references have been published which are mainly concerned with the design, operation, and waste treatment of nuclear reactors.

1. Abrams, C. S., W. J. Hahn, and J. R. LaPointe, "Performance of the radioactive waste disposal system at the Shippingport Station," *Chem. Eng. Progr.* **58**, 70 (1962).
2. Abrao, A., M. J. Nastasi, and A. A. Laranja, "Application of ion exchange resins to the separation of carrier-free fission products," Instituto de Energia Atomica, Report no. *IEA*-101, Sao Paulo; *Nucl. Sci. Abstr.* **20**, 35587 (1966).
3. Allemann, R. T., "Radiant spray calcination," U.S. Atomic Energy Commission, Report no. *TID*-7646, Division Reactor Development 1963.
4. Allemann, R. T., R. L. Moose, F. P. Roberts, and U. L. Upson, "Hot-cell studies on the solidification of Hanford high-level wastes by radiant heat spray and pot calcination," U.S. Atomic Energy Commission, Report no *HW-SA*-2877, Hanford Atomic Products Operation, January 1963.
5. Allemann, R. T., and B. M. Johnson, "Radiant-heat, spray-calcination process for the solidification of radioactive waste," *Ind. Eng. Chem., Process Design Develop.* **2**, 232 (1963).
6. Amarantos, S., and A. Hatzikakids, "The evaporation of radio-active wastes by solar energy," "Democratis" Nuclear Research Center, Athens, Report no. 2, Final Report no. *NP*-15874, 1965.
7. Amberson, C. B., et al., "Treatment of intermediate and low level radioactive wastes at the NRTS," Paper read at Symposium on Practices in the Treatment of Low- and Intermediate-Level Radioactive

Wastes, International Atomic Energy Agency, Vienna, Paper no. *SM*-71-42, December 1965.
8. Ames, L. L., Jr., "Mineral reaction work at Hartford," U.S. Atomic Energy Commission, Report no. *TID*-7644, Johns Hopkins University, January 1963.
9. Amphlett, C. B., "Treatment and disposal of radioactive wastes," *Roy. Soc. Health J.* **82**, 95 (1962).
10. Andelman, J. B., M. A. Shapiro, and I. M. Yu, "Competitive sorption of heavy metal radionuclides from aqueous solutions onto mineral surfaces," 1966.
11. Anderson, R. P., et al., "Ion exchange removal of Sr 90 from biologic wastes," U.S. Atomic Energy Commission, Report no. *UCD*-472-112, University of California, June 1965.
12. Anokin, U. L., and V. F. Zheverzheeva, "Extraction of fission products from water by foaming," *Gigiena i Sanit.* **3**, 64 (1966); *Nucl. Sci. Abstr.* **20**, 36934 (1966).
13. "Selected literature according to subject field, construction engineering, disposal of radioactive waste material, I and II," *Zentrestelle Atom Energie-Dokumentation, Gemein, Inst. Anorg. Chem. Grenzg.*, Report no. *AED-C*-01-12, Frankfurt am Main, 1965; *Nucl. Sci. Abstr.* **20**, 18153 (1966).
14. "Waste disposal," *Reactor Fuel Process.* **6**, 39 (1963).
15. "Waste disposal," *Reactor Fuel Process.* **6**, 41 (1963).
16. "Waste disposal," *Reactor Fuel Process.* **5**, 52 (1963).
17. "Fission product storage in glass," *Atomics* **16**, 19 (1963).
18. "Waste disposal," *Reactor Fuel Process.* **7**, 54 (1964).
19. "Waste disposal," *Reactor Fuel Process.* **7**, 200 (1964).
20. "Calcination of radioactive wastes," *Brit. Chem. Engr.* **8**, 701 (1963); *Nucl. Sci. Abstr.* **17**, 40418 (1963).
21. "Fission product storage in glass," *Atomics* **16**, 19 (1963).
22. "Separation of Cs^{137} from high-activity radioactive waste," *Bedriff En. Tech.* **18**, 31 (1963); *Nucl. Sci. Abstr.* **17**, 38581 (1963).
23. "Single-stage vertical evaporator engineering materials," U.S. Atomic Energy Commission, Report no. *CAPE*-859, Savannah River Laboratory, E. I. Dupont de Nemours Co., Aiken, S.C.
24. "Sulfur holds radioactive wastes," *Chem. Eng. News* **40**, 60 (1962).
25. Archambault, J., and J. Lemoine, "Underground storage conditions for radioactive wastes," *Colloq. Intern. Retention Migration Ions Radioactifs Sols, Saclay, France, 1962,* 1963, p. 21; *Nucl. Sci. Abstr.* **18**, 17340 (1964).
26. Arden, R. V., and J. Pilot, "Methods for the treatment of atomic wastes," *J. Inst. Public Health Engr.* **62**, 267 (1963); *Nucl. Sci. Abstr.* **18**, 43096 (1964).
27. Armstrong, A. A., Jr., et al., "Department of Chemical Engineering, Progress Report no. 1," U.S. Atomic Energy Commission, Report no. *TID*-19164, South Carolina University, 1963.
28. Arnold, W. D., et al., "Radium removal from uranium mill effluents with inorganic ion exchangers," *Ind. Eng. Chem., Process Design Develop.* **4**, 333 (1965).
29. Arsance, F. C., "Method for disposing of radioactive waste and resultant product," U.S. Patent no. 3093593, 11 June 1963.
30. "Atomic radiation. II: Monitoring, radiation protection, radioactive shipment, waste disposal," RCA Service Co., 1962; *Nucl. Sci. Abstr.* **16**, 15041 (1962).
31. Aurand, K., J. Arndt, and R. Wolter, "Behavior of radionuclides in sewer systems," Binw. F.-Boden Luft Hygiene, Bundesgesundheitsamt, Report, 1966.
32. Backman, G. E., "Control limits for the concentration of radioactive materials in aqueous and gaseous effluents from the plutonium recycle test reactor," *Nucl. Sci. Abstr.* **15**, 14569 (1961).
33. Backman, G. E., "Dispersion of 300 area liquid effluent in the Columbia river," U.S. Atomic Energy Commission, Report no. *HW*-73672, Hanford Atomic Products Operation, May 1962.
34. Baetsle, L., et al., "The use of inorganic ion exchangers in acid medium for the recovery of Cs and Sr from reprocessing solutions," Centre d'Etude de l'Energie Nucleaire, Brussels, Report no. *Blg*-267, August 1964; *Nucl. Sci. Abstr.* **19**, 7613 (1965).
35. Baestle, L., et al., "Treatment of highly active liquid wastes by mineral ion-exchanger," *Nucl. Sci. Abstr.* **17**, 33408 (1963).
36. Baker, B. L., et al., "Department of Chemical Engineering Progress Report no. 7," U.S. Atomic Energy Commission, Report no. *TID*-19187, South Carolina University, April 1962.
37. Baker, B. L., et al., "I. Inorganic sorbents, progress report no. 8," U.S. Atomic Energy, Report no. *TID*-19187, South Carolina University, June 1963.
38. Beetem, W. A., "Adsorption studies by the U.S. geological survey," in "The use of inorganic exchange materials for radioactive waste treatment. A working meeting held at Washington, D.C., Aug. 13–14, 1962," eds. D. K. Jameson, B. M. Kornegay,

W. A. Vaughan, and J. M. Morgan, Jr., U.S. Atomic Energy Commission, Report no. *TID*-7644, Johns Hopkins University, January 1963.

39. Belgian Patent no. 552692, "Procedure and apparatus for the treatment of liquid radioactive wastes," *Water Pollution Abstr.* **34**, p. 1511 (1961).

40. Belter, W. G., "Advances in radioactive waste management technology—its effect on the future U.S. nuclear power industry," U.S. Atomic Energy Commission, Report no. A/conf. 28-P-868; *Nucl. Sci. Abstr.* **18**, 33141 (1964).

41. Belter, W. G., "Ground disposal. Its role in the U.S. radioactive waste management operations," *Nucl. Sci. Abstr.* **18**, 17337 (1964).

42. Belter, W. G., "Recent developments in the processing and ultimate disposal of high-level radioactive waste," in Proceedings of 16th Industrial Waste Conference, Purdue University Engineering Extension Series, Bulletin no. 109, 1961, p. 29.

43. Belter, W. G., "Radioactive wastes," *Intern. Sci. Tech.* December 1962, pages 42 and 46.

44. Belter, W. G., "Treatment and storage of high-level radioactive wastes," in Proceedings of Symposium, October 1962, International Atomic Energy Agency, Vienna, February 1963.

45. Belter, W. G., *et al.*, "Radioactive waste research and development activities," *J. Water Pollution Control Federation* **37**, 316 (1965).

46. Belter, W. G., and H. Bernard, "Status of radioactive liquid waste management in the United States," *J. Water Pollution Control Federation* **35**, 168 (1963).

47. Bensen, D. W., "Crib disposal of NPR decontamination wastes," General Electric Co., Richland, Wash., Report no. *HW*-73482, May 1962; *Nucl. Sci. Abstr.* **16**, 30154 (1962).

48. Berdnikov, A. I., *et al.*, "On the possibility of using soils for the purification of low radioactivity sewer waters. I: Adsorption of Co^{60} by the soil of peat bogs," 12V, *Vysshikh Uchebn. Zavedenii, Khim., Tekhnol.* 1964, p. 7594; *Nucl. Sci. Abstr.* **19**, 22412 (1965).

49. Berryman, R. V., V. H. Clarke, and T. D. Wright, *A radioactive Effluent Filtration Plant*, U.K. Atomic Energy Authority, Harwell (1966).

50. Bhagat, S. K., and E. F. Gloyna, "Environmental radionuclides in municipal waste water," *Water Sewage Works* **110**, 205 (1963).

51. Bierman, S. R., "Concentration of liquid radioactive wastes by direct contact heat transfer," U.S. Atomic Energy Commission, Report no. *HW-SA*-2958, Hanford Atomic Products Operation, 1962.

52. Bixby, H. D., "Method and product for the disposal of radioactive wastes," U.S. Patent no. 3249555; *Nucl. Sci. Abstr.* **20**, 30951 (1966).

53. Blanco, R. E., "Pot calcination," U.S. Atomic Energy Commission, Report no. *TID*-7646, Division Reactor Development, 1963.

54. Blanco, R. E., *et al.*, "Waste treatment and disposal progress report for June–July, 1962," U.S. Atomic Energy Commission, Report no. *ORNL-TM*-396, Oak Ridge National Laboratory, December 1962.

55. Blanco, R. E., *et al.*, "Waste treatment and disposal progress report for April–May, 1962," U.S. Atomic Energy Commission, Report no. *ORNL-TM*-376, Oak Ridge National Laboratory, October 1962, p. 72.

56. Blanco, R. E., and F. L. Parker, "Waste treatment and disposal quarterly progress report, Aug.–Oct., 1962," U.S. Atomic Energy Commission, Report no. *ORNL-TM*-482, Oak Ridge National Laboratory, March 1963.

57. Blanco, R. E., and F. L. Parker, "Waste treatment and disposal semi-treatment annual progress report, July–Dec., 1965," U.S. Atomic Energy Commission, Report no. *ORNL-TM*-1465, Oak Ridge National Laboratory, 1966.

58. Blanco, R. E., and E. G. Struxness, "Waste treatment and disposal progress report for April–May, 1961," U.S. Atomic Energy Commission, Report no. *CF*-61-7-3, August 1961; *Nucl. Sci. Abstr.* **15**, 28846 (1961).

59. Blanco, R. E., *et al.*, "Recent developments in treatment of low and intermediate level radioactive waste in the United States," Symposium on Practices in the Treatment of Low- and Intermediate-Level Radioactive Wastes, Vienna, Paper no. *SM*-71-26, December 1965.

60. Blomeke, J. O., W. J. Boegly, Jr., and R. L. Bradshaw, "Disposal in natural salt formations," in "Health Physics Division annual progress report for period ending June 30, 1963," U.S. Atomic Energy Commission, Report no. *ORNL*-3492, Oak Ridge National Laboratory, September 1963, p. 19.

61. Blomeke, J. O., "Cold engineering development of the pot calcination process," Report of Third Working Meeting on Calcination and/or Fixation of High-Level Wastes in Stable, Solid Media at Wash-

ington, D.C., February 1962, p. 69, comp. and ed. W. H. Regan; U.S. Atomic Energy Commission, Report no. *TID*-7646, Division Reactor Development, 1962.
62. Blomeke, J. O., et al., "Disposal in natural salt formations," U.S. Atomic Energy Commission, Report no. *ORNL*-3492, Oak Ridge National Laboratory, September 1963.
63. Blomeke, J. O., et al., "Estimated costs for management of high activity power reactor processing wastes," U.S. Atomic Energy Commission, Report no. *ORNL-TM*-559, Oak Ridge National Laboratory, May 1963.
64. Blythe, H. J., "Disposal of solids and use of sludge for concentrating radioactive wastes," *J. Proc. Inst. Sewage Purif.*, Part 5 (1962), p. 452.
65. Boenzi, D., and G. Blanca, "Tuff as cation exchanger for the removal of Cs and Sr from aqueous solutions. I: Batch tests," *Energia Nucl.* **12**, 413 (1965); *Nucl. Sci. Abstr.* **19**, 38768 (1965).
66. Boenzi, D., Z. Dlouhy, and G. Denzi, "A study on the sorption properties of the natural tuffs occurring in the Lake Braciano region (Rome)," Report no. 2, Report *RT*-Prot/(65).
67. Bonhote, P. A., et al., "Low-level sludge concentration by solar evaporation," Symposium on Practices in the Treatment of Low and Intermediate Level Radioactive Wastes, Vienna, Paper no. *SM*-71-5, December 1965.
68. Bonner, W. P., H. A. Bevis, and J. J. Morgan, "Removal of strontium from water by activated alumina," *Health Phys.* **12**, 1691 (1966).
69. Bonniaud, R., et al., "Vitrification of concentrated solutions of fission products: study of classes and their characteristics," Institute of Atomic Energy Association Preprint no. *SM*-31-25; *Nucl. Sci. Abstr.* **17**, 10237 (1963).
70. Bogorov, V. G., and B. A. Tareen, "Ocean depths and the problem of burying radioactive wastes in them," *Izv. Akad. Nauk. SSSR, Ser. Geofiz.*, 1960, p. 3; *Transatom Bull.* **1**, 3886 (1961).
71. Bourdrez, J., "Evolution of the liquid waste problem in the centers in the Paris region," *Bull. Inform. Sci. Tech.*, February 1963, p. 11; *Nucl. Sci. Abstr.* **17**, 26936 (1963).
72. Bovard, P., and M. E. Gahinet, "The disposal of radioactive effluents," *Eau* **50**, 301 (1963); *Nucl. Sci. Abstr.* **18**, 6539 (1964).
73. Bouen, B. M., et al., "Radioactive waste disposal at the Georgia Nuclear Lab.," *Am. Ind. Hyg. Assoc. J.* **22**, 119 (1961).
74. British Patent no. 837967, "Method of handling radioactive waste solutions," *Nucl. Sci. Abstr.* **14**, 21181 (1960).
75. Bradshaw, R. L., "Disposal of high activity power reactor wastes in salt mines," *Nucl. Struct. Eng.* **2**, 438 (1965).
76. Bradshaw, R. L., J. J. Perona, and J. O. Blomeke, "Demonstration disposal of high level radioactive solids in Lyons, Kansas, salt mine: background and preliminary design of experimental aspects," U.S. Atomic Energy Commission, Report no. *ORNL-TM*-734, Oak Ridge National Laboratory, January 1964.
77. Bradshaw, R. L., J. J. Perona, J. T. Roberts, and J. O. Blomeke, "Evaluation of ultimate disposal method for liquid and solid radioactive wastes. I: interim liquid storage," U.S. Atomic Energy Commission, Report no. *ORNL*-3123, Oak Ridge National Laboratory, August 1961; *Nucl. Sci. Abstr.* **15**, 28853 (1961).
78. Branca, G., "Non-conventional methods for treatment of liquid radioactive waste (low activity) and elimination of wastes of high activity," *Ing. Sanit.* **4**, 138 (1962).
79. Branca, G., et al., "Experiment on the solidification of radioactive contaminated waste by means of inglobing in concrete blocks," *Ing. Sanit.* **3**, 1962, p. 12, *Nucl. Sci. Abstr.* **17**, 31792 (1963).
80. Branca, G., and G. Sternheim, "The processing plant for liquid radioactive waste at the Ispra Centro Comune di Richerche," *Nucl. Sci. Abstr.* **17**, 21377 (1963).
81. Bray, L. A., "Solvent extraction process for recovery of strontium, rare earths and cesium from radioactive waste solutions," U.S. Atomic Energy Commission, Report no. *HW-SA*-2982, Hanford Atomic Products Operation, June 1963.
82. Bray, L. A., et al., "Laboratory and hot-cell cesium-dipicrylamine solvent extraction studies," U.S. Atomic Energy Commission, Report no. *HW*-76222, Hanford Atomic Products Operation, January 1963.
83. Bray, L. A., et al., "Invention report—use of sugar to neutralize nitric acid waste liquors," U.S. Atomic Energy Commission, Report no. *HW*-75565, Hanford Atomic Products Operation, November 1962.
84. Brooksbank, R. E., et al., "Low-radioactive level waste treatment, II: Pilot plant demonstration—

scavenging—precipitation ion exchange process," U.S. Atomic Energy Commission, Report no. *ORNL*-3349, Oak Ridge National Laboratory, May 1963.

85. Brown, D. J., "Chemical effluents technology waste disposal investigations, (Jan.–Dec. 1964)," U.S. Atomic Energy Commission, Report no. *HW*-84549, Hanford Atomic Products Operation, 1964.

86. Brown, B. P., B. M. Legler, and L. T. Lakey, "Development of a fluidized bed calcination process for aluminum nitrate wastes in a two-foot-square pilot plant calciner. IV: Final process studies—runs 23 through 37," U.S. Atomic Energy Commission Report no. *IDO*-14627, Phillips Petroleum Co., June 1964.

87. Brown, J. M., Jr., J. F. Thompson, and H. L. Andrews, "Survival of waste containers at ocean depths," *Health Phys.* **7**, 227 (1962).

88. Brllue, O. H., "Underground storage and disposal of radioactive products," U.S. Patent no. 3286053,22 February 1966; *Nucl. Sci. Abstr.* **20**, 26585 (1966).

89. Buckham, J., "Fluidized bed calcination of high-level radioactive waste in plant scale facility," *Chem. Eng. Progr. Symp. Ser.* **62**, 52 (1966).

90. Buckham, J. A., "Design of a hot pilot plant facility for demonstration of the pot calcination process," in Proceedings of Treatment and Storage of High-level Radioactive Wastes Symposium, Vienna, October 1962, International Atomic Energy Agency, February 1963.

91. Bucham, J. A., L. T. Lakey, and J. A. McBridge, "Development of the fluidized bed waste calcination process," American Institute of Chemical Engineers, Preprint no. 6, 1963.

92. Buckham, J. A., and J. A. McBridge, "Pilot plant studies of the fluidized bed waste calcination process," in Proceedings of Treatment and Storage of High-Level Radioactive Wastes Symposium, Vienna, October 1962, International Atomic Energy Agency, February 1963.

93. Bunch, D. F., E. H. Markee, Jr., and D. K. Venson, "Environmental effects from operation of the waste calcination facilities," *Health Phys.* **12**, 1283 (1966).

94. Bunker, C. M., and W. A. Bradley, "Gamma radioactivity investigations related to waste disposal, Jackson Flats, Nevada, test site," U.S. Atomic Energy Commission, Report no. *TEI*-829, Geological Survey, September 1962.

95. Burns, R. H., "The safe disposal of radioactive wastes," *Roy. Soc. Health J.* **80**, 214 (1960).

96. Burns, R. H., et al., "Filtration and ion exchange plants in use at AERE, Harwell," Symposium on Practices in the Treatment of Low and Intermediate Level Radioactive Wastes, Vienna, Paper no. *SM*-71-15, December 1965.

97. Burns, R. H., et al., "Present practices in the treatment of liquid wastes at AERE, Harwell," Symposium on Practices in the Treatment of Low and Intermediate Level Radioactive Wastes, Vienna, Paper no. *SM*-71-58, December 1965.

98. "Calcining techniques to ease nuclear waste woes," *Chem. Eng.* **70**, 26 (1963).

99. Campbell, B. F., E. Doud, and R. E. Tomlinson, "Management of high-level wastes—current practice," New York Engineers' Joint Council, Report no. *HW-SA*-2478, Preprint no. 53, 1962; *Nucl. Sci. Abstr.* **16**, 30159 (1962).

100. Cardozo, R. L., "The decontamination of synthetic effluent by flotation," European Atomic Energy Community, Nuclear Research Centre, Ispra, Report no. *EUR*-262-e, 1963; *Nucl. Sci. Abstr.* **17**, 35240 (1963).

101. Carswell, D. J., et al., "Ion-exchange extraction of thorium from monazite," *Australian J. Appl. Sci.* **15**, 329 (1964).

102. Cartwright, A. C., "The treatment of low activity aqueous wastes," *Nucl. Eng.* **7**, 26 (1962).

103. Caudill, H. L., H. D. Haberman, and J. W. Kolb, "*CGC*–981 Process design engineering waste management program—phase II," U.S. Atomic Energy Commission, Report no. 77163, Hanford Atomic Products Operation, March 1963.

104. Cerrai, E., A. Scaroni, and C. Triulzi, "Some bentonite clays for decontaminating water containing radioactive cesium and strontium," *Energia Nucl.* **7**, 253 (1960).

105. Cerre, P., "Lyophilic processing of medium-activity liquid effluents for treatment and storage of high level radioactive wastes," in Proceedings of Symposium, October 1962, International Atomic Energy Agency, Vienna, February 1963.

106. Cerre, P., E. Mestre, and J. Courtault, "Lead decontamination," *Bull. Inform. Sci. Tech.*, February 1963, p. 49.

107. Christenson, C. W., "The ceramic sponge process," Report of Third Working Meeting on Calcination and/or Fixation of High-Level Wastes in Stable, Solid

Media at Washington, D.C., February 1962, p. 94, comp. and ed. W. H. Ryan, U.S. Atomic Energy Commission, Report no. *TID*-7646, Division Reactor Development, 1962.
108. Christenson, C. W., *et al.*, "Radioactive waste disposal. III," *Ceram. Age* **80**, 31 (1964).
109. Clark, H. J., *et al.*, "Purification of radioactive solvent with a flash vaporizer," U.S. Atomic Energy Commission, Report no. *DP*-848, E. I. Dupont de Nemours & Co., 1965.
110. Clark, W. E., *et al.*, "Investigations into the use of a centrifugal bed of vermiculite for the decontamination of radioactive effluents," U.K. Atomic Energy Authority, Research Group, Atomic Energy Research Establishment, Harwell, Report no. *AERE-R*-4314, May 1963.
111. Clark, W. E., *et al.*, "Development of processes for solidification of high-level radioactive waste," U.S. Atomic Energy Commission, Report no. *ORNL-TM*-1584, Oak Ridge National Laboratory, 1966.
112. Clark, W. E., and H. W. Godbee, "Laboratory development of processes for fixation of high-level radioactive wastes in glassy solids; wastes containing (1) aluminum nitrate, and (2) the nitrates of the constituents of stainless steel," U.S. Atomic Energy Commission, Report no. *ORNL*-3612, Oak Ridge National Laboratory, July 1964.
113. Clarke, J. H., T. D. Wright, and R. J. Berryman, "Investigation into the use of a centrifuged bed of vermiculite for the decontamination of radioactive effluent," U.K. Atomic Energy Authority, Research Group, Atomic Energy Research Establishment, Harwell, Report no. *AERE-R*-4314, May 1963.
114. Clarke, J. H., *et al.*, "Investigations into the use of ion exchange resins for the decontamination of radioactive effluent. II: Pilot plant trials," U.K. Atomic Energy Authority, Research Group, Atomic Energy Research Establishment, Harwell, Report no. *AERE-R*-4905, April 1965.
115. Clelland, D. W., "The concentration and storage of highly active wastes from the first stages of the U.K. civil nuclear power programme," in Proceedings of Symposium, October 1962, International Atomic Energy Agency, Vienna, February 1963.
116. Clelland, D. W., *et al.*, "The treatment of intermediate level radioactive liquid wastes by evaporation—the design performance of evaporation systems," Symposium on Practices in the Treatment of Low and Intermediate Level Radioactive Wastes, Vienna, Paper no. *SM*-71-70, December 1965.
117. Coleman, L. F., *et al.*, "Ruthenium removal from intermediate level radioactive wastes by electrodialysis methods," U.S. Atomic Energy Commission, Report no. *BNWL*-72, Batelle-Northwest Laboratory, April 1965.
118. Collins, S. C. (ed.), *Radioactive Wastes—Their Treatment and Disposal*, E. & F. N. Spon, Ltd., London, (1960).
119. Commander, R. E., G. E. Lohse, D. E. Black, and E. D. Cooper, "Operation of the waste calcining facility with highly radioactive aqueous waste," U.S. Atomic Energy Commission, Report no. *IDO*-14662, 1966.
120. Continental Oil Co., "Underground disposal of radioactive liquids or slurries," British Patent no. 926821, 22 May 1963.
121. Coppinger, E. A., "Pilot plant determination of Purex waste with sugar," U.S. Atomic Energy Commission, Report no. *HW*-7780, Hanford Atomic Products Operation, March 1963.
122. Cowser, K. E., and T. Tamura, "Significant results in low-level waste treatments at ORNL," *Health Phys.* **9**, 687 (1963).
123. Culp, R, L., "Will treatment plant processes remove radioactive wastes at the Savannah River Plant. Engineering considerations," *Nucl. Sci. Abstr.* **14**, 26509 (1960).
124. Cummings, R. L., "Radioactive waste systems for a sodium cooled reactor," *Chem. Eng. Progr.* **58**, 82 (1962).
125. Daniel, A. N., "Underground storage of low level radioactive wastes at the Savannah River Plant: engineering considerations," *Nucl. Sci. Abstr.* **14**, 26509 (1960).
126. Davis, M. W., Jr., "Calcination of radioactive wastes in molten sulfur," Report of Third Working Meeting on Calcination and/or Fixation of High-Level Wastes in Stable, Solid Media at Washington, D.C., February 1962, p. 20, comp. and ed. W. H. Regan, U.S. Atomic Energy Commission, Report no. *TID*-7646, Division Reactor Development, 1962.
127. Davis, T. F., "Fixation of fission products," *Nucl. Sci. Abstr.* **14**, 16 (1960).
128. DeLaguna, W., "Disposal of radioactive wastes by hydraulic fracturing. I and II," *Nucl. Eng. Design* **3**, 338 and 432 (1966).

129. DeLaguna, W., *et al.*, "Disposal by hydraulic fracturing," U.S. Atomic Energy Commission, Report no. *ORNL*-3492, Oak Ridge National Laboratory, September 1963.
130. Davids, J. A. G., and G. C. Van Dam, *Radioactive Waste Disposal and Investigation on Turbulent Diffusion in the Netherlands Coastal Areas*, International Atomic Energy Agency, Vienna (1966).
131. Davis, T. F. (comp.), "Radioactive waste processing and disposal—a literature search," U.S. Atomic Energy Commission, Report no. *TLD*-3555, Division Technical Information, June 1960.
132. Debiesse, J., "Natural radioactivity and the effluents at Soclay," *Bull. Inform. ATEN* **25**, 1 (1960).
133. Rejonghe, P., "Treatment installation for radioactive waters of the CEN at Mol (Belgium)," *Techn. Sci. Munic.* **58**, 429 (1963).
134. DeMier, W. V., "Fixation of radioactive fission products on inorganic ion exchange media," U.S. Atomic Energy Commission, Report no. *HW-SA*-3233, Hanford Atomic Products Operation, October 1963.
135. DeMier, W. V., *et al.*, "Development of radioactive waste fractionalization and packaging technique," U.S. Atomic Energy Commission, Report no. *HW-SA*-2786, Hanford Atomic Products Operation, January 1963.
136. "Design and construction of a radioactive waste treatment plant at JAERI," Japan Atomic Energy Research Institute, Tokyo, Report no. *JAERI*-1021, February 1962; *Nucl. Sci. Abstr.* **16**, 17334 (1962).
137. "Determination of caesium 137 in reactor fuel processing and effluent treatment plant solutions," U.K. Atomic Energy Commission Authority, Progress Report no. 202 (*W*); *J. Appl. Chem.* **11**, ii (1961).
138. "Determination of calcium 45 in cooling pond and effluent treatment plant solutions," U.K. Atomic Energy Authority, Production Group, Chemistry Services, Report no. *PG*-311 (*W*), 1962; *Anal. Abstr.* **9**, 2950 (1962).
139. Dohland, E., and T. S. Barendregt, "The processing of irradiated fission materials in the Eurochemic plant," *Atomwirtschaft* **6**, 149 (1961).
140. Domish, R. F., and L. P. Hatch, "Continuous calcination of high level radioactive wastes by means of a rotary ball kiln," U.S. Atomic Energy Commission, Report no. *BNL*-832, Brookhaven National Laboratory, November 1963.
141. Donato, S., *et al.*, "The decontamination of radioactive liquid wastes," *Acqua Ind.* **2**, 85 (1960).
142. Doud, E., H. W. Stivers, G. H. Bauer, G. W. Morrow, F. A. MacLean, and S. P. Robertson, "Process design engineering Purex essential waste routing system and 241-ax tank farm," U.S. Atomic Energy Commission, Report no. *HW*-72780, Hanford Atomic Products Operation, April 1962; *Nucl. Sci. Abstr.* **16**, 31343 (1962).
143. Duchaniel, F., "Resolution of problems concerned with treatment of radioactive wastes," *Bull. Inform. Sci. Tech.* February 1963, p. 69; *Nucl. Sci. Abstr.* **17**, 26934 (1963).
144. Edgerly, E., *et al.*, "Treatment of strontium-90-bearing organic waste by ion exchange," Paper presented at Meeting of American Chemists' Society, Philadelphia, April 1964; *Water Pollution Abstr.* **38**, 1018 (1965).
145. Eliassen, R., "Disposal of high level radioactive wastes," *J. Water Pollution Control Federation* **36**, 201 (1964).
146. Emility, L. A., *et al.*, "Disposal of AM^{241}-PV^{234} rapanate solutions by fixations with cement," U.S. Atomic Energy Commission, Report no. *LA*-3150, Los Alamos Science Laboratory, August 1964.
147. Emerson, F. M., *et al.*, "Demonstration of disposal of high level radioactive solids in salt," 2nd Symposium on Salt, Northern Ohio Geological Society, 1965, Vol. 1, p. 432.
148. Fernandez, N., "Effluents treatment plant. Its problems, operation, results," *Energie Nucl.* **5**, 282 (1963); *Water Pollution Abstr.* **38**, 1184 (1965).
149. Fiege, Y., *et al.*, "Analysis of waste disposal practice and control at ORNL," *Nucl. Sci. Abstr.* **15**, 1085 (1951).
150. Folsom, T. R., *et al.*, "A study of certain radioactive isotopes in selected wastewater treatment plants," *J. Water Pollution Control Federation* **35**, 304 (1963).
151. Forsman, R. C., *et al.*, "Formaldehyde treatment of Purex radioactive waste," *Chem. Eng. Progr.* **60**, 74 (1964).
152. Fozzy, P., "Atomic waste disposal," *Ind. Res.* **4**, 27 (1962).
153. Frank, W. H., *et al.*, "Investigation of the decontamination efficiency of slow sand filter action in ground water recharge," *Gas-Wasserfach.* **104**, 1400 (1964); *Publ. Health Eng. Abstr.* **45**, 1163 (1965).
154. Frost, C. R., "Decontamination of radioactive

aqueous effluent by the calcium-ferro phosphate process," *J. Am. Water Works Assoc.* **55**, 581 (1963).
155. Frysinger, G. R., and H. C. Thomas, "Fixation of radioactivity in solid media: stability of glasses and glazes," *Nature* **197**, 352 (1963).
156. Fujii, S., et al., "On the quality of radioisotopes in liquid wastes and some problems with regard to the storage tank of low level water," *Proc. Japan. Conf. Radioisotopes*, **5**, 37 (1963); *Nucl. Sci. Abstr.* **71**, 30071 (1963).
157. Gailledreau, C., "Study of the purification device based on mineral exchange with the view to discharging weakly radioactive residual liquids into the ground," *Nucl. Sci. Abstr.* **17**, 29872 (1963).
158. Galder, R., "Burying radioactive wastes," *New Scientist* **9**, 322 (1961).
159. Galea, V., E. T. Birsan, and I. Uray, "Decontamination of waters containing ^{90}Sr and ^{90}Y with gypsum and bentonites," *Rev. Chim.* **16**, 318 (1965); *Nucl. Sci. Abstr.* **20**, 29348 (1966).
160. Gardiner, D. A., and K. E. Cowser, "Optimization of radionuclide removal from low-level process wastes by the use of response surface methods," *Health Phys.* **5**, 70 (1961).
161. Gittens, G. J., et al., "Application of electrodialysis to demineralization," Institute of Chemical Engineers, New Chemical Engineering Problems in Utilization of Water, Paper no. 9.14, June 1965. p. 65.
162. Glauberman, H., and P. Laysen, "The use of commercial incinerators for the volume reduction of radioactivity contaminated combustible wastes," *Health Phys.* **10**, 237 (1964).
163. Gloyna, E. F., et al., "Radioactive waste disposal. I, II and III," *Ceram. Age* **80**, no. 6, p. 63, no. 7, p. 24, and no. 8, p. 31 (1964).
164. Godbee, H. W., and W. E. Clark, "The use of phosphite and hypophosphite to fix ruthenium from high activity wastes in solid media," U.S. Atomic Energy Commission, Report no. *ORNL-TM*-263, Oak Ridge National Laboratory, June 1962.
165. Goldman, M., et al., "The removal of strontium 90 from organic wastes," *Health Phys.* **9**, 847 (1963).
166. Goldsmith, W. A., and E. J. Middlebrooks, "Radioactive decontamination by slurrying with YAZOO and Zi/pha clays," *J. Am. Water Works Assoc.* **58**, 1052 (1966).
167. Goode, V. H., "Fixation of intermediate level radioactive waste in asphalt hot-cell tests," U.S. Atomic Energy Commission, Report no. *ORNL-TM*-1343, Oak Ridge National Laboratory, 1965.
168. Graham, R. J., E. L. Beard, and D. J. Kvam, "A decontamination process for low-level waste waters," Report no. *UCRL*-6641, U.S. Atomic Energy Commission, September 1961; *Nucl. Sci. Abstr.* **16**, 8440 (1962).
169. Grover, J. R., and B. E. Chidley, "Glasses suitable for the long-term storage of fission products," *J. Nucl. Energy: Pt. A B* **16**, 405 (1962).
170. Grune, W. N., et al., "Effects of carbon 14 and strontium 90 on anaerobic digestion," *J. Water Pollution Control Federation* **35**, 493 (1963).
171. Gyllander, C., *Water Exchange and Diffusion Processes in Tuaeren at Baltic Bay*, International Atomic Energy Agency, Vienna (1966); *Nucl. Sci. Abstr.* **21**, 15999 (1967).
172. Haas, P. A., et al., "Foam columns for countercurrent surface-liquid extraction of surface-active solutes," *Am. Inst. Chem. Engrs. J.* **11**, 319 (1965).
173. Haden, W. L., Jr., "Development of an effective alumino-silicate adsorbent for radioactive waste disposal," in "The use of inorganic exchange materials for radioactive waste treatment—a working meeting held at Washington, D.C., August 1962," eds. B. H. Kornegay, W. A. Vaughan, and J. M. Morgan, Jr., p. 201; U.S. Atomic Energy Commission, Report no. TID-7644, Johns Hopkins University, Jan. 1963.
174. Haefeli, R., "Glaciological introduction to the problems of discarding radioactive wastes in the great ice caps of Earth," *Schweiz. Z. Hydrol.* **23**, 253 (1961); *Chem. Abstr.* **56**, 8490 (1962).
175. Hancher, C. W., J. C. Suddath, and M. E. Whatley, "Engineering studies on pot calcination for ultimate disposal of nuclear waste: formaldehyde–treated Purex waste for 1965 (FTW-65)," U.S. Atomic Energy Control Commission, Report no. *ORNL-TM*-715, Oak Ridge National Laboratory, January 1964.
176. Hancher, C. W., and J. C. Suddath, "Pot calcination of simulated radioactive waste with continuous evaporation," *Chem. Eng. Progr. Symp. Ser.* **60**, 105 (1964).
177. Haltrich, W., "Elimination of both phosphorus isotopes P^{31} and P^{32} from domestic waste water," *Wasser, Luft* **29**, *Comit. Naz. Energia Nucl.*, Rome, Italy, 1965.
178. Haney, W. A., and J. F. Honstead, "A history and discussion of specific retention disposal of radio-

active liquid wastes in the 200 areas," *Nucl. Sci. Abstr.* **14**, 17636 (1960).
179. Hatch, L. B., "Ball kiln calcination," in Report of the Third Working Meeting on Calcination and/or Fixation of High-Level Wastes in Stable, Solid Media at Washington, D.C., February 1962, comp. and ed. W. H. Regan, p. 37; U.S. Atomic Energy Commission, Report no. *TID*-7646, Division Reactor Development, 1962.
180. Hatch, L. P., "Fixation of radioactive wastes in stable solids," Hearings on Industrial Radioactive Wastes Disposal, International Commission on Atomic Energy, 1959, Vol. 3, p. 1839.
181. Hauang, E. J., and T. S. Yang, "Treatment of liquid wastes containing ^{51}Cr," U.S. Atomic Energy Commission, Report no. *ORNL-TR*-916, Oak Ridge National Laboratory; translated from *Ho Tsu K'o Hsueh* **4**, 135 (1964); *Nucl. Sci. Abstr.* **20**, 10444 (1966).
182. Hawkins, D. B., "Removal of cobalt and chromium by precipitation and ion exchange on soil, lignite and clinoptilolite from citrate containing radioactive liquid waste," U.S. Atomic Energy Commission, Report no. *IDO*-12036, Idaho Operations Office, June 1964.
183. Hawkins, D. B., "A system for the evaluation of liquid waste disposal," U.S. Atomic Energy Commission, Report no. *IDO*-12052, Idaho Operations Office, 1966.
184. Hazegawa, K., "Removal of ^{89}Sr and ^{137}Cs on cellulose phosphate," *Rept. Fac. Sci., Shizuoka Univ.* **1**, 29 (1965).
185. Hedland, R., and A. Lindskog, "The radioactive waste management at Studsvick," Aktiebolaget Atomenergi, Stockholm, Report no. *AE*-225, 1966.
186. Hershey, H. C., R. D. Mitchell, and W. H. Webb, "Separation of cesium and strontium by electrodialysis," *J. Inorg. Nucl. Chem.* **28**, 645 (1966).
187. Higgins, I. R., "Ion-exchange treatment of low level radioactive water in which bicarbonate is the predominate anion," U.S. Atomic Energy Commission, Report no. *ORO*-603, Chemical Separation Corp., February 1963.
188. Higgins, I. R., et al., "Development of a process of continuous phenolic ion exchange treatment of alkaline intermediate radioactivity-level wastes," U.S. Atomic Energy Commission, Report no. *ORNL*-3322, Oak Ridge National Laboratory, July 1963.
189. Hoather, R. C., and R. F. Rackham, "Some observations on radon in waters and its removal by aeration," *J. Inst. Water Engrs.* **17**, 13 (1963).
190. Holcomb, R. R., "Low radioactivity waste level treatment. I: Development of a scavenging-precipitation ion exchange process for decontamination of process water wastes," U.S. Atomic Energy Commission, Report no. *ORNL*-3322, Oak Ridge National Laboratory, July 1963.
191. Holmes, J. T., et al., "Engineering development of a fluidized fluoride volatility process. II: Pilot scale studies," *Nucl. Appl.* **1**, 301 (1965).
192. Honstead, J. F., and J. L. Nelson, "Mineral reactions—a new waste decontamination process," *Health Phys.* **8**, 191 (1962).
193. Honstead, J. F., "The role of 'ground disposal' in radioactive waste management," *Colloq. Intern. Retention Migration Ions Radioact. Sols, Saclay* 1963, p. 11; *Nucl. Sci. Abstr.* **18**, 17338 (1964).
194. Horner, D. E., D. J. Crouse, K. B. Brown, and B. Weaver, "Fission product recovery from waste solutions by solvent extraction," U.S. Atomic Energy Commission, Report no. *CONF*-49–1, Oak Ridge National Laboratory, 1962.
195. Horski, J., "Removal of radioisotopes from aqueous solutions by silica gel," Institute of Nuclear Research, Warsaw, Report no. 636-XIII-*J*, 1965.
196. Hottenstein, E. R., "The disposal of radioactive waste." *Mech. Eng.* **84**, 60 (1962).
197. Howells, H., "Discharges of low-activity radioactive effluent from the Windscale works into the Irish Sea," in Proceedings of Symposium on Disposal of Radioactive Wastes into Seas, Oceans, and Surface Waters, International Atomic Energy Agency, Vienna, 1966, p. 769.
198. Hutchinson, J. P., and C. W. Christenson, "Treatment of radioactive wastes by chemical preparation and ion exchange at Los Alamos," *Nucl. Sci. Abstr.* **15**, 15320 (1961).
199. Hwang, Y. S., and A. I. Huang, "Treatment of radioactive waste," *Ho Tsu K'o Hsueh* **4**, 85 (1965); *Nucl. Sci. Abstr.* **20**, 14211 (1966).
200. Imaoka, M., "Biological treatment of radioactive waste water," *Koshu Eiseiin Kenyu Hokoku* **13**, 1964, p. 90; *Nucl. Sci. Abstr.* **19**, 24196 (1965).
201. Imaoka, M., "Treatment of radioactive waste water with activated sludge," *Koshu Eiseiin Kenkyu Hokoku* **13**, 96 (1964); *Nucl. Sci. Abstr.* **19**, 24197 (1965).
202. International Atomic Energy Agency, "Disposal

of radioactive wastes into fresh water," Safety Series no. 10, Vienna, March 1963.
203. International Atomic Energy Agency, "Management of radioactive wastes produced by radioisotope users," Safety Series no. 12, Vienna, 1965.
204. International Atomic Energy Agency, "Management of radioactive wastes produced by radioisotope users. Technical addendum," Safety Series no. 19, Vienna, 1966.
205. International Atomic Energy Agency, Proceedings of the Symposium on Disposal of Radioactive Wastes into Seas, Oceans, and Surface Waters, Vienna, 1966.
206. International Atomic Energy Agency, Proceedings of the Symposium on Practices in the Treatment of Low- and Intermediate-Level Radioactive Wastes, Vienna, 1966.
207. International Atomic Energy Commission, "Treatment and storage of high-level radioactive wastes," Vienna, February 1963.
208. Iranzo, E., "Decontamination of liquid waste from the regeneration of the ion-exchange units of a high-flux reactor," in Proceedings of Symposium on Practices in the Treatment of Low- and Intermediate-Level Radioactive Wastes, International Atomic Agency, Vienna, 1965, p. 255; *Nucl. Sci. Abstr.* **20**, 26584 (1966).
209. Irish, E. R. (ed.), "Fixation of radioactive residues. Quarterly progress report, Jan.–March, 1963," U.S. Atomic Energy Commission, Report no. *HW*-77299, Hanford Atomic Products Operation, April 1963.
210. Irish, E. R. (ed.), "Research and development activities. Fixation of radioactive residues quarterly progress report, July–Sept., 1962," U.S. Atomic Energy Commission, Report no. *HW*-75290, Hanford Atomic Products Operation, October 1962.
211. Ishiyama, T., *et al.*, "Decontamination of radioactive ruthenium by cellulose derivative bentonite flocculation. II," *Ann. Rep. Radiation Centre, Osaka Prefect.* **5**, 37 (1965); *Nucl. Sci. Abstr.* **19**, 33679 (1965).
212. Ishiyama, T., T. Matsumura, and T. Mamuro, "Decontamination of radioactive ruthenium by cellulose derivative bentonite flocculation. III," *Ann. Rep. Radiation Centre, Osaka Prefect.* **6**, 33 (1965); *Nucl. Sci. Abstr.* **20**, 36929 (1966).
213. Ito, M., M. Fukuda, and Y. Tanigawa, "Small-scale horizontal diffusion near the coast," in Proceedings of Symposium on Disposal of Radioactive Wastes into Seas, Oceans, and Surface Waters, International Atomic Energy Agency, Vienna, 1966, p. 471.
214. Ito, M., and M. Nishidoi, "Treatment of radioactive wastes with ion exchange membrane electrodialyzer," *Nucl. Sci. Abstr.* **15**, 4796 (1961).
215. Iwai, S., *et al.*, "Storage tanks packed with ion exchangers for treatment of radioactive wastes," *Proc. Japan. Conf. Radioisotopes* **5**, 43 (1963); *Nucl. Sci. Abstr.* **17**, 30073 (1963).
216. Jackson, J. D., H. A. Sorgenti, G. A. Wilcox, and R. S. Brodkey, "Nuclear waste disposal by fluidized calcination of simulated Al-type wastes," *Ind. Eng. Chem.* **52**, 795 (1960).
217. Jacobs, D. G., "An interpretation of cation exchange by alumina," *Health Phys.* **12**, 1565 (1966).
218. Jacobs, D. G., "Cesium exchange properties of vermiculite," *Nucl. Sci. Eng.* **12**, 485 (1962).
219. Jacobs, D. G., O. M. Sealand, and O. H. Myers, "Liquid injection into deep permeable formations," Health Physics Division Annual Progress Report, period ending 30 June 1963, p. 31; U.S. Atomic Energy Commission, Report no. *ORNL*-34092, Oak Ridge National Laboratory, September 1963.
220. Jacobs, D. G., "Mineral exchange work at ORNL," in "The use of inorganic exchange materials for radioactive waste treatment—a working meeting held at Washington D.C., Aug. 3–14, 1962," eds. D. K. Jamison, B. M. Kornegay, W. A. Vaughan, and J. M. Morgan, Jr., pp. 187–99; U.S. Atomic Energy Commission, Report no. *TID*-7644, Johns Hopkins University, January 1963.
221. Jaeger, T., "Design of storage tanks for highly radioactive waste products," *Kerntechnik* **3**, 307 (1961); *Nucl. Sci. Abstr.* **15**, 28845 (1961).
222. Jansen, G., Jr., W. E. Willingham, and W. V. Demier, "Buried radioactive waste storage tank temperatures and soil temperatures near leaks," U.S. Atomic Energy Commission, Report no. *BNWL*-181, Batelle-Northwest Laboratory, 1966.
223. Jansen, G., Jr., and S. V. DeMier, "Transient thermal behavior in fixation of radioactive materials," U.S. Atomic Energy Commission, Report no. *BNWL*-157, Batelle-Northwest Laboratory, 1966.
224. Jarman, L., and M. Matic, "A scheme for the rapid analysis of uranium–ore acid leach solution," *Talanta* **9**, 25 (1962).
225. Jenicek, A., and J. Kortus, "Plant treatment of radioactive wastes of Czechoslovak Institute of Nuclear Research," *Czech. Heavy Ind.* **3**, 1964, p. 9.

226. Jensen, H. W., "Contamination and decontamination," *Nucl. Sci. Abstr.* **16**, 32630 (1962).
227. Jessen, N. C., "Nature of wastes in fuel element manufacture," Hearings on Industrial Radioactive Wastes Disposal, International Commission on Atomic Energy, 1959, Vol. 1, p. 116.
228. Katz, D. L., "Liquid-waste storage in abnormal-pressure reservoirs," *Nucleonics* **20**, 71 (1962).
229. Katz, H., et al., "Decontamination of Savannah River waste supernate by ion exchange," U.S. Atomic Energy Commission, Report no. *BNL-853*, Batelle-Northwest Laboratory, 1964; *Water Pollution Abstr.* **38**, 712 (1965).
230. Karwat, H., "Radioactive gaseous wastes and protection of the environment," *Nucl. Sci. Abstr.* **16**, 5618 (1962).
231. Kaufman, W. J., "The containment of radioactive wastes in deep geological formations," in Proceedings of Conference on Disposal of Radioactive Wastes, Monaco, 1959, Vol. 2, p. 533.
232. Kavale, A. J., "Controlling radiation hazards in nuclear power plant," *Air Conditioning, Heating, Ventilating* **58**, 62 (1961).
233. Kawamoto, K., "Treatment of radioactive waste," *Nucl. Sci. Abstr.* **16**, 2813 (1962).
234. Keese, H., *The Decontamination of Radioactive Water with Chemical Precipitation*, Thesis, Technische Hochschule, Brunswick (1963); *Water Pollution Abstr.* **38**, 1 (1965).
235. Keese, H., "Contributions to the decontamination of radioactive contaminated water using chemical precipitation methods," Technische Hochschule, Brunswick, Report no. *BMwFBk-65-04*, 1965.
236. Kelly, P. N., "Thermal conductivity of alumina produced by the fluidized bed process," U.S. Atomic Energy Commission, Report no. *IDO-14592*, Phillips Petroleum Co., September 1962.
237. Kenny, A. W., R. N. Crooks, and J. R. W. Kerr, "Radium, radon and daughter products in certain drinking waters in Great Britain," *J. Inst. Water Engrs.* **20**, 123 (1966).
238. Kenny, A. W., "Sanitary engineering aspects of nuclear energy developments," *World Health Organ. Bull.* **26**, 475 (1962).
239. Kepak, F., "Sorption of the radioisotopes S, I, and Ru on hydrated oxides in laboratory columns," *Collection Czech. Chem. Commun.* **31**, 3500 (1966).
240. Keshishyan, G. O., P. F. Andrew, and L. J. Daniloo, "Extraction of thorium from dilute solutions by means of tannate of gelatin," *Nucl. Sci. Abstr.* **20**, 27067 (1966).
241. Kiedaisch, W., "Liquid radioactive waste handling," *Chem. Eng. Progr.* **58**, 79 (1962).
242. Kiefer, H., and R. Maushart, "Monitoring of radioactivity in waste water and off-gas," *Nucl. Sci. Abstr.* **15**, 29569 (1961).
243. King, L. J., and M. Ichikawa, "Pilot-plant demonstration of the decontamination of low-level process," U.S. Atomic Energy Commission, Report no. *ORNL-3863*, Oak Ridge National Laboratory, 1965.
244. Knoll, K. C., "Removal of cesium from redox alkaline supernatant wastes by ion exchange," U.S. Atomic Energy Commission, Report no. *HW-84105*, Hanford Atomic Products Operation, October 1964.
245. Knowles, D. J., "Liquid and gaseous-waste-effluent sampling and monitoring," *Nucl. Safety* **7**, 52 (1965).
246. Koch, H., "The decontamination of radioactive laboratory water," *Nucl. Sci. Abstr.* **17**, 2646 (1963).
247. Koda, Y., and S. Takagi, "Separation of ^{140}La from a ^{140}Ba–^{140}La mixture by coprecipitation with ferric hydroxide, using several organic bases as precipitants," *Nucl. Sci. Abstr.* **20**, 23202 (1966).
248. Kohrt, H. U., "Gas purification in nuclear establishments," *Chem. Ingr. Tech.* **33**, 135 (1961).
249. Kostyrko, A., "Study on removal of some radionuclide from sewage water solutions by sorption on peat," Institute of Nuclear Research, Warsaw, Report no. 609-XIX-D, 1965.
250. Kountz, R. R., "Radioactive waste water distillation data and costs," in Proceedings of 17th Industrial Waste Conference, Purdue University Engineering Extension Series, Bulletin no. 112, 1962, p. 409.
251. Kown, B. T., "The ion-exchange reactions of radioactive ions with soils and effects of organic compounds," University of Illinois, Department of Civil Engineering, *Sanit. Eng. Series* **39**, 1966.
252. Koyanoka, Y., "Studies on separation of fission products by flotation method, separation of Cs," *Radioisotopes* **15**, 77 (1966); *Nucl. Sci. Abstr.* **20**, 16392 (1966).
253. Kraus, H., and O. Nentwick, "The decontamination of the radioactive wastes of the Karlsruhe Nuclear Research Center from 1963 to 1965," *Kerntechnik* **8**, 105 (1966).
254. Krawczunshi, S., and B. Kanellahopulos, "Process for precipitation in radioactive waste waters with

different radiochemical composition," *Gas-Wasserfach.* **102**, 601 (1961).
255. Kupp, R. W., and S. M. Stoller, "The perpetual care of high-level radioactive liquid wastes in New York State, a technological and cost study," U.S. Atomic Energy Commission, Report no. NP-12484, S. M. Stoller Associates, October 1962.
256. Lacy, W. J., "A comparison of the composition of reactor waste solutions and radioactive fallout," *Am. Ind. Hyg. Assoc. J.* **21**, 334 (1960).
257. Lacy, W. J., "Laboratory work on certain aspects of the deep well disposal problem," *Health Phys.* **4**, 228 (1961).
258. Laguna, "Hydrologic analysis of postulated liquid waste releases," U.S. Geological Survey Bulletin no. 1156-E, 1966, 51 pages.
259. Langenhorst, W. T. J. P., M. Tels, J. C. Vlugter, and H. I. Waterman, "Cation exchangers on a sugar-beet pulp base. Application for decontaminating radioactive waste water," *J. Biochem. Microbiol. Technol. Eng.* **3**, 7 (1961); *Water Pollution Abstr.* **34**, 2136 (1961).
260. Lavatter, V. E., "An incinerator for wastes containing microcurie amounts of carbon 14," *Am. Ind. Hyg. Assoc. J.* **22**, 485 (1961).
261. Lawrence, C. H., *Removal of Low-Level Radioisotopes from Waste Water by Aerobic Methods of Treatment*, Thesis, University of Florida (1963).
262. Lawrence, C. H., et al., "Removal of low level isotopes from wastewater by aerobic treatment," *J. Water Pollution Control Federation* **37**, 1287 (1965).
263. Lazzarini, E., and G. Tognon, "Disposal of fission products in concrete," *Energia Nucl.* **10**, 117 (1963).
264. Ledbetter, J. O., "Environmental hazard of radio wastes," *J. San. Eng. Div. Am. Soc. Civil Engrs.* **91**, 59 (1965).
265. Lee, B. S., J. C. Chu, A. A. Jonke, and S. Lawroski, "Kinetics of particle growth in a fluidized calciner," *Am. Inst. Chem. Engrs. J.* **8**, 53 (1962).
266. LeGrand, H. E., "Geology and ground-water hydrology of the Atlantic and Gulf Coastal plain as related to disposal of radioactive wastes," *Nucl. Sci. Abstr.* **16**, 16024 (1962).
267. Leipunskii, O. I., "The physics of radiation protection," *Nucl. Sci. Abstr.* **16**, 15033 (1962).
268. Levi, H. W., "Fixation of radionuclides in titanium dioxide and titanates via coprecipitation," in Proceedings of Symposium on Radioactivity, October 1962, International Atomic Energy Agency, Vienna, February 1963.
269. Levi, H. W., "Chemical processes for the deactivation of effluent water," *Atomwirtschaft* **6**, 265 (1961).
270. Levi, H. W., "On the application of filter aids for the filtration of sludges from the decontamination of radioactive effluent," *Kerntechnik* **7**, 485 (1965).
271. Lewin, V. H., "Radioactivity and sewage treatment," *J. Proc. Inst. Sewage Purif.* (1965), p. 166; *Water Pollution Abstr.* **39**, 1531 (1965).
272. Lewis, R. E., T. A. Butler, and E. Lamb, "An aluminosilicate ion exchanger for recovery and transport of 137 Cs from fission-product wastes," *Nucl. Sci. Eng.* **24**, 118 (1966).
273. Lieberman, J. R., and D. C. Costello, "Safe discharge of low-level nuclear wastes to surface waters," *Waste Eng.* **31**, 531 (1960).
274. Linder, P., et al., "The evaporator used as the only treatment facility at research establishments," Symposium on Practices in Treatment of Low- and Intermediate-Level Radioactive Wastes, International Atomic Energy Agency, Vienna, Paper no. SM-71-19, December 1965.
275. Linderoth, C. E., and G. A. Little, "Contamination control at the Hanford laundry," *Nucl. Sci. Abstr.* **16**, 27463 (1962).
276. "Liquid-waste treatment by ion exchange and foam separation," *Reactor Fuel Process.* **6**, 39 (1963).
277. Lopez-Menchero, E., "Research and development work on the treatment of low- and medium-level wastes in the ENEA countries," Symposium on Practices in the Treatment of Low- and Intermediate-Level Radioactive Wastes, International Atomic Energy Agency, Vienna, Paper no. SM-71-62, December 1965.
278. Loeding, J. W., and A. A. Jonke, "Method of reducing gaseous radioactive nuclear wastes to solid form," *Nucl. Sci. Abstr.* **15**, 14046 (1961).
279. "Low-activity liquid wastes," *Nucl. Eng.* **6**, 329 (1961).
280. Lowman, F. G., et al., "Interactions of the environmental and biological factors on the distribution of trace elements in the marine environment," in Proceedings of Symposium on Disposal of Radioactive Wastes into Seas, Oceans and Surface Waters, International Atomic Energy Agency, Vienna, 1967.
281. Lyon, W. A., "Nature and control of radio wastes in Pennsylvania waters," *J. Am. Water Works Assoc.* **53**, 89 (1961).

282. Mackrle, V., O. Dracka, and J. Svec, "Hydrodynamics of the disposal of low-level liquid radioactive wastes in soil," Institute of Hydrodynamics, Czechoslovak Academy of Science, Report no. *Np-15859*, 1965.
283. Makowski, J., "Study of the use of mineral coal for purifying radioactive waste water," *Nucl. Sci. Abstr.* **18**, 38697 (1964).
284. Malasch, E., and O. Zach, "Low-activity waste treatment before permanent storage," *Nucl. Sci. Abstr.* **17**, 19716 (1963).
285. Mamuro, T., T. Matsumura, and T. Ishiyama, "A radioactive sludge concentrator," *Ann. Rep. Radiation Centre, Osaka Prefect.* **3**, 44 (1962).
286. Mamuro, T., et al., "Study on the dewatering treatment of radioactive sludges," *Nucl. Sci. Abstr.* **17**, 30072 (1963).
287. Mamuro, T., T. Ishigama, and G. Matsumura, "Volume reduction of radioactive sludges by freezing and thawing treatment," *Nucl. Sci. Abstr.* **18**, 6544 (1964).
288. Manneschmidt, J. F., and E. J. Witowski, "The disposal of radioactive liquid and gaseous waste at Oak Ridge National Laboratory," *Nucl. Sci. Abstr.* **16**, 28636 (1962).
289. Marcus, F. R., "Role of waste management at Eurochemic," *Atompraxis* **12**, 50 (1966).
290. Marillier, J. C., "Treatment of radioactive effluents," *Energie Nucl.* **4**, 320 (1962).
291. Marruse, H. R., "A licensed radioactive waste evacuation service," *Nucl. Sci. Abstr.* **14**, 17639 (1960).
292. Martenola, F., "Problems of decontamination with ion exchanges," *Centre Belge Etude Doc. Eaux Tribune* **16**, 459 (1963).
293. Matsumara, T., et al., "Humic acid as coagulation agent for calcium phosphate flocculation," *Ann. Rep. Radiation Centre, Osaka Prefect.* **4**, 28 (1963); *Water Pollution Abstr.* **38**, 307 (1965).
294. Matsumara, T., et al., "Decontamination of radioactive wastes water by humic acid," *Ann. Rep. Radiation Centre, Osaka Prefect.* **5**, 33 (1965); *Nucl. Sci. Abstr.* **19**, 33678 (1965).
295. Matsumura, T., T. Ishigama, and T. Mamuro, "On the decontamination of radioactive waste water by ferric hydroxide flocculation," *Ann. Rep. Radiation Centre, Osaka Prefect.* **3**, 32 (1964).
296. Mauchline, J., and W. L. Templeton, "Dispersion in the Irish Sea of the radioactive liquid effluent from Windscale Works of the U.K. Atomic Energy Authority," *Nature* **198**, 623 (1963).
297. Mawson, C. A., "Management of radioactive wastes at a nuclear power plant," in Proceedings of 8th Ontario Industrial Waste Conference, 1961, p. 18.
298. Mawson, C. A., "Nuclear waste management," *J. Water Pollution Control Federation* **35**, 1055 (1963).
299. McElfresh, A. J., "The effectiveness of a finned circulator-evaporator for concentrating radioactive waste," U.S. Atomic Energy Commission, Report no. *HW*-80056, Hanford Atomic Products Operation, 1963.
300. Mende, H., "Water contamination by nuclear radiation and its prevention," *Nucl. Sci. Abstr.* **16**, 13409 (1962).
301. Mende, H., "The decontamination of radioactive waste water and polluted air from radiological control areas," *Gesundheit. Ingr.* **84**, 90 (1963).
302. Michon, G., "The disposal of radioactive waste in irrigation water," Paper presented at Meeting of Health Physics Society, Cincinnati, Ohio, June 1964; *Nucl. Sci. Abstr.* **19**, 10666 (1965).
303. Mitchell, N. T., et al., "Removal of certain radioactive isotopes during the primary sedimentation of sewage," *J. Proc. Inst. Sewage Purif.* **6**, 581 (1963).
304. Mittag, I., and G. Sachse, "The purification of radioactive waste waters by contact precipitation. I: The principle of contact precipitation," *Nucl. Sci. Abstr.* **18**, 33169 (1964).
305. Moore, R. L., "Hot-cells studies of the spray and pot calcination of fully radioactive fuel reprocessing wastes," U.S. Atomic Energy Commission, Report no. *BNWL-SA*-569, Batelle-Northwest Laboratory, 1966.
306. Morgan, J. M., Jr., J. C. Geyer, and D. C. Costello, "Land burial of solid packaged radioactive wastes," *J. Sanit. Eng. Div. Am. Soc. Civil Engrs.* **88**, 139 (1962).
307. Murray, N. F., "Waste effluent evaporators (*CR*-10-1144)," Atomic Energy of Canada, Ltd., Chalk River, Ont., April 1963.
308. Napravnick, J., and F. Kepak, "The sorption of fission products on the suspension of the natural sorbents coagulated in an electric field," *Water Pollution Abstr.* **39**, 1194 (1966).
309. Nelson, B., "Thermal pollution," *Science* **158**, 755 (1967).

310. Nelson, J. L., R. W. Perkins, J. M. Nielsen, and W. L. Haushill, *Reactions of Radionuclides from the Hanford Reactors with Columbia River Sediments*, International Atomic Energy Agency, Vienna (1966).
311. Nelson, J. L., G. J. Alkire, and B. W. Mercer, "Inorganic ion exchange separation of calcium from Purex-type high-level radioactive wastes," *Water Pollution Abstr.* **37**, 849 (1964).
312. Nelson, R. W., and J. R. Eliason, "Prediction of water movement through soils—a first step in waste transport analysis," U.S. Atomic Energy Commission, Report no. *BNWL-SA*-678, Batelle-Northwest Laboratory, 1966.
313. Nelson, J. L., and B. W. Mercer, "Ion exchange separation of cesium from alkaline waste supernatant solutions," U.S. Atomic Energy Commission, Report no. *HW*-76449, Hanford Atomic Products Operation, Jan. 30, 1963, 31 pages.
314. Newton, T. D., "On the dispersion of fission products by ground water," *CRT-866*, November 1959; *Nucl. Sci. Abstr.* **14**, 7159 (1960).
315. Nodier, J., J. Scheidhauer, and M. Malabre, "Conditioning of radioactive waste by bitumen," *J. Appl. Chem.* **12**, page 350, April 1962.
316. Oak Ridge National Laboratory, "Fixation of fission products in aluminum nitrate wastes by ceramic masses," U.S. Atomic Energy Commission, Report no. *ORNL*-2611, October 1963.
317. Oak Ridge National Laboratory, "Waste treatment and disposal," U.S. Atomic Energy Commission, Report no. *ORNL*-3627, October 1964.
318. Oak Ridge National Laboratory, "Waste treatment and disposal," in Chemical Technology Division Annual Progress Report for Period Ending May 31, 1963, U.S. Atomic Energy Commission, Report no. *ORNL*-3452, September 1963.
319. Oglaza, J., and A. Siemaszko, "An investigation of the usability of some flocculants for treatment of radioactive waste water by mineral sorbents," *Nukleonika* **10**, 519 (1965).
320. Ojima, T., H. Toratoni, and H. Fujimoto, "Behavior of radioactive strontium, cesium and cobalt in marine water," *Ann. Rep. Radiation Centre, Osaka Prefect.* **6**, 93 (1965).
321. Otto, R., and P. Hecht, "Utilization of multiclone dust or decontaminating sewage waters containing 137 Cs and 90 Sr/90," *Kernenergie* **8**, 643 (1965).
322. Ozmidov, R. V., and N. I. Popov, "Some data on the diffusion of soluble contaminants in the ocean," in Proceedings of Symposium on Disposal of Radioactive Wastes into Seas, Oceans and Surface Waters, International Atomic Energy Agency, Vienna, 1966.
323. Paige, B. E., "Leachability of alumina calcine produced in the Idaho waste calcining facility," U.S. Atomic Energy Commission, Report no. *IN*-1011, Idaho Nuclear Corp., 1966.
324. Pasker, F. L., and R. E. Blanco, "Waste treatment and disposal progress report for Nov.–Dec. 1962, and Jan. 1963," U.S. Atomic Energy Commission, Report no. *ORNL-TM*-516, Oak Ridge National Laboratory, June 1963.
325. Parsons, P. J., "Migration from a disposal of radioactive liquid in sands," *Health Phys.* **9**, 333 (1963).
326. Parsons, P. J., "Movement of radioactive waste through soil. 5: The liquid disposal area," *Nucl. Sci. Abstr.* **16**, 28634 (1962).
326. Parsons, P. J., "Movement of radioactive waste through soil. 4: Migration from a single source of liquid waste deposited in porous media," *Nucl. Sci. Abstr.* **16**, 18703 (1962).
328. Pearce, D. W., "Radioactive waste fixation quarterly progress report for July–Sept., 1959," *Nucl. Sci. Abstr.* **14**, 13520 (1960).
329. Pearce, K. W., and M. J. Smyth, "Handling with filtration of radioactive sludges," *Filtration* **90**, September 1965.
330. Peleik, J., and J. Bar, "Adsorption of ruthenium on solid sorbents in acid aqueous solutions," *Jaderna Energie* **12**, 121 (1966).
331. Penney, E., "Radiation protection in the Atomic Energy Authority," *Brit. J. Radiol.* **35**, 510 (1962).
332. Pepper, D., "Ion removal and its application to radioactive effluent treatment," *Brit. Chem. Engr.* **9**, 510 (1964); *Water Pollution Abstr.* **38**, 309 (1965).
333. Perachylae, O., "Radioactive wastes and their handling," *Nucl. Sci. Abstr.* **17**, 15531 (1963); *Water Pollution Abstr.* **35**, 1404 (1962).
334. "Periodic characterization of radioactive waste disposal effluents, test evaluation," Duquesne Light Co., Report no. *DICS*-2180102, Shippingport, Pa., June 1962; *Nucl. Sci. Abstr.* **16**, 28635 (1962).
335. "Periodic waste disposal system material balance test, Core 1, seed 2," Duquesne Light Co., Report no. *DICS*-2390101, June 1961; *Nucl. Sci. Abstr.* **15**, 28847 (1961).

336. Perkins, R. W., "Radiochemical analysis of reactor effluent waste materials at Hanford," *Talanta* **6**, 117 (1960).
337. Perona, J. J., J. O. Blomehe, R. L. Bradshaw, and J. J. Roberts, "Evaluation of ultimate disposal methods for liquid and solid radioactive wastes. V: Effects of fission product removal on costs of waste management," U.S. Atomic Energy Commission, Report no. *ORNL*-3357, Oak Ridge National Laboratory, June 1963.
338. Petrie, J. C., et al., "Fluidized bed calcination of simulated zirconium fluoride waste in exploratory pilot plant test," U.S. Atomic Energy Commission, Report no. *IDO*-14653, Phillips Petroleum Co., July 1963.
339. Porcella, D. B., and A. G. Friend, "Field studies of specific radionuclides in fresh water," *J. Water Pollution Control Federation* **38**, 102 (1966).
340. Prechistenskii, S. A., "Radioactive waste in atmosphere (devices for purifying aerosols and gases)," *Nucl. Sci. Abstr.* **16**, 9065 (1962).
341. Pressman, M., D. C. Lindsten, and R. P. Schmitt, "Removal of nuclear bomb debris, strontium 90–yttrium 90, and cesium 137–barium 137 from water with corps of engineers mobile water-treating equipment," *Nucl. Sci. Abstr.* **15**, 27888 (1961).
342. Pritchard, D. W., A. Okubo, and H. H. Carter, "Observations and theory of eddy movement and diffusion of an introduced tracer material in the surface layers of the sea," in Proceedings of Symposium on Disposal of Radioactive Wastes into Seas, Oceans, and Surface Waters, International Atomic Energy Agency, Vienna, 1966.
343. "Process and waste characteristics and selected uranium mills," Robert A. Taft Sanitary Engineering Center, U.S. Public Health Service, Cincinnati, Ohio, Technical Report no. *W*62-17, 1962.
344. Proctor, J. F., "Geologic, hydrologic, and safety considerations in storage of radioactive wastes in a vault excavated in crystalline wastes," *Nucl. Sci. Eng.* **22**, 350 (1965).
345. Pushkarev, V. V., et al., "Clarifying and purifying of radioactive waste liquids by the flotation method," *At. Energya* **16**, 48 (1964); *Water Pollution Abstr.* **38**, 118 (1965).
346. "Radioactive effluents," *Chem. Trade J.* **148**, 1631 (1961); *Water Pollution Abstr.* **35**, 143 (1962).
347. "Radioactive waste concentrator," *Water Waste Treatment J.* **9**, 25 (1962).
348. "Radioactive waste handling in the nuclear power industry," *Reactor Fuel Process.* **4**, 62 (1961).
349. Rafferty, J. C., "Some notes on plant for the removal of radon by aeration," *J. Inst. Water Engrs.* **17**, 23 (1963).
350. Raggenbass, A., and J. Lefevre, "Fission product separation as a final solution to the problem of storing highly radioactive waste," in Proceedings of Symposium on Treatment and Storage of High-Level Radioactive Wastes, Vienna, October 1962, International Atomic Energy Agency, February 1963, p. 141.
351. Ralkova, J., and J. Saidl, "Incorporation of radioisotopes into melted silicates," International Atomic Energy Agency, Vienna, Preprint no. *SM*-31-20, 1962.
352. Rauzen, F. V., et al., "Udalenie radioaktivnykh izotopov iz sbrosnykh vod.," *At. Energia* **18**, 623 (1965); *Soviet J. At. Energy* **18**, 784 (1965).
353. Rauzen, F. V., S. Dudnik, and E. Gutin, "The use of electrodialysis with ion-exchange membranes for desalting and purifying low-level waste waters," in Proceedings of Symposium on Practices in the Treatment of Low- and Intermediate-Level Radioactive Wastes, International Atomic Energy Agency, Vienna, 1966.
354. Rettmer, R. S., et al., "The treatment of biologically-loaded radioactive sewage," *Atompraxis* **10**, 74 (1964); *Water Pollution Abstr.* **38**, 96 (1965).
355. Rhodes, D. W., "Adsorption by soil of strontium from 216-*S* crib waste," *Nucl. Sci. Abstr.* **14**, 17633 (1960).
356. Riech, H. G., "Pilot-scale aluminum bed decontamination of reactor effluent. II: Final report," U.S. Atomic Energy Commission, Report no. *HW*-72215, Hanford Atomic Products Operation, 1962.
357. Rimshaw, S. J., and D. C. Winkley, "Removal of Cs^{137}, Sr^{90}, and Ru^{106} from ORNL plant wastes by sorption on various minerals," *Nucl. Sci. Abstr.* **14**, 14604 (1960).
358. Roberts, J. T., and R. R. Halcomb, "A phenolic resin ion exchange process for decontaminating low-radioactivity-level process water wastes," *Nucl. Sci. Abstr.* **15**, 20474 (1961).
359. Robinson, B. P., "Ion-exchange minerals and disposal of radioactive wastes—a survey of literature," U.S. Geology Survey, Water Supply, Paper no.

1616, 1962; *Water Pollution Abstr.* **35**, 2196 (1962).
360. Robinson, L. R., Jr., "Monitoring streams for radioactive wastes," *Water Sewage Works* **112**, 152 (1965).
361. Rodger, W. A., "Radioactive waste disposal," *Batelle Tech. Rev.* **10**, 446 (1961).
362. Rodger, W. A., "Safety problems associated with disposal of radioactive wastes," *Nucl. Safety* **5**, 287 (1964).
363. Rodier, J., G. Lefillatre, and J. Scheidhauer, "Encasing radioactive sludge in bitumen," *Energie Nucl.* **6**, 81 (1964); *Nucl. Sci. Abstr.* **18**, 31042 (1964).
364. Rodier, J., R. Estournel, R. Court, and F. Gallisian, "Return of waste water to the environment. The experience of seven years' operation of the Marcoule Plutonium Production Center," in Proceedings of the Symposium on Disposal of Radioactive Wastes into Seas, Oceans, and Surface Waters, International Atomic Energy Agency, Vienna, 1966.
365. Rodier, J., *et al.*, "L'incineration des residues radioactifs et ses aspects industriels," *Energie Nucl.* **7**, 25 (1965).
366. Rubin, E., A. Reisser, and A. Karalis, "An evaporator-type radioactive liquid waste concentrator with disposable vessel," U.S. Atomic Energy Commission, Report no. *AD*-404279, Radiation Applications, Inc., 1963; *Nucl. Sci. Abstr.* **18**, 11498 (1964).
367. Rubin, E., "Foam separation, quarterly progress report, Jan. 1 to March 31, 1963," U.S. Atomic Energy Commission, Report no. *RAI*-113, Radiation Applications, Inc., 1963.
368. Rubin, E., E. Schonfeld, and R. Everett, Jr., "The removal of metallic ions by foaming agents and suspensions: laboratory and engineering studies, annual report, July 1, 1961, to June 30, 1962," U.S. Atomic Energy Commission, Report no. *RAI*-104, Radiation Applications, Inc., 1962.
369. Rudolf, A., "Improvements in and relating to the disposal of radioactive waste," British Patent no. 929863, 26 June 1963.
370. "RWD System incinerator tests, core 1, seed 3, test evaluation," U.S. Atomic Energy Commission, Report no. *TID*-14871, Feb. 7, 1962; *Nucl. Sci. Abstr.* **16**, 16025 (1962).
371. Rzekiecki, R., "Sorption of Sr^{90} on an aluminum oxide," Centre d'Etudes Nucleaires, Commissariat à l'Energie Atomique, Saclay, Report no. *CEA-R*-2851, 1965.

372. Sachse, G., and I. Mittag, "Feasibility of treatment of radioactive waste," *Kernenergie* **6**, 1 (1963).
373. Sachse, G., and R. Winkler, "Use of a peat-clay exchanger for the purification of radioactive solutions," Zentralinstitut fur Kernphysik, Dresden, Report no. *ZFA-RCH*-2, May 1963; *Nucl. Sci. Abstr.* **18**, 19449 (1964).
374. Saddington, K., "Waste treatment at some high-activity processing plants," in *Atomic Energy Waste. Its Nature, Use and Disposal*, Interscience Publishers, New York (1961), p. 257.
375. Saidl, J., "The disposal of high-specific activity wastes. II: Waste fixation into clay minerals," *Jaderna Energie* **8**, 314 (1962); *Nucl. Sci. Abstr.* **17**, 1164 (1963).
376. Saidl, J., and J. Ralkova, "Behavior of strontium and cesium during fixation of radioactive wastes in basalt," *Collection Czech. Chem. Commun.* **31**, 871 (1966); *Nucl. Sci. Abstr.* **20**, 24814 (1966).
377. Sakata, S., "Radioactive waste disposal in Europe," Japan Atomic Energy Research Institute, Tokyo, Report no. *JAERI*-4021, October 1962; *Nucl. Sci. Abstr.* **17**, 22961 (1963).
378. Sasahi, T., S. Watanabe, G. Oshelia, N. Okami, and M. Kajihara, "Studies on the container for disposing radioactive wastes into the sea," *J. Oceanog. Soc. Japan* **19**, 16 (1963); *Nucl. Sci. Abstr.* **17**, 29874 (1963).
379. Savage, D. J., *et al.*, "Liquid radioactive waste disposal at the Pretoria General Hospital," *S. African J. Radiol.* **2**, 11 (1964); *Water Pollution Abstr.* **38**, 306 (1965).
380. Schaffer, W. R., Jr., *et al.*, "Project salt vault: design and demonstration of equipment," U.S. Atomic Energy Commission, Report no. *ORNL-P*-1987, Oak Ridge National Laboratory, 1964.
381. Schanze, U. Q., "Determination of activity in radioactive waste waters," *Stadtehygiene* **12**, 28 (1961); *Water Pollution Abstr.* **35**, 692 (1962).
382. Scheidhauer, J., "Rapid method for the determination of plutonium in radioactive effluents," Service de Protection contre les Radiations, Centre de Marcoule, France; L. Messainguiral, *Chim. Anal.* **43**, 462 (1961) and *Anal. Abstr.* **9**, 1894 (1962).
383. Schneider, K. J., "Fluidized-bed calcination studies with simulated *ICPP* waste solution," U.S. Atomic Energy Commission, Report no. *HW*-65838-*RD*, Hanford Atomic Products Operation, June 1960.
384. Schoen, H. M., "Radium removal from uranium mill

waste water," *J. Water Pollution Control Federation* **34**, 1026 (1962).
385. Schoenbert, D., "Radioactive waste and its treatment," *Chem. Tech.* **14**, 28 (1962); *Nucl. Sci. Abstr.* **16**, 14369 (1962).
386. Schonfeld, E., and W. Davis, Jr., "Softening and decontaminating waste water by caustic carbonate precipitants in a slowly-stirred sludge blanket column," *Health Phys.* **12**, 407 (1966).
387. Schonfeld, E., et al., "The removal of Sr and Cs from nuclear waste solutions by foam separation: final report," *Nucl. Sci. Abstr.* **15**, 15319 (1961).
388. Schultz, N. B., "Removal of low-level radioactive wastes by a sanitary water treatment process," U.S. Atomic Energy Commission, Report no. *K*-1651, Oak Ridge Gaseous Diffusion Plant, 1966.
389. Schulze-Rettner, R., et al., "A small installation for the treatment of radioactive sewage," *Atompraxis* **6**, 320 (1965); *Nucl. Sci. Abstr.* **19**, 38288 (1965).
390. Schwarzer, W., "Protective measures against the effects of radiation from radioactive materials used in industry," *Kerntechnik* **4**, 56 (1962); *Nucl. Sci. Abstr.* **16**, 20894 (1962).
391. Sharpelos, J. M., "Treatment of a radioactive condensate waste," U.S. Atomic Energy Commission, Report no. *HW-SA*-2986, Hanford Atomic Products Operation, April 1963.
392. Shefcih, J. J., "Batch calcination studies. An interim report," U.S. Atomic Energy Commission, Report no. *HW*-74043, Hanford Atomic Products Operation, June 1962.
393. Silker, W. B., "Reducing radioisotope concentrations in reactor effluent by high-coagulant feed," *J. Am. Water Works Assoc.* **55**, 355 (1963).
394. Silker, W. B., "The effectiveness of high-coagulant feed in reducing Hanford reactor effluent radioisotope concentrations," U.S. Atomic Energy Commission, Report no. *HW-SA*-2192, Hanford Atomic Products Operation, July 1961.
395. Silker, W. B., "Separation of radioactive Zn from reactor cooling water by an isotope exchange method," *Anal. Chem.* **33**, 233 (1961).
398. Siple, G. E., "Geohydrology of storage of radioactive wastes in crystalline rocks at AEC Savannah River plant," U.S. Geological Survey, Professional Paper no. 501-C, 1964, p. 180.
399. Skarpelos, J. M., "Preliminary report on the characterization of Purex tank farm radioactive condensate waste," *Nucl. Sci. Abstr.* **15**, 14041 (1961).
400. Skarpelos, J. M., "Treatment of radioactive condensate wastes," *Oregon State Coll., Eng. Exp. Sta., Circ.* **29**, 110 (1963).
401. Smith, J. F. W., and R. H. A. Crawley, "Removing radioactive iodine from gaseous effluent," *Nucl. Eng.* **6**, 428 (1961).
402. Smith, J. V., "Separation of cesium from fission product wastes by ion exchange on ammonium molybdophosphate," *Ind. Eng. Chem., Process Design Develop.* **5**, 117 (1966).
403. Smit, J. V. R., and J. J. Jacobs, "Separation of cesium from fission-product wastes by ion exchange on ammonium molybdophosphate," *Ind. Eng. Chem., Process Design Develop.* **5**, 117 (1966).
404. Sobolev, I. A., and L. M. Khomchick, "Organization of a centralized removal of radioactive wastes from the sites of their accumulation," *Gigiena i Sanit.* **5**, 78 (1965); *Nucl. Sci. Abstr.* **20**, 1553 (1966).
405. Spragg, H. R., "Atomic radiation in sewage processing," *Water Sewage Works* **110**, 163 (1963).
406. Stevens, J. I., "Fluidized bed calcination," Report of Third Working Meeting on Calcination and/or Fixation of High-Level Wastes in Stable, Solid Media at Washington, D.C., February 1962, comp. and ed. W. H. Regan, p. 3; U.S. Atomic Energy Commission, Report no. *TID*-7646, Division Reactor Development, 1962.
407. Stevens, J. I., and J. A. McBridge, "What can we do with radioactive wastes?" *Mech. Eng.* **84**, 46 (1962).
408. Stevens, J. I., and J. A. McBridge, "Management of high-level waste—future possibilities," New York Engineers' Joint Council, Preprint Paper no. 78, (1962); *Nucl. Sci. Abstr.* **16**, 30160 (1962).
409. Straub, C. P., *Low-Level Radioactive Wastes, Their Handling, Treatment, and Disposal*, U.S. Government Printing Office, Washington, D.C. (1964).
410. Straub, C. P., et al., "Environmental implications of radioactive waste disposal as related to stream environments," in Proceedings of Second Conference on Disposal of Radioactive Wastes, Monaco, November 1959, p. 407.
411. Stevenson, C. E., "Current practices in handling and storing solid radioactive waste from reactors," *Nucl. Safety* **7**, 231 (1966).
412. Stricos, D. P., "A procedure for the quantitative recovery of Pa from irradiated thoria," U.S. Atomic Energy Commission, Report no. *KALP-M*-6554, Knolls Atomic Power Laboratory, 1966.

413. Struxness, E. G., R. J. Morton, and F. L. Parker, "Radioactive waste disposal health physics division annual progress report, 1962," U.S. Atomic Energy Commission, Report no. *ORNL*-3347, Oak Ridge National Laboratory, 1962.
414. Suchoza, B. P., "Shippingport atomic power station disposition of radioactive wastes from underground storage tanks," U.S. Atomic Energy Commission, Report no. *WAPD-T*-1851, Bettis Atomic Power Laboratory, 1965.
415. Szekacs, I., et al., "Decontamination of radioactive liquid wastes. I: Investigation of the adsorptive capacity of native bentonite," *Mag. Chem. Foly.* **65**, 218 (1959); *Water Pollution Abstr.* **36**, 1169 (1963).
416. Takizawa, M., K. Koda, S. Uehara, and A. Hirokawa, "Removal of Sr^{90} on Nigata montmorillonite clay by column technique," *Proc. Japan. Conf. Radioisotopes*, **5**, 46 (1963); *Nucl. Sci. Abstr.* **17**, 30074 (1963).
417. Tamise, M., "The treatment of radioactive effluents by evaporation," *Energie Nucl.* **5**, 85 (1962).
418. Tamura, T., "Disposal of radioactive wastes by hydraulic fracturing," *Nucl. Eng. Design* **5**, 477 (1967).
419. Tamura, T., "Improving cesium selectivity of bentonites by heat treatment," *Health Phys.* **5**, 149 (1961).
420. Tamura, T., and E. G. Struxness, "Removal of strontium from wastes," *Nucl. Sci. Abstr.* **15**, 5849 (1961).
421. Tamura, T., "Selective sorption reactions of strontium with soil minerals," *Nucl. Safety* **7**, 99 (1965).
422. Tamura, T., and E. G. Struxness, "Reactions affecting removal from radioactive wastes," *Health Phys.* **9**, 697 (1963).
423. Thiriet, L., P. Lesur, R. Giraud, and J. Wanlin, "Comparison of the economic aspects of the treatment and storage of fission products from installations processing irradiated natural uranium," U.S. Atomic Energy Commission, Report no. *STI-P4B*-110, 1966, p. 571.
424. Thomas, H. A., "Operations research in disposal of liquid radioactive wastes in streams," U.S. Atomic Energy Commission, Report no. *NYO*-10477, Harvard University, 1965.
425. Thomas, H. C., "Mineral exchange at the University of North Carolina," in "The use of inorganic exchange materials for radioactive waste treatment. A working meeting held at Washington, D.C., Aug. 13–14, 1962," eds. D. K. Jamison, B. M. Karnegay, W. A. Vaughan, and J. M. Morgan, p. 13; U.S. Atomic Energy Commission, Report no. *TID*-7644, Johns Hopkins University, January 1963.
426. Thorburn, R. C., and R. J. Chandler, "An incinerator for uranium contaminated wastes," *Ind. Water Wastes* **6**, 46 (1961).
427. Tomlinson, R. E., "The Hanford program for management of high-level wastes," American Institute of Chemical Engineers, Preprint no. *S*, 1963.
428. Touhill, C. J., Jr., *The Removal of Radionuclides from Water and Waste-water by Manganese Dioxide*, Rennselaer Polytechnic Institute, Troy, N.Y. (1964).
429. Tritremmel, C., et al., *Behavior of Radioisotopes Released to Streams*, International Atomic Energy Agency, Vienna (1966).
430. Tukai, R., *Fate of Some Radionuclides fixed on Ion-Exchange Resins in Seawater Medium*, International Atomic Energy Agency, Vienna (1966).
431. Tuthill, E. J., "Phosphate glass process for disposal of high-level radioactive wastes," *Ind. Eng. Chem., Process Design Develop.* **6**, 314 (1967).
432. Tuthill, E. J., et al., "Brookhaven National Laboratory process for the continuous conversion of high-level radiation wastes to phosphate glass," U.S. Atomic Energy Commission, Report no. *BNL*-, Brookhaven National Laboratory, 1965, p. 109.
433. Tuthill, E. J., "Continuous phosphate glass process," Report of Third Working Meeting on Calcination and/or Fixation of High-Level Wastes in Stable, Solid Media at Washington, D.C., February 1962, comp. and ed. W. H. Regan, p. 43; U.S. Atomic Energy Commission, Report no. *TID*-7646, Division Reactor Development, 1962.
434. "Treatment of radioactive effluents," *Energica Nucl.* **9**, 320 (1962).
435. "Treatment and storage of high-level radioactive wastes," in Proceedings of Symposium on Radioactivity, October 1962, International Atomic Energy Agency, Vienna, February 1963.
436. Tsivoglou, E. C., and R. L. O'Connell, "Waste guide for the uranium milling industry," Robert A. Taff Sanitary Engineering Center. U.S. Public Health Service, Technical Report no. *W*62–12, 1962.
437. U.K. Atomic Energy Authority, "Determination of iodine-131 in reactor fuel processing and effluent treatment plant solutions," Report no. 334(*W*), 1962; *Water Pollution Abstr.* **37**, 843 (1964).
438. "Ultrasonic leaching of simulated calcined wastes," *Nucl. Sci. Abstr.* **15**, 5850 (1961).
439. Upson, U. L., "Fixation of high-level radioactive wastes in phosphates glass; hot cell glass experiment,"

U.S. Atomic Energy Commission, Report no. *BNWL*-220, Batelle-Northwest Laboratory, 1966.
440. Upson, U. L., et al., "Fixation of high-level waste (radioactive) in phosphate glass," U.S. Atomic Energy Commission, Report no. *BNWC-SA*-564, Batelle-Northwest Laboratory, 1965.
441. Van deVloed, A., "Transportation of radioactive pollution by water to great distances," *Tech. Sci. Munic.* **58**, 55 (1963).
442. Verdnikov, A. E., et al., "Feasibility of using soils for low-radioactivity waste-water purification," *Nucl. Sci. Abstr.* **20**, 26550 (1966).
443. Verschraeghen, L., "Process for the destruction of the radioactive residues," *Nucl. Sci Abstr.* **15**, 30330 (1961).
444. Vesely, V., "Problems of radioactive wastes disposal in the nuclear research institute," *Nucl. Sci. Abstr.* **16**, 2812 (1962).
445. Vesely, V., A. Jenicek, and J. Drahogal, "Radioactive waste disposal plant in the Institute of Nuclear Research of the Czechoslovak Academy of Sciences," *Jaderna Energie* **9**, 3 (1963); *Nucl. Sci. Abstr.* **17**, 17621 (1963).
446. Vinaroo, I. V., et al., "Sorption and separation of zirconium and hafnium on native anionites," *Nucl. Sci. Abstr.* **20**, 29352 (1966).
447. Vol'khin, V. V., A. K. Shtol'ts, and E. M. Dosik, "Processing liquid laboratory waste containing radioactive isotope," *Nucl. Sci. Abstr.* **16**, 28639 (1962).
448. Voznesenskii, S. A., et al., "Use of flotation in treating radioactive effluents," *At. Energia* **9**, 208 (1960).
449. Waldichuk, M., "Sedimentation of radioactive wastes in the sea," *Nucl. Sci. Abstr.* **15**, 14043 (1961).
450. Wallace, deLaguna, "Engineering geology of radioactive waste disposal," *Rev. Eng. Geol.* **1**, 129 (1962).
451. Wamanacharya, G. A., "Separation of strontium and barium by cation exchange using polyaminepolycarbolic acids as eluants," U.S. Atomic Energy Commission, Report no. *ORNL-TM*-1342, Oak Ridge National Laboratory, 1966.
452. Watson, L. C., "Mineral exchange in Canada's waste treatment program," in "The use of inorganic exchange materials for radioactive waste treatment—a working meeting held at Washington, D.C., Aug. 13, 1962," U.S. Atomic Energy Commission, Report no. *TID*-7644, Johns Hopkins University, January 1963.
453. Weeren, H. O., "Disposal of radioactive wastes by hydraulic fracturing. III: Design of *ORNL*'s Shale Fracturing Plant," *Nucl. Eng. Design* **4**, 108 (1966).
454. West, P. J., "The treatment of radioactive effluent for discharge to inland waterways," Paper presented at Effluent and Water Treatment Convention, London, 1962.
455. West, S. W., "Disposal of uranium-mill effluent near Grants, New Mexico," U.S. Geological Survey, Professional Papers nos. 424-*D* and *D* 376–9, 1961; *Nucl. Sci. Abstr.* **16**, 5594 (1962).
456. Whatley, M. E., C. W. Hancher, and J. C. Suddath, "Engineering development of nuclear waste pot calcination," U.S. Atomic Energy Commission, Report no. *ORNL-TM*-549, Oak Ridge National Laboratory, April 1964.
457. Whatley, M. E., P. A. Haas, R. W. Horton, A. D. Ryon, J. C. Suddath, and C. D. Watson, "Unit operations section monthly progress report, July 1962," U.S. Atomic Energy Commission, Report no. *ORNL-TM*-343, Oak Ridge National Laboratory, January 1963.
458. Wheeler, B. R., J. A. Buckham, and J. A. McBridge, "A comparison of various calcination processes for processing high-level radioactive wastes," U.S. Atomic Energy Commission, Report no. *IDO*-14622, Phillips Petroleum Co., April 1964.
459. Winkler, R., "Purification of radioactive wastes with natural organic ion exchanges," Symposium Tech. Zagadniena Ochrony Promieniowaniem, Warsaw, 1962, p. 546; *Nucl. Sci. Abstr.* **18**, 40967 (1964).
460. Wlodarski, R., "Studies on the neutralization of wastes forming as the result of the sulfuric acid treatment on uranium ores," *Nukleonika* **7**, 494 (1962).
461. Wormser, G., "Problems presented by the treatment of weekly radioactive effluents in an atomic center," *Bull. Inform. Sci. Tech.* **69**, 43 (1963); *Nucl. Sci. Abstr.* **17**, 26940 (1963).
462. Wormser, G., *Treatment of Effluent at the Saclay Center for Nuclear Studies*, Centre d'Etudes Nucleaires, Commissariat a l'Energie Atomique, Saclay, 1960.
463. Yamamoto, Y., et al., "Chemical engineering studies on the processing of radioactive liquid wastes by evaporation," *J. Fac. Eng., Univ. Tokyo (1965); Nucl. Sci. Abstr.* **20**, 30946 (1966).
464. Zoch, O., and E. Malasek, "Permanent storage of low-activity radioactive wastes in Czechoslovakia," *Jaderna Energie* **8**, 231 (1962); *Chem. Abstr.* **57**, 14901 (1962).

INDEX

INDEX

Acid mine drainage wastes, 510
Acid neutralizing unit, 470
Acid waste, 4, 9, 70, 128, 468
Acid waste utilization in industrial proceses, 77
Acrylics, 497
Acrylonitrile, 5
Activated carbon, 101
Activated sludge treatment, 8, 9, 110, 113, 173
Adsorption, 137
Ad valorem tax, 157
Aeration, 518
   of equalizing basins, 80
Aerobic decomposition, 110, 111
Aerobic treatment (high rate), 110, 117
Aesthetic value, 50
Agricultural use, 12, 16
Air bubbles, 91
Alcoa-Bayer process, 413
Algae, 3, 103, 111
Algae (harvesting), 106
Algal photosynthesis, 112
Alkalinity, 6
Alkalis, 4, 9
Alkyl benzene sulfonate, 6
Allocation firm, 53
Allocation of costs, 158
Alpha oxidation of acids, 136
Aluminum, 413
Ammonia, 65
Anaerobic digestion, 8, 110, 122, 135, 137
Analyses of sewage effluents, 64
Animal glue manufacturing wastes, 463
Aquatic life, 4
Archimedes' principle, 90
Arsenic, 6

Assembly plants, 432
Atomic energy plants, 247
Atomized suspension, 135, 143
Atomized water, 515
Auto and truck crankcase oil, 453

Bacteria, 135, 228
Bacteriological quality, 231
Baking, 4
Batch or slug discharges, 65
Bathing water use, 11
Beer brewing, 4
Beet-sugar wastes, 350
Benefit measurements, 50, 51
Benefit-cost analysis, 50
Benefit-cost ratio, 50
Beta oxidation of acids, 136
Bio-disc system, 110, 130
Bio-filters, 119
Biological oxygen demand (BOD), 7
Biosorption, 116
Blast furnace, 398
Blaw-Knox-Ruthner process, 401
Blood recovery, 339
Blood wastes, 190, 341
Blowdown of cooling towers, 502
BOD, 92, 157
Boiler blowdown, 504
Brass and copper, 412
Brewery and distillery wastes, 329
Brownian movement, 97
Brush aeration, 110, 129
Bulking sludge, 114
Buoyancy, 90
Burial, 515, 546

By-product coke process, 398
By-product recovery, 68, 70, 339

Calcium sulfate, 3
Candle manufacturing wastes, 365, 465
Canned fruits, 4
Cannery wastes, 297
Capital costs of operation and maintenance, 187
Carbohydrates, 136
Carbon chloroform extract, 6
Carbon-dioxide treatment for alkaline wastes, 76
Carbonate, 3, 6
Carboxymethyl cellulose, 217
Caustic embrittlement, 4
Caustic-soda treatment for acid wastes, 74
Cavitation, 123
Cellulosics, 494
Centrifuging, 135, 144, 515
Cesium 137, 7
Changes in processes, 189
Changing production, 62
Chemical coagulation, 98
Chemical flocs, 98
Chloride ion, 3, 5, 6
Chlorides, 103
Chlorine, 6
Chrome plating, 199
Chromium-bearing plating, 418
Chromium (hexavalent), 5, 6
Churchill multiple-regression technique, 166
Churchill's method of multiple linear correlation, 26, 37
Circular tanks, 85
Citrus cannery waste, 299
Citrus fruits—oranges, lemons, 299
Classification of wastes, 61
Closed system, 61
Cloth and towels, 190
Coagulant aids, 87, 100
Coagulation, 514
Coal mining, 69
Coal preparation wastes, 509, 510
Coal storage, 502
Coal washeries, 509

Coil filter, 138
Coffee wastes, 355
Coliform, 238
Colloids, 97
Colloids, removal by adsorption, 101
Color, 5, 9
Color colloids, 100
Colored matter, 8
Combined treatment, 56, 192
Completely mixed systems, 118
Complete treatment, 193, 229
Composite waste analysis, 169
Composite waste sampling, 169
Composting, 149
Condensed and powdered milk, 316
Condenser waters, 5, 9
Conservation of waste water, 61
Contact stabilization, 110, 116
Cooling water, 11
Cooperation of industry and municipalities, 154, 191
Copper, 5, 6
Coprecipitation, 545
Cornstarch industry wastes, 473
Corrosion, 4
Corrosive effects, 157
Cost-effectiveness analysis, 50
Cost of radioactive waste treatment, 548
Cottage cheese, 316
Cotton, 255
Covering strip mines, 518
Curie, 532
Cyanide case hardening, 199
Cyanides, 6, 167, 415, 417

Dairy industry, 71
Dairy wastes, 316
Deep-well disposal, 518
Deep-well injection, 110, 124, 128
Demineralization (ion exchange), 106
Deoxygenation rate, 26, 219
Depth of tank, 83
Desize, 218
Detention time, 83

Detergents, 9, 65, 191
Dewatering, by lagoons, 180
    by vacuum filter and disposal by approved landfill, 180
    by vacuum filter–incineration (multiple-hearth), 180
Dewatering sludge, 137
Dialysis, 103, 263
Diesel trains, 453
Digesters, 8
Digestion of tannery–municipal solids, 183
Discharge of raw wastes, 206
Dispersed air flotation, 87
Dispersed growth aeration, 110, 114, 267
Dissolved air flotation, 87
Dissolved minerals, 103
Dissolved organic matter, 110
Dissolved oxygen, 4, 9, 212
Distribution and baffling, 79
Drainage from cinder and ash dumps, 502
Drinking water, 12
Dry cleaning, 515
Drying and incineration, 135, 143
Drying beds, 138
Drying time, 140
Dyeing and finishing mill, 188
Dyes, 8, 218

Economics of waste treatment, 48
Eddying, 85
Edible rendering, 340
Effect of river dilution, 220
Effluent standards, 11
Effluents, 61
Electrical-plating company, 189
Electrodialysis, 105
Electrostatic painting, 433
Elimination of batch or slug discharges, 61
Elutriation, 135, 138
Equalization of wastes, 68, 70, 79, 171
Equipment modifications, 68, 69
Evaporator blowdown, 502
Evaporation, 103, 140, 545
Explosives industry wastes, 484

Extended aeration, 129

Facultative anaerobic bacteria, 111
Facultative chemo-organotrophs, 112
Fair's classification, 28, 166
Fats, 8, 9, 136
Feathers, 9, 190
Federal quality guidelines, 16
Federal Register, 540
Feedlot wastes, 347
Fermentation, 4, 136
Filterability of either raw or digested solids, 183
Filter clog or "blind," 138
Filtration, 137
Fish, 4, 16
    diseases of, 211
Fish wastes, 358
Fishing use, 12
Fission, 532
Flammable or explosive liquids, 157
Flammable substances, 156
Flash distillation, 518
Floating materials, 8
Floating solids and liquids, 4
Flooding abandoned mines, 517
Flotation, 87
Flue gas for neutralization, 226
Fluid milk, 316
Fluoride, 6
Fly ash, 503
Foam phase separation, 110, 128
Foam-producing matter, 7
Food-processing industries, 297
Formaldehyde wastes, 489
Free market, 51
Froth flotation, 513, 515
Fruits, processing of peaches, tomatoes, 299
Fuel-element processing, 533
Fuel-oil wastes, 453
Fuel-processing wastes, 536

Gamma rays, 532
Garbage, 157
Gaseous radiation waste, 540

Gases, 9
Gasoline, 157
Gelatin, 4
Genossenschaften Ruhr Valley System, 52
Glass manufacturing wastes, 365, 459
Glue factories, 178
Glue manufacture, 4
Glue wastes, 365
Grease, 4, 8, 157, 190
Grease recovery, 341
Gross β-radiation concentration, 6
Gross national product, 52
Gob piles, 515
Gold, 413

Hard waters, 3
Heat-transfer coefficient (air flows parallel sludge), 144
Heated water, 5
Heavy metals, 194
Hemicelluloses, 99
Henry's law, 87
Hospitals, as contributors to radioactive waste, 533
Howard process, 377
Hydrolytic bacteria, 135

Ice cream and frozen desserts, 316
Incinerators, 144, 547
Industrial plant location, 12
Industrial production records, 171
Industrial water quality, 14, 16
Industries, separate contracts, 156
Inedible rendering, 340
Inflammables, 9
Inlet and outlet design, 87
Inorganic dissolved solids, 103
Inorganic salts, 3
In-plant changes, 195
Insecticides, 5
Intangible benefits, 50, 51
Interceptors for grease, oil, and sand, 157
Intermediate treatment, 98
Ion exchange, 103, 518
Iron, 3, 6

Iron-foundry wastes, 365, 434
Isoelectric point, 100

Joint disposal of untreated industrial waste and domestic sewage, 161
Joint treatment, 7, 153, 188

Kier wastes, 218
Koppers dephenolization process, 401
Kraus process, 114

Laboratory pilot plant studies, 121
Lagooning in oxidation ponds, 110
Lagooning of acid mine water, 517
Lagoons, 141
Laminar flow, 120
Landfill (sanitary), 135, 147
Lanolin, 264
Laundering of contaminated clothes, 533
Laundry wastes, 291
Lead, 6
Leather tanning, 4
Lime, 100
Limestone treatment for acid wastes, 74
Lime slurry treatment for acid wastes, 74
Local river basin, 51
Low-flow augmentation, 16
Lyophilic colloids, 97
Lyophobic colloids, 97

Magnesium, 6
Magnesium sulfate, 3
Manganese, 6
Maximum permissible concentration (MPC), 532
Meat-packing, 338
Mechanical aeration system, 110, 123
Mechanical agitation, 80
Menhaden fishing industry, 231
Metal cleaning, 4
Metal ions, 9
Metal plating wastes, 68, 70, 365, 415
Metal wastes, 365, 398
Methane fermentation, 112

INDEX 581

Microorganisms, 6
Microstraining, 93
Mixing wastes, 73
Modified aeration, 110
Monitoring waste streams, 68, 72
Monomolecular rate of decomposition of organic matter, 120
Motor industry wastes, 431
Municipalities' assistance to industry, 155
Municipal ordinance, 155, 156

Naval-stores wastes, 365, 459
Neutralization, 73, 517
Neutralizing overacidity, 73
New England Interstate Water Pollution Control Commission, 16
Nickel plating, 196
Nickel stripping, 199
Nitrate, 6, 103
Noxious gases, 157
Nylon wastes, 76

Ocean outfall (used to dispose of digested sludge), 147
Ohio River Sanitation Commission, 16
Oils, 4, 8
Oil fields, 435, 436
Oil refinery wastes, 91, 365, 435, 436
Organic matter, 4, 8, 65
Organic polymers, 98
Overalkalinity, 73
Overflowing rate, 84
Oxidation, of glucose, 115
Oxidation ditches, 129
Oxidation ponds, 107, 110, 235
Oxidation reduction, chemical reactions, 108
Oxygen-sag characteristics, 166
Oxygen-sag curve, 161, 217

Packing houses, 71, 338
Paint sludges, 144
Partially treated industrial wastes, 216
Paunch material and manure, 190, 340
Pearl starch, 217

Penfer gum, 217
Penicillin, 115
Pennsylvania effluent standards, 16, 24
Pesticide wastes, 101, 491
pH, 6, 157, 210
pH recorders, 190
Pharmaceutical houses, 71
Phenolic constituents, 455
Phenolic resins, 499
Phenols, 4, 6, 9
Phosphate industry wastes, 476
Phosphates, 6, 65, 103, 238
Phosphorus, 3, 166
Photographic wastes, 365, 397
Photosynthesis, 111
Photosynthetic pond, 236
Pickle processing, 206, 398
Pickle wastes, 360
Pit incineration, 144
Plant production study, 195
Plastic and resin wastes, 493
Plywood, 365
Plywood-plant glue wastes, 466
Pollution-carrying resources, 49, 52
Pollution index, 53, 54
Polyester and alkyd resins, 498
Polyolefins (polyethylenes), 494
Polypeptides, 99
Polystyrene resins and copolymers, 495
Poultry-plant waste, 188, 232, 338
Poultry-processing plant, 229, 339
Power-plant cooling waters, 533
Power-plant wastes, 505, 533
Precedent of joint treatment, 192
Pressure flotation, 87
Pretreatment, 190
Primary benefits, 49, 51
Process changes, 68
Process water, 12
Processing of uranium ore, 533
Producing carbon dioxide in alkaline waste, 76, 77
Proportioning of acid drainage to streams, 517
Proportioning of wastes, 68, 80, 81, 72
Protein–carbohydrate wastes, 116

Proteins, 8, 99, 136
Prototype, 177
Public water supplies, 16
Pulp and paper-mill waste standards, 16, 24
Pulp and paper-mill wastes, 365
Pumping raw or digested sludge, 147
Pumped storage, 16
Purex process, 537

Quality of water supply, 241

Radiation, 7, 518
Radioactive waste materials, 532
Radium 226, 6
Rag, rope, and jute-mill wastes, 9, 115, 372
Rate of heat transfer, 104
Raw sludge, 137
Reaeration rate, 26, 219
Receiving waters, 505
Recreation, 16
Rectangular basins, 85
Reducing wastes, 189
Refractory compounds, 101
Refractory solids, 229
Refueling stations, 453
Regeneration of ion-exchange softeners, 502
Regulation, 51
Rendering plant, 188, 338
Remelting, 548
Removal of nitrogen and phosphorus, 108
Research-laboratory wastes, 533
Reusing industrial and municipal effluents, 61, 62
Reverse osmosis, 103
Rice wastes, 358
River resource allocation boards, 52
River resources, 53
River studies, 218
Rubber reclaiming, 4
Rubber wastes, 365, 453
Ruhr River, 52

Salt concentration, 211
Salt water, salts, 3
Sampling and analysis program, 202

Saprophytes, 135
"Save-alls," 373
Schulze-Hardy rule, 99
Screening, 91
Sealing of abandoned mines, 517
Secondary benefits, 51
Secondary treatment, 50, 91
Sedimentation, 83
Segregation of wastes, 68, 69
Selenium, 6
Self-flocculation, 83
Semipermeable membrane, 97, 104
Settling ponds, 514
Settling velocity, 83, 84
Sewage treatment plant, 7
Sewer ordinance, 156
Sewer rental charges, 155, 157
Shellfishing, 12
Short-circuiting, 85, 87
Silver-plating plant, 71
Site selection, 241
Slaughterhouse, 338
Slaughterhouse and meat packing, 188
Slime growths in waste (zoogleal forms), 119
Sludge, 8, 86
  age of, 114
  barging of, 135, 147
  lagooning of, 141
  pumping of, 135, 147
Sludge concentration (flotation and thickening), 148
Sludge conditioners, 137
Sludge digestion, 173
Sludge drying beds, 138
Sludge float, 87
Sludge solids, 135
Sludge volume index, 114
Slug acid discharges, 190
Slug loads, 169
Soaps, 190, 264
Soap and detergent-industry wastes, 479
Soap manufacturing, 4
Social benefits, 16
Soda pulping waste, 372

Sodium hydroxide, 4
Soft-drink bottling wastes, 4, 361
Soft waters, 100
Solids handling, 183
Solid waste (apricots and clingstone peaches), 149
Soluble polymeric coagulation, chemicals, 98
Special assessments, 157
Special contracts, 157
Specialty paper mills, 71
Spray burning, 469
Spray irrigation, 110, 122
Spreading, 129
Stabilization in ponds, 110
Stamping plants, 431
Starches, 8, 68, 99, 218
State Quality Standards, 16
Steel mill wastes, 398
Steam power plants, 501
Stockyard waste, 339
Stokes' law, 90, 147
Storage, river, 220
Straining, 137
Stream aeration, 16
Stream classification, 11
Stream protection, 11
Stream quality, 11
Stream sampling, 43
Stream specialization, 16
Stream standards, 11
Stream survey, 162
Streeter-Phelps formulation and methods, 26, 166
Strength reduction, 68
Streptomycin wastes, 115
Strontium 90, 6, 7
Submarginal consumers, 52
Submerged combustion, 76
Substitution of soluble sizing, 224
Subsurface disposal, 110, 129
Sugars, 8
Sulfate, 6
Sulfur oxidizing bacterium, 510
Sulfuric acid, 4
Sulfuric acid treatment for alkaline wastes, 77
Sulfite liquors, 370

Sulfite waste-liquor by-products, 71
Surface area, 83
Suspended solids, 4, 8, 83
Sweco vibrating screens, 234
Synthetic fibers, 257

Tanneries, 161, 162, 178
Tannery wastes, 277
Tapered aeration, 114
Tax allowances, 56
Tax depreciation, 56
Technical external economy, 50
Temperature, 136, 211
Textile dyeing, 4, 266
Textile finishing mills, 68, 70
Textile industry, 104
Textile wastes, 255
Thermal efficiencies, 144
Thermal pollution, 501
Thickeners and hydrocyclones, 515
Thomas' Modification of Sag Curve Analyses, 26
Tidal dispersion characteristics of the receiving waters, 147
Total measurable benefits, 53
Total oxidation, 117
Total solids, 6, 65
Toxic chemicals, 5
Toxic limit for metals, 200
Toxic wastes, 157
Trickling filter plants, 9
Trickling filtration, 110, 119, 174
Trough separation, 516
Turbulence, 84
Two-stage digestion, 137
Tyndall effect, 97

Underground cavities, placement of wastes in, 129
Upflow limestone bed, 470
Urea and melamine resins, 498
U.S. Geological survey, 208

Vacuum filtration, 135, 137
Vacuum flotation, 87

Vegetable wastes, 299
Vinyl resins, 494
Volatile acid, 137
Volume, 9

Waste acids, 71
Waste boiler-flue gas, 76
Waste disposal, 241, 244
Waste oxidation basins, 112
Waste sludge, 117
Waste strength reduction, 68
Waste treatment required, 26
Wastes from manufacturing processes, 61
Wastes from sanitary uses, 61
Water treatment plants, 4
Water treatment plant wastes, 362
Water quality act of 1965, 15

Water quality goals, 50
Water quality management, 50
Water quality standards, 15
Water quality surveillance, 74
Waters used as cooling agents, 61
Wet air oxidation system, 183
Wet combustion, 110, 122, 135, 141
Wood fiber, 365
Wood-preserving wastes, 365, 463
Wool, 257

Zeolites, 105
Zeta potential, 98, 99
Zimpro, 183
Zimpro process, 141
Zinc, 6
Zinc plating, 197